A Companion to the
Philosophy of Science

Blackwell Companions to Philosophy

This outstanding student reference series offers a comprehensive and authoritative survey of philosophy as a whole. Written by today's leading philosophers, each volume provides lucid and engaging coverage of the key figures, terms, topics, and problems of the field. Taken together, the volumes provide the ideal basis for course use, representing an unparalleled work of reference for students and specialists alike.

Already published

1 The Blackwell Companion to Philosophy
   Edited by Nicholas Bunnin and Eric Tsui-James

2 A Companion to Ethics
   Edited by Peter Singer

3 A Companion to Aesthetics
   Edited by David Cooper

4 A Companion to Epistemology
   Edited by Jonathan Dancy and Ernest Sosa

5 A Companion to Contemporary Political Philosophy
   Edited by Robert E. Goodin and Philip Pettit

6 A Companion to the Philosophy of Mind
   Edited by Samuel Guttenplan

7 A Companion to Metaphysics
   Edited by Jaegwon Kim and Ernest Sosa

8 A Companion to Philosophy of Law and Legal Theory
   Edited by Dennis Patterson

9 A Companion to Philosophy of Religion
   Edited by Philip L. Quinn and Charles Taliaferro

10 A Companion to the Philosophy of Language
    Edited by Bob Hale and Crispin Wright

11 A Companion to World Philosophies
    Edited by Eliot Deutsch and Ron Bontekoe

12 A Companion to Continental Philosophy
    Edited by Simon Critchley and William Schroeder

13 A Companion to Feminist Philosophy
    Edited by Alison M. Jaggar and Iris Marion Young

14 A Companion to Cognitive Science
    Edited by William Bechtel and George Graham

15 A Companion to Bioethics
    Edited by Helga Kuhse and Peter Singer

16 A Companion to the Philosophers
    Edited by Robert L. Arrington

17 A Companion to Business Ethics
    Edited by Robert E. Frederick

18 A Companion to the Philosophy of Science
    Edited by W. H. Newton-Smith

*Blackwell
Companions to
Philosophy*

A Companion to the Philosophy of Science

Edited by

W. H. NEWTON-SMITH

Copyright © Blackwell Publishers Ltd 2000

First published 2000

2 4 6 8 10 9 7 5 3 1

Blackwell Publishers Inc.
350 Main Street
Malden, Massachusetts 02148
USA

Blackwell Publishers Ltd
108 Cowley Road
Oxford OX4 1JF
UK

Library of Congress Cataloging-in-Publication Data has been applied for.

ISBN 0-631-17024-3 (hardback)

British Library Cataloguing in Publication Data
A CIP catalogue record for this book is available from the British Library.

Typeset in 10 on 12½ pt Photina
by Graphicraft Limited, Hong Kong
Printed in Great Britain by MPG Books Ltd, Bodmin, Cornwall

This book is printed on acid-free paper.

Contents

List of Contributors   xi

Preface   xv

List of Logical Symbols   xvi

Introduction   1

1   **Axiomatization**   9
    FREDERICK SUPPE

2   **Berkeley**   12
    M. HUGHES

3   **Biology**   16
    PAUL THOMPSON

4   **Bohr**   26
    DUGALD MURDOCH

5   **Causation**   31
    PAUL HUMPHREYS

6   **Cognitive Approaches to Science**   41
    RONALD N. GIERE

7   **Computing**   44
    LESLIE BURKHOLDER

8   **Confirmation, Paradoxes of**   53
    J. D. TROUT

9   **Convention, Role of**   56
    LAWRENCE SKLAR

10 **Craig's Theorem**   65
    FREDERICK SUPPE

CONTENTS

11 **Darwin**   68
SAMIR OKASHA

12 **Definitions**   76
FREDERICK SUPPE

13 **Descartes**   79
TOM SORELL

14 **Discovery**   85
THOMAS NICKLES

15 **Dispositions and Powers**   97
ROM HARRÉ

16 **Einstein**   102
CHRISTOPHER RAY

17 **Evidence and Confirmation**   108
COLIN HOWSON

18 **Experiment**   117
DAVID C. GOODING

19 **Explanation**   127
W. H. NEWTON-SMITH

20 **Feminist Accounts of Science**   134
KATHLEEN OKRUHLIK

21 **Feyerabend**   143
JOHN PRESTON

22 **Galileo**   149
ROBERT E. BUTTS

23 **History, Role in the Philosophy of Science**   154
BRENDAN LARVOR

24 **Holism**   162
CHRISTOPHER HOOKWAY

25 **Hume**   165
W. H. NEWTON-SMITH

26 **Idealization**   169
YEMIMA BEN-MENAHEM

27 **Incommensurability**   172
   MUHAMMAD ALI KHALIDI

28 **Induction and the Uniformity of Nature**   181
   COLIN HOWSON

29 **Inference to the Best Explanation**   184
   PETER LIPTON

30 **Judgment, Role in Science**   194
   HAROLD I. BROWN

31 **Kuhn**   203
   RICHARD RORTY

32 **Lakatos**   207
   THOMAS NICKLES

33 **Laws of Nature**   213
   ROM HARRÉ

34 **Leibniz**   224
   WILLIAM SEAGER

35 **Locke**   229
   G. A. J. ROGERS

36 **Logical Empiricism**   233
   WESLEY C. SALMON

37 **Logical Positivism**   243
   CHRISTOPHER RAY

38 **Mach**   252
   GEREON WOLTERS

39 **Mathematics, Role in Science**   257
   JAMES ROBERT BROWN

40 **Measurement**   265
   J. D. TROUT

41 **Metaphor in Science**   277
   ELEONORA MONTUSCHI

42 **Metaphysics, Role in Science**   283
   WILLIAM SEAGER

CONTENTS

43 **Mill**   293
    GEOFFREY SCARRE

44 **Models and Analogies**   299
    MARY HESSE

45 **Naturalism**   308
    RONALD N. GIERE

46 **Natural Kinds**   311
    JOHN DUPRÉ

47 **Newton**   320
    RICHARD S. WESTFALL

48 **Observation and Theory**   325
    PETER ACHINSTEIN

49 **Peirce**   335
    CHERYL MISAK

50 **Physicalism**   340
    WILLIAM SEAGER

51 **Popper**   343
    JOHN WATKINS

52 **Pragmatic Factors in Theory Acceptance**   349
    JOHN WORRALL

53 **Probability**   358
    PHILIP PERCIVAL

54 **Qualities, Primary and Secondary**   373
    G. A. J. ROGERS

55 **Quantum Mechanics**   376
    RICHARD HEALEY

56 **Quine**   385
    LARS BERGSTRÖM

57 **Ramsey Sentences**   390
    FREDERICK SUPPE

58 **Realism and Instrumentalism**   393
    JARRETT LEPLIN

59 **Reductionism**   402
    JOHN DUPRÉ

60 **Relativism**   405
    JAMES W. MCALLISTER

61 **Russell**   408
    PAUL J. HAGER

62 **Scientific Change**   413
    DUDLEY SHAPERE

63 **Scientific Methodology**   423
    GARY GUTTING

64 **Simplicity**   433
    ELLIOTT SOBER

65 **Social Factors in Science**   442
    JAMES ROBERT BROWN

66 **Social Science, Philosophy of**   451
    ALEX ROSENBERG

67 **Space, Time, and Relativity**   461
    LAWRENCE SKLAR

68 **Statistical Explanation**   470
    CHRISTOPHER READ HITCHCOCK AND WESLEY C. SALMON

69 **Supervenience and Determination**   480
    WILLIAM SEAGER

70 **Technology, Philosophy of**   483
    MARY TILES

71 **Teleological Explanation**   492
    ANDREW WOODFIELD

72 **Theoretical Terms: Meaning and Reference**   495
    PHILIP PERCIVAL

73 **Theories**   515
    RONALD N. GIERE

74 **Theory Identity**   525
    FREDERICK SUPPE

CONTENTS

75 Thought Experiments   528
JAMES ROBERT BROWN

76 Underdetermination of Theory by Data   532
W. H. NEWTON-SMITH

77 Unification of Theories   537
JAMES W. MCALLISTER

78 The Unity of Science   540
C. A. HOOKER

79 Values in Science   550
ERNAN MCMULLIN

80 Verisimilitude   561
CHRIS BRINK

81 Whewell   564
JOHN WETTERSTEN

Index   568

List of Contributors

Peter Achinstein is Professor of Philosophy at Johns Hopkins University, Baltimore.

Yemima Ben-Menahem is in the Faculty of Humanities at the Hebrew University of Jerusalem.

Lars Bergström is Professor of Practical Philosophy and Head of the Department of Philosophy at Stockholm University.

Chris Brink is Professor of Mathematics and Pro Vice-Chancellor (Research) at the University of Wollongong, Australia, and Fellow of the Royal Society of South Africa.

Harold I. Brown is Professor of Philosophy at the University of Northern Illinois at De Kalb.

James R. Brown is Professor of Philosophy at the University of Toronto.

Leslie Burkholder is Professor of Philosophy at the University of British Columbia in Vancouver.

Robert E. Butts (died 1997) was Professor of Philosophy at the University of Western Ontario.

John Dupré is Professor of Philosophy at Birkbeck College of the University of London and Senior Research Fellow in the Department of Sociology at the University of Exeter.

Ronald N. Giere is Professor of Philosophy at the Minnesota Center for Philosophy of Science at the University of Minnesota.

David Gooding is Professor of History and Philosophy of Science and Director of the Science Studies Centre at the University of Bath.

Gary Gutting is Professor of Philosophy at the University of Notre Dame in Indiana.

Paul Hager is Associate Professor of Education at the University of Technology in Sydney, Australia.

Rom Harré is Emeritus Fellow of Linacre College of the University of Oxford and Professor of Psychology at Georgetown University.

Richard A. Healey is Professor of Philosophy and Director of Graduate Studies in the Department of Philosophy at the University of Arizona.

Mary Hesse is Emeritus Professor of History and Philosophy of Science at the University of Cambridge.

Christopher R. Hitchcock is Associate Professor of Philosophy at the Division of the Humanities and Social Sciences of the California Institute of Technology in Pasadena.

Cliff Hooker is Professor of Philosophy at the University of Newcastle in New South Wales, Australia.

Christopher Hookway is Professor of Philosophy at the University of Sheffield.

Colin Howson is Reader in Philosophy at the London School of Economics.

Martin Hughes is in the Department of Philosophy at the University of Durham.

Paul Humphreys is Professor of Philosophy in the Corcoran Department of Philosophy at the University of Virginia.

Muhammad Ali Khalidi is Assistant Professor and Chair of the Department of Philosophy at the American University of Beirut.

Brendan Larvor is Lecturer in Philosophy in the Department of Humanities at the University of Hertfordshire.

Jarrett Leplin is Professor of Philosophy at the University of Chicago.

Peter Lipton is Professor and Head of the Department of History and Philosophy of Science at the University of Cambridge.

James W. McAllister is a member of the Faculty of Philosophy at the University of Leiden.

Ernan McMullin is John Cardinal O'Hara Professor Emeritus of Philosophy at the University of Notre Dame in Indiana.

Cheryl Misak is Professor of Philosophy at the University of Toronto.

Eleonora Montuschi is in the Department of Philosophy at the London School of Economics.

Dugald Murdoch is a member of the Department of Philosophy at the University of Stockholm.

William H. Newton-Smith is Fairfax Fellow and Tutor in Philosophy at Balliol College, University of Oxford.

Thomas Nickles is Professor of Philosophy at the University of Nevada in Reno.

Samir Okasha is the Jacobsen Research Fellow in Philosophy at the London School of Economics.

Kathleen Okruhlik is Associate Professor of Philosophy and Dean of the Faculty of Arts at the University of Western Ontario.

Philip Percival is a member of the Faculty of Philosophy at the University of Glasgow.

John Preston is Senior Lecturer in the Department of Philosophy at the University of Reading.

Christopher Ray is Academic Senior Master at King's College School, London.

John Rogers is Professor of Philosophy at the University of Keele.

Richard Rorty is University Professor of Humanities at the University of Virginia.

Alex Rosenberg is Professor of Philosophy at the University of Georgia.

Wesley C. Salmon is University Professor of Philosophy, Professor of History and Philosophy of Science, and Fellow of the Center for Philosophy of Science at the University of Pittsburgh, and President of the International Union of History and Philosophy of Science, Division of Logic, Methodology, and Philosophy of Science.

Geoffrey Scarre is in the Department of Philosophy at the University of Durham.

William Seager is Professor of Philosophy at Scarborough College in the University of Toronto.

Dudley Shapere is Z. Smith Reynolds Professor of Philosophy and History of Science at Wake Forest University in Winston-Salem, North Carolina.

Lawrence Sklar is Professor of Philosophy at the University of Michigan.

Elliott Sober is Hans Reichenbach Professor of Philosophy and Vilas Research Professor in the Department of Philosophy at the University of Wisconsin, Madison.

Tom Sorell is Professor in the Department of Philosophy at the University of Essex.

Frederick Suppe is Professor of Philosophy and Chair of the History and Philosophy of Science Program at the University of Maryland.

Paul Thompson is Professor of Philosophy and Principal of the University of Toronto at Scarborough.

Mary Tiles is Professor of Philosophy at the University of Hawaii at Manoa, Honolulu.

J. D. Trout is Associate Professor of Philosophy and Adjunct Associate Professor of the Parmly Hearing Institute at Loyola University in Chicago.

John Watkins (died 1999) was Professor of Philosophy at the London School of Economics.

Richard S. Westfall (died 1996) was Professor of History and Philosophy of Science at Indiana University in Bloomington.

John Wettersten is in the Faculty of Social Sciences of Heuchelheim University, Germany.

Gereon Wolters is Professor of Philosophy at the University of Konstanz.

Andrew Woodfield is in the Department of Philosophy, Centre for Theories of Language and Learning, at the University of Bristol.

John Worrall is Professor of Philosophy at the London School of Economics.

Preface

My greatest debt of gratitude is to the patient contributors to this volume. I have been the barnacle on their ship of progress. I hope our readers will agree that this collection has been worth the wait.

In particular I want to thank for their able assistance Katie Jamieson, who launched the project, and Beata Ekert, who saw it through. I am also grateful to a series of editors at Blackwell Publishers. With the passage of time they have become too numerous to mention. I would single out the present editor, Beth Remmes, without whose gentle insistence nothing would have come from something. For help with the proofs I am grateful to James Driscoll.

There are major figures in the history of the philosophy of science who regrettably do not have entries. And there are significant topics that are not treated. It seemed better to advance with less than perfect coverage than to delay still further. I hope that the reader will regard this as a companion. And, like a good companion in life, one will enjoy it for what it provides and not regret what it does not have.

The volume is dedicated to my wife, the journalist Nancy Durham. The pleasure of her support, love, and companionship is greater than could be conceived if it were not for the fact that it is actual. Thank you JA.

W. H. Newton-Smith
Cefnperfedd Uchaf, 17 September 1999

Logical Symbols

In this work the following logical notation is used:

~	negation	"not"
∧	conjunction	"and"
∨	inclusive disjunction	"either — or — or both"
→	material conditional	truth functional "if — then —"
↔	material biconditional	truth functional "if and only"
(∃x)	existential quantifier	"There is . . ."
(x)	universal quantifier	"For all . . ."

Introduction

W. H. NEWTON-SMITH

We think that science is special: its products – technological spin-offs – dominate our lives. Sometimes it enriches our lives; sometimes it impoverishes them or even takes them away. For better or for worse, no institution has had more impact on the character of our existence this millennium than science. Penicillin, computers, atomic bombs make modern life *modern* life.

Science affects our lives profoundly in another way as well. These technological devices derive from sophisticated theories, some rudimentary understanding of which is the hallmark of the educated person. The theories – evolution, quantum mechanics, relativity, and so forth – tell us stories that are sometimes hard to believe or even to understand. But we do come to believe them. And in coming to believe them we form radically different pictures from our ancestors of ourselves, of the world, and of our place in it. The modern person is composed ultimately of subatomic particles with properties that defy comprehension. We trace our origins back to inanimate matter and live in a curved space-time created by a big bang. Nowhere is this hunger to know the scientific picture of the world more manifested than in the remarkable current demand for popular science writings. Stephen Hawking's *A Brief History of Time* (1988), on the list of best-selling books for well over a year, is but one of many attempts to paint the scientific world picture for the layperson.

We are so impressed with science that we give quite amazing credibility to any claim that is successfully represented as being scientific. A mundane example was provided by the Ben Johnson fiasco. Some years ago the entire Canadian population was convulsed with delight the night on which Johnson appeared to have won the Olympic gold medal in the 100-meter sprint. The Prime Minister declared him to be a new model for youth and a symbol around which to build the Canadian identity. At a press conference the next day it was revealed that a scientific test which was probably understood by only a handful of Canadians had shown on the basis of data available to only a couple of scientists that Johnson had been taking steroids. Notwithstanding the nation's desire to believe the contrary, within hours almost all of the 23,000,000 Canadians had changed their minds and concluded that Johnson had cheated: such is the power of science. No other institution in Western pluralistic societies has had such power to engender belief except possibly organized religion and the Communist Party. The Communist Party is finished. And while the influence of religion diminishes as educational levels go up, that of science increases.

Of course science has its detractors. But even they acknowledge its power. Indeed, occasionally science is attacked just because of its power. Michael Dummett (1981)

1

argued that it would have been better if all research in physics had been stopped permanently in 1900, on the grounds that the increase in our understanding of the universe was not worth the cost in terms of the creation of weapons of mass destruction. Even more outlandish critics such as Paul Feyerabend implicitly accorded science due respect. Feyerabend told us that science was a con game. Scientists had so successfully hoodwinked us into adopting their ideology that other equally legitimate forms of activity – alchemy, witchcraft, and magic – had unfairly lost out. He conjured up a vision of much enriched lives if only we could free ourselves from the domination of the "one true ideology" of science just as our ancestors freed us from the domination of the one true church (Feyerabend 1975, p. 307). He told us these things in Switzerland and in California, happily commuting between the two in the ubiquitous product of science – the airplane.

In light of the overwhelming success of science and its importance in our lives, it is not surprising that, at the millennium, when the vast majority of scientists who have ever lived are alive, the philosophy of science is a growth area. We want to understand what science is and how it works. Of course there were great nineteenth-century philosophers of science, Mill and Whewell. But within the discipline of philosophy they were a novelty among the hordes of epistemologists and metaphysicians. Today virtually all philosophy departments in the English-speaking world have their philosopher of science. The undergraduate course in philosophy of science is a staple. Books, conferences, journals, and professional associations abound.

But what is the philosophy of science? To answer this tricky question, perhaps we should turn to definitions of "philosophy" and of "science." But the characterization of philosophy is one of the most vexed of all philosophical questions. Periodically philosophy goes through a period of navel gazing in which this question is to the fore as it was in the heyday of Wittgenstein's influence. The profession eventually exhausts itself without reaching a consensus and moves more fruitfully back to focusing on particular philosophical questions. And what is science? Once upon a time it was fashionable to attempt neat answers to this one. The logical positivists defined science in terms of what was cognitively meaningful. Sentences other than definitions were cognitively meaningful just in case they could be verified by experience. Science is then coextensive with cognitively meaningful discourse! The discourses of ethics and aesthetics were not scientific. They were not even meaningful. And Popper defined a theory as scientific if it could be falsified. But neither of these definitions even fitted all of physics, which they took to be the paradigm science.

The dominant tendency at the moment is to reject the question (see Rorty 1991). Science has no essence. We have constituted our idea of science around a list of paradigm exemplars (including biology, chemistry, geology, medicine, physics, zoology) of particular disciplines. Let us call this our "shopping list" of sciences. There is debate over which other disciplines belong on the list. But it is best to start conservatively with a core on which we can agree. The disciplines on our list differ significantly. Rather than look for an essence that could be enshrined in a definition of science, we should appreciate the similarities and the differences between them. We can then consider similarities and differences with other disciplines – anthropology, economics, political science, sociology, and so on – and debate their significance. However, having had that debate, deciding just how far to extend the word "science" will not be a substantial matter.

2

The prospects of giving an enlightening characterization of the philosophy of science in terms of its constituent notions are bleak. It will be more efficient to consider what those who call themselves philosophers of science actually care about and do. There is no hiding the fact that they are an eclectic lot who do a diverse range of things, some of them strange. Still, a rough map of the terrain is possible. The map that is sketched in what follows also introduces many of the issues treated in this Companion. Given the cultural context in which philosophy of science is located, it is not surprising that a main point of departure for much of its activity is the special success of science. At what is science especially successful?

Manifestly the sciences on the shopping list give us the power over nature that Frances Bacon sought (Bacon 1905, p. 300). Is that all there is to it? Are theories just tools for predicting and manipulating the world? To treat them as just tools would mean that we could no longer cherish the stories of the scientist of particles that can be here and there at the same time, of fluctuations in a vacuum producing the big bang, of time stopping in black holes. We need to be told: are these stories just to help us predict or are they really true? Is it really like that? For the instrumentalist the answer is "no." For the realist the story that science tells about quarks and their charm is to be taken literally. What is more, claims the realist, we have good reasons to think that our best stories are at least approximately true.

This controversy about the aim of science takes us to explanation. Why should we care whether the stories of science are true? If they enable prediction and manipulation, what point would there be in knowing whether the stories are actually true? Because we have a deep-seated desire to understand. For this we seek explanations, which, to give real understanding, must be based on true or approximately true theories, or so one version of the story goes. Duhem, the French physicist and philosopher, famously thought that science could not achieve truth, and hence could not provide explanations. Rather touchingly, he advised readers seeking explanation to turn from physics to metaphysics (Duhem 1962, p. 10).

The plot thickens. Why does the realist think that there is truth in some of our theories? Typically, he argues for this by appeal to the power of the theory to explain. To take an example, the atomic theory was very controversial in the early part of the twentieth century. No one had seen an atom, so why should we think that there were any? Einstein argued that we should believe the atomic theory since positing the existence of atoms would explain Brownian motion. This is the apparently random motion we see of pollen particles immersed in water. If water were composed of atoms in motion, collisions between the atoms and the pollen particles would explain their motion. If it is explanatory power that is supposed to lead us to quarks and black holes, we had better know what an explanation is and how to assess its power. And, crucially, we need some reason for thinking that explanatory power really is a sign of truth. Our ancestors explained the adverse features of their lives by reference to the anger of the gods. What makes our stories any better? And could it be that radically different theories might be underdetermined in the sense that each could explain whatever the other can? Would that mean victory for the instrumentalist?

Interestingly, there are many realists but few instrumentalists. However, even if there were no instrumentalists, we would need to invent one. For the realist is asking us for such a major commitment that we need to be sure of our grounds. This debate

3

about the aim of science is one of the major issues in the philosophy of science today, and many of the contributions that follow bear directly or indirectly on it.

The realist and the instrumentalist agree that science delivers prediction and control. The realist sees success at this level as evidence for success at a deeper level, at the level of telling true stories about the underlying entities and structures responsible for what we observe. Consequently, they can set aside this disagreement to some extent and consider how science is able to achieve prediction and control. The classical answer is by using the scientific method. Scientists gather evidence, advance hypotheses, and bind themselves to using the scientific method in deciding between them. Unless we have some such special tool for making these decisions, how is it that science could make such impressive progress? Much work in the history of the philosophy of science has been expended attempting to characterize the scientific method.

In philosophical fiction there is a character, Algorithor, who searches for the One True Method: an algorithm for theory choice. Ernest Nagel claimed in the 1950s that our "basic trouble" was that "we do not possess at present a generally accepted, explicitly formulated, and fully comprehensive schema for weighing the evidence for any arbitrarily given hypothesis so that the logical worth of alternative conclusions relative to the evidence available for each can be compared" (Nagel 1953, p. 700). Algorithor sought a mechanical procedure for blowing away all but the most promising hypotheses. Even Algorithor did not aspire to a mechanical technique for generating hypotheses or theories in the first place. That was a matter of discovery: the province of conjecture, intuition, and hunch. His was the province of justification. Once hypotheses are advanced, he sorts them out.

Algorithor failed in his quest. The method has not been found in spite of the valiant efforts of Mill, Whewell, Popper, the logical positivists, and a host of others. The years of failure have taken their toll, and most philosophers of science would agree that the quest should be called off. If by "scientific method" we mean some such algorithm, there is no such thing. While many logical positivists searched for an algorithm, one of them, Neurath, noted the futility of this: "There is no scientific method. There are only scientific methods. And each of these is fragile; replaceable, indeed destined for replacement; contested from decade to decade, from discipline to discipline, even from lab to lab" (Cartwright et al. 1996, p. 253).

No philosopher of science has played a greater role in vanquishing Algorithor than Thomas Kuhn (1970). His *Structure of Scientific Revolutions* begins: "History, if viewed as a repository for more than anecdote or chronology, could produce a decisive transformation in the image of science by which we are now possessed" (Kuhn 1970, p. 1). Whether his model of science is tenable is much disputed. What is undisputed is that he did direct us to the history of science, which shows that theory choice is not the simple-minded affair pictured by Algorithor. The historical turn he effected in the philosophy of science can be seen on many of the pages that follow.

That we no longer seek an algorithm does not mean that there are no questions about scientific method. That topic is alive and well. Many of the chapters in this Companion touch on it directly or indirectly. Science attends to the evidence. But what is scientific evidence? Is it ultimately just observation and experimentation? Or does the evidence go beyond what we observe? Can we, for instance, justify one hypothesis over another on the grounds of its greater simplicity? Do we observe the world as it really is, or do our theories color what we see? Is observation the only way to find things out, or can we make discoveries about the world by doing thought experiments? What is the relation between discovery and justification?

These and a host of other issues are the legitimate legacy of Algorithor's abandoned quest. It is a messy but exciting business. It is messy because the range of considerations which scientists cite to justify their choices is wide and diverse. Kuhn offered a partial list, including accuracy, consistency, scope, simplicity, and fruitfulness. Sometimes choice may even come down to making judgment calls for which the scientist cannot articulate a reason. It is exciting because real and difficult questions abound. To take but one example, we prefer simple theories to complex ones. Does their simplicity give us a reason to think they are more likely to be true or just more likeable? How could we show that simplicity is a guide to the truth? We could if we could show that the world is ultimately simple. But how could we ever show that except by showing that simple theories tend to be true? And what, after all, is simplicity? Is there an objective way of determining whether one theory is simpler than another? Or is simplicity something that dwells in the beholder, not in the theory?

Much philosophy of science concerns issues of aim and methods. In the course of pursuing our aim (whatever it may be) using our rich range of methods, we have crafted descriptive and explanatory tools common to the different sciences. We talk of explanations, idealizations, laws, models, probabilities, theories, and so forth. Much of philosophy is like bicycle riding. Many of us can ride a bicycle. Few of us can give a verbal description of what it is we do to keep the bicycle upright. Just as we can use a bicycle, so we can use concepts. For instance, we use the concept of time all the time. We use tenses, we measure time, we assess the past, and we plan for the future. But if asked what time is, we are at a loss, like Augustine: "What then is time? If no one asks me, I know: if I wish to explain it to one that asketh, I know not" (Augustine 1907, p. 262). One of the tasks of philosophy is to articulate the context of familiar concepts. The philosophy of science is no exception. We can give examples of explanations. We would like to go beyond this and give a general characterization of what it is that makes our examples of explanations explanatory. Many contributors to this volume seek to cast light on these familiar tools of science.

This might seem a rather boring task more suited to the lexicographer. But, intriguingly, the attempt to explicate these and other concepts takes us into deep metaphysical and epistemological waters. Consider the notion of a law. Is a law just a true description of a regularity? There are regularities in the world. Bodies released near the Earth's surface fall. Lumps of gold near the Earth's surface are less than 5,000 kilograms in weight. What is the difference between these? The first regularity is a matter of law. It has to hold. The second could be broken. What do we mean by "has to hold"? What in the world confers this special status on that regularity? And if some regularities cannot be broken, how do we discover this? Some philosophers have found all answers to this particular question so unconvincing that they conclude that there is no content to the notion of a law of nature (van Fraassen 1989). It is merely a legacy from those like Newton who sought to discover laws, God's laws, in the workings of the world.

In addition to considering the aims, methods, and tools of science, the philosopher of science is interested in the products of science, the contents of its theories – in particular, those very general, powerful theories which offer pictures of the world radically at odds with our commonsense view of things. The scientist is interested in the challenges that a theory may present for deep-seated views about the world (does quantum indeterminacy show that causality is not operative?). He is also interested in the impact of theories on traditional philosophical issues. Leibniz and Newton disagreed as to whether space and time would exist if there were no bodies or events (see Alexander 1956). The theory of relativity replaces space and time as separate items by space-time. Would space-time exist if there were no matter and radiation? Many of these interpretative issues are specific to particular sciences and would receive more extensive treatment in Companions to particular sciences. However, in the face of their great importance, interest, and relevance to other issues, some are treated in this volume.

Science is a vast social institution governed by very tight social conventions. Until Kuhn, philosophers paid virtually no attention to the social dimension of science. In Popper and the logical positivists you find individual scientific heroes like Newton and Einstein, but never a society of them. Having recognized the social dimension at last, some writers (more commonly sociologists than philosophers) went to the extreme of treating science as just a social institution. Which theories triumphed in science were determined not by scientific methods conducive to the search for truth, but by social factors. In part this sociological turn was inspired by Kuhn's treatment of scientific revolutions as analogous to political revolutions. Theories before and after revolutions were incommensurable; that is, there was no common measure to be used in evaluating them. So, as in political revolutions, force, not rationality, determines the outcome. This extreme view (arguably not Kuhn's considered view, but certainly inspired by what he wrote) is not tenable. For it would be a total mystery why theories selected by the operation of social forces should continue to give us ever more power over nature. Several of the contributors seek to come to grips with the Kuhnian challenge and to explicate the proper role to be accorded to social factors in science.

Part of the explanation of the success of science lies in its institutional character (see Newton-Smith 1995). We have crafted an institution well designed to deliver the goods. We had to discover how to do this. Francis Bacon, writing in 1624 the seminal essay on the social character of science in the form of a fable, argued that progress required a social organization which he called the "College of Six Days Work" (Bacon 1905, p. 712). He got some things right and some things wrong. He argued that progress would require financial support from the state. But he also argued for secrecy. The members of the College were to take an oath not to reveal results to outsiders. We have discovered the importance of putting scientific work in the public domain and have developed a complex apparatus of rewards and promotions to encourage this. There is need for a constructive partnership between philosophy of science and the sociology of science. For if we really want to understand scientific progress, we need to investigate the contributions of social mechanisms. This volume reflects current discussion in the philosophy of science, and this particular topic, which Alvin Goldman has called "social epistemics," has yet to take its proper place within philosophy of science. I would urge the interested reader to look at his important book *Knowledge in a Social World* (Goldman 1999).

The philosophy of science seeks to understand the aims, methods, tools, and products of science. It focuses on the social and historical dimensions of science insofar as these are relevant to such concerns. What does it add up to? Scientists in particular sometimes express disappointment in philosophy of science. They had the false expectation that the philosophy of science would itself serve to advance the scientific process. Sometimes it does, as some articles illustrate. But the expectation that philosophy could provide major assistance to science probably makes sense only under the assumption that Algorithor's project was feasible. Philosophy of science aims to give us a certain kind of understanding without necessarily making us better at doing what it gives us an understanding of. An analogy will bring this out. A competent speaker of English can determine whether any sequence of words is a sentence of English or not. We recognize that this sequence of words is a sentence of English and that the following sequence is not. Tiger quark running green lovely. Linguists seek a system of mechanical rules, which would generate all and only those sequences of words that are sentences of English. The development of such a set of rules would be an exciting step in explaining our capacity to discriminate between sentences and nonsentences. It would not make us any better at doing this. We are competent at this. Philosophy of science has value through increasing our understanding of science without necessarily making us better scientists. That is not the game.

The craft of science gives rise to questions that science is powerless to answer. What is a theory? What is an explanation? Should other disciplines be added to our original minimal shopping list? Science is an awesome tool for answering scientific questions. But science gives rise to questions the answering of which requires techniques that a scientific training does not provide. There is no laboratory for investigating the nature of theories. We do not discover the nature of explanation by doing experiments. If we want answers to these questions, we have to turn to philosophy. Of course, answering some of our questions requires a partnership between philosophy and science. For instance, to what extent has determinism been refuted? Whether or not the reader finds the answers offered in this volume convincing, I hope that he or she finds the attempts interesting and stimulating. We might now venture an overall generalization of the philosophy of science. The philosophy of science seeks to answer those questions prompted through doing science for which scientific techniques are not adequate by themselves. Perhaps this characterization brings some historical and sociological issues within the purview of the philosophy of science, but that is no bad thing.

Whether or not the answers are, or ever can be, fully convincing (in the way that some particular scientific results might purport to be) is another matter. And there will always be those frustrated by the nebulous character of much philosophical writing who will argue that these questions are illegitimate. But at least the exploration of them is relatively harmless, and that is something very positive.

Our map of the philosophy of science – aims, methods, tools, and products – is illustrative rather than comprehensive, and there are issues discussed in this Companion that have not been touched on here. The topics in the philosophy of science are intimately connected. For that reason there is overlap in the discussions in the following chapters. I have let that stand, as contributors differ among themselves. And where they do not differ substantially, viewpoints are illuminated by being put in different ways.

There are other questions about which no one is thinking enough. Science is an expensive business. Its products are impressive, but often costly. Is our current level of commitment to science in terms of money and persons the most appropriate use of our resources? Scientists like to think of themselves as embodying a high degree of rationality. But it is hardly rational to pursue ends without considering costs. Philosophy of science has a duty to stimulate a debate on the crucial issue: should science continue to have the same place in our culture in the new millennium that it currently has? I hope that these articles will contribute to our understanding of science and encourage us to face this really big issue in a more informed way.

References

Alexander, H. G. (ed.) 1956: *The Leibniz–Clarke Correspondence* (Manchester: Manchester University Press).

Augustine 1907: *Confessions*, trans. E. B. Pusey (London: J. M. Dent & Sons).

Bacon, F. 1905: *The Philosophical Works of Francis Bacon*, ed. J. M. Robertson (London: George Routledge and Sons).

Cartwright, N., Cat, J., Fleck, L., and Uebel, T. E. 1996: *Otto Neurath: Philosophy between Science and Politics* (Cambridge: Cambridge University Press).

Duhem, P. 1962: *The Aim and Structure of Physical Theory* (New York: Atheneum).

Dummett, M. A. E. 1981: Ought research to be restricted? *Grazer Philosophische Studien*, 12/13, 281–98.

Feyerabend, P. 1975: *Against Method* (London: New Left Books).

Goldman, A. 1999: *Knowledge in a Social World* (Oxford: Clarendon Press).

Hawking, S. 1988: *A Brief History of Time* (New York: Bantam).

Kuhn, T. S. 1970: *The Structure of Scientific Revolutions*, 2nd enlarged edn (Chicago: University of Chicago Press).

Nagel, E. 1953: The logic of historical analysis. In *Readings in the Philosophy of Science*, ed. H. Feigl and M. Broadbeck (New York: Appleton-Century-Crafts), 688–700.

Newton-Smith, W. H. 1995: Popper, science and rationality. In *Karl Popper: Philosophy and Problems* (Cambridge: Cambridge University Press), 13–30.

Rorty, R. 1991: Is natural science a natural kind? In *Objectivity, Relativism and Truth* (Cambridge: Cambridge University Press), 46–63.

van Fraassen, B. 1989: *Laws and Symmetry* (Oxford: Oxford University Press).

1

Axiomatization

FREDERICK SUPPE

Axiomatization is a formal method for specifying the content of a theory wherein a set of *axioms* is given from which the remaining content of the theory can be derived deductively as *theorems*. The theory is identified with the set of axioms and its deductive consequences, which is known as the *closure* of the axiom set. The logic used to deduce theorems may be *informal*, as in the typical axiomatic presentation of Euclidean geometry; *semiformal*, as in reference to set theory or specified branches of mathematics; or *formal*, as when the axiomatization consists in augmenting the *logical axioms* for first-order predicate calculus by the *proper axioms* of the theory. Although Euclid distinguished *axioms* from *postulates*, today the terms are used interchangeably. The earlier demand that axioms be self-evident or basic truths gradually gave way to the idea that axioms were just assumptions, and later to the idea that axioms are just designated sentences used to specify the theory.

Axiomatization played a central role in positivistic philosophy of science, which analyzed theories as axiomatic systems. (See THEORIES.) Initially these systems were construed as purely *syntactical systems* which invoked no pre-axiomatic "meanings" in specifying the content of the theory. On this view the axiomatization *implicitly defined* the key terms of the axiomatization (see CRAIG'S THEOREM; DEFINITIONS). As positivists came to clearly distinguish syntax from semantics and to develop formal semantics (which specify referential or extensional meanings) for their logical systems, they would sometimes augment their syntactical axiomatizations with formal semantic conditions. For example, in their analysis of theories, a formal semantics would be specified for the observational vocabulary, but not the theoretical vocabulary.

Axiomatization is one of the main techniques used for philosophical analysis by *formalization*. It has become fashionable to distinguish between *semantic approaches* and *syntactical approaches* to formalization, with positivistic axiomatization being the syntactical paradigm. On semantical approaches one identifies an intended class C of systems or instances, then presents a formal structure S (a mathematical entity such as a configurated state space), and asserts a mapping relation R which holds between S and C. In effect, the analysis presents more precise structures S which, it is asserted, the less precisely specified C exemplify. Typically, S is presented as some standard kind of mathematical entity, such as a Hilbert or vector space or an algebra. Occasionally (e.g., Suppes 1967) S is specified axiomatically, indicating that axiomatization is not the exclusive possession of syntactical approaches.

On syntactical approaches to analyzing intended class C of systems or instances, one presents a set A of sentences (typically as the closure of one's axioms) and then asserts that *C* is among the *metamathematical models* (intended interpretations) which *satisfy* (instance) *A*. The difficulty is that the instances of *A* also include a number of wildly *unintended models* or systems, which, when discovered, are held up as counterexamples to the analysis of *C* provided by *A*. This could be precluded by producing a *representation theorem* proving that all the models of *A* are isomorphic to *C* – which is impossible if *C* is infinite (Löwenheim–Skolem theorem). Such representation theorems are virtually absent from syntactical philosophical analyses. Semantic approaches avoid this difficulty by starting with *C* and then relating it to *S*, thereby limiting consideration to just the intended systems *C*.

Many of the philosophical controversies surrounding positivistic axiomatic and other syntactical analyses of confirmation, laws, theories, etc. concern unintended models offered as counterexamples to the analyses. The susceptibility to these counterexamples is artifactual of the syntactical approach to formal analysis, and so such controversies tend to coalesce on matters tangential to the philosophical illumination or understanding of such notions as confirmation, laws, theories, and the like. The semantic approach preempts this species of artifactual diversion by starting with just the intended systems – a fact positivists themselves were beginning to realize towards the end, when, as in their notion of partial interpretation, they resorted to semantical notions to avoid unintended models (see Suppe 1974, sec. IV-C).

Formal methods within philosophy are controversial, but much of the controversy can be laid to rest, since it concerns the above shortcomings peculiar to syntactical approaches. But even the semantical approach can lead to wretched philosophical analyses. This will be so if the specifics of *S* are not rooted in careful sensitive informal philosophical analysis, textual exegesis, empirical facts, and the like. (These are required to establish that *S* in fact stands in the asserted mapping relation *R* to *C*.) It will also be so if the formal analysis consists in just a litany of formal definitions that lead to no new insights. As a rough rule of thumb, an adequately grounded formalization or axiomatization will be superfluous unless it is used to prove theorems that could not otherwise be obtained readily or to support claims that would not be convincing if defended only informally.

Turning to science itself, axiomatization occurs only infrequently, and then usually in foundational studies of well-developed theories. The classic example is von Neumann's axiomatization of quantum mechanics, which showed that wave and matrix mechanics were the same theory. (See THEORY IDENTITY.) Such scientific axiomatizations typically are informal or semiformal, and are best understood as instancing the semantic approach. Despite occasional successes (e.g., Williams 1973), attempts at predicate calculus axiomatizations of substantive scientific theories (e.g., Carnap 1958) have generally been unsuccessful, capturing only the more elementary parts of the theory before becoming excessively complex.

The rapidly increasing computerization of science provides new scope for axiomatization. The connections between digital computers and symbolic logic are sufficiently deep that computer programs can be construed as a form of axiomatization: Program *P* axiomatically specifies automaton system *S* which is embedded (mapped *R*) into computer *C* running the program. It is increasingly common in science to "axiomatically" present theories and models via programming. Fledgling efforts to do the same within philosophy of science are beginning. In all cases these efforts are best

10

understood as instancing a semantic approach to formalization, rather than a syntactical approach.

References and further reading

Carnap, R. 1958: *Introduction to Symbolic Logic and its Applications*, trans. W. H. Meyer and J. Wilkinson (1st pub. 1954; New York: Dover Publications), part II.

Hempel, C. G. 1974: Formulation and formalization of scientific theories. In *The Structure of Scientific Theories*, ed. F. Suppe (2nd edn 1977; Urbana, IL: University of Illinois Press), 244–65.

Henkin, L., Suppes, P., and Tarski, A. (eds) 1959: *The Axiomatic Method with Special Reference to Geometry and Physics* (Amsterdam: North-Holland).

Suppe, F. 1974: The search for philosophic understanding of scientific theories. In *The Structure of Scientific Theories*, ed. F. Suppe (Urbana, IL: University of Illinois Press; 2nd edn 1977), 3–241, secs. IV-C, F.

Suppes, P. 1957: *Introduction to Logic* (New York: Van Nostrand), part II, esp. ch. 12.

—— 1967: What is a scientific theory? In *Philosophy of Science Today*, ed. S. Morgenbesser (New York: Basic Books), 55–67.

—— 1968: The desirability of formalization in science. *Journal of Philosophy*, 65, 651–64.

van Fraassen, B. C. 1989: *Laws and Symmetry* (New York: Oxford University Press), ch. 9.

von Neumann, J. 1955: *Mathematical Foundations of Quantum Mechanics* (Princeton: Princeton University Press).

Williams, M. B. 1973: Falsifiable predictions of evolutionary theory. *Philosophy of Science*, 40, 518–37. (This and an earlier article of hers building on pioneering work by Woodger are widely regarded as among syntactical axiomatization's real successes.)

2

Berkeley

M. HUGHES

Berkeley was a bishop and a defender of orthodox Christianity in an age when science was beginning to be claimed as an ally by those who called themselves "freethinkers": people who wanted to modify religion and to discard awkward dogmas, or who might even be drawn towards atheism.

Berkeley's philosophy revolves around his attempt to prove the existence of God by a new argument. This argument starts by claiming that our perceptions exist only insofar as they are perceived, and being perceived is a passive state. Therefore perceptions have no power.

It may be supposed that perceptions are related to material objects; but this supposed relationship cannot be understood unless it involves resemblance or likeness between perception and object. But there is, he argues, no likeness except between perceptions; indeed, nothing could be more unlike than something perceived and something unperceived. Therefore, since there is no likeness, the supposed relationship cannot be understood at all, and hence does not exist. Thus the very concept of material objects is to be rejected; we cannot use any concept which cannot be related to our perceptions, the sole basis of our information and knowledge. Thus we cannot say that the power which causes our perceptions resides in material objects.

But some power causes our perceptions, since we have very little choice about what we perceive. If that power does not reside in matter or in the perceptions themselves, it must reside in a mind. Since we are minds and are active, we do at least know that minds exist and have powers. The power which creates our perceptions is the power which is supreme in our world and which, evidently, operates by steady and coherent rules (since our perceptions form a steady and coherent whole). Berkeley identifies this power with the God of Christendom.

No mistake has caused more harm in Western philosophy than the unexamined assumption that the culturally familiar idea of God is a coherent idea, instantly ready for philosophical use. Berkeley makes this mistake and, despite the fact that many of his arguments are brilliant and convincing, pays in the end an alarming price for making it.

This price is paid in several installments, as Berkeley is drawn into several controversies with a scientific aspect. His first published work, *Essay towards a New Theory of Vision*, superbly challenges Isaac Barrow and others in their theory of optics. Our ability to grasp the three-dimensionality of the world, Berkeley argues, is not just a mechanical process; it depends not only on receiving, but also on collating and interpreting, visual data. Here Berkeley lays the first foundations of his argument that the world is a meaningful whole, challenging us to interpret and understand its meaning.

But perhaps Berkeley willfully ignores the evidence which is on Barrow's side and which points to the fact that a merely mechanical difference – that is, the difference between seeing with two eyes and seeing with only one – has a significant impact on our depth vision, and thus on our ability to perceive the three-dimensional world. Berkeley challenges science on its own ground; in this respect, most philosophers have declined to follow him.

In his key text, *Principles of Human Knowledge*, Berkeley denies Locke's suggestion that there are two kinds of quality: primary and secondary. According to this suggestion, primary qualities, belonging to a material world, cause both genuine ideas of themselves in our minds – for instance, our perceptions of shape and texture – and also a secondary kind of ideas which do not belong to the material world in the same way.

Color, according to Locke, is a secondary quality: light is absorbed and reflected differently by differently textured surfaces; but we do not see this process, or the forms (wavelengths) of light which the process involves, strictly as they are. Instead, ideas of color, with each color corresponding to a form of light, are created in our minds.

The beautiful colors of the world, retorts Berkeley, are thus reduced to a "false imaginary glare." He argues that Locke is wrong, both because the primary and secondary qualities are connected with each other by inextricable logical bonds – we could not see shapes if we could not distinguish color masses – and because our perception of primary qualities is just as subject to mistakes and illusions as our perception of secondary qualities is. The two kinds of qualities are, Berkeley insists, manifestly alike.

But Locke's theory has a foundation in scientific ways of thinking, and therefore in most ways of thinking which belong to the modern world. There could hardly have been any modern investigation of colors had not the idea of the dependence of color on light wavelengths been accepted. Berkeley has, therefore, to say that the modern way of thinking is mistaken.

Berkeley is indeed convinced that efforts to modernize thought and to render it more scientific are a disastrous mistake, for they obscure the fact that God is immediately and graciously present in our lives. This great truth is hidden behind the petty details of modern learning, which he refers to contemptuously as "minute philosophy." The conflict between Christianity and "minute philosophy" is comprehensively treated in Berkeley's long dialogue *Alciphron*.

Berkeley believes that one profound mistake of modern philosophers is to say that color depends on wavelengths of light, or that heat depends on the motion of particles, or that the pain of an unpleasant burn depends on heat, as if all these were statements of cause and effect. His own basic argument prohibits this mistake: the world is a set of passive perceptions, not of active causes. Thus, he argues, there is only one true cause in the world, and that is God. The causes which we may think that we observe in the world are not true causes at all: they are merely "signs and prognostics" of future experiences. The heat of a flame warns us, under God's providence, that we may be burnt painfully, but the flame and its heat cannot cause the pain; only God can cause anything.

Here Berkeley pays a second installment of the price mentioned above: he asks people to move from the familiar Christian idea that God is the *supreme* cause to the strange idea that God is the *sole* cause. But as a reward, he gains powerful additions to his set of theories.

The first gain is the theory of science as a study of signs, a study validated not because it reveals hidden causes, but simply because it makes true predictions. We are accustomed to talk of "scientific realism" versus "instrumentalism," and to assign Berkeley to the "instrumentalist" sect. But to say that Berkeley saw science "merely as an instrument" is unfair.

The unfairness emerges if we consider that Berkeley saw the world both as real and as a set of signs. Indeed, the second gain from his theory of causation appears here: his theory, if valid, supports the contention, frequent throughout his writings, that the world is God's work of art, and that our perceptions are God's language. Thus the dependencies which exist among perceptions are the dependencies which exist among the different elements of a complex, meaningful statement. Thus science is not so much an instrument as a revelation: from science we learn that the nature of reality is to be significant and to communicate meaning. The richness and power of Berkeley's belief that the world is a text and that God is its author are beginning to be appreciated in modern commentary.

Berkeley's third gain is to offer a new philosophical exposition of ancient Christian ideas: his whole work is a kind of commentary on the Johannine claim (John 1: 1–2) that the Word was in the beginning with God and that without the Word "was nothing made."

Berkeley's theology drives him into conflict with Newton, who (following both Romans 4:17 and the Egyptian mystic Hermes Trismegistus) held to an idea of creation as the conferment of full reality on preexistent mathematical forms which are present both in space and in God's own being, where all forms are perfectly and fully realized. In this idea of creation we can see two roots of the conflict between Newton and Berkeley. First, Newton is committed to the idea of forms which are open to mathematical understanding but are not, or are not yet, physically real or perceptible. Second, God is not an author, intimately involved with his text, but an austere ruler, set apart by his unique perfection. Newton found this view of God in the Bible; thus he was drawn to the Arian heresy which denies the orthodox Christian view that God was incarnate in Jesus Christ. The "minute philosophers," who were (in Berkeley's view) distrustful of all religious tradition and authority, were pleased that someone with Newton's reputation for objectivity and accuracy had rejected orthodoxy.

In the *Principles*, *De Motu*, and elsewhere, Berkeley denies that there are forms of motion which can be understood only by reference to something not physically real or perceptible – the famous example is the "centrifugal" motion which occurs in a mass which is subject to rotation. It is not easy to explain this motion by reference to the relationship between the mass and its container; it is when the relative motion between the mass and its container is least that the mass is most deformed (as anyone can see by spinning a cup of water). Berkeley (along with Leibniz) founded a school of thought which says that centrifugal motion can still be treated as relative: it must be relative to some other mass, say that of the Earth or the fixed stars. We can stay well clear of Newton's world of unrealized mathematical forms.

In the *Principles*, *The Analyst*, and elsewhere, Berkeley insists that the ancient problem of squaring the circle has not been solved by reference to 'infinitesimals" – that is, by the idea that quantities can increase or decrease in amounts which are too small for perception but which can be understood by mathematical thought. If Newton had indeed solved this ancient problem, he had done so, Berkeley argues, simply by showing us how to balance errors. Berkeley argues that Newton does not proceed with complete accuracy and objectivity; instead, he adjusts his method to get a desired result.

Berkeley argues that all science requires a procedure for balancing errors: this is how knowledge progresses. Christian theology, he contends, is itself a science, a reading of the signs of the presence of God. Its elaborate dogmas resulted from a process in which errors were gradually and carefully balanced, just as in any science. Thus Berkeley achieves a theory of the power of our human minds: they are empowered to understand more and more of God's text. Minds are not only active, but also progressive: they gain in knowledge and in effective power.

Here Berkeley comes to pay the last and most alarming kind of installment of the price for his use of religion in philosophy. It is his view that every perception is directly caused by God: what, then, if someone is hurt or unjustly treated by another? What if scientific knowledge, the fruit of mental progress, is used for evil purposes? It seems either that the evil-purposed mind is controlled by God or else that it does, at least to some serious extent, have control over God. The first of these alternatives makes evil part of God's nature, the second opens that way not to science, but to magic: human control of a supernatural power.

It is in part the measure of Berkeley's greatness as a philosopher of science and in part the measure of his failure that he is drawn towards belief in magic. His last work, *Siris*, envisages philosophers finding a panacea for the cure of disease and generally breaking down the barriers of nature. He is really accepting the logic of his argument that the world is a text; but, as some modern thinkers have seen, the meaning of a text is influenced by interpreters as well as by its author, which means that God comes to share his power with us and in a way to depend on us. So Berkeley, like Newton, is drawn, partly by the influence of the magician Hermes Trismegistus, into a departure from orthodox Christianity.

Yet, in a way his whole philosophy depended on the idea that orthodox Christianity, the bedrock of English-speaking culture, provides a clear and acceptable idea of God and of divine power. In reality the idea of supreme or divine power is very complex and difficult, which means that Berkeley's philosophy contains difficulties that it never properly addresses. On the other hand, his later writings, for all their strangeness, are among the earliest to perceive that science, since it seems to offer progress without limit, suggests that we may advance beyond humanity and may "play God." The problems associated with this perception remain unsolved.

References

Works by Berkeley
1709: *Essay towards a New Theory of Vision.*
1710: *A Treatise Concerning the Principles of Human Knowledge.*
1713: *Alciphron in Three Dialogues between Hylas and Philonous.*

Works by other authors
Brook, R. J. 1973: *Berkeley's Philosophy of Science* (Amsterdam: Nijhoff).
Hughes, M. 1982: Absolute rotation. *British Journal for the Philosophy of Science*, 32.
—— 1992: Newton, Hermes and Berkeley. *British Journal for the Philosophy of Science*, 43.
Jesseph, D. M. 1993: *Berkeley's Philosophy of Mathematics* (London: University of Chicago Press).

3

Biology

PAUL THOMPSON

Philosophical reflection on biological phenomena and knowledge has a long history. Indeed, some topics of contemporary interest have a history dating back at least to Aristotle (e.g., teleology, the nature of laws, classification (taxonomy), and the nature and logic of explanation and causality). The modern period of philosophy of biology, however, was ushered in by Morton Beckner with the publication of his landmark book *The Biological Way of Thought*. In it, Beckner employs the analytic tools and methods of contemporary logic, epistemology, and metaphysics to examine many of the issues outlined above.

Two major events in the recent history of biology – that is, within the last 150 years – have given rise to new issues and new perspectives on perennial issues. The first of these was the introduction by Darwin of a conceptually rich theory of evolution. The key elements of Darwin's theory were variability (differences among organisms in a population), heredity, natural selection, and mutability of species (over time, organisms originally classified as belonging to one species could give rise to distant offspring that would be classified as belonging to another species). Darwin, in *The Origin of Species* (1859), provided extensive conceptual accounts of, and empirical evidence for, the latter two elements. Detailed accounts and evidence for the first two had to wait until the twentieth century.

In the early part of this century, Darwin's theory was developed by a number of individuals – most significantly, Ronald A. Fisher, Sewall Wright, and John B. S. Haldane – into the Modern Synthetic Theory of Evolution (MST). Two key issues that have emerged from MST are the concept of selection (and its relation to fitness) and the determination of the fundamental units of selection (populations, organisms, genes – and later molecules).

The second major event was the discovery, by James Watson and Francis Crick in 1953, of the chemical structure of DNA – the molecular basis of genetics. A topic that was invigorated by this discovery was reductionism (biology to chemistry, and population genetics to molecular genetics). Another issue flowing from this discovery was the appropriate definition of "gene" (should it be defined in terms of its function, its structure, or its effect?). As indicated in the last paragraph, this discovery also affected the debate over the units of selections.

Contemporary issues

The structure of theories: the received view

For most of this century the discussion of science has taken place within a framework that Hilary Putnam in 1962 dubbed "The Received View." The received view conception of scientific theories has its roots in logical positivism (see Suppe 1977 for an excellent exposition and historical account of this conception). Its influence on the philosophy of biology has been profound (see Ruse 1973; Hull 1974; Rosenberg 1985; Sober 1984a). The essence of this conception can be stated simply: a scientific theory is an axiomatic deductive structure which is partially interpreted in terms of definitions called "correspondence rules." Correspondence rules define the theoretical terms of the theory by reference to observation terms.

An axiomatic-deductive system consists of a set of deductively related statements (sentences), the structure of which is provided by mathematical logic. In the case of a scientific theory, the statements are generalizations (laws), a small subset of which are taken as the axioms of the theory. The axioms are laws of the highest generality within the theory. They constitute a consistent set, no one of which can be derived from any subset of the others. All laws of a theory, including the axioms, describe the behavior of phenomena. All laws except the axioms, in principle, can be derived from the axioms. Usually such deductions require numerous subsidiary assumptions.

This deductively related set of statements is given empirical meaning by definitions which *ultimately* link theoretical terms (e.g., population, fertile, disease, motility, polymorphonuclear, chemotaxis, gene, and so forth) to observations (e.g., a theoretical term like "fertile" is partially defined by reference to the outcomes of numerous sexual events of a specified kind under specified conditions). Some theoretical terms are defined by reference to one or more other theoretical terms. Ultimately, any chain of such definitions must end in theoretical terms that are defined by reference to observations. In this way, the theory as a whole is given empirical meaning. Because of this complex interconnection of theoretical terms, the meaning of any one term is seldom independent of the meaning of many, if not all, of the other terms of the theory. Hence, theories have a global meaning structure: changes to the meaning of one term will have consequences for the meaning of many, and usually all of the other terms of the theory.

This view is a powerful one, because it answers in a straightforward and compelling way numerous questions about the nature of the scientific enterprise. As a result, it has proved difficult to abandon it and difficult to challenge it. The strongest testimony to this is the fact that it is still the dominant view. Nonetheless, it has always had its critics. Recently, however, the ideas and investigations discussed below under "New directions" have significantly challenged the relevance of this view. It is, in part, for this reason that I hold that these new directions constitute a significant shift in philosophy of biology.

Laws in biology

J. J. C. Smart sparked considerable controversy in 1963 by claiming that biology has no laws. For Smart, a law had to be spatially and temporally unrestricted, and must

PAUL THOMPSON

not refer to specific entities. Biology, Smart argued, has no generalizations that meet these requirements. Within the received view of science, the most compelling response to Smart called into question his criteria.

Several criteria for distinguishing laws from contingent statements (accidental generalizations) have been advanced: spatiotemporal generality, capacity to support counterfactuals, and embeddedness in an accepted scientific theory. The first currently enjoys very little support. Even Newton's laws don't pass this test: they don't apply to objects at very high speeds or to subatomic particles. And, since no general theory of macro and micro phenomena is yet available, neither quantum theory nor relativity is general in the required sense. The second is widely regarded as correct, but is so because it is a consequence of the third. The truth of the counterfactual "If a ball bearing were released from a height of 10 meters on a planet with twice the mass of the earth, it would fall to the ground in 1.013 seconds" is supported by Galileo's law of free fall $(d = \frac{1}{2}gt^2)$. The reason why it is taken to support this counterfactual, however, is that it is derivable from the axioms of Newtonian mechanics. That is, it is embedded in a widely accepted theory (widely accepted as applicable to macro objects at relatively slow speeds). By comparison with laws, contingent statements stand outside the theory as mere statements of current fact. This view of laws is solidly rooted in the received view of theories. On this view of laws, biology has numerous laws: Mendel's laws, the Hardy–Weinberg law, the law of natural selection.

Explanation and prediction

"Cause" underlies much of the literature on explanation. To have explained an event is to have identified its cause. To have identified its cause is to have demonstrated that an event of kind x can, and in the specific instance did, produce an event of kind y. The joint requirement of establishing that x can produce y and that in a given case x did produce y led Hempel and Oppenheim in 1948 to formulate the logic of explanation as follows:

$$\frac{L_1, L_2, L_3, \ldots, L_n}{C_1, C_2, C_3, \ldots, C_m}$$
$$E$$

where E is the event to be explained, $L_1, L_2, L_3, \ldots, L_n$ are scientific laws, and $C_1, C_2, C_3, \ldots, C_m$ are events that are known to have occurred prior to the occurrence of E. The laws guarantee that the event could be produced by $C_1, C_2, C_3, \ldots, C_m$. The determination that $C_1, C_2, C_3, \ldots, C_m$ did occur in connection with the occurrence of E guarantees that $C_1, C_2, C_3, \ldots, C_m$ was the set event responsible for the occurrence of E. On this pattern of explanation, the laws and antecedent events conjointly are sufficient for the occurrence of E.

The essential difference between *causal* prediction and explanation is time. In explanation the event deduced has occurred. In the case of prediction, the C's are not known to have occurred, and E is not known to have occurred. Laws (as statements embedded in a theory) are a central part of this account of explanation. They link antecedent events and the event to be explained, and as a result bring the full power of a theory to bear on the explanation.

18

In the 25 years after the paper of Hempel and Oppenheim was published, philosophers debated vigorously the correctness of this model and its applicability to specific sciences. The best case for the applicability of this model to biology was made by Michael Ruse in *The Philosophy of Biology* (1973).

Reductionism

Is biology ultimately reducible to physics and chemistry? The discovery of the structure of DNA in 1953 provided significant support for an affirmative answer to this question. The discovery revealed the chemical basis of one of the fundamental phenomena of the biological world: heredity. The molecule DNA was shown to be the material basis for heredity. All genetic information was stored in the form of triplets of nucleotides. The triplets formed the rungs of a double helical structure. The way in which DNA stored and replicated information was elegant and simple.

The specific focus of much of the controversy was on whether population genetics could be reduced to molecular genetics (i.e., to a largely chemical theory). If this reduction was possible, then it made more plausible a claim that biology was reducible to physics and chemistry. Although, on the surface, the discovery of the structure of DNA seemed to tip the scales in favor of reductionism decidedly, the issue turns out to be extremely complex. In the end, the debate turned on the conditions which any successful reduction must satisfy. If population genetics (PG) is to be reduced to molecular genetics (MG), then:

- All the terms of PG must be translatable without loss to terms of MG.
- All the laws describing the behavior of entities in PG must be deducible from the laws of MG (or, perhaps more challenging, MG must be able to describe with equal explanatory, predictive, and unifying power all the phenomena described by PG).
- The laws and concepts of MG must be able to be substituted for those of PG in all other biological contexts.

Whatever one's initial intuitions might be, each of these proved to be a formidable task. Even the task of defining the concept of the gene in MG and PG such that a gene in PG was translatable into a gene in MG was complicated and not very successful. After 30 years, population genetics is alive and well as a theoretical framework and functions in ecology and evolutionary biology in a manner yet to be replaced by MG. This fact alone attests to the difficulty of bringing about the reduction.

Fitness and selection

Despite the perceived importance of the concept "fitness" to evolutionary theory, it continues to be attacked on the grounds that it is epistemologically vacuous. Fitness, the argument runs, is *defined* in terms of "reproductive survival," and "reproductive survival" is explained in terms of fitness. Hence, operationally, "fitness" is defined in terms of "fitness." The circularity and vacuity, however, are illusory. They are a function of inadequate appreciation of the richness of the contemporary concept of fitness. To claim that an organism is fit relative to an environment E is to claim: (1) that it has characteristics which some other members of the same population do not have; (2) that those characteristics have a genetic basis; and (3) that, statistically, organisms with those characteristics in an environment E leave more offspring than those without the characteristics. Each of these claims is empirical and far from vacuous. In effect, the concept of fitness is embedded in the framework of *selection*, and this framework is far from circular or vacuous.

The most sustained and penetrating examination of selection is to be found in Elliott Sober's *The Nature of Selection* (1984a). There, he argues that, in any given case of selection, one can construct at all levels of organization (organism, gene, molecule) a selection account of change. But there will be only one "real" level at which selection takes place: the accounts at other levels will be artifacts of the "real" selection account. The "real" level is determined by the level at which a *causal force* is present. This concept of "causal force" is central to Sober's account of selection. A second important contribution of Sober is the distinction he draws between "selection *for*" and "selection *of*." "Selection *for*" captures the causes which are operating to bring about evolutionary change, while "selection *of*" captures the effects of the operation of these causes.

New directions

In my opinion, what follows represents exciting new challenges to philosophers of biology. Indeed, these issues represent the cutting edge of the field.

The structure of scientific theories: the semantic conception

Characterized in nontechnical terms, a theory, on the semantic conception, is a mathematical structure that describes (models) the structure and behavior of a system. Contrary to the received view, a theory is not a linguistic structure consisting of statements.

Although not strictly accurate, theories, on the semantic conception, can be construed, for convenience, as abstract and idealized models of actual empirical systems. Whether the model applies to the empirical phenomena within its intended scope depends on whether an isomorphism (a sameness of structure and behavior) between the model and the phenomena can be established. Establishing the isomorphism is complex, involving numerous other theories and mathematical tools. The theory does not specify either the domain of its application or the methodology involved in establishing an isomorphism. If a theory is deemed to be isomorphic with the phenomena, then explaining and predicting outcomes within the model constitutes explaining and predicting outcomes in the empirical world. There are two prominent versions of this conception of a theory: a set-theoretical version and a state space version.

A set-theoretical version of the formulation of Mendelian genetics is as follows:

T:　　　　　A system $\beta = <P, A, f, g>$ is a Mendelian breeding system if and only if the following axioms are satisfied:

Axiom 1:　The sets P and A are finite and nonempty.

Axiom 2:　For any $a \in P$ and $l, m \in A$, $f(a, l) f(a, m)$ iff $l = m$.

Axiom 3:　For any $a, b, \in P$ and $l \in A$, $g(a, l) g(b, l)$ iff $a = b$.

Axiom 4:　For any $a, b, \in P$ and $l \in A$ such that $f(a, l)$ and $f(b, l)$, $g(a, l)$ is independent of $g(b, l)$.

Axiom 5:　For any $a, b \in P$ and $l, m \in L$ such that $f(a, l)$ and $f(b, m)$, $g(a, l)$ is independent of $g(b, m)$.

Where P and A are sets, and f and g are functions, P is the set of all alleles in the population, and A is the set of all loci in the population. If $a \in P$ and $l \in A$, then $f(a, l)$ is an assignment, in a diploid phase of a cell, of a to l (i.e., f is a function that assigns a as an alternative allele at locus l). If $a \in P$, and $l \in A$, then $g(a, l)$ is the gamete formed, by meiosis, with a being at l in the gamete (the haploid phase of the cell). Although more sophistication could be introduced into this example (to take account, for example, of meiotic drive, selection, linkage, crossing over, etc.), the example as it stands illustrates adequately the nature of a set-theoretical approach to the formalization of population genetic theory in its simple Mendelian system form.

Characterizing Mendelian genetics using a state space approach is more complicated. A theory on this view consists of the specification of a collection of mathematical entities (numbers, vectors, functions) used to represent states of the system within a state space (a topological structure), the behavior of which is specified by three functions. These functions are commonly called "laws," but are not the same as laws in the received view. In the received view, laws are statements describing the behavior of entities in the world. In the semantic view, laws are mathematical descriptions of the behavior of mathematical systems of mathematical entities. The three kinds of laws are: laws of coexistence (which specify the physically possible set of states of the system), laws of succession (which specify the possible histories of the system), and laws of interaction (which specify the behavior of the system under conditions of inputs from interaction with other systems). A theory also requires the specification of a set of measurable magnitudes (represented by a function defined on the state space). Statements which formulate propositions to the effect that a particular magnitude has a particular value at a particular time are elementary statements. A satisfaction function determines the sets of states which satisfy the assignment of a value to a physical magnitude.

For population genetic theory, the state space will be a Cartesian n-space, where n is a function of the number of possible pairs of alleles in the population. A law of coexistence to the effect that only alleles at the same locus can form pairs will select the class of physically possible pairs of alleles. States of the system (genotype frequencies of populations) are n-tuples of real numbers from zero to one, and are represented in the state space as points. These are the measurable magnitudes. An example of a satisfaction function for the elementary statement "genotype Aa occurs with a frequency of 0.5" would specify the set of states in the state space that satisfy the statement. In this case the set of states would be a Cartesian $(n-1)$-space, which is a subset of the state space. For population genetic theory, a central law of succession is the Hardy–Weinberg law.

Organization

Complex systems have properties that place constraints on organization and selection. These properties are both structural and dynamical and give rise to self-organization and to evolutionary change independently of natural selection.

Systems can be ordered or chaotic. At the interface – on the edge of chaos – lies a narrow band of unstable order. Complex systems occupy this narrow band, and self-organization and evolution take place within it. The behavior of systems in the band

21

at the edge of chaos is dramatically altered by small perturbations because they are unstable – not well ordered. Systems that are strongly ordered – far from the edge of chaos – respond only slightly, if at all, to minor perturbations. Major catastrophes are required to affect an alteration, and these most often completely destabilize the system, resulting in its extinction rather than alteration. For systems well within the chaotic zone minor perturbations have cascading effects which result in entirely random changes. Consequently, the *possibility of change* requires that the system not be well ordered, and the *possibility of control* over change requires that the system not be chaotic. These conditions apply at the edge of chaos. Interactions among systems can drive them to the edge of chaos and hold them there as they undergo dramatic change. Selection as a force on systems at the edge of chaos will yield evolution. However, the dynamics of the interaction of systems on the edge of chaos also results in change which is driven by the structural and dynamic properties of the systems – often without the aid of selection. Adaptation for complex systems is adaptation to the condition at the edge of chaos.

One prominent researcher of organization and self-organization is Stuart Kauffman. His book *Origins of Order: Self Organization and Selection in Evolution* (1992) provides a comprehensive account of research in this field. Kauffman's discussions of the origins of life provide the clearest analysis of the constraints on change and the self-organizing character of change.

Artificial life

A number of aspects of model construction and organization and self-organization come together in a field of inquiry known as artificial life. Artificial life is the study of simulations of carbon-based living organisms. Those simulations can be mechanical devices, computer-based models, conceptual mathematical models, or carbon-based entities. The only significant distinction between artificial life and "natural" life is that humans, rather than nature, are responsible for the existence and characteristics of the "organisms." By far the greatest attention currently is on computer and mathematical simulations. One of the benefits of artificial life is the enrichment it brings to theoretical biology. It extends the tools of theoretical biology beyond mathematical models to computer simulations, and by so doing enables the development of a richer theoretical understanding of the nature and processes of biological organisms.

Artificial life – as a field of inquiry – is based on several important assumptions. One of the major assumptions is that "life" is a property of the organization of matter, rather than a property of the matter itself. Another is that properties (behaviors) *emerge* at different levels of organization – and are solely a function of organization. A third and critical assumption is that there is no global control of a complex "living" system. Rather, behavior is distributively controlled: control is local. A fourth (which is a corollary of the third) is that complex behaviors are a function of a few elementary rules governing the behavior and interaction of entities at low levels of organization.

These assumptions, especially the first three, distinguish current research into artificial life from past attempts, and from much of the current research into artificial intelligence (a field of research that is closely allied to artificial life). A discussion of each of them takes us to the heart of artificial life.

The emphasis on organization follows the same line of thinking as was discussed in the previous section. To really understand organisms, one must concentrate on how they are organized, the constraints on that organization, and the dynamics of it. For example, in systems on the edge of chaos (those, for instance, far from thermodynamic equilibrium), the nature of the organization is such that dramatic reorganization which is solely a function of the original organization can take place. This self-organization is a function not of the properties of the material, but of the organization and interaction of systems. Although the material may place some constraints on the realizable forms of organization, it is the nature of the organization itself that determines the dynamics of the systems. In the case of a computer simulation of a complex behavior, the entire emphasis is on organization independently of the actual material that is so organized.

One of the achievements of artificial life is the demonstration that complex behaviors can be simulated on a computer screen by means of a few local rules of organization. One clear example of this is the computer simulation of flocking. Birds often move in flocks, in which the pattern of the flock – a result of the flying behavior of each bird – and the dispersion and reformation in the face of an obstacle are seen as a complex coordinated activity. The same is true of the behavior of schools of fish and the herd movements of cattle. Craig Reynolds (1987) has simulated flocking behavior on a computer screen. His entities (called "Boids," which are devoid of material relevance) behave in accordance with three rules of behavioral tendencies:

1   to maintain a minimum distance from other objects in the environment, including other Boids;
2   to match velocities with Boids in its neighborhood; and
3   to move towards the perceived center of mass of the Boids in its neighborhood.

These are rules governing individual Boids. They are rules of local control. There are no further rules for the aggregate: the flock of birds. Aggregate behavior *emerges* from the behavior of individuals governed by these rules. The result on the computer screen is that when a number of individual Boids are given a random starting position, they will come together as a flock and will "fly" with grace and naturalness around obstacles by breaking into sub-flocks and then regrouping into a full flock once around the object. The flock's actual behavior when confronted with an object *emerged* from rules that determined only the behavior of individuals. To watch the Boids on the screen is to watch a complex coordinated aggregate behavior.

This example illustrates all of the above outlined assumptions of artificial life. It illustrates the primacy of the organization of entities over the properties of the matter of which they consist. It illustrates that there are no rules governing the aggregate behavior – only rules which govern the behavior of all entities (rules that are local and distributed over the local domain of entities). The aggregate behavior *emerges* from the individual (uncoordinated) behavior of the entities. In some cases, several independent (from the point of view of potential organization) systems may interact to produce a higher-order complex behavior. One could view this higher-order system as a single larger system with a slightly larger set of rules, or as the interaction of several individually organized systems. Ultimately, the distinction is irrelevant as long as none of the rules under either description exercises global control.

The Boid example also illustrates the assumption that control is not global, but local. There are no rules of coordination for the aggregate. Coordination is a function of the rules of behavior of the individual entities. An important feature of local, distributed control is the importance of a neighborhood. The behavior of interacting entities is specified in terms of neighboring entities: their positions or states. This system of entities interacting according to local rules based on a neighborhood is, in effect, the heart of the concept of organization. And such systems can be described using precise mathematical models in terms of state spaces, as described above in the context of the semantic conception of a scientific theory. The emphasis in artificial life and in the semantic conception of theories is on the dynamics of systems. In both cases those dynamics are specified in terms of organization.

Finally, the Boids example illustrates the assumption that complex behavior is the outcome of a *few* local rules. The essential point of this assumption is that simple behaviors of interacting elements are the basis for high-level organizational complexity, and that the attempt to formulate rules at higher levels (globally) to describe high-level complex behaviors is wrongheaded. Chris Langton, a leading exponent of artificial life (1989), has claimed that the quest for global rather than local control mechanisms (rules) is the source of the failure of the entire program of modeling complex behaviors up to the present, including, especially, much of the work on artificial intelligence.

References and further reading

Auyang, S. Y. 1998: *Foundations of Complex-System Theories in Economics, Evolutionary Biology, and Statistical Physics* (Cambridge: Cambridge University Press).

Bechtel, W., and Richardson, R. C. 1993: *Discovering Complexity* (Princeton: Princeton University Press).

Beckner, M. 1959: *The Biological Way of Thought* (New York: Columbia University Press).

Darwin, C. 1859: *The Origin of Species* (London: John Murray). (Numerous modern versions are available. Six editions were issued, of which the first was in 1859.)

Depew, D. J., and Weber, B. H. 1995: *Darwinism Evolving: Systems Dynamics and the Genealogy of Natural Selection* (Cambridge, MA: MIT Press).

Hempel, C., and Oppenheim, P. 1948: Studies in the logic of explanation. *Philosophy of Science*, 15, 134–75.

Hull, D. L. 1974: *Philosophy of Biological Science* (Englewood Cliffs, NJ: Prentice-Hall).

Kauffman, S. A. 1992: *Origins of Order: Self-Organization and Selection in Evolution* (Oxford: Oxford University Press).

—— 1995: *At Home in the Universe: The Search for the Laws of Self-Organization and Complexity* (Oxford: Oxford University Press).

Langton, C. G. (ed.) 1989: *Artificial Life* (Redwood City, CA: Addison-Wesley).

Langton, C. G., Taylor, T., Farmer, J. D., and Rasmussen, S. (eds) 1992: *Artificial Life II* (Redwood City, CA: Addison-Wesley).

Levy, S. 1992: *Artificial Life: A Report from the Frontier where Computers Meet Biology* (New York: Vintage Books).

Lloyd, E. A. 1988: *The Structure and Confirmation of Evolutionary Theory* (Westport, CT: Greenwood Press).

Nagel, E. 1961: *The Structure of Science* (London: Routledge and Kegan Paul).

Putnam, H. 1962: What theories are not. In *Logic, Methodology and Philosophy of Science*, ed. E. Nagel, P. Suppes, and A. Tarski (Stanford, CA: Stanford University Press), 240–51.

Reynolds, C. W. 1987: Flocks, herds and schools: a distibuted behavioral model. *Computer Graphics*, 21, 25–34.

Rosenberg, A. 1985: *The Structure of Biological Science* (New York: Cambridge University Press).

Ruse, M. 1973: *The Philosophy of Biology* (London: Hutchinson & Co.).

—— 1999: *Mystery of Mysteries: Is Evolution a Social Construction?* (Cambridge, MA: Harvard University Press).

Schaffner, K. F. 1993: *Discovery and Explanation in Biology and Medicine* (Chicago: University of Chicago Press).

Smart, J. J. C. 1963: *Philosophy and Scientific Realism* (London: Routledge and Kegan Paul).

Sober, E. 1984a: *The Nature of Selection* (Cambridge, MA: MIT Press).

Sober, E. (ed.) 1984b: *Conceptual Issues in Evolutionary Biology* (Cambridge, MA: MIT Press).

Suppes, F. 1977: *The Structure of Scientific Theories*, 2nd edn (Urbana, IL: University of Illinois Press).

Thompson, P. 1989: *The Structure of Biological Theories* (Albany, NY: State University of New York Press).

4

Bohr

DUGALD MURDOCH

One of the most influential physicists of the twentieth century, Niels Bohr was born in Copenhagen on 7 October 1885, and died there on 18 November 1962. He came of a well-to-do, cultivated family, his father being Professor of Physiology at Copenhagen University, where Bohr himself received his education in physics. After taking his doctorate in 1911, Bohr went to Cambridge University to continue his research on the theory of electrons under Sir J. J. Thomson. After several months in Cambridge, he moved to Manchester to work under Ernest Rutherford, the world leader in the newly emerging field of atomic physics. It was in Rutherford's department that Bohr proposed his revolutionary quantum theory of the structure of the hydrogen atom in 1913.

In order to explain the surprisingly large angles at which alpha particles were scattered from atoms in a target, Rutherford had proposed that an atom consists of a positively charged nucleus, and negatively charged electrons which orbit the nucleus. According to classical electrodynamics, however, such a structure ought quickly to disintegrate, for the electrons ought rapidly to spiral down into the nucleus, giving off radiation on their way. Bohr tackled this problem, and solved it with a theory of extraordinary boldness. Blatantly contradicting the classical theory, he suggested that an atom can exist only in a finite number of discrete energy states, and while the atom is in such a state, the electrons remain in stable paths, and no radiation is emitted: radiation is emitted only when the atom 'jumps' from a higher state to a lower one. The hypothesis that an atom's energy is quantized was a brilliant extension of Plank's notion of the quantum of action, an idea that the physics community had only recently begun to take really seriously. Bohr's revolutionary theory, which was published in the *Philosophical Magazine* in 1913, launched the quantum theory of atomic structure.

Bohr was appointed Reader in Physics at Manchester University in 1914, and Professor of Theoretical Physics at Copenhagen University in 1916. Over the next ten years he built up his institute into a world center for quantum physics which, like a magnet, attracted many brilliant young researchers, such as Paul Dirac, Werner Heisenberg, and Wolfgang Pauli. Bohr himself continued to contribute to quantum physics. His correspondence principle – that, for states of increasingly high energies, the frequency of emitted radiation predicted by quantum theory tends to coincide asymptotically with that predicted by classical electrodynamics – was one of the guiding ideas in the development of quantum theory. Equally, if not more, important in the ensuing years was Bohr's contribution to the interpretation of the theory.

By 1926 two mathematically equivalent versions of a new mechanics, quantum mechanics, had been developed: namely, Heisenberg's matrix mechanics and Schrödinger's wave mechanics. This new theory enabled a vast range of empirical data to be explained and predicted with marvelous economy. Yet, in spite of its empirical success, it was far from clear what picture of physical reality, if any, the new mechanics provided. It did nothing to resolve wave–particle duality (i.e., the fact that radiation and matter appear in some experimental situations to behave like particles and in other situations like waves). Moreover, the theory was indeterministic, or probabilistic, in the sense that the state vector does not assign a determinate value to every observable (i.e., physical quantity), but only the probability of its having a determinate value, and, as Heisenberg discovered at Bohr's institute, for pairs of conjugate observables such as position and momentum, if the value of one observable (position, say) has a determinate value in a given quantum state, then the conjugate observable (momentum) has no determinate value in that state.

Bohr tried to make sense of this strange new theory with his notion of *complementarity*. He argued that the wave and the particle models, and conjugate observables such as position and momentum, are *complementary* in the sense that they are (a) equally indispensable for the interpretation of quantum mechanics and the experimental evidence to which it applies; (b) mutually exclusive, in that they cannot be applied or assigned to a system simultaneously in one and the same experimental situation. The complementarity of concepts such as position and momentum was, he argued, a consequence of the quantum of action, the discovery of which had brought to light a hitherto unrecognized presupposition of the unrestricted way in which such concepts were applied in classical physics: namely, that the interaction between a system and an instrument in the measurement process could be made negligibly small, or could be determined and taken into account. This presupposition, Bohr argued, was false, and its falsity entailed that concepts such as position and momentum could not be applied to a system simultaneously, as in classical mechanics, but only at different times and in different experimental contexts.

Bohr showed, by a meticulous analysis of thought experiments, that it is impossible to measure determinate values of complementary observables simultaneously (see THOUGHT EXPERIMENTS). Complementary observables call for mutually exclusive measurement procedures, and owing to the indeterminable measurement interaction, values obtained by such mutually exclusive measurement procedures cannot be assigned by extrapolation to one and the same time. A system and the experimental arrangement with which it has interacted constitute an *unanalyzable whole*, and cannot be described independently of such a whole.

He argued, moreover, that quantum indeterminacy ought not to be construed as being merely *epistemic*: i.e., as constituting an ineliminable ignorance about a value which in reality is perfectly determinate. The information about a system provided by the state vector at any given time is *complete*. He preferred, however, to construe quantum indeterminacy not in purely *ontic* terms – i.e., as implying that a system to which no determinate position can be assigned has an ontically indeterminate or fuzzy position – but rather in *semantic* terms – i.e., as implying that talk of determinate position in such a context is not well defined. Thus the question whether an observable to which no determinate value can be assigned has in reality a determinate, though

27

unknown, value is meaningless. Bohr's philosophical concern was primarily with semantic rather than ontological questions – i.e., with establishing meaning conditions for the use of theoretical concepts.

Bohr's interpretation of quantum mechanics has often been construed as being instrumentalist, in that theoretical concepts and statements do not describe physical reality and have no truth-value, but are merely tools to help us make predictions about our sensory experience. Some recent commentators, however, have argued that his position is basically realist (see REALISM AND INSTRUMENTALISM), in that he regarded theoretical concepts, such as electron, electromagnetic wave, momentum, etc., as purporting to denote real physical entities and properties. But his realism is of a very weak variety, for he did not regard our theoretical conception of microphysical reality as representing it as it is independently of our means of investigating it. Models, such as particle and wave models, were not so much faithful mirrors of microphysical reality as pictures painted with peculiarly human materials and from our peculiarly human perspective, and a measured value of an observable cannot be said to characterize the observable independently of the experimental arrangement by means of which it has been obtained. The purpose for which we construct theoretical concepts, then, is not strongly realist, to reveal reality as it is in itself, independently of human modes of inquiry, but rather is weakly realist or even pragmatist, to help us make sense of our sensory experience. Owing to the elusive style in which he wrote, however, the question whether Bohr's position is more realist than anti-realist is likely to remain a matter for debate (on this issue see Folse 1985; Murdoch 1987; and Faye 1991).

Bohr's interpretation of quantum mechanics, which was first presented in the fall of 1927, was quickly accepted by the majority of physicists and, as understood by Bohr's followers, came to be known as the *Copenhagen Interpretation* (see Bohr 1928). Yet there were dissenting voices, among them the greatest physicist of his time, Albert Einstein (see EINSTEIN).

Einstein was much more robustly realist in his attitude to quantum mechanics than was Bohr (though not nearly as robust as is usually made out), holding that all observables must in reality have determinate values at any given time, even though these values are beyond the ken of the theory. Thus he rejected Bohr's thesis that the quantum-mechanical description of a system is complete, and argued that quantum indeterminacy should be construed as being merely epistemic. He argued for this view in the famous 'EPR' paper (see Einstein, Podolsky, and Rosen 1935). Einstein's argument, which was presented more clearly in later papers, was essentially as follows. By making a measurement on a system A at time t, it is possible to assign to another, distant system B (which has previously interacted with A) at t either one (but not both) of two different state vectors ψ and ϕ belonging to two conjugate observables (depending on what measurement we chose to make on A), and we can do this without physically disturbing B. This being the case, we can hold that B is either in the same physical state at t whether we assign ψ to it or ϕ, or in a different physical state. In the former case, since one and the same physical state can be described by different state vectors ψ and ϕ, neither ψ nor ϕ gives a complete description of that physical state. In the latter case, the physical state of B at t must somehow depend upon the physical state of A at t. In other words, we must accept either the *incompleteness* of the quantum-mechanical state description or the *nonseparability* of certain spatially separated states. Einstein was unable to accept the latter.

Bohr's answer to this extremely ingenious argument was, in a nutshell, that Einstein was not justified in assuming that system B could meaningfully be said to be in the *same* physical state at time *t* regardless of which state vector we assign to it, for the notion of physical state which Einstein was employing here was not well defined: the concept of physical state could not be meaningfully applied independently of the *total* relevant experimental context, and in the EPR case this context included the experimental conditions prevailing at system A as well as at system B (see Bohr 1935 and 1949). Thus what we do at A may determine what we can meaningfully say about the physical state of B. Here is the crux of the disagreement between Bohr and Einstein: whereas Einstein was prepared to talk of physical reality independently of our conditions of observation, Bohr was not. To put the difference between them more precisely: Bohr held a verificationist view of the meaning of theoretical concepts, whereas Einstein's view was realist; that is to say, Bohr held that theoretical concepts are meaningfully applicable only in contexts in which the truth-value of statements applying them can be established by experiment, whereas Einstein held that all that was necessary was that we should have a coherent conception of their truth-value.

Bohr's verificationism may easily seem to be founded on positivism, for he holds that a theoretical concept is meaningfully applicable only in the context of a *phenomenon* (by which he means a system together with the well-defined experimental arrangement in which it appears), and hence only in a context in which the truth-value of a statement applying that concept can be established by experiment. There are, however, grounds for thinking that the basis of Bohr's view of the meaning of theoretical concepts was pragmatism rather than positivism, though, again, this is a question on which views may differ (see LOGICAL POSITIVISM).

It is generally held that in his great debate with Einstein, Bohr was on the side of the angels, since Bell's theorem implies that Bohr, as opposed to Einstein, made the right choice in opting for nonseparability rather than incompleteness. Whether nonseparability can be adequately understood in terms of Bohr's notion of unanalyzable wholeness is another question (see Murdoch 1993).

References and further reading

Works by Bohr

1928: The quantum postulate and the recent development of atomic theory. *Nature*, 121, 580–90; repr. in Niels Bohr, *Atomic Theory and the Description of Nature* (Cambridge: Cambridge University Press, 1961), 52–91.

1935: Can quantum-mechanical description of physical reality be considered complete? *Physical Review*, 48, 696–702.

1949: Discussion with Einstein on epistemological problems in atomic physics. In *Albert Einstein: Philosopher-Scientist*, ed. P. A. Schilpp (Evanston, IL: Northwestern University Press), 199–242; repr. in Niels Bohr, *Atomic Physics and Human Knowledge* (New York: J. Wiley & Sons, 1958), 32–66.

1963: *Essays 1958–1962 on Atomic Physics and Human Knowledge* (New York: J. Wiley & Sons).

Works by other authors

Einstein, A., Podolsky, B., and Rosen, N. 1935: Can quantum-mechanical description of physical reality be considered complete? *Physical Review*, 47, 777–80.

Faye, Jan 1991: *Niels Bohr: His Heritage and Legacy* (Dordrecht: Kluwer Academic Publishers).

Folse, Henry 1985: *The Philosophy of Niels Bohr: The Framework of Complementarity* (Amsterdam: North-Holland Publishing Company).

Honner, John 1987: *The Description of Nature: Niels Bohr and the Philosophy of Quantum Physics* (Oxford: Oxford University Press).

Murdoch, Dugald 1987: *Niels Bohr's Philosophy of Physics* (Cambridge: Cambridge University Press).

—— 1993: The Bohr–Einstein dispute. In *Niels Bohr and Contemporary Philosophy*, ed. Jan Faye and Henry Folse (Dordrecht: Kluwer Academic Publishers), 303–24.

Pais, Abraham 1991: *Niels Bohr's Times: In Physics, Philosophy, and Polity* (Oxford: Clarendon Press).

5

Causation

PAUL HUMPHREYS

Ordinary language is saturated with causal concepts, with talk of tendencies, consequences, mechanisms, and a host of other thinly disguised causal terms. But ordinary language is no reliable guide to ontology, and for scientific purposes we must ask whether advanced sciences need to refer to causes in their theories and methods. Then, if we find that they do need to make such causal references, we must ask what the nature of the causal relation is and how we can discover instances of it. Here we shall reverse that order of questioning, by first laying out some standard approaches to characterizing and discovering the causal relation, and then examining whether such relations are dispensable in certain parts of science.

Philosophical preliminaries

One's choice of causal ontology is crucial, for, like choosing a spouse, an initial error of judgment inevitably leads to later disaster. Of primary importance is the nature of the causal relata. Many kinds of things have been suggested to play the role of cause and effect: events, property instances, objects, variables, facts, states of affairs, propositions, events under a description, amongst others. Yet objects do not cause anything; it is the properties possessed by them that do the causing. Likewise, propositions themselves are not generally causally efficacious; rather, they are a mode of representing facts about the events that are the real causes and effects.

A sound choice for a basic ontology is to use property instances in the form of the exemplification of a quantitative property by a given system at a specific time. Then events can be construed as changes in such property instances and states of affairs as relatively static values of such properties. This choice also allows one to represent the kinds of events needed for science, because sciences of any degree of sophistication must use quantitative properties such as the degree of magnetization of a substance or the income level of an individual. By contrast, the simple events used in most philosophical accounts, such as striking a match, involve only qualitative properties that either do or do not occur. These are relatively uninteresting in scientific contexts, and in any case can easily be represented by binary-valued variables. It is important that events not be thought of as regions of space-time. For such regions contain numerous property instances, most of which have no causal relevance to a given effect and should thus be omitted on the grounds that they play no causal role. Moreover, allowing such irrelevant property instances as part of the complex spatiotemporal event leads to needless problems, such as allowing one to describe the same event as an acidification of a solution or as increasing the volume of the liquid by 2 cc. By contrast, the preferred choice of ontology enables us to avoid most ambiguities involving "events under a description," which usually, although not universally, trade on simultaneous property exemplifications in a region such as the one just cited. Many supposed examples of the intensionality of causal contexts – that is, the ability to change the truth-value of a causal claim by substituting a co-referring term for the original description of the cause or the effect – are a result of a simplistic "events under a description" view.

Also of philosophical importance is the distinction between causal generalizations and singular causal claims. Theories of the former, such as regularity accounts within which generalizations are primary, view singular causal claims as derivative from the causal laws that hold of our world. By contrast, singularist accounts view specific causal connections between individual property values as a basic feature of the world, and generalizations are then parasitic upon the singular claims. Although it is of course important which of these approaches is the correct one, there are not, in this author's view, absolutely conclusive reasons for preferring one to the other.

Finally, although linguistic analysis may be a useful rough guide to appropriate causal relata, it is dangerous to rely too heavily upon what one would ordinarily say about causation. The development of science often forces us to abandon cherished beliefs, and the nature of scientific causal relations is surely one that ought not to be constrained by mere common sense. Rather, we should attempt to construct a comprehensive theory that accounts for what we know about the scientific uses of causation.

Hume's account

It may seem odd that an argument first invented over 250 years ago should have any relevance to contemporary science; but a proper understanding of Hume's objections to causation enables us to see how his minimalism about causation continually recurs in modern treatments of causation. Straightforwardly, Hume's complaint is this: Suppose one is an empiricist for whom empirical evidence is the ultimate arbiter of what can legitimately be claimed to exist. Think now of what an observer could detect about an allegedly causal relation. This observer would detect that the cause occurred prior to its effect, that the two were either in spatial and temporal proximity to one another or else connected by a chain of events that were so related, and that events resembling the effect regularly occurred when events resembling the cause did. Of anything further peculiar to causation, in particular the supposed crucial feature of the cause necessitating the effect, nothing could be detected, *even by an ideal observer equipped with the best instruments imaginable*. It is important to add this italicized claim, because as long as one is willing to rule out observers who have peculiar abilities to detect natural necessity, a natural extrapolation of Hume's argument from what we can observe with the unaided human senses to what the most sophisticated scientific instrument can do leads naturally to this powerful conclusion. There is no natural necessity meter; nor is there a causal connection scanner (see HUME).

As with most skeptical arguments, Hume's relies on a combination of an under-determination argument together with a use of Occam's razor. To see this, compare two competing accounts. Account 1 says that there is only what Hume says there is: constant conjunction, spatiotemporal contiguity, and temporal priority. Account 2 says there is that plus a peculiarly causal connection. According to Hume, accounts 1 and 2 have exactly the same observable consequences, but the second postulates the existence of one more type of relation: a *sui generis* causal connection. Occam then advises us to prefer the first, ontologically economical, account.

Standard problems

Hume himself suggested that the feeling of necessitation associated with causation came from our psychological habit of expecting regular sequences to continue – effect after cause, effect after cause. One can criticize Hume's positive account on a number of grounds. I regularly awaken at 6:47 a.m., exactly three minutes before my otherwise silent alarm clock buzzes. The sequence contains all the elements of Hume's account: a regular succession of one event occurring before another, spatiotemporal contiguity, and an expectation on my part that the alarm will go off. But my awakening does not cause the alarm to go off; nor is the alarm the cause of my awakening. In fact the causal situation here is complex, involving amongst other things a biological timing mechanism the relationship of which to sidereal time and to electronic clock time is subtle and not obviously analyzable in terms of mere regularities. The situation is complicated by the fact that sometimes Humean regularities are the result of the joint action of a common cause, as with the regular succession of thunder after lightning (these both being the result of an electrical discharge); sometimes they are regularities that are not the result of direct or indirect causal connections (in the above clock example, the correlation between the time exhibited by a caesium clock and the time exhibited by a clock based on sidereal time is one induced at least partly by convention); and sometimes the earlier event would cause the later event only if some other event were not to occur. We shall call these the common cause problem, the coincidence problem, and the preemption problem. In addition, when the regular sequence is such that the first event is necessary as well as sufficient for the second (as Hume himself insisted), the only feature that marks out one event rather than the other as the cause is temporal priority. Yet some have been reluctant to use temporal priority as a distinguishing criterion of a cause, often because they hold that temporal ordering should be based on a causal ordering, rather than vice versa, or that causes could in principle occur after their effects. On such views, providing a direction of causation is a fourth primary problem. For more on this see Sklar 1974, ch. 4; also SPACE, TIME, AND RELATIVITY. In scientific contexts, the first, second, and fourth of these problems are of greatest importance.

Sufficiency accounts

Until quite recently, it was almost universally held that causation must be deterministic, in that any cause is sufficient to bring about its effect. This deterministic dogma has crumbled, but the influence of the prejudices behind it is still strong and underpins many sufficiency accounts. Modern versions of these come in two types: those that rely, as did Hume, on a regular succession of actual events as the basis of causal claims, and those that use a logical reconstruction of event sequences. A particularly influential version of the latter uses universally quantified conditionals asserting that "For all x, if Fx then Gx." This allows us, given a particular propositional representation of an event's occurrence, Fa, to deduce that Ga. Here, the universal generalization, Fa, and Ga are all propositional entities, and the conditional may be either a material conditional or a conditional of a logically stronger kind, such as a subjunctive conditional. In order to avoid the common cause and coincidence problems, it is generally required that the universal regularity be a law of nature (see LAWS OF NATURE).

The key feature of these logical reconstructions is that they replace a natural necessity holding between cause and effect with a relation of logical deducibility between propositions representing the cause and the effect. Thus, whereas Hume's account relocated the necessitation to our mental representation of the cause–effect relationship, modern regularity theories relocate it to our logical representations of that relationship. Such an approach underlay Hempel's seminal deductive-nomological account of explanation (see EXPLANATION).

Necessity accounts

With the recognition that indeterminism is a real feature of the microworld has come the accompanying realization that causation may operate in the absence of sufficiency. Even given determinism, we can ask whether it is the sufficiency of a cause or its necessity for the effect that gives it its causal influence. Overdetermination cases are often cited as a reason for denying that sufficiency is enough for a causal relation to hold. If a pane of glass shatters while being heated, but at the precise moment that it would have disintegrated a steel hammer hits it with a force sufficient to break the glass, was the cause of the breakage the heating, the hammer blow, both, or neither? Our inclination is to say that it was not the heating, precisely because if the heat had not been applied, the glass would have shattered anyway. And this is true, even though the heating was sufficient for the glass to break. A parallel claim can be made about the hammer blow. So, to put it sententiously, sufficiency is not sufficient for causation. One might say of these overdetermination arguments that they trade upon an imprecise description of the effect, and that were the effect described exactly, breaking by heating would differ in some way from breaking by a blow. This point is correct, and it is an important one, for surely scientific causal claims must be precise in a way that causal claims of everyday casual conversation generally are not. Nevertheless, the argument against sufficiency alone as a basis for causation is conceptual, not factual, and it has persuaded many that it is the failure of the putative cause to be necessary for the effect that prevents it from being a genuine cause.

It is hard to make sense of a cause being necessary for its effect in some universal fashion, because most effects can be brought about in a number of different ways. (This is what Mill termed the "multiplicity of causes.") Once again, we must be careful to acknowledge that superficially similar events might be of different kinds when described precisely, thus making the multiplicity of causes an artifact of imprecise description, or, alternatively, that underlying all the apparently different ways of producing a type of effect is a single type of causal mechanism. Still, the multiplicity of causation has led to singularist *sine qua non* accounts focusing on the idea of a cause being necessary in the circumstances for its effect. We then have that event A caused event B if and only if (1) A occurred; (2) B occurred; (3) the counterfactual "If A had not occurred, than B would not have occurred" is true (or can be asserted). This basic account needs to be refined in various ways to avoid both straightforward and subtle counterexamples, and perhaps the most detailed theory in this regard can be found in Lewis (1983a, 1983b), where it is claimed that three of the four major difficulties noted earlier are avoided. (His theory also, in fact, rules out coincidences.) The worth of such accounts depends almost entirely upon the truth conditions provided for the conditional in clause (3), and despite the sophistication of Lewis's position, to solve those problems requires one to agree to the truth of certain counterfactual condition- als that, in the view of this writer and many others, are simply false. Another seminal and eminently readable *sine qua non* theory can be found in Mackie (1974).

Probabilistic accounts

Although *sine qua non* accounts are compatible with indeterministic causation, they have usually been restricted in their application to deterministic contexts, because of the difficulty of providing truth conditions for the counterfactual when the world is irreducibly probabilistic. Probabilistic accounts are frequently motivated merely by appeal to examples, such as the fact that smoking increases the probability of lung cancer and (hence) is a cause of it. A better reason for urging the adoption of probabilistic theories of causation is the inability of the *sine qua non* approach to correctly deal with certain cases. Given that not only smoking but also exposure to industrial pollutants increases the probability of acquiring cancer of exactly the same kind, then neither smoking nor such exposure is necessary in the circumstances for that effect; yet one can argue that each is a contributing cause of the disease. It has to be said that at present there is no commonly agreed upon formulation of prob- abilistic causation, but a standard approach is to assert that A causes B if and only if A increases the probability of B invariantly across a wide range of contexts: that is, $P(B/AZ) > P(B/\neg AZ)$, where the exact nature of Z depends upon the account. At a minimum, Z should contain any common causes and preempting causes of B and A. By conditioning on these factors, B and A will be rendered probabilistically independent, thus avoiding the common cause problem and the preemption problem. It is also highly unlikely that a statistical association between factors would be invariant across many contexts, were it merely coincidental (although see below for a quantum-mechanical case). There are many issues of great scientific and philosophical interest in this area, not least of which is finding the right interpretation of probability to use in the causal theory (see PROBABILITY). The reader is referred to Salmon (1984) and to Humphreys (1989) for contrasting detailed accounts. A persistent worry about probabilistic theories of causation is that they seem to need supplementation by specifically causal knowledge in order to be applied correctly. One reason for this revolves around the interpretation of the probability in the above inequality. If relative frequencies are used, then Z can- not include causal consequences of B for which B is necessary; for if Z does, then both sides of the inequality will be equal to unity, and A could not be a cause of B. This prob- lem could be avoided by excluding from Z any events later than B, but the problem does not occur when a propensity interpretation of probability is used, for propensities are themselves temporally dependent, and conditioning on events after B always leaves the propensity value unchanged.

Discovery

The three approaches detailed above provide a good logical taxonomy of various ways of characterizing causation. It can easily be seen that many of the definitions already discussed, when they can solve the principal problems mentioned, can also serve as a means of discovering causal connections, at least in the sense that they can discriminate between those sequences of events that are genuinely causal and those that are not. More will be said below about matters involving discovery.

Skepticism about causation

The first question we raised was whether science needs causal concepts at all. The answer to this question depends upon one's position with regard to causal versions of realism and instrumentalism (see REALISM AND INSTRUMENTALISM). Generally speaking, empiricists have been suspicious of causation as a distinctive relation that is not reducible to other, empirically accessible concepts. For, unlike the relation of, say, "having a greater mass than," the relation "caused" is not open to direct measurement, nor, apparently, to indirect measurement. In consequence, when empiricist influences are particularly strong, as in the late nineteenth century and during the high tide of positivist influence between, roughly, 1925 and 1960, attempts are made to exorcise causation entirely from respectable science. This has proved to be remarkably difficult, and, in this author's view, there are good reasons to think that causation cannot be so eliminated from at least some of the sciences. Russell's famous remark that causation, like the monarchy, survives only because it is erroneously supposed to do no harm, is witty enough, but errs in not recognizing that most alternatives are even worse.

Why might one think that causal concepts are unnecessary in scientific practice? There are five, not completely distinct, lines of argument that are often used to establish this conclusion. The first is the general skeptical argument of Hume which we have already discussed. The second line is constructive, and argues that the causal relation is eliminable and replaceable by an account of the relationship that uses only noncausal concepts. This approach is characteristic of the regularity, *sine qua non*, and probabilistic accounts described above. The third line of argument proceeds by constructing a detailed scientific theory that does not employ causal concepts and by then arguing that all empirically confirmable consequences of that theory can also be deduced and tested without using causal notions. An important rider to this argument is that the theory in question provide a complete account of the represented phenomena. A particularly influential view here has been the idea that causal relations can be replaced by a dependency represented only by mathematical functions; i.e., that X is a cause of Y just in case for some function f, and some representations x,y of X and Y, $f(x) = y$. But this approach simply ignores all four of our problems, for joint effects of a common cause, coincidentally related variables, and variables in a preempted relationship can all be functionally related in this way. Moreover, when f is invertible, as any 1:1 function is, we have $x = f^{-1}(y)$, so no asymmetry of causation is provided. There are more persuasive versions of this third line, as we shall see. The fourth line of argument is a scientific version of the first, skeptical argument. It suggests that the methods employed by the science in question are essentially incapable of distinguishing between genuinely causal relations and mere associations. A fifth line of argument asserts that adding causal features to a theory would result in inconsistency with empirical data.

Scientific considerations

Two aspects of quantum mechanics are important for us (see QUANTUM MECHANICS). The first involves what are generally known as Bell-type experiments. Within these, two systems initially interact and are then separated. When a measurement is made on one system, say of the polarization of a photon in a given direction, the other system is invariably found to be in a state that is perfectly correlated or anti-correlated with the first. The obvious explanation for this is that the measurements on the two systems are determined by the values of some variable that is fixed when the two systems interact originally. Yet, under very weak assumptions, Bell proved that this obvious explanation is wrong. Assuming such "hidden variables" gives incorrect empirical results for the statistical distribution of other measurable variables. Such experiments thus provide clear examples of correlations that are not the result of any direct causal link; nor are they the result of a common cause. Unless some as yet undiscovered account is found for such correlations, they would have to be taken as examples of a brute fact about the world, as clear a case of Hume's position as one could get. Use of Bell's results is an example of the fifth line of skeptical argument noted earlier, and a clear discussion can be found in Hughes (1989, ch. 8).

A different, but equally strong, objection to the universal need for causal concepts in science comes from the almost complete absence of their use in nonrelativistic quantum theory. To perform calculations of the energy levels of molecular orbitals in hydrogen, for example, one needs no causal information at all. In general, the mathematical apparatus required for solving Schrödinger's equation (which is the canonical representation of the dynamics of the quantum state) requires no causal input either – even more so when more abstract approaches using self-adjoint operators on Hilbert spaces are used as the representation. It is thus commonly held that a purely instrumentalistic use of quantum formalism, free of any causal content, can be employed, and hence that causation is an unnecessary feature of quantum theory. This is an example of the third line of skeptical argument. Yet this view is quite misleading, because in order to bring the abstract representational apparatus to bear on specific physical systems, a physically motivated model is usually used, and such models frequently have a recognizably causal content. For example, in the representation of a two-dimensional Ising model of ferromagnetism, the interaction between adjacent lattice nodes that results in spin flips has a clear causal interpretation. Such models are admittedly crude, and of largely heuristic value, but the point remains that solving the necessary equations which represent a given physical system cannot often be done without justification of boundary conditions (which are often construed in terms of the absence of external – i.e., causal – factors) and some causally motivated model of the system. Thus, at best, this dispensability argument can be applied to particular systems, but not universally.

A second area within which skepticism about causation has been voiced is in the social and behavioral sciences. Here issues involving discovery tend to dominate. For example, one traditional scientific approach to causation insists that the key feature of a cause is that it could, at least in principle, be manipulated to produce a change in its effect, and this manipulability criterion is what differentiates between causes and their associated effects. Thus, in the case of our four problems, blocking the lightning flash with a polarized filter will not prevent the thunderclap, because the two phenomena are due to a common cause; changing the rate of the caesium clock will not affect sidereal time, or vice versa, and hence the one cannot be the cause of the other; and an increase in consumer confidence often precedes, and could cause, an end to an economic recession, but a manipulated decrease in interest rates is often the preempting cause of that effect. Moreover, it appears that manipulability has promise for providing a direction of causation, for changing the effect will not change the cause in general. These considerations underlie the manipulability or activity approach to causation.

Similarly, one of the traditional functions of controlled experiments in science has been to discover causal relationships (see EXPERIMENT). By virtue of holding constant, or eliminating entirely, all but one influence on a system and then varying that influence, not only the mere fact of a causal relation can be discovered, but its exact functional form (linear, exponential, etc.) can be isolated. Within many of the social sciences, especially economics and sociology, controlled experimentation is not possible, for either practical or ethical reasons. For a solid discussion of such issues, see Cook and Campbell (1979).

A common way to circumvent these problems is to use statistical analysis and so-called causal models. Regression analysis, analysis of covariance, structural equation models, factor analysis, path analysis, and an assortment of other methods attempt, by means of various statistical techniques, usually supplemented by substantive theoretical assumptions, to do in nonexperimental contexts what experiments do under controlled laboratory conditions. Those who subscribe to the manipulability approach to causation doubt that these statistical surrogates are an adequate substitute for genuine experimentation, and thus subscribe to the fourth line of skepticism we have discussed.

There is heated debate over whether these statistical methods so underdetermine the causal structure of the systems they model that they are essentially worthless or, at best, suggest via the third line of skepticism that one can do without causal concepts. There is further debate over whether in order to be employed, these statistical methods require a substantial amount of prior causal knowledge about the system. Recently, the development of computer search procedures on data has led to a distinctively different approach. Using directed graphs to represent causal structure, Judea Pearl and others successively test for statistical independence between pairs of variables conditional upon neighboring variables. If independence is found, the variables are considered to be causally independent, and the edge between the variables is eliminated from the graph. One of the central claims of this approach is that causal connections can be inferred without the need for substantive theoretical or causal knowledge. The full methodology is complex; one source is Spirtes, Glymour, and Scheines (1993). Critical responses can be found in Humphreys and Freedman (1996).

As a third positive example, causation plays a central role in many areas of philosophical psychology. A small selection should convey the flavor of the rest. Functionalism, regarded here in a narrow sense concerned with psychological states, asserts that such states can be characterized by the causal role they play with respect to other psychological states. So, depression might be characterized by its tendency to cause lassitude and loss of appetite, to be caused by psychologically painful events, and a tendency to be alleviated by sympathetic conversation. There could be many different kinds of physical instantiations of such causal relations in humans, automata, and, possibly, aliens; but being embedded in such a causal network is, according to functionalism, essential.

The status of mental causation is a second area of philosophical psychology in which the issue of whether or not causal relations exist is of great importance. For physicalists who hold that mental events are supervenient upon brain events, a powerful argument exists to the effect that there is no causal influence at all at the level of mental events. If the physical world is causally closed, in that everything physical that is caused is caused by something that is also physical (as physicalists hold), then mental events cannot cause physical events. Moreover, since every mental event is determined by a physical event via the supervenience relation, no mental event causes any other mental event. Hence mental events are causally impotent. This argument can in principle be generalized to show that only the most fundamental level of physics can contain any causal influence. Thus mental causation is cast into doubt, a worry reinforced by more traditional arguments that reasons cannot be causes, because a reason logically entails the corresponding action, whereas causes are only contingently tied to their effects. An examination of various arguments in this area may be found in Heil and Mele (1993) (see SUPERVENIENCE AND DETERMINATION).

Finally, in most of its formulations, Bayesian decision theory once avoided any reference to causal connections between states of nature, acts, and outcomes. When Newcomb's problem and more realistic variants of it were discovered, it became apparent that despite valiant efforts to formulate acausal Bayesian decision theories, it was essential to distinguish between genuine causal connections and noncausal connections, so that we can distinguish between mere evidential relations and the kind of causal relations upon which actions are in fact based. (For a good anthology on this and other issues, see Campbell and Sowden (1985).)

References

Campbell, R., and Sowden, L. (eds) 1985: *Paradoxes of Rationality and Cooperation: Prisoner's Dilemma and Newcomb's Problem* (Vancouver: University of British Columbia Press).

Cook, T., and Campbell, D. 1979: *Quasi-Experimentation: Design and Analysis for Field Settings* (Boston: Houghton-Mifflin).

Heil, J., and Mele, A. (eds) 1993: *Mental Causation* (Oxford: Clarendon Press).

Hughes, R. I. G. 1989: *The Structure and Interpretation of Quantum Mechanics* (Cambridge, MA: Harvard University Press).

Humphreys, P. 1989: *The Chances of Explanation* (Princeton: Princeton University Press).

Humphreys, P., and Freedman, D. 1996: The grand leap. *British Journal for the Philosophy of Science*, 47, 113–23.

Lewis, D. 1983a: Causation. In *Philosophical Papers*, vol. 2 (Oxford: Oxford University Press), 159–72.

—— 1983b: Postscripts to "Causation." In *Philosophical Papers*, vol. 2, 172–213.

Mackie, J. L. 1974: *The Cement of the Universe* (Oxford: Clarendon Press).

Salmon, W. C. 1984: *Scientific Explanation and the Causal Structure of the World* (Princeton: Princeton University Press).

Sklar, L. 1974: *Space, Time, and Spacetime* (Berkeley: University of California Press).

Spirtes, P., Glymour, C., and Scheines, R. 1993: *Causality, Prediction, and Search* (New York: Springer-Verlag).

6

Cognitive Approaches to Science

RONALD N. GIERE

Until very recently it could have been said that most approaches to the philosophy of science were "cognitive." This includes logical positivism (see LOGICAL POSITIVISM), as well as later, historically based philosophies of science, such as that of Imre Lakatos (see LAKATOS). Here the contrast is between the cognitive and the psychological or social dimensions of science. Central to all such "cognitive" approaches is a robust notion of rationality, or rational progress, in the evaluation of scientific theories or research programmes (see THEORIES). Carnap sought an inductive logic that would make the evaluation of hypotheses rational in the way that deductive inference is rational. Lakatos defined rational progress in terms of increasing empirical content. For both it was essential that the philosophy of science exhibit science as a *rational*, rather than a merely psychological or social, enterprise.

Today the idea of a *cognitive approach to the study of science* means something quite different – indeed, something antithetical to the earlier meaning (Giere 1988). A "cognitive approach" is now taken to be one that focuses on the cognitive structures and processes exhibited in the activities of individual scientists. The general nature of these structures and processes is the subject matter of the newly emerging cognitive sciences. A cognitive approach to the study of science appeals to specific features of such structures and processes to explain the models and choices of individual scientists. It is assumed that to explain the overall progress of science, one must ultimately also appeal to social factors (see SOCIAL FACTORS IN SCIENCE). So there remains a distinction between cognitive and social approaches, but not one in which the cognitive excludes the social. Both are required for an adequate understanding of science as the product of human activities.

What *is* excluded by the newer cognitive approach to the study of science is any appeal to a special definition of rationality which would make rationality a categorical or transcendent feature of science. Of course scientists have goals, both individual and collective, and they employ more or less effective means for achieving these goals. So one may invoke an "instrumental" or "hypothetical" notion of rationality in explaining the success or failure of various scientific enterprises. But what is at issue is just the effectiveness of various goal-directed activities, not rationality in any more exalted sense which could provide a demarcation criterion distinguishing science from other human activities, such as business or warfare. What distinguishes science is its particular goals and methods, not any special form of rationality. A cognitive approach to the study of science, then, is a species of naturalism in the philosophy of science (see

NATURALISM). It uses the cognitive sciences as a resource for understanding both the process of doing science and its products.

Three topics have dominated discussions among those pursuing a cognitive approach: (i) Representation: By what means, both internal and external, do scientists represent the world? What are scientific theories? (ii) Discovery and conceptual change: How do scientists use capacities like those for employing models and analogies to construct new representations? (See DISCOVERY and MODELS AND ANALOGIES.) (iii) Judgment: How do scientists judge the superiority of one representation over another? (See JUDG-MENT, ROLE IN SCIENCE.)

Among advocates of a cognitive approach there is near unanimity in rejecting the logical positivist ideal of scientific knowledge as being represented in the form of an interpreted, axiomatic system. But there the unanimity ends. Many (Giere 1988; Nersessian 1992) employ a "mental models" approach derived from the work of Johnson-Laird (1983). Others (Thagard 1988) favor "production rules" (if this, infer that), long used by researchers in computer science and artificial intelligence, while some (Churchland 1989) appeal to neural network representations.

The logical positivists are notorious for having restricted the philosophical study of science to the "context of justification," thus relegating questions of discovery and conceptual change to empirical psychology. A cognitive approach to the study of science naturally embraces these issues as of central concern. Here again there are differences. The pioneering treatment, inspired by the work of Herbert Simon (Langley et al. 1987), employed techniques from computer science and artificial intelligence to generate scientific laws from finite data. These methods have now been generalized in various directions (Shrager and Langley 1990). Thagard (1988) favors a more recent "spreading activation" model. Nersessian (1992) appeals to studies of analogical reasoning in cognitive psychology, while Gooding (1990) develops a cognitive model of experimental procedures. Both Nersessian and Gooding combine cognitive with historical methods, yielding what Nersessian calls a "cognitive-historical" approach. Most advocates of a cognitive approach to conceptual change are insistent that a proper cognitive understanding of conceptual change avoids the problem of incommensurability between old and new theories (see INCOMMENSURABILITY and the essays in Giere 1992, part I).

No one employing a cognitive approach to the study of science thinks that there could be an inductive logic which would pick out the uniquely rational choice among rival hypotheses. But some, such as Thagard (1991) think it possible to construct an algorithm that could be run on a computer which would show which of two theories is best. Others (Giere 1988, ch. 6) seek to model such judgments as decisions by individual scientists, whose various personal, professional, and social interests are necessarily reflected in the decision process. Here it is important to see how experimental design and the results of experiments may influence individual decisions as to which theory best represents the real world.

The major differences in approach among those who share a general cognitive approach to the study of science reflect differences in cognitive science itself. At present, "cognitive science" is not a unified field of study, but an amalgam of parts of several previously existing fields, especially artificial intelligence, cognitive psychology, and cognitive neuroscience. Linguistics, anthropology, and philosophy also contribute.

Which particular approach a person takes has typically been determined more by original training and later experiences than by the problem at hand. Progress in developing a cognitive approach may depend on looking past specific disciplinary differences and focusing on those cognitive aspects of science where the need for further understanding is greatest.

References

Churchland, P. M. 1989: *A Neurocomputational Perspective* (Cambridge, MA: MIT Press).

Giere, R. N. 1988: *Explaining Science: A Cognitive Approach* (Chicago: University of Chicago Press).

Giere, R. N. (ed.) 1992: *Minnesota Studies in the Philosophy of Science*, vol. 15: *Cognitive Models of Science* (Minneapolis: University of Minnesota Press).

Gooding, D. 1990: *Experiment and the Making of Meaning* (Dordrecht: Kluwer).

Johnson-Laird, P. N. 1983: *Mental Models* (Cambridge, MA: Harvard University Press).

Langley, P., Simon, H. A., Bradshaw, G. L., and Zytkow, J. M. 1987: *Scientific Discovery* (Cambridge, MA: MIT Press).

Nersessian, N. J. 1992: How do scientists think? In *Minnesota Studies in the Philosophy of Science*, vol. 15: *Cognitive Models of Science*, ed. R. N. Giere (Minneapolis: University of Minnesota Press), 3–44.

Shrager, J., and Langley, P. 1990: *Computational Models of Discovery and Theory Formation* (Palo Alto, CA: Morgan Kaufmann Publishers, Inc.).

Thagard, P. 1988: *Computational Philosophy of Science* (Cambridge, MA: MIT Press).

—— 1991: *Conceptual Revolutions* (Princeton: Princeton University Press).

7

Computing

LESLIE BURKHOLDER

Computing as a science is the study of computers, both the hardware and their programs, and all that goes with them (Newell and Simon 1976). The philosophy of computer science is concerned with problems of a philosophical kind raised by the discipline's goals, fundamental ideas, techniques or methods, and findings. It parallels other parts of philosophy – for example, the philosophy of economics or linguistics or biology – in primarily considering problems raised by one discipline, rather than issues raised by a group of disciplines, as does the philosophy of science in general.

One example of a philosophical problem in computing science is: What is a computer? This is a typically philosophical kind of question about one of the subject's fundamental notions. Careful inspection of texts in theoretical computer science shows that the question is not answered there; what are presented are various different models of computers and computation. Perhaps there is no one general answer to the question, or perhaps one has to be found outside computer science proper. Another example of a problem of a philosophical sort is: What kind of science is computer science? It certainly doesn't look like any of the paradigmatic empirical natural sciences such as physics or chemistry: there seem to be no laws or explanatory theories. One possibility is that it is a formal science, like mathematics or logic; another that it isn't a science at all. A third example of philosophical interest in computer science is the debunking of exaggerated claims about the discipline's goals or techniques or findings. It is so easy to find instances of this that it can seem to be the chief concern of the philosophy of computing science. One of the aims of this chapter is to demonstrate that this is not so.

It should be said before going on that the interest in computing for philosophers of science isn't restricted to consideration of computing as a science. An active area of research concerns computer programs that make scientific discoveries and decide between competing theories. This research has two goals. One is to show that there are procedures which can lead reliably from observational or laboratory data to true scientific laws like Ohm's law relating the voltage in an electrical circuit to its current and resistance. The idea is to demonstrate that there is, *contra* Popper and others, a logic or set of rules for scientific discovery (see POPPER). The other goal is sometimes to simulate the reasoning procedures which scientists like Kepler or Boyle or Wegener may have used in arriving at new laws or deciding between competing theories. The two goals need not coincide, but do so if successful scientists – for example, Kepler or Wegener – employed reliable discovery and theory change procedures. This topic cannot be considered in any detail here; recent surveys can be found in Glymour (1992) and Boden (1991, ch. 8).

Extravagant claims

Common targets for skepticism by philosophers are claims in and about artificial intelligence (AI), a sub-science of the science of computing. The goal in AI is to make computing machines do things that apparently require intelligence when done by people. One target is the idea that (1a) since this computer behaves – for example, it responds to queries – in ways not distinguishable at a rate better than chance from the ways intelligent people do, (2a) it thinks or has intelligence. This is the famous Turing imitation or indistinguishability test for the possession of intelligence (Moor 1992). Another target of skepticism is the claim or inference that (1b) since this computer solves problems or performs tasks typically believed to need thought or intelligence or ingenuity to solve, (2b) it thinks or has intelligence. This is the strong AI inference or thesis (Searle 1980). These two targets are not quite the same, although it is easy to run them together. Obviously (1a) and (1b) aren't the same: A computer might solve problems requiring thought, although its manner or ability to solve them is distinguishable from that of any human being: this is sometimes said of chess-playing computers.

The validity of the Turing imitation test is hotly disputed. But it does not seem to be treated as a serious canon of research in AI, so it won't be considered further here. The strong AI thesis or inference, on the other hand, is. For example, Herbert Simon, one of the founders of the discipline, announced to a group of his students back in 1956 that during the previous Christmas vacation he and Allen Newell had invented a program capable of making a computer think. The program was Logic Theorist, a program that by itself constructs formal proofs in a system for truth-functional propositional logic. This is a task that many people believe requires insight or cleverness or thought. Simon's announcement looks as if it assumed strong AI.

But does a machine running Logic Theorist think? Here is a variant of a famous imagination experiment due to the philosopher John Searle (1980) designed to show that it doesn't. Imagine the program written out so that a person could follow it. The instructions, for example, might say that if the task is to deduce a proposition of the form $A \rightarrow B$ from a proposition which is a substring of B, first assume A. The instructions would also say how to recognize that a proposition is of the form $A \rightarrow B$. Following the instructions, a person with no understanding of what was going on could produce proofs mechanically. No intelligence or ingenuity on the part of the program executor is needed. Contrary to Simon's announcement, there should be no thinking going on at all, just mindless mechanical following of instructions!

Perhaps computers can be made to think; but just having them execute programs that enable them to solve problems usually believed to need thinking isn't sufficient. Strong AI is a big mistake. Of course, researchers both within and outside AI have come to AI's defense: if strong AI is false, is anything interesting left of the subject's goals or methods? One possibility is that while computers executing programs like Logic Theorist don't think, they simulate the processes of real thinkers when they construct proofs by thinking or, to refer back to the earlier example of discovery programs, do scientific research (Pylyshyn 1989). Physics and biology increasingly include research done by computer simulation; so does the scientific study of the mind. Another possibility is this: Logic Theorist and its successors don't think, as the imagination experiment shows. That much is obvious. The right conclusion, then, is that doing proofs doesn't really require intelligence or ingenuity or thought, contrary to popular opinion. AI doesn't produce and study artificial intelligences or thinkers; put immodestly, it shows where thought and intelligence are, like the ether or phlogiston, unnecessary or mistaken hypotheses.

There are other targets for philosophical skepticism in less eye-catching parts of computer science than AI. For instance, some computer scientists believe, or at least appear to believe, that it is possible to mathematically prove or formally verify that programs and computer chips will operate correctly, exactly as intended. Such formally verified software or hardware presumably has a smaller probability of error when employed than similar but unverified software or hardware. So, cost and difficulty aside, it seems worthwhile to prove correctness. There are two caveats, however, against exaggerating what proofs of correctness achieve. One is that they don't guarantee zero probability of error when a piece of software or hardware is used (Fetzer 1988). Programs, for example, run under computer operating systems and on hardware and typically need to be interpreted or compiled. These can all be sources of errors when the program is executed. A correctness proof doesn't show that there aren't such sources. Similarly, the fabrication process for a computer chip can produce defective chips; manufacturers typically expect this. Almost all are caught by the inspection process, but some escape. A correctness proof doesn't show that this won't happen either. The mistake here, dubbed "the identification fallacy" by the mathematician and philosopher Jon Barwise (1989), is possible in other sciences too. An example in classical mechanics is to believe that truths provable about an abstract model of the simple pendulum must be exactly true of any physically realized one. The other problem is that proofs of correctness show only that a program or chip design meets explicitly stated engineering specifications concerning how the item is to work. These can easily not anticipate all possible contingencies, with the result that a computer behaves in ways not expected or intended. For example, the designers of America's computer-aided missile defense system failed to consider the reflection of radar signals back to the earth from the moon. The system could only interpret these reflected signals as a large Soviet missile attack. But for the intervention of people, this would have had the disastrous result of launching an American nuclear response. The result is hardly what was intended; but it was not anticipated in the system's behavior specifications.

What is a computer?

This question can seem to have an obvious answer, one provided by any computer science theory textbook or even by Turing's work back in the mid-1930s characterizing what is now known as the Turing machine computer.

A Turing machine consists of a control unit or head and an infinitely long tape divided into cells, each containing symbols from a finite alphabet or a blank. The control unit can read from and write on the tape and can move itself one section at a time to the left or right along the tape (or remain at the same location). It is limited to being in one of only a finite number of different states or conditions. What it will do at any instant, whether it reads or writes or moves, is governed by which of the states it is in and what it sees on the tape cell at its current location. A Turing machine program specifies how the control head is to behave. Typically, a program is presented as a state transition or flow diagram or table, although it is also possible to devise and use a programming language similar to Pascal or BASIC. The parts of a Turing computer can be matched with parts of a desktop computer. The central processing unit of a personal computer corresponds to the control unit of a Turing machine, and the memory devices in the desktop machine correspond to the Turing machine's tape. This matching can mislead. Turing machines have an infinite tape, so an infinite memory store external to the central processing unit. The memory in any desktop machine is only finite.

Is a computer always a Turing computer? If so, then the leading question, What is a computer?, is immediately answered. Unfortunately, as the remark at the end of the previous paragraph establishes, not every computer is a Turing machine computer: desktop personal computers are computers, but they aren't Turing machines. But suppose some idealization is allowed, as in theoretical physics. There, an idealized real spring has no mass; here, an idealized real computer will have infinite memory. Is every idealized computer a Turing machine? Most computer scientists assume that any function whose values can be calculated by some computer can be calculated by a Turing machine. This is one way of stating the Church–Turing thesis, discussed below. But this isn't the same as saying that all computers, even idealized, are Turing machines. Computer science theory textbooks, as well as presenting Turing machines, often describe other kinds of computing devices – for example, finite state and pushdown machines. These cannot calculate the values of some functions which Turing machines can; still, there is no doubt they are computers. Finally, there are computing machines with architectures quite different from all of these. Turing machine computers, pushdown machines, and so on are all von Neumann serial or sequential devices. They can make only one calculation at a time. Quite different are parallel machines, with more than one control head or central processing unit operating simultaneously. These certainly aren't Turing machine computers, even when they are equivalent in computing abilities to them. The upshot of all this is that neither computer science theory texts nor Turing's work gives a single general adequate answer to the question, What is a computer?

To answer it, the idea of an algorithm or effective method needs to be introduced. This is a procedure for correctly calculating the values of a function or solving a class of problems that can be executed in a finite time and mechanically – that is, without the exercise of intelligence or ingenuity or creativity. A familiar example of an algorithm is the truth-table method for calculating truth-functional logical validity.

A computer is anything that (when working normally) calculates the values of a function or solves a problem by following an algorithm or effective method. Two points are packed into this. First, a computer must have states which somehow contain representations of the input and output values of the functions or the problems and their solutions. These might be tape cell contents or chemical balances or switch settings or gauge pointers or any number of other things. Second, a computer must go through some algorithm or effective procedure-executing process in which the input- or problem-representing states regularly bring about the output- or solution-representing ones.

This answer to the question is generous enough to include all the different kinds of devices mentioned earlier and some others as well. But perhaps it is too generous.

47

Imagine that pairs of rabbits infallibly or nearly infallibly produced exactly two off-spring every breeding season and that none ever died. Every pair of rabbits is, by the account given, a special-purpose computer for powers of 2. At the start of things the pair represents 2, and at the end of n breeding seasons the offspring represents 2^n. Perhaps this device every once in a while, every billionth season or rabbit, breaks down, and an extra rabbit is produced. But, at least when working properly, it works as described. Is this really a computer? Yes; the only reason it seems odd is that ordinary rabbits don't reproduce with such regularity.

Findings

Computer science theory has some central results which are, or should be, of interest in much the same way that the results of quantum theory are often said to be of philosophical interest. One of these, the Church–Turing thesis, is very familiar. The other, sometimes known as the Turing complexity thesis, is less so.

The Church–Turing thesis concerns functions whose values can be calculated or problems whose solutions can be found by executing an algorithm or effective procedure. Researchers would often like to know of a particular function or problem whether there is an algorithm like the truth-table method for truth-functional logical validity, for example, for determining first-order logical validity. Unfortunately, the idea of an algorithm or effective method is not a rigorously defined one. What does it mean, for example, that a procedure needs no ingenuity or understanding or creativity to be executed? This informality makes it hard, in the absence of an obviously effective method for calculating the values of the function or solving the problem, to establish that a function is or isn't effectively calculable or that a problem is or isn't algorithmically solvable. The Church–Turing thesis is intended to make the idea of an effectively or algorithmically calculable function or an algorithmically solvable problem precise and rigorous by equating these informal notions with formally and rigorously defined ones.

There are various apparently different versions of the thesis. One is: Any function whose values are algorithmically or effectively calculable and any problems whose solutions can be found by execution of an algorithm or effective procedure can have their values or solutions determined by a Turing machine computer (and vice versa). Another is: Any effectively or algorithmically calculable function on positive integers is a general recursive one (and vice versa). Despite their difference in appearance, these two, and other versions of the thesis, all come down to the same thing. For example, on the one side of these two different versions there is really no difference between effectively calculable functions or algorithmically solvable problems in general and effectively or algorithmically calculable functions on positive integers. Functions and problems can all be coded as functions of positive integers. On the other side of the different versions, it turns out that general recursive functions are provably just the same as the Turing machine computable ones. This identity extends to yet other formally and rigorously defined types of functions not mentioned here (Kleene 1950, sec. 62).

The thesis has consequences which most philosophers will recognize. A theorem proved by Church says that the first-order validity function is not a general recursive one, or equivalently, not one whose values are computable by any Turing machine. The Church–Turing thesis then yields the conclusion that there is absolutely no algorithm or effective procedure like the truth-table method which will determine whether an arbitrary first-order proposition is logically valid. Some writers have further concluded that room is thus left even in first-order logic for human creativity and genius, at least if these are not algorithmically or effectively executable processes (Myhill 1952).

Two questions can be asked about the thesis. One concerns the standards for its evaluation. It is often said that it is not a theorem of a formal system and cannot be expected to be formally proved in the same way that, say, the equivalence of Turing machine computable functions and general recursive functions can be (Kleene 1950, sec. 62). These two are mathematically well defined, so can be the subject of formal proofs. But the idea of an algorithmically calculable function or effectively solvable problem isn't. So the claim of equivalence between algorithmically or effectively calculable function or problem, and, say, a Turing machine computable function or problem can't be formally established. How, then, should its truth be judged? The other question concerns evidence for its having met the standards. A sensible answer to the first question is that the thesis is to be judged by the standard which Tarski proposed to judge his definition of truth: the equivalence asserted has to be both materially or extensionally adequate and formally correct (Etchemendy 1992). The evidence for the thesis's being formally correct is within recursion or computation theory. What is required is that the formal ideas on the right side of the equivalence (for example, those of a Turing machine solvable problem or a general recursive function) be mathematically usable. The evidence for its extensional adequacy comes in two parts. First, there are the many cases in which a particular effectively calculable function or algorithmically solvable problem has been shown to be a general recursive function or solvable by a Turing machine (Kleene 1950, sec. 62). Second, as hinted above, there are many analyses like Turing's own of what seems to be essential in mechanical calculation and what is plainly not so. These analyses all seem to end up being provably equivalent, suggesting that the equivalences claimed by the different versions of the Church–Turing thesis leave nothing out, so are extensionally correct.

There are perhaps not so many different statements of the Turing complexity thesis as of the Church–Turing thesis. To understand them, a division among Turing machine computable functions or Turing machine solvable problems needs to be introduced. Consider the variety of Turing machine programs that might be devised, based on various different algorithms, for calculating the values of a Turing computable function. Some of these might be less efficient, take more Turing machine computation steps (e.g., move one tape cell to the left and change control head state), and hence more time to calculate values of the function, than others. In particular, some programs might take a time that in the worst case increases exponentially with the size of the function's input, while others do better, growing only polynomially with input size in the worst case. Now suppose that the fastest possible, not merely the fastest found so far, Turing machine program for calculating the values of some function has a worst-case time performance that grows exponentially in its demands. Functions with such requirements are termed exponential in their Turing machine time complexity, or merely, for shortness, exponential. For some other function the most time-efficient program may fortunately grow only polynomially in minimum requirements. Functions with these demands are termed polynomial in their Turing machine time complexity, or again for shortness, polynomial. (The common use here by computer scientists of the terms "exponential" and "polynomial" is a bit sloppy: factorials are to be counted as exponential.)

Versions of the Turing complexity thesis can now be stated. One is: If a function is exponential in its time-resource demands on a Turing machine, calculating its values will also be exponential in minimal demands on some resource (often, but not necessarily, time) on any other kind of computing device; if a function is merely polynomial in its time complexity on a Turing machine computer, then it will also be merely polynomial in its resource needs on every other kind of computer. Another version is: The functions whose values can be realistically or practically, as opposed to theoretically, computed are just the ones whose resource demands are merely polynomial; effectively or algorithmically calculable functions whose values can be mechanically calculated only with exponentially increasing amounts of resources are for all practical purposes uncomputable, even if they are in principle so.

Are these stylistic variants of the same thesis? Probably not. There are reasons for doubting that the functions whose values are computable practically are precisely the ones whose values can be computed with only polynomially increasing resource requirements. Suppose the resources devotable to computing the values of a function are fixed or increase in their productivity or availability at a rate smaller than the resource requirements for calculating values of the function. Then the function won't be practically computable, whether its resource needs grow exponentially or only polynomially. Conversely, suppose a function has worst-case requirements for computing its values that increase exponentially as the input size grows. It may still be usably calculable. As it happens, functions scheduling teachers and groups of students into school classrooms are typically exponential. But many schools employ programs run on micro-computers to successfully calculate values of these functions.

Computer scientists are confident that one interesting example of a function needing exponentially increasing amounts of at least some resource is the truth-functional logical validity function. It isn't difficult to arouse a suspicion, although not a proof, that this might be so. Familiar textbook methods for calculating truth-functional logical validity plainly take longer and longer to execute as the length of the proposition being tested grows, as any logic student can testify. These time demands on a serial or sequential processor, which is what a logic student executing one of these textbook procedures is, translate into resource demands on other kinds of computers. Imagine a machine with parallel processors and a costless way of breaking up a proposition to be tested for truth-functional logical validity into pieces, one piece for each processor. If time for calculating values of the function is to be kept constant or increase only polynomially with the length of the tested proposition, the number of processors needed will increase exponentially. Philosophers should find this interesting both on its own and because of its potential consequences. For example, common conceptions of ideal scientific rationality assume the ability to calculate truth-functional logical validity for arbitrarily long propositions. Cognitive scientist Chris Cherniak has argued that a consequence of the function's being exponential is that ideal rationality is practically speaking unattainable, hence never truly exhibited by anyone or anything (Cherniak 1984). Of course, as the considerations in the previous paragraph should suggest, even if the function is only polynomial in its resource needs, this consequence may result.

Computer science and other sciences

Is computer science a science at all, in the way that, for example, physics certainly is? On a common division among the sciences, if it is one, it must be either an empirical natural science like astronomy or biology, or a nonempirical science like pure mathematics or logic, or one of the social sciences (Hempel 1966, pp. 1–2). While some computer scientists are interested in the social effects of the widespread use of computers, this is certainly only a minor part of the discipline: computer science is not a social science. Nor again is it one of the nonempirical sciences. Some work in computer science is mathematical in nature: for example, the study of the mathematical properties of computer programs and algorithms. But not all of it is; the design of reliable and efficient information storage hardware or central processing units is certainly no more pure mathematics than the design of reliable and efficient automobile engines. Finally, computer science seems not to be like any of the empirical natural sciences. These use the experimental method to arrive at theories explaining facts about the part of the natural world studied. Computer science certainly seems not to study any part of the natural world. Although human beings and perhaps even honeybees compute, naturally occurring computation is not the specific object of study in computer science. And what theories or laws has it produced comparable to the theory of natural evolution in biology or any of the laws of Newtonian mechanics? What facts would such theories explain?

Computer science is an engineering science, an empirical science of the artificial rather than the natural world, so in a category not allowed for in the standard division (Newell and Simon 1976). Like other empirical engineering sciences, it conducts experiments and constructs theories to explain the results of them. Here is an easy-to-understand example. A binary tree is a widely used structure for storing orderable data. As new data comes in, the key by which it is retrieved is added to the bottom of the tree, to the left if it is prior in ordering to an existing data node, otherwise to the right. Suppose some data has been stored this way. How long, on average, will it take to retrieve the data by looking for its key? This is discoverable by experiment, by counting and averaging over randomly selected trees (Bentley 1991). Similarly, a function describing the relationship between the period of a simple pendulum and its length is discoverable by experiment, by counting and averaging over randomly selected pendulums. And just as the discovery about pendulums can be theoretically explained by classical mechanics, the finding about data retrieval times can be explained by a part of computer science, the theory of algorithms.

References

Barwise, J. 1989: Mathematical proofs of computer system correctness. *Notices of the American Mathematical Society*, 36, 844–51.

Bentley, J. L. 1991: Tools for experiments on algorithms. In *CMU Computer Science: A 25th Anniversary Commemorative*, ed. R. F. Rashid (New York: ACM Press), 99–124.

51

Boden, M. A. 1991: *The Creative Mind* (New York: Basic Books).

Cherniak, C. 1984: Computational complexity and the universal acceptance of logic. *Journal of Philosophy*, 81, 739–58.

Etchemendy, J. 1992: On the (mathematical) analysis of truth and logical truth (unpublished lecture, Boston).

Fetzer, J. H. 1988: Program verification: the very idea. *Communications of the ACM*, 31, 1048–63.

Glymour, G. 1992: Android epistemology: computation, artificial intelligence, and the philosophy of science. In *Introduction to the Philosophy of Science*, ed. M. H. Salmon et al. (Englewood Cliffs, NJ: Prentice-Hall), 364–403.

Hempel, C. G. 1966: *Philosophy of Natural Science* (Englewood Cliffs, NJ: Prentice-Hall).

Kleene, S. C. 1950: *Introduction to Metamathematics* (Princeton: D. van Nostrand).

Moor, J. H. 1992: Turing test. In *Encyclopedia of Artificial Intelligence*, ed. S. C. Shapiro, 2nd edn (New York: John Wiley & Sons), 1625–9.

Myhill, J. 1952: Some philosophical implications of mathematical logic. *Review of Metaphysics*, 6, 165–98.

Newell, A., and Simon, H. 1976: Computer science as empirical inquiry: symbols and search. *Communications of the ACM*, 19, 113–26.

Pylyshyn, Z. W. 1989: Computing in cognitive science. In *Foundations of Cognitive Science*, ed. M. I. Posner (Cambridge, MA: MIT Press), 49–92.

Searle, J. R. 1980: Minds, brains, and programs. *Behavioral and Brain Sciences*, 3, 417–24. [Followed at pp. 425–57 with replies, some defending strong AI.]

8

Confirmation, Paradoxes of

J. D. TROUT

The confirmation of scientific hypotheses has a *quantitative* and *qualitative* aspect. No empirical hypothesis can be confirmed conclusively, so philosophers of science have used the theory of probability to elucidate the quantitative component, which determines a degree of confirmation – that is, the extent to which the hypothesis is supported by the evidence (see PROBABILITY and EVIDENCE AND CONFIRMATION). By contrast, the qualitative feature of confirmation concerns the prior question of the nature of the *relation* between the hypothesis and the evidence if the hypothesis is to be confirmed by its instances. If a hypothesis is to be supported by a body of evidence, it must be related to the evidence in an appropriate way. The paradoxes of confirmation arise in the attempt to characterize in first-order logic the qualitative relation between hypothesis and evidence.

The most celebrated example – the raven paradox – begins with the simple hypothesis "All ravens are black," symbolized in first-order quantification as (x) (Rx → Bx). According to a natural and intuitive principle of evidential support, often called "Nicod's condition," a hypothesis is confirmed by its positive instances and disconfirmed by its negative ones. So, the hypothesis that all ravens are black, (x) (Rx → Bx), is confirmed by the observation statement "This is a raven and it is black" (Ra ∧ Ba), and disconfirmed by the observation statement 'This is a raven and it is not black' (Ra ∧ ~ Ba).

The central paradox arises from adherence to Nicod's criterion, in combination with a fundamental principle of confirmation known as the "equivalence condition." It proceeds from the observation that there are a number of logically equivalent ways of expressing any universal generalization. If one views the relation of logical implication as a species of the confirmation relation, then any two logically equivalent generalizations are equally confirmed or disconfirmed by the same evidence statement. Similarly, if a generalization is confirmed by some evidence statement e_1, it is confirmed by any other evidence statement logically equivalent to e_1.

In light of the equivalence condition, the logical equivalence of "All ravens are black" and "All non-black things are non-ravens" entails that these statements are equally supported by the same body of evidence *e*. Therefore (and here is the paradox), the observation of a non-black non-raven (~Ba ∧ ~Ra), such as a blue book, is a confirming instance of the hypothesis "All ravens are black." The equivalence condition has further consequences. The generalization (x) (Rx → Bx) is also logically equivalent to (x) [(Rx ∨ ~Rx) → (~Rx ∨ Bx)]. Here, the antecedent is a tautology, and thus any truth-value assignments that make the consequent true will confirm the hypothesis that all ravens are black. Since the consequent is a disjunct, and the extensional, deductive model of hypothesis testing requires only one of the disjuncts to be true for the entire (compound) proposition to be true, "All ravens are black" is confirmed by any observation of an object that is either black or a non-raven. Surprisingly, then, the discovery of either a hunk of tar or a boat confirms the generalization that all ravens are black. These results should be disturbing to anyone who desires an account of confirmation that bears some resemblance to what scientists actually do, or who ties the success of scientific methodology to routine judgments of evidential relevance.

Philosophers have handled the paradox in a variety of ways. Some rejected Nicod's condition, and others abandoned the equivalence condition. Hempel (1945) treats as a kind of psychological illusion the apparent irrelevance of an observation of a blue book to the hypothesis that all ravens are black. In point of logic, Hempel has argued, this hypothesis is as much about non-ravens as about ravens, black things as non-black things. One might argue that if this view is correct, then scientists' confirmational judgments must be seen as systematically defective, since they would not give equal weight to the observations, ignoring the blue book and attending to the black raven.

Against purely syntactic accounts of confirmation, Nelson Goodman has argued that for any empirical hypothesis, it is possible to construct an alternative hypothesis equally well supported by the evidence to date, so that it is not clear which hypothesis is confirmed. Goodman (1965) defines the predicate "grue" to express the property that an object has if and only if it is green at time t and blue thereafter. If confirmation is analyzed completely in terms of syntactic relations in a first-order observation language, then the universal generalizations "All emeralds are green" and "All emeralds are grue" are equally well supported by the evidence. We classify objects in terms of their greenness, not in terms of their color-plus-temporal properties, such as grueness. In Goodman's words, "green" is a *projectible* predicate, implicated in counterfactual-supporting, lawlike generalizations (see LAWS OF NATURE). "Grue" is not similarly projectible.

Why, then, do we use "green" rather than "grue"? Goodman explains this fact in terms of a pragmatic notion of "entrenchment" (see PRAGMATIC FACTORS IN THEORY ACCEPTANCE); "green" is more crucially implicated in our theoretical vocabulary and subsequent practice than "grue." Although it is surely true that "green" is more entrenched than "grue," many philosophers have found the concept of entrenchment to be merely a description of conditions making current usage possible, rather than an explanation of its prominence. Instead, they hold that the only adequate explanation for this entrenchment must appeal (in the typical case) to the reality of the properties in question. Although Goodman's merely pragmatic explanation for the persistence of certain "projectible" predicates may be unable to account for why certain entrenched predicates are abandoned in favor of others, his criticism is effective against the formal, purely syntactic account of confirmation.

The problem of grue arises from a conception of evidence favored by logical empiricists, according to which evidence must be exclusively observable. The paradoxes may be resolved if *theoretical* issues, such as explanatory power or unification (see UNIFICATION OF THEORIES), are regarded as evidential. But the paradoxes of confirmation have additional sources.

Although once central to discussion in the philosophy of science, the paradoxes of confirmation now receive less attention, for two closely related reasons. First, internal criticisms of the sort described here were ultimately decisive against purely syntactic or logical accounts of confirmation. Every technical maneuver designed to resurrect a first-order analysis of confirmation was met with an equally clever rejoinder. This led many to suppose that the difficulty resided not in genuine peculiarities of confirmation, but in frailties of general features associated with empiricist projects during the twentieth century, among them a purely syntactic conception of theories, a deductivist account of hypothesis generation and support, and an observational foundationalist understanding of theory testing and evidential support. Second, emerging naturalistic accounts of confirmation offered detailed alternatives to the aforementioned empiricist picture. The scientific hypotheses to be confirmed typically express causal relations. As earlier reductive analyses of theoretical disposition terms showed (see DISPOSITIONS AND POWERS; THEORIES; and THEORETICAL TERMS), causal notions cannot be captured by first-order operations, such as material implication. Drawing on causal theories of knowledge and evidence, naturalistic approaches to confirmation treat confirmation relations as themselves causal and subject to *a posteriori* study. Judgments of cumulative plausibility, then, have led most people to abandon efforts to resuscitate the particular qualitative conception of confirmation that gave rise to the paradoxes.

References and further reading

Brown, H. 1977: *Perception, Theory and Commitment* (Chicago: University of Chicago Press).

Carnap, R. 1962: *The Logical Foundations of Probability*, 2nd edn (Chicago: University of Chicago Press).

Goodman, N. 1965: *Fact, Fiction, and Forecast* (Indianapolis: Bobbs-Merrill); 4th edn (Cambridge, MA: Harvard University Press) 1983.

Hempel, C. 1945: Studies in the Logic of Confirmation. *Mind*, 54, 1–26, 97–121; repr. with a Postscript in *idem, Aspects of Scientific Explanation* (New York: Free Press, 1965), 3–51.

Sainsbury, M. 1988: *Paradoxes* (Cambridge: Cambridge University Press).

Salmon, W. 1973: Confirmation. *Scientific American*, 228(5), 75–83.

Stalker, D. 1994: *Gruel: The New Riddle of Induction* (Chicago: Open Court).

Swinburne, R. 1971: The Paradoxes of Confirmation – A Survey. *American Philosophical Quarterly*, 8, 318–30.

9

Convention, Role of

LAWRENCE SKLAR

The claim that some assertion is true "as a matter of convention" is likely to arise only in the circumstance that the assertion is allowed to be true, though an account of its believability as being warranted by its conformity to observable facts is taken to be inadequate.

One important realm where truth is alleged to be truth merely by convention is in the positivist account of the nature of logical truth (see LOGICAL POSITIVISM; LOGICAL EMPIRICISM). If one accepts some logicist reduction of all scientifically respectable a priori truth to the truths of logic, supplemented by the definitions needed to extend logical truth to analytical truth, one still has the problem of accounting for the truth of the truths of logic. The positivist program was to disarm the threat of this last domain of a priori knowledge by alleging that all logical truths themselves merely reflect convention or decision on our part to utilize our terms in a certain way. Thus the correct filling in of the truth table for a simple truth-functional logical connective was allegedly warranted by the merely definitional status of the truth table in "giving the meaning we assigned" to the connective in question.

Needless to say, the doctrine that all logical truths are true merely by convention is fraught with problematic aspects. While the anti-conventionalist will freely admit that the adoption of the English word "or" to stand for the disjunctive logical connective is, to be sure, a mere convention on our part, the facts about which logical systems are consistent, which derivational relations truth-preserving and, hence, worthy syntactical representatives of genuine logical entailment, and the other facts about logical truth are not in any sense "made true" merely by a decision on our part in the anti-conventionalist's view.

A quite different set of conventionalist claims arises, however, in the discussion of the alleged underdetermination of our theory of the world by the, in principle, possible totality of all observational data. It is to these issues of alleged conventionality that this discussion is directed.

Poincaré on geometry

The beginnings of the contemporary debate over alleged conventional aspects in physical theory is often traced back to H. Poincaré's discussion of the epistemology of geometry in his "Space and geometry." Having demonstrated the logical consistency (relative to Euclidean geometry) of non-Euclidean geometry, and having rejected the Kantian theory of a priori knowledge of geometry founded on transcendental idealism, Poincaré then took up the question of whether or not we could empirically establish

56

the truth of a claim to the effect that the geometry of the physical world obeyed the axioms of non-Euclidean geometry.

To illustrate how the world might appear non-Euclidean to the sentient beings in it, Poincaré outlined a Euclidean world bounded by a sphere in which all material objects shrink in a lawlike way as they are removed from the center of the sphere to its periphery. The world-ball is also filled with a medium whose index of refraction varies from center to boundary. If the geometry of this world is mapped out by the inhabitants in the usual ways, he claimed, they will posit it to be a Lobachevskian world of constant negative curvature and infinite extent. But while beings educated in such a world would posit it to be non-Euclidean, if we were transported to such a world, we would hold to our Euclidean geometry and instead posit the necessary "shrinking" fields and "index of refraction" fields that would reconcile the observational data with the posited Euclidean geometry.

Poincaré says: "Experiment guides us in this choice [of a geometry], which it does not impose upon us. It tells us not what is the truest, but what is the most convenient geometry" (1952, pp. 70–1). What are the crucial ingredients of this "conventionalist" claim?

Geometry and general theory

There are three major presuppositions behind the Poincaréan claim. First there is the assumption that the totality of our evidential basis for accepting or rejecting a geometric theory of the world is confined to a proper subset of the claims made by the theory in general. All that we can determine empirically are the relations of material objects with one another. But we have no direct empirical access to the structure of space itself. Thus, the path of a light ray is available to us empirically, but not what the actual shortest-distance paths in the space are. And only local relations, even among the material objects, are part of the observation basis. Thus we can empirically determine of two measuring rods whether or not their end points coincide when the rods are brought into contact. But we cannot tell directly of two separated rods whether or not they are of the same length.

Next is the claim that our theory is such that we can derive from it consequences about the limited part of the world of the directly observable only by invoking a multiplicity of parts of the structure of the theory. Thus to come to some conclusion about the intersections of light rays in our theory, we need to posit both a geometry for space and also a law connecting the paths of light rays in space to some geometric feature of the space. We might do this by assuming that light rays do, indeed, follow shortest-distance paths in the space. It is this need for multiple elements in the theory to be utilized to come to an observational conclusion that gives us the flexibility we need to be able to make more than one possible adjustment to the theory in the face of surprising data. If light rays don't behave the way we expect in our model of a Euclidean empty world, we can either posit a non-Euclidean geometry or else, alternatively, hold to Euclidean geometry and posit that space is filled with a medium with a variable index of refraction.

Finally, there is Poincaré's assumption about the role of physical theory in general. The purpose of theory, he believed, is to provide a set of lawlike regularities governing the order of the items that constitute the elements of our experience. Any theory that provides a complete and correct summary of these regularities has served its purpose. There can be nothing to choose, apart from matters of convenience, between two apparently distinct theories that serve the purpose equally well.

It is clear that Poincaré's claim is generalizable beyond claims of conventionality for our geometric theory of the world. Any sufficiently rich theory will, if these arguments are correct, be underdetermined by the totality of possible observational data relevant to its confirmation or disconfirmation. Geometry simply provides an ideal case for the claim of the conventionalist, for two reasons. First, a clearly delimited "observation basis" is available in the form of local relations among material objects. Second, the formalized nature of geometry provides us with a systematic way of actually characterizing the alternative theoretical possibilities.

Poincaré's argument preceded the actual invocation of non-Euclidean geometries in physics. His conventionalist arguments were subsequently applied both to the special theory of relativity (with arguments to the effect that, for example, the specification of distant simultaneity for events was purely conventional) and to general relativity in the form of arguments to the effect that the choice of a curved or a flat space-time in that theory was, once again, simply a conventional choice of "convenience" on our part.

Denying the problem

The problem of underdetermination takes for granted that there is at least a portion of our theoretical claims that is in principle immune to direct confrontation with observational data. But is this so? It has certainly been denied by some. Often the denial takes the following form: If we assume that the "observational" is "sense-data in the mind of the subject," then we can have a sharp observational/nonobservational distinction in our theories. But many philosophers deny the intelligibility of the "purely phenomenal" realm. Any distinction between observables and nonobservables must then be within the realm of the physical. But for physical objects there is no such hard-and-fast distinction, as "slippery slope" arguments will, for example, take us from tables to bacteria to viruses to molecules to atoms to quarks as all "observable in principle."

But one can, of course, hold fast to the older ideas of the observable as the realm of sense-data. Or one can observe that within much of foundational theoretical physics there is, built into the theories themselves, a presupposed distinguished class of observables. In relativistic theories these are the coincidences at a point of material events. In either case the possibility of empirical indistinguishability of theories remains a live issue.

Realist responses

One set of responses to the conventionalist thesis takes the semantics of that part of the theory that outruns the observable confirmational basis as being on a par with the semantics of the language referring to the observables (see REALISM AND INSTRUMENTALISM). On this view, referring terms have determinate denotations in objects not present in the observation basis, and the individual sentences of the theory describing the unobservables are to be taken as straightforwardly individually true or false.

A simple response to the underdetermination problem is just some version or other of skepticism. If the nature of the world outruns our ability to fix on one possible universe as the actual world in which we live, even if we have available to us the totality of possible observational facts, that is just too bad. A variant of this is to observe that even such underdetermination does not block us from fixing on a theory that is "empirically adequate" in that it correctly captures all the possible lawlike regularities among the observables. Then one can argue that science does not care about which theory ought to be believed, but only about which theory ought to be believed to be empirically adequate. The conventionality of theory choice is thus argued to be innocuous given what scientists really care about.

Alternatively, one can argue that although the alternative theories noted by Poincaré all have the same observational consequences, this does not prevent us from taking one of them as most rationally believable. If this were so, there would be no conventional choice to be made once all the elements of rational decision making are taken into account. It is often argued that we have grounds for rational belief in theories that outrun mere conformity of their observational consequences with the empirically accessible facts. Indeed, without such principles, we could not even engage in purely inductive generalization that takes us from what we have observed to lawlike claims over the observables themselves.

Of course, the skeptic will reply that at least in the inductive case an application of one of these inferential principles is subject to the test of further experiment. In induction, even if we do adopt the "simplest" hypothesis in conformity with the data, we could still be proved wrong in our guess by further experiment. But in the case of full underdetermination, we could not ever be proved wrong in choosing one of the empirically equivalent alternatives over the others.

Several principles for selecting among empirically equivalent theories have been proposed. Some are versions of methodological conservatism that tell us to choose the theory that makes a minimal change from the theory we antecedently accepted before considering the new empirical data. But the notion of minimal change can be ambiguous (minimal change in what respect?). And the rule suggests to the skeptic that we are just propagating earlier arbitrariness into the future. The skeptic also wonders why the mere fact that the theory is the most conservative choice should be an indication of its truth.

Another suggestion is to use "simplicity" to choose the appropriately believable theory (see SIMPLICITY). One version of this suggestion points out that one of the alternatives is often "ontologically" preferable to the others. For example, in the Poincaré parable case the Euclidean option assumes an undetectable "central point" of the universe to exist, but the Lobachevskian alternative has no such distinguished but empirically undeterminable point. Arguments of this kind play an important role in claiming for Einstein's special relativity the virtue of simplicity over its "aether frame plus compensating physical effects" alternatives, and for the general theory of relativity over some proposed "flat space-time plus gravity as force and metric field" alternatives.

Once again, however, the skeptic will wonder what entitles us to take simplicity as a mark of truth. An additional problem faced by advocates of choosing the ontologically thinnest theory that saves the observational phenomena is that it suggests moving to an anti-realist account in which we simply posit the lawlike relations among the observables themselves and invoke no unobservable explanatory structure. The realist who wants to avoid skepticism by invoking simplicity will need to tell us when we ought to stop opting for empirically equivalent alternatives with ever smaller ontologies.

The issue of the meaning of theoretical terms

An important response to the threat of skepticism presented by underdetermination is founded on a denial that the alternative theoretical accounts all apparently compatible with the same observational data really are alternative theoretical accounts at all. Rather, they are to be taken as fully equivalent theoretical descriptions of the world. Their equivalence, however, is masked by systematic ambiguity in the meaning of the terms in which they are expressed (see THEORY IDENTITY; THEORETICAL TERMS).

One version of this account, originating with A. Eddington (1920) and developed at length by M. Schlick (1963) and H. Reichenbach (1958), is based on the claim that each theoretical term of a theory has its meaning given by its association with some complex of observational terms in the form of a "coordinative definition." "Null geodesic," for example, in general relativity just means "path taken by an unimpeded light ray." In a flat space-time alternative to general relativity, however, light rays do not travel null geodesics. But all that means, according to this view, is that the term "null geodesic" is given a different meaning in the flat space-time rendition of the theory than it is in general relativity.

From this perspective, theories are fully translatable into the set of their observational consequences. Thus theories that have identical observational consequences are saying the same thing, although they frame what they say differently, using different meanings for the theoretical terms, terms which serve only to conveniently capture complexes of observational meaning. The only "conventionality" in our choice of theory is, according to this view, the usual "trivial semantic conventionality" that underlies our arbitrariness in choosing to mean what we wish by any term in our language.

This approach to theoretical meaning seems to have strongly irrealist consequences. If all a theory says about the world is completely contained in the set of observational assertions into which it can be translated by using the coordinative definitions to eliminate its theoretical terms, then it seems implausible to take apparent reference by theoretical terms as being anything more than hidden reference to the observables. If we are dealing with space-time theories, such an irrealism with regard to the ontology of "space-time itself" may, of course, be welcomed by space-time relationists. But applied more to theories in general, such an eliminationism with respect to the unobservable ontology of the world seems more disturbing. If one takes as the observables the phenomenal sense-data, in fact, this response to the skepticism latent in theoretical realism becomes full-blown phenomenalism.

Holism about meaning

The claim has often been made that, as far as confirmation goes, theories face experience as integrated wholes. We do not confirm or disconfirm the sentences of complex theories "one at a time." Rather, the theory as a whole faces the tribunal of experience. Such holistic claims have been made, for example, by P. Duhem and W. Quine (see QUINE).

A natural companion thesis to confirmational holism is semantic holism. According to this view, the meanings of the theoretical terms of a theory accrue to these terms solely by means of the role that the terms play in the theory. This claim is often accompanied by the additional claim that any partitioning of the assertions of a theory into "analytic definitions" of the theoretical terms and "synthetic" fact-stating assertions is artificial and without explanatory value. According to Quine, for example, the meaning-attribution role of a theory and its fact-stating function are both inextricably entwined in every assertion made by the theory.

How do the issues in the debate over conventionalism fare from this semantically holistic perspective? Insight into this question can be gained by looking some more at the issue of the relation between the observational equivalence of theories and their full equivalence (see UNDERDETERMINATION OF THEORY BY DATA).

Even the most ardent realist will agree that two apparently incompatible theories may very well be "saying the same thing," with their equivalence hidden by ambiguous use of terms. If one theory can be obtained from another by mere interchange of terms (for example, two electrostatic theories that are identical except that one speaks of negative charge where the other speaks of positive charge, and vice versa), all are likely to agree that the two accounts really do say the same thing (what one means by "negative," the other means by "positive," etc.). Realists will often assert that two theories are equivalent if they have a common definitional extension.

But what if the alternative theories cannot be reconciled to one another by some simple inter-translation scheme? The positivist can adapt the line espoused by Eddington, discussed above, to the holistic context by simply maintaining that having the same observational consequences – call it "observational equivalence" – is enough to guarantee full equivalence for two theories. Such a claim does not depend on the older demand for termwise translatability of theoretical terms into complexes of observational terms. Conventionality of theories then becomes, again, merely the conventional choice of how to express the identical cognitive content of the apparently alternative theories. Once again, however, it seems impossible to reconcile such an account with taking the theoretical ontology of the theories seriously.

Most realists would deny that mere observational equivalence is enough to establish the full equivalence of alternative theories. The realist will usually demand that, in addition to having the same observational consequences, the alternative accounts must be "structurally identical" at the theoretical level, in order for them to count as merely alternative ways of "saying the same thing." Some view of this sort seems to be the natural approach to trying to reconcile theoretical realism with a holistic role-in-theory account of the meaning of the theoretical terms. It has sometimes been claimed that the possibility of structurally unalike theories to save the same phenomena is clearly indicated by the possibilities for finding alternative models accounting for a given totality of observational consequences that differ from one another even in the cardinality of their theoretical ontologies.

Of course, the realist who takes such a line and who believes that there are, indeed, structurally unalike, observationally equivalent theories will once again have to confront the skeptical problem noted above, with its familiar responses in terms of nonempirical elements endemic in rational theory choice.

It is an important question as to whether, indeed, a realist interpretation of theories can be reconciled with a view of the meaning of theoretical terms which, even if holistic, takes the meaning of the terms to be fixed solely by the role that they play in the theory. The holistic role-in-theory view of the meaning of theoretical terms is closely associated with the important observation of Ramsey that one could think of an axiomatized theory as being given by its Ramsey sentence (see RAMSEY SENTENCES). This sentence is obtained by conjoining all the axioms of the theory in a single sentence, replacing the theoretical terms with second-order variables, and then prefacing the resulting form with existential quantifiers for each of these new second-order variables. The original theoretical terms are then just thought of as "placeholders" for the variables.

But the natural interpretation of a theory so construed is that it asserts the possibility of embedding the facts about the observable features of the world in an abstract structure. The components of this abstract structure and their interrelation, as well as their relation to the observables, are stated by the Ramsey sentence. But then it is misleading to think of the theory as positing some concrete physical structure over and above the observables. The theory is simply saying that the observable facts can be mapped by an embedding in an abstract structure. The assertion that such an embedding exists has as its consequence the full structure of lawlike relations among the observables which the theory was intended to capture. But, from this point of view, it ought not to be taken as genuinely extending the realm of concreta in the world beyond those already noted in the observation basis. From this perspective, the possibility of inequivalent theories saving the same phenomena, and the need to make a "conventional" choice among them, becomes just a conventional choice among "representing structures," and seems, like the choice as viewed by Eddington, an innocuous one.

Naturally the realist will resist such a representationalist reading of the theory. The question is, though, whether this resistance can be made coherent while maintaining a holistic role-in-theory account of the meaning of the theoretical terms. One version of realism does, in fact, drop that account of theoretical meaning, taking terms referring to the unobservables to have semantic content over and above that granted them by their role as placeholders in the theoretical structure. One version of such an account focuses on the fact that terms such as "particle" function at both the observational and the theoretical levels, and holds that such terms, when referring to unobservables, retain the meaning they acquired in the observational context.

Quine's approach

Quine has presented a subtle discussion of the issues of underdetermination of theories and allegations of conventionality in line with his views on the role of theory in general and the role of semantic theory in particular (see QUINE). Presented with two alternative theories which save the phenomena, we can sometimes reconcile them by mutual inter-interpretation or by the absorption of one by the other.

But what if these remedies fail? Quine suggests two options. In the "sectarian" option, we hold to our initial theory and bar the terms of the other theory from our language as "meaningless." We do this knowing that, had we started off with the other theory, we would have held to *it* in a sectarian manner and so disposed of the account we now take as true as "meaningless." Yet, we insist, "still we can plead that we have no higher access to truth than our evolving theory." Alternatively, we can adopt an "ecumenical" approach, couching the theories in distinct variables in a single language, asserting all the claims of both, and, understanding "true" in a disquotational manner, accepting both as true. Ultimately, Quine asserts, "the cosmic question of whether to call two such world systems true should simmer down, bathetically, to a question of words." Yet, "the rival theories describe one and the same world" (Quine 1990, pp. 100–1). Here one seems to have the permissiveness of representationalism, which is happy to have the phenomena representable by embedding in more than one distinct logical structure, with an insistence on the realistic interpretation of theories. Or, rather, the distinction between theories as delineating the world or as merely representing it in an abstract structure is denied. This is a radical conventionalism indeed.

Another concept of conventionality of geometry

The issues of conventionalism on which this chapter has focused take them to be those arising out of the alleged underdetermination of theory by its observational consequences and the skeptical challenges arising out of this epistemic situation.

It should be noted that there is another distinct understanding of what a claim of conventionality comes down to in the geometric case. This distinct set of issues originates in B. Riemann's famous inaugural lecture in which he presented the foundations of his general non-Euclidean geometry of arbitrary dimension. In this lecture Riemann noted the fact that the denseness of the spatial manifold prevents us from defining metric separation of points by the topological means of counting intervening points. It is this underdetermination of the metric of space by its topological structure that becomes the core of attributions of conventionality to geometry in the explorations of A. Grünbaum (1973).

Similarly, whereas we have noted a conventionalist issue in the theory of special relativity, one grounded on taking point coincidences as the observation basis and thus leaving simultaneity at a distance as underdetermined by the totality of directly observable facts, Grünbaum takes it that it is the alleged failure in the theory of special relativity of facts about the causal relations among events to fully fix simultaneity relations in the way in which they do this job in pre-relativistic physics that is the ground for attributing conventionality to the relativistic specification of distant simultaneity.

The detailed issues regarding the extent to which one kind of fact (topological, causal) fixes another kind (metric, simultaneity) in pre-relativistic and relativistic physics is one of some subtlety. There is much debate as well as to the relevance of such structural facts about the worlds described by these theories to allegations that some class of putative assertions about facts is merely a set of assertions that are true "conventionally."

63

References and further reading

Eddington, A. 1920: *Space, Time and Gravitation* (Cambridge: Cambridge University Press).

Friedman, M. 1983: *Foundations of Space-Time Theories* (Princeton: Princeton University Press).

Glymour, C. 1971: Theoretical realism and theoretical equivalence. *Boston Studies in the Philosophy of Science,* 8, 275–88.

Grünbaum, A. 1973: *Philosophical Problems of Space and Time* (Dordrecht: Reidel).

Poincaré, H. 1952: *Science and Hypothesis* (New York: Dover).

Putnam, H. 1975: What theories are not. In *Mathematics, Matter and Method* (Cambridge: Cambridge University Press), 215–27.

—— 1983: Models and reality. In *Realism and Reason* (Cambridge: Cambridge University Press), 1–25.

Quine, W. 1975: On empirically equivalent systems of the world. *Erkenntnis,* 9, 313–28.

—— 1990: *Pursuit of Truth* (Cambridge, MA: Harvard University Press).

Reichenbach, H. 1958: *The Philosophy of Space and Time* (New York: Dover).

Riemann, B. 1929: On the hypotheses which lie at the foundations of geometry. In *A Source Book in Mathematics,* ed. D. Smith (New York: McGraw-Hill), 411–25.

Schlick, M. 1963: *Space and Time in Contemporary Physics* (New York: Dover).

Sklar, L. 1974: *Space, Time, and Spacetime* (Berkeley: University of California Press).

—— 1985: *Philosophy and Spacetime Physics* (Berkeley: University of California Press).

Sneed, J. 1971: *The Logical Structure of Mathematical Physics* (New York: Humanities Press).

van Fraassen, B. 1980: *The Scientific Image* (Oxford: Clarendon Press).

10

Craig's Theorem

FREDERICK SUPPE

William Craig (1953) proved a theorem about theories in mathematical logic which has been utilized in philosophical attempts to analyze scientific theories.

Let ϕ be a sentence of mathematical logic containing names η and predicates π, where the η refer to individuals, and the π to attributes comprising some system M in such a manner that ϕ describes a possible configuration within M. If that configuration obtains in M, then M is a *model* of ϕ. If we let "$M(\phi)$" designate the class of all models M of ϕ, then ψ is a *valid consequence* of ϕ just in case $M(\psi) \subseteq M(\phi)$.

> *Craig–Lyndon interpolation theorem*: Let ϕ and ψ be formulae of first-order predicate calculus such that $M(\psi) \subseteq M(\phi)$. Then there exists a formula χ containing only those nonlogical symbols that occur in both ϕ and ψ such that $M(\chi) \subseteq M(\phi)$ and $M(\psi) \subseteq M(\chi)$.

χ is called the "interpolation formula."

Craig's theorem is used to prove two important results about the theories in first-order predicate calculus. (See AXIOMATIZATION.) For simplicity, we identify theories with their theorems.

> *Robinson's consistency theorem*: Let T_1 and T_2 be two theories. Then the theory $T_1 \cup T_2$ has a model if and only if there is no sentence in T_1 whose negation is in T_2.

Let $\tau, \beta_1, \ldots, \beta_n$ be distinct nonlogical terms of some theory T. Then τ is *implicitly definable from* $\beta_1, \ldots, \beta_n$ *in T* if and only if any two models $M(T)$ and $M'(T)$ that agree in what they assign to $\beta_1, \ldots, \beta_n$ also agree in what they assign to τ. If τ is not implicitly definable from $\beta_1, \ldots, \beta_n$ in T, then τ and the β are said to be *independent* (Padoa's principle). Confining attention to theorems of T containing at most $\tau, \beta_1, \ldots, \beta_n$ as nonlogical constants, let "T_τ" designate those theorems containing τ, and "T_β" designate those that do not. Then τ is *explicitly definable from* $\beta_1, \ldots, \beta_n$ *in T* if and only if there is some sentence $\delta(\tau, \beta_1, \ldots, \beta_n)$ in T containing only the nonlogical constants τ, $\beta_1, \ldots, \beta_n$ such that (1) for every sentence ϕ containing τ, there exists a sentence ψ not containing τ such that $M(\phi \leftrightarrow \psi) \subseteq M(T_\beta \cup \{\delta(\tau, \beta_1, \ldots, \beta_n)\})$, and (2) there is no formula ϕ not containing τ whose models include the $M(T_\beta \cup \{\delta(\tau, \beta_1, \ldots, \beta_n)\})$, but not all the $M(T_\beta)$. Condition (1) requires that τ could be *eliminated* from the T language without reducing its expressive power, and (2) that $\delta(\tau, \beta_1, \ldots, \beta_n)$ be *noncreative* in the sense that adding it to T_β introduces no theorems not already in T. When τ is explicitly definable by $\delta(\tau, \beta_1, \ldots, \beta_n)$, $M(T_\tau) \subseteq M(T_\beta \cup \{\delta(\tau, \beta_1, \ldots, \beta_n)\})$.

> *Beth's definability theorem*: τ is implicitly definable from $\beta_1, \ldots, \beta_n$ in T if and only if τ is explicitly definable from $\beta_1, \ldots, \beta_n$ in T.

(See DEFINITIONS.)

In their "received view" analysis, logical positivists analyzed scientific theories as being axiomatized in first-order predicate calculus (see THEORIES). Nonlogical or descriptive terms of the theory were bifurcated into an *observation vocabulary* V_O, of terms referring to directly observable entities or attributes, and a *theoretical vocabulary* V_T, of terms that do not. The epistemological and ontological statuses of theoretical terms were viewed as suspect, and some positivists maintained that they should be eliminable in principle from theories. (See REALISM AND INSTRUMENTALISM.)

Craig's theorem was thought to provide a formal proof that theoretical terms were not essential to specifying a theory's observational content. Consider a theory TC, with theoretical laws T involving terms from V_O but none from V_T and correspondence rules C involving terms from both V_O and V_T. Let O be the theorems derivable from TC that contain terms from V_O but none from V_T. Positivists took O to be the observational or testable content of the theory. Now take any observation sentence o in O as ψ, and TC as ϕ. Then, via Craig's theorem, there exists a sentence o^* in O such that o is derivable from o^*. Let O^* be the set of sentences o^* obtained by repeating the procedure for each o in O. Both TC and O^* have precisely the same observational consequences, O; but O^* does not contain terms from V_T.

Does this not vindicate the instrumentalist position that theoretical terms and laws are dispensable within scientific theories? Hempel (1958) and Maxwell (1962) argued not. Proofs of Craig's theorem are such that every sentence in O or a logical equivalent is made an axiom in O^*, and the set O^* will typically be neither finite nor finitely specifiable (except by applying the proof construction to TC). One crucial function of scientific theories is to systematize and illuminate the body O of observable phenomena, which requires the compact efficient formulations that TC, but not O^*, provides. Craig (1956, p. 49) himself pointed out that his theorem "fail[s] to simplify or to provide genuine insight" here. (See also RAMSEY SENTENCES.)

Logical positivists initially required that correspondence rules in theories provide explicit definitions of all the terms in V_T, and hence that such terms be eliminable and noncreative. However, Carnap (1936–7) argued that many legitimate theoretical terms in science were not so eliminable, and the requirement was loosened to allow creative definitions that provided *partial* observational *interpretations* of V_T terms. Controversy over coherence of the partial interpretation notion subsided after Suppe (1971) showed how Beth's theorem not only provides fundamental understanding of explicit definition, but that relaxing conditions on explicit definition enables understanding of Carnap's notion of partial interpretation: Whereas explicit definitions fully specify (implicitly define) the referents of theoretical terms within intended models of the theory, creative partial definitions serve only to restrict the range of objects in intended models that could be the referents of theoretical terms.

Robinson's theorem proves helpful in analyzing various notions of theory identity in science (see THEORY IDENTITY).

66

References and further reading

Boolos, G., and Jeffrey, R. 1974: *Computability and Logic* (Cambridge: Cambridge University Press), chs 23, 24.

Carnap, R. 1936–7: Testability and meaning. *Philosophy of Science*, 3, 420–68; 4, 1–40.

Craig, W. 1953: On axiomatizability with a system. *Journal of Symbolic Logic*, 18, 30–2.

—— 1956: Replacement of auxiliary expressions. *Philosophical Review*, 65, 38–55.

Hempel, C. G. 1958: The theoretician's dilemma. In *Minnesota Studies in the Philosophy of Science*, vol. 2, ed. H. Feigl, M. Scriven, and G. Maxwell (Minneapolis: University of Minnesota Press), 37–98; repr. in C. G. Hempel, *Aspects of Scientific Explanation and Other Essays in the Philosophy of Science* (New York: Free Press, 1965), 173–226.

Maxwell, G. 1962: The ontological status of theoretical entities. In *Current Issues in the Philosophy of Science*, ed. H. Feigl and G. Maxwell (New York: Holt, Rinehart, and Winston), 3–27.

Robinson, A. 1963: *Introduction to Model Theory and to the Metamathematics of Algebra* (Amsterdam: North Holland), ch. 5.

Suppe, F. 1971: On partial interpretation. *Journal of Philosophy*, 68, 57–76; repr. in *idem*, *The Semantic Conception of Theories and Scientific Realism* (Urbana, IL: University of Illinois Press, 1989), 38–54.

—— 1974: The search for philosophic understanding of scientific theories. In *The Structure of Scientific Theories*, ed. F. Suppe (Urbana, IL: University of Illinois Press; repr. 1977), 3–241. See esp. secs II-B and IV-C.

Suppes, P. 1957: *Introduction to Logic* (New York: Van Nostrand), ch. 8.

11

Darwin

SAMIR OKASHA

Discoverer of the theory of evolution by natural selection, and thus the founder of modern evolutionary biology, Charles Darwin is responsible for one of the most fundamental and far-reaching contributions to the modern scientific world view. Born in 1809 in Shrewsbury into a wealthy Victorian family, Darwin was educated at the universities of Edinburgh and Cambridge. Though his formal education was of little interest to him – "my time was wasted, as far as the academical studies were concerned" (1969, p. 58) – Cambridge did provide the opportunity to pursue his child-hood interest in natural history. Upon graduating in 1831, Darwin joined a charting expedition around South America on board *HMS Beagle*, which was to last for five years. During this time, he made lengthy expeditions inland, studying the flora, fauna, and geology of South America and the Pacific Islands, and accumulating a massive biological and geological collection. These years were critical for Darwin's intellectual development, as he admitted, sowing the seeds of his doubt in the fixity of individual species. Darwin was impressed by a large number of facts that could not be explained on the assumption of the fixity of species, including the close structural similarities between extinct and living species revealed by the fossil record. Most important of all were the tortoises and finches of the Galapagos Islands: each island was inhabited by very similar, yet distinct varieties – clear evidence for Darwin of common ancestry shaped by adaptation to local conditions. The creatures of the Galapagos, Darwin wrote, bring us close "to that great fact – that mystery of mysteries – the first appear-ance of new beings on this earth" (1839, p. 466).

On returning to England, Darwin published an account of his South American travels in a work generally referred to as the *Voyage of the Beagle* (Darwin 1839). Though primarily a naturalist's travel guide, a number of themes in the *Voyage*, notably the discussion of geographical distribution, extinction, and the utility of organs, were to play a major role in the *Origin of Species*. By 1837 Darwin had become convinced of the "mutability of species," as he put it, though the means by which one species might evolve into another was still unclear to him. His discovery of the missing mechanism – natural selection – was due to two factors: knowledge of the power of artificial selec-tion and his reading, in 1838, of T. R. Malthus's *Essay on the Principle of Population* (first published in 1798). Artificial selection was the process used by animal and plant breeders to improve their stock: dramatic changes in phenotype could be produced, in a few generations, by selecting animals or plants with a particular desirable trait and breeding from them alone. Darwin needed a natural equivalent of artificial selection to fuel the process of evolution. The answer came to him through reading Malthus.

Malthus had argued that human population growth would always outstrip food supply, drawing a pessimistic moral for the chances of improving human welfare. A key element in Malthus's theory was the "struggle for existence," a concept Darwin extrapolated to the natural world. Not all the organisms born in any generation could survive, so successive generations would contain a higher frequency of organisms whose traits best equipped them for survival and reproduction. Given sufficient variation among the members of a population, and presuming these variations to be heritable, it follows that future generations will exhibit different features from the original population: those successful in the struggle for existence will pass their features on to future generations. This process of "descent with modification," as Darwin called it, is capable over time of producing full-blown changes in a species, by the gradual accumulation of small differences. The theory of natural selection was born.

Though the core of Darwin's position was developed by the early 1840s, his major work, *On the Origin of Species by Means of Natural Selection*, was not published for nearly another 20 years, partly due to fear of an unfavorable reaction among his contemporaries. Darwin turned instead to a study of barnacles, which took ten years to complete and yielded four volumes. The *Origin*'s publication was spurred by Darwin's reading of a paper in 1858, sent to him by a fellow naturalist called Alfred Wallace, who had discovered the theory of natural selection independently. Darwin and Wallace did a joint presentation of their ideas at a Linnean Society meeting in 1858, and the *Origin* was to appear in the following year (Darwin and Wallace 1858; Darwin 1859).

Though Darwin regarded the *Origin* as a "sketch" of his position, a precursor to a longer work, the wealth of evidence and argumentation it contains made a compelling case for his central thesis: that today's species have evolved from common ancestors under the pressure of natural selection. The years following the *Origin*'s publication witnessed fierce public debate, marked by strong theological opposition; but the scientific community was converted to evolutionism very rapidly. The idea of evolution itself was not new with Darwin, though no scientist prior to him had come up with a plausible mechanism for why it might have occurred. Darwin's explanation – natural selection is developed at length in the early chapters of the *Origin*, through pursuing the artificial selection analogy and stressing the ubiquitous "struggle for existence." "Can we doubt," he wrote, "(remembering that many more individuals are born than can possibly survive) that individuals having any advantage, however slight, over others, would have the best chance of surviving and of procreating their kind? On the other hand, we may feel sure that any variation in the least degree injurious would be rapidly destroyed. This preservation of favourable variations and the rejection of injurious variations, I call Natural Selection" (1859, pp. 130–1).

Darwin offered a number of examples, both actual and hypothetical, to illustrate how natural selection works. Imagining a population of wolves with a limited supply of prey for food, he wrote: "I . . . see no reason to doubt that the swiftest and slimmest wolves would have the best chance of surviving, and so be preserved or selected" (1859, p. 138). He stressed the slowness and all-pervasiveness of natural selection – "natural selection is daily and hourly scrutinising, throughout the world, any variation, even the slightest" (1859, p. 133) – softening the reader to the idea that this mechanism could produce vast differences given enough time, and hence create new species. One particular kind of natural selection Darwin identified he called "sexual selection," a topic of considerable contemporary interest. This occurs in sexually reproducing species when males of a group fight for the females, or when females choose the most attractive males (or both). It is responsible for the evolution of certain fighting attributes used for intra-specific combat and features of mate attraction, such as the peacock's tail.

Having presented his basic thesis, Darwin produced a wealth of evidence, of various sorts, in support of evolution by natural selection, drawing on data from paleontology, biogeography, embryology, morphology, and other fields. The basic argument was simple: in each of these areas, there are phenomena that go unexplained on the supposition of immutable, individually created species, but are exactly what one would expect if evolution by natural selection were true. In modern jargon, Darwin argued for his theory by an inference to the best explanation (see INFERENCE TO THE BEST EXPLANATION). Though acknowledging the imperfection of the fossil record, Darwin nonetheless thought that it strongly favored evolution over creation. He stressed, among other things, the fact that similar fossils appear in succeeding strata, but not in widely separated ones – just what we would expect, on Darwin's theory, since once a species has become extinct, it can never return. The data from morphology and embryology are in many ways the most compelling. Darwin emphasized the significance of what are called "homologies" – structural similarities between very different organisms – e.g., the forelimbs of humans and dogs. These isomorphisms are clear evidence of common ancestry, but totally inexplicable if species were individually created. Similarly, why should the embryos of organisms from different species be so similar, unless those species have descended from a common ancestor? Darwin dwelt at length on the geographical distribution of organisms, which he thought was one of his strongest sources of evidence. Climatically and environmentally similar regions of the world are often inhabited by very different organisms, while the same forms often appear in radically different conditions. Both these facts are inexplicable if species were individually created by an intelligent designer to fit their environment, but easily explained by evolution and migration. Another class of facts anomalous without evolution concerns the inhabitants of islands: whole groups of organisms are often absent from islands, while the varieties that are there are often distinctive, as in the case of the Galapagos, tailored to suit local conditions by natural selection. The *Origin* is noticeably reticent on the subject of human evolution, due largely to Darwin's (unsuccessful) desire to avoid controversy. Though in no doubt about the applicability of his ideas to the human race, Darwin merely noted that in the future, "light will be thrown on the origin of man and his history" (1859, p. 458), postponing full treatment for a later work, *The Descent of Man* (Darwin 1871).

Darwin was quite open about the difficulties his theory faced, and they were many. However, subsequent scientific developments have provided convincing solutions to these difficulties, while retaining Darwin's core insight. An initial worry concerned the age of the Earth. Darwin was understandably troubled by Lord Kelvin's estimate in 1862, based on the cooling of the Earth's crust, of an absolute age of between 25 and 400 million years, as this was clearly insufficient for the evolution of today's species. However, the discovery of radioactive materials this century has led to a dramatic increase in the Earth's estimated age, defusing the problem. Darwin was candid about the imperfection of the fossil record, two features of which were particularly troubling to him: first, the absence of intermediate fossils, and second, the sudden appearance of sophisticated life forms in the period now known as the "Cambrian explosion" (about 600 million years ago). Both these features have received contemporary attention, particularly in the work of the controversial American paleontologist Stephen Jay Gould (Gould and Eldredge 1977; Gould 1989). Another puzzle for Darwin concerned the evolution of the "social insects" – sterile castes of ants, bees, and termites living together in colonies, serving a single fertile queen. Sterility obviously reduces an individual organism's fitness to zero, so how could it evolve by natural selection? The same problem arises for altruistic behavior generally. This puzzle was finally solved in the 1960s by William Hamilton's seminal work on "kin selection" (Hamilton 1964). Hamilton showed how seemingly altruistic behavior could evolve through natural selection: by helping genetically related relatives, an organism can pass on copies of its own genes to the next generation *in its relatives' offspring*. Under certain conditions, an organism can actually increase its genetic representation in future generations more efficiently by helping kin than by reproducing itself, so sterility can evolve.

Most troubling for Darwin were the problems of variation and heredity. Natural selection, in order to work, requires heritable variation in reproductive fitness; but where does this variation come from, and why should fitness-enhancing traits be inherited by future generations? The problem is an acute one, since the variation must be *continuous*, or else natural selection will quickly produce a homogeneous population, destroying the very variation on which its operation depends. Darwin noted that variation seemed to be widespread in the natural world, but admitted that "our ignorance of the laws of variation is profound" (1859, p. 202). He knew nothing about the mechanics of heredity, writing before the development of genetics, and actually believed, to some extent, in Lamarckism, the now discredited idea that characteristics of an organism acquired during its lifetime are inherited by offspring. He was therefore deeply concerned by Jenkin's objection that useful variations would not be transmitted to future generations but, rather, "blended away" unless both parents happened to possess the variant trait.

The problems of variation and heredity were solved by the rediscovery of Mendelian genetics at the start of the twentieth century. The tendency of offspring to inherit traits of their parents was explained by Mendel in terms of the transmission of *genes* – discrete hereditary particles – from generation to generation. In sexually reproducing organisms, each parent contributes one of the alleles at each locus in the offspring, governed by Mendel's two laws of transmission. This guarantees that each parent provides half the genetic material of each offspring. Occasionally, "mutations", – i.e., random genetic changes – occur, giving rise to new characteristics. Mutations are random, in that they are not biased towards improving the organism in which they occur, and the vast majority of mutations are actually harmful to an organism's fitness. Beneficial mutations do occur, however, and it is these that provide the raw material for natural selection to operate on: the mutant gene, and the new phenotypic characteristic it codes for, spread in frequency in the population. Genetic mutation, therefore, provides the continual source of variation that natural selection requires, but which Darwin himself was unable to account for.

The fusion of natural selection and Mendelian genetics began in the 1920s with the work of population geneticists Ronald Fisher, J. B. S. Haldane, and Sewall Wright. The Darwin–Mendel synthesis became known as the "neo-Darwininst synthesis" or the "synthetic theory of evolution" and was pursued with vigor in the following years by Dobzhansky, Mayr, Huxley, Stebbins, and others, in Britain, Europe, and the USA. Neo-Darwinism quickly acquired paradigmatic status in evolutionary theorizing, which it retains today. Since the 1950s, however, evolutionary studies have been increasingly influenced by molecular biology, after Watson and Crick's discovery of DNA in 1953. However, the relation between Mendelian and molecular genetics, though subtle, is essentially a harmonious one, and the extraordinary advances in molecular biology in the last four decades have not affected the status of the neo-Darwinist synthesis itself. Aside from the religiously motivated (and fundamentally misguided) attacks of the so-called scientific creationists in the USA, the major recent challenges to neo-Darwinism have come from Gould and his followers, in the 1970s and 1980s. Gould's attack was twofold. First, he challenged the standard interpretation of the fossil record, advancing instead the theory of "punctuated equilibrium" – roughly, the claim that evolution proceeded not gradually, but in leaps and bounds, periods of rapid evolutionary change interspersed by long periods of stasis. Gould argued that mass extinction and other historical contingencies played a more important role than traditional Darwinist selection. Second, he attacked the "pan-adaptationism" of many evolutionary theorists: their alleged tendency to insist that every major phenotypic and behavioral trait can be accounted for in terms of its adaptive value, in advance of specific evidence. These criticisms provoked a prolonged and heated controversy; consensus now appears to be growing that, while Gould raised some interesting and important points, these can easily be accommodated *within* the neo-Darwinist picture, rather than constituting a radical alternative, as Gould proclaimed (Gould 1980, 1989; Gould and Eldredge 1977, 1993; Gould and Lewontin 1979).

Inevitably, so radical a change in our world view as Darwin provided was bound to have far-reaching implications. Darwin himself wrote, in a notebook, "Origin of man now proved. Metaphysics must flourish. He who understands baboon would do more towards metaphysics than Locke" (Barrett et al. 1987, D 26, M 84). The impact of Darwinism on contemporary philosophy is multifaceted. In the first place, there is the philosophy of biology, which has seen a vigorous resurgence of late (see BIOLOGY). Most of the issues in this field have an evolutionary focus, due both to the centrality of evolutionary theory in biology itself and the plethora of conceptual questions that it raises. One of the earliest areas of philosophical interest was the concept of *biological fitness* and the related worry that the theory of natural selection was a tautology, as Popper once argued (Popper 1974). If the theory claims that the fittest tend to survive, and identifies the fittest with those who *do* survive, how can it avoid being vacuously true? The charge of unfalsifiability came to seem less pressing with the demise of Popper's views in the 1960s and 1970s, and has anyway been adequately answered. A more recent debate to which philosophers have contributed is the "units of selection" question. Do traits evolve because of the benefit they bestow on individual organisms, or on the genes that code for those traits, or on the species to which the individuals belong? What is the correct level of explanation? The currently popular "gene's eye" viewpoint, advocated by G. C. Williams and Richard Dawkins in particular, holds that selection is in the first place for individual genes, not organisms or groups; this involves key philosophical issues concerning reductionism, explanation, and causation (Dawkins 1976; Williams 1966). A third area of philosophical concern has been the recent controversy over creationism in the USA and the attempt to prevent the teaching of evolution in American classrooms. Philosophers of biology have risen eagerly to the challenge of uncovering the flawed logic behind "scientific creationist" arguments (Kitcher 1982; Sober 1993). An older – indeed, ancient – question in the philosophy of biology concerns systematics, or the theory of biological classification. Darwin's relevance is obvious here, for the mutability of species over time and the one-tree-of-life hypothesis bring evolutionary considerations to the fore in the question of how to classify organisms. This question and the related question of how to define a species have engaged philosophers' attention since Aristotle. A final topic worthy of mention is sociobiology, or the attempt to use evolutionary theory to explain social and behavioral traits of species, including humans. This program has a clear Darwinist pedigree (chapter 7 of the *Origin* was entitled "Instinct") and came to prominence with E. O. Wilson's controversial book *Sociobiology* in 1975 (though the sociobiological research program arguably began with William Hamilton's 1964 papers, discussed above) (Wilson 1975; Hamilton 1964). The ensuing debate, particularly about human sociobiology, caught the public attention, due partly to its level of bitterness and partly to its political overtones. Critics of sociobiology saw the program as the latest installment in evolutionary theory's notorious historical association with racist ideologies. Philosophers have done much to assess the methodological foundations of sociobiology and to explore its implications for ethics (Kitcher 1985; Richards 1987; Ruse 1986).

Philosophy of biology aside, the impact of Darwinism on philosophy can be felt in a number of ways. Most obviously, Darwin swept away one of the traditional arguments for the existence of God, the argument from design. According to this argument, the adaptation to the environment that is so ubiquitous in the living world is proof of an intelligent, benevolent designer. Though the design argument had been attacked before, notably by Hume, its ultimate demise required an alternative explanation of adaptation, and this Darwin provided. Darwin's views are often cited as contributing to the downfall of essentialism, the Aristotelian doctrine that the world is populated by fixed, immutable, eternal kinds (living and nonliving), each of which possesses an essence. Applied to biological species, the incompatibility of this view with Darwinism is obvious, for Darwin maintained precisely that species were mutable and *not* eternal, but rather evolving over time. However, it should be noted that Darwinism has done nothing to stop the resurgence in recent years of essentialist ideas, stemming from considerations in the philosophy of language. Recently, an increasing number of philosophers have tried to apply evolutionary ideas to traditional problems in epistemology, ethics, and elsewhere (Campbell 1974; Callebaut and Pinxten 1987; Richards 1986, 1987; Ruse 1986, 1995; Sober 1994). This trend is part of the broader "naturalist" turn in contemporary philosophy, which emphasizes the legitimacy (and importance) of appealing to science to solve philosophical problems. In the case of epistemology, the key idea is that the evolution of human intellectual capacities may admit of a Darwinist explanation; one optimistic thought is that this may help solve the problem of induction. However, evolutionary epistemology has met with an unfavorable reaction from many philosophers, probably partly due to the sparseness of

73

empirical evidence on the subject. The same is true of evolutionary ethics, often found guilty by association with the more speculative reaches of human sociobiology. It would be unwise to dismiss this approach, however, for sociobiology has undoubtedly shed important light on the evolution of altruism and cooperation, issues with obvious ethical implications. Most recently, certain philosophers (advocates of the "teleological theory of content") have attempted to shed light on the problem of intentionality, or meaning, through evolutionary considerations (Dennett 1987; Millikan 1984, 1993; Papineau 1987). All these attempts to expand the scope of Darwinist explanation are controversial and have met with serious objections; but they are testimony to the permanent appeal of adaptationist thinking, and ultimately to the power and originality of the new way of thinking about the natural world that Darwin taught us.

References

Works by Darwin

1839: *Journal of Researches into the Geology and Natural History of the Various Countries Visited by H. M. S. Beagle, etc.* (London: Colburn).

1859: *On the Origin of Species by Means of Natural Selection* (London: John Murray).

1871: *The Descent of Man, and Selection in Relation to Sex* (London: John Murray).

1969: *Autobiography*, ed. N. Barlow (New York: Norton).

and Wallace, A. R. 1858: On the tendency of species to form varieties; and on the perpetuation of varieties and species by natural means of selection. *Proceedings of the Linnean Society, Zoological Journal*, 3, 46–62; repr. in Darwin and Wallace, *Evolution by Natural Selection* (Cambridge: Cambridge University Press, 1958), 270–87.

Works by other authors

Barrett, P. H., Gautrey, P. J., Herbert, S., Kohn, D., and Smith, S. (eds) 1987: *Charles Darwin's Notebooks 1836–44* (Cambridge: British Museum (Natural History) and Cambridge University Press).

Callebaut, W., and Pinxten, R. (eds) 1987: *Evolutionary Epistemology: A Multi-paradigm Program* (Dordrecht: Reidel).

Campbell, D. T. 1974: Evolutionary epistemology. In *The Philosophy of Karl Popper*, ed. P. A. Schlipp (La Salle, IL: Open Court), 413–63.

Dawkins, R. 1976: *The Selfish Gene* (New York: Oxford University Press).

Dennett, D. 1987: *The Intentional Stance* (Cambridge, MA: MIT Press).

Gould, S. J. 1980: Is a new and general theory of evolution emerging? *Paleobiology*, 6, 119–30.

—— 1989: *Wonderful Life: The Burgess Shale and the Nature of History* (New York: Norton).

Gould, S. J., and Eldredge, N. 1977: Punctuated equilibria: the tempo and mode of evolution reconsidered. *Paleobiology*, 3, 115–51.

—— 1993: Punctuated equilibrium comes of age. *Nature*, 366, 223–7.

Gould, S. J., and Lewontin, R. C. 1979: The spandrels of San Marco and the Panglossian paradigm: a critique of the adaptationist programme. *Proceedings of the Royal Society of London*, B205, 581–98.

Hamilton, W. 1964: The genetical evolution of social behaviour, pts 1 and 2. *Journal of Theoretical Biology*, 7, 1–16, 17–52.

Kitcher, P. 1982: *Abusing Science: The Case against Creationism* (Cambridge, MA: MIT Press).

—— 1985: *Vaulting Ambition* (Cambridge, MA: MIT Press).

Malthus, T. R. 1798: *An Essay on the Principle of Population* (London: Everyman).

Millikan, R. 1984: *Language, Thought and Other Biological Categories* (Cambridge, MA: MIT Press).

—— 1993: *White Queen Psychology and Other Essays for Alice* (Cambridge, MA: MIT Press).

Papineau, D. 1987: *Reality and Representation* (Oxford: Blackwell).

Popper, K. 1974: Darwinism as a metaphysical research program. In *The Philosophy of Karl Popper*, ed. P. Schlipp (La Salle, IL: Open Court), 133–43.

Richards, R. J. 1986: A defense of evolutionary ethics. *Biology and Philosophy*, 1, 265–93.

—— 1987: *Darwin and the Emergence of Evolutionary Theories of Mind and Behaviour* (Chicago: University of Chicago Press).

Ruse, M. 1986: *Taking Darwin Seriously: A Naturalistic Approach to Philosophy* (Oxford: Blackwell).

—— 1995: *Evolutionary Naturalism* (London: Routledge).

Sober, E. 1993: *Philosophy of Biology* (Oxford: Oxford University Press).

—— 1994: Prospects for an evolutionary ethics. In *From a Biological Point of View* (Cambridge: Cambridge University Press), 93–113.

Williams, G. C. 1966: *Adaptation and Natural Selection* (Princeton, NJ: Princeton University Press).

Wilson, E. O. 1975: *Sociobiology: The New Synthesis* (Cambridge, MA: Harvard University Press).

12

Definitions

FREDERICK SUPPE

In the most fundamental scientific sense, to define is to delimit. Thus definitions serve to fix boundaries of phenomena or the range of applicability of terms or concepts. That whose range is to be delimited is called the *definiendum*, and that which delimits the *definiens*. In practice, the hard sciences tend to be more concerned with delimiting phenomena, and definitions are frequently *informal*, given on the fly, as in "Therefore, a layer of high rock strength, called the *lithosphere*, exists near the surface of planets." Social science practice tends to focus on specifying application of concepts through *formal* operational definitions. Philosophical discussions have concentrated almost exclusively on articulating *definitional forms* for terms.

Definitions are *full* if the *definiens* completely delimits the *definiendum*, and *partial* if it only brackets or circumscribes it. *Explicit definitions* are full definitions where the *definiendum* and the *definiens* are asserted to be equivalent. Examples are coined terms and stipulative definitions such as "For the purpose of this study the lithosphere will be taken as the upper 100 km of hard rock in the Earth's crust." Theories or models which are so rich in structure that sub-portions are functionally equivalent to explicit definitions are said to provide *implicit definitions*. In formal contexts our basic understanding of full definitions, including relations between explicit and implicit definitions, is provided by the *Beth definability theorem* discussed in a previous chapter (see CRAIG'S THEOREM). Partial definitions are illustrated by *reduction sentences* such as

When in circumstances C, *definiendum* D applies if situation S obtains,

which says nothing about the applicability of D outside C.

It is commonly supposed that definitions are analytic specifications of meaning. In some cases, such as stipulative definitions, this may be so. But some philosophers (e.g., Carnap (1928 and subsequently)) allow specifications of meanings to be synthetic. Reduction sentences are often descriptions of measurement apparatus specifying empirical correlations between detector output readings and values for parameters. These are synthetic and are rarely mere specifications of meaning. The larger point here is that specification of meanings is only one of many possible means for delimiting the *definiendum*. Specification of meanings seems tangential to the bulk of scientific definitional practices.

Definitions are said to be *creative* if their addition to a theory expands its content, and *noncreative* if they do not. More generally, we can say that definitions are creative whenever the *definiens* asserts contingent relations involving the *definiendum*. Thus definitions providing analytic specifications of meaning are noncreative. Most explicit definitions are noncreative, and hence *eliminable* from theories without loss of empirical content. (See CRAIG'S THEOREM for fuller discussion.) One could relativize the distinction so that definitions redundant of accepted theory or background belief in the scientific context are counted as noncreative. Either way, most other scientific definitions will be creative. Reduction sentences are almost always creative synthetic expressions of empirical correlation. Thus, for purposes of philosophical analysis, suppositions that definitions are either noncreative or meaning specifications demand explicit justification. Much of the literature concerning incommensurability and meaning change in science turns on uncritical acceptance of such suppositions (see INCOMMENSURABILITY).

Such confusions abound in scientific and philosophical discussions of *operational definitions*. The notion was first introduced by P. W. Bridgman (1938) with reference to noncreative explicit full definitions specifying meanings in terms of operations performed in the measurement process. Behaviorist social scientists expanded the notion to include creative partial definitions, and in practice most operational definitions can be cast as synthetic creative reduction sentences specifying empirical relations between measurement procedures and intervening variables or hypothetical constructs (the *definienda*). Thus, in practice, operational definitions are testable or subject to empirical evaluation. Yet, when their operational definitions are challenged, many social scientists respond that it's just a matter of quibbling over semantics – a response appropriate to Bridgman's sort of operational definitions but not to their own.

Many philosophers have been concerned with admissible *definitional forms*. Some require *real definitions* – a form of explicit definition in which the *definiens* equates the *definiendum* with an *essence* specified as a conjunction $A_1 \wedge \ldots \wedge A_n$ of attributes. (By contrast, *nominal definitions* use nonessential attributes.) The *Aristotelian definitional form* further requires that real definitions be hierarchical, where the species of a genus share $A_1, \ldots, A_{n-1}$, being differentiated only by the remaining essential attribute A_n. Such definitional forms are inadequate for evolving biological species whose essence may vary. *Disjunctive polytypic definitions* allow changing essences by equating the *definiendum* with a finite number of conjunctive essences. But future evolution may produce further new essences, so partially specified *potentially infinite disjunctive polytypic definitions* were proposed. Such "explicit definitions" fail to delimit the species, since they are incomplete. A superior alternative is to formulate reduction sentences for each essence encountered, which partially define the species but allow the addition of new reduction sentences for subsequently evolved essences.

Wittgenstein (1953) claimed that many natural kinds lack conjunctive essences; rather, their members stand only in a *family resemblance* to each other (see NATURAL KINDS). Philosophers of science have developed the idea in two ways. Achinstein (1968) resorted to cluster analysis, arguing that most scientific definitions (e.g., of *gold*) specify nonessential attributes of which a "goodly number" must be present for the *definiendum* to apply. Suppe (1989, ch. 7) urged that natural kinds were constituted by a single kind-making attribute (e.g., *being gold*), and that which patterns of correlation might obtain between the kind-making attribute and other diagnostic characteristics is a factual matter. Thus issues of appropriate definitional form (e.g., explicit, polytypic, or cluster) are empirical, not philosophical questions.

Definitions of concepts are closely related to *explications*, where imprecise concepts (*explicanda*) are replaced by more precise ones (*explicata*). The *explicandum* and *explicatum* are never equivalent. In an adequate explication the *explicatum* will accommodate all clear-cut instances of the *explicandum* and exclude all clear-cut noninstances. The *explicatum* decides what to do with cases where application of the *explicandum* is problematic. Explications are neither real nor nominal definitions and are generally creative. In many scientific cases, definitions function more as explications than as meaning specifications or real definitions.

References and further reading

Achinstein, P. 1968: *Concepts of Science* (Baltimore: Johns Hopkins University Press), chs 1–2.

Bridgman, P. W. 1938: Operational analysis. *Philosophy of Science*, 5, 114–31.

Carnap, R. 1928: *Der Logische Aufbau der Welt* (Berlin; 2nd edn, 1961); trans. R. George as *The Logical Structure of the World* (Berkeley: University of California Press, 1967).

—— 1936–7: Testability and meaning. *Philosophy of Science*, 3, 420–68; 4, 1–40.

Hempel, C. G. 1952: *Fundamentals of Concept Formation*, vol. 2, no. 7, of *International Encyclopedia of Unified Science*, ed. O. Neurath, R. Carnap, and C. Morris (Chicago: University of Chicago Press).

Hull, D. 1968: The Logic of Philogenetic Taxonomy (Ph.D. dissertation, Bloomington, Indiana University).

Kaplan, A. 1964: *The Conduct of Inquiry* (San Francisco: Chandler).

Suppe, F. 1974: The search for philosophic understanding of scientific theories. In *The Structure of Scientific Theories*, ed. F. Suppe (Urbana, IL: University of Illinois Press; 2nd edn 1977), 3–241. See esp. secs II-A; III; IV-B, C, E; V-2b.

—— 1989: *The Semantic Conception of Theories and Scientific Realism* (Urbana, IL: University of Illinois Press), chs 3, 6, 10.

Wittgenstein, L. 1953: *Philosophical Investigations*, trans. G. E. Anscombe (Oxford: Blackwell, 1953).

13

Descartes

TOM SORELL

Descartes has an unofficial, as well as an official, philosophy of science. The unofficial philosophy of science can be detected in his letters, in some of the essays that he presented as specimens of his method in 1637, in the closing pages of the *Discourse on Method* itself, in parts of the *Principles of Philosophy*, and in his physics treatise, *The World*. The official philosophy of science is to be found elsewhere: in the *Meditations*, for example, and in Parts 2 and 4 of the *Discourse*. There are also expressions of it in the 1647 preface to the *Principles*. The unofficial philosophy of science is that of a practicing scientist; the official one is that of a metaphysician trying to prepare the ground among theologians for the safe publication of his physics. The unofficial philosophy takes problem solving to be the leading kind of scientific activity. It allows for the legitimate use in science of hypotheses, experiments, and extensive observation. The official philosophy of science, on the other hand, is concerned with the conditions for the rigorous demonstration of a large number of effects from highly evident principles. While not ruling out a scientific use for observation and experiment, it stresses the importance of a certain kind of first principle in physics, the deduction of physical explanations of a very high order of generality, and the possibility of basing principles of physics and physical explanations on principles about the nature and knowability of God. The two philosophies of science have something in common. Both connect science with finding a preferred order in a series of things, or considerations bearing on a particular problem or thing to be demonstrated. The preferred order is always from the "simple" – in Descartes's sense of general and highly evident – to the less simple. Again, both philosophies of science identify scientific understanding with being able to take in a given ordering in a continuous mental survey – what Descartes calls a "deduction."

Philosophy of science in the *Discourse* and *Essays*

Indications of the two philosophies of science, and of the tensions between them, are to be found in Descartes's first full-scale publication: the *Discourse and Essays* of 1637, a set of four treatises published as a single book. Three of the four treatises were essays illustrating some of the results that could supposedly be achieved by a method of inquiry pioneered by Descartes. These treatises, the *Optics*, the *Meteorology*, and the *Geometry*, were prefaced by an autobiographical essay, the now famous *Discourse on the Method for Rightly Conducting Reason and Seeking Truth in the Sciences*, which, notwithstanding its title, contains no more than a brief description of the method itself. At the very end of the *Discourse*, there is a striking passage that cautions readers about what to expect when they pass from the account of the method to its applications:

> Should anyone be shocked at first by some of the statements I make at the beginning of the *Optics* and the *Meteorology* because I call them "suppositions" and do not seem to care about proving them, let him have the patience to read the whole book attentively, and I trust that he will be satisfied. For I take my reasonings to be so closely interconnected that just as the last are proved by the first, which are their causes, so the first are proved by the last, which are their effects. It must not be supposed that I am here committing the fallacy that logicians call "arguing in a circle." For as experience makes most of these quite certain, the causes from which I deduce them do not so much prove them as explain them; indeed, quite to the contrary, it is the causes which are proved by the effects. (CSM I, 150; AT VI, 76)[1]

These are not the views of an apriorist. Descartes is insisting that experience is a source of certainty about observed effects, and that effects can "prove" the principles that explain them. Is he entitled to these views, however, given other things he says in the *Discourse*, other things that belong to what I am calling his official philosophy of science?

In Part 2 of the *Discourse* he outlines the "method" or "logic" that is supposed to be the subject of the entire treatise. The very first precept of his method

> was never to accept anything as true if I did not have evident knowledge of its truth; that is, carefully to avoid precipitate conclusions and preconceptions, and to include nothing more in my judgements than what presented itself to my mind so clearly and distinctly that I had no occasion to doubt it. (CSM I, 120; AT VI, 18)

The context makes clear that his model of a clear and distinct perception was drawn from the highly evident mathematical sciences. Would the suppositions of the *Essays* provide material for such clear and distinct perceptions? It is not obvious that they would. At the beginning of the *Optics* Descartes invites the reader to

> consider the light in bodies we call "luminous" to be nothing other than a certain movement, or very rapid and lively action, which passes to our eyes through the medium of air and other transparent bodies, just as the movement or resistance of the bodies encountered by a blind man passes to his hand by means of the stick. (CSM I, 153; AT VI, 84)

This supposition is supposed to make other things intelligible, but some of these other things would have been hard for a reader to believe, so would have taken away from the credibility of the "supposition," certainly to the point where the supposition would not have seemed certain or undeniable. Thus, the supposition was supposed to make credible the denial of the idea

> that something material passes from objects to our eyes to make us see colours and light, [and] that there is something in the objects which resembles the ideas or sensations we have of them. (Ibid.)

80

But in other writings Descartes singles out this idea as one of the deeply entrenched "prejudices" we bring to the study of nature. Its denial would not have come easily; so neither would a supposition leading to its denial have seemed evident or "clear and distinct."

The use of suppositions at the beginning of the *Optics*, then, seems to flout Descartes's method of conducting scientific inquiry; either that, or it is inspired by a different, unstated method or approach to science, one that allows for reliance on unobvious, even speculative-seeming, hypotheses, if the explanatory rewards of this reliance are high enough. This second approach is the one announced in the passage at the end of Part 6 of the *Discourse*. It is also the approach followed in *The World*, the physics treatise that Descartes intended to release as his first book, but which he decided to suppress in 1633 out of fear of being condemned by the Roman Inquisition for teaching the movement of the earth. In *The World*, Descartes radically revises Aristotelian cosmology. For example, he jettisons Aristotle's qualitative theory of the "elements," putting in its place a theory that distinguishes the elements only by their mathematically measurable properties, and that reduces Aristotle's four to three. The first element – fire – has as its sole form the possession of "parts moving so rapidly and being so minute that there are no other bodies capable of stopping them" (CSM I, 89; AT XI, 26). The only reason that a reader with a traditional education would have had for believing in the existence of this element was that it would help to make intelligible phenomena like those observed in the heavens. The effects would "prove" the principles, as Descartes says at the end of Part 6 of the *Discourse*.

Philosophy of science in the *Meditations*

The suppression of *The World* in 1633 marked the beginning of the shift from the unofficial philosophy of science – the philosophy of science of a practicing physicist, physiologist, and biologist – to the official philosophy of science now widely associated with Descartes – a philosophy of science conceived by a metaphysician for a theological audience. The suppression of *The World* marked the beginning; the failure of the *Discourse and Essays* to win either universal scientific acclaim or theological endorsement marked a further stage on the path toward the official philosophy of science. The theologians would not endorse Descartes's philosophy until they had seen more of the metaphysical doctrine sketched in Part 4 of the *Discourse*; and Catholic teachers and other readers uncooperatively disputed Descartes's solutions to the mathematical and scientific problems in the *Essays*. Even the question of whether Descartes had "proved" anything in the *Essays* was raised. Descartes's friend Mersenne asked whether the account of refraction given in the *Optics* was a demonstration. Descartes replied:

> I think it is, in so far as one can be given in this field without a previous demonstration of the principles of physics by metaphysics – that is something I hope to do some day but it has not yet been done – and so far as it is possible to demonstrate the solution to any problem of mechanics, or optics, or astronomy, or anything else which is not pure geometry or arithmetic. But to ask for geometrical demonstrations in a field within the range of physics is to ask the impossible. (CSM III, 138; AT II, 134)

The *Meditations on First Philosophy* was precisely the "demonstration of the principles of physics by metaphysics" that was needed to turn the solutions to problems in the *Essays* into proofs.

It is hard to be sure exactly how many principles of physics are proved in the *Meditations*. One, undoubtedly, is that bodies exist. Another is that to be a body is essentially to be extended in space. A third is that bodies are not exactly as they are experienced to be by the use of the senses. Perhaps the list goes on to include the conclusions that there is no void, and that there are no atoms; or perhaps metaphysics stops short of these conclusions, and physics takes over. Since part of Descartes's point in the *Meditations* is that it is possible to order the truths of metaphysics and the most general truths of physics in such a way as to enable someone with a suitably disciplined mind to take them all in in a continuous mental sweep, we should not expect the borderline to be easy to spot. More worrying, perhaps, is the question of how principles like these are supposed to be made intelligible by the two first principles of metaphysics: "I am thinking, therefore I exist" and "God exists and is no deceiver." To someone who has not gone through the *Meditations*, the demonstration of the principles of physics from *these* principles of metaphysics may look hopeless. In fact, Descartes is not trying to win agreement to a complete *non sequitur*. The existence of a nondeceiving God is required to validate the rule that what is clear and distinct is true, and this rule certifies as true the principles of physics. As for "I am thinking, therefore I exist," this provides materials for a proof that God exists and is no deceiver – materials that are strongly independent of anything asserted or presupposed by physics.

At the same time as the *Meditations* proves the principles of physics by metaphysics, it makes the conditions for what it calls "perfect knowledge" or "science" (*scientia*) very exacting. An atheist cannot have perfect knowledge of anything in meteorology or optics even if he proposes explanations of the rainbow or of the nature of light exactly similar to Descartes's. The reason is that he cannot be sure that he is not being deceived even about what is most evident to him. He cannot be sure unless he has a proof of the existence of a nondeceiving God. Again, the *Meditations* makes it seem as if the senses and sensory experience are a positive obstacle in the way of the discovery of the truth, even the truth about matter, and as if the epistemically ideal condition is the near-solipsistic one of the isolated *res cogitans* or human soul that is the hero of the early *Meditations*. (It is the human soul that is capable of physics, according to Descartes, and only derivatively the embodied human being.) Both these implications, it seems, are by-products of the attempt to make the derivation of his metaphysical principles theologically impressive, and there is no doubt that, officially at least, Descartes's philosophy of science is the philosophy of *scientia*. Whether the official philosophy of science is more representative of Descartes's views than the philosophy of science which I am saying he leaves between the lines is another matter.

The two philosophies of science connected

To connect the official and unofficial philosophies of science, it is useful to bring in a pair of distinctions that Descartes used throughout his life in writing about the nature and methods of science. These are the distinctions between the simple and the complex and between intuition and deduction. Simple and complex relate to the order in which an inquiring mind arranges what it is thinking about for the purpose of solving a problem or demonstrating something. For a thing or consideration to be simple is for it to be the most, or among the most, intelligible things or considerations that a mind has analyzed a problem into, or that it has discerned as relevant to a conclusion to be demonstrated. Relative to the problem of showing that the human mind is capable of physics, the natures of the self and of God are simples, as are the truths corresponding to these: the *cogito* and the proposition that God exists and is no deceiver. Relative to the problems of optics, the nature of light is simple. Intuition and deduction are distinctions between kinds of mental vision that produce understanding. Intuition is instantaneous, vivid understanding to which nothing can be added without loss of vividness; deduction is the consciously prolonged mental vision of many different things or truths which, in the ideal case, are individually intuited (CSM I, 14–15; AT X, 368–70).

Now deductions are what the reader of the *Essays* is supposed to find, even though the dependence of these deductions on the principles of metaphysics is left unspecified. These deductions are self-sufficient in the sense of producing their own certainty as they unfold; but they nevertheless have a part in a larger deduction that is capable of producing even more certainty. The closest Descartes comes in his writings to presenting the inclusive deduction in one book is in *The Principles of Philosophy*. But the difference between inclusive deductions and noninclusive but self-sufficient deductions is not necessary for understanding Descartes's definition of science. We can say that for Descartes science is deduction from what can be intuited in a problem or intuited in a thing to be demonstrated; and we can take it that for Descartes the deduction will collect together in the mind a whole series of things or considerations – from simple to complex.

Note

1   All references are by volume and page number to *The Philosophical Writings of Descartes*, trans. J. Cottingham, R. Stoothoff, and D. Murdoch (Cambridge: Cambridge University Press, 1975), abbreviated "CSM," and to C. Adam and P. Tannery, *Oeuvres de Descartes* (Paris: Vrin, 1964–76), abbreviated "AT."

Further reading

Works by Descartes
The following are complete translations of works by Descartes only excerpted in CSM:

1965: *Discourse on Method, Optics, Geometry, and Meteorology*, trans. P. J. Olscamp (Indianapolis: Bobbs-Merrill).
1979: *Le Monde ou Traité de la Lumière*, trans. M. S. Mahoney (New York: Abaris Books).
1983: *Principles of Philosophy*, trans. R. Miller and R. P. Miller (Dordrecht: Reidel).

Works by other authors
Clarke, D. 1982: *Descartes's Philosophy of Science* (Manchester: Manchester University Press).
Des Chene, D. 1996: *Natural Philosophy in Late Aristotelian and Cartesian Thought* (Ithaca, NY: Cornell University Press).

Garber, D. 1992: *Descartes' Metaphysical Physics* (Chicago: University of Chicago Press).

Gaukroger, S. 1980: *Descartes: Philosophy, Mathematics and Physics* (Brighton: Harvester).

—— 1993: Descartes: methodology. In *The Renaissance and 17th Century Rationalism*, ed. G. H. R. Parkinson (London: Routledge), 167–200.

Shea, W. 1991: *The Magic of Numbers and Motion: The Scientific Career of René Descartes* (Canton, MA: Canton Publishing).

Voss, S. (ed.) 1993: *Essays on the Philosophy and Science of René Descartes* (Oxford: Oxford University Press).

14

Discovery

THOMAS NICKLES

We begin with some questions. What constitutes a scientific discovery? How do we tell when a discovery has been made and whom to credit? Is making a discovery (always) the same as solving a problem? Is it an individual psychological event (an aha! experience), or something more articulated such as a logical argument or a mathematical derivation? May discovery require a long, intricate social process? Could it be an experimental demonstration? How do we tell exactly *what* has been discovered, given that old discoveries are often recharacterized in very different ways by succeeding generations? What kinds of items can be discovered, and how? Is the discovery of a theory accomplished in much the same way as the discovery of a new comet, or is "discovery" an inhomogeneous domain of items or activities calling for quite diverse accounts? Must a discovery be both new and true? How is discovery related to (other?) forms of innovation, such as invention and social construction? Can there be a logic or method of discovery? Just one? Many? What could such a procedure be? How is it possible that an (a priori?) logic or method available now has so much future knowledge already packed into it? How could a logic of discovery itself be discovered? How general in scope must a method of discovery be? Must it apply to all sciences, independently of the subject matter (as we might expect of a "logic"), or might it apply only to problems of a certain type or depend on substantive scientific claims? How, if at all, is their discovery related to the justification of scientific claims? Is the manner in which scientists make discoveries at all similar to the way in which they test them? Is this justificatory "checkout" procedure really *part* of the larger discovery process rather than distinct from it? Can discoveries be explained rationally, or do they always contain irrational or nonrational elements, such as inspiration or blind luck? Are historians, sociologists, and psychologists better equipped than philosophers to explain scientific creativity? Can a methodology of discovery help to explain the explosion of scientific and technological progress since 1600? If there is no logic of discovery, and if discovery is irrelevant to justification, then why include the subject of discovery in the domain of philosophy (epistemology or methodology of science) at all? What could philosophers have to say about it? Are there historical patterns of discovery, for example, that tell us something about the rationality, if not the logic (in the strict sense), of the growth of scientific knowledge?

These are the central, overlapping clusters of questions that philosophers and methodologists raise about discovery. Some historical background will put in perspective some proposed answers to them.

In the seventeenth century, scientific discovery was the central problem of what we would now call scientific methodology, the study of scientific method (see SCIENTIFIC METHODOLOGY). In the twentieth century, we find a complete contrast: the first two major schools of professional philosophy of science (the logical positivists and the Popperians) both contended that discovery has no place at all in the logic or methodology of science. Nonetheless, since 1960 the topic has made something of a comeback, in historical philosophies of science, in artificial intelligence (AI) treatments of research, and even in the new science studies.

The Enlightenment was the heyday of the methodology of discovery. Plausible claims about the availability of methods of discovery supported the optimistic doctrine of progress. For discovery methods would be cornucopias of new results and would transform human life and society within a few generations. Then nineteenth-century scientific and Romantic attacks on the allegedly self-contradictory idea of a *method* for *creative* work, a *routine* for producing justified *novelty*, dealt discovery a blow from which it has never fully recovered. However, in AI today we find a neo-Enlightenment, or at least a post-Romantic, revival of interest in discovery. This revival is both supported and challenged by the new social studies of science. Science studies experts themselves focus on the process of scientific investigation and in this sense belong to the discovery movement; but they reject the idea of discovery as suggesting that science is a linear, cumulative revealing of natural reality, rather than a process of sociocultural construction of artifacts.

Francis Bacon, René Descartes, Isaac Newton, and their contemporaries largely invented the modern idea of a *method* of inquiry in the seventeenth century. They understood method as a method of *discovery*, as an orderly, systematic procedure for conducting inquiry that virtually guaranteed the discovery of abundant *new knowledge*. They placed much emphasis on the new, on novelty (as opposed to merely reformulating in rhetorically eloquent ways what people already knew), but an equal emphasis on knowledge. Genuine knowledge was *scientia*, the kind of certainty available in geometrical demonstration and nearly so in empirical demonstration. There was a puritanical, cleansing element to these methodologies of discovery. Each writer's touted method swept away most of what passed for knowledge, as mere opinion or the dogma of arbitrary authorities; and it claimed to provide a straight and narrow path toward empirical or intellectual truth. Hence, the method of discovery was also a method of justification: a claim was justified *because* it had been produced by the right method, by something akin to logical derivation. This integral method generated results that were both novel and true.

The fortunes of Baconian, Cartesian, and Newtonian methods (to mention the principal traditions) waxed and waned for various scientific and sociocultural reasons up through the end of the Enlightenment (roughly 1800). Then more Romantic notions of high creativity in poetry, music, and even the sciences emerged to cast doubt on the idea that a methodical procedure could produce genuinely novel and interesting results. Does not all creativity depend on insight and inspiration, rather than logic? And does not high creativity require flashes of genius? Thus was born the idea that discovery is a momentary, personal "aha!" experience, as opposed to sustained, logically systematic laboratory or theoretical work. The failure of Newtonian chemistry and optics, *inter alia*, reinforced the challenges to the old models of science. Increasing doubts about the certainty of scientific claims combined with analogies to recent mathematical work on successive approximations to suggest a new, more dynamic conception of science as self-correcting and hence progressive over time (yielding ever closer approximations to the truth), rather than as the static accumulation of definitively established theories. All these developments contributed to the decline of the old methodologies of discovery and to the ascendancy of their replacement, the method of hypothesis. Since Newton, hypotheses had been officially spurned as too conjectural, too remote from observable causes, to count as knowledge. However, the Newtonian method ruled out much hypothetical work that seemed fruitful. Accordingly, the hypothetical method itself came to look more promising.

Now, according to the hypothetico-deductive (H-D) method (as the hypotheticalist method is often called today), it does not matter how we discover hypotheses. They need not derive from vast tables of data, but may be the products of momentary inspiration. What matters is how we test them, once they have been proposed. A test consists in deducing (or otherwise deriving) an observable consequence (a prediction) from the claim being tested and then checking to see if the prediction is true. If so, the hypothesis is confirmed.

Here we find the historical and logical separation of justification from discovery – actually, more than a separation (Laudan 1981; Nickles 1987). For thoroughgoing hypotheticalists such as W. S. Jevons (1874) eventually *reversed* the logic of justification. Previously, a claim was warranted by the manner in which it was *generated* – roughly, by the soundness of the argument from which it was derived as a conclusion. On this view, a claim's mode of construction or discovery provides epistemic support for it. By contrast, the H-D method says that justification derives entirely from confirmed *consequences*, rather than antecedents (established premises), of the claim (the hypothesis). Following Laudan, let us call this second view *consequentialism*, and the first *generativism*. The old methodologies of science were generativist. The new, hypotheticalist methodologies were consequentialist. Consequentialism has dominated methodology to the present day.

In the twentieth century, many logical positivists (see LOGICAL POSITIVISM) and Popperians not only upheld the divorce of discovery from justification but also expelled the topic of discovery from epistemology. Popper's attack on inductivist methodology (which aspires to derive laws and theories from empirical facts) carried over naturally, if fallaciously, into an attack on all generative methodologies. Despite its title, Popper's *Logic of Scientific Discovery* (1959) completely rejected traditional logic of discovery (see POPPER).

Meanwhile, many positivists also denied that there could be a logic of discovery. Hans Reichenbach (1938) famously distinguished between "context of discovery" and "context of justification," a difference later misinterpreted as an invidious distinction denying the epistemological interest of the discovery process. Where possible, the logical positivists preferred to formulate their problems in abstract, formal terms. For them, a logic of discovery would be something like an algorithm for constructing interesting theories about the world. It was obvious to them that there could be no such logic.

Nonetheless, discovery has enjoyed a revival of interest since the 1960s. The three principal groups of "friends of discovery" are a "logical" group of AI experts, a "historical" group of philosophers of science who scarcely noticed the AI work until the 1980s, and the "science studies" group growing out of the new sociology of scientific knowledge. Historical philosophers such as N. R. Hanson (1958) took seriously the new history of science with its "internalist" accounts of important scientific discoveries. In the better accounts, discoveries were not momentary inspirations, but intellectual and experimental processes structured in time and by no means irrational. To the historical "friends of discovery" it was absurd to deny that discovery – the frontier where the growth of knowledge most obviously occurs – is relevant to the central problems of epistemology: What can we know, and how can we know it? How is inquiry possible? Also, historical philosophy of science seemed more relevant to science as practiced than the abstract, formal problems addressed by positivism and by much AI.

Yet these historical philosophers largely conceded that there is no logic or method of discovery in the old sense. Many "friends of discovery" retreated from defending the existence of a *logic* of discovery to defending the *rationality* of discovery in some sense. Dozens of detailed historical case studies portrayed scientific work as reasonable enough, but as so diverse and context-specific that there seemed no chance of finding anything like a master logic that could magically account for successes in all fields of research.

The historical case-study method faced criticism on several grounds. Just as one can appeal to the Bible to "prove" almost anything, so one can find a historical case to "establish" or "refute" practically any methodological claim. Paul Feyerabend (1975) could as easily use history as evidence "against method" as the so-called friends could use it as evidence *for* method (see FEYERABEND). Only the most careful use of history can provide general insights or genuine evidence for or against a methodological position (see Donovan et al. 1988).

Another criticism, this time from AI, is that historical "rationality of discovery" accounts are too weak and too vague to explain why the scientists proceeded in the ways they did and why those procedures were successful. What is needed is a constructive account sufficient to actually "compute" the discovery claims – i.e., a computer simulation sufficient to *rediscover* the results in question, given the information available at the time.

Interestingly, the history of AI work on discovery logics recapitulates the larger history of science and methodology. Just as the early, grand, universal, a priori, content-free logics of discovery gave way to more context-specific procedures laden with empirical and theoretical content, so the early attempts of Allen Newell and Herbert Simon (1972) to construct a General Problem Solver, using only very general, content-free rules, gradually gave way to knowledge-based and case-based "expert systems" in AI. Ironically, the Newell–Simon system of general rules was simply too weak to handle a diverse range of difficult problems. The problem domain of the sciences taken as a whole is too inhomogeneous to yield to a uniform treatment. According to Ira Goldstein and Seymour Papert, writing in 1977:

> Today there has been a shift in paradigm. The fundamental problem of understanding intelligence is not the identification of a few powerful techniques, but rather the question of how to represent large amounts of knowledge in a fashion that permits their effective use and interaction.... The current point of view is that the problem solver (whether man or machine) must know explicitly how to use its knowledge – with general techniques supplemented by domain-specific pragmatic knowhow. Thus, we see AI as having shifted from a *power-based* strategy for achieving intelligence to a *knowledge-based approach*. (1977, p. 85; emphasis original)

Already in 1971 Edward Feigenbaum had summarized the development thus: "There is a kind of 'law of nature' operating that relates problem solving generality (breadth of applicability) inversely to power (solution successes, efficiency, etc.) and power directly to specificity (task-specific information)."

By the 1980s, however, it became clear that the knowledge-based computation of standard AI was limited in its ability to handle complex problems. The standard approach, based on symbolic rules and symbolic representations of knowledge, worked best in well-structured domains and not so well for ill-structured problems at the frontier of knowledge. In any case, rule-based systems tend to be fragile, since it is difficult to verify the consistency of new rules with old rules. Accordingly, standard rule-based systems face difficulties precisely in the area of learning, including discovery. In addition there is the so-called knowledge elicitation problem, sometimes called "Feigenbaum's problem." Expert scientists often have difficulty articulating the rules they are supposedly using to solve problems, and they often violate rules they have previously proposed.

In the wake of these difficulties, many people in both AI and philosophy of science became fascinated with the promise of parallel distributed processing, or connectionism, with its neural-network-like models of various problem-solving processes. At first connectionism was touted as a clear alternative to standard, rule-based computation. However, given its own present shortcomings, there is a growing expectation today that more adequate systems will incorporate processes of both kinds. The same can be said for case-based reasoning, another alternative to purely rule-based reasoning that has emerged during this same period. Thagard (1992) is an example of a pluralistic approach to modeling innovative scientific reasoning.

The specific connection of this AI work on problem solving to scientific discovery was made already in the 1950s by Newell and Simon and associates. Their motivation was as follows: (1) Discovery is problem solving. Since we already know how to study problem solving, discovery, too, can be studied in the same, cognitive terms. (2) Problem solving usually amounts to search through "spaces" of possible solutions. (3) This search typically employs heuristic procedures – fallible rules or constraints that are often available when algorithms are not, and are often more efficient than algorithms anyway. Heuristic procedures can cut huge search "trees" or search spaces down to manageable size. Accordingly, we must broaden *logic* of discovery to include *heuristics*. Use of heuristics allows for insight without abandoning the logical space of reasons, arguments, and computations. (4) Problem solving is ubiquitous in science. *All* phases of research require problem solving; therefore, search tasks and hence discovery tasks are also ubiquitous. "Ubiquitous" embraces even the justification process – e.g., the search for novel predictions and for experimental designs to test such claims. (Connectionists largely retained these insights by treating heuristics as sub-symbolic constraints built into the very structure of networks or as patterns of connection weights among nodes.)

AI approaches have the twin virtues of reminding us how problem-centered scientific research is, and of revealing the naiveté of the old "stage" models of research used by philosophers. Following the divorce of discovery from justification, it became fashionable to speak of a "discovery stage" of research followed by a "justification stage." Actually, there was some confusion about whether these were temporal stages or logical stages, whether they were processes or reconstructed products, or whether discovery was the process and justification alone the logically reconstructed product. The two-stage model then gave way to a three-stage model, which sandwiched a "preliminary evaluation," "prior appraisal," or "pursuit" stage between discovery and "final" justification. The idea here was that some sort of assessment of the promise or fertility of the possible lines of investigation is needed even in the discovery process, some justification of the choice of one line over others. In this sense, justification, too, turns out to be ubiquitous – popping up here within the context of discovery. Concern with this middle stage did point up the need for more work on heuristic appraisal, a crucial and surprisingly neglected topic that straddles discovery and justification.

Since the 1980s, the historical and AI "friends of discovery" have interacted to a larger degree, as the historical philosophers have learned of the AI work and as AI experts have given more attention to historical cases. Some major differences of emphasis remain, however. AI advocates, deploring the low *logical* standards of the historical friends, impose on historical explanation the aforementioned rediscovery requirement that a fully adequate explanation be a "computational" explanation. Historicists reply by deploring the low *historical* standards of AI work. For is it not totally unrealistic to require that a successful historical explanation produce what is, in effect, a logic of discovery that computes or derives the discovery? We began by wondering whether a logic of discovery is even possible, and now we find that it is necessary, if we are to explain any discovery at all! It is surely false that every historical discovery was produced by, or can be explained by means of, a logic of discovery (a rule-based, problem-solving routine). For how were those discovery logics themselves discovered? The AI standard is reminiscent of Carl Hempel's famous old requirement that an adequate historical explanation be a deductive argument from covering laws as premises. Indeed, critics sometimes label the AI position "PC positivism" (personal computer positivism). The explanations attempted by Langley et al. (1987), for example, are clearly whiggish. They program in from the beginning the final, reconstructed conception of the problem and of the search space. In effect, our hindsight is attributed to the original discoverers (as foresight, prescience?). What the computer models is not the original, messy research, but only the sanitized, rationally reconstructed searches and solutions available a generation or two later in the textbooks. Accordingly, they skim over the most crucial phases of learning, including conceptual reorganization and acquisition of new skills. However, AI experts are aware of these problems. More recent AI programs, such as KEKADA, are capable of a sort of experimental exploration and are more tolerant of "noise." Today many experts have broadened the term "computational" to include nonsymbolic, connectionist processes and case-based reasoning, whereas it formerly referred to computation by means of symbolic rules and representations alone.

While AI experts fail to appreciate the degree of conceptual reconstruction and refinement of skill typical of scientific work, the same is true even of historical philosophers and sociologists! The discovery process is far more drawn out and structured than most methodologists appreciate. Theory construction, as exhibited in the original papers, is only the first stage of discovery, only the first "round" of innovation. To stop even at "'final justification" (empirical confirmation of the new claims) is to leave the task of describing/explaining discovery only half finished. For much or most of the innovation of every major discovery occurs in the successive technical refinements that occur in the decades *after* acceptance of the initial work into the literature – years after "the discovery" was supposed to have been made. Thomas Kuhn (1978) contends that Planck's famous discovery of the energy quantum in 1900 actually emerged over the next two or three decades, beginning with Einstein's and Ehrenfest's misreading of Planck in 1905. They attributed to Planck a solution that Planck never offered, and later explicitly repudiated, to a problem that Planck never entertained. And several basic physical theories developed since then have yielded new derivations of "Planck's radiation law." So virtually the entire discovery of energy quanta came years after Planck's papers of 1900. Meanwhile, Planck's own solution to his problem fell by the wayside. Similarly, Augustine Brannigan (1981) argues that Gregor Mendel's discovery of the basic laws of genetics was also a *post hoc* attribution of a solution Mendel never offered for a problem that did not emerge clearly until later.

For discovery to be relevant to epistemology, it must be subject to normative rules or standards of an epistemic nature; that is, it must be coupled to justification. But the just-described dissociation or divorce of what we may call "initial discovery" from "final discovery" – not to mention from final justification – casts doubt on the existence of such a coupling in actual research. Let us label this difficulty "the coupling problem." The Planck and Mendel examples suggest that the coupling may often be too tenuous to make discovery epistemically relevant. Or, on the contrary, do they suggest that the solution to the coupling problem will come from studying more closely the kinds of "rational reconstruction" by which scientists (not philosophers) transform or deform initial discovery into final discovery *cum* justification? Here there is plenty of room for controversy. Kuhn and Brannigan themselves offer very different sorts of accounts, yet both also reject the existence of a method of discovery.

Consider the following general argument from economy of research that coupling is necessary to achieve the goals of science, and hence that discovery is an essential topic for epistemology. The central problem of methodology is to show that the methods advocated have a reasonable chance of achieving the stated goals of the enterprise, that they are better than blind luck. The problem is to show how to achieve infinite aspiration ("Find the one true theory in an infinite domain of possibilities!") by finite means. Suppose that the goal of research is to find true laws, theories, models, and/or explanations. Of the infinite number of possible laws, theories, etc. possible for any scientific domain, scientists will, over time, actually formulate and consider only a finite set of candidates; or, at least, infinite subsets of the points in search space will go unnoticed. (For one thing, as the histories of deep conceptual change have taught us, we cannot now canvass possibilities that will only become available in future eras.) And these are precisely the law or theory candidates furnished by *discovery* procedures of whatever kind. Whatever its character, the discovery process filters out these few from the limitless set of potential laws, theories, or explanations. These are in turn poured through a second filter consisting of empirical testing and checks for compatibility with theoretical and methodological constraints. Thus research amounts to a two-stage filter or selections process. Unless a true (or sufficiently reliable) candidate is selected at the discovery stage (or developed during the transformation to "final discovery"), it has no chance at all of being selected at the second stage. In this sense at least, the discovery process is epistemically relevant. There must be some degree of coupling between the modes of generating theories and criteria of epistemic appraisal. Furthermore, since consequential testing obviously cannot govern the first-stage filter (since the candidates must already be selected before testing can commence), the economy argument establishes the necessity of a *generative* component of methodology, one that epistemically informs the initial selection/construction process.

The general argument from economy is conditional in form. *If* we are to efficiently realize our highest goals, *then* justification must be coupled to an epistemically constrained discovery process. But it may be simply impossible to achieve those aims. The universe does not fulfill our every desire! Note also that this argument for the epistemic importance of discovery does not entail the existence of a *logic* or *method* of discovery.

Now consider a powerful argument *against* the existence of methods of discovery. A naturalistic account of human inquiry, one that assumes that genuine learning is possible without commitment to nonnaturalistic faculties such as possession of fore-knowledge or a general criterion of truth, must treat inquiry at the frontier in a purely consequentialist way, as blind trial and error, as hazarding guesses followed by tests of their consequences. For at the outer frontier, by definition, knowledge already gained, including hitherto useful heuristics, no longer provides reliable guidance. And heuristics have epistemic force only insofar as they incorporate knowledge gained by previous inquiry. This is Donald Campbell's and Popper's position: that all human creativity, including scientific innovation, results, at bottom, from blind variation plus selective retention (BV + SR). Thus, in a sense, all scientific discoveries are products of serendipity (see Campbell 1974; Kantorovich 1993). They claim that this Darwinian model of inquiry is the only one compatible with what we know about our biology and psychology. Popper, Campbell, and followers such as Cziko (1995) go further to maintain that Darwinian BV + SR provides the only naturalistic solution to the Meno problem of how learning is possible. For any account that is not purely consequentialist, in allowing only selection by consequences, requires the assumption of a priori knowledge in the form of special, innate knowing capacities.

The Campbellians conclude, on this basis, that there can be no logic or method of discovery; for is not BV + SR the very antithesis of method? And commitment to non-Darwinian nonnaturalistic methods of discovery is tantamount to methodological creationism.

The irony is that precisely such BV + SR processes, when considered on a large enough (populational) scale, have recently become the basis of powerful problem-solving methods that can be mechanized. The field of genetic programming employs various versions of the so-called genetic algorithm. The idea is quite general. Begin with a population of candidate solutions to a problem (none of which need be at all adequate). Randomly recombine pieces of these solutions and test to determine which are more promising than others. Breed the population again in such a way that the better candidates have a higher probability of mating. Test again. Iterate this procedure. The process often breeds a solution in 30 to 100 generations. Koza (1992) furnishes many examples.

At first this idea sounds crazy as a problem-solving procedure. It reminds us of the old joke about monkeys at typewriters producing the works of Shakespeare. Yet it works. And when we stop to think about it, what process has been more creative than Darwinian evolution? The trouble with monkeys is that there are not enough of them, and they do no editing and no iterated retyping of chance recombinations of the more promising passages. The denial that an iterated chance process that proceeds by selection of consequences can produce interesting innovation is itself but the general form of the creationist objection to Darwin!

Is there a logic of discovery? Can there be? The answer depends on what counts as a logic and what counts as a discovery. The genetic algorithm is a promising candidate for a surprisingly general problem-solving method. It seems unlikely that we shall ever have a humanly usable method of quickly and routinely producing deep-structural theories of great conceptual novelty. However, there are standardized ways of "playing" with mathematical theories and models that can have profound, deep-structural consequences. For example, standard mathematical methods of generalizing and re-specifying claims are sufficient to derive the Lorentz transformation from the Galilean transformation, given the "empirical fact" that the velocity of light is finite. And consider the semi-automated modeling techniques now available. Routinized procedures and powerful heuristics can provide valuable support for work at the research frontier, making it far more efficient than it was a decade ago. If what is discovered is an empirically detectable object, such as a planet, a comet, a black hole, or a quasar, then it is again obvious that the degree of automation available today far exceeds anything of which Descartes or Newton ever dreamed. If what is discovered is an empirical regularity of some kind, consider the many available discovery logics for the collection and analysis of data, including statistical data. Factor analysis, path analysis, sampling methods, and many other forms of data analysis have been routinized in the form of computer programs (see, e.g., Glymour et al. 1987; Shrager and Langley 1990).

Ironically, some supposed enemies of discovery methodology have themselves developed discovery logics of the empirical variety. John Stuart Mill officially denied that there could be a logic of discovery; yet he refined previous strategies into "Mill's methods" of agreement, difference, concomitant variation, and residues – an inductive discovery toolkit. And Reichenbach's extensive work on inductive logic, including the so-called straight-rule of inductive inference, amounts to a logic of discovery (Laudan 1981).

Still, most "friends of discovery" would agree that there is no logic of discovery in the grand old sense of a single "logic" underlying scientific method. Rather, there exist many, diverse logics of discovery. These "logics" are not mere pipe dreams. They are in actual use and undergo frequent refinement. But neither are they *logics* in the strict, universal-formal sense. Rather, they are context-sensitive – problem-specific and often content-specific as well, laden with empirical and theoretical content of the subdisciplines that employ them. Many of them contain heuristic components.

Finally, it is important to realize that new discovery logics (e.g., new, standardized methods of problem formulation and solution) that are developed in times of major breakthroughs nearly always *postdate* the initial breakthroughs. Such a logic is not the cause or explanation of the corresponding initial discovery; rather, it is a *part* of the ongoing discovery itself, and fully articulated only as "final discovery." Typically, discovery logics are rational reconstructions of results arrived at by more haphazard routes. They are worked out by critical reflection on how the substantive problem solutions were originally achieved and how these methods might be streamlined or replaced by better ones. They are idealized discovery procedures – methods that *could* have been employed to make the original breakthroughs if (contrary to fact) we had known then what we know now. Accordingly, we might call them "discoverability logics." The objection raised above to the AI treatment of historical discoveries can now be restated in the new language. Simon and company are actually dealing with discoverability or final discovery rather than with initial discovery situations, and it is inappropriate for them to require that historians explain initial discovery in terms of search/construction procedures that would only become available as part of the "final discovery" as reported in standard textbooks. Their "rediscovery" requirement is actually a "discoverability" requirement. Kepler, Black, and Ohm did not have the textbook to help them! To reduce initial discovery to final discovery begs all the historical and epistemological questions.

Some additional objections (and brief replies) to the discovery program are these. (1) In his valuable book on formal learning theory, Kevin Kelly (1996), a fervent friend of discovery, contends that participants in the logic-of-discovery debate have taken history *too* seriously. The issue of the existence of a logic of discovery cannot be resolved by examining actual historical examples, for the question is one of logical existence in an abstract space of possible computer programs, not one of historical existence. Besides, the question is normative, not one of actual, historical practice. *Reply*: One *may* proceed purely formally in this way, but one traditional issue has been how to understand science as it has developed historically. And today we want an adequate, naturalistic account of how embodied human beings and human communities solve the Meno problem of gaining new knowledge. Criteria of historical relevance are just as valid for those interested in these issues as Kelly's dismissal of them for his purposes. Although Kelly is as critical of logical positivists as of historicists, any highly formal approach to knowledge is likely to present "relevance" difficulties analogous to those the positivists faced. (2) Scientists sometimes use insight in solving problems; they are not logic dopes. *Reply*: Granted. Not even Bacon and Descartes claimed that a logic of discovery is necessary to make discoveries, only that it is sufficient – and more efficient. Not even Campbell denies that powerful heuristics are sometimes available, and many investigators are simply lucky, in that their unjustified heuristics led them in directions that turned out to be fruitful. (3) "There is a method of discovery" implies "There is a scientific method." However, the diversity of the sciences and the variety of ways in which one may conduct scientific work show that the idea of method is too pretentious and inflexible. Some philosophers have claimed that there is no method of justification, any more than there is a method of discovery. The attack on method puts discovery and justification in the same boat, but the boat has sunk. *Reply*: The critics are right that there is no monolithic scientific method or logic of justification, anymore than there is a single logic of discovery. But why not allow a diversity of smaller-scale, context-sensitive methods in both cases? Specific, powerful, problem-solving routines are in daily use, after all. (4) Sociologists have also attacked methodologists' uses of the concepts of method and discovery. Method does little real work in science (they say); rather, talk about method (i.e., methodology) finds rhetorical employment in scientific disputes and provides *post hoc* rationalizations that make "internalist" histories of science seem appropriate, thus elevating science to a culturally transcendent status, above the play of "external" social forces. Methodology is a sort of secular theology (see Schuster and Yeo 1986). *Reply*: Well said, but we must tend our garden.

Sometimes method is mere rhetoric, but rhetoric, too, can play a crucial role in serious scientific work. The objection itself conceives of science too narrowly. (5) Once we replace the untenable, strongly realist notion of discovery by the concept of social construction or invention, there is no longer any reason to believe that there could be a logic of discovery. For social constructions are accomplished by tinkering with cultural resources in ways that are highly contingent, local, and diverse. The term "discover" naively suggests that scientists open a window on the world and gain direct access to the furniture of the world. Social-historical work has completely discredited this view (Schaffer 1986). *Reply*: These points are well taken, but overstated. By "discovery" many philosophers mean the discovery of problem solutions, not necessarily of absolute truths about the universe. In this sense, "discovery versus construction" is a false dichotomy. Routinized research procedures are rational reconstructions, and hence constructions. And among the humanly constructed items are some impressive problem-solving and data-analysis routines as well as automated laboratory equipment such as gene sequencers. Would not Bacon, Descartes, and Newton happily consider these logics of discovery?

References and further reading

Brannigan, A. 1981: *The Social Basis of Scientific Discoveries* (Cambridge: Cambridge University Press).

Campbell, D. T. 1974: Evolutionary epistemology. In *The Philosophy of Karl R. Popper*, ed. P. A. Schilpp (LaSalle, IL: Open Court), 413–63.

Cziko, G. 1995: *Without Miracles: Universal Selection Theory and the Second Darwinian Revolution* (Cambridge, MA: MIT Press).

Darden, L. 1991: *Theory Change in Science: Strategies from Mendelian Genetics* (New York: Oxford University Press).

Donovan, A., Laudan, L., and Laudan, R. (eds) 1988: *Scrutinizing Science: Empirical Studies of Scientific Change* (Dordrecht: Kluwer).

Feigenbaum, E. A., Buchanan, B., and Lederberg, J. 1971: On generality and problem solving: a case study using the DENDRAL program. *Machine Intelligence*, 7, 165–90.

Feyerabend, P. K. 1975: *Against Method* (London: New Left Books).

Glymour, C., Scheines, R., Spirtes, P., and Kelly, K. 1987: *Discovering Causal Structure: Artificial Intelligence, Philosophy of Science, and Statistical Modeling* (Orlando, FL: Academic Press).

Goldstein, I., and Papert, S. 1977: Artificial intelligence, language and the study of knowledge. *Cognitive Science*, 1, 84–123.

Hanson, N. R. 1958: The logic of discovery. *Journal of Philosophy*, 55, 1073–89.

Jevons, W. S. 1874: *The Principles of Science* (London: Macmillan).

Kantorovich, A. 1993: *Scientific Discovery: Logic and Tinkering* (Buffalo: SUNY Press).

Kelly, K. 1996: *The Logic of Reliable Inquiry* (Oxford: Oxford University Press).

Kleiner, S. 1993: *Scientific Discovery: A Theory of the Rationality of Scientific Research* (Dordrecht: Kluwer).

Koza, J. 1992: *Genetic Programming: On the Programming of Computers by Means of Natural Selection*, vol. 1 (Cambridge, MA: MIT Press).

Kuhn, T. S. 1978: *Black-Body Theory and the Quantum Discontinuity, 1894–1912* (Oxford: Oxford University Press).

Langley, P., Simon, H. A., Bradshaw, G., and Zytkow, J. 1987: *Scientific Discovery* (Cambridge, MA: MIT Press).

Laudan, L. 1981: *Science and Hypothesis* (Dordrecht: Reidel), esp. ch. 11.

Newell, A., and Simon, H. A. 1972: *Human Problem Solving* (Englewood Cliffs, NJ: Prentice-Hall). (A reconstructed summary of their work since the 1950s.)

Nickles, T. 1980a: *Scientific Discovery, Logic, and Rationality* (Dordrecht: Reidel).

—— 1980b: *Scientific Discovery: Case Studies* (Dordrecht: Reidel).

—— 1987: From natural philosophy to metaphilosophy of science. In *Kelvin's "Baltimore Lectures" and Modern Theoretical Physics*, ed. R. Kargon and P. Achinstein (Cambridge, MA: MIT Press), 507–41.

—— 1990: Discovery logics. *Philosophica*, 45, 7–32.

Popper, K. R. 1959: *The Logic of Scientific Discovery* (New York: Basic Books; trans. with revisions of *Logik der Forschung*, 1934).

Reichenbach, H. 1938: *Experience and Prediction* (Chicago: University of Chicago Press).

Schaffer, S. 1986: Scientific discoveries and the end of natural philosophy. *Social Studies of Science*, 16, 387–420.

Schuster, J., and Yeo, R. (eds) 1986: *The Politics and Rhetoric of Scientific Method* (Dordrecht: Reidel).

Shrager, J., and Langley, P. (eds) 1990: *Computational Models of Scientific Discovery and Theory Formation* (San Mateo, CA: Morgan Kaufmann).

Simon, H. A. 1977: *Models of Discovery* (Dordrecht: Reidel).

Thagard, P. 1992: *Conceptual Revolutions* (Princeton: Princeton University Press).

15

Dispositions and Powers

ROM HARRÉ

Dispositions: their forms and varieties

Conditional properties in general

The great variety of kinds of properties that are comprehended by such terms as "disposition," "power," "potential," "tendency," "capacity," "propensity," and "capability" share a common generic structure. They are ascribed to things and substances. However in all cases the basic structure of that attribution is conditional in form. To attribute a disposition to a thing or substance is to say that if certain conditions obtain, then that thing or substance will behave in a certain way, or bring about a certain effect – that is, that a certain outcome will occur. A fertile soil is one of which it is true that if seeds are planted in it and properly cared for, they will grow into flourishing plants. A negatively charged particle is one of which it is true that, if brought into proximity to another negatively charged particle, it will experience a force of repulsion.

All the variety we find in this class of properties can be expressed either by adding a qualifying clause to the basic form, or by variations in the modality of the conditional clauses, from "is" and "will be" to "were to be" and "would."

Bare and grounded dispositions

To say that chili peppers are "hot" is to say that if they are chewed (were to be chewed), one experiences (would experience) a tingling and burning sensation in the mouth. This simple conditional statement ascribes a bare disposition to these vegetables. However, one may raise two kinds of questions that lead to a natural elaboration of the matter. One may wish to know why chewing these seed pods and not others has (would have) this effect. And one may also wish to know how one can be justified in saying that the peppers *are* hot, when they are merely hanging from the kitchen ceiling, and not being chewed. Both these questions get the same answer: they have this effect because of their chemical constituents, and they possess these constituents even when not being chewed. The one question sets a problem for a biochemist, the other for an epistemologist. Their common answer is incorporated in a more elaborate formulation of dispositional ascriptions: namely, as grounded dispositions:

> If conditions C obtain, then effect E will occur by virtue of the nature of the things or substances involved, *ceteris paribus*.

The addendum clause describes the grounding of the disposition. The *ceteris paribus* qualification is required to take account of the fact that, in general, the conditions specified in the antecedent of the conditional clause are local and presume the stability of a more comprehensive environment, such as the atmosphere, the gravitational field, and so on. In many cases the logical form of the addendum is an existential quantification over properties. It would read, when expanded, something like this:

> There is some property, which we do not currently know, that is characteristic of the thing or substance involved, and the possession of that property is a necessary condition for the effect to occur in the defined circumstances.

It has been pointed out that the enrichment of a bare disposition to a grounded disposition is tantamount to the setting out of a scientific research program. The addendum asserts only that there is *some* grounding property, not which property it is. So it has the form of a hypothesis which could be further investigated, sometimes experimentally.

Affordances

The term was introduced by J. J. Gibson (1979) to distinguish those dispositions which we ascribe to material things, in which the conditional clause is expressed in terms of some human activity, requirement, and so on from all other kinds of dispositions. For instance, the declaration that the ice on a pond is safe means that it affords skating. A floor affords walking; scissors afford cutting; and a musical score affords performing. The notion has been generalized to include those dispositions which are ascribed to the material world on the basis of the reactions of a humanly constructed experimental apparatus. Thus a flow of current is that process in a conductor which affords a reaction of a galvanometer. Affordances are properties of the material world, but manifest themselves in circumstances devised and created by human beings. A central topic of concern to both physical and human scientists is whether the dispositions evoked by this or that experimental procedure are manifested in circumstances other than those devised by the investigator. Which affordances could be redefined as dispositions which could be manifested in circumstances other than those devised by human beings?

Liabilities and powers

From the very first uses of concepts that fit the "grounded disposition" schema in the seventeenth century, the distinction between passive dispositions, or liabilities, and active dispositions, or powers, has been important. Inertia is a passive disposition, or liability, since the capacity to resist acceleration becomes effective only when a body is subject to an impressed force. Weight is an active disposition, or power, since the tendency to accelerate towards the center of the Earth is continuously effective even when a body is prevented from falling, say, by a resting on a platform. This distinction, though of great importance in the way we structure explanations, is nevertheless relative. A material body has weight only when in a gravitational field, and that weight is proportional to the strength of the field. Relative to the gravitational field, weight is a liability. Elementary electric charges are counted among the fundamental entities of the universe, since, at least in orthodox electromagnetic theory, they depend for neither their existence nor their strength upon each other. They are, then, pure, or fundamental, powers. Considered with respect to other like charged bodies, these basic powers can also be seen to be liabilities, in that an elementary particle, by virtue of the charge, is acted upon by others. The question of whether any human dispositions, tendencies, capacities, and so on are active powers is still much debated. Considered as mere spectators of the working of cognitive mechanisms, people are denied active powers; while considered, according to the new discursive psychology, as active users of sign systems, people are taken to be original sources of activity.

The historical development of dispositional concepts

Locke's doctrine of qualities

The modern interest in dispositional concepts can be dated to their use in Locke's essay, famously developed out of the distinction between primary and secondary qualities. Locke thought that our sensory experience could be analyzed into simple ideas, such as the idea of square and the idea of yellow. These ideas, he supposed, must be caused in us by material things. How are the ideas in the mind related to the qualities which cause them in material bodies? According to Locke, the ideas of primary qualities, such as shape and number, resemble the qualities which cause them, while the ideas of secondary qualities, such as taste, warmth, color, and so on, do not. Yet these are caused, as are the primary ones. What are the corresponding secondary qualities? These, says Locke, are powers in the body to cause the ideas (see LOCKE; QUALITIES, PRIMARY AND SECONDARY).

It was quickly pointed out by both the friends and the opponents of natural powers that Locke's distinction between primary and secondary qualities and their corresponding ideas would not do. Greene (1727) pointed out that the argument for the powers analysis applied as much to primary as to secondary qualities. So the scientific realist, claiming to penetrate beyond the mere sensory appearances of things, must necessarily turn to an ontology of powers, as the real substance of the world. Reid, adopting a distinction made central by Berkeley, partitioned powers into active and passive, the active characterizing rational and sentient beings. Matter, Berkeley declared, is just a collocation of ideas, and, like the ideas that constituted it, was inert. Physics followed Greene, rather than Berkeley (see BERKELEY). There are, in nature, active causal powers, not associated with sentience or discretion.

The dynamicist metaphysics of physical science

If all the qualities of bodies, relevant to our observing them, are best taken as powers, what are we to make of these bodies themselves? Is there any place for material stuff in the ultimate constitution of the world? The dynamicists, of which party Leibniz, Greene, and Boscovich were enthusiastic members, declared matter to be redundant. The fundamental physical beings are point-centered fields of force. A field of force is characterized by a pattern of spatially distributed dispositions. The illusion of materiality arises from the way we perceive those surfaces in space at which the forces of attraction and repulsion are equal and opposite. A body will lie passively on a surface defined this way. The surface of forces in equilibrium will seem to enclose a volume of solid matter.

Dispositions and scientific realism

It can hardly be denied that we know the physical world, as scientists, through the reactions of our instruments. It is easy to slip into the positivist view that physical science is just the statistical study of the reactions of instruments, and our picture of the world a mere "as if." We cannot observe the states and processes that produce the reactions of instruments. "Yet the thing is not altogether desperate," as Newton once remarked (in the Scholium to Definition VIII), when faced with a similar impasse concerning our knowledge of space and time (see NEWTON). In this case, we know that the physical setup of world plus instrument *must* afford just those reactions. We can therefore ascribe a qualified disposition to the world. Our knowledge of the world as a system of powers is not wholly independent of human concepts and constructions, but is at least partially so. Our instruments do not behave in these ways unless "bolted to" the world.

Our confidence in this solution to the problem of defining an acceptable form of scientific realism rests on a historical observation about the development of a dispositionalist treatment of our knowledge of the physical world. Conceived in terms of dispositions, our knowledge presents itself in a hierarchical form, such that successive steps in the hierarchy are of epistemically different strengths; that is, as claims to knowledge. The dispositions of a substance to react in this or that way in appropriate circumstances are grounded in hypotheses about the constituents of that substance – for instance, in the arrangements of the elementary magnets that together constitute a bar magnet, whose active powers are revealed in the patterns it induces in a sea of iron filings. But the elementary magnets, revealed by metallurgists' microscopic techniques, are themselves bodies endowed with causal powers, which, in their turn, are grounded in molecular and submolecular constituents and their arrangements. The powers and dispositions of ions are grounded in structures of subatomic particles, each of which is endowed with its characteristic cluster of powers. The grounding of macro dispositions in the observable constituents of the material stuff in question lends inductive support to the next step: namely, grounding this level of disposition in the powers of further constituents, whose manipulability by a human agent gives weaker, but subtle, substantial inductive support to claims for their existence, and so on. The ultimate powers of matter are the elementary charges which define the nature of subatomic entities or beings of yet more subtle character. In this way the regress of dispositions and powers, structured by the intervening hypotheses as the groundings of the powers revealed or supposed at each level, confirms the general plan for ascribing dispositions, as real properties, even to the unobservable constituents of the universe.

The final step in setting up a dispositionalist ontology is to define a class of elementary beings, the only properties of which are their powers. The claim that any given kind of being is elementary is, of course, defeasible, by the discovery of some constituting structure of yet more elementary beings, properties of which ground the dispositions once taken as bedrock. The elementary beings that define the limits of a given ontology are its powerful particulars. In psychology these might be persons, and in physics certain classes of charged "particles."

100

References and further reading

Berkeley, G. 1985: *The Principles of Human Knowledge*, ed. G. J. Warnock (London: Fontana; 1st pub. 1710).

Boscovich, R. J. 1966: *A Theory of Natural Philosophy* (Cambridge, MA: MIT Press; 1st pub. 1763).

Cartwright, N. 1989: *Nature's Capacities and their Measurement* (Oxford: Clarendon Press).

Gibson, J. J. 1979: *The Ecological Approach to Visual Perception* (Boston: Houghton Mifflin).

Greene, R. 1727: *The Principles of the Philosophy of the Expansive and Contractive Forces* (Cambridge: Cambridge University Press).

Harré, R., and Madden, E. H. 1975: *Causal Powers* (Oxford: Blackwell).

Locke, J. 1961: *An Essay Concerning Human Understanding*, ed. J. Yolton (London: Dent; 1st pub. 1706), book II, ch. 21.

Reid, T. 1970: *An Inquiry into the Human Mind*, ed. T. Duggan (Chicago: University of Chicago Press: 1st pub. 1813).

Newton, Sir Isaac 1687: *Philosophiae Naturalis Principia Mathematica* (London: Royal Society).

16

Einstein

CHRISTOPHER RAY

In 1920, the eminent British astronomer and scientist Sir Arthur Eddington proclaimed that "Albert Einstein has provoked a revolution of thought in physical science" (p. vii). The preceding 15 years had seen historic scientific advances in three fields: quantum theory, relativistic kinematics, and gravitation. The genius of Einstein (1879–1955) had been perhaps the most important element in the early development of these fields. The year 1905 is often said to be Einstein's "annus mirabilis" – a truly miraculous year in which he published some 25 scientific articles including not only the landmark paper introducing the special theory of relativity, but also some quite remarkable contributions to quantum theory (see, e.g., Lanczos 1974, preface). By 1918, Einstein had extended the ideas involved in relativity to the problem of gravitation, and had laid the foundations for some astonishing discoveries in cosmology, with recent research into black holes and the big bang owing much to his pioneering work on the large-scale structure of the universe. Einstein had a tremendous and lasting influence on the scientific community; but he also captured the public imagination – and the attention of many leading twentieth-century philosophers.

Three episodes in Einstein's scientific life have attracted particularly close and sustained philosophical scrutiny: the birth of the special theory of relativity, with its robust challenge to the Newtonian conceptions of space and time (see SPACE, TIME, AND RELATIVITY); the development of the general theory of relativity, during which Einstein declared at least some allegiance to the positivism of Ernst Mach; and, in the 1920s and 1930s, the spirited discussions with the Danish physicist Niels Bohr, in which Einstein raised far-reaching questions about the character and implications of quantum theory (see MACH, BOHR, and QUANTUM MECHANICS).

The special theory of relativity

The essential ideas of the special theory of relativity are presented in Einstein's "On the electrodynamics of moving bodies," written in 1905. Although the special theory is invariably linked with the concept of space-time, this idea was not well formed until 1908, when Hermann Minkowski extended Einstein's basic ideas to produce an elegant four-dimensional union of space and time. Minkowski's geometrical characterization of special relativity is now the standard formulation in many textbooks. With special relativity, Einstein challenged the received "Newtonian" wisdom that we could rely on a background "absolute" frame of reference for space and for time. Within four-dimensional space-time, however, distant events could no longer be said to be "absolutely" simultaneous, since we are unable to transmit instantaneous signals from one location to another.

Einstein's 1905 paper begins by introducing two fundamental postulates: the so-called principle of relativity (the laws of physics should have the same form in all inertial frames) and the light postulate (the speed of light is independent of the motion of the emitting body). There is a long-standing debate about the status of these two postulates, with scientists and philosophers asking such questions as: Do the postulates by themselves determine the character of the special theory? Are they genuinely revolutionary, or can their "footprints" be found in pre-relativistic physics? Is the principle of relativity a theorem (rather than a foundational postulate) which may be deduced from the symmetrical character of special relativistic space-time? The emergence and rapid success of the special theory also provides an ideal case study for those who wish to understand how the scientific community "chooses" a new theory from competing possibilities. Mary Hesse (1974) argues that the choice is a rational one, and that Einstein's special theory was accepted initially because it had a far simpler form than its main rival, Lorentz's *ad hoc* modification of classical electrodynamics.

Although special relativity stresses the invariant nature of the speed of light, the theory does not explicitly rule out the idea of traveling faster than light. The structure of space-time in special relativity is such that any particle which travels faster than light might be able to travel backwards in time. However, such particles (usually called "tachyons") are unlikely to have any straightforward causal properties, so their existence would not lead to any of the paradoxes typically associated with time travel – for example, the possibility of traveling to the past to kill one's infant self.

The paradox which is most often associated with special relativity is the "clock paradox," or "twins paradox." Early commentators such as Henri Bergson were genuinely puzzled by Einstein's prediction that a moving clock would tick more slowly that a stationary clock; and that this would be true regardless of the nature of the clock – physical, chemical, or biological. In special relativity, the way that a clock behaves depends upon the space-time path along which it moves; and the path taken in turn depends upon the speed of the motion. The idea of the twins paradox derives from this fact. One twin is taken to stay at home "at rest," while the other undertakes a round-trip journey at a speed close to that of light. The twin who stays at home might age, say, ten years, but the other twin might age no more than a year or two. One of the consequences of the principle of relativity is that there is no way to identify whether or not an object is "really" moving unless it switches from one inertial frame of reference to another, and so experiences forces. Accordingly, for the greater part of the thought experiment, *each* twin might say that the other is the one in motion; and it might seem that we could describe the experiment in a neutral, symmetrical way: the twins move apart and then come together again. However, we immediately see that the symmetry is broken when we realize that only one twin will change frame of reference and experience "inertial" forces of deceleration and acceleration when the outwards motion stops and the twins begin to move towards each other. This has led some to suppose that the slowing of clocks is caused by the experience of forces. However, an extension of the twins thought experiment shows that special relativity predicts that moving clocks will run slow even when forces do not come into play (see Newton-Smith 1980).

Einstein's prediction about clocks is not a logical paradox, for no contradictions are involved. It may seem puzzling that the choice of route through space-time will determine the way a traveler ages; but force-free motion does seem to have real effects on objects. There is certainly no lack of empirical evidence to support the claim that motion with respect to a nonmaterial object – namely, space-time – has a material effect on the moving object. Only those who (positivistically) demand a material cause for the slowing down of clocks are likely to find the clock paradox genuinely paradoxical.

The general theory of relativity

The general theory of relativity is typically regarded as Einstein's masterpiece. In a TV interview on the BBC Horizon programme, John Archibald Wheeler, the renowned American physicist, once summed up Einstein's ideas on gravitation with remarkable elegance: "matter tells space how to curve; and space tells matter how to move." Einstein would have admired the economy with which Wheeler presents the general theory. For Einstein believed that simplicity is the key to the intelligibility of the physical world (see SIMPLICITY). He wrote in 1933: "Our experience up to date justifies us in feeling sure that in nature is actualized the ideal of mathematical simplicity" (p. 8). For many, Einstein's general theory encapsulates that ideal magnificently. The theory presents a view of gravitation in which the complexities of motion are deciphered by reference to local inertial frames. Even Newton had been suspicious of our tendency to rely on such global frames of reference as the distant "fixed stars." But he had replaced this material frame with another global frame, that of space itself (see NEWTON). Einstein's genius was revealed in his realization that local reference frames provide us with a new understanding of gravity and in his discovery that the fabric of space and time could have a dynamical role.

The general theory of relativity certainly owes much to the principles of equivalence and of general covariance. Einstein readily acknowledged the importance of these two principles in the development of the general theory. The first principle asks us to treat an accelerating object as physically "equivalent" to an object in a gravitational field. The second principle demands that the equations of relativity remain essentially the same when applied to any coordinate system. The status of these principles within relativity theory has been the subject of much philosophical attention (see Friedman 1983).

The general theory of relativity has frequently been regarded (erroneously) as the ultimate defeat for the "absolutist" belief that space exists as an independent, irreducible substantival entity – that is, for substantivalism. Einstein did hope, for some years, that the theory might incorporate at least the essentials of Ernst Mach's relationism – the idea of Mach's principle. In the years leading to his development of relativity, Einstein had been very much impressed by the work of Mach as well as by that of Henri Poincaré and David Hume (see Fösling 1999). Mach strongly believed that science should deal only in observable phenomena, and that our accounts of the physical world should be as economical as possible. He therefore saw no reason to embrace absolute space – a "substantival" entity which cannot be seen and which, in his opinion, was an unnecessary and extravagant element in our general account of motion and of gravitational and inertial forces. Einstein first referred to "Mach's Principle" explicitly in a paper of 1918, saying that he chose the name "because the principle implies a generalization of Mach's requirement according to which inertia should be reduced to the interaction of bodies" (p. 241n). In a paper written in 1917, Einstein found a static, spatially closed solution to his field equations – the "Einstein Universe," which he believed, for a short time, might be thoroughly Machian. But work by the Dutch physicist de Sitter demonstrated that Einstein's ideas were not fully consistent with Mach's principle.

Despite such setbacks, the philosopher Hans Reichenbach argued that the general theory was essentially Machian in character. Reichenbach's arguments in the influential *The Philosophy of Space and Time* (1957) convinced many members of the philosophical community for many years; but they failed to prevent mathematicians and physicists from developing "absolutist" models of general relativity in which space itself played an essential and irreducible role. Since the early 1970s, most philosophers have accepted that Einstein's theory seems to be an absolutist theory which does not explain dynamics in material terms alone (see Earman 1989). However, some philosophers argue that absolutism may involve an undesirable commitment to indeterminism (see ibid. and Ray 1991). Others claim that the models used to promote the absolutist case typically make unrealistic assumptions about the character of the actual universe (see Ray 1987). Hence, Einstein's dream of a thoroughly Machian theory is not quite dead.

Another of Einstein's ideas – the cosmological constant – also refuses to lie down and die. He introduced this constant into the field equations in 1917 in order to prevent the inwards collapse of a finite number of stars. The constant provided a long-range repulsive force which kept the stars apart. This allowed Einstein to promote the idea of a static universe used in the Einstein universe. However, he was forced to reconsider his views following Edwin Hubble's discovery in 1929 that the universe seemed to be far from static. Einstein recommended the removal of the cosmological constant in 1931, and later confessed that the constant was probably his greatest mistake. For the *ad hoc* addition of the constant to the field equations may have prevented him from predicting the possibility of an expanding universe. Ironically, modern cosmologists such as Alan Guth and Stephen Hawking see the constant as an essential element in theories of the big bang and the very early universe, acting as the powerhouse in an inflationary expansion (see Ray 1991).

Quantum mechanics and realism

The special and general theories of relativity are typically regarded as classical "deterministic" theories; and Einstein was very much attracted to physical determinism. However, quantum theory not only provided a tremendous challenge to Einstein's scientific ingenuity, it also tested his philosophical beliefs to their limits. He argued consistently against the claim, promoted by quantum theorists, that the fundamental laws of nature might be statistical. He explained his position in a letter to Max Born, written in 1926: "Quantum mechanics is certainly imposing. But an inner voice tells me it is not the real thing. The theory says a lot, but does not really bring us any closer to the secret of the 'old one'. I, at any rate, am convinced that *He* is not playing dice" (Born 1971, p. 90; emphasis original). Such views set Einstein on an ultimately fruitless search for new laws which might provide a basis for a unified physical theory – a theory of "everything" – consistent with his determinist prejudices.

What disturbed Einstein most about quantum theory was the fact that the theory, even in principle, did not seem to capture each and every aspect of physical reality. Einstein was profoundly skeptical about the "Copenhagen Interpretation" of quantum mechanics, as expressed in Niels Bohr's lecture to the International Physical Congress at Como in 1927. Einstein's response to Bohr later that year, during a series of meetings at the Solvay Institute in Brussels, was to set the scene for a debate which continued for over 20 years. The key disagreements in their debate were published in the *Physical Review* in 1935, with Bohr replying to a paper jointly written by Einstein, Podolsky, and Rosen. Einstein and his coauthors claimed that the description of the physical world embodied in quantum theory is incomplete – the basis of the celebrated EPR argument: there are elements of physical reality which do not have any counterparts in the theory – and, most critically, we are severely restricted in what we can say about the physical status of an undisturbed system. Bohr replied that the act of measurement, according to quantum mechanics, actually creates a physical state and does not reveal some preexisting reality (see MEASUREMENT). Their different points of view were taken up, clarified, and expanded in further papers, disputed at conferences, and, most notably, reexamined in Bohr's critical appraisal of Einstein's ideas in *Albert Einstein: Philosopher-Scientist* (Schilpp 1949). Elaborations of the two basic positions taken by Einstein and Bohr have formed the core of much further work in the philosophy of quantum mechanics, especially the problem of Bell's theorem (see QUANTUM MECHANICS, Healey 1989; Sklar 1992).

Einstein's philosophical perspective on quantum mechanics has often been described as essentially realist, contrasting sharply with what seemed to be his youthful enthusiasm for Machian positivism. Gerald Holton (1973) presents Einstein's philosophical development as a pilgrimage from a rather naive empiricism to a more sophisticated metaphysical realism. Arthur Fine claims that Einstein never abandoned his empiricist leanings. He argues that "Einstein's expressions of realism are presented in terms of motivations for the pursuit of science," rather than in terms of a firm set of beliefs about the nature of "reality" (Fine 1986, p. 7 and ch. 6). Fine sees Einstein as a motivational realist, concerned with the adequacy of empirical beliefs, driven by an almost religious desire to understand nature, but not especially preoccupied with scientific truth as such. It is easy to find support both for Holton and for Fine in Einstein's writings. Just as Einstein has provided entertainment for historians of science with some intriguing but contradictory statements about his knowledge of the work of other scientists, so too, his lack of philosophical consistency may aggravate philosophers of science for years to come.

References and further reading

Works by Einstein

1918: Principles of General Relativity. *Annalen der Physik, Leipzig,* 55, 214–45.
1933: *On the Method of Theoretical Physics* (Oxford: Oxford University Press); repr. in *Philosophy of Science,* 1, 162–85.
1945: *The Meaning of Relativity,* 2nd edn (Princeton: Princeton University Press).

1954: *Ideas and Opinions* (New York: Crown Publishing).

and Infeld, L. 1938: *The Evolution of Physics* (New York: Simon and Schuster).

and Lorentz, H. A., Weyl, H., and Minkowski, H. 1952: *The Principle of Relativity* (New York: Dover).

and Podolsky, B., and Rosen, N. 1935: Can quantum-mechanical description of physical reality be considered complete? *Physical Review*, 47, 777–80; with the reply by Bohr, *Physical Review*, 48 (1936), 696–702.

Works by other authors

Bernstein, J. 1973: *Einstein* (London: Fontana Modern Masters).

Bernstein, J., and Feinberg, G. (eds) 1986: *Cosmological Constants: Papers in Modern Cosmology* (New York: Columbia University Press) (Includes papers by Einstein, de Sitter, Hubble, and Guth.)

Born, M. (ed.) 1971: *The Born–Einstein Letters* (New York: Walker).

Earman, J. S. 1989: *World Enough and Space-Time* (Cambridge, MA: MIT Press).

Eddington, A. S. 1920: *Space, Time and Gravitation* (Cambridge: Cambridge University Press).

Fine, A. 1986: *The Shaky Game: Einstein, Realism and the Quantum Theory* (Chicago: University of Chicago Press).

Fösling, A. 1999: *Albert Einstein* (Harmondsworth: Penguin).

French, A. P. (ed.) 1979: *Einstein: A Centenary Volume* (London: Heinemann).

Friedman, M. 1983: *The Foundations of Spacetime Theories* (Princeton: Princeton University Press).

Healey, R. 1989: *The Philosophy of Quantum Mechanics* (Cambridge: Cambridge University Press).

Hesse, M. B. 1974: *The Structure of Scientific Inference* (London: Macmillan).

Holton, G. 1973: *Thematic Origins of Scientific Thought* (Cambridge, MA: Harvard University Press).

Hubble, E. P. 1936a: *The Realm of the Nebulae*.

—— 1936b: The luminosity function of nebulae. *Astrophysics Journal*, 84, 270–95.

Lanczos, C. 1974: *The Einstein Decade* (London: Elek Science).

Newton-Smith, W. H. 1980: *The Structure of Time* (London: Routledge).

Pais, A. 1982: *Subtle is the Lord* (Oxford: Oxford University Press).

Ray, C. 1987: *The Evolution of Relativity* (Bristol: Institute of Physics Press, Adam Hilger).

—— 1991: *Time, Space, and Philosophy* (London: Routledge).

Reichenbach, H. 1957: *The Philosophy of Space and Time* (New York: Dover).

Salmon, W. C. 1980: *Space, Time, and Motion*, 2nd edn (Minneapolis: University of Minnesota Press).

Schilpp, P. (ed.) 1949: *Albert Einstein: Philosopher-Scientist* (La Salle, IL: Open Court). (With autobiographical notes by Einstein, and critical discussions on Einstein by many philosophers and scientists, including Born, Bohr, Reichenbach, and Gödel.)

Sklar, L. 1974: *Space, Time, and Spacetime* (Berkeley: University of California Press).

—— 1992: *The Philosophy of Physics* (Oxford: Oxford University Press).

Stachel, J. et al. (eds) 1987– : *The Collected Papers of Einstein*, vols 1– (Princeton: Princeton University Press).

17

Evidence and Confirmation

COLIN HOWSON

Introduction

To say that a body of information is evidence in favor of a hypothesis is to say that the hypothesis receives some degree of *support* or *confirmation* from that information. What sorts of information confirm what hypotheses is a question which has long been controversial; it was discussed as avidly three centuries ago as it is today, when, under the heading of "confirmation theory," it is one of the central topics in contemporary philosophy of science. Its profound interest to philosophers is due to its intimate connection with the *philosophical problem of induction*, concerning what grounds, if any, observational data can give us for accepting as a basis for action and belief hypotheses whose content logically transcends the observational data. Presumably, if it could be shown that any such hypothesis is sufficiently well confirmed by the evidence, then that would be grounds for accepting it. If, then, it could be shown that observational evidence could confirm such a transcendent hypothesis at all, then that would go some way to solving the problem of induction.

That the problem of finding sound criteria for the confirmation of hypotheses which transcend the evidence is nontrivial is brought out neatly in the following simple example, due to Nelson Goodman (1954), although essentially the same point had been made earlier by Poincaré. Suppose h is the hypothesis that all emeralds are green. We might suppose that with enough evidence recording that all emeralds ever observed have been green, h would eventually be confirmed to the point where it would be regarded as more likely to be true than not. Let e be the statement that such very extensive evidence has been obtained. However, now consider the hypothesis h_0: "All emeralds are grue," where something is defined to be grue if it has been observed and found to be green, or it has not and is blue. Then h_0 stands in exactly the same logical relation to e as does h, for, according to e, all emeralds so far observed are also grue. So it seems that the evidence, however extensive, cannot discriminate between the two hypotheses. Yet it certainly cannot ever be the case that *both* h and h_0 are more likely to be true than not, since, given the existence of unobserved emeralds, they are mutually inconsistent. Indeed, the evidence seems incapable of discriminating between an *infinity* of mutually inconsistent hypotheses h_m, where $m = 0, 1, 2, \ldots$, of the form "All emeralds observed up to now + m minutes are green, and all others blue."

An easy answer to Goodman's riddle is that "grue" is a highly artificial predicate, presupposing the "natural" predicates "green" and "blue" for its definition, and that no evidence can support hypotheses incorporating such "cooked-up" predicates. But the answer is too easy. First, as Goodman pointed out, if we define "bleen" to mean "observed and

108

found to be blue or not yet observed and green," then the pairs {grue, bleen} and {green, blue} are interdefinable; and second, it remains to be shown why attributions of such allegedly artificial predicates are in principle empirically unconfirmable.

Goodman himself attempted to answer the second question with his theory of *projectibility*. He called a predicate "projectible" if the fact that it has so far been observed to characterize the members of some class of individuals inspires confidence that it will continue to do so. Projectible predicates, according to Goodman, are therefore likely to be just those already "entrenched" in our language: they are entrenched precisely because they have been found to be projectible. "Grue," unlike "green," is not entrenched, and is therefore not projectible.

There are two objections to Goodman's account. First, it fails to say why entrenched predicates should be thought to be likely to *continue* to project. Second, "scandium," "forward light cone," "momentum operator," "spin up," and "rest mass," for example, were entirely new, and hence highly nonentrenched, when the hypotheses incorporating them were strongly confirmed. So entrenchment is not necessary for projectibility, and "grue" is not, therefore, impugned by failing that criterion. Whatever the truth of the matter, entrenchment seems not to help us chart a course towards it.

Another puzzling case was discovered by Hempel (1945). Hempel considered the apparently plausible theory that positive instances of a general hypothesis "All A's are B's" support it (a positive instance is an individual which is both an A and a B). However, consider the following reasoning. If the observation of a green emerald supports the hypothesis "All emeralds are green," then, by symmetry, the observation of a non-green non-emerald supports "All non-green things are non-emeralds." But "All emeralds are green" is logically equivalent to "All non-green things are non-emeralds," and if evidence supports a hypothesis h, then it should presumably support anything equivalent to h (this is the so-called *equivalence condition*). So we seem impelled to the strongly counterintuitive conclusion that the observation of white handkerchiefs supports "All emeralds are green." Hempel's own example was the inessentially different hypothesis "All ravens are black," and for this reason the problem has become known as the "raven paradox."

The "grue" problem and the raven paradox are collectively known as the *paradoxes of confirmation* (see CONFIRMATION, PARADOXES OF). The hypotheses involved in them, "All emeralds are green" and "All ravens are black," are said to be *deterministic*, as opposed to *statistical*, hypotheses. The latter ascribe a statistical probability distribution to the outcomes randomly generated by some data source. Since such a hypothesis asserts not that any particular outcome will definitely occur in appropriate circumstances, but only that there is a particular statistical probability of its occurrence, then it is not obvious how the occurrence or nonoccurrence of any particular outcome bears upon the hypothesis. In what follows we shall use the problem of empirically evaluating statistical hypotheses, together with the paradoxes of confirmation, as a touchstone to illustrate the strengths and the weaknesses of each of the principal theories of confirmation.

Probabilistic theories of confirmation

A theory which commanded virtually universal assent in the eighteenth and nineteenth centuries is that confirming evidence for a hypothesis is evidence which increases its probability. What raised this intuitively compelling idea to the level of a nontrivial theory was that the probability in question was understood in the sense of the mathematical theory of probability, otherwise known as the *probability calculus*. Some basic principles of this calculus are the following; a and b are to be regarded as arbitrary sentences, which will function in appropriate contexts as hypotheses and evidence statements:

1   $0 \leq \text{prob(a)} \leq 1$.
2   prob(a or b) = prob(a) + prob(b), where a and b are mutually exclusive.
3   prob(a) = 1 if a is a necessary truth.
4   prob(a) = 0 if a is a necessary falsehood.
5   prob(a) = 1 − prob(not a).
6   if a entails the truth of b, then prob(a) ≤ prob(b).
7   prob(a) = prob(b) if a is logically equivalent to b.
8   prob(a | b) = prob(a and b) ÷ prob(b), where prob(b) > 0.
9   prob(a | b) = 1 if b entails the truth of a.
10 prob(a | b) = 0 if b entails the falsity of a.

These principles are not all independent; in fact, (1), (2), (3), and (8) generate the remainder (Howson and Urbach 1993, ch. 2). The function prob(a | b) is called *the conditional probability of a, given b*. It assumes central importance in a probabilistic theory of confirmation, because in such a theory evidence e is systematically related to a hypothesis h in terms of properties of the conditional probability, prob(h | e). If this is construed as the probability of h in the light of e, then the goal is to show that prob(h | e) > prob(h) for appropriate h and e.

Probabilities based on a "logical" metric

The probabilities involved in probabilistic theories of confirmation are *epistemic*, rather than statistical. It was at first hoped that these probabilities could be defined in some suitably objective way, making no appeal to assumptions that could not be justified on a priori grounds alone. This seemed to point to probabilities determined by some purely *logical* metric. The only seriously considered candidate, first identified by James Bernoulli in his posthumously published *Ars conjectandi* (1713), was based on the number k of those possibilities, in some appropriate partition Π of the logical space in which a sentence a is embedded, which make a true. If prob(a) is made proportional to k, it follows from the probability calculus that it is equal to k/n, where n is the number of elements in the partition. This is just the well-known ratio of favorable to possible cases which characterizes the so-called *classical definition* of probability.

The difficulties encountered within this theory are well documented. There are technical problems if Π is infinite, for example. But the fundamental objection is that Π can be chosen in various ways, and the probability is usually sensitive to that choice. For example, suppose Π is {red, not red}. The related probability that an arbitrary individual is red is $^1/_2$. Relative to $\Pi' = $ {red, green, neither red nor green}, however, it is $^1/_3$. But to say that one partition is a priori more "correct" than another introduces an arbitrariness quite at odds with the requirement of objectivity.

Subjective probabilities

Although Carnap attempted to salvage the idea of a theory of confirmation based on an allegedly logical probability (which in his 1950 book he called "probability"$_1$) in the middle years of this century, by the time of his death in 1970, it was generally agreed that an epistemically neutral logical metric is a chimera. However, in the twenties and thirties of this century the radical suggestion was made that everything of value in the probabilist program could be retained within a *subjective* probability framework. Laying the foundation for later work in decision theory, Frank Ramsey proved that, as long as they satisfy certain consistency conditions, an individual's preferences between gambles determine a utility function, determined up to the choice of zero and scaling constant, over the prizes and a unique probability distribution over the outcomes (Ramsey 1926). He showed that infringement of the conditions would involve the individual in accepting gambles that he or she was certain to lose. De Finetti (1937) gave a simple and elegant proof of this corollary, which has now become classic. Because a Dutch Book is a system of odds on which the bookmaker must lose overall, the corollary has come to be called the *Dutch Book theorem*.

Granted – which not everyone does – that one may interpret their mathematical results in the way in which de Finetii and Ramsey suggest, the criterion of confirmation in terms of increase of probability now becomes relativized to a particular individual's probability function. Indeed, the theory is usually called the *subjective Bayesian* theory of confirmation – "Bayesian" because of the central importance in it of a consequence of the probability calculus known as Bayes's theorem, after the eighteenth-century English clergyman and mathematician Thomas Bayes, who is (incorrectly) credited with first proving it. This theorem is the following identity:

$$\text{prob}(h\,|\,e) = [\text{prob}(e\,|\,h)\text{prob}(h)] \div \text{prob}(e)$$

and its importance is that it expresses the so-called *posterior probability* of h on evidence e (i.e., prob(h|e)) in terms of the product of its *likelihood* (prob(e|h)) and *prior probability* (prob(h)). Prob(e) can itself be expanded as a sum of similar products prob(e|h_i)prob(h_i), where the h_i are alternative explanations of e. The likelihood can usually be computed by appeal to deductive-logical considerations, if h is deterministic (cf. principles (9) and (10) above), or, if it is statistical, to the model it postulates.

This use of Bayes's theorem means that the subjectivity in the posterior probability is largely located in the priors, which function as undetermined parameters constrained only by the rules of the probability calculus. Some subjective Bayesians attempt to mitigate the subjectivism by pointing to certain consequences of Bayes's theorem that allegedly show that general criteria of confirmation follow from consideration of consistency alone. For example, if prob(h) > 0 and 0 < prob(e) < 1, and h predicts (entails) e, given suitable background information about initial conditions, etc. being satisfied, then prob(h|e) > prob(h). Also, with some manipulation, Bayes's theorem shows that prob(h|e) depends only on prob(h) and the *likelihood ratio* prob(e|h):prob(e|not h), and is an increasing function of both, tending to 1 as prob(e|not h) tends to 0. It follows that if e is regarded as very likely if h is true, and very unlikely if h is false, then prob(h|e) is close to the value of certainty.

111

These sorts of discriminations seem to endorse informal methodological principles. Some Bayesians also attempt to play down the role of the priors by pointing to the existence of mathematical theorems showing that asymptotically, and except on a set of evidence sequences of prior probability zero, an individual's posterior probability distribution over a class of mutually exclusive hypotheses is independent of the prior distribution. It has even been argued that the formally unconstrained priors are a strength, rather than a weakness, of the subjective theory, since a marked feature of the history of science is the frequent lack of agreement, even among experts with access to the same information, about the relative merits of rival hypotheses. Einstein, for example, always had severe doubts about quantum mechanics, while in our own time we see strong divergence over the status of string theory. In other words, the facts could be taken to suggest that there is a genuine indeterminacy of rationality at such points, which the subjective Bayesian theory makes due allowance for (Howson and Urbach (1993) contains a detailed discussion of the principal pros and cons of the subjective Bayesian theory).

Objective Bayesianism

The fact remains that many who are attracted to probabilistic accounts of confirmation find the absence of constraint in the choice of priors in the subjective approach unacceptable. So-called *objective Bayesians* look for further rules which, they hope, will determine the priors uniquely in appropriate cases. One idea, due originally to Jeffreys (1961), and taken up by E. T. Jaynes (1973), is that the context of a problem determines a group of transformations of the random variable x under which the prior probability density prob(x) should be invariant. In some, but by no means all, cases, the group allegedly so determined fixes prob(x) uniquely. Another proposal by Jaynes (1978) is to adopt that prior distribution prob(x), where there is one, which maximizes the *entropy* of the distribution (i.e., the expected value of −log prob(x), for an arbitrarily selected base), subject to the constraints imposed by the background information available. Finally, many Bayesians believe that *simplicity* affords an objective criterion for ordering the prior probabilities of hypotheses.

Each of these proposals faces problems. The first fails to give unique, or sometimes any, solutions in many cases, without a good deal of interpretation as regards the group of transformations "determined" by the problem. Maximum entropy distributions do not always exist either; while among those that do, there are some, called *improper priors*, which infringe the rule that probabilities cannot exceed unity. Simplicity turns out to be far from the univocal concept it appears at first sight; nor is there any convincing justification for identifying simpler hypotheses – in any sense – with more probable ones.

Grue, ravens, etc.

The purely formal characteristics of probabilistic theories of confirmation give them a promising line of attack on Goodman's "grue" problem. It follows from Bayes's theorem that the ratio prob(h|e):prob(h′|e) of the posterior probabilities of any two hypotheses h and h′ is equal to the ratio prob(h):prob(h′), when both h and h′ entail the evidence e. Hence each of the infinity of "grue" variants to the hypothesis h under test which we noted earlier has only to be assigned a smaller prior probability than h for it to have a smaller posterior probability. In particular, "All emeralds are green" can be given a posterior probability arbitrarily close to 1 with enough data of emeralds observed to be green. And here the subjective Bayesian theory enjoys an advantage over other probabilistic theories, identifying probabilities as it does with beliefs based on the sorts of considerations people actually regard as relevant. And people in fact rank the prior probability of "grue" hypotheses as very small (which is why we find the suggestion that they are as well supported as "All emeralds are green" counterintuitive), because they believe that there is no independent reason at all to think those hypotheses likely to be true: they are simply concocted to explain the data.

Probabilistic theories also provide an attractive diagnosis of the raven paradox. First, we should note that within these theories positive instances do *not* automatically confirm their respective hypotheses: the probability of the latter may increase, remain unaltered, or even decrease (it would only automatically increase if the evidence were entailed by the hypothesis, but "x is an A and a B" is not entailed by "All A's are B's"). And this is surely right. Rosenkrantz (1981) gives the following nice example of positive instances which decrease the probability (to 0). Suppose h is the hypothesis "Everybody in the room leaves with someone else's hat." Let e be evidence recording two positive instances of h, that a and b both leave with someone else's hat. But suppose we know that there are only three people in the room, a, b, and c, and that a has left with b's hat and b with a's hat. Clearly, given this information, e *refutes* h.

Second, probabilistic theories give a plausible explanation of why we regard randomly observed non-black non-ravens as irrelevant to the hypothesis h that "all ravens are black." Suppose we sample randomly from the set N of non-black things, and record whether what we find is a raven or not. Relative to the information that y is in N, h entails the prediction e': y is a non-raven (note that the equivalence condition follows from principle (7) above). However, the increase in h's probability is negligible, for by Bayes's theorem prob(h | e') = prob(h):prob(e'), and given that ravens are extremely sparse in the set of non-black things, prob(e') must be reckoned very close to 1, leaving prob(h | e') very close to prob(h).

Finally, Bayesian theory (subjective or objective) has no difficulty in principle in dealing with statistical hypotheses. A classic problem of statistical inference is to determine whether the value of a statistical parameter generates a sufficiently good fit between the sample data and the statistical model assumed for the data source. Bayesian theories tackle the problem by using a continuous form of Bayes's theorem to provide a posterior distribution over the infinite set of hypotheses specifying all possible parameter values.

Nonprobabilistic theories of confirmation

The Fisher–Popper theory

In the 1920s and 1930s R. A. Fisher and K. R. Popper proposed theories of confirmation which agree in claiming that (i) Bayesian probability distributions over hypotheses cannot be rendered both objective and nonarbitrary; (ii) the principal function of experimentally testing a hypothesis is to attempt to refute it; and (iii) the way in which evidence e discriminates between rival hypotheses h_i is determined by the likelihood function $\text{prob}(e \mid h_i)$. Further to (iii), it was Fisher who was responsible for calling $\text{prob}(e \mid h)$ the "likelihood" of h given e, and he explicitly regarded it as indicating the degree of rational belief in h merited by e (Fisher 1930). Popper's measure of what he called "the degree of corroboration" of h by e is proportional to $\text{prob}(e \mid h) - \text{prob}(e)$ (Popper 1959; in Popper's theory $\text{prob}(e)$ strictly cannot be calculated, and he seems to regard it as an informal qualitative estimate of the improbability of e relative to background information). Both Fisher and Popper emphasized that their measures of evidential support do not obey the laws of probability. Fisher, it is true, also developed a parallel theory of *fiducial probability*, which generates what look like posterior probability distributions in some cases; however, its ability also to generate inconsistencies, combined with an obscure theoretical foundation, makes it probably best regarded as a deviant part of the Fisher *oeuvre*.

Fisher proposed his theory in the context of statistical, and Popper in that of deterministic, hypotheses, but the structure of their accounts is otherwise so similar – excepting fiducial probability – that they can be regarded as different specializations of the same theory; Popper's own account of tests of statistical hypotheses is virtually identical to Fisher's celebrated theory of *significance tests*.

The raven paradox poses no threat in principle to this Fisher–Popper theory, since the mere observation of a positive instance is not regarded as sufficient to confirm the hypothesis concerned: confirmation comes only from experiments subjecting it to a stringent test. The "grue" problem is a very different matter, however, since for any hypothesis h, and test evidence e predicted by h, the infinite class of "grue" variants to h mentioned earlier all have the same likelihood, namely 1, and hence the same *maximum* corroboration, as h itself. Even if the test is a *crucial experiment*, deciding between the predictions of a rival h' and h, then it clearly also decides between h' and each of the grue alternatives h_0, h_1, h_2, etc.; if h is maximally corroborated, then, presumably, so should h_0, h_1, h_2, etc. The theory simply has no machinery for making extra-experimental discriminations (like prior probability distributions) between hypotheses which equally explain the evidence.

Ironically, since Fisher was a great statistician, and his account still forms a large component of the dominant statistical methodology, it nevertheless faces very grave difficulties with statistical hypotheses. Consider n repeated tosses of a coin. Let h be the hypothesis that the coin has a statistical probability of $\frac{1}{2}$ of landing heads at any toss, and that the tosses are independent. If e records the outcomes, heads or tails, or each toss, then $\text{prob}(e \mid h) = 2^{-n}$, which does not depend on the number of heads observed. It follows that the confirmation of h by e is the same for all such data sequences: none confirms, or disconfirms, the hypothesis more than any other. Yet, intuitively, those sequences which record a relative frequency of heads suitably near $\frac{1}{2}$ confirm the hypothesis, while those with a relative frequency suitably far from $\frac{1}{2}$ strongly disconfirm it.

Fisher appears to evade this sort of objection by allowing only certain functions of the "raw data" to represent evidence. However, not only do the criteria for admissible test statistics, formulated by him in a series of classic papers in the 1920s and 1930s, lack any convincing theoretical justification, but, as Neyman showed, they can be made to generate contradictory recommendations (Neyman 1952, pp. 43–54).

114

The Neyman–Pearson theory

Current statistical orthodoxy is an amalgam of Fisher's theory and one developed by Neyman, in collaboration with E. S. Pearson, which incorporates some of Fisher's ideas while attempting to avoid the counterexamples. The Neyman–Pearson theory is, unlike the others considered here, explicitly decision-theoretic, and is cast in the form of criteria for minimizing the chances of two types of error which may occur in a decision to accept or reject a statistical hypothesis after test. If h is the hypothesis, a *type 1 error* consists in rejecting h when it is true, and a *type 2 error* consists in accepting it when it is false. Neyman and Pearson identify the chance of making a type 1 error with the probability, according to h, with which the experimental outcome will lie in the rejection region, call it R, of a test statistic T (a statistic is a mathematical representation, or function, of the sample data). Suppose that some other hypothesis h' is actually true. Then if the complement R^c of R is taken to be the acceptance region for h, then the chance of making a type 2 error is identified with the probability which h' assigns to T's being in R^c. The *power* of the test of h against h' is defined to be 1 minus this probability, and Neyman and Pearson were able to show that for many of the standard distributions there exists a region R which, for any magnitude of type 1 error, maximizes power (for an extended discussion see Howson and Urbach 1993, ch. 13).

Despite its status, the Neyman–Pearson theory is open to powerful objections. First, there is no region in the range of a statistic which is uniformly most powerful against *all* the alternatives to h, even when these are restricted to hypotheses proposing different probability distributions over the values of T. (There is, of course, an infinity of distinct deterministic hypotheses which will explain any set of observations.) In fact, a uniformly most powerful test exists only when the class of alternatives to h is very severely restricted.

A further objection is that most empirical evaluations of scientific hypotheses, even in those exceptional cases, are not conducted in terms of a crude accept/reject decision. Nor do Neyman and Pearson provide a convincing explanation for why the magnitude of a statistical probability is relevant to a decision based on a single sample. They claim that in a sufficiently long run of such decisions the proportion of errors of either type will be equal to their probabilities; if the latter are minimized, so will the loss incurred. But the long run guarantees no such equality (the weak law of large numbers asserts merely that the probability of equality tends to 1); even if it did, nothing would follow about the rightness of the decision in any particular case, which, as Fisher, a strong opponent of the Neyman–Pearson theory, emphasized, should be the concern of any theory of inductive inference.

Conclusion

While the theories of confirmation discussed above dominate the field, they are far from being the only ones. Two of the more prominent alternative accounts developed in the last 20 years are Glymour's "bootstrap" theory (Glymour 1980), and the so-called Dempster–Shafer theory, based on Shafer's theory of belief functions and Dempster's rule for combining independent bodies of evidence (Shafer 1976). Glymour's theory investigated a feature of inductive reasoning which he believed to be not well explained by existing confirmation theories. This is the tendency of successful predictions to confirm, or disconfirm, just the parts of a general theory on which those predictions depend. Shafer's theory of nonadditive belief functions was constructed, among other reasons, to avoid the feature which additive measures like probability functions possess, of making the degree of belief in a hypothesis determine, and be determined by, the degree of belief in its negation. Both these theories initially commanded keen interest, which has since waned considerably under the impact of destructive criticism.

The development of artificial intelligence has stimulated practical interest in confirmation theory, and new theories, like the MYCIN theory of confidence factors, and others based on fuzzy logic and probability, have been developed by "knowledge engineers" themselves. Whether any of them will ever achieve the same degree of acceptance as the theories discussed above remains to be seen.

References

Bernoulli, J. 1713: *Ars conjectandi* (Basel).

Carnap, R. 1950: *Logical Foundations of Probability* (Chicago: University of Chicago Press).

de Finetti, B. 1937: Foresight: its logical laws, its subjective sources. In *Studies in Subjective Probability*, ed. H. Kyburg and H. Smokler (New York: Wiley, 1964, 93–159. Translation of "La Prévision: ses lois logiques, ses sources subjectives," *Annales de l'Institut Henri Poincaré*, 7, 1–68.

Fisher, R. A. 1930: Inverse probability. *Proceedings of the Cambridge Philosophical Society*, 26(4), 528–35.

Glymour, C. 1980: *Theory and Evidence* (Princeton: Princeton University Press).

Goodman, N. 1954: *Fact, Fiction and Forecast* (London: Athlone Press).

Hempel, C. G. 1945: Studies in the logic of confirmation. *Mind*, 54, 1–26, 97–121.

Howson, C., and Urbach, P. M. 1993: *Scientific Reasoning: The Bayesian Approach*, 2nd edn (Chicago: Open Court).

Jaynes, E. T. 1978: Where do we stand on maximum entropy? Repr. in Jaynes, Papers on *Probability, Statistics and Statistical Physics*, ed. R. D. Rosenkrantz (Dordrecht: Reidel, 1983), 210–314.

—— 1973: The well-posed problem. *Foundations of Physics*, 4, 477–93.

Jeffreys, H. 1961: *Theory of Probability*, 3rd edn (Oxford: Clarendon Press).

Neyman, J. 1952: *Lectures and Conferences on Mathematical Statistics and Probability*, 2nd edn (Washington: US Department of Agriculture).

Popper, K. R. 1959: *The Logic of Scientific Discovery* (New York: Basic Books).

Ramsey, F. P. 1926: Truth and probability. Repr. in *The Foundations of Mathematics and Other Essays* (London: Routledge & Kegan Paul, 1931), 156–99.

Rosenkrantz, R. 1981: *Foundations and Applications of Inductive Probability* (New York: Ridgeview Publishing Company).

Shafer, G. 1976: *A Mathematical Theory of Evidence* (Princeton: Princeton University Press).

18

Experiment

DAVID C. GOODING

Introduction

There have been many images of experiment. The contemplative narratives of Aristotle scrvcd to illustrate hypotheses and arguments. There was no expectation that they be performed. Even in Galileo's dialogues, the distinction between real experiments and imaginary ones is not sharp (see GALILEO). During the seventeenth century, performance and public description became essential to the probative power of experiment. These made its methods and procedures transparent, allowing any reader of the narrative to be a virtual witness of an active demonstration (Shapin and Schaffer 1985, pp. 22–79). Thus, by the end of the scientific revolution, illustrative narratives were distinguished from public accounts of experiment as the disciplined, systematic study of phenomena. Eventually this made the efficacy of thought experiments problematic (see Kuhn 1977, pp. 241–2, and THOUGHT EXPERIMENTS). As a source of experience of realms previously beyond the reach of the senses, real-world experiment contributed to the rise of objective scientific knowledge. Harvey, Galileo, Hooks, Boyle, Newton, and other proponents of experimental natural philosophy established an important new mode of argumentation.

As experimental argument gained authority, it displaced other modes of knowledge based on superstition, the authority of tradition, of ancient texts, or of religious dogma. When the point at issue involved reference to some feature of the world, arguments invoking experiments that could demonstrate mastery of the empirical domain were more convincing than arguments that could not. Experiment became the cutting edge of objectivity which enabled true ideas about nature to be distinguished from false ones. This heroic, idealized image of experiment remained intact for nearly three centuries, until post-empiricist critiques of the status of observation sentences undermined the independent status of experimental evidence, making experiment a mere handmaiden of theory.

Experiment as the handmaiden of theory

By the end of the seventeenth century, it was no longer sufficient to follow a process of reasoning about a possible world: for a fact to be established, actual performance must be seen to be done, and readers should act as witnesses. This was an important departure from the passive, Aristotelian notion of experiment prevalent until the scientific revolution (Tiles 1993). We think of publication as a means of disseminating results, but its primary role was to place results in the context of the methods that produced them. However, in the twentieth century philosophers separated the activity of experimentation from its results, reducing experiment to an invisible, unproblematic source of the observation statements that confirm or falsify theories. Thus, philosophers typically consider only observation, and they assume that observation sentences are known to be true or false. This assumption spares philosophers the difficult task of showing how scientists' experiments show observation statements to be true or false. But it proves to have been a hostage to fortune, because it has been left to historians and sociologists to provide answers. Their answers show that observations simply aren't what philosophers routinely take them to be.

According to the traditional account, experimentation is a method of placing beliefs before the tribunal of experience, in a more disciplined manner than the illustrative methods of Aristotelian and medieval science. Experiment was soon identified with the testing of theories; however, the program of artful "vexation" of nature which Bacon called for continued, for example, in the exploratory style of investigation of chemical, electrical, and optical phenomena suggested by Newton's *Opticks* (see Kuhn 1977, pp. 31–65, and NEWTON). As I will show below, the exploratory, nondemonstrative use of experiment continues to be important in contemporary science. Yet philosophers identify experiments with observed results, and these with the testing of theory. They assume that observation provides an open window for the mind onto a world of natural facts and regularities, and that the main problem for the scientist is to establish the uniqueness or the independence of a theoretical interpretation. Experiments merely enable the production of (true) observation statements. Shared, replicable observations are the basis for scientific consensus about an objective reality.

The philosophical imperative for narrowing the empirical base to observation statements was Carnap's belief that science should be modeled on logic (see also LOGICAL POSITIVISM). The sort of logic underlying scientific method was, and remains, a fiercely debated matter (see SCIENTIFIC METHODOLOGY, also EVIDENCE AND CONFIRMATION and POPPER). Inductive methods of confirmation advocated by Carnap require the accumulation of observational evidence. Hypothetico-deductive methods involve deduction of hypotheses or predictions from theories, to be tested by experiment. Subjecting beliefs about nature to empirical tests that can be repeated anywhere distinguishes science from other activities.

Both approaches reduce science to a set of operations on propositions. The latter must conform to logical rules, and some of the propositions must contain terms referring to observable states of affairs. Observation statements thus support or contradict theories as structures of propositions. The meaning or content of experimental results are given by the theory that defines the terms in it or by the procedures that link so-called observational predicates ostensively with states of affairs in the world, such as pointer readings on instruments.

The presumption that propositional knowledge is the only kind that matters restricts philosophy of science to theory. Experiment is defined in terms of the needs of theory; its only role is to deliver propositions about the world, for only these could count as grounds for accepting or rejecting theories. From a logical point of view, fallibilist methodologies (such as falsificationism) and inductive methodologies look very different; nevertheless, they share the same assumption that theories cannot interact with observation, or with an inchoate world of experimental practice, but only with observation statements.

The narrow view of experiment as the handmaiden of theory is not wholly false. The caricature considers only observations in the form of statements, and dismisses the processes of eliciting and producing observable phenomena. But there is an element of truth in this caricature, and here the element of truth is that experiment is important. The importance of the processes of making things observable, countable, and measurable becomes evident as soon as we consider how scientists establish the truth of observation statements.

Though philosophers such as Putnam claim that theories have a pragmatic role as guides to experimental practice, theories are rarely specific enough about the minutiae of engaging the world. They cannot show, for example, how to get equipment to work; nor can they anticipate precisely how an experiment needs to be varied, as the following account of immunology research illustrates:

> The initial question – In Hodgkin's disease are lymphocytes trapped in the spleen? – might sound simple enough, but from then on we would be in a maze all the way. This was not just a simple matter of counting lymphocytes. The leitmotiv was complexity: complexity of painstaking experimental and surgical techniques; of repetitive tests, each of which would reveal one small facet of this living microcosm; of time-consuming procedures . . . of machines for counting, spinning, blending and seeing which must not fail; of reagents and substances that must be pure; of temperatures that must not change; of figures that must be accurately assessed; of points of a graph precisely plotted, and curves correctly outlined; of deductions that must be cautiously drawn . . . (Goodfield 1982, p. 47)

The result is a complex constellation of practices, developed and refined over time. Hacking (1992) describes this as a "self-vindicating" feature of the laboratory sciences, since practices – methods, procedures, materials, instruments – are likely to be altered by an unexpected or unbelievable result more than theoretical propositions. Anomalies can be explained away by theories and *ad hoc* hypotheses. However, to identify a result as an anomaly rather than an artifact shows that theoretical expectations have been outweighed by experimenters' confidence in the material technologies and instrumental practices they have built up. It is sometimes possible to identify the point at which experimental results, however unexpected or anomalous, cease to be dismissed as artifacts, as Gooding has shown (1990, pp. 195–201). Theories appear to be underdetermined by the little bit of knowledge that philosophers recognize as data; but when the larger body of knowledge actually used is taken into account, it turns out that theories are not underdetermined at all (see Pickering 1989; Hacking 1992, pp. 29–31).

Philosophies that restrict knowledge to what can be expressed in propositions cast science back into the passive, purely literary form that the practical experimental philosophy of Bacon, Galileo, Boyle, and Newton superseded. According to their critics, mainstream philosophies of science fail to grasp the nature of experiment's contribution to science. They have "mummified" science (Hacking 1983, p. 1), treating scientists as disembodied reasoners with 20–20 vision and a strong sense of logical propriety, but few other cognitive capacities (Gooding 1990, pp. 3–4, 203–9; Giere 1988, pp. 109–10; see also COGNITIVE APPROACHES TO SCIENCE).

Do experiments test theories?

Two sorts of criticism have been made of the view that experiment provides only observation statements that test theories: philosophical critiques and those arising from the observation that philosophical models of science bear little resemblance to the history and current practice of science. The most serious philosophical objection is that logical positivism and post-positivist philosophies of science have undermined experiment's traditional, privileged position as a source of empirical knowledge. Philosophers of science in this century, notably Duhen and Quine, argued that experiment cannot provide observational evidence for a theory independently of some assumptions of that theory. The problem arises from the fact that observations must be interpreted, and that interpretation always invokes theoretical assumptions about phenomena or data. Since these are usually part of the theory to be tested, such results cannot provide independent evidence for or against that theory. In Quine's terms, the beliefs of scientists face the tribunal of experience as a whole, and not as Popper and other positivist philosophers maintained, via hypotheses that implicate discrete units of theory. Theory-laden observations cannot decide the truth or falsity of theories (see QUINE and UNDERDETERMINATION OF THEORY BY DATA).

This consequence is incompatible with the findings of studies of scientific practice (see, e.g., Gooding et al. 1989 and Hacking 1992). In practice, it is never the case that theories interact with experiment through observation statements. Hacking (1983) criticized this "theory-dominated" view of experiment for ignoring the extent to which experimenters' strategies can vindicate observational judgments independently of particular theories, and for failing to notice that experiment has "a life of its own" – that is, experiments often stimulate inquiry independently of particular theories. Nor is experiment an unproblematic source of stable, transparent observations. As Hacking comments: "Noting and reporting readings of dials . . . is nothing. Another kind of observation is what counts: the uncanny ability to pick out what is odd, wrong, instructive or distorted in the antics of one's equipment" (1983, p. 230). I have identified some philosophical reasons why philosophers have neglected this aspect of science and will consider others below.

The neglect of experiment (Franklin 1986) involves the neglect of discovery, as well as of the processes of validating experimental results. The neglect of discovery means that an important fact about science is ignored: testing and matching of predicted and observed values comes towards the very end of the process of developing and defending empirical theories. In a seminal paper Kuhn argued that "large amounts of qualitative work have usually been prerequisite to fruitful quantification in the physical sciences" (1977, pp. 183–224, at pp. 180, 213) (see MEASUREMENT). Neglect of this work explains why philosophers find theory so much more important than experiment: "It is only because significant quantitative comparison of theories with nature comes at such a late stage in the development of a science that theory has seemed to have so decisive a lead" (Kuhn 1977, p. 201). Moreover, the narrow image of science as testing predictions by observations is largely derived from just one source: "To a very much greater extent than we ordinarily realise, our image of physical science and of measurement is conditioned by science texts" (ibid.). One purpose of science texts is to make the connections between theory and evidence as strong as possible;

so their accounts of experiment must be as logically transparent as possible, and they must conform to accepted methodological standards. This means editing out what went into learning how to achieve stable, demonstrable results. To see how the language of science comes to describe natural phenomena, we have to go behind the reconstructions of textbooks, journals, and even of private accounts (see Goodfield 1982, pp. 217ff and Gooding 1990, pp. 4–8), because the processes of achieving fit between theory and data are written out of most texts and research reports (see also Hacking 1992). I turn to a philosophical implication of this in the final section.

Observation and instrumental practice

Theoretical philosophy of science remains resistant to the fact that seventeenth-century experimental natural philosophy did away with observation as looking and seeing. The nature of observation is not fixed; it is transformed by new representational techniques and by experimental technologies. Observation was transformed first by the invention of sense-extending instruments such as the microscope and the telescope, then by a new synthesis of mathematical and manipulative methods associated principally with Galileo, Descartes, Leibniz, and Newton, and by new instruments designed specifically as investigative tools for establishment of new natural facts. It became a complex activity involving sophisticated skills. Some instruments, such as Boyle's air pump or C. T. R. Wilson's cloud chamber, created special situations found only in laboratories. Others, such as balances, clocks, thermometers, barometers, and calorimeters, selected new quantities, bringing them into the realm of things that can be counted, measured, and represented mathematically. It is pertinent to note that Locke, Berkeley, Hume, and Kant developed new theories of knowledge, largely in response to the new sorts of knowledge being produced by science (see LOCKE; BERKELEY; and HUME).

The qualities and quantities that experiments make observable are not found, ready-made, therefore they cannot be passively observed. Rather, they are created with special laboratory techniques in special circumstances, without which they would not otherwise exist. Results are further constituted as public facts through the process of writing and publishing experimental narratives (Shapin and Schaffer 1985), so their existence depends on institutionalized publication practices (not to mention printing and other technologies of dissemination).

Sociological writers like Latour and Woolgar (1979) argued that such laboratory phenomena are constructs, because they depend wholly upon the culture of representational practices and technologies that produces and disseminates them. Hacking, Gooding, and others counter that phenomena produced by human artifice are not necessarily artifacts: those that change the world, or require us to act differently with regard to the world, may be taken to be real.

The experimenter's regress

We saw earlier that the theory-ladenness of observation means that experiment cannot – as logical positivists supposed – access any aspect of the world "as it actually is." Historians and sociologists have argued that if theory is underdetermined by

121

observational data, interests and other extra-scientific factors provide the additional constraints needed to bring about consensus. Kuhn and sociologists such as Collins (1985) and Latour and Woolgar (1979) extend Duhem's critique to the validation of experimental methods (see KUHN). They point out that these must be open to criticism, and that disagreement about the adequacy of methods, procedures, and interpretations cannot be resolved by appeal to empirical evidence or a theoretically "correct" result. Writing of experiments to detect the existence of gravity waves (a consequence of Einstein's general theory of relativity), Collins argues that:

> What is the correct outcome depends on whether there are gravity waves hitting the Earth in detectable fluxes. To find this out we must build a good gravity wave detector and have a look. But we won't know if we have built a good detector until we have tried it and obtained the correct outcome! But we don't know what the correct outcome is until . . . and so on *ad infinitum*. (1985, p. 84)

Here, the theoretical beliefs of scientists do not face the tribunal of experience directly at all. The experimenter's regress, like the Duhem–Quine thesis, shows that they cannot do so. To end the regress, an independent criterion of evaluation must be provided. The experimenter's regress states that the epistemological warrant of an experiment must reside outside the experiment's outcomes. Therefore it cannot be captured in a logical relationship between an observation statement and a theory. As for the argument that experiments can be crucial turning points in the rise and fall of competing theories, such cruciality must also reside outside the mere propositions that report their outcomes.

The epistemic force of the result of an experiment is grounded in consensus about what is the case. This is obvious while such judgments remain controversial, before scientists' agreement defines "the right answer." This is ultimately decided by a group of experts, members of a scientific community who propose, criticize, and evaluate experiments and resolve disagreements about them by argument. These arguments appeal to matters such as practitioners' competence, their status, or that of the laboratory, to an established, authoritative tradition of interpretation, analysis, or instrumentation, and also to political or moral imperatives and to religious or metaphysical precepts.

Such factors are no more empirical than those used to settle disputes prior to the rise of experimental science in the scientific revolution (see SOCIAL FACTORS IN SCIENCE). This compromises the dual image of experiment as a means of empirical access to a world that is not constructed by its investigators and as an arbiter of the truth of theoretical claims about that world. Philosophers, historians, and sociologists thus seem to agree (though for different reasons) that experiment does not have – and could never have had – the decisive, epistemological function once attributed to it, or the direct, probative force that science texts still give to it for pedagogical reasons.

Experiment and knowledge

Experiment seems to be an epistemological football – essential to the game, but of no intrinsic philosophical interest. As a knowledge-producing activity, experiment engages the inchoate, the practical, and the particular. The disorderly, inchoate, and personal character of scientific discovery and the complexity of experimental work needed to elicit meaning from phenomenological disorder have persuaded many that there is nothing philosophically interesting to recover (see DISCOVERY). Thus, creative, exploratory, and constructive aspects of experimentation are largely neglected by philosophers. Disdain for mundane practice is an obstacle to philosophical understanding of how a language – and the arguments formulated in it – comes to grips both with a material, phenomenologically complex world and with the intellectual and social world of scientists, who are the primary audience for such arguments. Philosophical theses about experiments' contribution to scientific knowledge have been stimulated primarily by work outside the mainstream of Anglo-American philosophy of science. Besides the work of historians and sociologists (some of whom are cited in this chapter), students of experiment have drawn on philosophies as diverse as those of Bachelard, Foucault, Ryle, Polanyi, and Wittgenstein. Recently, writers such as Collins, Franklin, Galison, Goodifield, Gooding, Hacking, Pickering, and Pinch have addressed the variety of complex strategies that experimentalists use to make result stable and believable. Franklin's (1986) and Galison's (1987 and 1997) studies are substantial contributions to understanding how the experimentalists of modern physical science deal with the ever-present problem of determining which phenomena and data are indicative of natural processes rather than artifacts of their complex instrumentation.

An essential role of experimentation is to provide new information about how to investigate the world: in other words, instrumental knowledge about the world under investigation. Many of the strategies which scientists use to test their methods and win confidence in their experiments are learned through the process of experimentation itself. Gooding's and Pickering's studies of what makes successful discoveries and demonstrations of natural facts show that learning to make experiments is a crucial source of information about both natural and social conditions of experiment (Gooding 1985, pp. 106–7; Pickering 1989). The studies by Franklin and Galison cited above show how teams of scientists design and learn to use the complex technologies of experimentation of twentieth-century physics.

As remarked earlier, experiment is informative in ways that theory cannot be. Experiments change our view of the world, as Harré (1981) puts it, but they do so through what experimenters discover they must do to change the world. Novelty – whether presented in experience or prompted by new questions – stretches interpretative and manipulative skills, technologies of observation and the representational capacity of practitioners' language. Much experimentation is about developing and extending the representational capabilities of a science in order to bring new features of the world into the domain of discourse and argument. To state new results often involves inventing new representational techniques, as well as new manipulative ones.

On this view, the practical observational and manipulative skills of Harvey, Hooke, Davy, Faraday, or Darwin show as much ingenuity – and have contributed as much to the progress of science – as the more celebrated theoretical accomplishments of Maxwell, Bohr, Einstein, Schrödinger, or Feynman (see EINSTEIN). The study of experiment, therefore, has a role analogous to the critical role of experiment in the sciences: it can expose some philosophical blind spots about the sorts of knowledge that scientists produce.

123

Some implications of experiment

As we saw earlier, the distinction between theory and observation created puzzles such as underdetermination (observation must be theory-laden, when in fact it is practice-laden) and falsification (where the whole burden of proof falls on simple observation statements, when in practice it involves the whole context of experiment). Another problem is to show how two things as different as theory (propositional knowledge) and experience can interact. Dissatisfied with methodological prescriptions, some philosophers have recognized the more sophisticated nature of the interaction between theory and observation. Thus Franklin argues that epistemological strategies provide rational grounds for believing their results to be true, and denies that validation depends ultimately on culturally accepted practices (see Franklin 1986, pp. 166–225; Gooding et al. 1989, pp. 458–9).

However, the historical and sociological studies cited here suggest a different view. The many new technologies of experience we refer to loosely as "experimentation" are among the most significant innovations of Western culture. Experimentation not only transforms the way we perceive the world; it alters the world as well. On this view, the two transformations are inextricably linked. This denies that theory and observation belong to ontologically distinct categories, dispensing with the need for a model of interaction based on logical operations on propositions. The denial is supported by an explanation of how the apparently self-evident distinction between representations and the things that they denote comes about in scientific practice. The explanation draws on the work of Latour and Woolgar, Hacking, Pickering, and others (see Gooding 1990, chs 7–8). Theory and observation are not brought into interaction as scientists design and conduct tests. Rather, representations and their objects are brought into correspondence by the work that scientists do. Hacking puts it succinctly: "Our preserved theories and the world fit together so snugly less because we have found out how the world is than because we have tailored each to the other" (1992, p. 31). On this view, establishing a matter of fact is inseparable from establishing the adequacy of what represents it, and this is a practical matter. Of course, once this has been established, it seems manifestly the case that the representation exists – and has always existed – independently of the facts that it purports to represent (for example, that little blobs near Jupiter are moons, rather than faults in a glass lens).

A radical implication of this view is that the independence of representations and objects is an illusion, fostered by the fact that the experimental work that accomplishes this is systematically written out of research reports. This editing is necessary for effective communication, pedagogy, and argument. However, once we have identified the process, the disappearance of the human contribution to the meaningfulness of the terms that describe the world *explains* the apparent independence of representations from the world, as something accomplished by human agency: "the thing and the statement correspond for the simple reason that they come from the same source. Their separation is only the final stage in the process of their construction" (Latour and Woolgar 1979, p. 183).

Philosophers have wrongly inferred that because the laws of nature are universal and timeless, the results of experiment have these same qualities. Most of the time, the opposite is true. Experimental practice is particular and local, and so are its results.

The world implicated by work in one laboratory or field may be subject to disturbing effects of a kind absent in another. Different groups testing the same prediction invent different technologies (thus, Fairbanks believed he had found quarks with niobium spheres; Morpurgo believed he did not find them using graphite grains (Pickering 1989)). These differences are often tacit, based on skilled responses to particularities that may not be present in other laboratories. This is why replication is so difficult, and usually requires the transfer of skills. Sometimes differences are exposed only by close analysis of a failure to transfer successful techniques from one laboratory to another, as in the case of TEA laser technology (Collins 1985, pp. 51–78).

This is because, as remarked earlier, the guidance which theory gives to material practice is much looser than the guidance it can give to intellectual operations. Actual testing just isn't like the textbooks say. Research scientists engage a world that is much more complex than the idealized world of theory, so theory cannot anticipate the specifics of engaging the actual world. As Galison puts it:

> The world is far too complex to be parcelled into a finite list of all possible backgrounds. Consequently there is no *strictly logical* termination point inherent in the experimental sciences. Nor, given the heterogeneous contexts of experimentation, does it seem productive to search after a universal formula for discovery, or an after the fact reconstruction based on an inductive logic. (1987, p. 3)

Nevertheless, scientists can and do identify and eliminate these local or contingent factors. What can be manipulated in any laboratory, when all the relevant variables are controlled, is taken to be natural, or "out there." The important point is that scientists work to achieve the "out-thereness" of a phenomenon (Latour and Woolgar 1979, pp. 181–3) and the demonstrative, "golden-event" status of a result (Galison 1987, pp. 18–19).

This process dissociates results from particular individuals or laboratories, and it complements the mastery and dissemination of skills. Latour calls it the demodalization of factual claims, because references to particular persons, practices, and places (the "modalities") are gradually dropped from scientists' talk. This makes facts context-free (Latour and Woolgar 1979, pp. 151ff, 236). Though such transportable effects are often named after individuals, titles such as "Stokes' law" or the "Hall effect" do not function as modalities. Stripping the modalities from particular results confers upon them the status of observations that support or contradict theoretical claims, but it does not diminish the significance of the local, particular character of experimentation. That work, which justifies both the result and the decision that the work is sufficient to establish a result, must be reproduced for analysis and criticism whenever the status of a result is challenged.

This explanation also turns the deductive model on its head: insofar as it is empirically constrained, a theory is tested not directly through a set of (defeasible) observations, but through the myriad of small, instrumental findings that establish the practices that in turn give confidence in the result. The specifics of a test are worked out through experimentation. What scientists call "recalcitrance" in experiment helps to identify just those aspects of the world that are actually implicated in a particular laboratory at a particular time. Different laboratories will encounter different recalcitrances. This is why, although simulations based on numerical methods have been used for decades in high energy physics (see Galison 1987, pp. 189–93, 265–6), it remains very difficult to design realistic discovery programs (see COMPUTING). Without real experiments that develop the skills and technologies of observation – and methods to disseminate them – there can be no demodalization, no generality, no observation statements.

The regress into experimental practice (see above) turns out to be the experimenter's redress. It is possible to explain the epistemic force of experiment in the discourse of science. To do so, we need to understand both that experiments are made into arguments and how scientists actually accomplish this transition from the exploration of expectations to the demonstration of results.

References

Collins, H. M. 1985: *Changing Order: Replication and Induction in Scientific Practice* (Beverly Hills, CA, and London: Sage).

Franklin, A. 1986: *The Neglect of Experiment* (Cambridge: Cambridge University Press).

Galison, P. 1987: *How Experiments End* (Chicago: University of Chicago Press).

——— 1997: *Image and Logic: A Material Culture of Microphysics* (Chicago and London: University of Chicago Press).

Giere, R. N. 1988: *Explaining Science: A Cognitive Approach* (Chicago: University of Chicago Press).

Goodfield, J. 1982: *An Imagined World: A Story of Scientific Discovery* (Harmondsworth: Penguin; previously published New York: Harper & Row, 1981).

Gooding, D. C. 1985: "In Nature's School": Faraday as an experimentalist. In *Faraday Rediscovered: Essays on the Life & Work of Michael Faraday, 1791–1867*, ed. D. C. Gooding and F. James (London: Macmillan; also New York: American Institute of Physics, 1989), 105–35.

——— 1990: *Experiment and the Making of Meaning: Human Agency in Scientific Observation and Experiment* (Dordrecht and Boston: Kluwer Academic Publishers).

Gooding, D., Pinch, T. J., and Schaffer, S. (eds) 1989: *The Uses of Experiment: Studies in the Natural Sciences* (Cambridge: Cambridge University Press).

Hacking, I. 1983: *Representing and Intervening: Introductory Topics in the Philosophy of Natural Science* (Cambridge: Cambridge University Press).

——— 1992: The self-vindication of the laboratory sciences. In *Science as Practice and Culture*, ed. A. Pickering (Chicago: University of Chicago Press), 29–64.

Harré, H. R. 1981: *Great Scientific Experiments: Twenty Experiments that Changed our View of the World* (Oxford: Phaidon).

Kuhn, T. S. 1977: *The Essential Tension: Selected Studies in Scientific Tradition and Change* (Chicago: University of Chicago Press).

Latour, B., and Woolgar, S. 1979: *Laboratory Life: The Social Construction of Scientific Facts* (Beverly Hills, CA, and London: Sage).

Pickering, A. 1989: Living in the material world. In Gooding et al., 275–97.

Shapin, S., and Schaffer, S. 1985: *Leviathan and the Air-Pump: Hobbes, Boyle and the Experimental Life* (Princeton: Princeton University Press).

Tiles, J. E. 1993: Experiment as intervention. *British Journal for the Philosophy of Science*, 44, 463–75.

19

Explanation

W. H. NEWTON-SMITH

The point of departure for all discussions of nonstatistical explanation in the philosophy of science has been the deductive-nomological or covering law model of explanation that was given its most influential exposition by Carl Hempel, who was also the pioneering figure in the discussion of statistical explanation (see STATISTICAL EXPLANATION). On this account, to explain a particular event, we cite other particular events together with a general law or laws which "cover" what we want to explain. For example, we might explain why a balloon expanded when placed on a radiator, by citing the fact that the radiator was hot together with the relevant laws relating heat to expansion.

More precisely, on the deductive-nomological or D-N model, an explanation of an event is a valid deductive *argument* of the following form:

$$\frac{\begin{array}{c} C_1, C_2, \ldots, C_k \\ L_1, L_2, \ldots, L_r \end{array}}{E}$$

$C_1, C_2, \ldots, C_k$ are statements describing various particular facts involved, the initial conditions; $L_1, L_2, \ldots, L_r$ are general laws; and the conclusion E is a statement describing the event to be explained. An argument of this form constitutes an explanation if and only if it is deductively valid, the statements describing the initial conditions are true, the statements $L_1, L_2, \ldots, L_r$ are true and express genuine laws, and enter essentially into the derivation.

This model captures an important intuitive feature of explanation. We feel we understand why something happened if we see that, given certain features in the situation, it had to happen. And that condition is captured if we have a valid deductive argument from true statements of the initial conditions and true statements of laws. It is also attractive in that, apart from the notion of a law (see LAWS OF NATURE), it explains explanation in terms of relatively clear and simple notions.

Very few explanations actually encountered in everyday life or in science have this precise form. Hempel intended his model to characterize a theoretical ideal. Deviations of actual explanations from the norm were to be explained away. According to Hempel, sometimes we offer as an explanation an *elliptical* formulation of the full argument that we have not bothered to elaborate. For instance, we might explain that the butter is melting by just citing the fact that the pan was hot. We do not bother to spell out the relevant laws. In other cases what we give is only an *explanatory sketch*; that is, we sketch part of a story which we are betting could be elaborated so as to incorporate appropriate laws, given further empirical research.

We not only explain particular events; we also explain laws. Newton used his laws of motion together with his universal law of gravitation to explain Kepler's laws of planetary motion. It might seem that the D-N model is ideally suited to cover the explanation of laws by other laws, as well as the explanation of particular events. For the explanation of a law, we simply derive that law from other laws without the introduction of initial conditions. However, as Hempel himself noted, this gives rise to a difficulty. Not only can we derive Kepler's laws from the Newtonian laws, we can also derive Kepler's laws from the conjunction of Kepler's laws with, say, Boyle's gas laws. But that would not be counted as an explanation of Kepler's laws. As Hempel saw no resolution of this difficulty, he focused on the explanation of events.

The model was attractive. But even when restricted to the explanation of particular events, it met insurmountable objections. Explanation is asymmetric. That is, if A explains B, B cannot explain A. No model of explanation can be acceptable unless it meets this condition. But consider a flagpole of a certain height that casts a shadow of a certain length. The height of the flagpole explains the length of the shadow, given the position of the Sun. This can be represented in D-N form. From the position of the Sun and the height of the flagpole, we can derive the length of the shadow using elementary laws of optics with the help of a little mathematics. Unfortunately for the D-N model, we can equally well derive the height of the flagpole from the length of the shadow. But the length of the shadow does not explain the height of the flagpole! The D-N modeled is fundamentally flawed in that it does not guarantee that explanations will satisfy the condition of explanatory asymmetry.

Explanations are sensitive things. Adding a simple truth to a good explanation may destroy it. In a certain context I might explain why a particular stick looks bent by reference to the fact that, unknown to you, it is partially submerged in a container of water. Depending on how much you know and want to know, I may need to add an account of the refractory properties of water on light. Let us suppose that a priest has blessed the water in the container. If I had said instead that the stick is submerged in water blessed by the priest, I would not have provided a satisfactory explanation. However, if we can cast my original story in the appropriate D-N form, we can equally replace "The stick is partially submerged in water" by the equally true "The stick is partially submerged in water that has been blessed by a priest." The derivation will still be valid. But we would not accept as explanatory the claim that it looks bent because it is in water that has had the attentions of a priest. That is simply irrelevant. As developed by Hempel, the D-N model is powerless to prevent the generation of countless similar counterexamples. To save the model, we would need to add a requirement that all premises used in the argument are relevant. Explicating an appropriate notion of relevance will not be easy, and if such a condition is added, the model will lose in part its particularly attractive feature of explaining explanation in terms of notions that are relatively unproblematic.

Laws are often important in explanations, and D-N models give prominence to this fact. But are laws always required for explanation? Michael Scriven (1962) argued to the contrary. He claimed that fully satisfactory, complete explanations can be given without even tacit appeal to laws. The explanation of the black stain on Professor

Jones's carpet is the fact that he accidentally bumped his desk on which there was an open bottle of ink. No explicit or implicit appeal to laws is needed according to Scriven to make the explanation work. (See CAUSATION.)

In this regard it is instructive to consider the motivating example drawn from Dewey which Hempel used in his seminal paper (Hempel 1965). Dewey, washing his dishes, noticed that on taking glass tumblers from the hot soapy water and placing them to drain upside down, soap bubbles emerged from under the rims, grew for a while, and then receded inside the tumblers. Dewey explained this phenomenon to himself by reflecting that the cool air trapped in the glasses was warmed by the heat of the glasses. The warming air increased its pressure, which forced the bubbles under the rims. As the air cooled, the pressure dropped, and the bubbles receded. Hempel used the example to develop the D-N model, remarking that laws not explicitly mentioned by Dewey would need to be added, such those governing the elastic behavior of bubbles. Hempel is certainly right that Dewey possessed a good explanation of the phenomenon. But there is no reason to think that Dewey knew the laws of elasticity of bubbles. Nor, more seriously, is there any reason to think that those laws which are no doubt of a highly mathematical character can be married with the initial conditions as known to Dewey ("The water is pretty hot and quite soapy") to produce the required deductively valid argument. And it is not clear that a D-N story, if it could be elaborated, would have further advanced Dewey's understanding. Ironically, then, Hempel's own motivating example itself prompts a suspicion as to the adequacy of his model.

The literature abounds in counterexamples to the D-N model. Many of these, including those given above, gave rise to the suspicion that the D-N modeled error in not giving due attention to causation. Causation is asymmetric. If A causes B, B does not cause A. In the case of the flagpole, we can explain the length of the shadow by reference to the height of the flagpole, and not vice versa, because the flagpole causes or produces the shadow. The shadow does not produce the flagpole. And in the case of the partially submerged stick, that the water has been blessed is causally irrelevant to the fact that it looks bent. Perhaps, then, to explain a specific event is to cite an event causally relevant to its production. Professor Jones's bumping his desk explains the stain, as it is one of its causes. Dewey understood his soap bubbles when he was able to grasp some of the causal mechanism in operation, and not because someone might be able to recast his reflections in the form of a particular kind of deductive argument.

Reflections on the role of causation in explanation have given rise to the *causal relevance* model of explanation, hereafter cited as C-R. On this model explanations are not arguments. It is the causal features of the world, aspects of the causal mechanisms in the world, which explain. In giving explanations we draw attention to what in the world is in part at least causally responsible for what we want to explain. This model fits our actual explanatory practices in science and ordinary life much better than the D-N model. If I am asked why my car stopped and I cite the fact that it has run out of gas, that is the explanation. I have drawn attention to a causally relevant feature of the world: no gas. I do not need to suppose that it is an explanation in virtue of the fact that it might be possible to transform this into an argument of the D-N form. In many cases we are more confident that we have given an explanation than that we or anyone else could provide the transformation.

129

For a very wide class of explanations, both in science and in everyday life, this looks like a promising approach. However, there are difficulties. First, we are in danger of explaining the obscure, explanation, in terms of the equally if not more obscure, causation. Ever since at least Hume, philosophers have been aware that causation is a problematic notion. Hume himself attempts a demystifying analysis of causation in terms of constant conjunction; roughly, A causes B just in case events similar to A are constantly conjoined with events similar to B. However, if we are to develop an adequate causal relevance model of explanation we will need a stronger notion of causality than Hume's (see HUME and CAUSATION). For many of the counterexamples to the D-N model are equally counterexamples to the Humean account of causation. The difficulties in providing a satisfactory non-Humean account of causality have inclined some to resist the C-R approach. The proponents of the C-R model argue that we need a non-Humean account of causation in any event and see the attraction of their model as providing an additional incentive for redoubling our efforts to achieve this.

The C-R model fits at best an important class of explanations in science. There are a variety of types of respectable noncausal explanations. One of these is explanation by identification. It was an important scientific discovery that temperature is mean molecular motion. The fact that temperature is mean molecular motion explains why increasing the temperature of a gas increases pressure (increasing temperature is increasing the mean motion of the molecules of the gas, which then strike the walls of the container with greater force). Certainly there are causal factors involved, but the real explanatory work is done by the identity between temperature and mean molecular motion, and that is not a causal relation (see Achinstein 1983; Ruben 1990). Furthermore, various authors, including Cartwright (1986), Glymour (1982) and Kitcher (1985) have offered additional examples of respectable noncausal scientific explanations. The situation is further complicated by the fact that in everyday life and in science we also explain things using models and analogies (see METAPHOR IN SCIENCE; MODELS AND ANALOGIES).

A good explanation increases our understanding of the world. And clearly a convincing causal story can do this. But we have also achieved great increases in our understanding of the world through unification. Newton was able to unify a wide range of phenomena by using his three laws of motion together with his universal law of gravitation. Among other things he was able to account for Kepler's three laws of planetary motion, the tides, the motion of comets, projectile motion, and pendulums. Friedman (1974) and Kitcher (1981, 1989, 1993) have elaborated this idea that one form of explanation is unification. Friedman's idea is that we increase our understanding of the world as we decrease the number of independently acceptable hypotheses needed to account for the phenomena of the world. For Kitcher it is matter of reducing the number of argument patterns. It is hard to deny that unification is in some sense one aspect of explanation. And it can be expected that this approach to explanation will receive further elaboration and refinement. We need to see how it relates to other forms of explanation, such as causal relevance. Salmon (1998) has suggested that these two approaches may be complementary rather than competitive.

We have an embarrassment of riches. We have explanations by reference to causation, to identities, to analogies, to unification, and possibly to other factors. Philosophically we would like to find some deeper theory that explained what it was about each of these apparently diverse forms of explanation that makes them explanatory. This we lack at the moment. Dictionary definitions typically explicate the notion of explanation in terms of understanding: an explanation is something that gives understanding or renders something intelligible. Perhaps this is the unifying notion. The different types of explanation are all types of explanation in virtue of their power to give understanding. While certainly an explanation must be capable of giving an appropriately tutored person a psychological sense of understanding, this is not likely to be a fruitful way forward. For there is virtually no limit to what has been taken to give understanding. Once upon a time, many thought that the fact that there were seven virtues and seven orifices of the human head gave them an understanding of why there were (allegedly) only seven planets. We need to distinguish between real and spurious understanding. And for that we need a philosophical theory of explanation that will give us the hallmark of a good explanation.

In recent years there has been a growing awareness of the pragmatic aspect of explanation. What counts as a satisfactory explanation depends on features of the context in which the explanation is sought. Willy Sutton, the notorious bank robber, is alleged to have answered a priest's question "Why do you rob banks?" by saying "That is where the money is." We need to look at the context to be clear about for what exactly an explanation is being sought. Typically we are seeking to explain why something is the case rather than something else. The question which Willy's priest probably had in mind was: "Why do you rob banks rather than have a socially worthwhile job?" and not the question "Why do you rob banks rather than churches?" We also need to attend to the background information possessed by the questioner. If we are asked why a certain bird has a long beak, it is no use answering (as the D-N approach might seem to license) that the bird is an Aleutian tern and all Aleutian terns have long beaks if the questioner already knows that it is an Aleutian tern. A satisfactory answer typically provides new information. In this case the speaker may be looking for some evolutionary account of why that species has evolved long beaks. Similarly, we need to attend to the level of sophistication in the answer to be given. We do not provide the same explanation of some chemical phenomena to a school child as to a student of quantum chemistry.

Attention to the pragmatic aspect of explanation has been a further useful corrective to the Hempelian approach. Hempel's goal was a nonpragmatic conception "which does not require relativization with respect to questioning individuals any more than does the concept of mathematical proof" (Hempel 1965, p. 426). This has now been seen to be a hopeless project. Important as it is to note the pragmatic aspect of explanation, this does not provide us in itself with a theory of explanation. For once we have specified the relevant aspects of the context in which an explanation is sought, we need an account of what it is appropriate to offer. The C-R model (and even the D-N model) can be regarded as seeking to provide objective conditions that any answer must satisfy in order to be satisfactory once all the relevant features of the context have been specified.

Van Fraassen whose work has been crucially important in drawing attention to the pragmatic aspects of explanation has gone further in advocating a purely pragmatic theory of explanation (van Fraassen 1980). A crucial feature of his approach is a notion of relevance. Explanatory answers to "why" questions must be relevant, but relevance itself is a function of the context for van Fraassen. For that reason he has denied that it even makes sense to talk of *the* explanatory power of a theory. However, his critics (Kitcher and Salmon 1987) point out that his notion of relevance is unconstrained, with the consequence that anything can explain anything! This *reductio* can be avoided only by developing constraints on the relation of relevance, constraints that will not be a function of context and hence take us away from a purely pragmatic approach to explanation.

This chapter provides a rough sketch of the tip of a very large iceberg. The literature abounds with further models of explanation. And there are many alternative ways of elaborating the models discussed above (see Ruben 1994). The prospect of any one of these models being developed to cover all good scientific explanations (let alone good explanations in general) are dim. Perhaps we should opt for pluralism. Perhaps there are several types of explanation, each with its appropriate model. Philosophical reflection may winnow the field a little. The time has come, for instance, to acknowledge that the D-N model is historically important, but not at all acceptable – though we have yet to explain satisfactorily what it was that made it once seen so attractive. But pluralism in philosophy lacks the attraction it may have in social and political affairs. We are left not understanding what it is that knits these different types of stories together as explanatory stories. The matter is too serious to take refuge in any Wittgensteinian notion of family resemblance. It is too serious, because scientists as part of their day-to-day craft judge theories and hypotheses in terms of their abilities to explain – period. They do not make judgments in terms of their abilities to explain on different scales – on the causal scale, on the unification scale, etc. For the moment we should preserve the pursuit of a single model or an underlying principle that links the divergent models together.

The current situation is an embarrassment for the philosophy of science. Indeed, one might go so far as to say that it is the sort of scandal to philosophy of science that Kant thought skepticism was to epistemology. While we have insightful studies of explanation, we are a very long way from having this single unifying theory of explanation. Why should we want this? As noted above, we would like to be able to explain what it is that leads us to count different explanations as explanatory. This task is made all the more pressing as most philosophers of science hold that *a* main task, if not *the* main task, of science is to provide explanation, whatever that may be. And it is hard to see how we will be able to adjudicate the substantial claims about the relation of explanation to epistemology without such a unifying account. Realists, for instance, typically claim that the greater a theory's explanatory power, the greater its likely truth or approximate truth (see INFERENCE TO THE BEST EXPLANATION). Without the backing of a unifying account of explanation, this claim is suspect. Of course, that is not to imply that a unifying account will sanction the claim. But it would at least allow it to be assessed.

References and further reading

Achinstein, P. 1983: *The Nature of Explanation* (Oxford: Oxford University Press).
Cartwright, N. 1986: *How the Laws of Physics Lie* (Oxford: Clarendon Press).
Friedman, M. 1974: Explanation and scientific understanding. *Journal of Philosophy*, 71, 5–19.

Glymour, C. 1982: Causal inference and causal explanation. In *What? Where? When? Why?*, ed. R. McLaughlin (Dordrecht: Reidel), 179–91.

Hempel, C. G. 1965: *Aspects of Scientific Explanation* (New York: Free Press).

—— 1966: *The Philosophy of Natural Science* (Englewood Cliffs, NJ: Prentice-Hall).

Humphreys, P. 1989: *The Chances of Explanation* (Princeton: Princeton University Press).

Kitcher, P. 1981: Explanatory unification. *Philosophy of Science*, 48, 507–31.

—— 1985: Salmon on explanation and causality: two approaches to explanation. *Journal of Philosophy*, 82, 632–9.

—— 1989: Explanatory unification and the causal structure of the world. In Kitcher and Salmon, 410–505.

—— 1993: *The Advancement of Science* (New York: Oxford University Press).

Kitcher, P., and Salmon, W. C. 1987: Van Fraassen on explanation. *Journal of Philosophy*, 84, 315–30.

Kitcher, P., and Salmon, W. C. (eds) 1989: *Scientific Explanation, Minnesota Studies in the Philosophy of Science*, vol. 13 (Minneapolis: University of Minnesota Press).

Ruben, D.-H. 1990: *Explaining Explanation* (London: Routledge).

Ruben, D.-H. (ed.) 1994: *Explanation* (Oxford: Oxford University Press).

Salmon, W. C. 1984: *Explanation and the Causal Structure of the World* (Princeton: Princeton University Press).

—— 1998: *Causality and Explanation* (Oxford: Oxford University Press).

Scriven, M. 1962: Explanations, predictions, laws. In *Minnesota Studies in the Philosophy of Science*, vol. 3 (Minneapolis: University of Minnesota Press), 172–230.

van Fraassen, B. 1980: *The Scientific Image* (Oxford: Oxford University Press).

20

Feminist Accounts of Science

KATHLEEN OKRUHLIK

Feminist accounts of science expose the ways in which the various sciences exhibit androcentric bias in their theories, practices, and presuppositions. Some, but not all, of these accounts also raise questions about the extent to which our understanding of what it is to be rational, objective, and scientific is itself gender-laden. The analyses are wide-ranging and diverse, reflecting a broad range of commitments within philosophy of science and within feminist theory. It is a mistake to treat feminist critiques of science as constituting a monolithic body of literature, since doing so leads to caricature and to the inevitable repression of crucial issues. Moreover, one of the chief lessons to be gleaned from the accumulating research in this area is that the role played by gender in science is exceedingly complex and variable. Popular presentations and offhand references that downplay this complexity and variability misrepresent feminist research, and contribute to maintaining the unfortunate gulf between it and most "mainstream" philosophy of science.

Not all of the vast literature on "gender and science" can be touched on here. Related bodies of literature that will not be dealt with include equity studies, efforts to reform science education, research on women scientists, and the literature on women and technology. The focus instead is on scholarship that addresses directly questions relating to the content, methodology, and epistemology of science.

Feminist critiques of science, even in this somewhat restricted sense, are very widely dispersed, due to their diverse origins. Some arise from within the sciences themselves, in response to particular instantiations of androcentric theory and practice. Particularly in the biological and social sciences, feminist researchers have dramatically and effectively presented case studies that show how the omission or misrepresentation of women and gender has led to work that is deeply flawed and demonstrably unbalanced. Much of this work is to be found in journals and anthologies specific to the disciplines in question.

Other feminist analyses are more general in their orientation. They represent extensions of feminist epistemological projects and sometimes extensions of other genres of science criticism, including contextualist, sociological, and relativist approaches. Despite some similarities to other forms of science criticism, feminist critiques are distinguished by their emphasis on the power differences that are embodied in gender relations and the way in which these power relations are reflected in the processes and products of science. In order to provide a reasonable overview of the field, examples of both first-order, discipline-specific research and second-order, epistemological reflections will be cited.

Some of the most powerful and accessible feminist critiques are in the form of case studies demonstrating the ways in which gender-related biases have affected the content of the biological and social sciences. A good introduction to this genre is *Myths of Gender: Biological Theories about Women and Men* (1985) by Anne Fausto-Sterling. Although the author was trained as a scientist rather than a philosopher, the book is philosophically sensitive, and consistently addresses questions of methodology and occasionally epistemology. Fausto-Sterling examines attempts to provide biological explanations of alleged cognitive differences between the sexes, genetic accounts of behavior, and hormonal explanations of aggression as well as other phenomena. She also discusses evolutionary accounts that purport to "explain" why it is natural for women to function in socially subordinate roles, why men are smarter and more aggressive than women, why women are destined to be homebodies, and why men rape. Evidence for each claim is examined; experimental designs are scrutinized; and methodological issues are raised. Fausto-Sterling draws attention consistently to the ways in which some evidence is ignored, some questions not asked, some hypotheses never considered, and some experimental controls never instituted. She points to the tenacity of biological explanations for women's socially inferior role, often in the light of extremely recalcitrant evidence. A good example is her treatment of hypotheses about spatial ability in women and men. It has been suggested that spatial ability is X-linked, and therefore exhibited more frequently in males than in females; that high levels of prenatal androgen increase intelligence; that low levels of estrogen lead to superior male ability at "restructuring" tasks. Some have held that female brains are more lateralized than male brains, and that more lateralization interferes with spatial functions. Others have argued that female brains are *less* lateralized than male brains, and that less lateralization interferes with spatial ability. Some have attempted to save the hypothesis of X-linked spatial ability by suggesting that the sex-linked spatial gene can be expressed only in the presence of testosterone. Others have argued that males are smarter because they have more uric acid than females.

None of these hypotheses is well supported by the evidence, and most seem to be clearly refuted. Yet, for many researchers, the one element of the theoretical network that they are unwilling to surrender in the face of recalcitrant evidence is the assumption that there must be predominantly *biological* reasons for inferior intellectual achievement in women.

Another useful and accessible text in the same genre is *The Politics of Women's Biology* (1990), by Ruth Hubbard, who has migrated from research in photobiology to feminist critiques of science. One of the chapters is a version of her influential article "Have only men evolved?," in which she examines some of the biases and blind spots of evolutionary theory. She cites passages from *The Origin of Species* in which Darwin attributes evolutionary development in human beings almost exclusively to male activity. In defending women and children, capturing wild animals, and making weapons, men have constantly had to draw on their higher cognitive faculties. Since these faculties are constantly being tested and selected for, men have become superior in intelligence to women. Darwin concludes that it is indeed fortunate that fathers pass on their brains to their daughters, because "otherwise it is probable that man would have become as superior in mental endowment to woman, as the peacock is in ornamental plumage to the peahen."

135

Hubbard also presents examples of gender bias in more recent evolutionary biology, exemplifying the sort of circular argument that is often used to "prove" that behaviors and social roles are biologically determined. A sexist stereotype derived from twentieth-century gender relations among human beings is imported (without independent evidence) into the animal world, and then the animal "evidence" is cited to justify the human gender relations. The circularity is particularly breathtaking when it involves creatures as different from us as algae. Yet even strands of algae are identified as male or female according to whether they are active or passive. One and the same strand of algae is identified as masculine when it takes the active role in sexual relations, feminine when it assumes the passive role. There is no independent evidence for these attributions. Yet they are then cited as part of the evidence for the claim that throughout the animal world, it is males who are active and engage in goal-directed behavior.

The influence of Darwin's androcentric bias has not been limited to evolutionary biology, because that theory functions as an auxiliary hypothesis in many other disciplines, especially in the social sciences. Anthropology is a good example. If one holds the view that man-the-hunter is chiefly responsible for human evolutionary development, one interprets fossil evidence in light of the changing behavior of males. Helen Longino and Ruth Doell, for example, in a 1983 article entitled "Body, bias, and behaviour: a comparative analysis of reasoning in two areas of biological science," trace the ways in which the androcentric account attributes the development of tool use to male hunting behavior. Longino and Doell point out that some recent research attributes up to 80 percent of the subsistence diet of "hunter-gatherer" societies to female gatherers. If that is the background theory informing one's interpretation of the evidence, then quite a different account of the same fossil evidence emerges.

The gynecentric story explains the development of tool use as a function of female behavior, portraying women as innovators who contributed more to the development of human intelligence and flexibility than did males. It emphasizes the importance of tools made from organic materials such as sticks and reeds, which are said to have been developed by women to defend against predators while gathering, as well as for carrying, digging, and food preparation. These tools are supposed to antedate the stone tools attributed to male hunters.

What matters here is not that the gynecentric hypothesis be true, but rather, that it makes obvious the extent to which the standard interpretation of the anthropological evidence has been colored by androcentric assumptions. It highlights the distance that sometimes exists between evidence and hypothesis and the difficulty of bridging that distance with independently corroborated auxiliaries.

Donna Haraway, in a series of articles and, more recently, in her book *Primate Visions: Gender, Race, and Nature in the World of Modern Science* (1989), has suggested that in primatology (at least) that distance cannot be bridged, and that what we are faced with is in fact a variety of irreducible narratives about our origins. These narratives were fashioned to serve a variety of political needs, and choices among them are based on similar considerations. Primatology is "politics by other means."

During the same period that these case studies were developed, other authors were looking at the scientific revolution itself for manifestations of gender ideology. One of the most influential and widely cited works of this type is *The Death of Nature: Women, Ecology, and the Scientific Revolution* (1980) by Carolyn Merchant. Merchant links the mechanization of the world picture to the demise of an older, feminine-nurturant vision of the Earth as a loving provider. This older vision embodied, according to Merchant, a system of values that promoted harmonious cooperation with the environment and a holistic approach to understanding nature. She argues that the replacement of this view by the mechanical world view has had disastrous effects for the environment and for women. Merchant also argues that the metaphors employed by Francis Bacon and others at this time reveal that the new philosophy implicitly embodied the view of a male knower who manipulates, dominates, and exploits the object of his knowledge.

In a series of essays written during the late 1970s and early 1980s and collected in a volume called *Reflections on Gender and Science* (1985), Evelyn Fox Keller also examines the events of the scientific revolution (and other topics). She argues that the seventeenth century witnessed a contest between masculine and feminine principles: head versus heart, purified versus erotic stances toward knowledge, attitudes of domination of the object versus merging with the object of knowledge. The victory of masculine over feminine forces has meant, according to Keller, that even today a "masculine" cognitive stance is required of all working scientists: one that emphasizes autonomy, separation, and distance between subject and object. The defeat of feminine principles in the seventeenth century meant the banishment of "sympathetic understanding" from the methodology of science. In making this argument, Keller draws heavily upon psychoanalytic theory, especially object relations theory, to explain how concepts of objectivity and masculinity have become intimately connected and mutually reinforcing. The theory, very roughly speaking, is that since the primary caregiver for very young children is almost invariably the mother, little girls and little boys are differently positioned with respect to the project of self-definition. Because little girls share the sex of the caregiver, they do not have to break so abruptly with her in order to establish a suitable gender identity and sense of self. The connection with the mother, the strong sense of relatedness and interdependence, can survive relatively unscathed. The little boy, however, if he is to establish a suitable masculine identity, must define himself in opposition to his mother – by separating from her, establishing distance, drawing boundaries between himself and her. In this way, he asserts his autonomy. Keller (as well as some other feminist epistemologists) maintains that this separation, distancing, delineation of boundaries, and emphasis on autonomy is central not only to our concepts of masculinity, but also to our notions of objectivity and scientific rationality. A circular process of definition is set up, so that it is not simply that our notions of rationality are masculine, but also that the prestige of science affects our concepts of masculinity. It is important to note that if arguments of this sort are successful, they would apply not just to the biological and social sciences, with their gendered ontologies, but to all the sciences, including physics.

A more recent historical treatment of the scientific revolution is Londa Schiebinger's excellent book *The Mind Has No Sex? Women in the Origins of Modern Science* (1989).

In light of these case studies (and hundreds of others like them), a number of questions arise. Chief among them is: What ought to be concluded about the nature of science? One approach is simply to maintain that insofar as these episodes do in fact exhibit androcentric bias, to that extent they fail to be scientific. Gender bias is not characteristic of science on this account, but only of *bad* science. Another response is to acknowledge that science-as-usual (and not just bad science) is gender-laden, and

137

then to consider the consequences of this acknowledgment. In fact, the stances adopted by feminist critics have been sufficiently various and the volume of literature sufficiently large that considerable effort was expended during the 1980s in an effort to develop typologies of feminist critiques of science. Most influential was a taxonomy developed by Sandra Harding in her very important book *The Science Question in Feminism* (1986). Harding's taxonomic categories are feminist empiricism, feminist standpoint epistemology, and feminist postmodernism.

Very roughly speaking, feminist empiricists believe that gender bias in the sciences reflects a failure to live up to their own epistemological ideals, and that a more rigorous and thoroughgoing application of scientific methodology would eliminate this bias, thus producing better science. The epistemic commitments and methodologies of science are not called into question by feminist empiricists, and the underlying assumption is that the sex of the knower would be irrelevant if science were done properly. It is this assumption that is repudiated by feminist standpoint theorists. They argue that the standpoint of the knower is epistemically relevant, and that a kind of epistemic privilege is available to women (or to feminists, depending on the account). Just as Hegel's slave could know things the master did not, so women (or feminists) are in a position not only to effectively criticize masculinist science, but also to produce a feminist successor science that is epistemically superior to what went before. Two kinds of standpoint theory that have been particularly influential are varieties informed by Marxism and object relations theory respectively. It is important to note that although feminist empiricist and feminist standpoint theorists disagree about the adequacy of current conceptions of scientific method and the relevance of the standpoint of the knower, they both espouse "successor science projects" in the sense that both strive for epistemic progress, for *better* science. This goal is not shared by feminist postmodernism, which eschews the very idea of a successor science and aims instead for "a permanent multiplicity of partial narratives."

An interesting characteristic of this taxonomy of feminist critiques is that the three categories are presented in a way that sometimes suggests that they represent successive *stages* in feminist inquiry, each stage being developed in response to tensions and inadequacies in the preceding stage. So Harding herself, a leading developer and proponent of feminist standpoint epistemology, appeared in 1986 to be moving toward a postmodern position in response to criticisms of standpoint theory. Although these criticisms have been numerous and diverse, the one that most affected Harding (and many other feminist theorists) was the insistence that there is no single feminist standpoint. Just as the standpoint of women differs from that of men, so the standpoint of women of color differs from that of white women, the standpoint of poor women from that of rich women, the standpoint of lesbians from that of heterosexual women, and so on. Fractured identities lead to fractured standpoints and, so it might seem, to the permanent multiplicity of partial narratives espoused by postmodernists. Feminist standpoint epistemology appeared to presuppose a kind of gender essentialism that was found to be no longer supportable. Although Harding insisted that all three types of feminist critique serve useful purposes, she seemed to suggest in 1986 that feminist postmodernism was the most sophisticated and theoretically adequate of the three.

This is the position she partially disavows in *Whose Science? Whose Knowledge?* (1991). Postmodern approaches have been unwelcome in some feminist circles for a variety of reasons. Perhaps chief among these is the belief that both feminist theory and feminist action require a fairly robust notion of objectivity. One wants to be able to say, for example, that masculinist accounts of female roles are *false* and ought to be replaced by accounts that are *objectively better*, not simply that there are many narratives available for a variety of purposes. Similarly, Harding argues in her most recent book, we must recognize that although science is politics by other means, it also generates reliable information about the empirical world. It has both progressive and regressive tendencies. An adequate feminist epistemology will have to take into account not just the political dimensions of science, but also its empirical successes: it will have to develop strategies for promoting the progressive tendencies of science while blocking its regressive tendencies. At the same time, it must never lose sight of the crucial point that the observer and the observed are in the same causal plane. The challenge for feminist epistemology is to articulate how it is that scientific knowledge is, in every respect, socially situated, without denying its considerable empirical success. One of the chief strategies employed by Harding to achieve this end is the reconceptualization of the relationship between the natural and the social sciences. Instead of seeing the social sciences as derivative from, and potentially reducible to, physics, she urges us to treat physics as a social science.

Rather than repudiate the notion of objectivity altogether, Harding criticizes the traditional conception for being too weak and calls for its replacement with the notion of *strong objectivity*. Although she recognizes and accepts descriptive relativism (it is true that different people have different belief systems), she rejects judgmental relativism as being simply the flip side of weak objectivity. The strong objectivity that Harding embraces extends the notion of scientific research to include systematic examination of cultural agendas and other powerful background beliefs that inform the scientific enterprise itself. It is in this context that the social sciences become paradigmatic; physics is just one human social activity among many others, and is amenable to investigation in the same way as other social activities.

Strong objectivity requires that we not only take into account the standpoint of the knower, but that we constantly question and analyze the assumptions that inform the standpoint that confers epistemic privilege. Harding's brush with postmodern feminism has left its mark. In fact, she describes her current position as a "postmodernist standpoint approach," where "postmodernist" (with lower-case *p*) describes any approach that fundamentally challenges the assumptions of Enlightenment epistemology, rather than a specific set of views about epistemology. She acknowledges that feminist standpoint theories tend to stress gender differences at the expense of ignoring other important differences, and she acknowledges that such theories contain an essentializing tendency. But, Harding argues, the logic of the standpoint approaches also contains the resources to combat these very same tendencies. To ground claims in women's lives is to ground them in differences "within women," as well as between women and men. And the same kinds of considerations that compelled us to theorize from the perspective of women's lives will also compel us to see the importance of theories created from the perspective of poor people, people of color, and others not represented in the current knowledge establishment. When we center the lives of lesbians, for example, we learn things we would not have learned otherwise.

139

The difficulty, of course, is to come up with an integrated theory; for this is clearly what Harding wants, particularly an account that successfully integrates gender, class, and race. And it is at this point that one may wonder in just what sense Harding's new position is correctly characterized as a standpoint theory after all. For we have now a multiplicity of standpoints, from which we must fashion an integrated theoretical account, one that strives toward strong objectivity. The standpoints are now just starting points; and no one of them possesses any ultimate epistemological privilege. So the sorting out, the adjudicating, the integrating of theories arising from radically different standpoints will have to proceed along lines not dictated by any one of these standpoints. In any event, it is clear that the vigor with which feminist epistemological projects are being pursued has caused the boundary of the old taxonomy (however useful it was) to give way. Harding's work has not only reflected, but also precipitated, much of the fast-moving debate in this arena.

Another prominent theorist whose views have recently undergone something of a shift is Evelyn Fox Keller, whose earlier work was sketched above. In *Secrets of Life, Secrets of Death* (1992), she relates that, although she stands by her earlier work, she finds it strategically impossible to proceed with her psychodynamic explorations of scientific postures. She also wishes to shift her focus away from the question of how science *represents* nature to an examination of the *force and efficacy* of its representations, not just with respect to gender, but more generally. In this transition, she has been largely influenced by such "interventionists" as Ian Hacking and Nancy Cartwright. Keller wishes now to focus on the constitutive role of language, studying the way in which it both reflects and guides the development of scientific models and methods. In doing so, she hopes to be able to shed some light on the logical and empirical constraints that make scientific claims so compelling. She is not tempted to join certain other science studies types in dethroning science; instead, she feels the need to explain its special efficacy, particularly in the case of physics.

In addition to Harding and Keller, the third prominent figure whose views deserve special mention is Helen Longino, author of *Science as Social Knowledge: Values and Objectivity in Scientific Inquiry* (1990). Longino's goal in this book is to develop an analysis of scientific knowledge that reconciles the objectivity of science with the role of contextual values in its social and cultural construction. Whereas *constitutive* values are generated from an understanding of the goals of science, and determine what constitutes acceptable scientific method and practice, *contextual* values belong to the larger social and cultural environment within which science is done. The task which Longino sets herself is to show how contextual values can play an important role even in "good" science, without thereby undermining the objectivity of the scientific enterprise. She calls her view "contextual empiricism," because it is empiricist in treating experience as the basis of scientific knowledge claims, and contextual in its insistence upon the relevance of contextual values to the construction of knowledge.

Longino stakes out her own position by contrasting it with the positions of positivists and holists. Although she agrees with the positivists that data can be specified independently of the hypotheses and theories for which they have evidential relevance, she takes issue with the positivist understanding of the nature of that evidential relationship. The crucial link in the argument is Longino's stress on the role played by background assumptions and beliefs in mediating the relationship between hypotheses and evidence. Given appropriately differing background beliefs, the same state of affairs can be taken as evidence for differing and even conflicting hypotheses. So two parties can *rationally* infer different conclusions from the same evidence. Furthermore, these mediating background assumptions will often introduce contextual values which cannot be eliminated without introducing constraints far too restrictive for the analysis of evidential relations in the actual practice of science.

The position does not deteriorate into holism, according to Longino, and there is no need to embrace incommensurability, because, however difficult it may be to ferret out background assumptions, they are articulable. And once articulated, they can be subjected to criticism. The critical function of scientific inquiry is heavily emphasized, because it is central to the argument that a form of scientific objectivity can be defended even when we recognize that contextual values permeate scientific inference.

Longino's strategy in outlining a modified account of objectivity is to treat scientific inquiry as a set of necessarily *social* practices, rather than as the disembodied application of a set of rules, or even as the mere sum of individual practices. Objectivity becomes on this account a characteristic of the community's practice of science, rather than a product of abstract methodology or a property of individual practice. One of the chief requirements of the community is that it attempts to articulate background assumptions and subject them to criticism. Because these background assumptions will typically incorporate nonempirical elements (including contextual values), it is essential that the critical function of the scientific community exhibit a *conceptual* as well as an empirical dimension.

What gives this book particular value, however, is Longino's ability to flesh out her argument by drawing on extensive case studies developed by herself and other feminist critics of science. These case studies give texture and substance to the foregoing rather abstract analysis. They don't simply illustrate the position; they constitute the best argument in its favor. They also begin to bridge the regrettable gap between "mainstream" philosophy of science and the feminist literature.

The case studies are chiefly meant to show how contextual values can affect the description of data, local background assumptions in a specific area of inquiry, and the global assumptions that set up a framework of inquiry. Two of the chief case studies involve, respectively, human evolutionary studies and behavioral endocrinology. The former is a further development of some of Longino's earlier work with Doell sketched above. Throughout her discussion, Longino is at pains to insist that she is not dismissing evolutionary theory or behavioral neuroendocrinology as "bad science" (in the sense of silly, sloppy, or fraudulent science). Instead, she is attempting to show how even "good" science may be permeated by contextual values.

In light of this, what should be the attitude of feminists toward science? What happens to the notion of a "feminist science"? Longino believes that if we focus on science as practice rather than content, "we can reach the idea of feminist science through that of doing science as a feminist" (1990, p. 188). This requires us to deliberately use background assumptions appropriately at variance with those of mainstream science. If, however, oppositional science is to be successful, it must always be *local*; and it must be respectful of some of the standards of the specific scientific community in question. Wholesale replacement of existing science by a "feminist paradigm" is not on the agenda proposed here.

141

The selection of Harding, Keller, and Longino to represent the second-order epistemological analyses of feminist research on science is not entirely arbitrary. Not only are they individually important, but collectively they represent a good part of the very wide range of feminist analyses of science.

References and further reading

Fausto-Sterling, A. 1985: *Myths of Gender: Biological Theories about Women and Men* (New York: Basic Books).

Haraway, D. 1989: *Primate Visions: Gender, Race, and Nature in the World of Modern Science* (New York: Routledge).

Harding, S. 1986: *The Science Question in Feminism* (Ithaca, NY: Cornell University Press).

—— 1991: *Whose Science? Whose Knowledge?* (Ithaca, NY: Cornell University Press).

Hubbard, R. 1990: *The Politics of Women's Biology* (New Brunswick, NJ and London: Rutgers University Press).

Keller, E. F. 1985: *Reflections on Gender and Science* (New Haven and London: Yale University Press).

—— 1992: *Secrets of Life, Secrets of Death* (London and New York: Routledge).

Longino, H. 1990: *Science as Social Knowledge: Values and Objectivity in Scientific Inquiry* (Princeton: Princeton University Press).

Longino, H., and Doell, R. 1983: Body, bias, and behaviour: a comparative analysis of reasoning in two areas of biological science. (This article and many others that are now considered classics first appeared in *Signs: Journal of Women in Culture and Society* between 1975 and 1987. They are conveniently collected in an anthology entitled *Sex and Scientific Inquiry*, ed. S. Harding and J. F. O'Barr (Chicago: University of Chicago Press, 1987).)

Merchant, C. 1980: *The Death of Nature: Women, Ecology, and the Scientific Revolution* (San Francisco: Harper & Row).

Okruhlik, K. 1992: Birth of a new physics of death of nature? In *Women and Reason*, ed. E. Harvey and K. Okruhlik (Ann Arbor, MI: University of Michigan Press).

—— 1994: Gender and the biological sciences. *Biology and Society, Canadian Journal of Philosophy*, Suppl. vol. 20, 21–42. Repr. in *Philosophy of Science: The Central Issues*, ed. M. Curd and J. Cover (New York: Norton), 192–208.

Schiebinger, L. 1989: *The Mind Has No Sex? Women in the Origins of Modern Science* (Cambridge, MA: Harvard University Press).

Wylie, A., Okruhlik, K., Morton, S., and Thielen-Wilson, L. 1990: Philosophical feminism: a bibliographic guide to critiques of science. *Resources for Feminist Research*, 19(2), 2–36.

21

Feyerabend

JOHN PRESTON

Paul K. Feyerabend (1924–94) was an imaginative maverick philosopher of science, a critic of positivism, as well as, more recently, falsificationism, philosophy of science itself, and of "rationalist" attempts to lay down or discover rules of scientific method.

In the 1950s, influenced by Karl Popper and Ludwig Wittgenstein, Feyerabend began a vigorous critique of logical empiricist philosophy of science, conducted through a study of observation and theory (see POPPER; LOGICAL EMPIRICISM; LOGICAL POSITIVISM; and OBSERVATION AND THEORY). He applied to the dispute over the interpretation of scientific theories (see THEORIES; also REALISM AND INSTRUMENTALISM) a strong measure of Popperian conventionalism (see CONVENTION, ROLE OF), arguing that this dispute is not a factual issue, but a matter of *choice*. We can choose to see theories either as descriptions of reality (scientific realism) or as instruments of prediction (instrumentalism), depending on what ideals of scientific knowledge we aspire to. These competing ideals (high factual content and sense certainty) are to be judged by their consequences. Stressing that philosophical theories have not merely reflected science but have *changed* it, Feyerabend held further that the form of our knowledge can be altered to fit our ideals. So we can have certainty, and theories that merely summarize experience, if we wish. But, mobilizing the equation between empirical content and testability, he urged that we should decisively reject certainty and opt instead for theories which go beyond experience and say something about reality itself.

He therefore argued that the idea, common to positivists, that the interpretation of observation terms doesn't depend upon the status of our theoretical knowledge has consequences undesirable to positivists. One of these is that "every positivistic observation language is based upon a metaphysical ontology" (Feyerabend 1981a, p. 21). Another follows from the thesis, which he relished, that the theories we hold influence our language and perceptions. This implies that as long as we use only one empirically adequate theory, we will be unable to imagine alternative accounts of reality. If we also accept the positivist view that our theories are summaries of experience, those theories will be void of empirical content and untestable, and hence there will be a diminution in the critical, argumentative function of our language. Just as purely transcendent metaphysical theories are unfalsifiable, so too what began as an all-embracing scientific theory offering certainty will, under these circumstances, have become an irrefutable dogma, a *myth*.

Feyerabend defended a realism according to which "the interpretation of a scientific theory depends upon nothing but the state of affairs it describes" (1981a, p. 42). At the same time he claimed to find in Wittgenstein's *Philosophical Investigations* a contextual theory of meaning according to which the meaning of terms is determined not by their use, nor by their connection with experience, but by the role they play in the wider context of a theory or an explanation. The key proposition of Feyerabend's early work, that "the interpretation of an observation language is determined by the theories which we use to explain what we observe, and it changes as soon as those theories change" (1981a, p. 31), is supposed to encapsulate both the contextual theory of meaning and scientific realism. Only realism, by insisting on interpreting theories in their most vulnerable form as universally quantified statements, leads to scientific progress, rather than stagnation, he argued. Only realism allows us to live up to the highest intellectual ideals of critical attitude, honesty, and testability.

Unlike positivism, which conflicts with science by taking experiences as unanalyzable building blocks, realism treats experiences as *analyzable*, explaining them as the results of processes not immediately accessible to observation. Experiences and observation statements are thus revealed as more complex and more structured than positivism had realized. Feyerabend overextended the contextual theory of meaning to apply not only to theoretical terms but to observation terms too, arguing that there is no special "problem" of theoretical entities, and that the distinction between observation terms and theoretical terms is a purely pragmatic one (see THEORETICAL TERMS). If, as the contextual theory also implies, observation statements depend on theoretical principles, any inadequacy in these principles will be transmitted to the observation statements, whence our beliefs about what is observed may be in error, and even our experiences can be criticized for giving only an approximate account of what is going on in reality. *All* our statements, beliefs, and experiences are "hypothetical." Observations and experiments need *interpretation*, different interpretations being supplied by different theories. If existing meanings embody theoretical principles, then instead of passively accepting observation statements, we should attempt to find and test the theoretical principles implicit in them, which may require us to *change* those meanings.

Feyerabend therefore idolized semantic *instability*, arguing that the semantic stability presupposed by positivist accounts of reduction, explanation, and confirmation has been, and should be, violated if we want progress in science. If meaning is determined by theory, terms in very different theories can't share the same meaning. Any attempt to derive the principles of an old theory from those of a new one must either be unsuccessful or must effect a change in the meaning of the terms of the old theory. The "theoretical reduction" beloved of empiricists is therefore actually more like replacement of one theory and its ontology by another. Feyerabend concluded that semantic instability precluded any formal account of explanation, reduction, or confirmation.

His most important argument for realism was methodological: realism is desirable because it demands the proliferation of new and incompatible theories. This leads to scientific progress, because it results in each theory having more empirical content than it would otherwise have, since a theory's testability is proportional to the number of potential falsifiers it has, and the production of alternative theories is the only way to ensure the existence of potential falsifiers. So scientific progress comes through *theoretical pluralism*, through allowing a plurality of incompatible theories, each of which will contribute by competition to maintaining and enhancing the testability, and hence the empirical content, of the others. According to Feyerabend's pluralistic criterion, theories are tested against one another. He thus idealized what T. S. Kuhn called

144

"pre-paradigm" periods and scientific revolutions, when there are many incompatible theories, all forced to develop through their competition with each other (see KUHN and SCIENTIFIC CHANGE). But he downplayed the fact that theories are compared with one another *primarily* for their ability to account for the results of observation and experiment.

Thus far, the argument for theoretical pluralism follows that of John Stuart Mill (see MILL). But Feyerabend went on to try to demonstrate a *mechanism* whereby theories can augment their empirical content. According to this part of the argument, theories may face difficulties which can be discovered only with the help of alternative theories. A theory can be incorrect without our being able to discover this in a direct way: sometimes the construction of new experimental methods and instruments which would reveal the incorrectness is excluded by laws of nature; sometimes the discrepancy (were it to be discovered) might be regarded as an oddity, and might never be given its correct interpretation. Circumstances can thus conspire to hide from us the infirmities of our theory. The "principle of testability" demands that we develop alternative theories incompatible with the existing theory, and develop them in their strongest form, as descriptions of reality, not mere instruments of prediction. Instead of waiting until the current theory gets into difficulties, and only then starting to look for alternatives, we ought vigorously to proliferate theories and tenaciously defend them in the hope that they may afford us an *indirect refutation* of our existing theory. Only theories which are empirically adequate will thus contribute to raising the empirical content of their fellows. But Feyerabend argues that *any* theory, no matter how weak, may become empirically adequate, and so may contribute to this process. To be a realist, he therefore suggests, involves demanding support for any theory, including implausible conjectures with no independent empirical support which are inconsistent with data and well-confirmed laws. We should retain theories that are in trouble, and invent and develop theories that contradict the observed phenomena, just because in doing so we will be respecting the intellectual ideal of testability.

In thus appealing to the "principle of testability" as the supreme methodological maxim, Feyerabend forgot that testability must be traded off against other theoretical virtues. Only his pathological fear of theories losing their empirical content and becoming myths led him to want to maximize testability and embrace an unrestricted principle of proliferation. He also disregarded historical evidence that anti-realist approaches can be just as pluralistic as realism.

Another consequence of the contextual approach to meaning that emerged gradually was the thesis of incommensurability, developed in harmony with Kuhn (see INCOMMENSURABILITY). In Feyerabend's version, the semantic principles of construction underpinning a theory in its realist interpretation can be violated or "suspended" by another theory. As a result, theories cannot always be compared with respect to their content, as "rationalists" would like. This opens the door to relativism, the thesis that there is no objective way of choosing between theories or traditions (see RELATIVISM).

By the late 1960s, Feyerabend was ready to fly the falsificationist coop and expound his own perspective on scientific methodology. His *tour de force*, the 1975 book *Against Method*, which got him branded an "irrationalist", contained most of the themes mentioned so far, sprinkled into a case study of the transition from geocentric to heliocentric astronomy. He emphasized that older scientific theories, like Aristotle's theory of motion, had powerful empirical and argumentative support, and he stressed, correlatively, that the heroes of the scientific revolution, such as Galileo, were not as scrupulous as they were sometimes represented to be (see GALILEO). He also sought further to downgrade the importance of empirical arguments by suggesting that aesthetic criteria, personal whims, and social factors have a far more decisive role in the history of science than rationalist or empiricist historiography would indicate (see SOCIAL FACTORS IN SCIENCE).

Against Method explicitly drew the "epistemological anarchist" conclusion that there are no useful and exceptionless methodological rules governing the progress of science or the growth of knowledge. The history of science is so complex that if we insist on a general methodology which will not inhibit progress, the only "rule" it will contain will be "Anything goes." In particular, logical empiricist methodologies and Popper's critical rationalism would inhibit scientific progress by enforcing restrictive conditions on new theories. The "methodology of scientific research programmes" developed by Imre Lakatos either contains ungrounded value judgments about what constitutes good science, or is reasonable only because it is epistemological anarchism in disguise (see LAKATOS). The phenomenon of incommensurability renders the standards which these "rationalists" use for comparing theories inapplicable.

Feyerabend thus saw himself as having undermined the arguments for science's privileged position within culture, and much of his later work was a critique of the position of science within Western societies. Because there is no scientific method, we can't justify science as the best way of acquiring knowledge. And the *results* of science don't prove its excellence, since these results have often depended on the presence of nonscientific elements; science prevails only because "the show has been rigged in its favour" (Feyerabend 1978, p. 102), and other traditions, despite their achievements, have never been given a chance. The truth, he suggested, is that

> science is much closer to myth than a scientific philosophy is prepared to admit. It is one of the many forms of thought that have been developed by man, and not necessarily the best. It is conspicuous, noisy, and impudent, but it is inherently superior only for those who have already decided in favour of a certain ideology, or who have accepted it without ever having examined its advantages and its limits. (Feyerabend 1975, p. 295)

The separation of church and state should therefore be supplemented by the separation of science and state, in order for us to achieve the humanity of which we are capable. Setting up the ideal of a free society as "a society in which all traditions have equal rights and equal access to the centres of power" (Feyerabend 1978, p. 9), Feyerabend argues that science is a threat to democracy. To defend society against science, we should place science under democratic control, and be intensely skeptical about scientific "experts," consulting them only if they are controlled democratically by juries of laypeople.

Although the focus of philosophy of science has moved away from interest in scientific methodology in recent years, this may not be due to the acceptance of Feyerabend's argument. That argument has been criticized as resting on a failure to appreciate that since methodological rules are normative, one can't show their nonexistence by demonstrating that they have occasionally been profitably violated (Newton-Smith 1981, ch. 6). It's not simply ironic that one who once made so much appeal to methodo-logical principles and ideals should later seek to show that methodology doesn't exist: Feyerabend's skeptical conclusions bear the mark of his initial misconceptions about methodological rules, ideals, and the epistemology of science.

Feyerabend came to be seen as a leading cultural relativist, not just because he stressed that some theories are incommensurable, but also because he defended rela-tivism in politics as well as in epistemology. Throughout the 1980s, for example, he espoused what he called "Democratic relativism," the normative ethical view that all traditions should be given equal rights and equal access to power. In his last work, however, while remaining enamored of what he perceived as the tolerant spirit of relativism, he repudiated the "conversion philosophy" with which he had become associated, expressing reservations about the relativist presupposition that theories and cultures are "closed" domains. These works also saw him working towards a social constructivist alternative to the metaphysics of scientific realism. Nevertheless, his lengthy critique of "the rise of Western Rationalism" still issued in the conclusion that orthodox accounts of the triumph of science and reason involved a con trick, since reason triumphed by taking advantage of preexisting social processes, not because its arguments were good enough to refute previous world views.

His denunciations of Western imperialism, his critique of science itself, his conclu-sion that "objectively" there may be nothing to choose between the claims of science and those of astrology, voodoo, and alternative medicine, as well as his concern for environmental issues, ensured that Paul Feyerabend has become a hero of the anti-technological counterculture.

References and further reading

Works by Feyerabend
1965: Problems of empiricism. In *Beyond the Edge of Certainty: Essays in Contemporary Science and Philosophy*, ed. R. G. Colodny (Englewood Cliffs, NJ: Prentice-Hall), 145–260.
1975: *Against Method* (London: Verso).
1978: *Science in a Free Society* (London: New Left Books).
1981a: *Philosophical Papers*, vol. 1: *Realism, Rationalism, and Scientific Method* (Cambridge: Cambridge University Press).
1981b: *Philosophical Papers*, vol. 2: *Problems of Empiricism* (Cambridge: Cambridge University Press).
1987: *Farewell to Reason* (London: Verso/New Left Books).
1991: *Three Dialogues on Knowledge* (Oxford: Basil Blackwell).
1993: *Against Method*, 3rd edn (London: Verso).
1995: *Killing Time: The Autobiography of Paul Feyerabend* (Chicago: University of Chicago Press).
1999: *Philosophical Papers*, vol. 3: *Knowledge, Science and Relativism*, ed. J. Preston (Cambridge: Cambridge University Press).
2000: *The Conquest of Abundance* (Chicago: University of Chicago Press).

Works by other authors
Burian, R. M. 1984: Scientific realism and incommensurability: some criticisms of Kuhn and Feyerabend. In *Methodology, Metaphysics and the History of Science*, ed. R. S. Cohen and M. W. Wartofsky (Dordrecht: D. Reidel), 1–31.

Churchland, P. M. 1979: *Scientific Realism and the Plasticity of Mind* (Cambridge: Cambridge University Press).

Couvalis, S. G. 1989: *Feyerabend's Critique of Foundationalism* (Aldershot: Avebury Press).

Machamer, P. K. 1973: Feyerabend and Galileo: the interaction of theories, and the reinterpretation of experience. *Studies in History and Philosophy of Science*, 4, 1–46.

Maia Neto, J. R. 1991: Feyerabend's scepticism. *Studies in History and Philosophy of Science*, 22, 543–55.

McEvoy, J. G. 1975: A "revolutionary" philosophy of science: Feyerabend and the degeneration of critical rationalism into sceptical fallibilism. *Philosophy of Science*, 42, 49–66.

Munévar, G. (ed.) 1991: *Beyond Reason: Essays on the Philosophy of Paul Feyerabend* (Dordrecht: Kluwer).

Newton-Smith, W. H. 1981: *The Rationality of Science* (London: Routledge and Kegan Paul).

Preston, J. M. 1997: *Feyerabend: Philosophy, Science and Society* (Cambridge: Polity Press).

Preston, J. M., Munévar, G., and Lamb, D. 2000: *The Worst Enemy of Science? Essays in Memory of Paul Feyerabend* (New York: Oxford University Press).

Zahar, E. 1982: Feyerabend on observation and empirical content. *British Journal for the Philosophy of Science*, 33, 397–409.

22

Galileo

ROBERT E. BUTTS

Galileo Galilei was born at Pisa in Italy on 18 February 1564 and died at Arcetri, near Florence, on 8 January 1642. He excelled in observational and theoretical astronomy, natural philosophy, and applied science. An outstanding theoretical and experimental physicist, he is perhaps best known for his defense of the Copernican heliocentric theory in astronomy, and for his humiliating treatment at the hands of the Catholic Inquisition, following the papal condemnation (23 February 1616) of heliocentrism as heretical and at odds with biblical teaching. Forced to recant his Copernican convictions, Galileo spent the last years of his life under house arrest in Arcetri. Even though completely blind and continually harassed by his enemies, in his last years he completed his *Discorsi e Dimonstrazioni matematiche intorno a due nouve scienze* (1638), a work that created modern mechanics.

The philosophical importance of Galileo's science rests largely upon the following closely related achievements: (1) his stunningly successful arguments against Aristotelian science; (2) his proofs that mathematics is applicable to the real world; (3) his conceptually powerful use of experiments, both actual and employed regulatively; (4) his treatment of causality, replacing appeal to hypothesized natural ends with a quest for efficient causes; and (5) his unwavering confidence in the new style of theorizing that would come to be known as mechanical explanation.

As a student, Galileo was confronted by a medieval Aristotelian cosmology consisting of an interwoven fabric of conceptually comforting speculative hypotheses. In time, especially with the appearance of his *Dialogo sopra i due Massimi Sistemi del Mondo* (1632), he would tear that fabric to shreds. The received cosmology had something for everybody: the magic of the motions of the heavenly bodies and the assurance – for humans – that God's in his heaven and all's well with the world. The Earth, the heaviest stone in the universe, is in its natural place at the center. Earthlike objects moving near the surface of the Earth endeavor to get as close to the center as possible. Why? They seek their natural places, much as the lover seeks to be one with his beloved; thus do Aristotelians explain accelerated motions. Above, the fixed stars give eternal guidance, and the heavenly bodies are constituted of a subtle matter that yields celestial motions as distinct in kind from terrestrial ones.

The universe thus conceived is created for the pleasure and temptation of human beings. It was also accorded official status by papal decree. The tentativeness of Copernicus and the fate of Bruno are easy to comprehend. Galileo's courage stands out as an emblem of the nobility of the scientist. Even before 1632, his disenchantment with Aristotelianism began to take form. In a letter to Kepler (1597) he forthrightly stated his acceptance of Copernicanism. In *Sidereus Nuncius* (1610) he recorded his telescopic observations: the mountains on the Moon, numerous stars not seen by the naked eye, four of Jupiter's satellites (the Medicean stars), the form of the Milky Way. Later, he discovered the phases of Venus, the existence of sunspots, and the composite structure of Saturn. Perhaps the telescopic observations alone constituted sufficient reason to abandon the distinction between terrestrial and celestial motions, the concept of fixed stars, and the geocentric model of the universe.

However, these observational results alone could not convince the Aristotelians, whose trust of naked-eye perceptions is a key feature of their epistemology. A new instrument might, after all, provide only a magic show; and where, indeed, are the images that one sees through a telescope? Seeing may be believing; but, perhaps paradoxically, what one believes is not necessarily what one sees. (Some who saw the mountains on the moon through Galileo's telescope believed that the image they perceived was *in* the instrument.) This is why the *Dialogo* is such an important part of the program of Galileo, for it is in this work that he provides his arguments in support of the new cosmology and the new ways of seeing. The arguments are for the most part attempts to reduce the main tenets of medieval Aristotelianism to absurdity by disclosing hitherto undetected logical flaws. Galileo's logic may not be persuasive, but as propaganda, his arguments are breathtaking (Butts 1978; Feyerabend 1970).

The *Dialogo* is dominated by argumentative (often sophistical) exchanges between Salviati (who represents Galileo) and Simplicio (who represents the prevailing Aristotelianism). The main effort of the work is to demonstrate that mathematics is true not only formally and abstractly, as Aristotle held, but is also true when applied to the material world. At one place Salviati challenges Simplicio by insisting that the geometrical truth – a plane struck on a tangent with a sphere touches that sphere in only one point – is not just true abstractly, but is a truth about actual physical objects. Simplicio, trained in the commonsense method that relies heavily on naked-eye observation, replies that the theorem may be true in geometry, but it cannot be true of real spheres and planes, because a bronze sphere and a steel plate will touch not in only one point, but along several points of their surfaces. Salviati now makes an argumentative move that will become one of the mainstays of his methodology.

Suppose, he says, that we begin with a geometrically perfect sphere and plane. If, when they touch, they do so at more than a single point, they cannot, during the operation of touching, have remained perfect. But this is true both of physical objects and abstract, geometrically defined objects. For if, even considered abstractly, the sphere and the plane are not perfect, they will touch at more than one point (Galileo 1953, p. 207). Now the twist of epistemological irony: these considerations ought not to lead us to conclude that the geometrical theorem is false, for what we do to realize the truth of the theorem in application to imperfect objects is to find ways to render those objects perfect. Later physicists and philosophers of science will repeat this point, but in more precise language, when they urge that any geometry may be preserved by making suitable adjustments in relevantly associated physical laws (e.g., see Grünbaum 1973, pp. 131–2).

What we do to preserve the geometry, thought Galileo, is "deduct the material hindrances" (1953, pp. 207–10). Removing the impediments preserves the geometry, proving that what is true in geometry is true in the material world, and securing the applicability of mathematics to that world. In the case of the theorem at issue, we would presumably have to find experimental ways of discounting the impediments, ways of accommodating the respects in which a certain material sphere and a certain material plane deviate from these kinds of objects as defined in the geometry. Since we have an enormous amount of direct evidence that objects and events in nature deviate in striking ways from what is true in geometry, the thesis of deducting the material hindrances becomes a central postulate of Galileo's methodology, one that is applied by him repeatedly.

Perhaps the most striking example of Galileo's employment of the postulate of deducting the material hindrances is his discussion of the law of free fall in the *Discorsi*. The law is now well known. All physical bodies in a state of free fall moving in the direction of the surface of the Earth accelerate at the same rate. This is true regardless of differences in weight, the property which the Aristotelians had emphasized; yet pennies are seen to get to the surface of the Earth faster than feathers. Just as the geometrical theorems can be held to be true by deducting the material hindrances, so also the law of equal acceleration, despite naked-eye evidence to the contrary, can be shown to be true by removing physical impediments. The composition and form of an object, as well as the medium through which it moves, suffer hindrances of several kinds: resistance, specific weight, shape of the moving body, and contact between the surface of the moving body and the fluid medium. The effects of these and other hindrances in given cases should have led Galileo to the conclusion that the law of equal acceleration is false. Instead, he argued as follows, employing the postulate of deducting the material hindrances.

He observed that as the medium through which a physical object moves becomes less dense, the more the movement of the object conforms to the law of free fall. Thus, given two physical objects of different weights and a medium through which they are moving whose density approximates zero, the two objects will accelerate at the same rate. A penny and a feather get to the bottom of an evacuated vacuum tube at exactly the same time. Galileo, characteristically, accepted the law of equal acceleration even though he thought a vacuum impossible, and had no means of direct experimental confirmation of the law. His confidence in the applicability of mathematics and in the complete reliability of the procedures by means of which we can deduct the material hindrances led him to accept conclusions only confirmed by later scientific work.

Galileo's science totally abandons the *details* of Aristotelian cosmology, but not the Aristotelian ideal of science as demonstrative (McMullin 1978). Nevertheless, we are no longer to regard the senses as providing reliable information. Experimentation is introduced to confirm or refute what has been deduced by geometry. This may seem to commit Galileo to acceptance of the hypothetico-deductive method pure and simple. The acceptance may be pure, but it is certainly not simple. In his polemical work of 1623, *Il Saggiatore*, Galileo had distinguished between those qualities – colors, tastes, pains, in short, sensations – whose very existence depends upon receptor organs and upon consciousness, and those qualities – shape, place, motion, in short, mathematical properties having determinable and exact characters – which are the causes of the sensory qualities. The realities of the universe are thus numbers, or, more precisely, physical objects having exactly determinable numerical characteristics. The hypothesized laws in Galileo's hypothetico-deductive explanations will ideally be propositions derived from the reduction of certain kinds of events and objects to mathematized physical objects (which may in some explanations be unperceived or unperceivable). The propositions will also refer to causes. Galileo despised appeals to occult causes, and he had no room for teleological considerations in his physics. But he certainly sought causes, efficient causes, causes productive of the bringing about of events and objects, causes that function in mechanical explanations: pushes and pulls.

We have seen, however, that in his use of the postulate of deducting the material hindrances, Galileo was not so much arguing for the success of an experimental confirmation of the law of equal acceleration as he was experimentally *preparing* the way for testing by showing how accelerated motions can be accommodated in mathematics. William Shea (1972, pp. 159–63) refers to this second use of experiment in Galileo as a *regulative* use. Experimentation in this sense does not involve repetition of various conditions in the effort to produce (inductively) theories. Rather, such experimentation (if often only in thought) reveals principles useful in a physical (mathematical) *interpretation* of nature. Regulative experiments provide conceptual context for framing hypotheses in ways exact enough to allow the application of the hypotheses to actual instances. Experimental testing, in the ordinary sense, results from hypothesizing based upon principles of mathematization, chief among them the postulate of deducting the material hindrances.

Galileo's use of the hypothetico-deductive method should be understood as supplemented by deeper metaphysical commitments, especially his commitment to the necessary truths of mathematized scientific experience. What legitimizes the metaphysical regulation of nature which Galileo proposed? Not the word of God; for Galileo, there is no theology of science. Not a fully articulated metaphysical system; for that, we will have to wait for Descartes, Leibniz, Kant, and others. Galileo's methodological practice discovered some exact laws of nature and some technological applications as well. What better credentials can a scientist ask for? The proof of the program is its success.

References and further reading

Works by Galileo
1953: *Dialogo sopra i due Massimi Sistemi del Mondo* (Florence, 1632); trans. by Stillman Drake as *Dialogue Concerning the Two Chief World Systems* (Berkeley and Los Angeles: University of California Press).

1954: *Discorsi e Dimonstrazioni matematiche intorno a due nuove scienze* (Leiden, 1638); trans. by Henry Crew and Alfonso de Salvio as *Dialogues Concerning Two New Sciences* (New York: Dover).

1957a: *Il Saggiatore* (Rome, 1623); trans. by Stillman Drake as "The assayer." In *Discoveries and Opinions of Galileo* (New York: Doubleday), 229–80.

1957b: *Sidereus Nuncius* (Venice, 1610); trans. by Stillman Drake as "The starry messenger." In *Discoveries and Opinions of Galileo* (New York: Doubleday), 21–58.

Works by other authors
Butts, R. E. 1978: Some tactics in Galileo's propaganda for the mathematization of scientific experience. In *New Perspectives on Galileo*, ed. R. E. Butts and J. C. Pitt, University of Western Ontario Series in Philosophy of Science, 14 (Dordrecht: D. Reidel), 59–85.

Feyerabend, P. 1970: Problems of empiricism, part II. In *The Nature and Function of Scientific Theories*, ed. R. G. Colodny (Pittsburgh: University of Pittsburgh Press), 275–353.

Grünbaum, A. 1973: *Philosophical Problems of Space and Time*, 2nd enlarged edn, Boston Series in the Philosophy of Science, 12 (Dordrecht: D. Reidel).

Koyré, A. 1939: *Études galiléennes* (Paris: Hermann).

McMullin, E. 1978: The conception of science in Galileo's work. In *New Perspectives on Galileo*, ed. R. E. Butts and J. C. Pitt, University of Western Ontario Series in Philosophy of Science, 14 (Dordrecht: D. Reidel), 209–57.

Shea, W. R. 1972: *Galileo's Intellectual Revolution* (New York: Science History Publications).

23

History, Role in the Philosophy of Science

BRENDAN LARVOR

The leading philosophers of science of the first half of the twentieth century had little use for the history of science. There are several possible explanations for this. One is that philosophers of science sometimes (knowingly or not) mimic the methodological habits and values of scientists. Many philosophers of science are motivated by admiration for the perceived rigor and intellectual hygiene of the exact sciences. Historical sense is not normally a cardinal virtue among physicists. Hence, those philosophers who take their methodological cues from their scientific heroes are unlikely to think of philosophy as a historical discipline. Indeed, the idea of a close relationship between history and philosophy may have fallen into disrepute as a consequence of its association with idealist philosophers, of whom Hegel is the most notorious. Then there is Feyerabend's explanation, which is that the philosophy of science was transformed by the spectacular advances in formal logic which took place in the late nineteenth century. The development of modern formal logic made possible a range of hitherto inconceivable projects in the philosophy of science. The most radical of these was the positivist attempt to replace philosophy with the logical analysis of scientific language. Equally dependent on the new logic was Popper's attempt to explain the workings of empirical science using only deductive patterns of inference. According to Feyerabend, the philosophical projects engendered by the advances in formal logic came to dominate so thoroughly that everything else was marginalized (see FEYERABEND). It was not necessary for philosophers to know anything about the history of science, but they did have to be competent logicians: "As always maturity in a narrow domain means illiteracy elsewhere" (Feyerabend 1981, p. 20).

In the period since the Second World War, philosophers have gradually abandoned their indifference to the history of science. The most dramatic moment of this reversal was the publication in 1962 of Thomas Kuhn's *The Structure of Scientific Revolutions* (see KUHN). Kuhn seemed to claim that most scientists most of the time are unthinking drones, mindlessly copying the latest paradigm as closely as possible. Kuhnian "normal science" seemed like painting by numbers. Moreover, Kuhn suggested that the change from one paradigm to the next is not a reasoned transition, but is rather a "Gestalt switch" induced by a crisis of confidence. Worse still, Kuhn robbed science of its moral authority. Many philosophers had been attracted by Popper's description of the scientific community as an open society of rational mutual critics. In Kuhn's version, the scientific establishment constantly rewrites the textbooks so that history always appears to lead directly to the present orthodoxies, even as those orthodoxies change. He readily acknowledged that his work suggested that "the member of a mature scientific community is, like the typical member of Orwell's *1984*, the victim of a history rewritten by the powers that be" (1970, p. 167). Many philosophers were outraged by what they (mistakenly) took to be Kuhn's authoritarian relativism, and hastened to defend science against it. The task was particularly urgent, as Kuhn's book had been seized upon by philosophers and critics anxious to debunk the alleged objectivity of science. Unfortunately for the defenders of science, Kuhn could not be dismissed as a crank. He was a trained physicist turned historian with a respectable record of publications in the history of the empirical sciences. *The Structure of Scientific Revolutions* is full of historical illustrations which philosophers ignorant of the history of science were unable to challenge. Philosophers had to take an interest in history if they hoped to defeat Kuhn's heresy (see KUHN).

The impact of *The Structure of Scientific Revolutions* was magnified by the fact that it ran into a body of literature generated by philosophical attempts to articulate the essence of science. Kuhn's work was a contribution to this "demarcation problem," but it seemed to contest the presuppositions of most other contributions – namely, that mature science is the crowning achievement of human rationality and is, morally, a good thing. The Popperian school in particular understood the demarcation problem to be that of finding a principled distinction between rational, virtuous science and irrational dogma sustained by brute force. The relationship between the history of science and the demarcation question was most thoroughly addressed as a theoretical question by one of Popper's ablest students and sharpest critics, Imre Lakatos (see POPPER and LAKATOS).

The simplest model of this relationship is one in which the history of science serves as a source of counterexamples with which to test putative characterizations of science. For example, if Popper is right about the nature of science, then its history should be full of bold conjectures and "crucial experiments." So we can check Popper's theory against the archives. However, the relationship between theory and evidence is always complex and delicate. As Lakatos pointed out, this is as true of philosophical theories and historical evidence as it is of physical theories and empirical evidence. Philosophers (including Popper) abandoned simple falsificationism as a logic of science, because they realized that any decently rich theory can be kept free from falsification by deft logical footwork. The central claim of the theory can be protected by means of carefully chosen "auxiliary hypotheses." For example, a theory which calculates the mass of the universe to be ten times greater than the figure suggested by observation might seem to be decisively refuted. All its defenders need do, however, is to postulate the existence of "dark matter." With the aid of this auxiliary hypothesis, the theory is saved – for the moment. The same point holds true for philosophical characterizations of science and the relevant historical evidence. Philosophers can save their theories from refutation by selecting auxiliary hypotheses to explain away recalcitrant evidence. In particular, they can exploit the fact that textual records are the chief form of evidence for historians of science. Shades of meaning can always be distinguished. In a contest between interpretations, it is usually possible to force a draw because the evidence (the books, diaries, notes, and so on) is finite, whereas distinctions of meaning can be sliced ever finer and finer.

The fact that philosophical characterizations of science can usually be saved from direct historical refutation by the use of auxiliary hypotheses does not imply that they are all of equal value. For Lakatos, each characterization of science should be treated as the organizing principle of a historical "research program." Popperians should write Popperian histories of science, inductivists should write inductivist histories, and so on. These philosophically inspired histories may then be evaluated and compared for their plausibility and coherence. The original philosophical theories then inherit the evaluation placed upon their associated accounts of the history of science. However, in addition to the usual standards for judging histories, there is a further comparison, which concerns the fact that a putative solution to the demarcation question is also a criterion of rationality. Suppose (for example) that a falsificationist philosopher of science, armed with her preferred solution to the demarcation question, is writing a history of science. Suppose that she encounters a historical episode which does not, at first sight, meet her criterion of the scientific. For a falsificationist, such an episode would be one in which a historical scientist stuck with a theory which he knew to be refuted. But the philosopher has a 'choice. She could reinterpret the episode so that it did meet her standards after all. She could (for example) argue that the scientist did not stick with his refuted theory, in spite of appearances, because the meanings of his theoretical terms changed. He seems to have ignored the refutation because he went on using the same words, but they expressed a different theory after the refutation than they did before. Hence the historical scientist is not guilty of irrationality by falsificationist lights, and falsificationism survives as a criterion of the scientific.

Such careful rereadings of the archives in order to protect a philosophical thesis vary in their plausibility, and the philosopher should beware. She does her theory no favor if she tries to bolster it with strained and unlikely reinterpretations of hitherto unproblematic texts. If the episode cannot be plausibly reinterpreted to fit the philosopher's template, then she should admit as much. She still has options, one of which is to appeal to the normativity of her theory. That is, she can argue that the demarcation question calls for a characterization of science as an ideal, not as an existing institution. Falsificationism, inductivism, and similar theories are prescriptions for rationality. They attempt to explain how people *ought* to behave if they wish to be properly scientific. Methodologists need not claim that all scientists always behave scientifically. However, the danger with this move is that the episode in question may be an illustrious scientific success. It would be hubris for a philosopher to argue that Newton was not doing science properly when he produced his most famous results. Note that if our philosopher appealed to normativity, her argument would not be that Newton had false methodological beliefs. It may often be plausible to claim that great scientists misunderstand the logic of what they are doing. Rather, she would have to assert that Newton's actual practice was mistaken. But that cannot be correct. He must have been doing something right to produce such a successful theory.

If some historical episode does not conform to the philosopher's normative description of good science, then that episode is irrational by her lights. If the episode in question is a great moment in scientific history, then the fact that her theory condemns it counts against the philosophical theory. Thus, according to Lakatos, philosophers should try to produce characterizations of science which accommodate as many of the agreed successes of science as possible. In other words, every solution to the demarcation problem divides the history of science into rational, scientific episodes and irrational, unscientific ones. In developing a criterion of the scientific, the philosopher should maximize the number of agreed scientific success stories which are judged to be rational by her lights. Every theory of scientific rationality suffers some exceptions, for Lakatos. Therefore, every philosophically inspired history of science which explains events by fitting them into a chosen logical form (such as inductive generalization or Popperian falsification) must be complemented by a causal account of what went wrong in those episodes which fail to conform to the philosopher's preferred pattern.

Lakatos's account of the relationship between the demarcation question and the history of science required a change in philosophical methodology. Philosophy was not, for him, an a priori exercise in conceptual analysis or the discernment of necessary truths. On his account, the testing of a philosophical theory requires empirical historical research. Philosophy (or, at least, the demarcation question) had become a matter of theorizing the established judgments of scientific success and failure. Lakatos's philosophy was intended to be a reasoned exchange between methodological theory and the best scientific practice. The traffic was not all one way, however. History had something to learn from philosophy. Lakatos claimed that "all historians of science *who hold that the progress of science is progress in objective knowledge*, use, willy-nilly, some rational reconstruction" (1978, p. 192; emphasis original). A "rational reconstruction" is a historical narrative in which events are explained by reference to some methodological pattern (such as falsification or induction). The argument is that all historical writing has some theoretical bias – somehow historians must identify periods, problems, and explanatory motifs. A historian of science who regards the history of science as largely a story of advancing knowledge (rather than as a chain of esoteric texts with no special epistemic status) must be committed to some normative conception of the scientific. Her theoretical bias must include some (probably implicit) answer to the demarcation problem.

If Lakatos is right, it sounds as if the history and philosophy of science are two aspects of a single subject. All it needs for this to become explicit is for philosophers to learn to do philosophy by writing case studies in the history of science, and for historians to become more self-aware regarding their theoretical commitments. It will then be possible for the history and philosophy of science to become a single seamless discipline, with no hard distinction between historical fact and philosophical theory (just as there is no hard distinction between theory and observation in physics). Indeed, there are already departments, societies, and journals for the history and philosophy of science. The thought that history and philosophy are at bottom one and the same is not new; but is it true? And even if it is true, would an institutional union of history and philosophy be well advised?

In the opening chapter of *The Essential Tension* (1977), Kuhn suggested that the answer to both questions is "No." He argued the point by reporting his experience of leading seminars on classic works in science and philosophy attended by graduate students of history and philosophy. The philosophers and historians were equally diligent; but, said Kuhn, "it was often difficult to believe that both had been engaged with the same texts" (1977, p. 6). The philosophy students tended to produce readings which depended on subtle distinctions which were invisible to the history students for the very good reason that they were not present in the text. The introduction of these subtleties exposed logical gaps in the original argument which the philosophers filled without realizing it. The philosophy students were often surprised when the historians showed them that neither the gap-exposing distinctions nor the gap-filling improvements were present in the original. As a result, "The Galileo or Descartes who appeared in the philosopher's papers was a better scientist or philosopher but a less plausible seventeenth-century figure than the figure presented by the historians" (1977, p. 7). Kuhn noticed other differences too: the historians drew on a wider range of supplementary material, and their essays were longer, but less tightly argued, than those of the philosophers.

The explanation for these differences is clear: philosophy and history have different goals. Historians want to understand how this individual (or group) came to believe that specific theory. Philosophers tend to be interested in the merits of the theory as such, so it is entirely natural for them to construct a rigorous, decontextualized version of whatever they are asked to discuss. In other words, both historians and philosophers focus on "the essentials" of the text in hand, but their judgments of salience differ widely. Kuhn maintained that the differing goals of history and philosophy require mutually exclusive intellectual habits and attitudes. One may alternate between history and philosophy, but one cannot do both at once. Students educated in a single discipline – historico-philosophical studies – would acquire neither the analytical acumen of the philosopher nor the narrative skills of the historian. Consequently, said Kuhn, history and philosophy should cooperate, but not coalesce.

Kuhn first delivered that chapter as a paper in 1968, and it is likely that philosophers of science are less analytic and unhistorical now than they were then, but the differing approaches and results of history and philosophy are still recognizable from his description. One of the most striking differences is that philosophers tend to regard all intellectual products as theories, or, failing that, as evidence for theories. That is the origin of Lakatos's claim that all history is written with some theoretical bias. He is half right: all history is written with some bias, but it need not be a *theoretical* bias. By way of analogy, consider another case of the philosopher's tendency to see all thought as theory, that of so-called folk psychology. "Folk psychology" is the everyday understanding of other people that we use to navigate social life. Some philosophers like to compare it unfavorably with psychological and social theory. To see the mistake in that comparison, consider the following argument: If you make use of a certain conception of human nature in order to explain one behavioral episode, then all your future explanations should be consistent with that conception because, by using it, you commit yourself to it. This is a good argument which most people quite properly ignore most of the time. The reason is that coping with other people is far more important than maintaining a consistent theoretical base. Eclecticism is a small price to pay for a successful social life.

Most historians take a similar attitude. Other things being equal, they would of course prefer their work to have a defensible historiographic basis, but such metahistorical concerns have a very low priority compared with their primary business of producing plausible narrative explanations. Any historian who seeks to raise the priority of historiographic consistency is in danger of abandoning history in favor of philosophy. Lakatos is right to say that a historian of science as knowledge must be informed by some normative notion of the scientific; but this notion need not be a latent theory of scientific rationality. Lakatos convinced himself otherwise by treating historians of science in just the way that Kuhn's philosophy students treated the intellectual giants of the early modern period. Lakatos's claim is not true of many real historians, but it may be true of philosophically idealized historians.

These observations explain why historians are normally unimpressed by philosophers' attempts to write history. The scorn of historians often leaves the philosophers bemused, because they imagine that good history writing means not fiddling with the evidence or ignoring salient facts. Basic intellectual honesty is of course a necessary condition, but it is hardly sufficient. What prevents most philosophers from writing good history is their inability to judge a narrative on its own terms, without asking what philosophical lessons can be drawn from it. Few philosophers can resist the temptation to raid the archives for evidence and illustrations of their favorite philosophical theses. Sometimes, philosopher-historians present their philosophical doctrines as hypotheses to be established by appeal to history. When the history is actually written, however, the philosophical doctrines undergo a subtle change of status. They cease to be theses awaiting corroboration, and become the organizing principles of the historical narrative. Historians then accuse the philosopher of writing bad history, because they see that, while there may be no lies or omissions in it, his narrative is driven by a purpose other than that of finding the best explanation for the events in question.

In view of this, historians may be tempted to ban philosophers from the archives altogether; but a better response would be to recognize the philosophically motivated historical narrative as a distinct genre, with its own standards of evaluation. These standards should reflect the fact that the philosophy of science has an interest in the history of science which is distinct from that of historians. Narrative is a legitimate mode of argument, which should be available to philosophers of science, but it will remain so only if philosophical argument by narrative is sharply distinguished from history proper. Sometimes this distinction is clear from the text: no one imagines that Plato's dialogues are accurate records of real conversations. One device is to write a fiction which is sufficiently similar in structure to the real history to be philosophically instructive. That was the strategy of Lakatos's *Proofs and Refutations* (1976), in which a highly tendentious "distilled history" was presented as a dialogue, with footnotes to spell out its connection with real history. If nothing else, it is necessary to maintain this distinction in order to prevent any particular philosophically driven reading of the past from becoming *the* reading.

What, then, is the role of the history of science in philosophy? Or rather, what are its roles? First of all, history is a source of philosophical problems and questions. Philosophers can find useful employment addressing such issues as the identify of theories over time, the evaluation of competing interpretations of antique scientific documents, and the relationship between scientists and the science they produce. Second, the history of science can work in the way that Lakatos suggested, as a standard against which to test philosophical theories. Lakatos's historiography was an account of the relationship between the history of science and the demarcation question. Since then, the demarcation question has dropped out of favor – in its original form, at least. This may be because none of the most popular answers have fared well before the tribunal of history. None of the methodologies proposed by philosophers have been remotely adequate to explain the development of the sciences. Those philosophical characterizations of science which retain any plausibility (including the models offered by Kuhn and Lakatos) do so only because they are highly schematic. Nevertheless, the demarcation question resurfaces from time to time. Sometimes it appears outside philosophy – for example, in the arguments over whether creationism should be taught in American schools. At other times it emerges from within philosophy, as in the debate over whether so-called folk psychology should be evaluated as a science, or as some other sort of discourse. As long as the question continues to be asked, it will be necessary to deploy history against the appeal of simple answers.

History need not work only negatively, to debunk philosophical myths, however. It can also play a part in the solution of philosophical riddles. One example is the discovery that the important features of a scientific theory are not only its logical properties (such as consistency or complexity), but also how well it does over time: whether it progresses or degenerates. If the dark matter hypothesis, for example, provokes a lot of fruitful questions and leads to unexpected discoveries, it can be written up as a good move. If, on the other hand, a century of searching passes with no sign of any dark matter, then it will appear to posterity as an *ad hoc* device. This insight has consequences for methodology and for the development of a sophisticated reply to relativism. Naturally, it was not formulated by any historian. Rather, it is a philosopher's abstraction of a certain kind of historical explanation. Historians often want to know why this or that theory fell out of favor after a period of popularity. One answer is that theories have progressive and degenerative phases. In the hands of a philosopher, this historical thought easily becomes the basis for a theory of scientific rationality. Of course, answers generate new questions. We will want to know what is meant by "progress" and "degeneration," and these terms cannot be elucidated without reference to the history of science.

There is the germ of an answer to the demarcation question here too: if something is a science, then it must make sense to ask whether it is progressing. Astrology and folk psychology are therefore not sciences, because they do not have leading problems or outstanding questions on which they could make progress (or fail to make progress, for that matter). This remark about what is and is not science may or may not be true, but its genesis is instructive. It was not arrived at by searching through the history books for paradigmatic cases of "science," and then abstracting their common properties. It is the result of a long, complex exchange between history and philosophy which cannot be summarized in some simple formula.

Finally, the history of science provides the raw materials for philosophical narratives. This is its most important role for those philosophers who think that science is best understood and explored by writing philosophically inspired histories of its development. Even philosophers who avoid narrative argument are informed by some version of the story of modern science. Current philosophy of science takes place in a context in which science is hugely successful and has become arguably the most prestigious of all forms of knowledge. It was different for early modern figures such as Bacon, for example, who were the theorists and champions of something new and undeveloped. At that time, the existing technology was largely a product of the craft traditions, and science had had little effect on the lives of most people. It had yet to deliver penicillin and portable televisions, and skepticism about science was far more tenable than it is now. Other forms of knowledge could still compete for the authority that attends possession of *the* truth. Things are rather different today. All philosophers nowadays work in the shadow of the massive presence of modern science. The sheer power of our technology renders skepticism absurd, and raises new questions. Is our technology always good for us? What are the characteristics of a mature science? What is the relationship between basic research and technological advancement? Indeed, the question for this chapter only makes sense because science is as big, as old, and as successful as it is. In Bacon's day, science had no history. In short, the range of intelligible philosophical questions and plausible answers is conditioned by the historical context in which philosophy takes place. If philosophers hope to understand the effect of science on their own thought, they must position themselves in a story which includes both science and philosophy.

All philosophy deals in idealizations. When philosophy takes the form of idealized history, it is possible (through comparison with ordinary history) to appreciate the philosophical argument and simultaneously recognize the idealization that made it possible. Moreover, it is through history that the philosophy of science has been able to correct what was its most glaring flaw: that it had little to do with actual science. A knowledge of the history of science refines one's intuitions regarding the plausibility of philosophical theses. This could be achieved by having philosophers learn more contemporary science. However, learning about the current state of science has the drawback that one gets only a snapshot of the present moment in its development. It is hard to infer movement from a single frame of film (a "still"), and it is hard to develop a philosophical understanding of how and why science changes just by looking at it as it is now. There is also the danger that philosophers can assimilate too well the myths and values of the institution they hope to study. History creates critical distance simply by presenting a variety of forms of scientific life. Suddenly, the practices and norms of today's science lose the appearance of immutable laws. Hence it is only through history that philosophers can address the general question of change in scientific theory, scientific method, and the self-understanding of scientific disciplines. Without history, these issues are worse than intractable – they are invisible.

References

Feyerabend, P. 1981: *Philosophical Papers*, vol. 2: *Problems of Empiricism* (Cambridge: Cambridge University Press).

Kuhn, T. S. 1970: *The Structure of Scientific Revolutions*, 2nd edn (Chicago: University of Chicago Press; 1st pub. 1962).

—— 1977: *The Essential Tension* (Chicago: University of Chicago Press).

Lakatos, I. 1978: Why did Copernicus's programme supersede Ptolemy's? *Philosophical Papers*, vol. 1: *The Methodology of Scientific Research Programmes*, ed. J. Worrall and G. Currie (Cambridge: Cambridge University Press), 102–38; originally published in *Boston Studies in the Philosophy of Science*, vol. 8, 1971.

—— 1976: *Proofs and Refutations*, ed. J. Worrall and E. Zahar (Cambridge: Cambridge University Press).

24

Holism

CHRISTOPHER HOOKWAY

The term "holism" refers to a variety of positions which have in common a resistance to understanding larger unities as merely the sum of their parts, and an insistence that we cannot explain or understand the parts without treating them as belonging to such larger wholes. Some of these issues concern explanation (see EXPLANATION). It is argued, for example, that facts about social classes are not reducible to facts about the beliefs and actions of the agents who belong to them; or it is claimed that we only understand the actions of individuals by locating them in social roles or systems of social meanings.

The most discussed forms of holism in recent philosophy of science have concerned epistemological and semantic issues which have sometimes appeared to threaten reasonable assumptions about rationality. When we make predictions in order to test hypotheses, we rely upon extensive background knowledge: from the reliability of our senses to the other theories presupposed by our experimental techniques, from information about the context of observation to techniques drawn from logic and mathematics, and so on. If our prediction is disappointed, this shows only that something is wrong somewhere: but judgment is needed to conclude that the best overall response is to reject the hypothesis we were testing. In principle, we could reject the embarrassing observations, place some of our background knowledge in doubt, or even question parts of logic or mathematics. Holistic theories of confirmation hold that the rules we follow in making such adjustments are concerned with the overall virtues of whole systems of beliefs, rather than with the support of particular claims. For example, they enjoin us conservatively to minimize change or to seek overall simplicity.

Holistic approaches were defended by Duhem, Poincaré, and others. A classic but sketchy version is in Quine's "Two dogmas of empiricism" (1953, ch. 2): "our statements about the external world face the tribunal of sense experience not individually but only as a corporate body" (p. 41); "total science is like a field of force whose boundary conditions are experience" (p. 42). And one consequence Quine draws is that "the total field is so underdetermined by its boundary conditions, experience, that there is much latitude of choice as to what statements to reevaluate in the light of any singular contrary experience" (p. 43). His more recent "moderate holism" is less extreme, holding that "more modest chunks" of theory are at stake in each observation (1981, p. 71). But he still denies that confirmation focuses on particular claims (see QUINE).

Two apparent consequences of such positions seem particularly disturbing. First, once attention is drawn to the virtues of overall systems of beliefs, or the comparative merits of different revisions in our beliefs, some common assumptions about rationality seem questionable. Objectivity seems to require that the bearing of evidence on our beliefs and theories should be a matter of consensus among all rational inquirers, and it is supposed that relatively specific rules guide us in evaluating our opinions. The holistic picture suggests that different revisions in our beliefs may be equally good in the light of the available evidence, or that different "total theories" may be equally satisfactory in the face of all possible evidence. There need be no fact of the matter concerning what scientific rationality requires of us. An instrumentalist view of theories beckons.

The second consequence concerns meaning. Logical empiricists held that our understanding of a proposition was fixed by a set of analytic truths specifying which experiential results were compatible with its truth (see LOGICAL EMPIRICISM). These analytic truths provided standards of rationality: we should reject the hypothesis if the predictions they license are disappointed. Holism undermines this picture of meaning and analyticity. No distinction is to be drawn between meaning-constitutive analytic propositions and synthetic ones, and any opinion could (in principle) be abandoned in order to make sense of surprising experiences. Our account of what is involved in understanding a proposition must instead talk of the varying degrees to which beliefs are embedded in our overall system of opinions: some could be abandoned with little disruption to the rest of the corpus, while others could not. Understanding, like confirmation, is thus holistic. Our understanding of theoretical propositions is a function of their relations to all of the rest of our knowledge. In principle, any of our opinions could influence how we evaluate the proposition in question. If we endorse the empiricist project of defining meaning in terms of confirmation, the unit of meaning becomes "the whole of science." Either we need a holistic theory of meaning (cf. Dummett 1975), or we conclude, skeptically, that the concept of meaning has little philosophical role.

These corollaries of holism depend upon an empiricist understanding of confirmation and meaning. They may be avoidable if this understanding is abandoned. Suppose that we emphasize the role of activities such as experimentation in scientific testing, and note that this involves a causal interaction with theoretical entities. We may then be able to explain how the *reference* of such terms is not fixed wholly by our current views about how to test statements involving them. As our knowledge progresses, we try to preserve reference, although our concepts and conceptions evolve (cf. Putnam 1990; Hacking 1983). This may enable us to defend realism, and to make better sense of how we decide which revisions to our beliefs are appropriate.

Analogous problems emerge in the philosophy of mind. Granted that our actions and predictions reflect the total pattern of our beliefs and desires, observations of how people act and what they expect may be compatible with very different ascriptions of beliefs and desires to them. And the contents of beliefs and desires will depend upon their place in this entire network of attitudes. Holism about meaning and content can lead to skepticism about whether propositional attitudes can have any role in psychological explanation (Fodor and Lepore 1992). This can be used to support the eliminativist view that the proper vocabulary for psychological research is physiological, or to argue that explanations in terms of propositional attitudes do not form part of an inchoate or informal *science* of mind.

References

Dummett, M. A. E. 1975: What is a theory of meaning? In *Mind and Language*, ed. S. Guttenplan (Oxford: Oxford University Press), 97–138.

Fodor, J. A., and Lepore, E. 1992: *Holism: A Shopper's Guide* (Oxford: Blackwell).

Hacking, I. 1983: *Representing and Intervening* (Cambridge: Cambridge University Press).

Putnam, H. 1990: Meaning holism. In *Realism with a Human Face* (Cambridge, MA: Harvard University Press), 278–302.

Quine, W. V. O. 1953: *From a Logical Point of View* (Cambridge, MA: Harvard University Press).

—— 1981: *Theories and Things* (Cambridge, MA: Harvard University Press).

25

Hume

W. H. NEWTON-SMITH

David Hume is the greatest figure in the empiricist tradition in philosophy and was a particular source of inspiration for the logical positivists (see LOGICAL POSITIVISM). Hume was born in 1711 and entered Edinburgh University at the age of 12. After graduating, he had a varied career in commerce, diplomacy, as a librarian, and as a writer of history. Twice he was secretary to General St Clair and on one occasion set off with him on an expedition to drive the French out of Canada. Forced back by the wind, it was decided instead to make a brief incursion on the coast of Brittany. This failed, for the expedition had maps of Canada, but none of Brittany. During a three-year stay in France he wrote his major philosophical work *Treatise of Human Nature*. It appeared in 1739 and, in his own words, "fell dead-born from the Press." His *Inquiry Concerning Human Understanding*, which was intended to be a more popular version of book I of his *Treatise*, was published in 1748 but was only a little more favorably received. His six-volume *History of England*, on the other hand, was a great success. He died in 1776, never having succeeded in his ambition to be appointed to a professorship of philosophy in Scotland, due to his religious skepticism.

Hume aimed to develop a science of man in his acting, feeling, and thinking aspects. His hope of achieving for the study of mind what Newton had achieved for the study of the natural world is conveyed in the subtitle of his *Treatise*: *being an Attempt to Introduce the Experimental Method of Reasoning into Moral Subjects*. And in his *Inquiry* he mused that philosophy "if cultivated with care and encouraged by the attention of the public may carry its researches further and discover the secret springs and principles by which the human mind is activated in its operations" (Hume 1999, p. 93). There are formal similarities between Hume's system and that of Newton. The ultimate building blocks of the mind for Hume are atoms of experience (impressions), on which the mind operates with three laws of "association." Hume did not succeed in this venture and is remembered primarily for his philosophical skepticism rather than for any positive theory of the mind.

For Hume, experience is the sole source of all meaning and knowledge. This is enshrined in his cardinal principle: all our ideas are derived from impressions. Impressions are experiences. Some ideas – simple ideas such as, for example, the idea of green – have content in virtue of being produced by the experience of seeing, say, grass. The only other way an idea can have content is by being a concatenation of ideas produced by experience. Thus, to use one of Hume's more fanciful examples, I can have the idea of a virtuous horse by dint of having had ideas derived from experience of horses and of virtue, which I concatenate together. His philosophical method is to determine the real content of an idea by breaking it down into its constituent simple ideas that derive directly from simple experiences. Unless we can trace an idea back to experience in this way, it is to be condemned as bogus. Perhaps the most dramatic use of this principle of significance arises when Hume seeks the origin of his idea of himself. He could find nothing in experience from which this idea could be derived. Finding on introspection only a confusion of impressions of other things and no impression corresponding to himself, he concluded that he was nothing but a "bundle of impressions" (Hume 1978, p. 252).

In contemporary terminology, Hume's principle of significance amounts to the claim that a word is meaningless unless it can be defined in terms of words standing for types of simple experiences. The discourse of the metaphysician is thus utterly without meaning. Exciting as this doctrine is, it clearly will not do. For even simple words like "and" or "not" would turn out to be meaningless. However, this general orientation of tying meaning to experience was to resurface with the logical positivists (see LOGICAL POSITIVISM), who avoided some of the most obvious problems facing Hume by focusing on sentences rather than words. Their doctrine that the meaning of a sentence is given by its method of verification in experience leads similarly to a rejection of all metaphysics.

Another crucially important feature of Hume's philosophical method is his rigid adherence to a dichotomy between *relations of ideas* and *matters of fact*. Relations of ideas correspond to analytical propositions whose truth can be discovered a priori by reflecting on the meanings of the words involved and whose denial is self-contradictory. Matters of fact correspond to contingent, empirical truths whose denial is not self-contradictory and whose truth is to be discovered in experience. This dichotomy which is now much disputed is known as "Hume's fork."

In addition to his general contribution to the elaboration of empiricism, his specific views on causation and induction have had particular and continuing importance in the philosophy of science. In his treatment of causation Hume puts his fork and his principle of significance to work as follows. Imagine billiard ball A striking billiard ball B, causing B to move. That A caused B to move is not something we can discover as a matter of the relations of ideas. For there is no contradiction in the assumption that A hits B without B moving off. Causation, then, is something to be discovered by experience. But what do we experience? All we see is the event of A striking B conjoined with the event of B's moving off. All we find in experience corresponding to what we call "causation" is constant conjunction. For Hume, all that we can mean by "A caused B" is "events similar to A are constantly conjoined with events similar to B." To this he adds the requirements that there is spatiotemporal continuity between A and B and that A occurs before B. There can be nothing more to causation than that. To assume that there is some tighter connection between A and B would be to illegitimately posit a relation between A and B which could not be explained in terms of experience.

Hume is well aware that we feel that there is something more going on, that we feel there is some element of necessity when A causes B that is not involved in the idea of events similar to A being merely constantly conjoined with events similar to B. Of this Hume offers a psychological explanation. Having experienced many cases of an A followed by a B, our minds form the habit of passing to the thought of a B whenever we think of an A. This is a propensity in ourselves that we erroneously project onto the world. In the world apart from ourselves there is only constant conjunction. There is no causal glue cementing the events of the world together.

This is the classic empiricist account of causation. All we can ever experience is constant conjunction. Therefore, given the principle of significance, we can have no legitimate idea of causation that goes beyond that. There are a host of difficulties with Hume's account (Strawson 1989), but it remains the source of inspiration for those who seek to demystify causation (Mackie 1974).

Hume was a skeptic about induction. He argued that I have no reason whatsoever to think that bread will nourish me tomorrow, in spite of the fact that it has always nourished me in the past. The argument begins with his fork. We can establish relations of ideas by a priori reasoning, or we can seek to establish contingent matters of fact by reasoning from experience. I can imagine that the bread tomorrow will not nourish me. So there is no contradiction in the assumption that it will not. Consequently, it cannot be established by a priori reasoning that the bread will nourish me. But neither can we establish that the bread will nourish me by appeal to experience. The fact that bread has nourished me in the past only gives me a reason to think that it will nourish me tomorrow if I have a reason to think that nature is uniform, that the future will be like the past. Could I have a reason for thinking that? I only have reasons for thinking that the future will be like the past if I have reasons to think that, among other things, bread will continue to nourish. I can only have a reason for thinking that a particular regularity will continue to hold (that bread will continue to nourish) if I have a reason to think that the future will be like the past. But I can only have a reason for that general claim if I have a reason for thinking that that particular regularity among all others will continue to hold. Thus any attempt to justify any inductive conclusion about the future will involve an argument which runs in a vicious circle.

Hume did not argue that we should not continue to act and reason inductively. For it is part of our nature to do so. We are creatures of custom and habit, a notion of the greatest importance for Hume. And we will continue to act on the assumption that induction is a reliable process. It is simply a bemusing fact of the human condition that what we do does not have the sanction of reason. As humans, we cannot help but use induction, but as philosophers, we have to recognize that there is no possible justification for it. Hume is sometimes taken as making the simple error of asking for a deductive justification for induction. If his argument is mistaken, this is not the mistake. His claim is rather that any justification of induction is unconvincing because it is circular. His conclusion is radical. It is not just that we cannot have certain knowledge that the bread will nourish; we cannot have any reason for thinking that it is more likely than not to nourish. This is an ironic conclusion for one whose aim was to produce for the study of man what Newton had achieved in his study of the natural world.

Hume's argument has generated an enormous literature (see INDUCTION AND THE UNIFORMITY OF NATURE). There have been attempts to provide a justification, or at least a vindication of induction. Others have sought to dissolve the problem by arguing that no justification is called for. Popper, on the contrary, accepts that Hume has shown that induction has no justification. Contrary to Hume, he argues that we (or at least good scientists) do not in fact use induction (see POPPER). Unlike Popper, probably most philosophers hold that induction is justified. However, few of them would be confident that they can adequately answer Hume's skeptical challenge.

Given his assumption that causation is constant conjunction, this means that for Hume we can never have any reason for thinking that anything caused anything! For to justify the claim that a particular event A caused a particular event B, we would have to justify the claim that all events similar to A have been and will be followed by events similar to B. Given his skepticism about induction, no reason at all, however weak, can be provided for such a claim.

Hume's skeptical strategy that he applied initially to induction about the future can be generalized. There is no noncircular argument for any factual claim that goes beyond claims about my present experiences. At the end of the day, for Hume, all we can know are the contents of our mind.

While these and other particular philosophical claims have had great influence in the history of the philosophy of science, Hume's greatest impact was the vision he gave of linking everything to experience. This is stirringly conveyed in his rallying call at the end of the *Inquiry*:

> When we run over libraries, persuaded of these principles, what havoc must we make? If we take in our hand any volume; of divinity or school metaphysics, for instance; let us ask, Does it contain any abstract reasoning concerning quantity or number? No. Does it contain any experimental reasoning concerning matter of fact and existence? No. Commit it then to the flames: For it can contain nothing but sophistry and illusion.

References and further reading

Works by Hume

1754–62: *History of England* (Edinburgh: Hamilton, Balfour and Neil).
1978: *A Treatise of Human Nature* (Oxford: Oxford University Press).
1990: *Dialogues Concerning Natural Religion* (Harmondsworth: Penguin).
1999: *An Inquiry Concerning Human Understanding* (Oxford: Oxford University Press).

Works by other authors

Flew, A. 1997: *Hume's Philosophy of Belief: A Study of His First Inquiry* (Bristol: Thoemmes).
Mackie, J. L. 1974: *The Cement of the Universe* (Oxford: Clarendon Press).
Mossner, E. C. 1954: *The Life of David Hume* (Oxford: Clarendon Press; repr. 1970).
Norton, D. F. (ed.) 1993: *The Cambridge Companion to Hume* (Cambridge: Cambridge University Press).
Pears, D. 1991: *Hume's System: An Examination of the First Book of His Treatise* (Oxford: Oxford University Press).
Strawson, G. 1989: *The Secret Connexion* (Oxford: Oxford University Press).
Stroud, B. 1981: *Hume* (London: Routledge).

26

Idealization

YEMIMA BEN-MENAHEM

When Sadi Carnot carried out the pioneering work on heat engines which led to the second law of thermodynamics, he contemplated an ideal heat engine, one that was completely reversible. Carnot's use of idealization was particularly successful – while the ideal engine cannot actually be constructed, the conclusions he derived for the ideal engine hold *a fortiori* for actual heat engines. For example, the greater the temperature difference between the two heat reservoirs, the higher the engine's efficiency. But this is not always the case. What holds true in the ideal limit may be false in reality. Nevertheless, the ideal case, simple and tractable, can be expected to shed light on actual cases, the precise treatment of which is impossible or impractical. Idealizations abound in science: ideal gases, closed systems, perfectly rational agents, and evolutionarily stable strategies. Indeed, thinkers as diverse in their outlook as Edmund Husserl and Albert Einstein have pointed to idealizations as the hallmark of modern science.

The use of idealizations raises a number of problems for the philosopher of science. One such problem is that of confirmation. On the deductive nomological model of scientific theories (see CONFIRMATION, PARADOXES OF, and EXPLANATION), a theory is a deductive scheme which uses laws and initial conditions to derive predictions of events or lower-level laws. A deductive scheme should yield true consequences when the premises are true. However, if idealizations are admitted as premises, the premises are, strictly speaking, false, and the conclusions need not be true even when the argument is valid. But if a theory does not purport to yield true conclusions, how can it be tested? Intuitively, a theory can be confirmed even in such a case if its predictions deviate from the truth no more than is "justified" by the deviation of the input, and otherwise disconfirmed. Ronald Laymon (1987) has introduced the requirement of *monotony* to capture this intuition. A theory is monotonous if better approximations of input yield better approximations of output, and confirmed if monotonous. Not all theories, however, are of this kind. A typical feature of chaos is the sensitivity of its equations to initial conditions and their failure to meet the requirement of monotony.

The logic of approximation developed by Michel Katz suggests another way of handling the problem of confirmation. This logic extends the standard two-valued notion of deduction to real-valued logic. An argument is valid when, if the premises are true enough, the conclusions are very nearly true. Thus, under the extended sense of deduction, a set of formulas Γ can be said to yield another set Δ if, for every $\varepsilon > 0$, there is a $\delta > 0$ such that if the maximal truth-value among the formulas of Γ is smaller than δ, the minimal truth-value among the formulas of Δ is smaller than ε.

(In this logic the truth-value of a formula is its degree of error, 0 representing absolute truth and 1 absolute falsehood.) Since extended deduction preserves important features of standard deduction, it seems reasonable to suggest that standard deduction be replaced by extended deduction in the deductive nomological model. Note that on this suggestion the standard DN model itself functions as an idealization in the philosophy of science in the same sense that frictionless motion is an idealization in physics.

A closely related problem concerns the relation between successive theories, such as that between Kepler's and Galileo's laws and Newtonian mechanics. According to the ideal picture of "convergence," comprehensive theories entail lower-level laws. This picture has come under attack: it has been argued that, strictly speaking, Kepler's and Galileo's laws are refuted by Newton's theory, not entailed by it. If this case is typical, there is neither convergence nor progress in science. The logic of approximation can again rescue the ideal picture. It can be shown that in many cases higher-level theories entail lower-level laws in the extended sense, though not when theories completely replace each other, as Aristotelian mechanics was replaced by Newtonian mechanics. "Convergence" is therefore saved precisely for cases in which it is intuitively attractive. Saving the idea of "convergence" has far-reaching implications for the problem of incommensurability and the controversy over realism (see INCOMMENSURABILITY and REALISM AND INSTRUMENTALISM).

A salient feature of an idealization is the way it approximates the actual case. By contrast, fictions, which are also frequently used in science, lack this characteristic verisimilitude (see VERISIMILITUDE). In solid state physics, a "hole" in the "sea of electrons" is treated as a positive particle. A real force is sometimes treated as if it were composed of fictitious component forces. The idea here is that the fictitious case is analogous to the real one. But, unlike the ideal case, a fiction cannot be approached gradually; nor is it considered even approximately true to the facts. A good example of a fiction outside natural science is the notion of the social contract. Most social contract theorists do not claim that an actual contract was ever drawn up. Rather, they claim that political arrangements are, or should be, *as if* a contract had been drawn up.

Recourse to idealizations in science has led several thinkers to be concerned about the relation between science and reality. Husserl (1954) saw the ideal nature of modern science as the root of what he called the crisis of European science – the gulf between science and the world we live in (*Lebenswelt*). Recently, from a very different perspective, Nancy Cartwright (1983) has also drawn attention to the ideal character of the fundamental laws of physics. Not only do these laws deviate from the truth, but there is, on her view, a trade-off between truth-value and explanatory power: the better the explanation, the lower the likelihood of its truth. The above considerations do not support this thesis, for on the suggested criteria for adequacy for idealizations, such as monotony, truth-value and explanatory power are congruent. The metaphysical conception underlying Cartwright's provocative claim is Aristotelian: nature is diverse, and uniformities are imposed by the observer. Galileo, the first to turn idealization into a fundamental scientific tool, subscribed to a diametrically opposed metaphysics. He believed that the "book of nature" was written in mathematical language. The argument from the indispensability of idealization to a particular metaphysics seems inconclusive.

References and further reading

Ben-Menahem, Y. 1988: Models of science: fictions or idealizations? *Science in Context*, 2, 163–75.

Cartwright, N. 1983: *How the Laws of Physics Lie* (Oxford: Clarendon Press).

Funkenstein, A. 1986: *Theology and the Scientific Imagination* (Princeton: Princeton University Press). [Ch. 3 traces the theological roots of ideal laws of nature.]

Husserl, E. 1954: *The Crisis of European Sciences and Transcendental Phenomenology*, trans. D. Carr (Evanston, IL: Northwestern University Press; repr. 1970).

Katz, M. 1982: Real-valued models with metric equality and uniformly continuous predicates. *Journal of Symbolic Logic*, 47, 772–92.

Laymon, R. 1987: Using Scott domains to explicate the notions of approximate and idealised data. *Philosophy of Science*, 54, 194–221.

Vaihinger, H. 1924: *The Philosophy of "As If"*, trans. C. K. Ogden (London: Routledge and Kegan Paul; 1st pub. 1911). (A classic work on the utility of fictions in science and other disciplines.)

27

Incommensurability

MUHAMMAD ALI KHALIDI

Along with "paradigm" and "scientific revolution," "incommensurability" is one of the three most influential expressions associated with the "new philosophy of science" first articulated in the early 1960s by Thomas Kuhn and Paul Feyerabend (see KUHN and FEYERABEND). But, despite the fact that it has been widely discussed, opinions still differ widely as to the content and significance of the claim of incommensurability. What is uncontroversial is that the term "incommensurability" was borrowed from mathematics, where it can be used, for example, to apply to the relation between the side of a square and its diagonal. Since the side of a square is measured by a rational number, and its diagonal by an irrational number, and since an irrational number cannot be represented by a point on the rational number line, the two quantities are said to have no common measure; they are literally *incommensurable*. Kuhn and Feyerabend adapted this term and applied it to some pairs of rival scientific theories, to indicate that such theories also had no common measure, or, in some sense to be determined, could not be compared directly. Both Kuhn and Feyerabend agree that they hit upon the term independently and used it in print for the first time in 1962, in *The Structure of Scientific Revolutions* and "Explanation, reduction, and empiricism," respectively. But the two writers explicate the claim and argue for it rather differently. After tackling the concept of incommensurability as it appears in the works of each of these authors in turn, some reactions and responses will be sampled.

Kuhn's notion of incommensurability

One common and natural interpretation of the idea that there is no common measure among rival scientific theories is that they cannot be phrased in a common set of linguistic terms, or, to put it more simply, that they cannot both be translated into a single language. That is, the claim of incommensurability can be taken to be about the impossibility of the linguistic mode of comparison in the first instance. This interpretation is confirmed by some, though by no means all, of Kuhn's early articulations of the concept. In *The Structure of Scientific Revolutions* Kuhn often puts incommensurability in terms of change of meaning, but he sometimes suggests that translation *is* possible between two incommensurable theories or paradigms. He writes that "the physical referents of these Einsteinian concepts [space, time, and mass] are by no means identical with those of the Newtonian concepts that bear the same name" (1970a, p. 102). He continues: "This need to change the meaning of established and familiar concepts is central to the revolutionary impact of Einstein's theory" (ibid.). Kuhn refers to this revolutionary change from classical to relativistic mechanics as a "displacement of the conceptual network" (ibid.). In the "Postscript" to the text, he reiterates the view that incommensurability involves differences in meaning between two agents espousing incommensurable theories: "Two men who perceive the same situation differently but nevertheless employ the same vocabulary in its discussion must be using words differently. They speak, that is, from what I have called incommensurable viewpoints" (1970a, p. 200). However, he then goes on to say that "what the participants in a communication breakdown can do is recognize each other as members of different language communities and then become translators," resorting to "shared everyday vocabularies" in doing so (1970a, p. 202). If this is carried out successfully, Kuhn thinks, then "Each will have learned to translate the other's theory and its consequences into his own language and simultaneously to describe in his language the world to which that theory applies. This is what the historian of science regularly does (or should [do]) when dealing with out-of-date scientific theories" (ibid.).

Since Kuhn sometimes suggests that translation is indeed possible between two incommensurable scientific theories, how are we to understand the claim of incommensurability? At some points in the "Postscript," he hints that it is a claim about the impossibility of a more general assessment of two scientific theories. This second construal of the notion of incommensurability – that it precludes a neutral way of appraising scientific theories – seems to rest on a different claim: namely, that scientific theories or paradigms contain within themselves their own standards for success or criteria of appraisal. Not only do scientific paradigms differ "about the population of the universe and about that population's behavior"; Kuhn writes that they are also "the source of the methods, problem-field, and standards of solution accepted by any mature scientific community at any given time" (1970a, p. 103). These "non-substantive differences" are an integral part of incommensurability, which is demonstrated by the fact that adherents of two scientific paradigms "will inevitably talk through each other when debating the relative merits of their respective paradigms," since "each paradigm will be shown to satisfy more or less the criteria that it dictates for itself and to fall short of a few of those dictated by its opponent" (1970a, pp. 109–10).

However, in later developments of Kuhn's view, less emphasis is placed on what might be called "evaluative incommensurability," and more on "linguistic incommensurability." Indeed, by 1983, Kuhn appeared to have moved away from evaluative incommensurability entirely, by saying that speaking of differences in "methods, problem-field, and standards" is "something I would no longer do except to the considerable extent that the latter differences are necessary consequences of the language-learning process" (1983, p. 684, n. 3). And in a 1990 article, Kuhn states quite baldly: "Incommensurability thus equals untranslatability" (p. 299). In a footnote, he writes: "My original discussion described nonlinguistic as well as linguistic forms of incommensurability. That I now take to have been an overextension resulting from my failure to recognize how large a part of the apparently nonlinguistic component was acquired with language during the learning process" (1990, p. 315, n. 4). Not only does Kuhn in his later work take incommensurability to be more explicitly the denial of translatability, he also states that this version of the claim is the same as the "original version" of the incommensurability thesis, which he characterized as follows: "The claim that two theories are incommensurable is then the claim that there is no language, neutral or otherwise, into which both theories, conceived as sets of sentences, can be translated without residue or loss" (1983, p. 670). Therefore, if incommensurability equals untranslatability, what is it about scientific paradigms that precludes translation into a single common language, so that their claims can be set side by side and their points of agreement and disagreement isolated? Moreover, how does this claim square with Kuhn's earlier claim (in the "Postscript") that historians of science can and do translate out-of-date scientific theories? (Some commentators on Kuhn have regarded this as the supreme irony of his work, that he denies translatability while at the same time serving as an articulate expositor of historical scientific theories.)

The resolution of this tension lies in what Kuhn says after equating incommensurability with untranslatability: "what incommensurability bars is not quite the activity of professional translators. Rather, it is a quasi-mechanical activity governed in full by a manual that specifies, as a function of context, which string in one language may, *salva veritate*, be substituted for a given string in the other" (1990, p. 299). Such a quasi-mechanical translation cannot be effected because of certain concrete problems posed by the translation of a scientific theory by a translator who does not share that theory. Kuhn claims that the problems of translating a scientific text into a foreign language or a later version of the same language are very similar to the problems of translating literature (1990, p. 300). In an illuminating passage which is worth quoting in full, he comments on the translational difficulties which are shared by literary and scientific discourse:

> In both cases the translator repeatedly encounters sentences that can be rendered in several alternative ways, none of which captures them completely. Difficult decisions must then be made about which aspects of the original it is most important to preserve. Different translators may differ, and the same translator may make different choices in different places, even though the term involved is in neither language ambiguous. Such choices are governed by standards of responsibility, but they are not determined by them. In these matters there is no such thing as being merely right or wrong. The preservation of truth values when translating scientific prose is as delicate a task as the preservation of resonance and emotional tone in the translation of literature. Neither can be fully achieved; even responsible approximation requires the greatest tact and taste. In the scientific case, these generalizations apply, not only to passages that make explicit use of theory, but also and more significantly to those their authors took to be merely descriptive. (1990, pp. 300–1)

In this passage, Kuhn does not clarify the specific translational difficulties involved, but in other works, certain specific obstacles emerge. Although Kuhn does not always distinguish them clearly, two can be singled out for special attention.

The first kind of translational difficulty implicated in incommensurability is the problem of *clusters of interdefined terms*. Kuhn uses the example of the eighteenth-century chemical term "phlogiston" to illustrate his point. He says that the term cannot be translated into terms of later chemical theory because of its relation to a number of other terms in the phlogiston theory, like "principle" and "element." "Together with 'phlogiston'," Kuhn explains, "they constitute an interrelated or interdefined set that must be acquired together, as a whole, before any of them can be used, applied to natural phenomena" (1983, p. 676). He acknowledges that one can introduce a neologism for a term from a previous scientific theory which is no longer part of the current scientific vocabulary. However, he suggests that when there are whole clusters of such interrelated terms, translation is no longer possible, presumably because each neologism needs to be explicated in terms of the extant vocabulary, making whole clusters of them resistant to such explication.

Another translational problem is that of *conceptual disparity* among terms. Kuhn brings this out by adverting to an example drawn from nonscientific discourse. He explains that the French word *doux* does not correspond to any single word in English. It "can be applied, *inter alia*, to honey ('sweet'), to underseasoned soup ('bland'), to a memory ('tender'), or to a slope or a wind ('gentle'). These are not cases of ambiguity, but of conceptual disparity between French and English" (1983, pp. 679–80). He emphasizes that *doux* is a unitary concept for French speakers, and that English speakers have no single equivalent. English paraphrases for this French term provide no substitute because of their clumsiness, and because the term must be learned together with other parts of the French vocabulary (1983, p. 685, n. 12). While he acknowledges that a translation manual is adequate to deal with cases of straightforward ambiguity, he argues that the examples he uses are not to be seen in this light, and should be distinguished from standard examples of ambiguous words, such as "bank" or "cape." The reason seems to be that it is crucial for French speakers, as opposed to English speakers, that there is a single concept at play, rather than a single term which happens to stand for a number of distinct concepts. Thus, a translation which substituted a different English word for *doux* depending on context would be misleading. Though he does not explicitly say so, a scientific example of this phenomenon might be found in Kuhn's discussion of Aristotle's concept of speed, which he says contains "two disparate criteria," the first giving rise to our concept "average speed," the second to our concept "instantaneous velocity" (1977, pp. 246–7). However, Aristotle himself never made the distinction but employed what he would have considered to be a unitary concept.

Therefore, according to Kuhn's mature view, it is not possible to phrase all the claims of two scientific theories in a single language so that they can be put side by side and their exact points of difference pinpointed. Kuhn thereby denies the possibility of what is perhaps the most direct and natural method of comparing two scientific theories. As a result, choices between scientific theories are not based on a point-by-point comparison. Scientists who learn a new theory do not merely translate the new terms into the old terms; rather, they begin from scratch in the way that learners of a first language do. A language-learner, Kuhn states, will not always "be able to translate from his newly acquired language to the one with which he was raised" (1990, p. 300).

Kuhn never says that incommensurable theories can never be compared at all. Since the mismatches between incommensurable theories are local, we should expect that certain comparisons *can* be effected. Often such comparisons will involve concrete measurements of phenomena, presumably ones described in terms shared by the two theories. He states that "proponents of different theories can exhibit to each other, not always easily, the concrete technical results achievable by those who practice within each theory" (1977, p. 339). Although the Ptolemaic theory and the Copernican theory were incommensurable because of such problematic terms as "planet," "The quantitative superiority of Kepler's Rudolphine tables to all those computed from the Ptolemaic theory was a major factor in the conversion of astronomers to Copernicanism" (1970a, p. 154). But there are also other criteria for comparison; for example, "there are arguments . . . that appeal to the individual's sense of the appropriate or aesthetic – the new theory is said to be 'neater', 'more suitable', or 'simpler' than the old" (1970a, p. 155). Hence, many grounds for comparison remain despite incommensurability, including "accuracy, scope, simplicity, fruitfulness, and the like" (1970b, p. 261).

Feyerabend's notion of incommensurability

Feyerabend is more consistent than Kuhn in giving a linguistic characterization of incommensurability, and there seems to be more continuity in his usage over time. He generally frames the incommensurability claim in terms of language, but the precise reasons he cites for incommensurability are different from Kuhn's. One of Feyerabend's most detailed attempts to illustrate the concept of incommensurability involves the medieval European impetus theory and Newtonian classical mechanics. He claims that "the concept of impetus, as fixed by the usage established in the impetus theory, cannot be defined in a reasonable way within Newton's theory" (1981a, p. 66). On the basis of this and other considerations, he concludes:

> [W]hat happens when a transition is made from a restricted theory T′ to a wider theory T (which is capable of covering all the phenomena which have been covered by T′) is something much more radical than incorporation of the unchanged theory T′ into the wider context of T. It is rather a replacement of the ontology of T′ by the ontology of T, and a corresponding change in the meanings of all descriptive terms of T′ (provided these terms are still employed). (1981a, p. 68)

On several occasions Feyerabend explains the reasons for incommensurability by saying that there are certain "universal rules" or "principles of construction" which govern the terms of one theory and which are violated by the other theory. Since the second theory violates such rules, any attempt to state the claims of that theory in terms of the first will be rendered futile. "We have a point of view (theory, framework, cosmos, mode of representation) whose elements (concepts, 'facts', pictures) are built up in accordance with certain principles of construction. The principles involve something like a 'closure': there are things that cannot be said, or 'discovered', without violating the principles (which does *not* mean contradicting them)" (1975, p. 269). After terming such principles "universal," he states: "[L]et us call a discovery, or a statement, or an attitude *incommensurable* with the cosmos (the theory, the framework) if it suspends some of its universal principles" (ibid.). As an example of this phenomenon, consider two theories T and T′, where T is classical celestial mechanics, including the space-time framework, and T′ is general relativity theory. About these theories, Feyerabend claims:

> The classical, or absolute idea of mass, or of distance, cannot be defined within T′. Any such definition must assume the absence of an upper limit for signal velocities and cannot therefore be give within T′. *Not a single primitive descriptive term of T can be incorporated into*

T' ... the meanings of all descriptive terms of the two theories, primitive as well as defined terms, will be different: T and T' are *incommensurable theories*. (1981c, p. 115; emphasis original)

Such principles as the absence of an upper limit for signal velocities govern all the terms in celestial mechanics, and these terms cannot be expressed at all once such principles are violated, as they will be by general relativity theory.

The reason that these universal rules affect the meanings of all the terms of the theory which contains them is to be found in Feyerabend's theory of meaning, which he calls a "contextual theory of meaning." He uses this contextual theory to define "strong alternatives" to a given scientific theory: theories which can be considered true competitors to a dominant theory, as opposed to those which are mere variants. He explains that "One of the main properties of strong alternatives is that they disagree everywhere if they disagree at a finite number of points" (ibid.). In other words, one sign that a theory is substantively different from another is that the differences between them affect the meanings of all terms; otherwise, Feyerabend implies, the rival theory is not a genuine alternative, but a mere variant. All such strong alternatives are incommensurable. Elsewhere, he writes that the meaning of a term is not an intrinsic property of it, but is dependent on the way in which the term has been incorporated into a theory (1981a, p. 74). This is the gist of what Feyerabend calls a "contextual theory of meaning." It also accords with his ridicule of what he calls the "hole theory," or the "Swiss cheese theory" of meaning, which holds that the conceptual cavities in a theory or language can be plugged without displacing the meanings of any of the existing terms. "According to the hole theory every cosmology (every language, every mode of perception) has sizeable lacunae which can be filled, *leaving everything else unchanged*" (1975, p. 266). The idea seems to be that the meaning of every term is affected by the general principles governing the theory, and that the principles change with every substantial theoretical change, so that the meaning of every term also changes. But even Feyerabend concedes that large parts of our *total* theory of the world remain constant across some scientific theory changes. "It may be readily admitted," he writes, "that the transition from T to T' will not lead to new methods for estimating the size of an egg at the grocery store" (1981b, p. 100). And he say that the transition from Newtonian mechanics to the general theory of relativity has left the arts, ordinary language, and perception unchanged (1975, p. 271).

Comparison of Kuhn and Feyerabend

Feyerabend's differences with Kuhn can be reduced to two basic ones. The first is that Feyerabend's variety of incommensurability is more global, and cannot be localized in the vicinity of a single problematic term or even a cluster of terms. That is, Feyerabend holds that fundamental changes of theory lead to changes in the meanings of all the terms in a particular theory. The other significant difference concerns the reasons for incommensurability. Whereas Kuhn thinks that incommensurability stems from specific translational difficulties involving problematic terms, Feyerabend's variety of incommensurability seems to result from a kind of extreme holism about the nature of meaning itself.

One significant point of agreement between Kuhn and Feyerabend is that neither thinks that incommensurability is incomparability *tout court*. Both countenance, and indeed recommend, alternative modes of comparison. Feyerabend says that "the use of incommensurable theories for the purpose of criticism must be based on methods which do not depend on the comparison of statements with identical constituents. Such methods are readily available" (1981c, p. 115). But although he mentions a number of methods, he does not explicate them in full. For example, he says that theories can be compared using the "pragmatic theory of observation," according to which you attend to causes of the production of a certain observational sentence, rather than the meaning of that sentence (1981a, p. 93). He does not elaborate further, but this claim is difficult to uphold given his insistence that even the meanings of "descriptive terms" are different in incommensurable theories. He also argues that "when making a comparative evaluation of classical physics and of general relativity we do not compare meanings; we investigate the conditions under which a structural similarity can be obtained" (1981b, pp. 102–3). And he insists that "there may be empirical evidence against one [theory], and for another theory without any need for similarity of meanings" (1981c, p. 116). On a more sarcastic, though revealing, note, Feyerabend states; "Of course, *some* kind of comparison is *always* possible (for example, one physical theory may sound more melodious when read aloud to the accompaniment of a guitar than another physical theory)" (1975, p. 32; emphasis original). At any rate, he insists that "it *is* possible to use incommensurable theories for the purpose of mutual criticism," adding that this removes "one of the main 'paradoxes' of the approach" that he suggests (1981c, p. 117; emphasis original). Finally, he uses the same analogy that Kuhn uses to explain a scientist's ability to learn a new theory, that of a child learning a new language. Rather than translating between languages, "[w]e can learn a language or a culture from scratch, as a child learns them, without detour through our native tongue" (1987, p. 266).

Responses to incommensurability

Responses to incommensurability have been profuse in the philosophy of science, and only a small fraction can be sampled here. Two main trends may be distinguished. The first denies some aspect of the claim, and suggests a method of forging a linguistic comparison among theories, while the second, though not necessarily accepting the claim of linguistic incommensurability, proceeds to develop other ways of comparing scientific theories.

In the first camp are those who have argued that at least one component of meaning is unaffected by untranslatability: namely, reference. Israel Scheffler (1982) enunciates this influential idea in response to incommensurability, but he does not supply a theory of reference to demonstrate how the reference of terms from different theories can be compared. Later writers seem to be aware of the need for a full-blown theory of reference to make this response successful. Hilary Putnam (1975) argues that the causal theory of reference can be used to give an account of the meaning of natural kind terms, and suggests that the same can be done for scientific terms in general (see NATURAL KINDS). But the causal theory was first proposed as a theory of reference for proper names, and there are serious problems with the attempt to apply it to science. An entirely different language-based response to the incommensurability claim is found in Donald Davidson (1985). Davidson contends against Kuhn that all putative conceptual schemes, presumably including the scientific theories embedded within them, are intertranslatable. The argument is a powerful one, but it proceeds at a purely general level. Davidson does not show in practice how specific scientific theories can be expressed in the same terms.

The second kind of response to incommensurability proceeds to look for nonlinguistic ways of making a comparison between scientific theories. Among these responses, one can distinguish two main approaches. One approach advocates expressing theories in model-theoretic terms, thus espousing a mathematical mode of comparison. This position has been advocated by writers such as Joseph Sneed and Wolfgang Stegmüller, who have shown how to discern certain structural similarities among theories in mathematical physics. But the methods of this "structuralist approach" do not seem applicable to any but the most highly mathematized scientific theories. Moreover, some advocates of this approach have claimed that it lends support to a model-theoretic analogue of Kuhn's incommensurability claim. Another trend which has emerged more recently involves the so-called cognitive approach to science, which takes scientific theories to be entities in the minds or brains of scientists, and regards them as amenable to the techniques of recent cognitive science; proponents include Paul Churchland, Ronald Giere, and Paul Thagard. Thagard's (1992) is perhaps the most sustained cognitivist attempt to reply to incommensurability. He uses techniques derived from the connectionist research program in artificial intelligence, but relies crucially on a linguistic mode of representing scientific theories without articulating the theory of meaning presupposed. Interestingly, another cognitivist who urges using connectionist methods to represent scientific theories, Churchland (1992), argues that connectionist models vindicate Feyerabend's version of incommensurability.

The issue of incommensurability remains a live one. It does not arise just for a logical empiricist account of scientific theories, but for any account that allows for the linguistic representation of theories. Discussions of linguistic meaning cannot be banished from the philosophical analysis of science, simply because language figures prominently in the daily work of science itself, and its place is not about to be taken over by any other representational medium. Therefore, the challenge facing anyone who holds that the scientific enterprise sometimes requires us to make a point-by-point linguistic comparison of rival theories is to respond to the specific semantic problems raised by Kuhn and Feyerabend. However, if one does not think that such a piecemeal comparison of theories is necessary, then the challenge is to articulate another way of putting scientific theories in the balance and weighing them against one another.

References and further reading

Churchland, P. M. 1992: A deeper unity: some Feyerabendian themes in neurocomputational form. In *Minnesota Studies in the Philosophy of Science*, vol. 15: *Cognitive Models of Science*, ed. Ronald Giere (Minneapolis: University of Minnesota Press), 341–63.

Davidson, D. 1985: On the very idea of a conceptual scheme. In *Inquiries into Truth and Interpretation* (Oxford: Oxford University Press), 183–98 (1st pub. 1974).

Feyerabend, P. K. 1970: Consolations for the specialist. In *Criticism and the Growth of Knowledge*, ed. I. Lakatos and A. Musgrave (Cambridge: Cambridge University Press), 197–230.

—— 1975: *Against Method* (London: Verso).

—— 1981a: Explanation, reduction, and empiricism. In *Philosophical Papers*, vol. 1: *Realism, Rationalism and Scientific Method* (Cambridge: Cambridge University Press), 44–96 (1st pub. 1962).

—— 1981b: On the "meaning" of scientific terms. In *Philosophical Papers*, vol. 1: *Realism, Rationalism and Scientific Method* (Cambridge: Cambridge University Press), 97–103 (1st pub. 1965).

—— 1981c: Reply to criticism: comments on Smart, Sellars, and Putnam. In *Philosophical Papers*, vol. 1: *Realism, Rationalism and Scientific Method* (Cambridge: Cambridge University Press), 104–31 (1st pub. 1965).

—— 1987: Putnam on incommensurability. In *Farewell to Reason* (London: Verso), 265–72.

Kuhn, T. S. 1970a: *The Structure of Scientific Revolutions*, 2nd edn (Chicago: University of Chicago Press; 1st pub. 1962).

—— 1970b: Reflections on my critics. In *Criticism and the Growth of Knowledge*, ed. I. Lakatos and A. Musgrave (Cambridge: Cambridge University Press), 231–78.

—— 1976: Theory-change as structure-change: comments on the Sneed formalism. *Erkenntnis*, 10, 179–99.

—— 1977: *The Essential Tension* (Chicago: University of Chicago Press).

—— 1983: Commensurability, comparability, communicability. In *PSA 1982: Proceedings of the 1982 Biennial Meeting of the Philosophy of Science Association*, vol. 2, ed. P. D. Asquith and T. Nickles (East Lansing, MI: Philosophy of Science Association), 669–88.

—— 1990: Dubbing and redubbing: the vulnerability of rigid designation. In *Minnesota Studies in the Philosophy of Science*, vol. 14: *Scientific Theories*, ed. C. Wade Savage (Minneapolis: University of Minnesota Press), 298–318.

Putnam, H. 1975: Is semantics possible? In *Philosophical Papers*, vol. 2: *Mind, Language and Reality* (Cambridge: Cambridge University Press), 139–52 (1st pub. 1970).

Scheffler, I. 1982: *Science and Subjectivity*, 2nd edn (Indianapolis: Hackett; 1st pub. Indianapolis: Bobbs-Merrill, 1967).

Thagard, P. 1992: *Conceptual Revolutions* (Princeton: Princeton University Press).

28

Induction and the Uniformity of Nature

COLIN HOWSON

The problem of induction is one of the oldest, and one of the most intractable, of philosophical problems. Possibly its clearest formulation occurs in a celebrated discussion by David Hume, where it is posed as the question of whether there is anything *"in any object, considered in itself, which can afford us a reason for drawing a conclusion beyond it."* Hume's answer, famously, is that there is not: *"we have no reason to draw any inference concerning any object beyond those of which we have had experience,"* even after the observation of their "frequent or constant conjunction" (1739, bk 1, pt III, sec. XII; italics original). However extensive the observational evidence, there is, according to Hume, no legitimate inference to the truth or even the probability of any hypothesis whose logical content transcends that evidence; what today we call *ampliative*, or *inductive*, inference is for Hume no species of reasoning at all, merely a psychological propensity (see HUME).

If this is true, then science stands on no surer evidential foundation than the crudest superstition. The uncongenial nature of this conclusion prompted generations of philosophers to try to find some flaw in his reasoning. This has proved far from easy, and many have reluctantly concluded that it cannot be successfully rebutted. First, Hume points out, extrapolations from experience, in whatever way they might be made, are not *deductive*: it is not contradictory to affirm both that bread has always nourished and that it will cease to do so tomorrow. What other reason might there be to infer, from the evidence that it has always nourished, that it will continue to nourish, or even that it probably will? One might adduce the additional fact that inferences from a numerous sample to future instances have proved successful in the past. But, as Hume pointed out, since it is precisely the validity of an inference from past to future which is at issue, to cite in its support a further fact about the past does not advance the argument.

Some people, like Russell (1959, p. 33), have reluctantly conjectured that the validity of inductive inferences can be secured only by adopting an independent postulate, or *inductive principle*, that observed regularities can be extrapolated into the future with a probability that approaches certainty with sufficiently extensive evidence. A variant proposal of Strawson's (1952), which amounts to much the same thing, is to regard the principle not as a major premise transforming a deductively invalid inference into a deductively valid one, but as a valid *rule of inductive inference*.

Whether as rule of inference or independent postulate, the rule can be viewed as giving methodological expression to a synthetic *principle of the uniformity of Nature*. Clearly, there can be no a posteriori justification for this principle, for it was proposed precisely in order to validate a posteriori reasoning. A fact perhaps less widely appreciated is that unless restricted in some more or less *ad hoc* way, any rule licensing the extrapolation of observed regularities is also unsound, however large the sample is stipulated to be. For one property which characterizes without exception the members of any sample is, of course, that of belonging to that sample. But it is clearly false that this property will belong to any individuals not yet sampled.

It will not help to require that the sample property in question be capable in principle of being possessed by any individual whatever. To take a famous example of Nelson Goodman's (1954), suppose that all emeralds so far observed are green. Call something "grue" if it has already been observed and found to be green, or it has not been observed so far and is blue. It follows that all emeralds observed so far are also grue. But the next emerald to be observed, let alone all those as yet unobserved, cannot be both green and grue. Goodman's example is a modern version of the older curve-fitting problem: any finite set of points which lie on one curve also lie on infinitely many others, and so by themselves afford no basis for distinguishing the true curvilinear hypothesis, if there is one.

Some writers, including Goodman, have responded by proposing criteria for discriminating properties which are *projectible* beyond the sample from those, like "grue," which allegedly are not. Goodman himself suggested *entrenchment* (see EVIDENCE AND CONFIRMATION). Others, like the Bayesian, Jeffreys (1961, p. 47), have proposed *simplicity*, determined in Jeffreys's theory by the number of independent adjustable parameters involved in the definition of whatever property characterizes members of the sample. Jeffreys's *simplicity postulate* assigns the hypothesis "All emeralds are grue" a smaller a priori probability than "All emeralds are green," since the former incorporates a time parameter determined by the sample (though Goodman has shown that by adopting different basic predicates, that ordering can be reversed). Any proposed criterion, however, returns us to the original and fundamental objection: that the inductive rule incorporating it will necessarily beg the question of its own reliability.

The conclusion appears to be an unresolvable dilemma. On the one hand, inductive inferences seem to require an inductive principle to warrant their validity. In some systems of probabilistic inductive logic, like that of Carnap (1950), the inductive principle is implicit in some initial distribution of probabilities over a partition of the logical space of a language. In Carnap's so-called λ-continuum (1952), it is represented by λ itself, which determines the degree to which the probability function is "prepared" to extrapolate sample uniformities.

On the other hand, any such principle inevitably begs the question of its own authenticity. Popper has tried to avoid the impasse by denying that a rule of inductive inference is necessary at all. In the light of the grue example, this position seems rather obviously untenable, and indeed Popper does, in his theory of corroboration, in effect provide such a rule (1959, appendix *ix). C. D. Broad observed that induction was the glory of science and the scandal of philosophy; so it seems set to remain (1952, p. 143).

References

Broad, C. D. 1952: The philosophy of Francis Bacon. In *Ethics and the History of Philosophy* (London: Routledge and Kegan Paul).

Carnap, R. 1950: *Logical Foundations of Probability* (Chicago: University of Chicago Press).

—— 1952: *The Continuum of Inductive Methods* (Chicago: University of Chicago Press).

Goodman, N. 1954: *Fact, Fiction and Forecast* (London: Athlone Press).

Hume, D. 1739: *A Treatise of Human Nature*, Books 1 and 2 (repr. London: Fontana Library).

Jeffreys, H. 1961: *Theory of Probability* (Oxford: Clarendon Press).

Popper, K. R. 1959: *Logic of Scientific Discovery* (New York: Basic Books).

Russell, B. A. W. 1959: *The Problems of Philosophy* (Oxford: Oxford University Press).

Strawson, P. F. 1952: *Introduction to Logical Theory* (London: Methuen).

29

Inference to the Best Explanation

PETER LIPTON

Science depends on judgments of the bearing of evidence on theory. Scientists must judge whether an observation or the result of an experiment supports, disconfirms, or is simply irrelevant to a given hypothesis. Similarly, scientists may judge that, given all the available evidence, a hypothesis ought to be accepted as correct or nearly so, rejected as false, or neither. Occasionally, these evidential judgments can be make on deductive grounds. If an experimental result strictly contradicts a hypothesis, then the truth of the evidence deductively entails the falsity of the hypothesis. In the great majority of cases, however, the connection between evidence and hypothesis is nondemonstrative or inductive. In particular, this is so whenever a general hypothesis is inferred to be correct on the basis of the available data, since the truth of the data will not deductively entail the truth of the hypothesis. It always remains possible that the hypothesis is false even though the data are correct.

One of the central aims of the philosophy of science is to give a principled account of these judgments and inferences connecting evidence to theory. In the deductive case, this project is well advanced, thanks to a productive stream of research into the structure of deductive argument that stretches back to antiquity. The same cannot be said for inductive inferences. Although some of the central problems were presented incisively by David Hume in the eighteenth century, our current understanding of inductive reasoning remains remarkably poor, in spite of the intense efforts of numerous epistemologists and philosophers of science.

The model of Inference to the Best Explanation is designed to give a partial account of many inductive inferences, both in science and in ordinary life. One version of the model was developed under the name "abduction" by Charles Sanders Peirce early in the twentieth century, and the model has been considerably developed and discussed over the last 25 years (see PEIRCE). Its governing idea is that explanatory considerations are a guide to inference, that scientists infer from the available evidence to the hypothesis which would, if correct, best explain that evidence. Many inferences are naturally described in this way. Darwin inferred the hypothesis of natural selection because, although it was not entailed by his biological evidence, natural selection would provide the best explanation of that evidence. When an astronomer infers that a star is receding from the Earth with a specified velocity, she does this because the recession would be the best explanation of the observed red shift of the star's characteristic spectrum. When a detective infers that it was Moriarty who committed the crime, he does so because this hypothesis would best explain the fingerprints, blood stains, and other forensic evidence. Sherlock Holmes to the contrary, this is not a matter of deduction. The evidence will not entail that Moriarty is to blame, since it always remains possible that someone else was the perpetrator. Nevertheless, Holmes is right to make his inference, since Moriarty's guilt would provide a better explanation of the evidence than would anyone else's.

Inference to the Best Explanation can be seen as an extension of the idea of "self-evidencing" explanations, where the phenomenon that is explained in turn provides an essential part of the reason for believing that the explanation is correct. For example, a star's speed of recession explains why its characteristic spectrum is red-shifted by a specified amount, but the observed red shift may be an essential part of the reason the astronomer has for believing that the star is receding at that speed. Self-evidencing explanations exhibit a curious circularity, but this circularity is benign. The recession is used to explain the red shift, and the red shift is used to confirm the recession; yet the recession hypothesis may be both explanatory and well supported. According to Inference to the Best Explanation, this is a common situation in science: hypotheses are supported by the very observations they are supposed to explain. Moreover, on this model, the observations support the hypothesis precisely because it would explain them. Inference to the Best Explanation thus partially inverts an otherwise natural view of the relationship between inference and explanation. According to that natural view, inference is prior to explanation. First, the scientist must decide which hypotheses to accept; then, when called upon to explain some observation, she will draw from her pool of accepted hypotheses. According to Inference to the Best Explanation, by contrast, it is only by asking how well various hypotheses would explain the available evidence that she can determine which hypotheses merit acceptance. In this sense, Inference to the Best Explanation has it that explanation is prior to inference.

There are two different problems that an account of induction in science might purport to solve. The problem of description is to give an account of the principles that govern the way in which scientists weigh evidence and make inferences. The problem of justification is to show that those principles are sound or rational – for example, by showing that they tend to lead scientists to accept hypotheses that are true and to reject those that are false. Inference to the Best Explanation has been applied to both problems.

The difficulties of the descriptive problem are sometimes underrated, because it is supposed that inductive reasoning follows a simple pattern of extrapolation, with "More of the Same" as its fundamental principle. Thus we predict that the sun will rise tomorrow because it has risen every day in the past, or that all ravens are black because all observed ravens are black. This model of "enumerative induction" has been shown, however, to be strikingly inadequate as an account of inference in science. On the one hand, a series of formal arguments, most notably the so-called raven paradox and the new riddle of induction (see CONFIRMATION, PARADOXES OF), have shown that the enumerative model is wildly over-permissive, treating virtually any observation as if it were evidence for any hypothesis. On the other hand, the model is much too restrictive to account for most scientific inferences. Scientific hypotheses typically appeal to entities and processes not mentioned in the evidence that supports them and often themselves unobservable, not merely unobserved, so the principle of More of the Same does not apply. For example, while the enumerative model may account for the inference that a scientist makes from the observation that the light from one star is red-shifted to the conclusion that the light from another star will be red-shifted as well, it will not account for the inference from observed red shift to unobserved recession.

The best-known attempt to account for these "vertical" inferences that scientists make from observations to hypotheses about the often unobservable reality that stands behind them is the hypothetico-deductive model. According to it, scientists deduce predictions from a hypothesis (along with various other "auxiliary premises") and then determine whether those predictions are correct. If some of them are not, the hypothesis is disconfirmed; if all of them are, the hypothesis is confirmed and may eventually be inferred. Unfortunately, while this model does make room for vertical inferences, it remains, like the enumerative model, far too permissive, counting data which are in fact totally irrelevant to it as confirming a hypothesis. For example, since a hypothesis (H) entails the disjunction of itself and any prediction whatever (H or P), and the truth of the prediction establishes the truth of the disjunction (since P also entails (H or P)), any successful prediction will count as confirming any hypothesis, even if P is the prediction that the sun will rise tomorrow and H the hypothesis that all ravens are black.

What is wanted is an account that permits vertical inference without permitting absolutely everything, and Inference to the Best Explanation promises to fill the bill. Inference to the Best Explanation sanctions vertical inferences, because an explanation of some observed phenomenon may appeal to entities and processes not themselves observed; but it does not sanction just any vertical inference, since obviously a particular scientific hypothesis would not, if true, explain just any observation. A hypothesis about raven coloration will not, for example, explain why the sun rises tomorrow. Moreover, Inference to the Best Explanation discriminates between different hypotheses all of which would explain the evidence, since the model sanctions only an inference to the hypothesis which would best explain it.

Inference to the Best Explanation thus has the advantages of giving a natural account of many inferences and of avoiding some of the limitations and excesses of other familiar accounts of nondemonstrative inference. If it is to provide a serious model of induction, however, Inference to the Best Explanation needs to be developed and articulated, and this has not proved an easy thing to do. More needs to be said, for example, about the conditions under which a hypothesis explains an observation. Explanation is itself a major research topic in the philosophy of science, but the standard models of explanation yield disappointing results when they are plugged into Inference to the Best Explanation. For example, the best-known account of scientific explanation is the deductive-nomological model, according to which an event is explained when its description can be deduced from a set of premises that essentially includes at least one law. This model has many flaws (see EXPLANATION). Moreover, it is isomorphic to the hypothetico-deductive model of confirmation, so it would disappointingly reduce Inference to the Best Explanation to a version of hypothetico-deductivism.

The difficulty of articulating Inference to the Best Explanation is compounded when we turn to the question of what makes one explanation better than another. To begin with, the model suggests that inference is a matter of choosing the best from among those explanatory hypotheses that have been proposed at a given time, but this seems to entail that at any time scientists will infer one and only one explanation for any set of data. Yet scientists are sometimes agnostic, unwilling to infer any of the available hypotheses, and they are also sometimes happy to infer more than one explanation, when the explanations are compatible. "Inference to the Best Explanation" must thus be glossed by the more accurate but less memorable phrase, "inference to the best of the available competing explanations, when the best one is sufficiently good." But under what conditions is this complex condition satisfied? How good is "sufficiently good"? Even more fundamentally, what are the factors that make one explanation better than another? Standard models of explanation are virtually silent on this point. This does not suggest that Inference to the Best Explanation is incorrect, but, unless we can say more about explanation, the model will remain relatively uninformative.

Fortunately, some progress has been made in analyzing the relevant notion of the best explanation. We may begin by considering a basic question about the sense of "best" that the model requires. Does it mean the most probable explanation or, rather, the explanation that would, if correct, provide the greatest degree of understanding? In short, should Inference to the Best Explanation be construed as inference to the *likeliest* explanation, or as inference to the *loveliest* explanation? A particular explanation may be both likely and lovely, but the notions are distinct. For example, if one says that smoking opium tends to put people to sleep because opium has a "dormative power," one is giving an explanation that is very likely to be correct, but not at all lovely: it provides very little understanding. At first glance, it may appear that likeliness is the notion that Inference to the Best Explanation ought to employ, since scientists presumably infer only the likeliest of the competing hypotheses they consider. This is probably the wrong choice, however, since it would severely reduce the interest of the model by pushing it towards triviality. Scientists do infer what they judge to be the likeliest hypothesis, but the main point of a model of inference is precisely to say how these judgments are reached, to give what scientists take to be the symptoms of likeliness. To say that scientists infer the likeliest explanations is perilously like saying that great chefs prepare the tastiest meals: true perhaps, but not very informative if one wants to know the secrets of their success. Like the dormative power explanation of the effects of opium, "Inference to the Likeliest Explanation" would itself be an explanation of scientific practice which provides but little understanding.

The model should thus be construed as "Inference to the Loveliest Explanation." Its central claim is that scientists take loveliness as a guide to likeliness, that the explanation that would, if correct, provide the most understanding is the explanation that is judged likeliest to be correct. This, at least, is not a trivial claim, but it poses at least three challenges. The first is to identify the explanatory virtues, the features of explanations that contribute to the degree of understanding they provide. The second is to show that these aspects of loveliness match judgments of likeliness, that the loveliest explanations tend also to be those that are judged likeliest to be correct. The third challenge is to show that, granting the match between loveliness and judgments of likeliness, the former is in fact the scientists' guide to the latter.

To begin with the challenge of identification, there are a number of plausible candidates for the explanatory virtues, including scope, precision, mechanism, unification, and simplicity. Better explanations explain more types of phenomena, explain them with greater precision, provide more information about underlying mechanisms, unify apparently disparate phenomena, or simplify our overall picture of the world. Some of these features, however, have proved surprisingly difficult to analyze. There is, for example, no uncontroversial analysis of unification or simplicity, and some have even questioned whether these are genuine features of scientific hypotheses, rather than mere artifacts of the way they happen to be formulated, so that the same explanation will count as simple if formulated in one way, but complex if formulated in another.

A different, but complementary, approach to the problem of identifying some of the explanatory virtues focuses on the contrastive structure of many why-questions. A request for the explanation of some phenomenon often takes a contrastive form: one asks not simply "Why P?," but "Why P *rather than* Q?" What counts as a good explanation depends not just on fact P, but also on the foil Q. Thus an increase in temperature might be a good explanation of why the mercury in a thermometer rose rather than fell, but would not be a good explanation of why it rose rather than breaking the glass. Accordingly, it is possible to develop a partial account of what makes one explanation of a given phenomenon better than another by specifying how the choice of a foil determines the adequacy of contrastive explanations. Although many explanations both in science and in ordinary life specify some of the putative causes of the phenomenon in question, the structure of contrastive explanation shows why not just any causes will do. Roughly speaking, a good explanation requires a cause that "made the difference" between the fact and the foil. Thus the fact that Smith had untreated syphilis may explain why he, rather than Jones, contracted paresis (a form of partial paralysis), if Jones did not have syphilis; but it will not explain why Smith rather than Doe contracted paresis, if Doe also had untreated syphilis. Not all causes provide lovely explanation, and an account of contrastive explanation helps to identify which do and which do not.

Assuming that a reasonable account of the explanatory virtues is forthcoming, the second challenge to Inference to the Best Explanation concerns the extent of the match between loveliness and judgments of likeliness. If Inference to the Best Explanation is along the right lines, then the lovelier explanations ought also in general to be judged likelier. Here the situation looks promising, since the features we have tentatively identified as explanatory virtues seem also to be inferential virtues – that is, features that lend support to a hypothesis. Hypotheses that explain many observed phenomena to a high degree of accuracy tend to be better supported than hypotheses that do not. The same seems to hold for hypotheses that specify a mechanism, that unify, and that are simple. The overlap between explanatory and inferential virtues is certainly not total, but at least some cases of hypotheses that are likely but not lovely, or conversely, do not pose a particular threat to Inference to the Best Explanation. As we have already seen, the dormative power explanation of opium's soporific effect is very likely, but not at all lovely; but this is not a threat to the model, properly construed. There surely are deeper explanations for the effect of smoking opium, in terms of molecular structure and neurophysiology, but these explanations will not compete with the banal account, so the scientist may infer both without violating the precepts of Inference to the Best Explanation.

The structure of contrastive explanation also helps to meet this matching challenge, because contrasts in why-questions often correspond to contrasts in the available evidence. A good illustration of this is provided by Ignaz Semmelweiss's nineteenth-century investigation of the causes of childbed fever, an often fatal disease contracted by women who gave birth in the hospital where Semmelweiss did his research. Semmelweiss considered many possible explanations. Perhaps the fever was caused by "epidemic influences" affecting the districts around the hospital, or perhaps it was caused by some condition in the hospital itself, such as overcrowding, poor diet, or rough treatment. What Semmelweiss noticed, however, was that almost all of the women who contracted the fever were in one of the hospital's two maternity wards, and this led him to ask the obvious contrastive question, and then to rule out those hypotheses which, though logically compatible with his evidence, did not mark a difference between the wards. It also led him to infer an explanation that would explain the contrast between the wards: namely, that women were inadvertently being infected by medical students who went directly from performing autopsies to obstetrical examinations, but only examined women in the first ward. This hypothesis was confirmed by a further contrastive procedure, when Semmelweiss had the medics disinfect their hands before entering the ward: the infection hypothesis was now seen also to explain not just why women in the first rather than the second ward contracted childbed fever, but also why women in the first ward contracted the fever before, but not after, the regime of disinfection was introduced. This general pattern of argument, which seeks explanations that account not only for a given effect, but also for particular contrasts between cases where the effect occurs and cases where it is absent, is very common in science – for example, wherever use is made of controlled experiments.

This leaves the challenge of guiding. Even if it is possible to give an account of explanatory loveliness (the challenge of identification) and to show that the explanatory and inferential virtues coincide (the challenge of matching), it remains to be argued that scientists judge that a hypothesis is likely to be correct *because* it is lovely, as Inference to the Best Explanation claims. Thus a critic of the model might concede that likely explanations tend also to be lovely, but argue that inference is based on other considerations, which have nothing to do with explanation. For example, one might argue that inferences from contrastive data are really applications of Mill's method of difference (see MILL), which makes no explicit appeal to explanation, or that precision is a virtue because more precise predictions have a lower prior probability and so provide stronger support as an elementary consequence of the probability calculus (see PROBABILITY).

The defender of Inference to the Best Explanation is here in a delicate position. In the course of showing that explanatory and inferential virtues match up, he will also inevitably show that explanatory virtues match some of those other features that competing accounts of inference cite as the real guides to inference. The defender thus exposes himself to the charge that it is those other features, rather than the explanatory virtues, that do the real inferential work. Meeting the matching challenge will thus exacerbate the guiding challenge. The situation is not hopeless, however, since there are at least two ways to argue that loveliness is a guide to judgments of likeliness. As we have seen, at least some of the competing accounts of inference are fraught with difficulties, inapplicable to many scientific inferences and incorrect about others. If it is shown that Inference to the Best Explanation would give a better account of more inferences than any other available account, this is a powerful reason for supposing that loveliness is indeed a guide to likeliness. Second, if there is a good match between loveliness and likeliness, as the guiding challenge grants, this is presumably not a coincidence, so itself calls for an explanation. Why should it be that the hypotheses that scientists judge likeliest to be correct are also those that would provide the most understanding if they were correct? Inference to the Best Explanation gives a very natural answer to this question, similar in structure to the Darwinian explanation of the fact that organisms tend to be well suited to their environments. If scientists select hypotheses on the basis of their explanatory virtues, the match between loveliness and judgments of likeliness follows as a matter of course. Unless the opponents of the model can give a better account of the match, the challenge has been met.

We have been considering the prospects of Inference to the Best Explanation as a partial solution to the problem of describing the structure of scientific inferences, but the model has also been applied to problems of justification. The most fundamental problem of inductive justification is due to David Hume, who argued that there can be no good reason to believe that our inductive practices are even moderately reliable, tending to take us from true observations to true hypotheses or predictions (see HUME). According to Hume, to justify induction, we would have to produce a cogent argument whose conclusion is that induction is generally reliable, and whose premises are not themselves inductively based. The only such premises are reports of past observations and the demonstrative truths of logic and mathematics. All cogent arguments are either deductive or inductive. Now we face a dilemma. There can be no cogent deductive argument for the reliability of induction, since no number of past observations (along with demonstrative truths) deductively guarantees that induction is generally reliable. In particular, past observations will never entail that induction will be reliable in the future. Neither is there a cogent inductive argument for induction, since any such argument presupposes the very practice it is supposed to justify. For example, to argue that induction is likely to be reliable in future on the grounds that it has been reliable in the past would beg the question, even if it were granted that the past reliability of induction could itself be known on the basis of observation. Hence our inductive practices are unjustifiable.

If Hume's argument is sound, there is no reason whatever to believe any scientific claim that goes beyond what has been directly observed, which is, at the very least, to say that there is no reason to believe any scientific prediction, hypothesis, or theory. This is incredible; but the skeptical argument has proved extraordinarily resilient, and there is still no generally accepted answer to it. For all of Hume's sophistication in presenting the problem of justification, however, his solution to the problem of description is rather primitive. He seems to have accepted a version of the simple enumerative "More of the Same" model of induction discussed above. Consequently, one might hope that a more sophisticated and accurate account of inductive practice would make it possible to avoid or rebut Hume's skeptical argument. In particular, it is sometimes supposed that Inference to the Best Explanation provides such an account.

Unfortunately, Inference to the Best Explanation does not solve Hume's problem. The description he gave of induction was incorrect, but his skeptical argument does not depend on it. Indeed, the argument seems to depend on little more than the undeniable fact that inductive arguments are not deductively valid. Reports of past observations will never entail that future inferences to the best explanations will in fact select true hypotheses; and any argument that the reliability of Inference to the Best Explanation would itself be the best explanation of what we have observed begs the question. It might even be claimed that Inference to the Best Explanation exacerbates the problem of justification, since it is quite unclear why the hypothesis that would, if correct, provide the deepest understanding is also in fact likeliest to be correct. Why should we suppose that ours is the loveliest of all possible worlds? This additional worry may be an overreaction, however, since what Hume's skeptical argument suggests is that the success of any other method of induction would be equally mysterious.

Inference to the Best Explanation has also been invoked to solve more modest problems of inductive justification. Even if the model is of no avail against a complete inductive skeptic, it might have a role to play in the defense of scientific realism, according to which there are good reasons to believe that well-supported theories are likely to be at least approximately true, as against positions such as constructive empiricism, according to which we can have reason to believe only that our best theories are empirically adequate, that their observable consequences are true. (Constructive empiricism has been developed in detail by Bas van Fraassen, who is also a vigorous critic of Inference to the Best Explanation.) The constructive empiricist is no inductive skeptic, since to say that all the observable consequences of a theory are true is a much stronger claim than to say merely that its observed consequences are true; but the realist goes further by sanctioning, in addition, vertical inferences to the truth of a theory's claims about unobservable entities and processes.

Perhaps the best-known example of this application of Inference to the Best Explanation in defense of scientific realism is the so-called miracle argument, discussed by Hilary Putnam. He takes it that the model provides a good solution to the descriptive problem, and proposes that philosophers may themselves make an inference to the best explanation in defense of scientific realism. Suppose that all the many and varied predictions derived from a particular scientific theory are found to be correct: what is the best explanation of this predictive success? According to Putnam, the best explanation is that the theory itself is true. If the theory were true, then the truth of its deductive consequences would follow as a matter of course; but if the hypothesis were false, it would be a "miracle" that all its observed consequences were found to be correct. So, by a philosophical application of Inference to the Best Explanation, we are entitled to infer that the theory is true, since the "truth explanation" is the best explanation of the theory's predictive success. This higher-level inference is supposed to be distinct from the first-order inferences that scientists make, but of the same form.

This justificatory application of Inference to the Best Explanation has considerable intuitive appeal, but it faces three objections. The first is that the truth explanation for the predictive success of a theory is not really distinct from the substantive scientific explanation that the theory provides and on the basis of which it was inferred by scientists in the first place. If this is so, then the miracle argument provides no additional reason to believe that the hypothesis is correct: it is merely a repetition of the scientific inference it was supposed to justify. This objection can be answered, however, by observing that the two sorts of explanation have a different structure. The scientific explanations that a theory provides are typically causal, whereas the truth explanation is logical. The truth of a theory does not physically cause its consequences to be true; the explanatory connection is rather that a valid argument with true premises must also have a true conclusion.

The second objection to the miracle argument is that, even if the truth explanation is distinct from the scientific explanations, the inference to the truth of the theory is vitiated by the same sort of circularity that Hume appealed to in his skeptical argument. In effect, the miracle argument is an attempt to use an inference to the best explanation to justify scientific inferences to the best explanation; so, the objector will claim, such an argument must beg the question of the reliability of this form of inference. In particular, the constructive empiricist may insist that, although he will allow the legitimacy of some forms of induction, inferences to the truth of theories that traffic in unobservables are precisely those that are at issue. One possible response to the circularity objection is to argue that the circle is broken in virtue of the difference between inferences to causal and to logical explanations; but the objection has considerable force.

The third objection to the miracle argument is that truth is simply not the best explanation of predictive success, so the argument fails on its own terms. The obvious way to flesh out this objection is to give another explanation that is at least as good. For example, the constructive empiricist may claim that we can explain the predictive success of a theory by supposing that it is empirically adequate, that all its observable consequences are true, whether or not the theory is true as a whole. In this case, however, the defender of the miracle argument has two ready replies. First, it is far from clear that the explanation in terms of empirical adequacy is as lovely as the truth explanation, since it is dangerously close to saying that the consequences of the theory are true because they are true, an extremely unlovely explanation, reminiscent of the appeal to opium's dormative power. Moreover, even if, as in the opium case, we infer this explanation, it does not preclude an inference to the truth explanation, since the two explanations are compatible: a theory may be both empirically adequate and true. The third objection to the miracle argument can be made more pressing, however, through a better choice of alternative explanations. For, given any set of successful predictions, there are always in principle many theories incompatible with the original one which nevertheless share those consequences (see UNDERDETERMINATION OF THEORY BY DATA). The truth of any of the competing theories would also explain the predictive success they share with the original theory, and it is unclear that these alternative truth explanations would be any less lovely than the original. The inference to the truth of the original theory may thus be blocked.

Neither of the justificatory applications of Inference to the Best Explanation we have considered appears promising. If the model can help to solve problems of inductive justification, these are likely to concern more specific aspects of scientists' inductive practices. For example, the model has been plausibly applied in an argument to show why it is rational for scientists to put greater weight on data that a hypothesis correctly predicts than on data that was available when the hypothesis was formulated and which it was constructed to accommodate. Whatever the justificatory potential of Inference to the Best Explanation, however, the model may be counted a philosophical success if it can be shown to give an illuminating description of some of the general inferential principles that guide scientific practice.

Further reading

Garfinkel, A. 1981: *Forms of Explanation* (New Haven: Yale University Press).

Harman, G. 1965: The inference to the best explanation. *Philosophical Review*, 74, 88–95.

Hempel, C. G. 1965: *Aspects of Scientific Explanation* (New York: Free Press).

Hume, D. 1975: *An Enquiry Concerning Human Understanding* (Oxford: Clarendon Press; 1st pub. London, 1777), secs 4, 5.

Lipton, P. 1991: *Inference to the Best Explanation* (London: Routledge).

Peirce, C. S. 1931: *Collected Papers*, ed. C. Hartshorn and P. Weiss (Cambridge, MA: Harvard University Press), 5. 180–9.

Putnam, H. 1978: *Meaning and the Moral Sciences* (London: Hutchinson), 18–22.

Thagard, P. 1978: The best explanation: criteria for theory choice. *Journal of Philosophy*, 75, 76–92.

van Fraassen, B. 1980: *The Scientific Image* (Oxford: Oxford University Press).

30

Judgment, Role in Science

HAROLD I. BROWN

Introduction

According to a widely held view of science, scientific hypotheses are evaluated on the basis of observational data in accordance with the rules of inductive logic. Inductive logic, like deductive logic, is supposed to consist of a set of formal rules. These rules abstract from any details of the specific hypothesis under examination, the context in which the evaluation is taking place, and the individuals who carry out the evaluation. Observational data are also independent of the context or the observer in the following sense: there may be individual and cultural limitations on what observations are undertaken, but the outcome of an observation procedure is presumably independent of the observer's preferences or peculiarities. Now, given impersonal data and a set of formal rules, the assessment of a hypothesis will be completely impersonal. The fact that a scientist thinks of a hypothesis and undertakes to evaluate that hypothesis may depend on individual features of that scientist and on the current state of science, but the resulting evaluation will be free of individual, historical, or cultural factors. Any two scientists who evaluate a hypothesis on the basis of the same observational evidence must arrive at the same evaluation of that hypothesis.

We shall see in this chapter that this ideal does not come close to describing how science actually works. Rather, the process of evaluating a hypothesis requires multiple decisions that must be made by individual scientists or by an organized scientific community without benefit of formal rules. We shall also see that the reliability of science depends on the reliability of these decisions. If this claim is true, a long tradition will tempt many to draw a skeptical conclusion about science. According to this tradition, if the evaluation of a hypothesis depends on human decisions, rather than on the dictates of rigorous rules and impersonal data, scientific beliefs are subjective and, ultimately, arbitrary. We shall see below that the skeptical conclusion should not be accepted. It will be argued that as scientists learn their craft, they develop an ability to exercise scientific judgment which yields evaluations that are not carried out in accordance with rules, but that are more reliable than arbitrary choices. In addition, science is organized so as to marshal scientific judgment in a way that improves this reliability – although the element of fallibility is never eliminated. Our first concern in developing this view is to consider what is meant by *judgment*.

Judgment

Judgment should be thought of as a cognitive skill that is analogous to physical skills. Examples of physical skills include the ability to ride a bicycle, hit or catch a baseball, and use specific tools. The key point about these skills is that they are not learned by explicitly learning a set of rules, which are then followed in exercising the skill. There are many well-developed skills whose practitioners cannot formulate any set of rules that guide their behavior; in some cases, no one can formulate such rules. Rather, skills are learned by practice – usually, but not always, under the guidance of someone who has already mastered that skill. Skill learning exhibits the phenomenon of talent: some people learn a particular skill more readily than others, and some achieve a higher level of performance than others, even given comparable opportunities. Skills tend to deteriorate with lack of practice, although resumption of practice will typically result in relearning at a more rapid rate than was required for the original learning. Finally, the exercise of skills is fallible: even the most adept may fail on some occasions. But the fallibility of skills does not provide any grounds for denying that skills are learned and exercised.

Cognitive skills can be introduced by considering the construction of deductive proofs in a typical system of formal logic. Such systems provide a set of explicit rules, and each step in a proof must be justified by appeal to one of those rules. But the process of constructing a proof – that is, the process of deciding which rule to apply at each stage – is not governed by any comparable set of rules. Rather, people learn to construct proofs much as they learn to use physical tools – by practice under the guidance of someone who has already learned the skill. Most people can learn to construct proofs, and their skill improves with practice. Constructing proofs exhibits the phenomena of talent and of "becoming rusty," as well as the fallibility that characterizes physical skills. Moreover, no one can provide an algorithm for constructing formal proofs. As a second example, consider the writing of computer programs. This is another skill that is learned by practice and pursued without following established rules, even though the exercise of this skill results in the production of an algorithm.

These examples of cognitive skills also provide instances of the exercise of judgment: they require individuals to decide how to proceed in carrying out a cognitive activity, and to do so without following an algorithm. We develop the ability to exercise judgment as we master a particular subject matter. For example, people develop logical judgment, medical judgment, legal judgment, or engineering judgment. The ability to exercise judgment is a result of experience, and is specific to definite subject matters; there is no general ability to judge on every subject whatsoever.

The main thesis of this chapter is that scientists learn to exercise judgment in their particular specialty or sub-specialty as they master available knowledge and techniques in that field. Decisions such as whether to accept or reject a hypothesis, to carry out an experiment, or to devote individual or communal resources to the further development of a research program are judgment calls. It must be emphasized that observations and formal techniques are not irrelevant – they are vital, and scientific judgment cannot proceed in their absence. But observation and formal techniques are not sufficient for dictating these decisions. Let us consider some of the major points at which judgment is required in evaluating scientific claims.

Inductive confirmation

We begin by considering reasons for accepting a universal generalization; we shall examine two types of situations (see EVIDENCE AND CONFIRMATION). The first concerns isolated generalizations such as "All ravens are black." Here the evidence will consist of a large number of observations of black ravens and no observations of ravens that are not black. The second case concerns scientific theories which consist of a set of generalizations that cannot be evaluated separately – such as Newton's theory of gravitation, which melds his laws of motion and principle of gravitation (see NEWTON). In this case empirical support is developed by deducing observable results from the theory and confirming these results.

Assume, for the moment, that observation provides impersonal data. In each of our cases there may exist a large body of observations that support the hypothesis, but no body of observations can *prove* the hypothesis true, because universal hypotheses make predictions that go beyond any body of evidence that we may have accumulated at any point in time. It is this point that yields the key distinction between deductive and inductive logic. The conclusion of a deductive argument makes explicit information that was built into the premises; thus we have a guarantee that if the premises are true, then the conclusion is also true. Inductive arguments seek to establish conclusions that go beyond the evidence provided by observation. An inductive evaluation of a hypothesis always leaves open the possibility that the conclusion is false, even though it is supported by all available evidence. This feature of the logic of induction leads directly to the classical problem of induction, first formulated by Hume (see HUME).

Hume's central point is deceptively simple. It consists of noting, for example, that even though every raven observed so far has been black, it is possible to conceive of a bird that is like a raven in every respect except that its feathers are not black. A classic example is provided by the fate of the hypothesis that all swans are white. This hypothesis had long served as a standard example of a universal generalization that had been proved by induction. But the hypothesis was show to be false when black swans were discovered in Australia.

We can read Hume as underlining the point that we should not expect deductive certainty from inductive logic. But once we acknowledge this conclusion, we must ask whether scientists should ever accept any universal generalization, given the intrinsic fallibility of such acceptance. Yet it is clear that science cannot exist at all without accepting a substantial body of such generalizations. In addition to an intrinsic interest in universal generalization, scientists implicitly assume many such generalizations whenever they accept the reliability of, say, a microscope or a calculator.

One might be tempted to conclude that our acceptance of universal generalizations lacks a rational basis; but we need not jump to this conclusion. Scientists have good reasons for accepting many universal generalizations, even though these reasons are fallible. But these reasons require more than just observational data and formal relations. Rather, generalizations are accepted as a result of judgments made by skilled individuals who reflect on the information and the alternatives that are available in the field of their expertise. To take but one example, a biologist working in the context of late twentieth-century genetics and ecology would not be likely to conclude that all swans ever to be observed will be white, even if massive numbers of swans were observed and all were white. Too many factors are now understood to be involved in determining the outer color of an animal for this simple inductive generalization to be justified. As the example illustrates, accepting a generalization requires weighing a number of factors in addition to the available observational evidence. Such judgments will be highly dependent on the scientific context in which they must be made. Moreover, the ability to make such judgments will require much information that can be acquired only by studying the particular subject matter in question, as well as the skills that are developed through experience in that field. The development of judgment is a problem of a totally different kind from that involved in examining formal relations between a hypothesis and a body of data.

There is a second problem involved in accepting universal generalizations, which Goodman (1965) has labeled "the new riddle of induction." This problem arises because as long as we restrict ourselves to formal relations, a given body of evidence provides equal support for an unlimited number of mutually incompatible hypotheses. An artificial example will illustrate the problem. Suppose we have the following data on the relationship between an independent variable, X, and a dependent variable, Y:

X	1	2	3	4	5	6	7	8	9	10
Y	2	4	6	8	10	12	14	16	18	20

We now have two concerns: to predict values of Y for further values of X and to find a general relation between X and Y. The thought that $X = 11$ and $X = 12$ would yield $Y = 22$ and $Y = 24$ respectively is virtually irresistible; so is the thought that the general relation is $Y = 2X$. But a bit of cleverness will show that many other possibilities are equally supported by the data. Consider, for example, $Y = (X - 1)(X - 2) \ldots (X - 10) + 2X$, where the ellipses indicated that there are ten terms in the initial product. This relation generates the above data, but yields $Y = 3,638,822$ for $X = 11$ and $Y = 39,916,824$ when $X = 12$.

Such possibilities occur in real science. Indeed, they are easier to arrive at in real science, because one requires that a formula yield predictions only within the limits of accuracy of the observational data. So, Galileo concluded that objects fall to the Earth with constant acceleration after measuring falls of only a few feet, and he generalized this result to all distances of fall – he even used it to compute how long it would take a stone to fall back to the Earth from the Moon (see GALILEO). Newton's more complex inverse square law yields Galileo's data within the limits of accuracy of Galileo's measurements, but very different results when we consider longer falls. Moreover, Newton (in effect) generalized from data on low-velocity motions to high-velocity motions (see NEWTON). This step was corrected by Einstein's vastly more complex account – an account which yields the Newtonian and Galilean results within the limits of precision available to those earlier scientists (see EINSTEIN).

One more example will show that this variety of possible generalizations is not limited to quantitative cases. Both Locke and Hume pointed out that someone who lived only in a tropical climate would probably conclude that water remains liquid as the temperature drops, and would be shocked at the suggestion that at some sufficiently low temperature water would suddenly solidify.

These examples introduce another cluster of judgments that is required for scientific research. Given any set of data, the decision to opt for one of the possible hypotheses cannot be dictated solely by the data and some formal rules. Even if we are only selecting hypotheses for further testing, we must still decide which hypotheses are worth pursuing. Many considerations, such as simplicity or coherence with other beliefs, may enter into this decision. But, as Kuhn (1977) has argued, scientists must decide which additional factors to consider and how strongly they shall be weighted (see KUHN). Moreover, as science develops, our understanding of the range of possibilities and their relative importance changes. Sitting at the end of the twentieth century and reflecting on the tale of Galileo, Newton, and Einstein, we might be reluctant to assume that the mathematically simplest hypothesis is likely to be sustained. We should also be more concerned than Galileo was with the accuracy of our measurements, since we have superior technology and relevant historical experience: using Atwood's machine, we can show that Galileo's law of fall is wrong even for short falls. In other words, the range of factors that should enter into scientific judgments changes as the historical context changes.

There are many different respects in which one can be said to "accept" a hypothesis. These include considering it worthy of further evaluation, which requires a commitment of individual and (often) communal resources; using it in the development and interpretation of instrumentation; assuming it when constructing theories in related domains; and using it as the basis for technological applications. The reliability of any of these acceptance decisions rests ultimately on the judgment of those who must make the decisions.

Let us consider one further approach to inductive confirmation which begins by evaluating the probability that a hypothesis is true. We will focus our attention on Bayesian confirmation theory, an important current attempt to develop this program. This approach is based on a theorem of probability calculus. Suppose we are considering a set of competing hypotheses. For each hypothesis, H_i, we have an initial estimate of the probability that the hypothesis is true, $P(H_i)$. We seek to adjust these estimates on the basis of an evidence statement, E. For each hypothesis we determine the probability that E is true if that hypothesis is true, $P(E/H_i)$, and multiply this term by the corresponding $P(H_i)$. Let S be the sum of these terms for all the hypotheses under consideration. Then the adjusted probability for a given hypothesis is:

$$P(H_i/E) = P(H_i) \times P(E/H_i)/S.$$

Note that we are now seeking to evaluate the probability that a hypothesis is true solely on the basis of evidence and a formal algorithm.

But again, our evaluation of H_i will depend on a large set of judgments. For example, our results will depend on the set of hypotheses we judge worthy of consideration. Probability theory requires that the sum of the initial probabilities not exceed one. Thus we do not just assign probabilities to individual hypotheses; rather, we distribute a finite pool of initial probability among a set of hypotheses. Bayes's law can be applied only *after* this initial distribution has been decided; the law then yields a comparative evaluation among the hypotheses under consideration. As a result, if two groups of researchers compare different sets of hypotheses, Bayes's law and the available evidence may lead each group to assign maximum probability to a hypothesis that the other group has not even considered. Yet Bayes's law and the evidence can never *require* that we add a new hypothesis to the mix. Any decision to enlarge the set of hypotheses to include those considered by both groups will have to be arrived at by means other than evidence and the algorithm. Indeed, if we once assign a probability of zero to a hypothesis, Bayes's law and evidence can never change this value. New evidence may lead us to reconsider a hypothesis that we once rejected completely, but this too will have to occur by means not included in Bayes's law.

These examples illustrate some of the ways in which the probabilities we arrive at by applying Bayes's law depend on judgments. In addition, once we arrive at a probability value for a hypothesis, we must still decide whether to accept or reject that hypothesis in all the respects noted above. In other words, Bayes's algorithm is a useful formal tool, but deciding how to apply the algorithm and what to make of its outputs depends on judgment.

Falsification

We have been considering confirmation, but the case in which evidence contradicts a hypothesis appears to be logically simpler (see POPPER). A single non-white swan disproves the hypothesis "All swans are white," while the deduction of a single false evidence statement from a theory shows that there is something wrong with that theory. But the need for judgment reappears dramatically when a theory is falsified. A false consequence deduced from a set of premises proves that at least one of those premises is false, but does not give a clue as to how many are false or which ones these are. Moreover, significant theoretical predictions typically require a large number of premises. Sometimes several theories may be involved; there will also be auxiliary hypotheses stating specific conditions required for the prediction; and deductions from complex mathematical theories will typically involve a number of approximations. Any or all of these may be subject to reconsideration when observation and theory conflict. In addition, the details of the procedure that yields the troubling observation may also be questioned. There are no general rules that dictate what should be done in all these cases.

For example, at the time of writing, an experiment that measures the rate at which neutrinos from the Sun arrive at the Earth has been providing data since 1968, and has consistently yielded results much lower than had been predicted on the basis of available theory. Something is wrong, surely, but the prediction and observation involve a large number of hypotheses. These include the theories of how stars produce energy and of how neutrinos behave, plus hypotheses about the Sun's composition, magnetic field, and more. In addition, the detection procedure for neutrinos involves some complex chemistry, along with our understanding of radioactive decays and the instruments by which these decays are measured. The attempt to determine what is wrong has spawned much further research. New theories have been proposed, and some of these have been subjected to new experimental tests; accepted parameters describing the Sun have been reexamined; multiple evaluations of the detector have been carried out; and, because there are intrinsic limits to the detector, new, more accurate (and more expensive) detectors have been built. There is no possibility that all conceivable options and combinations of options can be considered, and any decision to pursue some option involves a commitment of limited resources that might have been used elsewhere. Even the original decision to carry out the solar neutrino experiment was arrived at only after considerable deliberation, since, at the time it was proposed, there was no serious doubt about the theory of stellar energy production that the experimenters sought to test. But this confidence had to be balanced against the fact that this particular test had never been done before (it was less than a decade since the first experimental detection of neutrinos) and our general understanding of the fallibility of all scientific results.

Observation

As the last example suggests, scientific observation is far from straightforward (see OBSERVATION AND THEORY). Indeed, multiple judgments are often involved in accepting a report as accurate. Consider some examples.

When Galileo turned his telescope on the heavens, he made several observations that were in direct conflict with naked-eye observations. He could not just report his observations; he had to assess which were to be accepted and provided grounds for believing that at least some telescopic observations were more reliable than naked-eye observations.

Eighteenth-century astronomers regularly measured the time at which a moving object reached a specific point by setting the telescope's cross hair on that point and watching, while listening to a clock that ticked off seconds. Since the object would rarely reach the cross hair on a tick, fractions of a second had to be estimated. This complex judgment was considered highly reliable until Bessel discovered systematic differences among astronomers. The problem was ameliorated by the introduction of photography and electronic equipment, but accepting the output of this equipment requires an assessment of the reliability of the devices used. As science has developed, the technology has become steadily more complex, and the body of science that must already be accepted if we are to accept a particular observational result has become progressively richer. The judgment that an observational result is correct depends on all the judgments that went into the acceptance of this prior science.

A particularly important situation has been studied by Galison (1987). Every experiment involves a number of "backgrounds": interfering factors that can confuse the outcome. For example, the solar neutrino detector is in a deep underground cavern. The earth above the detector filters out many cosmic particles that would confuse the result, but a small number of other troublesome particles are produced by the materials making up the walls of the cavern. Experimenters must assess the possible backgrounds, and use additional experimental or calculational techniques to eliminate their effect. There is no algorithm for carrying out this assessment. Rather, the decision that the backgrounds have been adequately dealt with depends on the judgment of the scientists involved, a judgment that may be challenged by other members of the scientific community.

Consider one more example. An important body of data that initially supported Newton's law of fall did not come from more precise laboratory measurement, but from Newton's hypothesis that the Moon continually falls towards the Earth under the influence of the same force that causes a stone to fall. In effect, Newton enlarged the body of available observations by considering two cases which had previously seemed quite disparate to be relevantly similar. This was the result of a judgment on Newton's part.

Conclusion

Let me summarize the main points at which judgment enters into scientific decisions. Consider, first, the decision that a hypothesis is worthy of further consideration. Several hypotheses may occur to various scientists who examine a body of data, or reflect on the current state of a science, or engage in a variety of other reflections. These hypotheses cannot all be subjected to detailed analysis and observational testing; judgments must be made as to which are worthy of further pursuit. Second, once it is decided that a hypothesis shall be pursued, the exact nature of this pursuit must be considered. Hypotheses must be explored for internal consistency, as well as consistency with other hypotheses, and testable consequences must be derived. These are largely problems of deductive logic, but solving them requires the skills needed for any construction of a deductive proof. In addition, if actual testing is to ensue, one must derive conclusions that are in fact testable under existing technological – and perhaps social and economic – conditions. Once a testable result is arrived at, means of carrying out that test must be developed and shown to be reliable. At every stage there will typically be many alternative routes that can be taken without violating any principles of logic or established scientific practice.

Now suppose the test is carried out. Backgrounds must be examined, and an assessment made as to whether a result worthy of being reported has been produced. If so, we may consider three further possibilities. First, the observational result may not have any impact on the hypothesis. Here are two ways in which this can occur. Observational errors may be too large for this to count as a significant test under prevailing conditions. This is not always a straightforward matter. In the solar neutrino case, as it became clear that the experimental results were much lower than expected, the effects of particles produced in the surrounding rock became more significant. Alternatively, the result may fall in a range that fails to distinguish between competing hypotheses.

Second, the outcome may support the hypothesis being tested. Now scientists must decide what attitude to take to the hypothesis, and this may depend on several other factors. It might have to be decided whether further testing is in order. Or it might have to be decided whether the hypothesis is sufficiently well confirmed to be used as the basis for designing an instrument that will then be used to test a quite different hypothesis. Or there might be a question of a technological application that could affect the economics of a company or the lives and safety of people.

Third, suppose the test is negative. Here, all the options noted above come into play: should the hypothesis be rejected, or some auxiliary hypothesis be reconsidered, or the experimental equipment be reassessed, or the approximations made at any of a number of stages be rethought? And in all the cases there remains the decisions as to when enough has been done and a conclusion can be considered definitive – until new challenges arise.

The picture we have arrived at is a long way from the traditional account of scientific knowledge established by the application of a rigorous methodology. But that picture was naive, and it would be a mistake to conclude that without a rigorous method there is no basis for accepting scientific results. The thesis of this chapter is that the accomplishments of science depend on the ability to exercise judgment, an ability that scientists develop as they learn and practice their craft. Scientists develop judgment in the specific fields that they have mastered, and this results in the existence of a set of cognitive skills in the scientific community. These skills are fallible, but their fallibility does not make them epistemically worthless. Rather, cognitive skills provide genuine grounds for evaluating proposals, and all our epistemic accomplishments ultimately rest on our ability to exercise these skills.

References and further reading

Brown, H. I. 1985: Galileo on the telescope and the eye. *Journal of the History of Ideas*, 46, 487–501.
—— 1987: *Observation and Objectivity* (New York: Oxford University Press).
—— 1988: *Rationality* (London: Routledge).
—— 1994: Judgment and reason. *Electronic Journal of Analytic Philosophy*, 2, 5.
Earman, J. 1992: *Bayes or Bust? A Critical Examination of Bayesian Confirmation Theory* (Cambridge, MA: MIT Press).
Galison, P. 1987: *How Experiments End* (Chicago: University of Chicago Press).
Goodman, N. 1965: *Fact, Fiction and Forecast*, 2nd edn (Indianapolis: Bobbs-Merrill).
Hempel, C. G. 1965: Studies in the logic of confirmation. In *Aspects of Scientific Explanation* (New York: Free Press), 3–51.
Horgan, T. E., and Tienson, J. L. 1989: Representation without rules. *Philosophical Topics*, 17, 147–74.
Kuhn, T. S. 1977: Objectivity, value judgment and theory choice. In *The Essential Tension* (Chicago: University of Chicago Press), 320–51.
Newton-Smith, W. H. 1981: *The Rationality of Science* (London: Routledge and Kegan Paul).
Polanyi, M. 1958: *Personal Knowledge: Towards a Post-Critical Philosophy* (New York: Harper & Row).
Popper, K. 1968: *The Logic of Scientific Discovery*, 2nd English edn (New York: Harper & Row).
Salmon, W. C. 1990: Rationality and objectivity in science: Tom Kuhn meets Tom Bayes. In *Scientific Theories*, Minnesota Studies in the Philosophy of Science, 14, ed. C. Wade Savage (Minneapolis: University of Minnesota Press), 175–204.
Sankey, H. 1997: Judgment and rational theory choice. In *Rationality, Relativism and Incommensurability* (Aldershot: Ashgate), 135–46.
Suppes, P. 1984: *Probabilistic Metaphysics* (Oxford: Oxford University Press).
Wartofsky, M. 1980: Scientific judgment: creativity and discovery in scientific thought. In *Scientific Discovery: Case Studies*, ed. T. Nickles (Dordrecht: D. Reidel), 1–20.

31

Kuhn

RICHARD RORTY

Thomas S. Kuhn, historian and philosopher of science, was born on 18 July 1922 in Cincinnati, Ohio, and died 17 June 1996 in Cambridge, Massachusetts. He entered Harvard in 1939 and remained there until 1956, receiving a Ph.D. in physics in 1949. For three years he was a member of the Harvard Society of Fellows, and then began teaching in James Bryant Conant's recently established General Education Program. Conant used a historical approach to communicate the nature of science to undergraduates; working with Conant helped shift Kuhn's interests from physics to the history of science. After leaving Harvard, Kuhn taught at Berkeley for 9 years, at Princeton for 15, and at the Massachusetts Institute of Technology for 12. He retired from teaching in 1991.

After publishing two articles in the *Physical Review* and one in the *Quarterly of Applied Mathematics*, in 1951 Kuhn began publishing in *Isis*, the journal of the history of science edited by George Sarton. In 1957 his first book, *The Copernican Revolution*, appeared. Following up on the work of Alexandre Koyré and others, this book spelled out in detail the gradual breakdown of attempts to reconcile Aristotelian ways of describing physical processes, and Aristotelian ways of thinking about the methods and function of scientific inquiry, with Copernican astronomy and Galilean mechanics. (See GALILEO.) It made clear that the traditional account of the New Science as a victory of "reason" over prejudice and superstition was much too simple, and showed why a revolutionary mechanics and a revolutionary astronomy both required the other and a revolution in the philosophers' account of the nature of science, before either could be fully accepted.

In his second book, *The Structure of Scientific Revolutions* (1962), Kuhn made explicit the philosophical moral of this historical story. The late 1950s and early 1960s were a time of ferment in philosophy of science, for writers such as Michael Polanyi, Imre Lakatos, Stephen Toulmin, Paul Feyerabend, and Norwood Russell Hanson had begun to challenge the picture of scientific inquiry which had been sketched by Rudolph Carnap, Karl Popper, Carl Hempel, and others associated with logical positivism (see LAKATOS; FEYERABEND; POPPER; LOGICAL EMPIRICISM; and LOGICAL POSITIVISM). This picture had taken for granted the idea of an observation language, neutral between alternative scientific theories, in which the explananda of all such theories might be formulated. The logical positivists had tended to assume that there must be a quasi-algorithimic logic of justification, producing rational choices among alternative theories on the basis of such neutrally formulated data – a logic which could be studied without reference to the history of science. Although Hempel, Goodman, and others had pointed out various difficulties faced by attempts to construct such a logic, most philosophers of science in the 1950s still took the idea of such a logic for granted.

Its radical and thoroughgoing repudiation of the idea of such a logic, and of that of a neutral observation language, made *The Structure of Scientific Revolutions* the most widely read, and most influential, work of philosophy written in English since the Second World War. Dozens of books have been written in response to it. Constantly assigned in undergraduate as well as graduate courses, in almost every academic department, it has altered the self-image of many disciplines, from philosophy through the social sciences to the so-called hard natural sciences. For more than two centuries, up through the heyday of logical positivism, practitioners of many disciplines had wondered if they were being "sufficiently scientific" – a term which they used almost interchangeably with "sufficiently rational" and "sufficiently objective." The "hard" sciences – physics in particular – were viewed as models which other disciplines should imitate. Kuhn's book suggested that decisions between physical theories are no more algorithmically made than are decisions between alternative political policies. This suggestion was greeted with sighs of relief by some, who felt relieved of their previous methodological worries, and with consternation (and even anger) by others, who interpreted Kuhn as denying science's claim to objective knowledge. (See PRAGMATIC FACTORS IN THEORY ACCEPTANCE; JUDGMENT, ROLE IN SCIENCE; SCIENTIFIC CHANGE).

A host of critics gathered to defend science's rationality and objectivity against Kuhn. As defenses of Kuhn against these critics proliferated, new battle lines were drawn which gradually transformed the philosophy of science, and which brought history of science into ever more fruitful interaction with philosophy of science. (See HISTORY, ROLE IN THE PHILOSOPHY OF SCIENCE.) The resulting controversies interlocked with broader philosophical controversies about the nature of rationality itself, and in particular with the debate between the atomistic accounts of language and thought familiar from the tradition of British Empiricism and more holistic accounts offered by Quine (see QUINE), the later Wittgenstein, Donald Davidson, and Putnam. Kuhn's work thus became central to the development of post-positivistic analytic philosophy. Reactions to his book produced an enormous increase in the amount and sophistication of philosophical discussion of meaning change, and of the distinction (if any) between observational and theoretical terms. (See OBSERVATION AND THEORY and THEORETICAL TERMS).

Much discussion of Kuhn's work has focused on the question of whether either tables or electrons can be said to exist independently of human thought. Even though Kuhn for many years explicitly characterized himself as a realist, he was often accused of lacking a sufficiently robust sense of mind-independent reality, and of lending aid and comfort to anti-realism: the view that there is no fact of the matter about which of two scientific theories is true (see UNDERDETERMINATION). He also insisted that he had no intention of breaking down the distinction between science and nonscience, but merely wished to demythologize scientific practice by setting aside a simplistic picture of scientific practice as the patient accumulation of "hard facts." He clarified his position considerably in a postscript to the second edition of *Structure* and also in various further explanations and replies to criticisms (collected in Kuhn 1977).

Kuhn turned away from philosophy to history for a time, while preparing a history of the origins of quantum mechanics (Kuhn 1978). But since the publication of that book the bulk of his work consisted in detailed defenses of the claim that there is no language-independent reality, no single "Way The World Is" (a claim first defended, in those terms, by Nelson Goodman) (see NATURAL KINDS). He subsequently said: "I aim to deny all meaning to claims that successive scientific beliefs become more and more probable or better and better approximations to the truth and simultaneously to suggest that the subject of truth claims cannot be a relation between beliefs and a putatively mind-independent or 'external' world" (Kuhn 1993, p. 330).

Kuhn then defended his much-discussed thesis that Aristotle lived in a different world from Galileo (Kuhn 1962, ch. 10) by an analogy between the evolution of scientific ideas and that of biological species: "Like a practice and its world, a species and its niche are interdefined; neither component of either pair can be known without the other" (Kuhn 1993, p. 337). On this view, you can no more identify the world to which a statement or a theory corresponds, or which it accurately represents, without a knowledge of the language in which the statement or theory is framed, than you can identify a biological niche without knowledge of the behavior of the species which inhabits, or inhabited, that niche.

Kuhn's critics continued to press on the question of whether this line of thought can be reconciled with his claim that science produces genuine *knowledge* of nature. These critics insist that, if we drop the notion of a language-neutral reality to be accurately represented, we endanger the distinction between increasing knowledge of nature and mere pragmatic adjustments in response to novel stimuli. Kuhn's response consisted in denying that the "objective of scientific research" is accuracy of representation. Rather, "whether or not individual practitioners are aware of it, they are trained to and rewarded for solving intricate puzzles – be they instrumental, theoretical, logical or mathematical – at the interface between their phenomenal world and their community's beliefs about it" (ibid., p. 338). The principal question raised, though not yet resolved, by Kuhn's work is: Can the link between representation and knowledge, a link still taken for granted by most post-empiricist analytic philosophers, be broken without abandoning the distinction between rational and irrational human practices? Kuhn clearly thought that it could. He ended a response to his critics with the sentence: "Those who proclaim that no interest-driven practice can properly be identified as the rational pursuit of knowledge make a profound and consequential mistake" (ibid. p. 339).

References and further reading

Works by Kuhn

1957: *The Copernican Revolution: Planetary Astronomy in the Development of Western Thought* (Cambridge, MA: Harvard University Press).

1962: *The Structure of Scientific Revolutions* (Chicago: University of Chicago Press). (A second, enlarged edition published in 1970 included an important Postscript.)

1977: *The Essential Tension: Selected Studies in Scientific Tradition and Change* (Chicago: University of Chicago Press). (Contains Kuhn's much-discussed essay "Objectivity, value, judgment and theory choice.")

1978: *Black-Body Theory and the Quantum Discontinuity, 1894–1912* (Oxford: Clarendon Press).

1993: Afterwords. In *World Changes: Thomas Kuhn and the Nature of Science*, ed. Paul Horwich (Cambridge, MA: MIT Press), 311–41. (This book contains important critical appraisals of Kuhn's work, to which he responds in "Afterwords.")

Works by other authors

Barnes, B. 1982: *T. S. Kuhn and Social Science* (New York: Columbia University Press).

Goodman, N. 1978: *Ways of Worldmaking* (Indianapolis: Hackett).

Gutting, G. (ed.) 1980: *Applications and Appraisals of Thomas Kuhn's Philosophy of Science* (Notre Dame, IN: University of Notre Dame Press).

Hoyningen-Huene, P. 1993: *Reconstructing Scientific Revolutions: Thomas S. Kuhn's Philosophy of Science*, with a Foreword by Thomas S. Kuhn (Chicago: University of Chicago Press). (Trans. by Alexander T. Levine from *Die Wissenschaftsphilosophie Thomas S. Kuhns: Rekonstruktion and Grundlagenprobleme* (Braunschweig: Friedrich Vieweg, 1989).)

Stegmüller, W. 1973: *Probleme und Resultate der Wissenschaftstheorie und Analytischen Philosophie*, vol. 2: *Theorie un Erfahrung*. part 2: *Theoriensktruktur und Theoriendynamik* (Berlin: Springer).

32

Lakatos

THOMAS NICKLES

Imre Lakatos (9 November 1922–2 February 1974) is the most important philosopher of mathematics and one of the most influential philosophers of science since the mid-twentieth century. A Hungarian, Lakatos changed his name from Lipschitz to Molnar during the Nazi era and then to Lakatos ("locksmith"). After the war he remained politically active, as secretary in the Hungarian Ministry of Education. Later he was imprisoned as a dissident, and escaped to the West during the revolt of 1956. He studied at Budapest, Moscow, and Cambridge (Ph.D., 1958). During the 15 years preceding his death, he taught at the London School of Economics and Political Science, where he became Professor of Logic in 1969. He was a lively teacher, discussant, and social critic. An inspired circle of friends and colleagues gathered around him within the Popperian stronghold at the London School of Economics.

Lakatos was a major contributor to twentieth-century debates about the nature and status of scientific methodology. During the 1960s battle among the great theories of scientific change, he worked out his own distinctive position, "the methodology of scientific research programmes" (MSRP) in debate with other prominent methodologists and anti-methodologists such as Karl Popper, Thomas Kuhn, Michael Polanyi, Paul Feyerabend, and Stephen Toulmin. (See SCIENTIFIC CHANGE; POPPER; KUHN; and FEYERABEND.)

The seeds of MSRP are evident already in Lakatos's doctoral research, eventually issuing in *Proofs and Refutations: The Logic of Mathematical Discovery*, a highly original investigation of creative problem solving and the growth of knowledge in mathematics. The four-part article from which the book grew is a philosophical case study, a "rational reconstruction," of the history of a mathematical problem and its solutions: namely, the Descartes–Euler conjecture that the relative numbers of vertices (V), edges (E), and faces (F) of all polyhedra is $V - E + F = 2$. Lakatos shows rather convincingly that at least this sort of mathematical research belies the stereotype of mathematics as a dry, a priori, purely formal enterprise of discovering unchallengeable deductive proofs. Such "formalist" accounts, he insisted, fail to address even the problem of the growth of mathematical knowledge – that is, fail to admit a "logic of mathematical discovery" in Popper's sense, much less in Lakatos's richer sense of the term. On his account, much mathematical work is informal and possesses a strong heuristic component. And on the frontier of research, where knowledge most evidently grows, nearly all work must be informal and heuristic.

Accordingly, Lakatos rejected the twentieth-century formalist metamathematicians' reduction of mathematics to formalized mathematics. He also blurred the logical positivists' sharp distinction(s) of (philosophy of) mathematics from (philosophy of) empirical science on the grounds that mathematical work is more heuristic, more critical, and even more empirical-looking than formalists and positivists would have us believe (see LOGICAL EMPIRICISM; LOGICAL POSITIVISM). (Here Lakatos acknowledged major debts to mathematician Georg Pólya's work on heuristic problem solving in mathematics and to philosopher Karl Popper's critical approach to empirical science, according to which science consists of criticizing conjectures in an attempt to refute them.) Mathematical thinking, he said, is responsive to the "situational logic" of ongoing debates between competing research programs rather than consisting of self-evident and unassailable definitions, axioms, and proofs that somehow descend from heaven. Hence Lakatos found it natural to write in dialogue form. The very title of the work, *Proofs and Refutations* is oxymoronic for formalists, who retain the "dogmatic" (uncritical) image of genuine mathematics as "authoritative, infallible, irrefutable" (Lakatos 1976, p. 5). For how could a proof possibly be refuted?

Lakatos turned his attention increasingly to empirical science, and explicitly developed MSRP in his famous article, "Falsification and the methodology of scientific research programmes" (in Lakatos and Musgrave 1970). This and ensuing papers reveal deep debts to, but also increasing differences from, Popper. According to MSRP, the relevant units of and for analysis are not individual theories or conjectures, but entire research programs. A research program is defined by a "hard core" of unrevisable principles plus a "protective belt" of auxiliary assumptions that are to be revised whenever necessary to deflect criticism away from the hard core, thus protecting it from falsification (see UNDERDETERMINATION OF THEORY BY DATA). This tolerance of apparent falsification is Lakatos's compromise between Popper's negativist epistemology and Kuhn's vision of scientific development, in which paradigms face anomalies at all times. Both the hard core and the protective belt provide substantial heuristic guidance to theory construction. The negative heuristic warns, "Do not violate the hard core." The positive heuristic exhorts, "Do develop the program, stage by specified stage, by constructing an ordered series of more sophisticated, protective-belt theories consistent with the world picture that the program attempts to articulate." The core principles may in fact be violated by the deliberately oversimplified theories and models developed in the early stages of the program, but these internal problems must be solved by later stages. Thus internal criticism has a positive, heuristic-constructive role. (See the detailed Newtonian mechanics example in Lakatos 1970.) By avoiding immediate falsification, with an eye to long-term growth, MSRP makes scientific inquiry a more ordered and connected affair than Popper's herky-jerky multiplicity of conjectures and refutations. MSRP appraises connected *series* of theories, rather than isolated theories (Lakatos 1970, p. 118). How is this done?

If a research program, via changes in its auxiliary assumptions, generates theory changes (new theories in the series) that yield novel predictions, then MSRP evaluates the program as "theoretically progressive," and we may speak of a "progressive problem shift." If some of these predictions are confirmed, the program is "empirically progressive" as well. Heuristically unmotivated changes that merely protect the hard core from falsification and yield no novel predictions are "ad hoc" and produce a "degenerative problem shift" (ibid., p. 133). A research program whose "theoretical growth" (production of novel predictions) outpaces its empirical growth is progressive, while a program whose theoretical growth lags behind its empirical growth (in accommodating new information only in a *post hoc* manner) is "degenerating" or "stagnating" (Lakatos 1971, §d).

Although he sometimes wavered on the point, Lakatos refused to provide rules for determining *when* a degenerative program should be abandoned. He distinguished "methodological appraisal" from "heuristic advice." It is not irrational for you to stick with a degenerating program if you recognize the risk of its not becoming progressive again in the future – and the unlikelihood of finding grant support. But, contrary to Kuhn on commitment to a paradigm, neither are you bound to stick with your current research program until you think its heuristic power is completely exhausted. Something more promising may come along. There are no rules for deciding here and now whether your decision is "rational." Here we find a Hegelian strain in Lakatos's attack upon "instant rationality" and his advocacy of a historical methodology of science (1970, pp. 154f). There can be no rule that instantly justifies a decision to retain or abandon a research program, for a program that is progressive now may degenerate in the future, relative to a competitor. We cannot know now what we shall only know later (Popper). "We can only be "wise" after the event" (Lakatos 1971, §d). For this reason, scientific rationality is largely a matter of *post hoc* rational reconstruction of long-term developments. For the same reason, a "crucial experiment" is usually a *post hoc* reconstruction.

Against the logical positivists, Popper insisted that not all metaphysics is bad, for theoretical research is often driven or motivated by metaphysical world views (e.g., a mechanistic world picture, wave picture, particle picture, germ theory of disease). Lakatos's heuristics is an attempt to dignify and articulate this metaphysical guidance as an essential part of scientific methodology, rather than to dismiss it as merely a matter of aesthetic taste or psychological motivation. For Lakatos, the stronger the heuristic, constructive resources – the forward drive – of a program, the better. Ideally, a program will never be surprised by new facts, but will anticipate all future developments. Insofar as it does not, the program must rely on *ad hoc* accommodation to the facts. (Zahar (1983) develops this idea in detail.)

Thanks to its constructive component, MSRP can offer a fuller account of the growth of scientific knowledge than Popper's falsificationism, which says that methodology begins not with discovery, but only with the critical testing of conjectures already "on the table." MSRP provides a quasi-rational account of how candidate theories got on the table in the first place. In this respect MSRP is reminiscent of seventeenth- and eighteenth-century methodologies, which included an account of discovery as well as an account of justification – indeed, as *part* of an account of justification (see DISCOVERY). Thus it is surprising that Lakatos repeatedly denies that the heuristic-constructive component of a program provides any epistemic justification for its conclusions. He insisted that *"The term 'normative' no longer means rules for arriving at solutions, but merely directions for the appraisal of solutions already there. Some philosophers are still not aware of this problem shift"* (1974, §1; Worrall and Currie 1977a, p. 140; emphasis original). So, after all, Lakatos leaves heuristics with the epistemic status of metaphysics: it offers methodological stimulation, but has no epistemic teeth. How, then, can Lakatos claim to demarcate good from bad programs on the basis of his heuristic *ad hoc*-ness criterion?

The difficulty is that Lakatos officially retained Popper's epistemic position that only novel predictions carry epistemic weight. Lakatosians agree with Popper that information used to construct a theory can have no epistemic significance. Nature is allowed to speak only through novel predictions, the new predictive consequences of theories under consideration. This ardent consequentialism distances Lakatos's project from methodologies of discovery (ranging from Bacon and Descartes to contemporary inductive-statistical methods), and leaves it unclear why heuristic theory construction is important to methodology, given its lack of epistemic significance (Nickles 1987).

We may conclude by briefly examining some other major problems raised for Lakatos's work. In some cases I indicate the lines of Lakatos's presumed answers, without claiming that these are necessarily satisfactory.

1. Lakatos weakens normative methodology excessively by providing no rules for when a stagnating program should be abandoned in favor of a more progressive one, or when it is rational to abandon a modestly progressive program in favor of a competitor. How long must we wait? How much failure must we endure? *Reply*: Strict rules are incompatible with Lakatos's conception of creative research as informal (not rigidly rule-governed), with his distinction between appraisal (for which he does provide rules) and advice (the realm of free, informed decisions), and with his attack on the "instant rationality" that has spoiled most traditional methodology. Feyerabend (1975, p. 2) lauds Lakatos's ability to combine "strict criticism with free decision, historical accident with rules of reason. It is one of the most important achievements of twentieth-century philosophy."

2. The MSRP is an incoherent compromise between Popper and Kuhn, whose projects are too different to amalgamate. (Similarly, Lakatos has been viewed as an attempt to bridge Polanyi and Popper.) Kuhn demonstrated how far science as historically practiced differs from Popperian philosophy. Lakatos wants to have his cake and eat it, too. His research programs are an obvious surrogate for Kuhn's paradigms, and he attempts to historicize methodology and even rationality; but, unlike Kuhn, these latter steps presuppose that he still wants to defend the project of a rational yet critical and undogmatic *methodology* of science. *Reply*: Lakatos's work lends some credence to this charge, since his own later work moves increasingly away from Popper.

3. Both Kuhn and Lakatos are correct to find more than Popper's scatter-gun pattern of isolated conjectures and refutations in the history of scientific research. However, both impose a single, oversimple pattern upon history. And as in the case of Kuhnian revolutions and paradigm shifts, the patterning, the historical coherence, achieved by MSRP is deceptive, for Lakatosian research programs tend to be collapsed histories, retrospective reconstructions. The pattern is whiggishly imposed with the benefit of reconstructive hindsight, and is then said to have been guiding the work from the beginning. *Reply*: Again, there seems to be some truth to this charge; however, Lakatos is correct that programmatic elements are often important in the history of science.

4. Lakatos's attempt to historicize methodology has been criticized by historians and sociologists of knowledge, as well as by philosophers. Disagreeing with Feyerabend, some philosophers have complained that Lakatos does not adequately combine, or reconcile, the demands of logic (or methodology) and history. For example, if the best methodology is the one that makes more of the history of science rational than any

competitor, will there not exist a methodology that trivially accommodates all of history? *Reply*: Such a methodology must be heuristically powerful, and thus could not be trivial.

Historians object vehemently that Lakatos's "rational reconstructions" terribly deform history (e.g., Pearce Williams 1975). Lakatosian history is whig, or presentist, history. Lakatos relegates "real" history to the footnotes of his rational reconstructions: "The history of science is always richer than its rational reconstruction. *But rational reconstruction or internal history is primary, external history only secondary, since the most important problems of external history are defined by internal history*" (1971, §e; Worrall and Currie 1977a, p. 118; emphasis original) (see HISTORY, ROLE OF IN THE PHILOSOPHY OF SCIENCE). Sociologists of a social constructivist bent complain about Lakatos's use of methodology to retain an untenable distinction between internal and external factors in research. David Bloor (1978), answered by John Worrall, adds that Lakatos's own case studies in *Proofs and Refutations* would have been richer and more explanatorily revealing had he factored in the sociopolitical circumstances of the combatants.

References and further reading

Works by Lakatos

1970: Falsification and the methodology of scientific research programmes. In Lakatos and Musgrave (1970), 91–196; repr. in Worrall and Currie (1977a), ch. 1.

1971: History of science and its rational reconstructions. In *PSA 1970*, ed. R. C. Buck and R. S. Cohen, 91–135; repr. in Worrall and Currie (1977a), ch. 2.

1974: Popper on demarcation and induction. In *The Philosophy of Karl R. Popper*, ed. P. A. Schilpp (La Salle, IL: Open Court); repr. in Worrall and Currie (1977a), ch. 3.

1976: *Proofs and Refutations: The Logic of Mathematical Discovery*, rev. edn, ed. J. Worrall and E. Zahar (Cambridge: Cambridge University Press). (Series of articles in *British Journal for the Philosophy of Science*, 1963–4.)

1977a: *Philosophical Papers*, vol. 1: *The Methodology of Scientific Research Programmes*, ed. J. Worrall and G. P. Currie (Cambridge: Cambridge University Press). (Includes a bibliography of Lakatos's works.)

1977b: *Philosophical Papers*, vol. 2: *Mathematics, Science and Epistemology*, ed. J. Worrall and G. P. Currie (Cambridge: Cambridge University Press).

(ed.) 1967: *Problems in the Philosophy of Mathematics* (Amsterdam: North Holland).

1968: *The Problem of Inductive Logic* (Amsterdam: North Holland).

—— and Musgrave, A. (eds) 1968: *Problems in the Philosophy of Science* (Amsterdam: North Holland).

1970: *Criticism and the Growth of Knowledge* (Cambridge: Cambridge University Press). (This is the most important confrontation of the major theorists of scientific change.)

Works by other authors

Bloor, D. 1978: Polyhedra and the abominations of Leviticus. *British Journal of the History of Science*, 11, 245–72. (A reply by, and response to, J. Worrall occurs in the 1980 issue.)

Cohen, R. S., Feyerabend, P. K., and Wartofsky, M. W. (eds) 1976: *Essays in Memory of Imre Lakatos* (Dordrecht: Reidel). (Contains a bibliography of Lakatos's works.)

Feyerabend, P. K. 1975: Imre Lakatos. *British Journal of the Philosophy of Science*, 26, 1–18.

Howson, C. (ed.) 1976: *Method and Appraisal in the Physical Sciences* (Cambridge: Cambridge University Press). (Case studies that use and appraise Lakatos's work.)

Koetsier, T. 1991: *Lakatos' Philosophy of Mathematics: A Historical Approach* (Amsterdam: North Holland).

Kuhn, T. S. 1970: *The Structure of Scientific Revolutions*, rev. edn (Chicago: University of Chicago Press).

Latsis, S. (ed.) 1976: *Method and Appraisal in Economics* (Cambridge: Cambridge University Press).

Nickles, T. 1987: Lakatosian heuristics and epistemic support. *British Journal of the Philosophy of Science*, 38, 181–205.

Pearce Williams, L. 1975: Should philosophers be allowed to write history? *British Journal of the Philosophy of Science*, 26, 241–53.

Zahar, E. 1983: Logic of discovery or psychology of invention? *British Journal of the Philosophy of Science*, 34, 243–61.

33

Laws of Nature

ROM HARRÉ

Introduction

From the very beginnings of science there was the realization that amidst the apparent diversity of patterns to be observed in nature there are some which regularly repeat themselves. There are natural regularities. It was also realized that behind much that appeared irregular and chaotic there are deeper regularities. At one time it was thought that these regularities existed because there were Laws of Nature, in the sense in which behind certain regularities in human conduct there are the Laws of the Land. Even as late as the eighteenth century the two senses of law were knowingly conflated by philosophers. In 1710 Berkeley proposed to account for the fact that there are all sorts of regularities in the world as we observe it by reference to God's role in engendering our ideas. From his theology he took the notion of the rules that God prescribes for himself in thinking the world and thinking us and our experiences. It is these rules that account for, and are reflected in, whatever regularities we perceive in nature (see BERKELEY).

In recent times philosophers of science have taken the Laws of Nature, as they appear in the physical sciences, as descriptions of tendencies and regularities that preexist our attempts to describe them. Most philosophers now believe that the laws play no part in the genesis of natural regularities or the natural tendencies that are displayed in them. However, from time to time philosophers have tried to combine both points of view. It has been argued that our beliefs have some role in what we are able to observe of the many-faceted face of Nature. Perhaps the very organization of primitive sensory experience into patterns owes something, perhaps a great deal, to our prior beliefs and the inbuilt patterning tendencies of our minds.

Sometimes philosophers have used the expression "Laws of Nature" both for the regularities described and for the statements that describe them. Despite the seal of antiquity, this usage is undesirable, in that it glosses over certain important considerations that come from the relation between the characteristics of what is described and those of the descriptions of it.

What are the main characteristics of those statements which we dignify by the title "Laws of Nature"? First of all, they are taken to hold universally. The patterns they describe are supposed to be found wherever phenomena of the sort they cover occur. They express concomitance or regularities – that is, repeated patterns of event sequences and of sets of coexisting properties. They are also taken to be in some way necessary – that is, the universal regularities that they describe could not be other than they are. Throughout the universe material bodies attract each other according to the law

$$F = G\ m_1.m_2/d^2$$

This law expresses a kind of necessity. Drop an apple in the proximity of a planet, and not only does it always fall in a manner close to the regular way described by the above law; but, we believe, everything else being equal, this pattern of events could not be otherwise.

In discussing the philosophical problems that the idea of Laws of Nature seems to involve, we must account for both these characteristics. But in doing so we must pay attention to a possible ambiguity in the application of the idea of necessity. Is the necessity of Laws of Nature a logical or conceptual necessity, a feature of the statement of laws? Or is it also a feature of the processes and property clusters that law statements describe? If it is the latter, what are we to make of the idea of necessity in nature? In the terminology of logic, are the necessities we encounter in science *de dicto*, of propositions, or *de re*, of things in the world?

But there is another problem. There is a gap between the somewhat messy rough regularities we observe in nature and what the Laws of Nature seem to be saying about the world. In all cases laws seem to describe a more perfect world than that with which we are acquainted. For example, Newton's first law, that a body continues at rest or in uniform motion in a straight line unless acted upon by an impressed force, seems to describe a phenomenon which we never actually come across in this complex, fragmented world of ours. No material thing is ever in the kind of stripped-down circumstances in which it could display pure inertial motion. Perhaps our initial thought, that Laws of Nature describe natural regularities, is not quite right. We could say that Laws of Nature are abstracted from descriptions of our world, or that they describe models of aspects of this world. We shall see that there are reasons for favoring the latter resolution.

Looked at from the point of view of the philosophy of science, the major problems that the Laws of Nature present are the following:

How do we justify the implicit claim for the universality of laws?
How do we account for the apparent necessity that each law ascribes to the patterns it describes?

We shall use these questions to probe a number of treatments of the Laws of Nature, particularly in relation to what those treatments take the subject matter of the laws to be.

There are three main views as to what the Laws of Nature are about. In Aristotle's philosophy of science, there is no discussion of Laws of Nature as such, but rather of the definitions which play a similar explanatory role for him as laws do for us. His discussion of definitions treats them as expressing relations between concepts definitive of the essences of material substances. Necessity for Aristotle is primarily a conceptual matter. Hume's views on causality will serve as an exemplar of a widely held view that the Laws of Nature are summaries of sensory experience, a view developed in detail by Mach (1894) (see MACH). Hume offers a psychological explanation of natural necessity in terms of habits of expectation, which are "projected" onto bare sequences of elementary sensory impressions. The third view, that Laws of Nature are descriptions of real natural tendencies or powers has a long history. Locke presented it in a half-hearted way, but it has come into its own only in contemporary philosophy.

Tendency theorists analyze natural necessity in terms of the real essences of material things and substances that account for the powers of those things to manifest themselves in the ways we observe in particular circumstances. In the history of philosophy we find a great many variations, but these three types of account are the recurring themes.

Running in parallel, so to say, with discussions of these questions has been a sustained effort to give a satisfactory account of the logical form of law statements. The primary intuition upon which much of the discussion rests is that genuine laws support "counterfactuals," while mere reports of accidental generalization do not (Goodman 1965). For example, if it is a law that reducing agents are electron donors, then we are entitled to say that if this substance were to be (had been, should prove to be . . .) a reducing agent, then it would be (would have been, should . . .) be an electron donor. We shall see how our three main accounts of Laws of Nature "shape up" to interpreting and justifying this intuition.

Laws of nature as expressing relations among concepts

Aristotle: essences and definitions

In Aristotle's philosophy of science (Aristotle 1981) we have the first major attempt to develop an account of what we would now call the Laws of Nature – that is, propositions describing the regularities to be observed in sublunary nature. We begin with his oft-repeated claim that there is no scientific knowledge of individuals, only of species, types, or sorts. When we seem to be making a claim about what must be true of some individual, it is only *qua* that individual's membership of a species that we can make it.

Demonstration for Aristotle is a pattern of reasoning which, so to say, works backwards from that which we wish to explain to fundamental and indisputable grounds which are primitive and necessary. What is this necessity? It is manifested in definitions of kinds. A statement which is used to define a kind assigns an essence to that kind; for example, the essence of thunder is "a noise in the clouds due to the quenching of fire."[1] This is what thunder is, in itself, *per se*. The definition is like a modern Law of Nature, in that with it we explain the existence of particular instances of thunder.

In summary, then, Aristotle traces back the necessity of a Law of Nature, as we would call it, to the necessity of the premises of a demonstration. And that necessity is identified with, or grounded in, essence – that is, in how properties relate to one another in species, genera, and the like. These property relations are given or immediate, and stand in no need of further accounting. Are these relations in the world, or in the conceptual structures with which we describe the world? Are they *de re* or *de dicto*? Is there necessity in nature? Scholars have differed in their readings of Aristotle, but the consensus seems to be that though Aristotle accepts both applications for the nation of necessity, his primary meaning is conceptual, *de dicto*.

Laws of Nature as "grammatical rules"

Consider a law such as F = ma, Newton's second law, which specifies a relation between the force applied to a body of a certain mass and the acceleration thereby produced relative to some frame of reference. As Mach (1942) pointed out, there is no way of making an independent measure of the force in this equation other than via the determination of some other kinematic variable, such as velocity – for instance, by using the law $Fs = \frac{1}{2}mv^2 - \frac{1}{2}mu^2$ to make a calculation of the value of F. The second law cannot then be a summary of observed correlations. There is another possibility. The law is a definition of the concept of mechanical force.

Wittgenstein (1953) used the metaphor of frame and picture to bring out the different role that such definitions play in discourse from those which report observations or the results of experiments. The frame propositions specify the language in which the descriptive propositions are couched. As such, they are not subject to empirical criteria, and are neither true nor false. In a very broad sense, they specify the "grammar" of a certain kind of discourse. Wittgenstein pointed out that such frame propositions, laying down the boundaries of what is to make sense in a certain discourse, have a characteristic logical property. The negation of such a proposition is not false, but meaningless. It is not false to say that something is both red and green all over, but it makes no sense if the proposition "Nothing can be red and green all over at once" is being used to specify the way the words "red" and "green" are to be used in a color discourse. Likewise, it is not false to say that the force applied is not equal to the product of mass and acceleration, but meaningless, if the Newtonian second law is being used to specify the way the concepts of "force," "mass," and "acceleration" are to be used.

The conceptualist analysis of Laws of Nature proves an unproblematic justification for asserting the corresponding counterfactual statements to each law. Since counterfactuals refer to possible states of affairs, and conceptual relations cover all cases, possible and actual, there is a logical link between them. Counterfactuals are justified exactly by the conceptual relations the corresponding law expresses.

Laws of Nature as summaries of experience

Hume's account of Laws of Nature

Hume's famous discussion appears in both his major philosophical works, the *Treatise* (1739) and the *Enquiry* (1777). The discussion is couched in terms of the concept of causality, so that where we are accustomed to talk of laws, Hume talks of causal statements (see HUME). The notion of causation between events, Hume contends, involves three root ideas (see CAUSATION):

1   that there should be a regular concomitance between events of the type of the cause and those of the type of the effect;
2   that the cause event should be contiguous with the effect event;
3   that the cause event should necessitate the effect event.

(1) and (2) occasion no difficulty for Hume, since he believes that there are patterns of sensory impressions unproblematically related to the ideas of regular concomitance and of contiguity. But the third requirement is deeply problematic, in that the idea of necessity that figures in it seems to have no sensory impression correlated with it. However carefully and attentively we scrutinize a causal process, we do not seem to observe anything that might be the observed correlate of the idea of necessity. We do

not observe any kind of activity, power, or necessitation. All we ever observe is one event following another, which is logically independent of it. Nor is this necessity logical, since, as Hume observes, one can jointly assert the existence of the cause and a denial of the existence of the effect, as specified in the causal statement (Law of Nature) without contradiction. What, then, are we to make of the seemingly central notion of necessity that is deeply embedded in the very idea of causation, or lawfulness? To this query Hume gives an ingenious and telling answer. There is an impression corresponding to the idea of causal necessity, but it is a psychological phenomenon: our expectation that an event similar to those we have already observed to be correlated with the cause type of event will come to be in this case too. Where does that impression come from? It is created as a kind of mental habit by the repeated experience of regular concomitance between events of the type of the effect and the occurring of events of the type of the cause. And this is the impression that corresponds to the idea of regular concomitance, (1) above. A Law of Nature then asserts nothing but the existence of the regular concomitance.

Can we give any proof of the universality of such a law? Since there is no contradiction between our observation that an event of the type of a cause has occurred and our belief that an event of the type of the effect will not occur, no matter how often we observe suitable pairs of concomitant events, there is no guarantee that "the course of nature will remain always the same." Anything might happen, as far as we can tell solely by reasoning from the phenomena we experience. Justifying the universality of Laws of Nature as conceived in the manner of Hume or Mach, as generalizations of observed cases of patterns in sequential events or in coexisting properties, how could one justify an induction from what is seen to be the case in some local region of space-time to all regions? Hume is credited with the clearest formulation of the problem of inductive skepticism, though he was not the first to notice the gap between what we can be sure of and the scope we wish to ascribe to laws. Looked at positively, the best we can say for a putative Law of Nature is that it is probable or likely to be true universally. But, strictly, we can draw no logical conclusion from locally true premises. For all we know, the way the world goes outside our little region of space-time may be very different. Only if we could justifiably claim that Nature is uniform could we recruit logic to support the induction we want. But that route is closed to us, since that very principle is in need of justification, and we cannot use induction to achieve it, on pain of circularity.

However, there is another line of thought in philosophy of science, the tradition of negative or eliminative induction. From Francis Bacon (1620) and in modern times K. R. Popper (1963) we have the idea of using logic to bring falsifying evidence to bear on hypotheses about what must universally be the case (see POPPER). Putative Laws of Nature are not confirmed by finding instances in which the patterns they describe do hold, but false conjectures are eliminated by the discovery of cases in which, though expected to hold, they do not. Knowledge grows by the continuous testing of our beliefs. Unfortunately, this approach, which looks so promising at first sight, suffers from the same defect as the more positive approach of inferring universality of application from a few positive cases. When we eliminate a putative Law of Nature, we want to be sure that though it does not hold in this region of space-time, there may not be other regions in which it does hold. For example, though false today, Nature may so change as to make that conjecture true from tomorrow on. The Principle of the Uniformity of Nature is involved in both the positive and the negative uses of evidence to support the claim of each and every Law of Nature to universality.

An account of laws as summaries of what did happen, is happening, and will happen has great difficulty in accommodating the intuition that laws support counterfactuals. This is partly because it has great difficulty with any conception of natural necessity other than Hume's psychological projection view. Possibilities, real possibilities, which are the subject matter of counterfactuals, simply find no place in the Humean/Machian catalog of the actual.

Laws as descriptions of natural tendencies

In the accounts we have culled from Aristotle and from Hume we have identified what one might call the "poles" of the discussion. For Aristotle there is a necessity in causal relations that comes from the fact that they derive from the essential characteristics of material things. For Hume there is no necessity in events or things. There are just, as a matter of fact, sequences of like pairs of elementary events. Most philosophers of science have been unwilling to rest content with either pole. There surely is natural necessity, expressed in Laws of Nature. There is some sense to be given to the "must" in statements like "If released in a gravitational field, an unsupported body must fall." Yet that necessity does not seem to be logical. We can imagine a body that does not fall when released – for example, a balloon filled with just the right amount of helium! Nor does natural necessity seem to be just a projection of a psychological state onto the world. There are some concomitances which, however long and however reliably they have been observed to occur, would never tempt anyone to think the relation causal, even so common a concomitance as the invariable succession of night and day.

Philosophers of science differ over what they take to be the subject matter of Laws of Nature. Actualists, such as Hume and Mach, take the Laws of Nature to be about what can actually be observed, or, if phenomenalists, about the experiences an observer is actually having (see MACH). Dispositionalists, such as Harré and Madden (1977), Bhaskar (1975), and Cartwright (1989), take the Laws of Nature to be about the powers, dispositions, or tendencies of natural systems to bring about observable phenomena. In the real world, according to dispositionalists, there are huge numbers of tendencies and powers acting simultaneously in open systems, the joint upshot of which is what we observe. Experimental procedures, by creating closed systems, enable investigators to isolate tendencies and to study their effects when acting singly. In nature there are no closed systems. Our Laws of Nature, abstracted from real processes in open systems, are true, not of the world, but only of abstract and simplified models of aspects of that world. Thus, each time we apply a Law of Nature to the solution of a problem in the real world, we must state the law *ceteris paribus* – that is, as applying, all else being equal.

Laws as singular statements about universals

The Hume/Mach treatment of the Laws of Nature simply takes for granted that law statements, whatever their status, are universal statements about particulars, the repeated event patterns they describe. However, there are many cases in which we seem to be dealing not with events, but with properties. It is a Law of Nature that the property of being a diamond is regularly and reliably associated with being a conductor of electricity. Of course, this association is manifested from time to time in events, such as subjecting a diamond to a difference in electrical potential, and finding that it is conducting electricity. But the law is about properties, not about the events in which they are manifested. Dretske (1977) has suggested that this sort of case be adopted as the archetype of natural lawfulness. There are many cases of laws which could be formulated in terms of events but are best set out as singular statements about patterns of properties. But there are also many cases in which the shift to the property formulation looks strained. For instance, laws of kinematics and elementary dynamics seem more properly analyzed in the event mode. The law

$$s = ut + \frac{1}{2}at^2$$

seems to be about the sequential happenings as a body accelerates under a uniform force field. After 2 seconds the body is moving with velocity 9 meters per second and is 15 meters from its starting point, and so on. This could be said to describe a pattern of properties of the system, but then at each moment the system would be exemplifying different properties, and that seems to call for an event interpretation.

What sort of properties are they that figure in Laws of Nature expressed in the property mode? Properties are universals in traditional philosophical terminology. The law, if one can so dignify it, that rubies are red is not a statement about this or that ruby's redness, but about the general property, or universal "redness," which is instantiated in innumerable cases. It is to this feature of properties that Dretske draws attention in his brief sketch of this position. This interpretation goes along with the distinction between "occurrent" and "dispositional," the kind of properties upon which dispositionalists base their account of the Laws of Nature. The property "red" displayed on a particular occasion to a particular observer of rubies is an occurrent property, manifested then and there. But the general property of redness which that ruby reliably possesses through time and space is not just a summary of its appearances. We want to say that it possesses the property even when not displaying it to anyone or to any test equipment. Such a property is a power or disposition. In general, when we adopt the Dretske way of setting out a Law of Nature, the properties of which such a law speaks are dispositional.

We can now see how superficial is the Hume/Mach treatment. A Law like $V = IR$ is not a summary of the innumerable particular voltages and currents to be measured in this or that concrete, particular circuit, but a statement about the relation of dispositions to affect voltmeters and ammeters, dispositions that are permanent and reliable properties of any circuit. As we shall see, the Dretske account of laws, developed through a dispositionalist treatment of properties, is not the whole story. We have still to consider the question of what it is about material systems that makes possible their possession of dispositions which could figure in Laws of Nature.

The world as a hierarchy of powers and natures

Necessity in nature is the idea that in a given set of circumstances not only is there no alternative to what exists or comes to pass, but that nothing other than what does exist or occurs could have. What grounds could the scientific community have for assuming that a statement about a real tendency expresses a natural necessity? There must be a reason other than the observed correlations of properties or concomitance of types of events for believing in the efficacy of the natural tendency or power to act. The physical sciences are organized hierarchically, in that the observed powers and tendencies of material things are explained by the workings of natural mechanisms; for instance, the chemical phenomena produced in test tubes by isolating the powers and tendencies of elements and compounds are explained by reference to unobserved mechanisms of molecular and ionic exchanges. So we do have a reason for thinking that not only does carbon reduce mercuric oxide, but that it must do so, given the molecular structures and processes that are occurring in the reaction. But these very processes are themselves described in Laws of Nature. To what can we attribute their natural necessity? Physical scientists proceed by repeating the move we have just described, adverting to "deeper" and more recondite processes and to the micro-entities through which they occur. At each level of the natural hierarchy, our belief in the natural necessity of the laws for that level is grounded in our beliefs about the nature of the causal mechanisms which operate at the level "below."

But though natural hierarchies could go on forever, there is a strong presumption that they do close, that the world is not infinitely complex. At the lowest level we encounter the ontological foundations of science, in which the most basic powers and tendencies of material substances must be grounded. The basic Laws of Nature can be grounded only in the properties of the universe as a whole, in the deep symmetries that find expression in those laws which are covariant through all transformations of frames of reference and which describe the tendencies of the most primitive kinds of beings we can conceive.

Something is still missing from this account. The natural mechanisms, which at higher levels of the hierarchy of laws are used explanatorily, are definitive of the nature of the substances involved. We explain the tendency of sodium to react with water by reference to its electronic architecture, and at the same time we define sodium by that very architecture. We can now see how hierarchy and necessity are linked. Sodium must react with water (*ceteris paribus*) because if this sample does not do so, it is not sodium; that is, it does not have that electronic architecture upon which its natural tendency to react with water depends. This is not logical necessity, because it is conceivable that sodium, as specified by its behavior alone, might have had a different architecture – indeed, that a quite different account of the nature of the chemical elements than the familiar proton, neutron, electron story should have turned out to be the right (or best) one. There is a further advantage to making the connection between natural tendencies and defining natures of the things and substances that possess them. The same account will hold good of the most basic beings in the ontology of a science. Poles and charges are the most elementary physical beings. As they are now understood, we assign them no internal complexity. They are located at mere points in space-time. But they are specified by the powers and dispositions they possess to act throughout their fields with other beings of the same elementary sorts. It is these point-centered powers that the basic vector and tensor matrices of physics describe. If a being fails to act as we expect, it is not that kind of being. Thus at every level laws express natural necessities through the conceptual necessities that define what it is to be a magnetic pole and what it is to be acetic acid, and so on. To put the point in another terminology, there is a close relation between what we take to be the Laws of Nature and what we recognize as natural kinds.

At first thought we might be inclined to say that if we have necessity, then we have universality for free, so to speak. Whatever is necessary must also be found to hold universally. But the matter is more complicated. We must attend to Bhaskar's (1975) distinction between what is actual and what it real. The actual world – the world as it manifested to us prior to any attempts at isolating tendencies by conducting experiments – is open. There are numerous tendencies acting at any point in space-time, and what can actually be observed to happen is the joint or resultant effect of all of them. Universality of Laws of Nature is a more subtle notion than the inductivists like Mach and Hume managed to grasp. We must be able to say that a tendency is universal but never manifested as such. We must also be able to say what it would bring about, were it to act alone. Neither its universality nor its necessity is a property of what actually takes place. This is quite alien to Mach's definition of a Law of Nature as "the mnemonic reproduction of facts in thought," a device to save us the trouble of remembering all the instances.

What account do dispositionalists give of counterfactuals? Natural tendencies tell us the possibilities inherent in the natures of things. And what type of thing we are dealing with is itself specified in terms of the dispositions that define it. Types and laws are intimately and mutually involved. As a description of the dispositions and natural tendencies of things, laws tell us what things would do, *ceteris paribus*. They do not, and indeed could not, tell us what actually happens. Since counterfactuals express what is possible for things of a certain kind, as do descriptions of dispositions and tendencies, laws which describe such dispositions and tendencies must entail counterfactuals (Aronson et al. 1994, ch. 7).

Conclusion

In our examination of the various ways in which philosophers have accounted for the universality, reliability, and necessity of Laws of Nature, we have encountered three major positions. Empiricists are inclined to interpret laws as summaries of observations. Realists are inclined to interpret laws as tendency statements grounded in a hierarchy of assumptions about the natures of the physical systems which possess them. Yet other philosophers are inclined to interpret at least some Laws of Nature as grammatical rules, specifying the way in which certain concepts are to be used. Which account should we prefer?

In this case, as in many other such situations in philosophy, the answer seems to be that all three accounts have merit. The general term "Law of Nature" comprehends more than one kind of proposition, each having its appropriate context of application. There is a family resemblance between the various cases in which we would use the term "Law of Nature," but there is no one common feature which marks out all and only such laws, other than the bare formal requirement that they be universal in application and necessary in force. A Law like $PV = RT$, the general gas law, seems best interpreted as a generalized and idealized summary of the way gases behave under moderate experimental conditions. Our granting of some measure of necessity to this law can be justified by reference to the molecular theory of the constitution of gases, which provides an account of the unobserved processes which are manifested in the relations the law expresses. A law like Le Chatelier's principle describes the tendency of a solution to behave in a certain way, *ceteris paribus*, and in the conditions created in a certain experimental setup. In the absence of an experimental or contrived closure of a chemical system, the phenomena the Principle describes are not observed, though the tendency is operative in bringing about that very result. Finally in a law like the relativistic mass law, $m = m_0/\sqrt{1 - v^2/c^2}$, we have neither a summary of observations nor a tendency statement, but a frame proposition expressing a "grammatical" rule for the use of "m," the concept of relativistic mass.

What should we conclude from this discussion? We have seen that the necessity of laws is related to the existence of stable mechanisms that generate the phenomena they describe. We must test and see if such mechanisms exist, and we must do this over and above any collection that we might make of observed instances in which the law under consideration is found to hold. So a claim that a certain statement is a Law of Nature, and so should be deemed necessary, is vulnerable to the discovery that the generative mechanism which we thought existed and which we had supposed we had represented correctly in some explanatory model at the heart of the relevant theory does not. Equally, the claim that a statement is a Law of Nature, and so universal in application, is vulnerable to the turning up of instances in which, though expected to hold, it does not. The claim that a statement is a Law of Nature boils down to the claim that it is both universal in application and necessary. It seems that all such claims, though worth making, must always be to some extent tentative. In a similar way, if we ground our belief in the universality and necessity of a statement as a Law of Nature in our definitions of the natures of the things, substances, and so on that are involved in some phenomenon, this belief is vulnerable to discoveries which show that we were wrong about what we thought were the definitive properties of the relevant substances. It may look as if the lawfulness of "Pure water boils at 100°C at NTP" is indisputable, since, if a liquid does not consistently behave this way, we should say that it is not water. But nothing obliges us to take this route. We could say that while most water obeys this law, some does not. The temperature at which water boils has then ceased to be criterial for whether the substance in question is water. Of course, we could make this statement only if we had adopted some other property as criterial of water of both "boiling" kinds.

Note

1   I am greatly indebted in this section to Richard Sorabji (1981).

References and further reading

Aristotle 1981: *Posterior analytics*, trans. H. C. Apostle (Gimmell: Peripatetic Press).
Aronson, J., Harré, R., and Way, E. C. 1994: *Realism Rescued* (London: Duckworth).
Bacon, F. 1620: *Novum Organon* (London: Lee).
Berkeley, G. 1957: *A Treatise Concerning the Principles of Human Knowledge* (Indianapolis: Indianapolis: Bobbs-Merrill; 1st pub. Dublin, 1710).
Bhaskar, R. 1975: *A Realist Theory of Science* (Leeds: Leeds Books).

Cartwright, N. 1989: *Natural Capacities and their Measurement* (Oxford: Clarendon Press).

Dretske, F. 1977: The laws of nature. *Philosophy of Science*, 44, 248–68.

Goodman, N. 1965: *Fact, Fiction and Forecast* (Indianapolis: Bobbs-Merrill).

Harré, R., and Madden, E. H. 1977: *Causal Powers* (Oxford: Blackwell).

Hume, D. 1957: *An Enquiry Concerning Human Understanding* (Oxford: Clarendon Press; 1st pub. 1777).

—— 1958: *A Treatise of Human Nature* (Oxford: Clarendon Press; 1st pub. 1739).

Locke, J. 1975: *At Essay Concerning Human Understanding* (Oxford: Clarendon Press; 1st pub. 1690).

Mach, E. 1890: *The Analysis of Sensations* (New York: Dover).

—— 1894: *Popular Scientific Lectures*, trans. T. J. McCormack (Chicago: Open Court).

—— 1942: *The Science of Mechanics* (La Salle, IL: Open Court; 1st pub. 1883).

Popper, K. R. 1963: *Conjectures and Refutations* (London: Routledge and Kegan Paul).

Sorabji, R. 1981: Definitions: why necessary and in what way? In *Aristotle on Science: The Posterior Analytics*, ed. E. Berti (Padua: Editrice Antenore), 205–44.

Wittgenstein, L. 1953: *Philosophical Investigations*, trans. G. E. M. Anscombe (Oxford: Blackwell).

34

Leibniz

WILLIAM SEAGER

Although one of the most important and prolific thinkers of all time, Gottfried Wilhelm Leibniz (1646–1716) spent his life as a courtier, wasting time in diplomatic business or preparing documents to shore up claims of lineage or territory for his patrons. He also spent a good deal of time on practical matters of engineering, such as his dreams of a system of windmills that would have ameliorated the chronic flooding of the Harz silver mines, and on his visionary mechanical calculators. Most of his working life was spent in Hannover, pursuing philosophy when he could spare the time from his professional activities, but nonetheless maintaining an astonishing pace of philosophical, mathematical, and scientific writing (of which very little was published in his own lifetime), as well as a relentless correspondence with all the important intellectual figures of his day. When Leibniz died in 1716, his funeral was scandalously insignificant, but time's revenge has been sweet.

Contributions to science

Leibniz's scientific legacy is vast, but perhaps his most important contribution was the differential and integral calculus. Of course, Leibniz was not the sole inventor, but the bitter battle over priority with Newton has ended in the recognition that both men independently discovered the fundamentals of the calculus (see NEWTON). Leibniz's open attitude towards his work and the approach he took to the calculus led to its rapid, widespread use and acceptance, at least throughout continental Europe. Although Leibniz adapted techniques that were well known to contemporary mathematicians, he was the first (with Newton) to see that differentiation and integration were inverse operations ("the fundamental theorem of analysis") of a fully generalizable system of calculation, and he based his work on a clear idea of *mathematical function*. Unlike Newton, Leibniz celebrated the appeal to infinity and infinitesimals. This was mathematically suspect, but scientifically fruitful. Leibniz's formalism, now standard, was designed to facilitate what appeared to be familiar mathematical manipulations. For example, by choosing to write the derivative of a function, f, as df/dx instead of Newton's $\dot{f}$, Leibniz gave the derivative the appearance of a simple ratio. For example, the chain rule, $dg/dx = dg/df \times df/dx$ (where g is a function of the function f), appears obviously valid – a simple case of dividing through by df. One can do a lot of working calculus by appeal to such operations, and Leibniz's notation aids (or abets) this. There were foundational problems, of course, but Leibniz had little patience with them. Such problems are exemplified by the cavalier way in which differentials are neglected whenever this appears productive. An elementary example stems from Leibniz's doubts about the correct value of $d(fg)$. At first, Leibniz thought this might simply be $dfdg$. He obtained the correct value by the method of differences, letting f go to $f + df$, g to $g + dg$, then subtracting fg from $(f + df)(g + dg)$. One almost obtains the right answer this way: $fdg + gdf + dfdg$. Leibniz eliminates the extra term by noting that it is infinitely smaller than the other elements of the sum. Also, Leibniz did not fully appreciate the problems of convergence of the infinite series appealed to in the calculus. He goes so far as to say that "error is impossible in this calculus . . . because it contains its own demonstration." In fact, the calculus was fully workable as presented by Leibniz, and was crucial for the development of mathematical science.

The calculus grew out of Leibniz's stay in Paris, as did another notable contribution which emerged from Leibniz's extended tussle with the Cartesians. More was at stake here than just the correct vision of the physical world, for Leibniz used Cartesian physics to illustrate the interpenetration of science and metaphysics. An outstanding example is Leibniz's criticism of Descartes's dynamics (see DESCARTES) by appeal to the metaphysical principle of continuity (difficult to spell out, but well summed up by Leibniz's own semi-technical dictum: "when two conditions . . . continuously approach each other until the one passes into the other, then the results sought for must also approach each other continuously"). Leibniz shows that, according to Descartes, as the mass of a body which impacts on another at rest increases continuously, there will be a discontinuous change in the result of impact when the masses become equal. The principle of continuity has an obvious relation to the calculus, which requires this kind of continuity for its safe use. Metaphysical considerations here make the world safe for mathematics, as well as revealing physical law.

Leibniz's investigation of Cartesian physics also led to a famous controversy about what is conserved in physical interactions, the so-called *vis viva* controversy. Descartes held that the quantity of motion (roughly, what we would call momentum, mv) was the conserved quantity, Leibniz that it was what he called "living force" (what we would call kinetic energy, $\frac{1}{2}mv^2$). By considering the amount of work done by falling bodies, Leibniz could show that it was living force, rather than quantity of motion, that was conserved. As usual, Leibniz drew more than a physical lesson from this controversy, seeing in the need to posit living force (nowadays, energy) which was "different than size, figure and motion" a metaphysical reality "grounding" the material world. Thus it is possible, with some strained squinting, to see Leibniz as a precursor of the modern view of matter and energy as interchangeable, with energy being the more fundamental.

In the final year of his life he entered a debate with Samuel Clarke, mouthpiece for the reticent Newton, ostensibly about the theological dangers of the new philosophy, but remembered for the dispute about the nature of space and time. Newton held that space and time are absolute, independently real "containers" of the world's objects and events, Leibniz that space and time are relative, dependent upon the relations among objects and events. As Leibniz elegantly puts his view: "space is nothing but the *order* of coexistents . . . time is the *order* of inconsistent possibilities."

Leibniz's grounds for relationalism were deeply metaphysical. For Leibniz, the physical world is at bottom phenomenal. The true reality is the infinite set of what he called "monads," each one of which is a perceiving soul (though only relatively few attain self-consciousness, and only one, God, complete clarity of perception). Each is a causally independent substance, indestructible, programmed to run through a certain sequence of perceptions for all eternity; and each one is a mirror of the entire universe. Each monad perceives the world from its own point of view, and its perceptions vary in clarity and distinctness. Every mind, Leibniz says, is omniscient but confused. The monads each represent a physical world, and *reality* is defined as what all the monads agree on, *illusions* as perceptions which represent to one monad a state of the world but to the others merely a state of consciousness. Here, perhaps, we see a foreshadowing of the modern doctrine that the real is what is invariant across different frames of reference. Returning to the main issue, let us imagine a "change" in the spatial arrangement of the world, such as Leibniz's example of God moving the entire universe one foot to the left, which is imperceptible. At the level of the monads, this is no change at all; *nothing* has been effected by this putative action. Anything real has to be reflected in monadic perception, and thus space (similarly, time) is not real.

Clarke *assumed* the intelligibility of God's moving the universe as a whole, but he adverted to an argument that Newton had often used in defense of absolute space. This is that acceleration is a kind of motion that has effects independent of relational configurations. There is a difference between the state of two *motionless* spheres connected by a cord and their state when rotating about their centre of gravity: namely, a tension in the cord and a very great difference in dispositional properties. This difference would be "relationally invisible" if, say, the linked spheres were the sole occupants of space. Where Leibniz argues for the equivalence of circular accelerated and linear constant motion, his argument appears to rest on ideas drawn from the calculus, for he avers that curvilinear motion arises from a composition of rectilinear motions, and hence that circular motion will be relative, since rectilinear motion is.

Leibniz independently invented binary arithmetic, the basis of modern computing. Characteristically, Leibniz saw in this a deeper truth – namely, a theological metaphor: from unity and nothing comes everything. He is also well known for his pioneering research on symbolic logic and artificial computational languages. Finally, mention should be made of his contribution to the culture of science. Leibniz helped to establish the modern pattern of scientific activity: free exchange of scientific results, scientific journals, and scientific societies and academies. His sense of optimism about what science could accomplish and his recognition of the communal nature of science, his openness and willingness to appreciate and build upon the work of others, were significant forces in the creation of both real and ideal modern science.

Philosophy of science

In the philosophical pantheon Leibniz resides among the rationalists, whose philosophy of science is often denigrated as aprioristic and anti-empirical. There is some truth here; Leibniz was convinced of the possibility of a priori demonstrations of scientific law, but he was never anti-empirical. The a priori strain comes, naturally, from considerations of metaphysics and theology. Since the world's all-perfect creator was constrained by a set of accessible metaphysical principles (e.g., continuity, sufficient reason, plenitude, perfection, and noncontradiction), it should be possible to re-create, as it were, the conditions of creation, and from that to deduce the world's structure. In this, Leibniz does not differ in method very much from modern theoretical cosmologists, though the latter tend to avoid any overtly theological underpinning (though some of the uses of the so-called anthropic principle might be noted – a notion that Leibniz would have been keenly interested in). The generality of Leibniz's principles allowed for a much broader application than would be acceptable today. At one famous extreme, Leibniz shows that there cannot be two qualitatively indiscernible physical objects – to use his example, two leaves distinct but otherwise identical (for suppose otherwise, then God could not decide where to place the two leaves in the world, for any reason would apply equally to both, and *everything* has a sufficient reason). This does not strike us as "scientific thought." But Leibniz also used his principles to verify Snell's law of refraction by what he called the method of final causes: the principle of perfection guarantees that light will take the "simplest" path, where the simplest path is taken to mean one that preserves identity of a ratio no matter what the angle of incidence. This is not so unlike appeals to simplicity or symmetry that are common in modern science. Modern philosophy (but not all modern scientists) would say that this was all merely within the "context of discovery"; that it makes good sense to try out the simplest hypotheses first, but that nature cannot be dictated to a priori. Leibniz agrees with this in fact, if not in spirit. A priori knowledge of nature is merely hypothetical, for finite minds cannot see the full interconnection of *all* things which determines how the world will satisfy the metaphysical principles. The complex interplay of empirical research and metaphysical principle that makes up Leibniz's philosophy of science cannot be examined here, but perhaps the appended chart (fig. 34.1), which

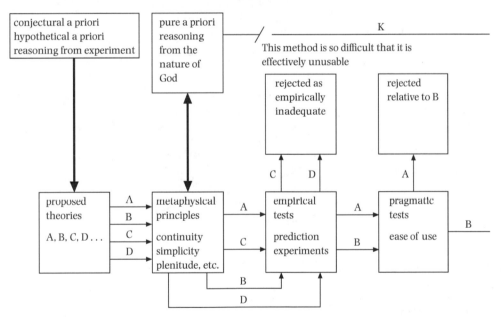

Figure 34.1 Theories are created by a variety of methods, empirical or semi-empirical (*top left*) or pure (*top centre*). The absolutely true theory, K, cannot be discovered in practice, due to the finitude of the human mind. Theories D and B conflict with metaphysics, but proceed to empirical testing. In this case the metaphysically false theory, B, turns out to be the most useful, as compared to the still possibly true theory, A. Thus B will remain in use for the present. Note that thick lines represent conceptual connection, thin lines the paths theories follow during testing. From Seager 1981, p. 493.

traces the path a hypothesis may take to acceptance, along with the above remarks, will give some sense of the depth of Leibniz's views on the nature of science.

Leibniz was an optimist about science. It was to be the most important force for human good and even cultural harmony. It would reveal the truth about the world far beyond the reach of the senses, and point to the metaphysical under pinnings that supported the material world. Through the combination of experiment and theory, science will unravel nature's subtlety, for, as Leibniz beautifully puts it, "a corpuscle hundreds of thousands of times smaller than any bit of dust which flies through the air, together with other corpuscles of the same subtlety, can be dealt with by reason as easily as can a ball by the hand of a player."

References and further reading

Principle collections

(Leibniz's writings remain in an unsatisfactory state. The *Sämtliche Schriften und Briefe* of the Deutsche Akademie der Wissenschaften will provide a complete critical edition, but so far relatively few volumes have been published, although work began in 1923. For now, the Gerhardt collections are the most comprehensive and accessible.)

Couturat, L. 1903: *Opuscules et fragments inédits de Leibniz* (Paris; repr. Hildesheim: Olms, 1961) (Primarily works on logic and methodology.)

Gerhardt, C. I. (ed.) 1971: *Die Philosophischen Schriften von Gottfried Wilhelm Leibniz*, 7 vols (Hildesheim: Olms; 1st pub. Berlin, 1875–90). (For this collection there exists a concordance, R. Finster et al., *The Leibniz Lexicon* (Hildesheim: Olms, 1988).)

—— 1971: *Die Mathematische Schriften von Gottfried Wilhelm Leibniz*, 7 vols (Hildesheim: Olms; 1st pub. Berlin and Halle, 1849–55).

Translations

Alexander, H. G. (ed.) 1998: *The Leibniz–Clarke Correspondence: Together with Extracts from Newton's* Principia *and* Optics (Manchester: Manchester University Press).

Ariew, R., and Garber, D. (eds and trans.) 1989: *G. W. Leibniz: Philosophical Essays* (Indianapolis: Hackett).

Loemker, L. (ed. and trans.) 1969: *Gottfried Wilhelm Leibniz: Philosophical Papers and Letters* (Dordrecht: Reidel). (Still the most comprehensive collection.)

Parkinson, G. H. R. (ed. and trans.) 1961: *Logical Papers* (Oxford: Oxford University Press).

Remnant, P., and Bennett, J. (eds and trans.) 1996: *New Essays on Human Understanding*, 2nd edn (Cambridge: Cambridge University Press).

Works by other authors

Mates, B. 1986: *The Philosophy of Leibniz: Metaphysics and Language* (Oxford: Oxford University Press).

Meli, Domenico Bertoloni 1993: *Equivalence and Priority: Newton versus Leibniz* (Oxford: Oxford University Press).

Okruhlik, K., and Brown, J. (eds) 1985: *The Natural Philosophy of Leibniz* (Dordrecht: Reidel).

Rescher, N. 1981: *Leibniz's Philosophy of Nature* (Dordrecht: Reidel).

Seager, W. 1981: The principle of continuity and the evaluation of theories. *Dialogue*, 20, 485–95.

Woolhouse, R. S. (ed.) 1981: *Leibniz: Metaphysics and Philosophy of Science* (Oxford: Oxford University Press).

35

Locke

G. A. J. ROGERS

Locke was born in Wrington, Somerset, on 29 August 1632. After the Civil War he was sent to Westminster School, and in 1652 to Christ Church, Oxford. A feature of the university in Locke's early years was growing interest in the natural sciences, fostered by, amongst others, Robert Boyle, John Wilkins, and Robert Hooke. After graduating, Locke was much attracted to the work of these men, and soon he was engaged in medical research with Robert Boyle. He remained in Oxford until 1667, when a chance meeting with Anthony Ashley Cooper, later Lord Shaftesbury, led to his joining Shaftesbury's London household, which from then on became Locke's usual residence, and where he acted as Shaftesbury's personal physician, conducted research with Thomas Sydenham, and became a Fellow of the Royal Society.

In 1671 Locke drafted a paper as a basis for discussion with friends, which in the next 18 years was to evolve into his great philosophical work. In the meantime he was engaged not only with medicine but also with politics and travel, especially in France and, in particular, at the medical school in Montpellier.

Locke fled to Holland in 1683, returning to England only after the Revolution of 1688. In the following year his two most important works, *Two Treatises of Government* and the *Essay Concerning Human Understanding*, were published, the first anonymously. The latter soon established him as the leading English philosopher of the age. The remaining years of his life were devoted to revising and defending his published works, writing new ones, and in scientific, philosophical, and theological discussion with his friends, including Isaac Newton. He died at High Laver in Essex on 28 October 1704.

The *Essay Concerning Human Understanding* is the first modern systematic presentation of an empiricist epistemology, and as such had important implications for the natural sciences and for philosophy of science generally. It was written from entirely within the new scientific community associated with the names of Bacon, Galileo, Descartes, and Newton and is one of the most influential works of epistemology ever to be published (see GALILEO; DESCARTES; NEWTON).

Like his predecessor, Descartes, Locke began his account of knowledge from the conscious mind aware of ideas. Unlike Descartes, however, he was concerned not to build a system based on certainty, but to identify the mind's scope and limits. The premise upon which Locke built his account, including his account of the natural sciences, is that the ideas which furnish the mind are all derived from experience. He thus totally rejected any kind of innate knowledge. In this he was consciously opposing Descartes, who had argued that it is possible to come to knowledge of fundamental truths about the natural world through reason alone. Descartes had argued, for example, that we can come to know the essential nature of both mind and matter by pure reason. Locke accepted Descartes's criterion of clear and distinct ideas as the basis for knowledge, but denied any source for them other than experience. It was information that came in via the five senses (ideas of sensation) and ideas engendered from our inner experiences (ideas of reflection) that were the building blocks of the understanding.

Locke combined his commitment to "the new way of ideas" with a tentative espousal of the "corpuscular philosophy" of Robert Boyle. This, in essence, was an acceptance of a revised, more sophisticated account of matter and its properties that had been advocated by the ancient atomists and recently supported by Galileo and Gassendi. Boyle argued from theory and experiment that there were powerful reasons to justify some kind of corpuscular account of matter and its properties. He called the latter qualities, which he distinguished as primary and secondary (see QUALITIES). Locke, too, accepted this account, arguing that the ideas we have of the primary qualities of bodies resemble those qualities as they are in the object, whereas the ideas of the secondary qualities, such as color, taste, and smell, do not resemble their causes in the object.

There was no strong connection between acceptance of the primary–secondary quality distinction and Locke's empiricism, and Descartes had also argued strongly for it. But it did fit closely with the new account of matter that was rapidly gaining almost universal acceptance by natural philosophers, and Locke embraced it within his more comprehensive empirical philosophy. But Locke's empiricism did have major implications for the natural sciences, as he well realized. His account begins with an analysis of experience. All ideas, he argues, are either simple or complex. Simple ideas are those like the red of a particular rose or the roundness of a snowball. Complex ideas, our ideas of the rose or the snowball, are combinations of simple ideas. We may create new complex ideas in our imagination – a dragon, for example. But simple ideas can never be created by us: we just have them or not, and characteristically they are caused by, for example, the impact on our senses of rays of light or vibrations of sound in the air coming from a particular physical object.

Since we cannot create simple ideas, and they are determined by our experience, our knowledge is in a very strict and uncompromising way limited. Furthermore, our experiences are always of the particular, never of the general. It is *this* particular simple idea or *that* particular complex idea that we apprehend. We never in that sense apprehend a universal truth about the natural world, but only particular instances. It follows from this that all claims to generality about that world – for example, all claims to identify what were then beginning to be called laws of nature – must to that extent go beyond our experience and thus be less than certain (see LAWS OF NATURE).

There was another important limitation on knowledge. Since it can never extend beyond experience, it must follow that suppositions about unobserved or unobservable entities and their properties can have no higher status than that of hypotheses. Locke explained the implications which this has for knowledge of nature by distinguishing between the real and the nominal essence of things. We believe that all material substances have an inner nature, an object's real essence. But all we can ever observe are the outward properties of that thing, not its internal constitution. Our idea of the essence of any particular thing is thus limited: we have no way of knowing whether it matches its inner constitution. The situation here stands in marked contrast with our knowledge of mathematical objects, such as a square or a triangle. Here we know precisely what the real essence is: namely, its definition as given in Euclidean geometry. Thus we know – we do not merely suppose – that a square is bounded by four straight lines which are equal in length and which intersect at right angles, and this is a general truth that applies to all possible squares. This contrasts with, say, gold, where, while we can suppose, we can never know, that its essence consists of the set of properties which scientists have so far identified in all its samples. Locke expressed the difference between these by saying that in the case of mathematics the real and the nominal essence of the objects concerned coincide, whereas in the case of substances, or as we might say today, natural kinds, we can never know this, and often discover that they do not (see NATURAL KINDS).

Locke realized that such considerations have major implications for our systems of classification for natural substances. In book III of the *Essay*, "Of Words," Locke offers the first sustained modern discussion of *meaning*, and some of his account has particular significance for the natural sciences. Words, Locke says, in their immediate signification stand for the ideas in the mind of the person who uses them. In their mediate signification they stand for the objects which the words pick out. In identifying natural kinds, such as gold, water, or cassowaries, we operate with a classificatory system which enables us to identify individuals as members of these natural kinds. But this classificatory system is determined not by the essences of things, but by our complex ideas of them, which are often quite different. As Locke, with his own experience in the laboratory wistfully tells us, "bodies of the same species, having the same nominal essence, under the same name . . . often upon severe ways of examination, betray qualities so different one from another, as to frustrate the expectations and labour of very wary chymists" (*Essay*, III. vi. 8). More generally, the imperfections of words are likely to mislead us in matters of reasoning and scientific investigation. Thus Locke tells us of a dispute among learned physicians, to which he was a party, as to whether any liquor passed through the filaments of the nerves. He tells us that he was able to defuse the debate when he showed that the disagreement turned on differing conceptions or "complex ideas" for which the word "liquor" stood in the minds of the disputants (*Essay*, III. ix. 16).

Locke's empiricist philosophy thus excluded the possibility of ever coming to know not only general truths about the natural world a priori, but also any such a posteriori knowledge as well. To this extent his position anticipates a central claim of the logical positivists: that the only certain general propositions are verbal, and the remainder are but conjectures, beliefs, or suppositions (see LOGICAL POSITIVISM). Locke, of course, did not wish to deny that we do and must make use of general propositions, and that when they are well supported by evidence, we are entitled to do so; but we must be careful not to be deceived into accepting as "an unquestionable truth, which is really at best but a very doubtful conjecture, such as are most (I had almost said all) of the hypotheses in natural philosophy" (*Essay*, IV. xii. 13).

If certainty was not to be a feature of our expectations about the natural world, Locke did not see this as in any sense placing a crippling burden on the human intellect. Where knowledge was not possible, assessment of probability, which he defined as "likeliness to be true" (*Essay*, IV. xv. 3), often was. In his account of the degrees of assent, he distinguishes different kinds of evidence, including testimony, which we may have for a claim. He also distinguishes between two different kinds of proposition: namely, those, on the one hand, where there is the possibility of some relevant empirical evidence, and, on the other, where no such evidence is possible. The latter include hypotheses about all unobservables – for example, the inner structure of minute particles. In these cases, Locke argues, all we can do is to use analogical argument, and assess things as more or less probable, "only as they more or less agree to truths that are established in our minds, and as they hold proportion to other parts of our knowledge and observation" (*Essay*, IV. xvi. 12). There is nothing wrong in using analogical argument, Locke says, so long as we are well aware of its limitations – in particular, the uncertain nature of conclusions drawn on the basis of it.

Locke's philosophy, while evidently of great relevance to the natural sciences and how they might be pursued, also had major implications for the development of the social sciences. The most obvious of these was that his account implied a model of the human mind and a science of its investigation. In that sense, Locke's philosophy provided a program for psychology and was often seen as such by his eighteenth-century readers – Voltaire, for example, who saw him as having written a natural history of the human soul to go alongside Newton's account of the physical world. And the associationist psychology inspired by a late chapter in the *Essay* is particularly linked to David Hartley's *Observations on Man, His Frame, His Duty, and His Expectations* (1749). Locke's work also had important implications for other social sciences. The rejection of innate ideas in book I of the *Essay* encouraged an emphasis on the empirical study of human societies, to discover just what explained their variety, and thus towards the establishment of the science of social anthropology.

In the eighteenth century, Locke's empiricism and the science of Newton were, with reason, combined in people's eyes to provide a paradigm of rational inquiry that, arguably, has never been entirely displaced. It emphasized the very limited scope for absolute certainties in the natural and the social sciences, and more generally underlined the boundaries to certain knowledge that arise from our limited capacities for observation and reasoning. To that extent it provided an important foil to the exaggerated claims sometimes made for the natural sciences in the wake of Newton's achievements in mathematical physics.

References and further reading

Works by Locke
1689: *Two Treatises of Government* (London).
1690: *Essay Concerning Human Understanding* (London).

Works by other authors
Ayers, M. 1991: *Locke* (London: Routledge).
Hartley, D. 1749: *Observations on Man, His Frame, His Duty, and His Expectations* (London).
Rogers, G. A. J. 1998: *Locke's Enlightenment* (Hildesheim: Olms).
Voltaire 1733: *Letters Concerning the English Nation*.
Woolhouse, R. S. 1971: *Locke's Philosophy of Science and Knowledge* (Oxford: Blackwell).

36

Logical Empiricism

WESLEY C. SALMON

The fundamental tenet of logical empiricism is that the warrant for all scientific knowledge rests upon empirical evidence in conjunction with logic, where logic is taken to include induction or confirmation, as well as mathematics and formal logic (see EVIDENCE AND CONFIRMATION). This appears to conflict strongly with Thomas Kuhn's famous statement that scientific theory choice depends on considerations that go beyond observation and logic, even when logic is construed so as to include confirmation (see KUHN and PRAGMATIC FACTORS IN THEORY ACCEPTANCE). Logical empiricists deny the possibility of synthetic a priori knowledge – that is, substantive knowledge of the world based on pure reason. Those who, with W. V. Quine, reject the analytic/synthetic distinction, would, I suppose, question the possibility of a priori knowledge altogether (see QUINE). Contemporary logical empiricists disagree, however, about such basic issues as the nature of empirical evidence, the status and structure of confirmation or inductive inference, the nature of scientific explanation, and the character of scientific theories, to name but a few examples (see THEORIES; EXPLANATION; CONFIRMATION, PARADOXES OF).

Historical background

The roots of logical empiricism are so intimately entwined with those of logical positivism (see LOGICAL POSITIVISM) that the two movements are often mistakenly identified with one another. Both arose in reaction to the post-Kantianism of the nineteenth century; both insisted upon empiricism in epistemology; both emphasized the importance of modern logic; both looked to the special sciences for inspiration; and both completely rejected speculative metaphysics (see METAPHYSICS, ROLE IN SCIENCE). In spite of this, as we will see, the logical empiricists later reacted against logical positivism because of fundamental philosophical differences. Logical positivism arose and flourished in Vienna in the 1920s and early 1930s, whereas Berlin was the center of logical empiricism. During that period there was a good deal of communication and friendly cooperation between the two groups. For example, *Erkenntnis*, the leading journal in scientific philosophy, was co-founded and co-edited by Rudolf Carnap (Vienna) and Hans Reichenbach (Berlin). Both groups suffered severely because of Nazism, and both groups dispersed with its rise in the German-speaking world.

Despite these common roots, logical empiricism continued to be a vital movement in philosophy in the second half of the twentieth century, while logical positivism had, by mid-century, ceased to be a significant philosophical force. In many contexts "logical positivist" now functions chiefly as a term of abuse, while "post-positivist" has become a widely used term of approbation. As Herbert Feigl, who had been a member of the Vienna Circle, later quipped, by that time (roughly mid-century) all of the positivists had changed either their views or their names (i.e., the designation of their philosophical affiliations). Among those early members of the Vienna Circle who began as logical positivists, but evolved into logical empiricists, Carnap, Feigl, and Carl G. Hempel have been the most influential.

Carnap's *Der Logische Aufbau der Welt* (*The Logical Structure of the World*) (1928) was the epitome of the program of logical positivism. As an introductory motto, Carnap (1967, p. 5) selected a famous quotation from Bertrand Russell: "The supreme maxim in scientific philosophizing is this: Whenever possible, logical constructions are to be substituted for inferred entities" (1929, p. 155). In the realm of mathematics, for instance, Russell applied this maxim when he defined cardinal numbers as of sets of sets (see RUSSELL). On this definition, numbers do not enjoy independent existence, but are constructs from sets. Carnap's *Aufbau* was a monumental effort to apply this maxim to *all* domains of scientific knowledge, and to "construct" the natural world as we know it in a precise manner, using a single individual's experiences as substantive content, and employing the most powerful tools of symbolic logic to carry out the construction. Carnap emphasized that other constructions on other bases were possible and legitimate, but he maintained that the phenomenalistic basis is epistemologically privileged, because it arises out of direct experience. In this work Carnap was attempting to carry out in detail a project Russell had sketched but never developed.

In 1933 Reichenbach published a largely laudatory review of this book. He begins by saying, "This extensive work by Rudolf Carnap represents a massive compilation of his ideas on logic and epistemology. Yet it is not Carnap's ideas alone [as Carnap explicitly declares] but also those set forth by the Vienna Circle as a scientific conception of the world which have been fully presented for the first time in Carnap's major work" (1978, p. 405). Reichenbach expresses just one reservation:

> But it seems to me to be at least doubtful whether this reduction to perceptual reports and pure logic exhausts everything we mean to include in our assertions about reality. These doubts are principally aroused when we consider the use of the concept of probability in the natural sciences, for if we accept Carnap's reduction of scientific assertions, we forfeit the indisputable basic principle that such assertions are not merely reports of past perceptual experiences, but are also invariably predictions of future perceptual experiences. It is a puzzle to me just how logical neo-positivism [as opposed to the earlier positivism of Ernst Mach] proposes to include assertions of probability in its system, and I am under the impression that this is not possible without an essential violation of its basic principles. (1978, p. 407)

The problems concerning prediction and probability, which are inseparably linked in Reichenbach's thinking, constituted the opening wedge for the split between logical positivism and logical empiricism (see PROBABILITY and MACH).

In 1933 Hitler took power in Germany. Reichenbach fled to Turkey, where he taught at the University of Istanbul until 1938, at which time he moved to the University of California at Los Angeles, where he remained for the rest of his life. Carnap, who was in Prague, remained until 1935, at which time he emigrated to America, taking a position at the University of Chicago. After Reichenbach's death in 1953, Carnap became his successor at the University of California at Los Angeles. Feigl, who had come to America in 1930, was an early emissary of logical positivism. He taught first at the University of Iowa and subsequently at the University of Minnesota, where he was the founder and director of the Minnesota Center for the Philosophy of Science until his retirement. Hempel, who was a member of both the Vienna and the Berlin groups, moved to Belgium in 1934. Three years later he went to the University of Chicago as Carnap's assistant for a year; later he taught at Yale and Princeton universities (among others), and, after retirement from Princeton, for 10 years at the University of Pittsburgh.

In 1936, A. J. Ayer published the first edition of *Language, Truth, and Logic*, which introduced logical positivism to the English-speaking world, and has remained the most influential text on positivism in the English language. This work takes up the central views that we find Reichenbach disputing in his 1938 book. The second edition of Ayer's book, unrevised except for the addition of a substantial introduction, was issued in 1946.

In "Logistic empiricism in Germany and the present state of its problems" (1936) – an ironic title, since there was no such thing in Germany by that time – Reichenbach expanded on the problem raised in his review of Carnap.

> An analysis of meaning [according to positivists] was reached according to which any proposition of science contains nothing but a repetition of "report propositions." Since every report consists of statements about the *immediate present*, science states nothing but relations existing between present phenomena. This conclusion, however, is in sharp contrast to the actual practice of science, for scientific propositions make assertions about the *future*. Indeed, there is no scientific law which does not involve a prediction about the occurrence of future events; for it is of the very essence of a scientific law to assure us that under certain given conditions, certain phenomena will occur. (p. 152)

He goes on to declare: "This was the precise reason why the Berlin group could not accept positivism" (ibid.).

Reichenbach's fullest account of his differences with the positivists came in his major epistemological treatise, *Experience and Prediction* (1938), which he explicitly characterized as his *refutation* of logical positivism. He takes issue with the positivists on three major points: (1) phenomenalism, (2) the verifiability theory (or criterion) of cognitive meaning, and (3) scientific realism (see REALISM AND INSTRUMENTALISM). Each of these ideas is directly connected to the concept of probability, the topic of Reichenbach's dissertation (1915) and his treatise *Wahrscheinlichkeitslehre* (*Theory of Probability*) (1935).

1. Reichenbach rejected the version of phenomenalism used by Carnap in the *Aufbau*, as well as the sensationalism or neutral monism of Mach and Russell. He adopted physicalism as a basis for his epistemology; that is, he claimed that we observe middle-sized physical objects, and these observations constitute our empirical data (see PHYSICALISM). Sense impressions are not the given data of experience; they are theoretical constructs of psychology (see THEORETICAL TERMS). He was, of course, fully aware that our observations of middle-sized physical objects are fallible. Thus he rejects the

positivistic search for certainty in the basis of all knowledge. He maintained that we can probabilistically distinguish veridical perceptions from dreams, illusions, and hallucinations, through our attempts to establish regularities in nature. This amounts to a coherence criterion of veridicality, but not a coherence theory of truth. Indeed, in a famous 1952 symposium he argued that even phenomenal reports are not absolutely certain (Reichenbach 1952).

On the issue of phenomenalism, the position of the logical empiricists seems thoroughly vindicated. Nelson Goodman's *The Structure of Appearance* (1951) showed convincingly that Carnap's efforts at construction in the *Aufbau* not only failed, but failed spectacularly. Carnap continued to maintain, until his death in 1970, that we can choose a phenomenalistic language if we wish, on the basis of such pragmatic factors as convenience and utility, but by mid-century it was clear that a phenomenalistic language is hopelessly inadequate even for the analysis of everyday discourse about ordinary material objects – to say nothing of the needs of science. To the best of my knowledge, not one single statement about such an object has ever been successfully translated into a phenomenalistic language.

2. The early positivists had maintained that no statement is cognitively meaningful unless it is capable in principle of verification by sensory experience. Some had declared that the meaning of a statement *is* the means of its verification. It was recognized that any such criterion is too strict to admit universal generalizations into the realm of meaningful statements; in response, such sentences were classified as rules of inference rather than statements capable of being true or false.

In *Experience and Prediction* Reichenbach declined to identify any such abstract entity as *the meaning* of a sentence; instead, he offered criteria to classify sentences as cognitively meaningful or meaningless and to determine when two sentences are equivalent in meaning. According to his criteria, a sentence is cognitively meaningful if and only if it is possible in principle to find empirical evidence to support or undermine it probabilistically. Moreover, two sentences are equi-significant if and only if both are supported or undermined to the same extent by any possible empirical evidence.

It should be noted that some positivists – for example, Ayer – also recognized that strict verifiability is too stringent a criterion, and in *Language, Truth, and Logic* he formulated a criterion in terms of confirmability to some degree, rather than complete and conclusive verifiability. His first formulation was defective inasmuch as it allowed any sentence whatever to qualify easily as confirmable in principle. In the introduction to the second edition, he attempted a much more careful formulation; but it, too, fell victim to the same criticism, as Alonzo Church (1949) showed in an elegant review. To many philosophers, I think, this failure sounded the death knell for the verifiability or confirmability criterion. It should be emphatically noted, however, that the defects in both the first and the second formulations were in Ayer's superficial explication of confirmability; *they had nothing whatever to do with confirmability as a criterion of cognitive meaningfulness.* In contrast to Ayer, Reichenbach earlier and Carnap later both had well-developed theories of induction, confirmation, and probability. In "Testability and meaning" (1936–7), Carnap had also relinquished the requirement of strict verifiability, and in (1956) he offered an even more ample criterion of empirical meaning, though it, too, was unsuccessful. Contemporary logical empiricists seem, by and large, to have given up on criteria of cognitive significance, though some see considerable value in a criterion of demarcation, such as Karl R. Popper's, between genuine science and pseudo-science (see POPPER).

3. In *Experience and Prediction* Reichenbach distinguishes between *reduction* and *projection*. Reduction is essentially a definitional relation – just the sort of relation found in the constructions of Carnap's *Aufbau*. Projection is not a definitional relation; it is a probabilistic relation that involves inductive inferences. Reichenbach tried to explain his position in terms of a rather puzzling analogy. Imagine a "cubical world" with translucent walls, in which human-like beings are confined for their entire lifetimes. By means of a complex set of lights and mirrors outside the cube, shadows of birds, also outside the cube, are cast on the ceiling and one of the walls. An observer, irremediably confined to the interior of the cube, noting correlations between shadow behavior on the ceiling and wall, infers the existence of unobservable birds outside the cube. A physicist would make this kind of inference, according to Reichenbach, and it is a legitimate probabilistic inference. It is an example of projection. A logical positivist, says Reichenbach, using reduction, would have to say that the entire reality consists of the shadows and their correlations; the positivist is not entitled to affirm the existence of the unobservable entities. If positivists were to refer to the birds, they would have to say that the birds are nothing more or less than these shadows and their correlations.

Unfortunately, Reichenbach (1938) does not explain how the inference of the physicist lends probability to the independent external existence of the unobservable birds. He says that probabilistic inferences enable us to extend our knowledge from one domain (the interior of the cubical world) to another (the birds outside the cube), but he provides hardly any argument. In my opinion, he first proffered a justification for this thesis in *The Direction of Time* (1956), where he presented and analyzed *the principle of the common cause* (see STATISTICAL EXPLANATION). Though he did not mention his cubical world analogy in that context, I believe he had this principle implicitly in mind when he wrote *Experience and Prediction*. In any case, regardless of the adequacy of his justification, Reichenbach unequivocally embraced scientific realism. Feigl and Hempel also advocated scientific realism, but for reasons quite different from Reichenbach's. Carnap, I believe, never did.

Current trends

In attempting to characterize contemporary logical empiricism, I shall enumerate some pertinent issues, but I do not intend to lay down a "party line" for logical empiricists. As I said at the outset, logical empiricists differ on a number of fundamental problems, and there is room for considerable latitude within the movement. For example, logical empiricists need not insist on a sharp distinction between the observable and the unobservable; moreover, if a sharp distinction is affirmed, they need not agree on how to draw the line. In addition, logical empiricists need not agree on whether scientific theories are partially interpreted axiomatic systems or some sort of model-theoretic construction. Other examples follow.

If Carnap's *Aufbau* represented the zenith of logical positivism, Hempel's *Aspects of Scientific Explanation and Other Essays in the Philosophy of Science* (1965) is the most representative expression of the logical empiricist approach. I am not suggesting that he provided definitive answers to the many problems he raises; indeed, to a large extent I strongly disagree with him, and so do many others. But the selection of issues and the methods employed in dealing with them are strongly indicative of the spirit of logical empiricism.

One major part of this book takes up *confirmation, induction, and rational belief*. As we saw, Reichenbach criticized the Carnap of the *Aufbau* for failure to take account of probability. In the 1940s Carnap began work on a project that occupied most of his attention for the rest of his life: to develop a precisely articulated system of inductive logic and confirmation in which two concepts of probability were clearly explicated (see EVIDENCE AND CONFIRMATION). One concept is *degree of confirmation* (often known as inductive probability); the other is *relative frequency* (often known as statistical probability). To this task he brought powerful formal techniques in the hope of developing an inductive logic as precise and as well founded as the systems of deductive logic available at the time.

Reichenbach strongly disagreed with Carnap, holding that the *frequency interpretation* is the only legitimate interpretation of the probability calculus (see PROBABILITY). Reichenbach believed that, with the aid of Bayes's theorem, it was possible to give a precise account of the probability of scientific theories. Although I believe that this view has at least a grain of truth, I must add that Reichenbach never spelled it out clearly, and most philosophers, including Carnap and Hempel, are completely skeptical.

More recently, a strong Bayesian movement has emerged in philosophy and in statistics; most of its adherents adopt a *subjective* or *personalistic interpretation* of probability as degree of rational belief. Carnap's system of inductive logic was also Bayesian, in the sense of accepting Bayes's theorem as a schema for confirmation, though he did not accept a thoroughgoing subjective interpretation. But whatever interpretation or interpretations of probability are adopted, investigators in all these categories regard the mathematical calculus of probability as a formal guide, and all parties to the controversies pursue logical and mathematical precision. A philosopher holding any of these views could qualify as a logical empiricist; moreover, various non-Bayesian approaches are also advocated by philosophers of this persuasion – for example, by those who adopt *bootstrapping* or *orthodox statistical approaches*.

Hempel's classic essay "The theoretician's dilemma," along with "A logical appraisal of operationism," both reprinted in the 1965 volume, squarely address the problems that led positivists to an anti-realistic view of theories that appear to refer to unobservable entities. He concludes that, for purposes of *deductive systematization*, it is logically possible (though heuristically undesirable) to dispense with "theoretical terms." For purposes of *inductive systematization*, he argues, they are indispensable. Two issues are involved: namely, the existence of unobservable entities and the meanings of theoretical terms. Like Carnap in his post-positivist phase, he admits that theoretical terms cannot be fully defined; they can nevertheless be meaningfully embedded in nomological networks. Although Hempel's reasons differ from Reichenbach's, he supports the same general realistic interpretation of scientific theories that Reichenbach offered in opposition to the logical positivists.

Another major section of Hempel's book deals with scientific explanation (see EXPLANATION), including statistical explanation. This represents another basic contrast between logical positivism (see LOGICAL POSITIVISM) and logical empiricism. Positivists were by and large convinced that there is no such thing as *scientific* explanation; the business of science is to describe and predict natural phenomena, and to systematize our knowledge of them. If one wants explanations – answers to why-questions – it is necessary to appeal to metaphysics or theology. Hempel and other logical empiricists believe, by contrast, that explanation is within the scope of science; indeed, it is one of the fruits, if not *the most important* fruit, of modern science. Moreover, they hold that philosophers can explicate the concept of scientific explanation with reasonable precision.

Significant discussion of scientific explanation in the latter half of the twentieth century takes as its point of departure the famous 1948 article by Hempel and Paul Oppenheim, which is reprinted in *Aspects*. This article contains a precise formal explication of the deductive-nomological model of explanation of particular facts; it also acknowledges (without attempting explications) that science contains deductive-nomological explanations of general laws, as well as statistical or probabilistic explanations. The article "Aspects of scientific explanation" in *Aspects* deals with all these types; for about two decades thereafter, this constituted a "received view" of the nature of legitimate scientific explanation. Although many logical empiricists, myself included, have severely criticized Hempel's models – with the result that they are no longer "received" – their effort should be regarded as an attempt to extend and improve Hempel's constructions, not to undermine the enterprise.

In one respect logical empiricism extends far beyond the topics taken up by Hempel in *Aspects*: namely, in the investigation of foundational issues in particular fields of scientific investigation. In the beginning, this tradition is most conspicuously exhibited by Reichenbach, who, before his flight to Turkey, worked closely with Albert Einstein in Berlin (see EINSTEIN). Well-versed in relativity theory, Reichenbach wrote extensively on problems involving space and time (see SPACE, TIME, AND RELATIVITY). He also contributed significantly to the philosophy of quantum mechanics, and to the foundations of thermodynamics and statistical mechanics (see QUANTUM MECHANICS). These topics, pertaining to areas of tremendous progress in twentieth-century physics, are subjects of intense philosophical activity today. In addition, extensive work addresses philosophy of biology, psychology and psychoanalysis, and the social sciences (see BIOLOGY and SOCIAL SCIENCE, PHILOSOPHY OF). In all cases, the philosophical investigations are conducted against a background of detailed mastery of the scientific subject matter.

The opposition

In its early stages, as we have seen, logical empiricism made common cause with logical positivism against the idealistic metaphysics that was strongly influential early in the twentieth century. However, a fundamental split occurred, initially on issues connected directly or indirectly with probability, and an opposition developed between these two movements.

During the middle decades of the twentieth century, as debates between logical positivism and logical empiricism waned, a strong opposition developed between logical empiricists and the practitioners of "ordinary language analysis," as exhibited, for example, in the spirited debates between Hempel and Michael Scriven. The fundamental disagreement seemed to hinge on formalization; Scriven and others attacked Hempel's models of explanation largely because of their degree of formality and their consequent failure to conform to ordinary usage. Another area of opposition centered on questions regarding probability and induction. Against Peter Strawson and many others, I claimed that the ordinary language *dissolution* of the problem of the justification of induction was seriously inadequate (Salmon 1967); I still hold that view, though I have serious misgivings about other approaches as well.

In his monumental *Logical Foundations of Probability* (1950) Carnap, correctly I think, placed great emphasis upon "clarification of the explicandum" – a process of informal analysis designed to make sure that we are reasonably clear about the concept we are trying to explicate before giving a formal and precise explication. At this crucial preliminary stage ordinary language analysis is valuable, especially if it contains analysis of the ordinary discourse of scientists. To the logical empiricist, such linguistic analysis thus becomes an important prolegomenon to precise explication.

In more recent decades, especially since Kuhn's *The Structure of Scientific Revolutions* (1962), the chief opposition has been between logical empiricism and a school of philosophy strongly committed to a historical approach. This opposition should not be overdrawn, however, inasmuch as many logical empiricists see the history of science as an essential source of enlightenment for the philosophy of science. Kuhn's book was the last monograph published in *The International Encyclopedia of Unified Science*, an enterprise originated and pursued by logical positivists and logical empiricists. Carnap was one of the editors. In "Did Kuhn kill logical empiricism?" Reisch (1991) reveals Carnap's strong support and extensive agreement with Kuhn. Careful study of Kuhn's work reveals, I believe, that much of the opposition between Kuhn and logical empiricism has been based on serious misunderstanding of his views – especially the notion, vehemently denied by Kuhn, that he holds an irrationalist view of the nature of science.

In my 1990 article I suggest that a good deal of the controversy between Kuhn and the logical empiricists has also hinged on a serious misconception on Kuhn's part regarding the nature of scientific confirmation. Consider, for example, Kuhn's rejection of the distinction, emphasized by Reichenbach (1938), between the *context of discovery* and the *context of justification*. Kuhn was apparently thinking of confirmation in terms of the traditional hypothetico-deductive method. Plausibility considerations, however, play a crucial role in scientific theory preference. Since such factors have no place in the hypothetico-deductive method, to Kuhn it appears that they must fall outside the context of justification. But if one adopts a Bayesian logic of confirmation, plausibility considerations enter directly in the form of prior probabilities, and thus play, not only an admissible role, but also an obligatory one, in the *logic* of scientific confirmation, and consequently in the context of justification. Again, when Kuhn claims that scientific theory choice depends on factors that go beyond observation and logic, it is reasonable to suppose that he has plausibility considerations in mind. But plausibility considerations do not go beyond the logic of confirmation when confirmation is construed in the Bayesian way.

Between the logical empiricists and those who, *unlike Kuhn*, do hold that science is irrational, the gulf is immense. Logical empiricists are well aware that social, political, economic, and psychological factors strongly influence scientific activity. Nevertheless, the fundamental aims of science, on their view, relate to the description and understanding of an objectively knowable real world. To those who say that the chief activity of scientists is puzzle solving, the logical empiricist would respond, "Why bother unless the solution contributes to our understanding of the world?"

Logical empiricism has developed dramatically since its beginnings in Berlin. Comparing the early writings of logical positivists and logical empiricists with Hempel's *Aspects*, one readily notices a diminution in dogmatism and an increased spirit of self-criticism. Moreover, in the years since 1965 we have seen a decrease in the rigidity of philosophical stance. For example, in Hempel's work we are told that genuine scientific explanations must meet certain criteria of adequacy; more recent work places less emphasis upon preliminary criteria of adequacy and more emphasis upon trying to find explanatory patterns used in scientific practice. Because of various philosophical scruples, Hempel denies causation a fundamental role in explanation (see CAUSATION); I am convinced that causation is an indispensable concept in many explanatory contexts (Salmon 1984). I believe, however, that the philosophical scruples are of utmost importance, and that, consequently, we must provide an account of causation that does not shrink from David Hume's classic challenges (see HUME). The causation issue has a direct bearing, for example, on the nature of functional explanations (see TELEOLOGICAL EXPLANATION). Hempel relegated them to a status inferior to genuine explanations; nevertheless, they are considered acceptable in such fields as sociology, anthropology, and biology. It seems to me that they can be understood as an important type of causal explanation.

Undeniably, many philosophers today consider logical empiricism wrongheaded or passé; often this results from a confusion of logical empiricism with logical positivism. For this reason, as well as historical accuracy, it is important to recognize the fundamental differences between the two movements. Not only in America – where several important figures found refuge from Hitler – but throughout the world, many philosophers continue to find the approach philosophically rewarding.

References and further reading

Ayer, A. J. 1936: *Language, Truth, and Logic* (London and New York; 2nd edn, New York: Dover, 1946).

Carnap, R. 1936–7: Testability and meaning. *Philosophy of Science*, 3, 419–71; 4, 1–40.

—— 1950: *Logical Foundations of Probability* (Chicago: University of Chicago Press; 2nd edn, 1962).

—— 1956: The methodological character of theoretical concepts. In *The Foundations of Science and the Concepts of Psychology and Psychoanalysis*, Minnesota Studies in the Philosophy of Science, 1, ed. H. Feigl and M. Scriven (Minneapolis: University of Minnesota Press), 38–76.

—— 1967: *The Logical Structure of the World* (Berkeley and Los Angeles: University of California Press), a translation by R. A. George of *Der Logische Aufbau der Welt* (Berlin-Schlachtensee: Weltkreis-Verlag, 1928).

Church, A. 1949: Review of Ayer's *Language, Truth and Logic*. *Journal of Symbolic Logic*, 14, 52–3.

Goodman, N. 1951: *The Structure of Appearance* (Cambridge, MA: Harvard University Press).

Hempel, C. G. 1965: *Aspects of Scientific Explanation and Other Essays in the Philosophy of Science* (New York: Free Press).

Kuhn, T. 1962: *The Structure of Scientific Revolutions* (Chicago: University of Chicago Press; 2nd edn, 1970).

Reichenbach, H. 1915: *Der Begriff der Wahrscheinlichkeit für die mathematische Darstellung der Wirklichkeit* (Leipzig: Barth).

—— 1933: Review of Carnap, *Der Logische Aufbau der Welt. Kantstudien*, 38, 199–201.

—— 1935: *Wahrscheinlichkeitslehre: Eine Untersuchung über die Logischen und Mathematischen Grundlagen der Wahrscheinlichkeitsrechnung* (Leiden: A. W. Sijthoff's Uitgeversmaatschappij).

—— 1936: Logistic empiricism in Germany and the present state of its problems. *Journal of Philosophy*, 33, 141–60.

—— 1938: *Experience and Prediction* (Chicago: University of Chicago Press).

—— 1952: Are phenomenal reports absolutely certain? *Philosophical Review*, 61, 147–59.

—— 1956: *The Direction of Time* (Berkeley and Los Angeles: University of California Press).

—— 1978: Carnap's *Logical Structure of the World*. In *Hans Reichenbach: Selected Writings, 1909–1953*, vol. 1, ed. Maria Reichenbach and Robert S. Cohen (Dordrecht: Reidel).

Reisch, G. A. 1991: Did Kuhn kill logical empiricism? *Philosophy of Science*, 58, 264–77.

Russell, B. 1929: *Mysticism and Logic* (New York: W. W. Norton).

Salmon, W. C. 1967: *The Foundations of Scientific Inference* (Pittsburgh: University of Pittsburgh Press).

—— 1984: *Scientific Explanation and the Causal Structure of the World* (Princeton: University of Princeton Press).

—— 1990: Rationality and objectivity in science *or* Tom Kuhn meets Tom Bayes. In *Scientific Theories*, Minnesota Studies in the Philosophy of Science, 14, ed. C. W. Savage (Minneapolis: University of Minnesota Press), 175–204.

—— 1999: The spirit of logical empiricism: Carl G. Hempel's role in twentieth-century philosophy of science. *Philosophy of Science*, 66.

37

Logical Positivism

CHRISTOPHER RAY

The Vienna Circle

Logical positivism and the Vienna Circle are almost synonymous. The Vienna Circle grew in strength throughout the 1920s, attracting philosophers such as Rudolf Carnap, Friedrich Waismann, and Otto Neurath and mathematicians and scientists such as Kurt Gödel and Hans Hahn. It started as an intellectual club (initially known as the Ernst Mach Society), with Moritz Schlick, Professor of Philosophy at the University of Vienna, as its leading light. As the club debated and discussed problems in science, logic, and philosophy, a definite consensus emerged. The members of the Vienna Circle were bound, initially at least, by commitments, to:

- the development of the positivistic heritage of David Hume and Ernst Mach, whose disdain for metaphysics and whose focus upon empirical investigation were echoed by the Circle again and again;
- the promotion of scientific inquiry as *the* model for all intellectual investigations;
- the conviction that physics was not simply a model for other sciences – but that all sciences, including psychological and social sciences, might one day be unified and reduced to common, fundamental physical terms;
- the systematic use of logical analysis to reduce complex statements to elementary propositions, so that the "high-level" scientific statements of a given theory might be unpacked into "low-level" (and directly verifiable) claims about observation and experience; here the members of the Circle were inspired by the foundational work of such mathematicians as Frege and by the logical atomism of Bertrand Russell and Ludwig Wittgenstein.

Logical positivism emerged quite naturally from the philosophical preoccupations and scientific bias of the Vienna Circle. Schlick and his comrades typically believed that science delivers knowledge through careful, direct observational and/or experimental verification. Given that the practices of the empirical sciences provide an exceptionally high, privileged status for scientific knowledge, the methods of science therefore provide the yardstick against which all other claims to knowledge must be measured. Logical positivism came to be associated with the distinctive slogan "The meaning of a statement is the method of its verification" – the so-called Verification Principle, with its origins in Wittgenstein's *Tractatus Logico-Philosophicus* and enthusiastically promoted by Waismann and Schlick (see Hanfling 1981, pp. 5 and 329). Indeed, questions concerning both meaning and the distinction between meaningful and meaningless statements became the chief preoccupation of logical positivists, distinguishing them from positivists and empiricists in general.

During the 1920s, the Vienna Circle gained in strength, forming an alliance with the "Berlin School," which included Hans Reichenbach and Richard von Mises. With the publication, in 1929, of their manifesto "The Vienna Circle: Its Scientific Outlook," logical positivists formalized their increasingly political role as a *movement* in science and philosophy. That role was developed further during an international congress held in Prague in the same year. In 1930, the journal *Erkenntnis*, edited by Carnap and Reichenbach, became the flagship for the positivist program. Books and monographs followed, and international conferences promoted the positivists' views to an enthusiastic and growing audience of philosophers, scientists, logicians, and mathematicians. Many young philosophers traveled to Vienna and to various congresses to learn at first hand of ideas which appeared to challenge what they regarded as established philosophical dogma, with its outmoded focus upon irrelevant metaphysical problems. A. J. Ayer visited Vienna in 1933, revealing his excitement in a letter to Isaiah Berlin: "Philosophy is grammar. Where you would talk about laws, they talk about rules of grammar. All philosophical questions are purely linguistical . . . Altogether a set of men after my own heart" (see Ignatieff 1998, p. 82). Ayer returned to England to write the enormously influential *Language, Truth and Logic*, the first edition of which was published in 1936. Many of Ayer's colleagues in Oxford, such as J. L. Austin, Stuart Hampshire, and Isaiah Berlin, were very much taken with the new ways of thinking – at least in these early days. Other visitors to Vienna included a young logician from the United States, Willard van Orman Quine, who was later to challenge much that the logical positivists had to say about meaning and analysis, and Carl Hempel, whose later work on the covering-law (or deductive-nomological) model of explanation owed much to his early enthusiasm for logical positivism (see Hempel 1965 and EXPLANATION).

During the 1930s, logical positivism maintained its momentum, attracting admirers and critical interest, as well as hostile reviews. Despite the death of Hahn and the murder of Schlick by one of his students in Prague, many were working hard to clarify and promote the ideas of logical positivism. For example, Carnap's *The Logical Syntax of Language* (1937) was a bold, if severely flawed, attempt to establish the rules and connections which link language and experience. However, with the traumatic onset of the Second World War, logical positivism as a formal movement, with the Vienna Circle as its locus, was scattered across the world. Despite this, in continental Europe, in England, and perhaps above all in the United States, the influence of individual logical positivists persisted into the 1960s. Carnap, Hempel, Nagel, Reichenbach, and many kindred spirits continued to expound the ideas of logical positivism. Some, like Charles Morris and Philipp Frank, preferred to call themselves "logical empiricists," but the essential message was the same, at least in these early days. However, the failure of Carnap and others to complete their positivistic projects, in the face of fierce opposition and often damning criticism, increasingly limited the influence of the movement.

From the late 1920s, Karl Popper, associated with, but still separate from, the Vienna Circle, had been wary of the commitment shown by its members to the role of verification; but in time he became rather more ferocious in his assertions that they had failed to provide an adequate demarcation between science and metaphysics (see Popper's tribute to Carnap, reproduced in Popper 1963, p. 263). In fact, from 1950 onwards, the onslaught grew apace. Some, like Quine in his "Two dogmas of empiricism" (1953), tackled logical positivism head-on, questioning some of the positivists' most cherished beliefs: Quine found it hard to accept their claim that individual experiential statements include distinct factual and linguistic components, for this requires a sharp and, he argued, erroneous distinction between synthetic and analytic; he also could not accept that each higher-level scientific statement could at least in principle be reduced to basic observational terms thereby rendering itself susceptible to verification. Others, like N. R. Hanson in his *Patterns of Discovery* (1958), undermined the foundations of logical positivism without making a single direct reference to the Vienna Circle. Hanson (1958, p. 19) maintained that "seeing is a 'theory-laden' undertaking. Observation of x is shaped by our prior knowledge of x." Perhaps the most ironical rebuke came from Thomas Kuhn, whose attacks on the epistemological purity of observation (see OBSERVATION AND THEORY), in chapters 6 and 10 of *The Structure of Scientific Revolutions* (1962), were published as part of the *International Encyclopedia of Unified Science*, a project started by the Vienna Circle in the 1930s, with Neurath himself as its editor-in-chief. One modern empiricist, Bas van Fraassen, makes an accurate assessment of the logical positivists, when he says that there was "the cavalier euphoria of being involved in a philosophical programme all of whose problems were conceived of as certain to be solved some time later on . . . (however) the culmination of Carnap's, Reichenbach's, and Hempel's attempts, which is found in Ernest Nagel's (1961) *The Structure of Science*, was still strangely inconclusive" (1989, p. 38). However, to understand logical positivism and why it had such influence on a whole generation of philosophers, one must explore earlier and perhaps less refined empiricist beliefs.

Origins

The scientist and philosopher Ernst Mach is frequently regarded as the father of logical positivism, as well as the chief architect of what might be called "scientific" positivism, a philosophy of science which regards the possibility of observational and/or experimental verification as the defining characteristic of all scientific statements (see MACH). His empiricist polemic reinforced the views of earlier philosophers such as Berkeley and Hume (see BERKELEY and HUME). Mach's influence in the scientific world was far-reaching. Mach claimed, in his book *The Conservation of Energy*, that only the objects of sense experience have any role in science: the task of physics is "the discovery of the laws of the connection of sensations (perceptions)"; and "the intuition of space is bound up with the organisation of the senses . . . (so that) we are not justified in ascribing spatial properties to things which are not perceived by the senses" (1911, p. 87). Thus, for Mach, our knowledge of the physical world is derived entirely from sense experience, and the content of science is entirely characterized by the relationships among the data of our experience.

Many scientists, including Einstein, admired the *spirit* of Mach's ferocious attack on Newton's "metaphysical" conception of absolute space (see EINSTEIN). Yet, even those in the scientific community who sympathized with Machian sentiments did not wholeheartedly embrace Mach's radically sensationalistic empiricism. When extending the central ideas of the general theory of relativity, Einstein tried (and failed) to incorporate Mach's basic intuition – what Einstein called Mach's Principle – that the source of all dynamical forces should be (observable) matter. Discouraged, perhaps in part by his failure to reconcile the general theory with Mach's Principle and in part by his growing conviction that physical theories could not be entirely tied to observation, Einstein increasingly distanced himself from Mach. In particular, he challenged the Machian notion that the aim of science is only to present the facts (given in observation). Einstein observed (in 1922) that if Mach had his way, science would look more like a mere catalog than a creative, integrated system of ideas (see Ray 1987). Other prominent scientists with positivistic leanings, including Niels Bohr, also kept a respectable distance from Mach's extreme positivistic views (see BOHR). Mach had repeatedly claimed that the concept of an atom is merely of instrumental value, being no more than a shorthand reference to observational data. In 1924, as the members of the Vienna Circle were developing their own positivistic philosophy, Bohr wrote to the American physicist Michelson proclaiming his deep commitment to the essential reality of atoms (if not electrons) in the quantum theory – a theory which took atomic physics into new and strange territory.

Despite the reservations expressed by such scientists, the Machian spirit flourished within the more philosophical atmosphere of Vienna. Circle members looked for ways in which to give rigorous expression to Mach's phenomenalistic philosophy of science. Three of Mach's ideas played a key role in the development of their positivist program:

1   We should regard sense experience as the only admissible guarantor of our physical descriptions; hence statements involving an essential reference to theoretical or unobservable entities may have at best an instrumental status in our accounts of the world.
2   Our knowledge about the world may only be regarded as secure if it can be checked against observation and experiment.
3   We should not seek anything more than complete descriptive powers in our accounts of the physical world; accordingly, "fundamental" explanations, particularly those involving supposed causal connections or metaphysical entities, should have no place in science.

Ernst Mach was not the only source for positivistic ideas. David Hume's strident empiricism, with his celebrated outburst demanding that we commit metaphysical books to the flames, was much appreciated by the logical positivists (see HUME). They also welcomed the ideas of the nineteenth-century French thinker Auguste Comte, who wrote the *Course of Positive Philosophy* between 1830 and 1842. Comte argued that science was on the threshold of breaking free from the religious and metaphysical prejudices which had too often hampered genuine and positive scientific progress. He claimed that even social thought could be treated scientifically and so looked forward to a final "positive" phase in our intellectual development in which science ruled supreme. The Vienna Circle shared with Comte this high regard for science and viewed physical science as a paradigm, not just for scientific thought in general, but for all human inquiry.

The shift from positivism to logical positivism was promoted by the work of Frege, Russell, Wittgenstein, and other mathematicians and logicians. Frege, writing in the late nineteenth century, had helped to bring about a revolution in logic, which now regarded propositions, rather than thoughts or ideas, as the fundamental units of meaning. Russell, like Frege, promoted the development of symbolic logic as a tool for the analysis of propositions (see RUSSELL). Wittgenstein believed that we could use logic to clarify propositions and so clarify our beliefs about the world. Schlick and the Vienna Circle were excited by the insights into the nature of logic achieved by these thinkers. The logical positivists focused strongly upon logic as they tried to develop a new, scientific method of philosophizing which consisted in the logical analysis of the statements and concepts of empirical science; indeed, Carnap regarded earlier, speculative philosophies as meaningless "concept-poetry." Philosophy was now to be a servant of the empirical sciences, used to clarify and analyze scientific statements. Logical positivism was to be a methodological strategy, saying nothing about the reality of the world as such, but aiming to relate *all* scientific statements – physical, material, or psychological – to what is given in sense experience by means of logical analysis. This, in turn, would lead to the development of a *unified science* with one logical method and one epistemological foundation – namely, the content of immediate experience.

Verification, meaning, and truth

The driving force behind the central ideas of logical positivism is the Principle of Verification. Taking their cue from Wittgenstein, Waismann maintained that we understand a statement when we know how it is to be verified, and Schlick said that, in order to specify the meaning of some proposition, we must spell out how the proposition would be verified or falsified. With such assertions, Schlick and his followers were clearly committed to the construction of a theory of meaning – a monumental task which ultimately was to prove fruitless. A. J. Ayer focused international attention upon logical positivism and the problem of verification with the publication of the first edition of *Language, Truth and Logic* in 1936. He avoided being drawn into any discussion about meaning as such by limiting himself to the general problem of verification, asking how we can pick out genuinely factual statements and distinguish them from meaningless statements with no empirical content. Following the lead set by the Vienna Circle, Ayer suggested that we might do so using a criterion of verification: "it is the mark of a genuine factual proposition . . . that some experiential proposition can be deduced from it in conjunction with certain other premises without being deducible from those other premises alone" (Ayer 1946, pp. 38–9). Ayer argued that all metaphysical statements – such as "God is omnipotent" – are ruled out as meaningless by his criterion. Factual claims, such as "The effective temperature at the surface of the Sun is 5770 K," are said to be meaningful. Even though we cannot check this temperature directly, we can check other observational statements which are entailed by this claim.

In a celebrated retort, Isaiah Berlin demonstrated that Ayer's criterion was far too liberal. Berlin asked us to consider the following argument, which begins with a "nonsense" statement to be tested by the criterion:

> This logical problem is a bright shade of green.
> I dislike all shades of green.
> Therefore I dislike this logical problem.

The argument is logically valid and grammatically correct. And an experiential statement has been deduced from the first statement. Hence, according to Ayer's criterion, this first statement is empirically meaningful. Ayer accepted Berlin's objection, for it became clear that his verification criterion would confer meaningfulness upon any indicative statement at all. For verification to play any vital part in the positivist program, Ayer realized that the criterion had to be improved (see Ayer 1946 and Berlin 1939). Unfortunately, he was unable to suggest any alternative which did not fall to further serious objections (see, e.g., Church 1949). Even so, the very idea of the principle attracted considerable interest and attention.

A further problem for Ayer and for all logical positivists was how to spell out what is to count as an experiential statement. Saying that an actual or possible observational claim fits the bill only pushes the difficulty one step away. For we must still say what is to count as an observational claim. Logical positivists and their supporters tried two principle maneuvers. Some, like Schlick, took Mach's phenomenalism as their inspiration: what is observable is what is "given" in immediate sense experience and hence incorrigible. Consequently, true experiential statements correspond in some straightforward way to the world: they are synthetic statements which express the facts with clarity and simplicity. Logic is then used to analyze the connection via analytic rules of statements in ordinary language to the statements of a more elementary language pitched at the sensory level. The task for philosophy is to articulate the rules which connect the two types of language. The rules are analytic, since they have nothing to say about the world itself, and so all can agree upon them. Hence, the approach here is underpinned by an analytic/synthetic distinction. Initially, the Vienna Circle was drawn towards the simple idea that we might build knowledge of the world from fundamental sentences which cash out our individual sensory experiences. However, their hope was forlorn. No convincing strategy was developed to analyze ordinary language in terms of sense experience (see Quine 1953).

Although Schlick remained faithful to this first approach until his death, Neurath and Carnap became increasingly dissatisfied with its naïveté, arguing that the supposed incorrigibility of our private experience could not underwrite the intersubjective domain of scientific knowledge. They were further discouraged by developments in logic. Frege and Russell had given the logical positivists confidence that there is a single, "true" logic which could be used as the common foundation for all logical analysis. However, the development of nonstandard logical systems (such as intuitionist logic) produced a dilemma: which logic should be chosen to underpin the positivist dream? So Neurath and Carnap advocated a second, more sophisticated approach in which elementary or "protocol" sentences, as they were called, referred to public physical experience and cashed out, in formal terms, the content of that experience. Protocol sentences are supposed to provide a precise record (or protocol) of a scientist's experience. These sentences are set within a given linguistic framework and their terms are given meaning by "meaning postulates" – analytic rules connecting the terms to the synthetic observational claims made by the sentences. Carnap suggested that a primitive protocol recording (say) an observed electrical spark across the air gap between two wires might be: "Arrangement of experiment: at such and such positions are objects of such and such kinds . . . here now pointer at 5, simultaneously spark and explosion, then smell of ozone there" (see Carnap in Hanfling 1981, p. 153). In order to elucidate the idea of protocol sentences, Carnap introduced a distinction between "formal" and "material" modes of speech. His distinction was designed to avoid any confusion between the private content of subjective experience (expressed through the material mode) and the public basis for intersubjective agreement about experiences (expressed through a formal, linguistic mode). Carnap asks us to avoid such "material" questions as "What objects are the elements of given, direct experience?" and to embrace the "formal" "What kinds of word occur in protocol sentences?" (ibid., p. 154) He says that protocol sentences do not refer materially to physical events as such. They are merely formal records of the event stated in a neutral, nonmaterial way.

Once a scientist decides to work within a given framework, intersubjective agreement can be delivered, but at a considerable price: the logical principles which provide the essential structure for the framework, and indeed the framework itself, are chosen, Carnap tells us, on pragmatic grounds. Two rival groups of scientists may use two quite separate frameworks with distinct logics. They no longer aim to describe the world, at least in any straightforward manner. Instead, the judgments which they make are relative to the framework itself: what is "true" or "false" depends upon internal consistency within each separate framework. This resonates with Carnap's view that the whole system of physics must be taken into account when judgments are made. In this he is following Pierre Duhem's assertion that, although "agreement with experiment is the sole criterion of truth for a physical theory . . . the physicist can never subject an isolated hypothesis to experimental test, but only a whole group of hypotheses" (Duhem 1954, pp. 21 and 187). What Carnap adds to Duhem is the concession that more than one physics is possible, at least in principle. Schlick and Ayer found all this rather too much to bear – it seemed to them that the empirical foundations of logical positivism were being undermined. Not only did the requirement that protocol sentences be expressed formally appear to drive a wedge between science and the world, but they were told that experience could not help us decide between rival scientific theories. Hempel, in defence of Carnap, argued that in practice there is only one scientific system, and that is the one accepted by the prevailing scientific culture (see Hempel 1935). However, this failed to calm Ayer and Schlick, who saw the breach between science and experience widening. Ayer argued that each rival and incompatible system might include the proposition that it was the only acceptable system. Schlick, rather more emotively, said: "If all the scientists in the world told me that under certain experimental conditions I must see three black spots, and if under those conditions I saw only one spot, no power in the universe could induce me to think that the statement 'there is only one black spot in the field of vision' is false . . . the only ultimate reason why I accept any proposition as true is to be found in . . . simple experiences" (Schlick in Hanfling 1981, p. 201).

A third way forward was provided by the Harvard physicist P. W. Bridgman, who developed, independently of the Vienna Circle, a variant of positivism called "operationism" (sometimes referred to as "operationalism"). Observational claims, he argued, should be based on "operational" procedures used in the laboratory. "In general, we mean by any concept nothing more than a set of operations; the concept is synonymous with the corresponding set of operations" (Bridgman 1927, p. 5). Hence, the concept of length may be determined fully by precisely those operations involved in the measurement of length: length involves no more and no less than the way we make such measurements. Some believed that Bridgman's strategy offered a way out of the vagueness and subjectivism associated with sense experience. Others were less convinced. For example, since length can be measured in an apparently unlimited number of ways, pinning the concept down does not prove quite as easy as Bridgman suggests (see Papineau 1979).

The unity of science was also a key element in the program of the logical positivists. Carnap forcefully expressed the Vienna Circle's commitment to unity in the first volume of the *International Encyclopedia of Unified Science*: although "it is obvious that, at the present time, laws of psychology and social science cannot be derived from those of biology and physics . . . no scientific reason is known for the assumption that such a derivation should be in principle and forever impossible" (Carnap in Hanfling 1981, p. 128). Here, as elsewhere, positivists demonstrated their faith in the future. What gave Carnap and others hope was their belief that there is a unity of language in science – through a reductionist program. Physicists and psychologists might not share the same laws, but so long as they restrict themselves to phenomenal talk in their respective domains (with such words as "large," "cold," and "red"), Carnap believed that they share the same language. This, it was argued, would provide a basis for the eventual reduction of the social and psychological sciences either to the physical sciences (Neurath) or to a general phenomenological science (Carnap). The commitment to reductionism is evident. The core of their reductionist faith rests on the assumption that it will always be possible to reduce all empirical statements to more basic statements with clear-cut observational consequences. Despite considerable efforts by Carnap and others, this assumption was not given any convincing support. Here too, logical positivists hoped for much but achieved little themselves.

Post-Second World War verdicts on logical positivism have been numerous: some have been sympathetic, but others have been rather less so, as we have already seen. Quine has been more understanding than most; in his (1953) challenge to some of the core beliefs of logical positivism (the analytic/synthetic distinction and the faith in reductionism), his aim is to provide firmer empirical foundations for our view of science (see Hookway 1988). Like many modern commentators, Ian Hacking finds "the success of the verification principle amazing . . . for no one has succeeded in stating it!" (1975, p. 95). John Earman is less charitable, stating baldly that "the extreme forms of positivism and operationalism are not a suitable basis on which to found an adequate epistemology" (1970, p. 298).

References and further reading

Ayer, A. J. 1936: *Language, Truth and Logic*, 1st edn (London: Victor Gollancz; 2nd edn 1946).
Ayer, A. J. (ed.) 1959: *Logical Positivism* (New York: Free Press). (This book includes many key source essays and articles by logical positivists, including Bertrand Russell on logical atomism, Schlick and Carnap on meaning, Carnap and Hahn on logic, and Neurath on protocol sentences.)

Berlin, I. 1939: Verification. *Proceedings of the Aristotelian Society*, 39, 225–48.

Bridgman, P. W. 1927: *The Logic of Modern Physics* (London: Macmillan).

Carnap, R. 1936: Testability and meaning. *Philosophy of Science*, 3, 419–71; 4 (1937), 1–40.

—— 1937: *The Logical Syntax of Language* (New York: Harcourt, Brace and World Inc.).

Church, A. 1949: Review of *Language, Truth and Logic*. *Journal of Symbolic Logic*, 14, 52–3.

Duhem, P. 1954: *The Aim and Structure of Physical Theory* (Princeton: Princeton University Press).

Earman, J. 1970: Who's afraid of absolute space? *Australasian Journal of Philosophy*, 48, 287–319.

Feigl, H., and Brodbeck, M. (eds) 1953: *Readings in the Philosophy of Science* (New York: Apple-Century-Crofts).

Hacking, I. 1975: *Why does Language Matter to Philosophy?* (Cambridge: Cambridge University Press).

—— 1983: *Representing and Intervening* (Cambridge: Cambridge University Press).

Hanfling, O. (ed.) 1981: *Essential Readings in Logical Positivism* (Oxford: Blackwell). (This book includes many key source essays and articles by logical positivists, including Waismann, Rynin, and Reichenbach, as well as Carnap, Schlick, Neurath and Ayer.)

Hanson, N. R. 1958: *Patterns of Discovery* (Cambridge: Cambridge University Press).

Hempel C. G. 1935: On the logical positivist's theory of truth. *Analysis*, 2/4, 10–14.

—— 1965: *Aspects of Scientific Explanation* (New York: Free Press).

Holton, G. 1993: *Science and Anti-Science* (Cambridge, MA: Harvard University Press).

Hookway, C. 1988: *Quine* (Cambridge: Polity Press).

Ignatieff, M. 1998: *A Life: Isaiah Berlin* (London: Chatto and Windus).

Kuhn, T. S. 1962: *The Structure of Scientific Revolutions*, 1st edn (Chicago: Chicago University Press; 2nd edn 1970).

Mach, E. 1911: *The Conservation of Energy* (LaSalle, IL: Open Court).

—— 1959: *The Analysis of Sensations* (New York: Dover).

Nagel, E. 1961: *The Structure of Science* (New York: Harcourt, Brace and World Inc.).

Neurath, O., Carnap, R., and Morris, C. (eds) 1938: *International Encyclopedia of Unified Science* (Chicago: Chicago University Press). Later published as *Foundations of the Unity of Science* (1969).

Papineau, D. 1979: *Theory and Meaning* (Oxford: Oxford University Press).

Parkinson, G. H. R. (ed.) 1968: *The Theory of Meaning* (Oxford: Oxford University Press).

Passmore, J. 1957: *A Hundred Years of Philosophy* (London: Duckworth).

Popper, K. R. 1959: *The Logic of Scientific Discovery* (London: Hutchinson).

—— 1963: *Conjectures and Refutations* (London: Routledge).

Quine, W. V. O. 1953: *From a Logical Point of View* (Cambridge, MA: Harvard University Press).

Ray, C. 1987: *The Evolution of Relativity* (Bristol: Adam Hilger).

Reichenbach, H. 1951: *The Rise of Scientific Philosophy* (Berkeley: University of California Press).

van Fraassen, B. C. 1989: *Laws and Symmetry* (Oxford: Oxford University Press).

Wittgenstein, L. 1961: *Tractatus Logico-Philosophicus* (London: Routledge).

38

Mach

GEREON WOLTERS

Ernst Waldfried Josef Wenzel Mach was born 18 February 1838 in the Moravian village of Chrlice (near Brno), at that time part of the Austrian Monarchy, now the Czech Republic, and died 19 February 1916 in Vaterstetten (near Munich). He enjoyed a very successful career as an experimental physicist (the unit for the velocity of sound has been named after him). His importance for the philosophy of science derives mainly from his "historico-critical" writings (Mach 1872, 1883, 1896b, 1921). Mach studied mathematics and physics at the University of Vienna (1855–60, doctorate in physics 1860, his "Habilitation" (i.e., qualification to become a university professor) 1861) and his subsequent work was in the physiology of the senses. In 1864 he became professor first of mathematics and then (1866) of physics at Graz University; from 1867 to 1895 he was professor of experimental physics at Prague University; and in 1895 he took a chair in "Philosophy, especially the History and Theory of the Inductive Sciences" at Vienna University. In 1898 a stroke ended Mach's university teaching, but he was able to continue scientific work to a certain degree.

Mach's philosophical activities can be subsumed under the general heading of "anti-metaphysics." This means the attempt to make philosophy (i.e., epistemology) more scientific and science more philosophical by dismissing from ontology everything that cannot be shown to be empirically significant.

The anti-metaphysical reform of epistemology led Mach to a sort of phenomenalism with so-called neutral elements as the irreducible basis of all knowledge. Examples of elements are memories, imaginations, etc., as well as colors, sounds, heats, pressures, spaces, times, etc. They "are interconnected in manifold ways" (Mach 1886, p. 2) to complexes or clusters. Only these complexes, not the elements they consist of, are the objects of unreflected awareness. Those clusters of elements that display a certain stability may be called "things" or "bodies" for the sake of convenience. For the same reason they receive a proper name or predicate. Among the "things" one also finds one's own body. It is distinguished from other things particularly by the fact that the elements that constitute it are closely (mostly functionally) interconnected with elements like volitions, feelings, memories, etc. Because of its continuity, the "I" is the relatively stable complex of the elements that constitute one's body and the volitions, memories, etc. functionally connected to it. There is no strict borderline between one's "I" and the bodies, because bodylike complexes of elements too may vary according to their functional relationships to I-like elements; for example, a stick partly immersed in water is crooked when seen and straight when touched (ibid., p. 10). For Mach it makes no sense to ask what the stick *really* is.

Mach's approach contradicts realistic conceptions that conceive of elements as causally generated by "things"; it asserts just the reverse: that things are clusters of elements. Only those elements of thinglike complexes of elements that are regarded in their functional dependence on elements that constitute our own body may be called "sensations." So "a color is a *physical object*, as soon as we pay attention to its dependence on the illuminating source of light (other colors, heats, spaces, etc.). But if we pay attention to its dependence on the retina (or other bodily elements), the same color is a *psychological object*, a *sensation*" (ibid. p. 17).

On the other hand, Mach contradicts the idealistic project of constituting the world of objects out of subjective sensations. For Mach's elements are neither objective nor subjective. They are just there. These neutral elements are the "given" of Mach's positivism. What is called "objective" or "subjective" in the traditional sense is only a special type of functional relationship between neutral elements: a "subjective" relationship expresses a connection between "I-like" and bodylike complexes of elements, whereas an "objective" relationship refers to dependencies among those bodylike complexes themselves.

From Mach's epistemological "neutral monism," three important consequences are derived: (1) causality is nothing more than a functional dependence between elements; (2) there is no "substance" as carrier of properties, but only elements in more or less stable complexes; (3) the mind–body problem is a *pseudo-problem*, because there are no generic differences between elements. Only according to the type of the functional dependency of its elements might a complex of elements be called "physical" or "psychological."

Mach emphasizes (addition 1 of the 5th–9th German editions of Mach 1886) that working physicists may easily dispense with his epistemology. It is indispensable only in research on the psychophysical relationship. Accordingly, Mach's methodology is systematically independent of his epistemology, although it can be regarded as an application of it.

For Mach, science has two central features: (a) its "biological" function for humans, and (b) its essentially "historical" nature – i.e., the transience of its respective outlooks. Both features reveal the anti-metaphysical thrust of Mach's thinking.

Anti-metaphysics in Mach's biological conception of science consists in restricting science to the *description of facts*, for only facts provide the orientational stability needed for acting with differential survival value. But a totally descriptive science is only the ideal, but unattainable, final goal of science. For the time being one has to rely on hypotheses and theories ("indirect descriptions"), that, with scientific progress, should gradually be replaced by "direct descriptions." Note that Mach does not advocate sensualism; for not only observations qualify as facts, but also not directly observable items like phases of sound waves, the law of propagation of heat, or, most important, theoretical "principles" (e.g., the energy principle, the principle of inertia). Principles are not observed in nature, but "intuited" by imaginative power on the basis of intimacy with natural phenomena. They are selected according to their "economic" value (cf. below); they are "conventions," as Mach agrees with H. Poincaré (see CONVENTION, ROLE OF) (Mach 1883, p. 306).

Mach presents – again with anti-metaphysical intention – two fundamental rules of concept formation in empirical science: (1) distrust all concepts that do not actually have observable referents; (2) exclude all concepts from science that in principle cannot have observable reference. From these rules follows a fundamental critique of all attempts to reduce empirically adequate conceptions to allegedly "deeper" theories whose concepts fail to have any observable referents in the domain in question. This leads Mach to a strict, *anti-mechanistic position* in physics. In this vein he ontologically rejected the existence of atoms and other invisible particles, and attributed, at best, instrumental value to mechanistic models of nonmechanical phenomena (e.g., the kinetic theory of heat). Only towards the end of his life does Mach seem to have given up his anti-mechanism (see Wolters, in Haller and Stadler 1988).

There is one more reason to consider science a "biological" endeavor: science is basically nothing else than a professionalized continuation of a particular form of everyday human survival activity – namely, observing nature and craftsmanship. This kind of activity has existed even since the dawn of of human cultural evolution.

The biological characterization of science has a variety of consequences. It follows, according to Mach, that we should adopt *theoretical instrumentalism*. The primary aim of science is not to tell us what the world as such is like, but rather to give us a successful explanatory and prognostic orientation. Only in a secondary sense does reliable orientation require correspondence to facts. It also follows that science correlates observables, and is thus based on, and restricted to, *empirical* quantities. The consequences of scientific theories have to match observations. In addition, for Mach, science is not only part of human cultural evolution, but also an activity that has itself to be described in *evolutionary terms*. Mach characterizes science (1905, ch. 10) as (a) "adaptation of thoughts to facts" (i.e., "observation") and (b) "adaptation of thoughts to each other" (i.e., "theory"). But he does not foreshadow the observation–theory dichotomy of logical empiricism, because he already emphasizes the theory-ladenness of observation (ibid., p. 120) as well as (in his "adaptation of thoughts to each other") a holistic theory conception (see LOGICAL EMPIRICISM and HOLISM). But not only the conceptual core of Mach's conception of science is evolutionary. The development of science, too, has to be described in evolutionary terms. Theories "fight their struggle for life no differently than the ichthyosaurus, the Brahman, and the horse" (Mach 1896a, p. 40 (dt.)). Finally, Mach's famous principle of economy is part of the biological characterization of science: first, in the rather external sense, that science saves experiences "by the reproduction and anticipation of facts in thought" (Mach 1883, p. 577). Internally, the principle of economy allows us to concentrate on selected features of the facts and requires their "completest possible presentment . . . with the least expenditure of thought" (ibid., p. 586). So simplicity and range become for Mach criteria for the assessment of theories (see SIMPLICITY).

History reveals science as (1) "unfinished, variable" (Mach 1872, p. 17). History is (2) of greatest value, because the study of the origin and development of ideas renders them familar to us in a similar way as if we ourselves had found and developed them. At the same time the understanding of origins (3) makes us more open to scientific progress, because a view whose origin and development we know "is never invested with that immobility and authority which those ideas possess that are imparted to us ready formed. We change our personally acquired views far more easily" (Mach 1896b, p. 5).

Although Mach has no recipe for bringing about scientific progress, the study of history offers a number of successful heuristic procedures: for example, (1) *analogy* between different domains (e.g., the understanding of light waves as analogous to sound waves); (2) the *"principle of continuity"* (Mach 1883, p. 167), as the attempt to retain under varied circumstances, as much as possible, an idea derived from a special case (e.g., Galileo's discovery of the law of free fall by "continuing" the regularities observed with the inclined plane); (3) *"abstraction"* – that is, elimination of nonrelevant aspects in the case under question; and (4) "paradoxes" as strong incentives to bring a theoretical system into harmony once again.

Mach's thought has exerted great influence in both science and philosophy. His anti-mechanism, as well as his rules of concept formation (particularly the critique of "absolute space") stimulated Einstein in his theories of special as well as general relativity (see EINSTEIN). Posthumously published texts ascribed to Mach that reject relativity were almost certainly forged (see Wolters 1987). In recent years too, Mach's principle in cosmology, which had fallen into disregard already in the 1920s, has been successfully revived in a new interpretation. Mach's strict empiricism was instrumental for the Copenhagen Interpretation of quantum mechanics (see QUANTUM MECHANICS).

In philosophy, logical empiricism saw itself, as far as its empiricism was concerned, as continuing the work of Mach. R. Carnap's phenomenalistic constitutional system in his *Der logische Aufbau der Welt* is directly influenced by Mach's positivism. Mach's anti-metaphysics played an important motivational role for the anti-metaphysics of the Vienna Circle. Its external, educational activities were carried out by the officially registered Ernst Mach Society.

But Mach's philosophy of science, with its emphasis on the biological function of science and the transient historical character of all theorizing, with its insight into the theory-ladenness of observation as well as its holism, seems to be less close to mainstream logical empiricism (with the exception of O. Neurath) than it is to the critics of logical empiricism since the 1960s.

References and further reading

Works by Mach

1872: *Die Geschichte und die Wurzel des Satzes von der Erhaltung der Arbeit*; 2nd edn, Leipzig, 1902; repr. in Ernst Mach: *Abhandlungen*, ed. J. Thiele (Amsterdam: E. J. Bonset, 1969); trans. P. E. B. Jourdain as *History and Root of the Principle of the Conservation of Energy* (Chicago: Open Court, 1911).

1883: *Die Mechanik in ihrer Entwicklung historisch-kritisch dargestellt*; 9th edn, 1933; repr. with an introduction by G. Wolters (Darmstadt: Wissenschaftliche Buchgesellschaft, 1991); trans. T. J. McCormack as *The Science of Mechanics: A Critical and Historical Account of its Development*, 6th edn (La Salle, IL: Open Court, 1974).

1886: *Die Analyse der Empfindungen und des Verhältnis des Physischen zum Psychischen*; 9th edn, 1922; repr. with an introduction by G. Wolters (Darmstadt: Wissenschaftliche Buchgesell-schaft, 1991); trans. C. M. Williams and S. Walerlow, with an introduction by T. S. Szasz, as *The Analysis of Sensations and the Relation of the Physical to the Psychical* (New York: Dover, 1959).

1896a: *Populär-wissenschaftliche Vorlesungen*; 5th edn, 1923; repr. with an introduction by A. Hohenester (Vienna: Böhlau, 1987); trans. T. J. McCormack as *Popular Scientific Lectures* (Chicago: Open Court, 1895, repr. with an introduction by J. Bernstein, 1986).

1896b: *Die Principien der Wärmelehre: Historisch-kritisch entwickelt*; 4th edn (1923; repr. Frankfurt: Minerva, 1981); trans. T. J. McCormack, P. E. B. Jourdain, and A. E. Heath, with an introduction by M. J. Klein, as *Principles of the Theory of Heat: Historically and Critically Elucidated* (Dordrecht: Reidel, 1986).

1905: *Erkenntnis und Irrtum: Skizzen zur Psychologie der Forschung*, 5th edn (1926; repr. Darmstadt: Wissenschaftliche Buchgesellschaft, 1991); trans. T. J. McCormack and P. Foulkes, with an introduction by E. Hiebert, as *Knowledge and Error: Sketches on the Psychology of Enquiry* (Dordrecht: Reidel, 1976).

1921: *Die Prinzipien der physikalischen Optik: Historisch und erkenntnispsychologisch entwickelt* (repr. Frankfurt: Minerva, 1982); trans. J. S. Anderson and A. F. A. Young as *The Principles of Physical Optics: An Historical and Philosophical Treatment* (London: Methuen & Co, 1926).

Works by other authors

Barbour, J. B. 1995: *Mach's Principle: From Newton's Bucket to Quantum Gravity* (Boston: Birkhaueser).

Blackmore, J. 1972: *Ernst Mach: His Work, Life, and Influence* (Berkeley: University of California Press).

Cohen, R. S., and Seeger R. J. (eds) 1970: *Ernst Mach: Physicist and Philosopher* (Dordrecht: Reidel).

Haller, R., and Stadler, F. (eds) 1988: *Ernst Mach – Werk und Wirkung* (*Ernst Mach – Work and Influence*) (Vienna: Hölder-Pichler-Tempsky).

Sommer, M. 1996: *Evidenz im Augenblick: eine Phänomenologie der reinen Empfindung*, 2nd rev. edn (Frankfurt: Suhrkamp).

Wolters, G. 1987: *Mach I, Mach II, Einstein und die Relativitätstheorie: Eine Fälschung und ihre Folgen* (*Mach I, Mach II, Einstein, and Relativity Theory: A Forgery and its Consequences*) (Berlin: de Gruyter).

39

Mathematics, Role in Science

JAMES ROBERT BROWN

We count apples and divide a cake so that each guest gets an equal piece; we weigh galaxies and use Hilbert spaces to make amazingly accurate predictions about spectral lines. It would seem that we have no difficulty in applying mathematics to the world; yet the role of mathematics in its various applications is surprisingly elusive. Eugene Wigner has gone so far as to say that "the enormous usefulness of mathematics in the natural sciences is something bordering on the mysterious and that there is no rational explanation for it" (1960, p. 223). The issue is not much discussed under the heading "applied mathematics," yet it is pivotal to several philosophical debates. In recent years three rather general questions have been central: (1) Just how does mathematics "hook onto" the world? This is the main concern of a rather technical branch of philosophy of science known as measurement theory (see MEASUREMENT). (2) Are some of the objects referred to in various theories merely mathematical objects, or do they have some other status? This problem often comes up in the philosophy of the special sciences. For example, do space-time and the quantum state exist in their own right, separate from their mathematical representations; or are they nothing but mathematical entities? (3) Is mathematics essential for science? Following Hartry Field's work, this has become a focal point in the debate between realists and nominalists in the philosophy of mathematics.

I shall take each of these topics up in order; but there is considerable overlap among the issues involved, so answers to any one question are likely to have some impact on the others.

Let us begin by asking how mathematics is applied. The common view in measurement theory begins by assuming two distinct realms: one is a mathematical realm, which is rich enough to represent the other, a distinct nonmathematical realm (see MEASUREMENT). (The ontological status of the mathematical realm is not at issue here; Platonists can ask how *numbers* hook onto the world, while nominalists can ask about *numerals*.) We then pick out some part or aspect of the world, and find a similar mathematical structure to represent it. For example, *weight* is represented on a numerical scale. The main physical relations among objects that have weight (determined, say, by a balance beam) are that some have more weight than others, and that when combined, the weight increases. Weight can then be represented by any mathematical structure (such as the nonnegative real numbers) in which there is a *greater-than* relation matching the physical greater-than relation and an *addition* relation matching the physical addition or combination relation.

More generally, a mathematical representation of a nonmathematical realm occurs when there is a homomorphism between a relational system P and a mathematical system M. P will consist of a domain D and relations R_1, R_2, . . . defined on that domain; M similarly consists of a domain D* and relations R_1^*, R_2^*, . . . on its domain. A homomorphism is a mapping from D to D* that preserves the structure in the appropriate way.

To make this a bit more precise, consider a simplified example. Let D be a set of bodies with weight, let D* = $\mathbb{R}$ (the real numbers), and let $\leqslant$ and $\oplus$ be the relations of *physically weighs the same or less than* and *physical addition*. ($\leq$ and + are the usual relations on real numbers of *equal or less than* and *addition*.) The two systems, then, are P = $\langle D, \leqslant, \oplus \rangle$ and M = $\langle \mathbb{R}, \leq, + \rangle$. Numbers are then associated with the bodies (a, b, . . .) in D by the homomorphism $\phi : D \rightarrow \mathbb{R}$ which satisfies the two conditions:

$$a \leqslant b \rightarrow \phi(a) \leq \phi(b)$$

$$\phi(a \oplus b) = \phi(a) + \phi(b).$$

In other words, the relations that hold among physical bodies get encoded into the mathematical realm, and are there represented by relations among real numbers. One of the objects can be singled out to serve as the unit, u, so that $\phi(u) = 1$. (An isomorphism is a special case where the two systems have the same structure; in general, a homomorphism is weaker; the structure gets preserved from D to D*, but not necessarily the other way.)

I must add a caveat to the assumption of two distinct realms, the mathematical and the nonmathematical. Since they are linked by the embedding homomorphism, ϕ, which is a function defined on D, there must be sets of nonmathematical objects as well as pure sets. This means that we start with the usual set theory, including *urelements*. Among *urelements* are physical objects, of course, but also abstract and fictional objects (faith, hope, and charity are three virtues; Santa's sleigh is pulled by eight reindeer). Having sets, sets of sets, etc. of *urelements* is just a start. The difficult part in setting up, or discovering, an association between the physical system P and some mathematical system M usually consists in finding the right set of physical relations. Much of the focus of current measurement theory is in psychology and the social sciences, where attempts to quantify such concepts as utility, desirability, IQ, degree of belief, intensity of pain, etc. are exceptionally difficult. Not only are there natural difficulties, but the air of rigor brought by mathematics often tends itself, in the hands of the inept or unscrupulous, to the production of pseudo-science.

Measurement theory often classifies different types of scale. *Ordinal* measurements are the simplest. The Mohs scale of hardness, for example, uses the numbers 1 to 10 in ranking the physical relation of "scratches"; talc is 1 and diamond is 10. The only property of the numbers used is their order; addition, for example, plays no role. By contrast, addition is crucial in *extensive* measurements, such as for weight. (In this case the physical combination of two bodies is represented by the addition of two real numbers. But the embedding homomorphism isn't always so simple as it is in the weight case; the relativistic addition of two velocities, for example, is constrained by an upper limit on their joint velocity.) An *interval* measurement uses the *greater-than* relation between real numbers, but does not employ addition. (Temperature and (perhaps) subjective probability are examples. Two bodies at 50° each do not combine to make one at 100°.)

In passing, it should be noted that the mathematical representation of the world need not be with numbers. From the Greeks to Galileo and after, geometrical objects did the representing. The increasing speed of a falling body, for example, was represented by Galileo by a sequence of increasing areas of geometrical figures. Newton's *Principia* was written in this geometrical style; but thereafter the tremendous power of the calculus has made analysis dominant. The geometric spirit, however, is far from dead. (See, e.g., the "visual" book by Abraham and Shaw (1983).) Graphs are geometrical, of course, but they tend to depict numerical results; that is, they are representations of representations of the world.

Color, beauty, and other such things are not readily mathematizable. But the alleged subjectivity of these properties has nothing to do with it; felt warmth and pain intensity are subjective, but have the appropriate structure to be mathematized. The reason for the nonmathematizability of color may have more to do with its internal features; it does not have the same structure as mass, length, temperature, or other so-called extensive magnitudes which would make it easy to associate with the real numbers.

So far I have spoken loosely of numbers hooking onto objects. Perhaps, instead of *objects*, numbers are associated with *properties of objects*. From a practical point of view, there isn't much difference, but philosophically the divergence is considerable. The former view is strongly empiricist and dominant today (Nagel 1932; Krantz et al. 1971–90); the latter is somewhat Platonistic, and has had notable support, too (Russell 1903; Campbell 1920; Mundy 1987; Swoyer 1987). The natural languages for these accounts are first- and second-order logic, respectively. To say that the weight of a and b combined is such and such, is to say, according to the first-order theory of measurement, that there is an object c which equals the weight of a and b combined (understood in a somewhat operationalist way, with c balancing a and b on a scale). This is physically unrealistic, and at best an idealization. However, it is not a problem for the second-order theory, since it is not objects, but properties that are assigned numbers. The property *weight* is postulated to be continuous and unbounded; there need not be exemplars of any particular weight in order to talk meaningfully about it.

These two accounts of measurement tie into rival accounts of laws of nature. The relations that hold in the (nonmathematical) relational structure are presumably laws of nature. The empiricist-motivated regularity theory fits harmoniously with the first-order theory. The more realist account of some philosophers which takes a law of nature to be a relation between universals (i.e., properties) fits very naturally with the second-order version. So the question, Does mathematics hook onto objects or onto properties of objects?, may have a bearing on the metaphysical issue of the nature of scientific laws (see LAWS OF NATURE).

The second concern with the role of mathematics in the sciences involves the possible presence of mathematical artifacts. Measurement theory is somewhat farfetched in assuming that we can *first* discern relations among nonmathematical objects and then *later* pick out mathematical structures to represent them. In reality, of course, mathematics plays an enormous role in theory construction. Because of this, it is sometimes difficult to distinguish the mathematics proper from its physical counterparts.

259

For example, the average family has $2\frac{1}{2}$ children. Of course, there is no family with that many children; the "average family" is a mathematical artifact. No one is likely to be fooled by this example, but many of the things that physicists regularly talk about have a contentious status: Are they physically real, merely mathematical, or what?

When Maxwell introduced classical electrodynamics, his electrodynamic field was thought by many to be just a mathematical entity. In terms of measurement theory, this is to say that the domain of the physical theory consisted of charged particles, but no field points. This relational structure would then be embedded within a mathematical structure of a vector field. So the only "field" is the mathematical one. The following argument tipped things the other way. Consider two separated charged particles. If one is wiggled, the other jiggles at a *later* time. Energy is conserved. Before and after the motions of the two particles, all energy can be located in the particles themselves, but not at intermediate times. Energy must be located somewhere. Thus, it must be in the field; so the field is physically real. (Note that this argument would not apply to the gravitational field of Newton; action is instantaneous in that theory, so energy can always be located in particles.) The consequence of this is that the electromagnetic field, though it is represented by a mathematical vector field which is isomorphic to it, is a distinct, physically real entity, not a mere mathematical artifact.

Similar problems about how to interpret the mathematical apparatus arise in quantum mechanics and in space and time (see QUANTUM MECHANICS and SPACE, TIME, AND RELATIVITY). Quantum mechanics makes heavy use of a notion of *state*, represented by a vector, ψ, in a Hilbert space. The mathematics of ψ is reasonably well understood; the same cannot be said about the state. One view says that there is nothing to the state other than the mathematical vector, ψ, itself. (Texts use the same symbol for both, making this seem natural.) At the other extreme, ψ might be a real field (e.g., Bohm's quantum potential). So, much of the problem of interpreting quantum mechanics amounts to determining how mathematics hooks onto a quantum system: Is the mathematical vector, ψ, associated with the electron, or with the *state* of the electron?

The modern debate between absolutists and relationalists in space-time concerns the status of the space-time manifold (Friedman 1983). Undisputed is the reality of events. Absolutists hold that actual events are the occupied points of a larger space-time manifold, which is taken to be physically real. (Some prefer to think of space-time points as abstract entities. Whether physical or abstract, however, the main claim is that they are real and distinct from their mathematical representation.) The space-time manifold is then associated with the mathematical structure, $\mathbb{R}^4$. By contrast, relationalists hold that the set of events is directly associated with $\mathbb{R}^4$ (bypassing the manifold). So once again, a major philosophical issue turns on the question of how mathematics is applied to the world.

Let us now take up the third question: Is mathematics necessary for science? The answer may be "yes," but it is not obviously so. The statement "There are two apples in the basket" seems to make essential use of numbers; yet we can capture its content without appeal to any mathematics at all by recasting it as: $\exists x \exists y \forall z (Ax \wedge Ay \wedge (Az \rightarrow z = x \vee z = y))$, where "A" means "is an apple in the basket." Hartry Field (1980) maintains that all science can be done in principle in the spirit of this simple example, without the use of numbers. Of course, there's no denying that mathematics

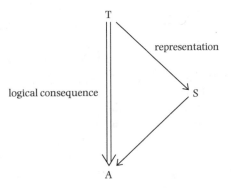

Figure 39.1

is heuristically powerful, and perhaps even psychologically essential, for doing the physics that has been done to date; but, according to him, it is not necessary in any deep ontological sense.

Field is mainly interested in combating a view of Quine and Putnam (1971) that since mathematics is essential for science, it must be true; and since it's true, there must exist objects such as sets, functions, numbers, etc. (see QUINE). Against this Platonism, Field upholds a brand of nominalism, claiming that mathematics is not essential, but merely provides an extremely useful shortcut. In particular, he claims that the role played by mathematics is quite different from that played by other theoretical entities, such as electrons. Field is surely right about this last point: mathematics works, as we noted above, by providing models in which the world (or some part of it) is *represented*. (But this does not mean that Field is right in his nominalism. These are many other – much better – reasons for mathematical realism than the one he attacks.) In this representing capacity, says Field, mathematics is *conservative*. His principle result is this: If A is a consequence of $T + S$ (where T is a nominalistically acceptable theory and S is a mathematical theory), then A is a consequence of T alone (schematically represented in fig. 39.1). The conservativism claim is then used by Field to justify his view that mathematics is not essential for science, since the consequences of the theory exist independently of mathematics.

Field's commentators have been largely skeptical (e.g., Irvine (1990), Malament (1982), Shapiro (1983), Tiles (1984)). Among the objections have been these: The notion of *logical consequence* that is needed is that of second-order logic. But second-order logic is not recursively axiomatizable, which means that the notion of consequence must be semantical. Syntactical consequence (i.e., a derivation) is perhaps nominalistically acceptable, but surely not semantic consequence, since this involves the idea of being *true in all models*, a set-theoretic idea if ever there was one. (Quine famously holds that second-order logic is really just set theory in disguise.)

It would seem, moreover, that we need mathematics to make sense of some crucial notions like *determinism*. In doing physics, we talk not only about how things are, but about what is or is not possible. For example, determinism is defined as follows: a theory is *deterministic* if all of its models with the same initial conditions have the same final conditions; it is *indeterministic* when two of its models with the same initial conditions have different final conditions. For this we need the notion of a model, obviously; but that is the kind of abstract entity provided by set theory that seems to give the nominalist indigestion. (Perhaps concern with determinism is not really part of science proper, but is instead a philosophical issue. In that case, mathematics would seem to be essential for metaphysics.)

Related to this sort of consideration, though less precise, is the role of mathematics in methodology. In the past, empiricists have often maintained that the meaning of a theoretical term (electron, gene) must be given via observation terms. Most philosophers today have abandoned this view, leaving it something of a mystery how we do manage to understand highly theoretical notions. Mathematics may provide the answer, since it would seem to provide a framework for thinking about the world. Highly theoretical concepts (symmetries, resonances, etc. in particle physics) which have no hope of being tied to empirical concepts can often be explicated and understood via mathematics. This would seem to make mathematics not just heuristically useful in drawing consequences from our scientific theories (as Field readily grants), but methodologically *essential* in the very creation and comprehension of those theories.

Finally, it may be a mistake to think that theories are nominalistically acceptable, independently of the mathematics. The example that Field develops in detail is Newtonian gravitation theory. This involves massive bodies and space-time points. These are probably acceptable for a nominalist, though some critics have objected that even space-time points are abstract. However, some theories employ abstract entities right from the start. As mentioned above, the quantum state, ψ, for instance, is arguably not just a mathematical entity, but a real (though abstract) object with something like causal powers of its own.

Field (1989) has replied to some of these objections. He adopts a modal logic, for instance, to cover the idea of logical consequence and to handle determinism. Whether this will resolve the difficulties in a way satisfactory to a nominalist remains an open question.

At the outset I made the assumption that there are two quite distinct realms: the mathematical and the nonmathematical, and that the former *represents* the latter. This isn't the only way to view the situation. Perhaps mathematics *describes* the world. The Pythagoreans, for example, thought that the world *is* mathematical. And John Stuart Mill held that numbers are a kind of very general property that objects possess. A four-legged, blue, wooden chair has the property four, just as it has the properties blue and wooden. Philip Kitcher has proposed (1983) an updated version of Mill. Elementary arithmetic, for example, stems from our ordinary experience; such statements as $2 + 3 = 5$ are not truths about a separate mathematical realm, but are, rather, general truths about the physical world. More sophisticated mathematics is created by an "ideal agent" who can carry out infinitely many operations. The application of mathematics to the world, consequently, is no more mysterious than is the applicability of "Red and yellow mixed together make orange." Like Mill, Kitcher goes far in explaining how mathematics is applied to the world. To be fair, though, many of the sophisticated uses of mathematics don't seem to fit this view. The *properties* of quantum systems are associated with the eigenvalues of linear operators defined on a Hilbert space. It is wholly implausible to see this as an extension from everyday experience exemplified by counting bananas. Perhaps the greatest weakness, as it was for Mill, is in failing to do justice to mathematics itself. Kitcher's "ideal agent," for example, has been repeatedly criticized for being an assumption just as strong as any made by Platonism, and a good deal more obscure.

A view that has been gaining ground in recent years is *structuralism* (Resnik 1997; Shapiro 1997). On this view, there are underlying structures which may be common both to the physical world and to mathematical systems. An infinite string of stars, for example, has the same underlying *structure* as an infinite sequence of moments of time, or as an infinite string of strokes, | | | | | | |. . . . It is easy to see why on this view mathematics is applicable to the nonmathematical realm: mathematics describes the structure or pattern, and the structure is present in the physical system itself. Mathematics is the science of structures or patterns. Structures – not objects – are primary on this view. An object is just a place in a structure; a star, a temporal moment, or a stroke could exemplify the number 27 just by being at the appropriate place. There is no number 27 over and above any of these places in the structure.

Like traditional Platonism, structuralism is a realist view of mathematics. The difference when it comes to applied mathematics is that structuralism sees structures right in the world itself, making mathematics descriptive, while Platonism sees the mathematical world as transcendent, making mathematics representative.

Looking back on the debate, Field versus Quine and Putnam, we can see it as implicitly a debate about whether mathematics represents (Field) or describes (Putnam and Quine). Field has explicitly used the results of measurement theory as done in the representational way. My reason for saying that Quine and Putnam see mathematics as describing the world stems from their various remarks about the possibility of revising mathematics and logic in the face of experience. On the representationalist view of applied mathematics, this would be absurd; an empirical upset would simply make us look for a different mathematical model to represent things; it would not lead us to change our mathematical theories themselves. In this regard, it must be said that the history of mathematics strongly supports the autonomy of mathematics, and hence the representationalist account.

Returning to Wigner, "The miracle of the appropriateness of the language of mathematics for the formulation of the laws of physics is a wonderful gift," he says, "which we neither understand nor deserve" (1960, p. 237, see also Steiner 1998). We may not grasp it fully, but we're partway; and what remains unexplained is not a hopeless mystery.

References and further reading

Abraham, R. H., and Shaw, C. D. 1983: *Dynamics: The Geometry of Behavior*, 3 vols (Santa Cruz, CA: Aerial Press).

Benacerraf, P., and Putnam, H. (eds) 1983: *Philosophy of Mathematics*, 2nd edn (Cambridge: Cambridge University Press).

Brown, J. R. 1990: *Philosophy of Mathematics: An Introduction to the World of Proofs and Pictures* (London and New York: Routledge).

Campbell, N. 1920: *Physics: The Elements* (Cambridge, MA: Harvard University Press); repr. as *Foundations of Science* (New York: Dover, 1957).

Field, H. 1980: *Science without Numbers* (Princeton: Princeton University Press).

—— 1989: *Realism, Mathematics, and Modality* (Oxford: Blackwell).

Friedman, M. 1983: *Foundations of Spacetime Theories* (Princeton: Princeton University Press).

Helmholtz, H. 1887: An epistemological analysis of counting and measurement. In *Selected Writings of Herman von Helmholtz*, ed. R. Kahl (Middletown, CT: Wesleyan University Press, 1971).

Irvine, A. (ed.) 1990: *Physicalism in Mathematics* (Dordrecht: Kluwer).

Kitcher, P. 1983: *The Nature of Mathematical Knowledge* (Oxford: Oxford University Press).

Krantz, D. H., Luce, R. D., Suppes, P., and Tversky, A. 1971–90: *Foundations of Measurement*, 3 vols (New York: Academic Press).

Malament, D. 1982: Review of Field, *Science without Numbers*. *Journal of Philosophy*, 79(9), 523–34.

Mundy, B. 1987: Faithful representation, physical extensive measurement theory and Archimedean axioms. *Synthese*, 373–400.

Nagel, E. 1932: Measurement. *Erkenntnis*, 313–33.

Putnam, H. 1971: *The Philosophy of Logic* (New York: Harper & Row). (Also repr. as a chapter in the author's *Philosophical Papers*, vol. 1, 2nd edn) (Cambridge: Cambridge University Press), 1–31.

Resnik, M. 1997: *Mathematics as a Science of Patterns* (Oxford: Oxford University Press).

Russell, B. 1903/37: *The Principles of Mathematics* (Cambridge: Cambridge University Press).

Savage, C. W., and Ehrlich, P. (eds) 1992: *Philosophical and Foundational Issues in Measurement Theory* (Hillsdale, NJ: Lawrence Erlbaum).

Shapiro, S. 1983: Conservativeness and incompleteness. *Journal of Philosophy*, 80(19), 521–31.

—— 1997: *Philosophy of Mathematics* (Oxford: Oxford University Press).

Steiner, M. 1998: *The Applicability of Mathematics as a Philosophical Problem* (Cambridge, MA: Harvard University Press).

Swoyer, C. 1987: The metaphysics of measurement. In *Measurement, Realism, and Objectivity*, ed. J. Forge (Dordrecht: Reidel), 235–90.

Tiles, M. 1984: Mathematics: the language of science? *Monist*, 67(1), 3–17.

Wigner, E. 1960: The unreasonable effectiveness of mathematics in the natural sciences. Repr. in *Symmetries and Reflections* (Cambridge, MA: MIT Press, 1967), 222–37.

40

Measurement

J. D. TROUT

Measurement – a central epistemic activity in science – relates a number and a quantity in an effort to estimate the magnitude of that quantity. A quantity is typically a property of a physical configuration, such as length or weight, and determines a function that applies to a domain or class of objects. At this high level of abstraction, the description of the purpose and relation of measurement is metaphysically neutral, leaving open the question of whether the domain is observable (empirical) or unobservable (nonempirical).

We can determine and express the value of a quantity as long as we can describe the quantitative relationships that obtain between two or more objects. The familiar mathematical relationships of "greater than," "less than," and "equal to" are the ones most commonly used to express quantitative relations; these relations provide a basis for the articulation of other quantitative relationships such as "farther than," "shorter than," or "same heaviness as." It is instruments that most commonly allow us to discover and formulate the relationships expressed.

Measurement practice has a long history, primarily occupied with astronomical inquiry and engineering concerns of volume, density, and speed, and associated with the most notable figures in the history of science. Historically, measurements were made with laboratory instruments and with telescopic and navigational instruments; but in the nineteenth century probabilistic methods found their way into the estimation of population characteristics in the areas of demography, mortality rates, annuities, and epidemiology. In the middle of the nineteenth century, Quetelet applied the theory of errors to an array of social and biological statistics, attempting to construct a kind of "social physics," as it was sometimes called. Shortly thereafter, Fechner and others pioneered a psychophysical theory designed to measure sensory magnitudes. Later, Spearman introduced factor analysis, and other psychologists such as Stevens (1951), Cattell, and Thurstone developed techniques for the systematic estimation of the magnitude of psychological properties. The twentieth century witnessed vigorous advances in the area of physical measurement, owing to sophisticated laboratory instrumentation. The development of a statistical theory of experimental design, proposed by R. A. Fisher, allowed statistical and laboratory instruments to be used with greater power.

Contemporary interest in measurement has generated two bodies of research. The first is primarily mathematical, concerned with the formal representation of empirical (i.e., observable) structures and the deductive consequences thereof. The second is primarily philosophical, being concerned with the epistemological and metaphysical assumptions and lessons of the practice of measurement. The former we might call "measurement theory," and the latter, "philosophical theories of measurement." This chapter covers both.

Measurement theory

The mathematical theory of measurement, though in principle a metaphysically neutral description of the measurement process, has been developed in a way that displays the distinct and deep influence of empiricism. Although measurement theory is compatible with the realist view that some of the objects of measurement are unobservable, contemporary treatments of measurement theory typically import the additional stricture that the domain be observational or empirical. The following discussion will reflect that empiricist orientation.

Representation of the domain

In order to estimate the magnitude of a quantity – mass, charge, cell firing rate, cognitive dissonance, etc. – we must have some systematic way of representing the relations between that quantity and others in a domain. (For a clear and thorough introduction to the technical apparatus used in measurement theory, see Wartofsky 1968, ch. 6.) Often, the objects, events, processes, properties, and states (hereafter, simply "objects") that we want to measure are *unobservable*, or at least *unobserved*. If we are to draw out the order among objects in a class, we must first *represent* them in some way. (The most common and convenient way of representing qualitative order in the objects of a class is numerically. This, by itself, does not render the data themselves quantitative, even if quantitative methods are used to analyze them.) When the objects in question are *observable*, their representation serves the purpose of keeping track of other complex and dynamic systems.

The process of measurement demands that we set up certain correspondences between a representational (typically numerical) system and an empirical, observational domain. These correspondences are fixed by relations of specific sorts and, depending upon the nature of the measurement relation in question, we want to arrive at a mapping that preserves that relation. A mapping from one relational system to another which preserves all the relations and operations of the system is called a *homomorphism*. If there is a one-to-one homomorphism between the representational system and the domain, the relation is an *isomorphism*.

According to standard measurement theory (Krantz et al. 1971), we begin with a system U of *observed* or *empirical* relations, and try to arrive at a mapping to a numerical relational system B which preserves all the relevant empirical relations and structures of U. In the case of the measurement of temperature, for example, we attempt to find an assignment of numbers that allows us to preserve the relation "warmer than," so that the numbers assigned increase as the temperature increases, etc.

If measurement is the representation of quantitative relations by numerical relations, what conditions must be satisfied if there are to be scales of various types? The most common sort of feature in terms of which objects are represented is a *relation*. Relations can be binary, triadic, quadratic, to n-adic. Binary relations are determined by ordered couples <x,y>, triadic relations by ordered triples <x,y,z>, and so on.

The first condition on the measurement of a quantity is that the objects in the domain can be *ordered* according to the relation chosen. Domains are often depicted as classes, or sets. The relation "same length as," for example, can be determined by ordered pairs of material objects. But, in order to compare the lengths of two objects in a class – that is, to depict the relative presence of some property or magnitude (in this case, length) – we must first define an *equivalence class* of that property or magnitude, such as all those objects with length. We then estimate the relative presence of that magnitude or property in two objects by defining *equivalence classes* of a unit. A unit of measurement defines an *equivalence class* of that magnitude. Objects in the equivalence class may be compared by an *equivalence relation*. Two classes of objects, *a* and *b*, can bear the equivalence relation to each other with respect to a specific property. The unit of one gram defines the class of all those things that are one gram, one meter the class of all those things that are one meter.

Measurement theorists differ in the notation that they use, but the expressions below illustrate common uses. With respect to some class, a binary relation R is an equivalence relation if and only if it is *transitive, symmetric,* and *reflexive*. More formally, two objects, x and y, bear the relation R of "same length as" if and only if: (1) for every x, y, and z in the class, if $R(x,y)$ and $R(y,z)$, then $R(x,z)$ [transitivity], (2) for every x and y in the class, if $R(x,y)$ then $R(y,x)$ [symmetry], and (3) for every x in the class, $R(x,x)$ [reflexivity]. Any two objects compared for length that bear these three relations to one another will be of equal length.

Other orders can be determined by further sets of ordered pairs. With respect to some class, a binary relation R is:

intransitive on a class if and only if, for every x, y, and z in the class, if $R(x,y)$ and $R(y,z)$, then not $R(x,z)$ – being the mother of

asymmetric on a class if and only if, for every x and y in the class, if $R(x,y)$, then not $R(y,x)$ – being the father of

anti-symmetric on a class if and only if, for every x and y in the class, if $R(x,y)$ and $R(y,x)$, then x = y (i.e., x is identical to y) – being at least as great as on the real numbers

irreflexive on a class if and only if, for every x in the class, not $R(x,x)$ – being the brother of

strongly connected on a class if and only if, for every x and y in the class, either $R(x,y)$ or $R(y,x)$ – being at least as great as on the natural numbers

connected on a class if and only if, for every x and y in the class such that x ≠ y, either $R(x,y)$ or $R(y,x)$ – being less than upon the natural numbers.

In the hands of many modern measurement theorists, the further empiricist demand is introduced that the homomorphism satisfies a function between a number and a formal *representation* of a domain that is *empirical or observable*. No additional epistemological argument has been offered that the domain in question must be observable, and therefore it is at this stage quite incorrect to understand the relevant formalism as an anti-metaphysical feature of the theory of measurement.

It might be thought that this empiricist assumption of the representationalist-formalist approach is not just inadequately defended, but false. The attempt to treat formalism as an anti-metaphysical feature of a theory has failed for related projects. The representationalist-formalist approach is often criticized on the grounds that, despite its formal appearance, it does not divest itself of assumptions concerning the specific nature of the domain being measured. Measurement employs scales, and the grounds for use of those scales include substantial assumptions about the objects, observable and unobservable. According to standard criticisms of projects which attempt to eliminate appeal to unobserved causes, the anti-realist goal is not achieved until the causal direction represented in the models of the empirical substructures (as well as other features of the model) can be determined without theoretical commitment. A similar difficulty arises on the syntactic conception of theories. There, the effort to eliminate theoretical terms was itself parasitic upon the complete articulation of that theoretical system. According to the representationalist-formalist view, the goal of the theory of measurement is the construction of an observable representation of the consequences of measurement axioms, a Ramsey sentence if you will (see RAMSEY SENTENCES). It is this feature that ties the formalist-representationalist approach to the empiricist tradition. For philosophers more aggressively empiricist, the function of measurement theory is eliminative; by providing an observable representation of the objects and relations in the domain, the model-theoretic, semantic approach allows one to appeal to embedded substructures as rendering superfluous the commitment to nonempirical components of the theory. Craig's theorem attempted to execute a similarly eliminative project for the syntactic approach to scientific theories (see CRAIG'S THEOREM). In such a case, this eliminativist project loses its original rationale. The initial motivation concerned the possibility of exclusive reliance on the observation statements of the theory. However, if the theoretical terms can be eliminated from the system only *after* the system has been developed, then theoretical terms appear to isolate an essential portion of a theory's content. It has been argued that, as a result, the eliminativist goals of Craig's theorem and Ramsey sentences foundered on this general difficulty.

The second conception of measurement might be called the causal (or sometimes, interactionist) approach. I shall have more to say about this approach in the discussion of philosophical theories of measurement.

Representation of order and scale

To achieve a representation of relations that is systematic, we can use *scales*. There are four basic scales, corresponding to four levels of measurement: nominal, ordinal, interval, and ratio. As we ascend from nominal to ratio scales, the scales increase in their power to represent characteristics of the data, while preserving the ability to represent the data in accordance with the preceding level. The process of measurement attempts to estimate magnitudes, the amounts of the attributes or properties under investigation.

Not all scales represent magnitude, or even order; some merely classify data. *Nominal scales*, for example, sort observations into different categories or classes. The classification of people into gender groups of male and female is an instance of nominal measurement, as is the categorization of subject responses into "Yes" and

"No." Nominal scales cannot represent many important properties of objects in the class, but the importance of the nominal classification of objects into stable categories should not be ignored. *Ordinal scales* order objects along a dimension – or "rank order" – but do not indicate how great the difference is (along that dimension) between any of the objects. Such scales are common in surveys where choices range from "Strongly disagree" to "Strongly agree." Although this scale depicts order, it does not capture information about magnitude; we don't know by how much "Strongly disagree" differs from "Disagree." Such scales do not have the power to honor the principle of equal intervals, a feature of interval scales. *Interval scales* allow us to infer specific differences among objects from differences in scale points. The most common example here is that of temperature; any 10-point difference on the scale has the same meaning. *Ratio scales* allow us to infer from numerical proportions among the representations of objects that one magnitude is twice or three times another. Ratio scales have a true zero point, unlike the other scales we have seen thus far. The octave scale in music, based on frequencies, is a ratio scale. The zero point in this case occurs when there are no cycles per second. One can say that middle C is half the frequency of the C an octave above.

There are two kinds of theorems in measurement theory, *representation theorems* and *uniqueness theorems*. Representation theorems state the conditions under which a numerical representation can be found for empirical structures; they thus state the existence of certain types of scales given that the nonnumerical relational structure of a domain satisfies certain conditions. Uniqueness theorems tell us whether the resulting scale is unique, or whether there are permissible transformations from one scale to another. (What we are trying to establish is not uniqueness proper, but uniqueness in some relevant sense.) The characterization of these transformations is important, because they specify invariances among scales, and these invariances are thought to reflect important features of a property in the domain.

Providing an *interpretation* of the scale is a necessary condition for measuring the domain. Aiding this interpretation was the theory of error, one of the most important contributions to the theory and practice of measurement. In light of certain statistical assumptions or axioms – most often the Kolmogorov axioms – the effects of measurement error could be estimated. Several of the axioms are expressed as follows, in terminology more approachable than that used by Kolmogorov. Where P(A) is the probability of event A:

Axiom I: $0 \leq P(A) \leq 1$.
Axiom II: $P(A) = 1$, if A must be true (or is certain).
Axiom III: If A and B are incompatible ("mutually exclusive"), then $P(A \text{ or } B) = P(A) + P(B)$.

The first axiom might be thought of as the axiom of positiveness, stating that the probability assigned to any event is positive or zero. The second axiom might be called the axiom of certainty, and says that the probability of the entire set of events is 1. The third axiom might be called the axiom of unions or addition, and states that if event A and event B are mutually exclusive, then the probability of A or B is the sum of their independent probabilities. These axioms can be used to calculate the probability that a certain distribution of measurement values could be expected by chance. In order to do so, we need some method for estimating variation. Therefore, we can estimate this probability by calculating the variance and the standard deviation. We begin with the sum of the differences between measurements and the mean value. The variance is the sum of the squares of these differences, and the standard deviation is the square root of the variance. Once supplemented with the theory of error, measurement theory might be understood as the effort to explore the deductive consequences of the Kolmogorov axioms of probability.

The estimation of the magnitude and direction of error is an important condition of valid and reliable measurement, and makes explicit the routine inexactness of measurement. A measuring instrument estimates the value of a quantity, yielding a number, but does so only subject to a certain range of error. Therefore, the accuracy of the measuring instrument, be it a pH meter or a statistical design, can be assessed only in light of a theory of error.

Philosophical theories of measurement

There have been three main philosophical approaches to measurement: operationalism, conventionalism, and realism. Directly or indirectly, these approaches attempt to address the apparently realist character of measurement; measurement procedures are formulated in light of, and to operate on, quantities that are theoretical, or unobservable. The thesis that measurement or its outcome is typically theoretical – in short, the theory dependence of measurement – did not sit well with the empiricist epistemology that motivated the operationalist account with which this history begins. In light of their diverse epistemological, metaphysical, and semantic commitments, these three approaches attempt to explain away this apparent theory dependence of measurement, to concede it but limit its significance, or to defend its literal interpretation and its consequences.

Operationalism

Operationalism was an early logical empiricist approach to measurement that attempted to cleanse unreconstructed appeal to theoretical quantities, and to define theoretical magnitudes exclusively in terms of observables (see LOGICAL EMPIRICISM). The resulting "correspondence rules" or "operational definitions" formed identity statements, linking each observable measurement and detection procedure with a single type of unobservable quantity. Empiricists eventually rejected this account of theoretical measurement as incoherent. The problem arises as follows. According to unreconstructed scientific practice, a single theoretical quantity can be measured in a variety of ways. But, according to operationalism, the particular procedure is *definitive* of the theoretical quantity, and thus one would have to postulate a different type of quantity for each different procedure that was (perhaps even could be) used.

Consider improvements in the measurement of pH. Because these revisions constitute changes in operations, according to the operationalist empiricist, they must reflect a change in what (in this case) "pH" refers to (see THEORETICAL TERMS) – for each new type of procedure, a different type of quantity. But scientists clearly take themselves to be measuring the same quantity, no matter how many different types of procedures they use; and the substantial continuity of their measuring procedures appears to vindicate this supposition. So, when applied to an instrument as familiar and well understood as a pH meter, the operationalist account of measurement, squarely within the logical empiricist tradition, supplied a disappointingly misleading account of scientific practice. Nevertheless, it was precisely because empiricists recognized and respected the fact that the same theoretical quantity persisted under diverse measurement procedures that they rejected early operationalist accounts of theoretical definition.

In describing this objection to the operationalist conception of measurement, another important feature of operationalism emerges. Because the operationalist holds that the particular procedure is *definitive* of the theoretical term, in this respect operationalism is an early conventionalist doctrine. Because the meaning of "pH" is associated with a certain convention – that we take the specific measurement procedure as definitive of the term – the term "pH" couldn't fail to refer (see CONVENTION, ROLE OF).

Moreover, it became clear that any single measurement procedure tacitly depended upon a variety of auxiliary hypotheses in the design of the instrument. The pH meter employs a collateral theory of electricity, for example. The accuracy of the pH meter was therefore partly dependent upon the accuracy of the auxiliary electrical theory. The recognition of this fact about routine measurement led to the two most influential features of the critique of operationalism, features which set the terms of the dispute after the demise of operationalism. The first is the *holistic* character of measurement (see HOLISM); revisions and improvements in measurement procedures are made in light of background or auxiliary theories. The second is the *realistic* assumptions underlying those revisions and improvements. Increasingly successful measurement could be accounted for only in terms of the accuracy of our *theoretical* knowledge.

Many empiricists acknowledged the holistic aspect of measurement, but were not prepared to accept the metaphysical consequences that define the realist program. For such empiricists, measurement is guided, if not dominated, by convention.

Conventionalism

Conventionalism about measurement states that the interpretation of measurement procedures reflects our conventions. Alternatively, measurement procedures do not provide evidence of quantities that exist independently of our efforts to measure. The conventional aspect of measurement is most forcefully illustrated in cases where two or more scales equally well represent the empirical order. In such cases, only pragmatic factors can determine our choice of a scale, factors such as the simplicity of the numerical laws (see PRAGMATIC FACTORS IN THEORY ACCEPTANCE and SIMPLICITY). Indeed, on this view, the simplicity of our laws would be inexplicable unless we suppose it to result from our selection of a theoretical framework. Unless we take this stance, the conventionalist argues, we must suppose that the laws of nature, with all the complex details of their observational consequences, could be revealed to us through diverse experimental strategies and sundry instrumentation. The conventionalist argues that increasing success in measurement is not a reason for thinking that the measurement reflects accurate causal (typically theoretical) information; it is only a reason for thinking that the measurement procedure or instrument exhibits the empirical order in a way we find simple or otherwise aesthetically pleasing.

Conventions are employed in the introduction of a unit, as in the choice to use the familiar meter bar located near Paris as the standard meter. Conventions operate at a later stage as well, in the application of those units and scales (see Ellis 1966). According to conventionalists such as Reichenbach, we must adopt certain conventions if we are to maintain simple laws. One such convention is that a measuring rod remains rigid when transported through space (to measure distances or objects at remote locations). In light of the underdetermination of theory by (observable) evidence, one might suppose that there are forces that operate in such a way that the rod *changes* length when transported (see UNDERDETERMINATION OF THEORY BY DATA). But such suppositions would generate less simple laws. A specific convention is therefore adopted – that forces operate uniformly on the measuring rod – to rule out such complicating possibilities.

At the same time, critics of conventionalism interpreted these conventions not as harmless stipulations, but rather as substantial theoretical assumptions. After all, judgments of simplicity, elegance, parsimony, or convenience, it has been charged, are themselves theory-dependent judgments, which cannot be made on the basis of empirical or observational considerations alone. According to this criticism, a numerical law counts as simple only in light of certain theoretical considerations; so the conventionalist, like the operationalist, cannot avoid the epistemic (as opposed to merely pragmatic) function of theoretical commitment in the selection of a scale or measurement procedure. For the realist, this theory describes causally important dimensions of the world.

Realism and causal analysis

A realist account of measurement treats the act of measurement as a product of a causal relation between an instrument (broadly interpreted) and a magnitude. The relation is one of *estimation*. These magnitudes or quantities (properties, processes, states, events, etc.) exist independently of attempts to measure them, and are sometimes too small to detect with the unaided senses. Mean kinetic energy is one such theoretical magnitude.

From the realist (see REALISM AND INSTRUMENTALISM) perspective of unreconstructed scientific practice, a pH meter is thought to measure an unobservable property: the concentration of hydrogen ion in a solution. The term "pH" is then thought to refer to this unobservable quantity. This ungarnished account is openly metaphysical. The realist account that replaced operationalism treated the introduction of new procedures in exactly the way that scientists seemed to – as new ways of measuring the same quantities identified by earlier procedures, now improved in response to the instrument's increasing sensitivity to unobservable features of magnitudes (Byerly and Lazara 1973). The impressive pace and character of the developments in the natural sciences, particularly the fashioning of sophisticated instruments, seemed to warrant the realist claim of *theoretical improvement* in measurement procedures. This argument for a realist interpretation of successful measurement is based upon an inference to the best explanation; the accuracy of our theoretical knowledge provides the best explanation for the improvement of our measurement procedures (see INFERENCE TO THE BEST EXPLANATION).

According to realists, there is a core set of reasons, common to lab and to life, that provide powerful general grounds for a realist rather than an empiricist conception of measurement. The first reason we have already seen. Early empiricist accounts, most notably operationalism, cannot explain the routine use of diverse procedures for the measurement of (what we regard as) the same quantity.

Second, the interpretation of measurement error, compulsorily reported in nearly all behavioral and social science journals, is rendered obscure on an empiricist account of measurement. It is misleading to describe a measurement as *inaccurate*, or as *in error*, if there is no difference between measured and real (or unmeasured) value. It would seem that the only natural way of expressing the incorrectness of a measurement is in terms of its difference from a correct measurement. There are other, less natural ways that the empiricist might reconstruct the initial concept of a "correct" measurement – perhaps in terms of ideal measurements, infinite samplings, or the measured value at the limit of inquiry – but none of these alternatives has the kind of grounding in experience required by the empiricist. And were these notions invoked to explain any practice less central to science than measurement, they would be anathema to the empiricist. By contrast, the realist holds that the typical quantities have a real value, independent of attempted measurements; error is the distance between the real and the measured value, distance produced by limitations of knowledge, instrument design, and noise. So the realist can provide a consistent rationale for the estimation of measurement error. (The realist does not presume to know the objective value, but merely states that there is one. The most familiar estimators of the true value of a quantity are *biased estimators*, such as the sample mean.)

Finally, early techniques for the measurement of certain kinds of magnitudes, such as pH, have been corrected and improved upon by later ones. For example, early methods of pH measurement did not correct for the fact that the same solution would yield different pH readings owing to differences in temperatures of the solution. More recent methods make more accurate estimates by accounting and correcting for the contribution of temperature in the measurement of pH. This fact of increasing accuracy is difficult to explain without supposing, first, that both the early and the more recent measurements are of the same quantity, and, second, that it is at least plausible to talk about the measured item as having an objective value, toward which successive measurement procedures are converging.

Something like the following account captures part of our unreconstructed conception of measurement:

X has some property P of magnitude M, and instrument *i* measures P if and only if it reports the actual value of M.

As stated, however, this simple account of measurement would be inadequate. In the first place, it would count coincidentally correct reports by defective instruments as instances of measurement, allowing that a stopped clock measured the time simply because the hands indicate the correct time twice a day, or a thermometer stuck at 100.4° will report the person's temperature correctly whenever that person has a fever of 100.4°, etc. So we need to add a further condition in order to rule out coincidentally correct reports by defective instruments:

> X has some property P of magnitude M, and instrument *i* measures P if and only if it reports the actual value of M, and *i* would not have reported that value unless M had that value.

But now it would seem that the account of measurement is too strong, ruling out as measurement cases in which the reported value is incorrect, but within an acceptable margin of error. So we need to add a clause that tolerates some amount of systematic and random error:

> X has some property P of magnitude M, and instrument *i* measures P if and only if it reports the actual value of M, and *i* would not have reported that value unless M, within certain well-understood parameters of error, had that value.

Spelling this distinction out, however, yields an analysis that treats measurement as a relation of causal dependence. This relation can be expressed by following the form of the counterfactual analysis of causation:

> *The Counterfactual Analysis of Causation*: P causes Q if and only if (1) P obtains, (2) Q obtains, and (3) Q would not be the case if P were not the case.

> *The Counterfactual Analysis of Measurement*: Instrument *i* measures P if and only if *i* reports the approximate value of M, and *i* would not have reported that value unless M, within certain well-understood parameters of error, had that value.

The above analysis of measurement does not require that the real value of M be known in order for the parameters to be well understood. For example, calibration occurs against a standard, and that standard is devised in light of our best theory (like a standard used to calibrate a pH meter). A regress is avoided, because we have diverse and increasingly better ways of measuring pH, time, etc.

Difficulties remain for the realist account of measurement. In the first place, it is not clear that a counterfactual analysis of causation requires realism. (For a counterfactual analysis of causation that is more in the Humean spirit, see Lewis 1973a.) Second, it has been claimed that counterfactual analyses founder because, though causation may be transitive, counterfactual dependence is not (see Lewis 1973b, pp. 31–6). More ambitious critics charge that causation itself is not transitive. It is not clear, however, whether the standard inferences concerning transitivity appear to fail only because of contextual features of the examples used.

The generic realist account of measurement states that quantitative procedures, routinely applied in light of our best theories, exploit the relation between (a) theoretically important dimensions of the population and (b) the products of measurement procedures and instrumentation, according to which the former regulate the latter. First, when a procedure or instrument accurately measures a quantity, the values yielded by that procedure or instrument are conditionally dependent on the real values of the population dimensions. Therefore, it might be thought that the most natural way to understand these dependencies is causal. Second, some of the theoretically important population dimensions form equivalence classes of quantities that are unobservable, and measurement procedures are refined with respect to these unobservable phenomena. The causal aspect of measurement not only explains the standard asymmetry of explanation, but also accounts for (i) the apparent dependence of the measured value on the real value in valid measurement systems, (ii) the counterfactual dependence evident in the statement of measurement relations, and (when interpreted realistically) (iii) the indifference shown by scientists about whether the postulated causal factor is an unobservable or just an unobserved observable.

Realists regard the typical quantities measured in science as natural kinds (see NATURAL KINDS). For this reason, measurement outcomes are thought to represent real, enduring, or stable features of the population. We can measure objects that are perhaps not eternal, but stable enough to support generalizations; such objects include quantities in psychology and social science (see SOCIAL SCIENCE, PHILOSOPHY OF). (For a realist account of measurement with special application to the psychological sciences, see Trout 1998.) Even so, philosophical analyses of measurement have always been devoted to special features of *physical* measurement. These early analyses, however, were profoundly influenced by empiricist attempts to syntactically reduce higher- to lower-level theories and to produce reductive definitions in terms of observation. Freed from these special philosophical concerns, a generic realist account of measurement covers the social and psychological, as well as the physical, sciences.

References and further reading

Blalock, H. 1961: *Causal Inferences in Nonexperimental Research* (Chapel Hill, NC: University of North Carolina Press).

Bridgman, P. W. 1927: *The Logic of Modern Physics* (New York: Macmillan), 1–32; repr. in *The Philosophy of Science*, ed. R. Boyd, P. Gasper, and J. D. Trout (Cambridge, MA: MIT Press/ Bradford Books, 1991), 57–69.

Byerly, H., and Lazara, V. 1973: Realist foundations of measurement. *Philosophy of Science*, 40, 10–28.

Carnap, R. 1966: *Philosophical Foundations of Physics: An Introduction to the Philosophy of Science* (New York: Basic Books).

Ellis, B. 1966: *Basic Concepts of Measurement* (Cambridge: Cambridge University Press).

Hempel, C. G. 1970: Fundamentals of concept formation in empirical science. In *Foundations of the Unity of Science*, vol. 2, nos 1–9, ed. O. Neurath, R. Carnap, and C. Morris (Chicago: University of Chicago Press), 653–745.

Krantz, D., Luce, R. D., Suppes, P., and Tversky, A. 1971: *Foundations of Measurement*, vol. 1 (New York: Academic Press). (The classic work on measurement theory.)

Kyburg, H. 1984: *Theory and Measurement* (Cambridge: Cambridge University Press).

Lewis, D. 1973a: Causation. *Journal of Philosophy*, 70, 556–67; repr. in *Causation*, ed. E. Sosa and M. Tooley (Oxford: Oxford University Press, 1993), 193–204.

Lewis, D. 1973b: *Counterfactuals* (Cambridge, MA: Harvard University Press).

Luce, R. Duncan, and Krumhansl, C. 1988: Measurement, scaling, and psychophysics. In *Stevens' Handbook of Experimental Psychology*, vol. 1; *Perception and Motivation*, ed. R. Atkinson, R. Herrnstein, G. Lindzey, and R. Duncan Luce (New York: John Wiley & Sons), 3–74.

Luce, R. Duncan, and Narens, L. 1987: Theory of measurement. In *The New Palgrave: A Dictionary of Economics*, vol. 3 (London: Macmillan), 428–32.

Roberts, F. S. 1979: *Measurement Theory: With Applications to Decisionmaking, Utility, and the Social Sciences* (Reading, MA: Addison-Wesley).

Savage, C. W., and Ehrlich, P. (eds) 1992: *Philosophical and Foundational Issues in Measurement Theory* (Hillsdale, NJ: Erlbaum). (A collection of recent articles on measurement, all by major figures in the field, and includes a fine editorial introduction.)

Spirtes, P., Glymour, C., and Scheines, R. 2000: *Causation, Prediction, and Search*, 2nd edn (Cambridge, MA: MIT Press).

Stevens, S. S. 1951: Mathematics, measurement, and psychophysics. In *Handbook of Experimental Psychology*, ed. S. S. Stevens (New York: Wilcy), 1–49.

Swoyer, C. 1987: The metaphysics of measurement. In *Measurement, Realism and Objectivity*, ed. J. Forge (Dordrecht: Kluwer Academic Publishers), 235–90.

Trout, J. D. 1998: *Measuring the Intentional World: Realism, Naturalism, and Quantitative Methods in the Behavioral Sciences* (New York: Oxford University Press, 1998). (Ch. 2, "Population-Guided Estimation," advances a systematic account of measurement. Ch. 4, "Measured Realism," introduces a related ontological and epistemological view.)

Wartofsky, M. 1968: *Conceptual Foundations of Scientific Thought* (New York: Macmillan). (Ch. 6 contains what is still the clearest and most thorough introduction to measurement and the associated technical apparatus.)

41

Metaphor in Science

ELEONORA MONTUSCHI

It is widely acknowledged that metaphors are used in science. Great scientists, such as Darwin and Einstein, believed that the use of metaphors is vital to the development of scientific ideas. The history of science is full of examples of scientific metaphors as tools at the forefront of discoveries of new facts and new concepts.

The question which concerns the philosopher of science is how to understand the role of metaphor in science – namely, what theoretical or methodological function is to be ascribed to metaphors in scientific language. A short answer to this question, I believe, can be found, emblematically, in the title to a well-known article by Richard Boyd (1979): "Metaphor and theory change: what is 'metaphor' a metaphor for?"

This title points up the fact that metaphor has been used by the philosopher of science as an instructive paradigm for the description of certain constructs in science. In particular, metaphor has entered the vocabulary of the philosophy of science to illustrate how models work in relation to scientific theories (see MODELS AND ANALOGIES), and how theoretical terminology is introduced into scientific language and its meaning developed (see THEORETICAL TERMS). In other words, metaphor (its linguistic mechanism) is mainly used in the philosophy of science as a "metaphor for" models and theoretical terms.

It is interesting to note that each tradition in the philosophy of science has employed a different "paradigm" of metaphor to fulfill the descriptive task.

The comparison view

In the formalist (logicist) tradition (see LOGICAL POSITIVISM), a scientific theory is conceived of as a deductive system to be represented, symbolically, by means of a calculus, and interpreted, empirically, by means of a reduction to the observable. Symbols and observational terms constitute the "literal language" of the theory, while models and theoretical terms (T-terms) – which are also means for interpreting the deductive system – are "like metaphors" of the theory. That is, they are useful "paraphrases" of the theory, introduced for illustrative purposes, or for reasons of descriptive economy, but devoid of any epistemic value. In fact, neither models nor T-terms have anything special to add to the theory, nothing which the theory itself cannot display or provide by an appeal either to its calculus or to observation (see OBSERVATION AND THEORY). Therefore, they are both dispensable vis-à-vis scientific theory.

Implicit in the formalist perspective on models and T-terms is a comparison view of metaphor (see Black's (1962) typology of metaphor). According to this view, metaphorical expressions are said to be built upon literal statements of comparison between sets of similar properties (Richard is a lion = Richard is like a lion – in being brave). This would explain why metaphors can always be rephrased in terms of a literal comparison: metaphor is nothing but an elliptical form of simile. The reason why a metaphorical expression is used instead of some equivalent, literal expression, is normally stylistic, or ornamental. In other words, with a metaphor we might be able to say something "better"; but we do not say anything "more" than the corresponding literal expression.

The interactive view

The critics of the formalist tradition (the so-called post-positivist philosophers of science) were particularly interested in showing (1) that theories are not simply formal or logical structures; (2) that the language they are built upon is not "static," but changes and transforms itself under the pressure of new cognitive acquisitions; and (3) that new discoveries in science cannot always be reduced to a stable realm of observable facts or dealt with in logical terms.

With these projects in mind, an interest arose in modeling as a form of theory construction, and in the mechanisms of scientific language (taken as a whole, and regardless of any artificial split between observational and theoretical terminology), not so much as representing, but rather as shaping, new concepts and new meanings. Metaphor was looked upon as a rather congenial descriptive device. A different paradigm of metaphor, however, was appealed to: the interactive view, as proposed by Max Black (1962).

According to this view, in a metaphor two subjects "interact" in such a way that a principal subject, or focus, is "seen through" a subsidiary subject, or frame (as in the widely exploited example: "Man is a wolf"). This means that features, implications, and commonplaces normally associated with the subsidiary subject are displaced, or transferred onto the principal subject. The transfer is selective; that is, it works as a "filter" which produces new meaning implications for the principal subject.

Due to this "filtering" mechanism, as Black himself claimed, metaphors *create* analogies, rather than spelling out preexisting ones (Black 1962, p. 37). In this respect, the interactive view differs from the comparative view: metaphors are not the result of independently established similarity relations. For this reason they cannot be translated into a literal comparison without loss of meaning.

So, among the post-positivist philosophers of science, the relativists (see RELATIVISM and KUHN) found in interactive metaphors a way to express the idea that, according to the language (theory, paradigm) used, the world changes. Metaphorical descriptions of the world are not simply alternative descriptions of some purportedly literal description of it. In Kuhn's words, we can live in different worlds, because – as metaphor reminds us – "the world" can be cut at different joints by another language (Kuhn 1979, p. 414). Instead, the realists (see REALISM AND INSTRUMENTALISM) found in metaphor a means to assert that science has ways to produce approximate descriptions of the basic constituents of the world, and to use these descriptions to explain what the world is like in its deeper strata.

Mary Hesse used interactive metaphors to describe the working of scientific models in relation to theory. In particular, Hesse was concerned both with a modification of the Hempelian model of explanation (see EXPLANATION) by means of a theory of the metaphorical function of models, and with the problem of concept formation in scientific theories.

Black had pointed out that the use of models in science resembles the use of interactive metaphors. What is required for both uses is, according to him, an "analogical transfer of a vocabulary." Both metaphor and model making are "attempts to pour new content in old bottles" (Black 1962, pp. 238–9). Hesse adopted and articulated this resemblance in the following way: the principal subject (or primary system) of the metaphor is to be related to the domain of the *explanandum*, and expressed in observational terms; the secondary subject (or system) is related to the *explanans*, and it is expressed either in observational terms or in the language of a familiar theory. The model "takes off" from the latter system: by virtue of a principle of assimilation, the two systems interact in such a way that features of the secondary system are "transferred" (selectively) to the primary, to such an extent that the latter becomes describable through the "frame" provided by the former (Hesse 1966). Examples of the relation between the two systems are of the following kind: "*Sound* (primary system) propagates via *undulatory motion* (secondary system)"; "*gases* are *collections of particles with random movement*"; etc.

By rethinking the Hempelian view of theory and explanation through this picture, we are forced to acknowledge that the relation between *explanans* and *explanandum* is not strictly deductive, but rather a relation of approximation, of mutual adaptation. Moreover, we must accept that the language of the *explanans* is modifiable. The distinction between the observational and the theoretical, in the case of scientific language, as well as the distinction between the literal and the metaphorical, in the case of language in general, can only be a relative one (Arbib and Hesse 1986). This does not necessarily mean that all language and all languages are metaphorical. But it does mean that metaphorical processes are more rooted in, and relevant to, the study of language, meaning, and forms of categorization than they are often, *prima facie*, taken to be (Hesse 1988; Rosche and Lloyd 1978). Metaphorical expressions are not to be taken as "parasitic" on literal language: as a matter of fact, we can look at the functioning of metaphors and learn something about "literal" meaning or, better still, about meaning in general (Hesse 1985–6).

The problem of the introduction of new concepts and new meanings into the language of science has been studied by Richard Boyd (1979). Once again the "paradigm" for this study is the Blackian conception of metaphor.

Boyd claims that the mechanism of interactive metaphors displays a strong "parallelism" to the procedure by means of which theoretical terms are introduced and used in scientific discourse. This parallelism is derived from the fact that both metaphors and theoretical terms provide "epistemic access" to the identification of possible, or purported, referents. This identification amounts to acknowledging the role which a certain term (metaphorical or theoretical) plays in a social setup (a conversation, a poem, the implementations of a scientific research project).

Without having to be "definite descriptions," theoretical terms, like metaphors, can still insure some kind of continuity in the use of a certain expression in "approximating" a certain, not yet identified referent. The expression can be taken as a nondefinitional strategy of "reference fixing," by means of which we accommodate our language to the world. So, for example, "gene," or "genetic code," before being or becoming a term for a real entity (gene), is a linguistic, tentative, and provisional identification (i.e., relative to our degree of cognitive awareness) of whatever the real entity will turn out to be.

Nonetheless, both Hesse and Boyd accept that there are substantial differences between scientific metaphors (i.e., models and theoretical terms) and interactive metaphors in common or literary languages. These differences can be explained either by the specific epistemic constraints of scientific languages, which ordinary metaphors cannot cope with (Hesse and Boyd's option), or instead by reference to some inadequacies in the paradigm of metaphor adopted by the philosophy of science (Aronson et al. 1994). In neither case are the differences taken as an argument against the use of metaphors in science (as sometimes Black himself seems to suggest). These two lines of inquiry will now be explored.

Scientific and literary metaphors contrasted

Hesse tells us that "metaphorical" models should be based on some kind of preexisting similarity, or analogy. Whatever may be the case with poetic uses, "the suggestion that *any* scientific model can be imposed a priori on *any* explanandum and function fruitfully in its explanation must be resisted. Such a view would imply that theoretical models are irrefutable" (Hesse 1966, p. 161). In order to work at an intersubjective level, interactive models are not meant to shock, or to surprise, by being unexpected, striking, or unrepeatable. A constitutive trait of literary metaphors is that they are often intentionally *imperfect*. On the contrary, scientific models ideally aim at becoming "perfect metaphors" – that is, constructs which might eventually, or virtually, become literal interpretations (ibid., p. 170).

A literal interpretation – even as an ideal – entails a belief in the "truth" of the description offered of the world. Here a further distinction emerges between poetic and scientific metaphor: contradiction is inextricable from interaction in poetic metaphors, even up to the point of paradox. Scientific metaphors, instead, are not "peculiarly subject to formal contradictoriness," and "their truth criteria, although not rigorously formalizable, are at least much clearer than in the case of poetic metaphor" (ibid., p. 169).

Boyd is even more emphatic. Scientific metaphors (T-terms), in order to function as theory-constitutive, must, first of all, trade their "open texture," which is suggestive of new connotations and of redescriptions, for some criteria of precision for their use. So, for instance, we read in Boyd that these "theoretical" metaphors can be *clarified*, as far as their analogical sources are concerned, in terms of the degrees of explanatory success achieved by them in the domain of a theory. This need for clarity with scientific metaphors is interpreted by Boyd as a sign of the existence of two different sorts of metaphorical "open-endedness": one pertains to theory-constitutive metaphors and can be called "inductive," while the other, labeled "conceptual," pertains to literary metaphors. The function of the latter, concisely put, "is not typically to send the reader out on a research programme" – which is precisely the function of inductive open-endedness. It is on the basis of this distinction that Boyd is inclined to conclude that scientific metaphors are, after all, "highly atypical" (Boyd 1979, pp. 361–3).

Could the mismatch between scientific and other metaphors be resolved so as to allow us to talk of scientific metaphors as "typical" constructs? A suggested solution takes into account type hierarchies (a notion borrowed from developments in artificial intelligence) as the clue to representing the interactive mechanism at work both in metaphors and in scientific models.

Interactive metaphors and type hierarchies

A type hierarchy is a particular kind of semantic network, organized according to levels of generality (concepts become more abstract as one moves up the hierarchy and more concrete as one moves down). Within this structure, properties and relations of any type can be "inherited" by all its sub-types. Inheritance proves, then, to be a nonarbitrary way of structuring the hierarchy – that is, "according to whether or not the subtypes can take on the meta-properties of the supertype" (Aronson et al. 1994, p. 38). This would explain how similarities among systems occur, and how relevant similarities can be distinguished from irrelevant ones. For example, the atom and the solar system can be placed under a common super-type (a central force field system), and this would make both of them inherit the meta-properties of central force fields in general. In this way, only the "positive" analogies will be selected, since the "negative" ones will not be represented at the level of the super-type.

This aspect is essential to both models and metaphors. The interactive view had left the problem of selecting relevant analogies practically unsolved, by simply suggesting the rather vague image of the "filter." The inheritance mechanism explains how the filtering process determines and controls the selection of the appropriate similarity relations.

The inheritance mechanism also develops and specifies Black's intuition that metaphor creates the similarity, rather than formulating similarities that exist antecedently. In fact, the mechanism of "interaction" can be represented by the type hierarchy structure in the following way: "the tenor or subject of the metaphor is redescribed in terms of a new hierarchy brought into play by the ontology of the vehicle or modifier" (Aronson et al., p. 102). By so doing, we explore new possibilities and extend the meaning of our concepts, but not arbitrarily – which is precisely what models are all about.

It must be emphasized that, by representing models and metaphors through type hierarchies, it can be proved that, pace Black, no "absurd conjunction of words" is involved in the process of creating analogies. The relevant similarities are selected according to the combinations allowed by the type hierarchy involved and by the interaction with the other type hierarchies brought into play by the metaphor. In this sense, then, metaphorical "creativity" also applies to scientific models without fear of producing paradoxical results.

References and further reading

Arbib, M., and Hesse, M. B. 1986: *The Construction of Reality* (Cambridge: Cambridge University Press).

Aronson, J. L., Harré, R., and Way, E. 1994: *Realism Rescued* (London: Duckworth).

Black, M. 1962: *Models and Metaphors* (Ithaca, NY: Cornell University Press).

—— 1979: More about metaphor. In *Metaphor and Thought*, ed. A. Ortony (Cambridge: Cambridge University Press; 2nd edn, 1993), 19–43.

Boyd, R. 1979: Metaphor and theory change: what is "metaphor" a metaphor for? In *Metaphor and Thought*, ed. A. Ortony (Cambridge: Cambridge University Press; 2nd edn, 1993), 356–408.

Hesse, M. B. 1966: *Models and Analogies in Science* (Notre Dame, IN: University of Notre Dame Press).

—— 1985–6: Texts without types and lumps without laws. *New Literary History*, 17, 31–48.

—— 1988: Theories, family resemblances and analogy. In *Analogical Reasoning*, ed. D. H. Helman (Dordrecht: Kluwer Academic Publishers), 317–40.

Kuhn, T. S. 1979: Metaphor in science. In *Metaphor and Thought*, ed. A. Ortony (Cambridge: Cambridge University Press; 2nd edn, 1993), 403–19.

Rosche, E., and Lloyd, B. B. (eds) 1978: *Cognition and Categorization* (New York: Wiley).

42

Metaphysics, Role in Science

WILLIAM SEAGER

We must begin with the admission that the term "metaphysics" does not have a very precise or agreed upon meaning (no more does "science"). In current philosophy of science, "metaphysics" is, by and large, a pejorative term applied to whatever is regarded as illicitly nonempirical. Traditionally, metaphysics is regarded as the study of what lies behind the world of appearance – perhaps constitutes that world, but is itself the only *true* reality. Obviously, a great many people would regard science, or at least the more basic sciences such as physics, chemistry, and perhaps astronomy, as fitting this description. Consider, for instance, Eddington's famous description of the scientific as opposed to the "commonsense" table:

> [The scientific table] does not belong to the world . . . which spontaneously appears around me when I open my eyes. . . . My scientific table is mostly emptiness. Sparsely scattered in that emptiness are numerous electric charges rushing about with great speed; but their combined bulk amounts to less than a billionth of the bulk of the table itself. . . . I need not tell you that modern physics has by delicate tests and remorseless logic assured me that my second scientific table is the only one which is really there. (Eddington 1928, pp. xii, xiv)

Eddington's view of science is not only realist, but *confrontational*, and while few would want to stress so much the supposed opposition between the scientific and the manifest images of the world, many would embrace the claim that science reveals an underlying reality behind appearances. So science *is* metaphysics, according to a common notion of that discipline.

It will be objected that metaphysics is supposed to deal with the reality behind the physical or material world, while science must remain snared within the world of becoming, no matter how subtle and complex science might find that world to be. At most, this would show that science was not the whole of metaphysics, which no one would ever claim, though some might want to say that science is the whole of acceptable metaphysics. This last would itself appear to be a metaphysical hypothesis which goes beyond the bounds of science, thus corroborating the modest claim of the last sentence. Anyway, the idea that there are nonphysical components to reality is metaphysically contentious; it may be that modern science shows, in some sense, that the correct metaphysics is one of physicalism (see PHYSICALISM), and perhaps even one of despair, judging by Weinberg: "the more the universe seems comprehensible, the more it also seems pointless" (1988, p. 154).

Others might object that what is truly characteristic and deplorable about metaphysics is that it goes beyond what can be established *empirically*. Unless this is interpreted as quite a radical thesis, science once again turns out to be metaphysics.

Is our choice of theories strictly determined by the empirical evidence? Surely not, for theories go beyond present evidence and contain theoretical material (see UNDERDETERMINATION OF THEORY BY DATA). But theories are confirmed by empirical evidence. Is it clear that metaphysical theses are not? One strand of anti-metaphysics stems from the infamous verificationism of the logical positivists (see LOGICAL POSITIV-ISM), which asserts that the meaning of a claim is, in one way or another, a function of its method of verification, so that if a claim cannot be verified empirically, it is simply nonsense. But what counts as empirical verification? Some recent physical theories (the so-called superstring theories) appear to differ empirically from standard models only at energies attainable during the creation of a universe, energies unlikely to be reproduced in any laboratory. More down to earth, some theories of the genetics of color vision suggest that about 0.14 percent of males have reversed red and green color perception, something which, it is admitted, would be difficult or impossible to prove (I owe this example to Martine Nida-Rümelin). Such scientific, though unverifiable, theses are elements of theories which are *otherwise* verifiable. Bad meta-physics, then, must be absolutely immune to empirical consideration, either directly or indirectly. Very few such theses have ever been propounded, and very few, if any, metaphysicians are utterly careless of empirical data.

In order to see whether it is possible or advisable to distinguish rigorously between science and metaphysics, I want to examine their history from a viewpoint limited to the *attitudes* taken up towards them, and from there consider the transmutation of these attitudes in modern philosophy of science. Finally, I would like to make some remarks about the proper role of metaphysics in science.

A history of attitudes

Western metaphysics and science can be traced back to the birth of philosophy with the pre-Socratic philosophers of the sixth and fifth centuries BC. They could make no distinction between science and metaphysics, but the crucial change in attitude comes with the idea that the world is an independent entity obeying its own set of internal principles, and from which capricious supernatural intervention is banished. The older attitude persisted as well: in his histories, Herodotus unblushingly allows the gods to intervene at strategic moments (e.g., he reports that Poseidon thwarted the Persians by commanding the sea to flood a swamp at a crucial juncture). Against this traditional backdrop, the ideas of the pre-Socratics stand out vividly, no matter how "unscientific" they may appear to us, for they are fundamentally naturalistic; natural events are the result of natural processes (e.g., consider Heraclitus's view that the luminous heavenly bodies are bowls of fire, and that eclipses occur when the open side of the bowl turns away from us). Furthermore, there developed the idea that the irregular complexity of the observed natural world was mere surface phenomena obscuring a simple system of primary elements and principles. We recall the doctrine of the four elements (earth, air, water, and fire), which was further reduced to a single element by certain advanced thinkers already seduced by the idea of a Grand Unified Theory (take your pick: water for Thales, air for Anaximenes, etc.). This magnificent and crucial idea – that the world's fulsome complexity and variety rest on a few simple elements and principles – prepares the way for both science and metaphysics, which, in their turn, develop this idea in their distinctive (though I would say overlapping) ways. Another aspect of this idea worth mentioning is the Pythagorean insistence that what lies beneath the world of appearances is, somehow, a purely mathematical universe. The interpretation of this doctrine is unclear, but it at least introduced the idea that the reality beneath nature might be amenable to mathematical analysis.

There is no basis for differentiating a scientific, as opposed to a metaphysical, sense of "underlying reality" until Plato (*c*.427–347 BC), in whose work it appears quite clearly. Plato's doctrine of the Forms is what most would call a metaphysical thesis. The Forms are not objects of empirical study; they are not material objects at all, and can be known only by a direct intellectual "grasp" issuing from what Plato calls *dialectical* reasoning. But they also inform – literally – the sensible world, which is, somehow, constituted of imperfect copies of the ideal Forms. However, according to the *Timaeus*, this sensible world is itself constructed out of simpler substructures which are not Forms (nor are they exactly material), but pure geometrical objects – specifically, triangles (since these form the five regular solids). Plato tells us that the doctrines of the *Timaeus* are merely likely hypotheses, as befits their status as elements of the second-class material world, whereas the apprehension of the Forms provides a kind of certainty that transcends even pure mathematical knowledge (i.e., knowledge of mathematical *theorems*, as opposed to hypotheses about the fundamental components of the sensible world).

Here is another historically important feature of metaphysics: it is a study which aspires to absolute certainty based on pure reason (this aspect more than any other draws, perhaps deservedly, modern scorn). Science may well posit hidden principles and structures, but our knowledge of these is hypothetical and tentative, dependent upon empirical observation and experimentation. The demand for certainty might also lead one to suppose that metaphysics concerns a realm beyond the sensory if, for example, one's epistemological presuppositions entail that certainty about the sensible world is unattainable. In fact, this is one of Plato's grounds for positing the supra-sensible Forms: their perfect stability allows them to be the objects of knowledge, whereas the sensible realm supports, at best, opinion (or likelihood). All very well, but we cannot forget that Plato adduces much empirical evidence in support of the doctrine of the Forms, and that, correlatively, this doctrine is invoked to *explain* features of the empirical world (such as categorical grouping and similarity). Thus, although we have a *logical* distinction between metaphysics and science, in fact the two remain inextricably entangled, mutually supporting and shaping each other.

This is still more evident in Aristotle (*c*.384–322 BC), who characteristically provides a succinct definition of "metaphysics" as the study of being *qua* being. Aristotle's metaphysics is a curious mixture of what appears to us as at least quasi-scientific and as more traditionally metaphysical. For example, we have the doctrines that motion requires a mover (eventually overthrown by the scientific hypothesis of inertia, and surely a *rival* of a scientific proposition is a scientific proposition), the claim that a vacuum is impossible, and the claim that there is a rational creator of the universe characterized as the "unmoved mover," proved, in part, on the basis of Aristotle's quasi-scientific account of motion.

I think we receive from the ancients two guiding ideas: first, that there are underlying explanatory structures in the world which might or might not be "part" of the physical world, and second, that there is a special way of knowing this underlying structure which transcends empirical investigation without being utterly independent of it. These ideas cannot flower fully until there is a large body of truly scientific work which can stand in clear contrast to metaphysical doctrine. Such a body of work arrives with the scientific revolution; there we see an initial vibrant interaction between science and metaphysics, then a reactive revolt against metaphysics in favor of "empirical purity."

Descartes's (1596–1650) writings are full of appeal to metaphysical principles to establish scientific fact (see DESCARTES). One example is Descartes's "proof," given in his *Principles of Philosophy*, of the principle of inertia from a metaphysical assumption: the immutability of God. Descartes goes on to prove a great number of other principles of motion and impact from the same assumption. He also frequently notes how experience confirms the laws of inertia and motion. On the other hand, the metaphysical principles are to be established a priori, either by argument based on pure premises or directly, by the "light of nature." Of course, it is far from clear whether Descartes's procedure is entirely cogent. More important, despite Descartes's claims, it is not clear whether his scientific picture of the world really depends upon metaphysics. Certainly, the law of inertia was not really deduced from metaphysical considerations in anything like the straightforward way that Descartes would suggest. But the overarching framework in which all of Descartes's scientific work resides is that of the mechanical philosophy – nothing will be admitted as a causal agent within the world save matter and its motions. Descartes says, "The only principles which I accept, or require, in physics are those of geometry and pure mathematics; these principles explain all natural phenomena" (1644, 2, §64). This is a metaphysical article of faith which is not given the support we expect – that is, a priori demonstration (and, as Descartes must admit, it is, strictly speaking, false, for the human mind introduces nonnatural causes into nature). At best, it seems to stand on an induction from past success at explaining natural phenomena. Nonetheless, Descartes unashamedly appeals to it, as well as to many other metaphysical doctrines, in the development of his scientific picture of the world, and explicitly maintains that all knowledge rests on metaphysical knowledge (of the nature of God and the innate knowledge which he has implanted within us).

The zenith of mindful metaphysical influence on science is reached in the work of Leibniz (1646–1716), who is crystal clear about the nature of his reasoning (see LEIBNIZ). Leibniz appeals to a set of metaphysical principles from which he can deduce a vast range of empirical facts, some more, some less what we would regard as scientific. As examples, he appeals to the principle of continuity to refute certain of Descartes's laws of impact, to the principle of plenitude to refute physical atomism, and to the principle of perfection to solve problems of descending bodies. The striking change in Leibniz's attitude toward the metaphysical principles is that he regards them as hypothetical, explicitly subject to confirmation in experience (save, of course, for purely logical principles). This is because we cannot be certain that our finite minds have appreciated all the forces of reason which move the infinite mind of God; for the metaphysical principles are, at bottom, reflections of his perfect intelligence. Furthermore, the world being infinitely complex, we cannot be certain that our applications of the metaphysical principles, even granting that we have correctly apprehended them, will lead to the truth, for we cannot see the complete interconnection of all things which determines God's choice of the actual world from the infinite range of the merely possible. Thus we see that, for Leibniz, all metaphysical principles are grounded in pure reason, which explains why we can hope to have some access to them. But our finitude demands that we seek confirmation of both the metaphysical principles and their applications in the material world to which they apply (for a recent view of science reminiscent of Leibniz's "Platonic fallibilism," see Brown 1991).

The dizzying heights of metaphysical speculation which philosophers like Leibniz reached, and the apparent impossibility of turning speculation into fact, as evidenced by the ceaseless controversies of the metaphysicians save where empirical confirmation could rein in "pure reason," eventually led to a fervent backlash against all metaphysics. Hume (1711–76) is the champion skeptic (see HUME) with his ringing declaration:

> When we run over libraries, persuaded of these [i.e., Hume's empiricist] principles, what havoc must we make? If we take in our hand any volume; of divinity or school metaphysics, for instance; let us ask, Does it contain any abstract reasoning concerning quantity or number? No. Does it contain any experimental reasoning concerning matter of fact and existence? No. Commit it then to the flames: For it can contain nothing but sophistry and illusion. (1748, §12)

Hume's grounds for the rejection of metaphysics are strangely similar to Leibniz's for demanding empirical validation of the use of metaphysical principles: namely, the abject finitude of the human mind. To quote Hume again:

> Here indeed lies the justest and most plausible objection against a considerable part of metaphysics, that they are not properly a science; but arise . . . from the fruitless efforts of human vanity, which would penetrate into subjects utterly inaccessible to the understanding. (1748, §1)

It is true that the metaphysical principles transcend Hume's epistemology, for they are neither provable by logic alone nor directly observable. However, as has been stressed repeatedly, they are not utterly remote from empirical considerations. Thus it is not surprising to find that Hume also employs a great number of metaphysical principles to buttress his anti-metaphysical philosophy. One example: Hume demands that the relation of cause and effect hold only between events that are contiguous and successive (see, e.g., 1739, bk 1, pt 3, §2). Now, Hume knows of the moon's influence on the tides, apparently failing his test, yet he explicitly admits gravity as a kind of cause (see 1748, §4, pt 1). The issue of action at a distance was extremely contentious during the seventeenth and eighteenth centuries; but at that time it was a *metaphysical* controversy, not subject to empirical investigation (Newton denied the intelligibility of action at a distance – clearly a metaphysical scruple). Presumably, the requirements of succession and contiguity stem from Hume's empiricist theory of meaning, which demands that all concepts be grounded in experience *plus* the notion that the experience which grounds our concept of cause is that of observing successive, contiguous events. This just seems wrong as a story of the genesis of this concept; but, more

287

important, the demand reveals a further, if unspoken, appeal to metaphysics: a metaphysics of nature which Hume assumes, but which itself goes beyond empiricism.
I believe that Hume was aware of this tension in his philosophy, and would respond
by appeal to more naturalism – this time the natural attributes of the human mind,
which espouses certain principles, possesses certain passions, and obeys certain laws
simply as a matter of empirical fact. Thus Hume will not deny that we appeal to
metaphysical principles, but he will deny that these are epistemologically privileged –
they are merely parochial elements of the human mind (whose genesis must remain
a mystery). Hume's tactic is dangerous, however, for it leaves science itself in a precarious state, apparently subject to many of the same attacks leveled against the
metaphysicians (e.g., the problems of induction and causality). The bath of metaphysics in which the philosophers lolled is perhaps too indulgently warm and relaxing,
but we don't wish to throw out the baby of science with it.

If Hume wanted to dethrone the "Queen of the Sciences," Immanuel Kant (1724–
1804) wished merely to strip her of all power, while leaving her in possession of a
barren crown. In his *Critique of Pure Reason* (1781/7) Kant argues that any speculation beyond the bounds of empirical experience can never become knowledge. The
reason is that knowledge is dependent upon certain conditions which turn out to
be the limits of experience. These conditions are what the old rationalists called
metaphysical principles (e.g., every event must have a cause; Euclidean geometry
necessarily applies to physical space). No more are they metaphysical but rather
quasi-psychological principles which outline the structure of the mind insofar as
it *constructs* the world of experience. The validity of these principles is established by
an argument to show that the possibility of a coherent world of experience, whose
existence is taken as evident, requires their applicability (the method of transcendental
argument). Note how this goes beyond Hume. Science is secure, (a) because it is limited
to the world of experience (by definition) and (b) because the background principles
required for its epistemic security are built into that world. The price of a secure
science is the death of metaphysics, for once the intellect sets sail beyond the bounds of
experience to the "world within the world," it cannot safely trust principles which are
known to hold only in the world of experience. Abandoning such principles leaves
reason floundering, bereft of the premises needed to generate knowledge of the underlying world of, as Kant calls it, things-in-themselves. Failure to accept the limits of
reason leads to absurdity: the so-called antinomies of pure reason by which Kant
sought to show that reason freed from the bounds of experience could provide apparently sound proofs of contradictory metaphysical conclusions (e.g., that the world
both did and did not have a beginning in time).

Metaphysics in modern science

Though to modern eyes Kant's general framework and method appear paradigm
cases of the metaphysical, his and Hume's lessons were taken to heart by modern
philosophers of science. Metaphysics transcends the empirical and is, *therefore*, unknowable or unverifiable; metaphysics is mere empty speculation. A progression can be discerned from Hume to Kant. Roughly speaking, Hume claims only that metaphysical
speculation goes beyond that for which we could have evidence; Kant adds the claim

that such speculation transgresses another boundary, beyond which we have no right even to *apply* the principles normally used to generate knowledge. But for Kant, as for Hume, metaphysical speculation remains intelligible or meaningful (indeed, for Kant the mere intelligibility of certain esoteric doctrines is sufficient and crucial for his system). The final step in the degradation of metaphysics was taken by the logical positivists of the twentieth century, who declared that metaphysical statements were actually meaningless nonsense. A. J. Ayer in his famous *Language, Truth, and Logic* says:

> we may . . . define a metaphysical sentence as a sentence which purports to express a genuine proposition, but does, in fact, express neither a tautology nor an empirical hypothesis. And as tautologies and empirical hypotheses form the entire class of significant propositions, we are justified in concluding that all metaphysical assertions are nonsensical. (1946, p. 41)

This being a proof from a definition, it remains to show that there are any metaphysical statements as defined. This is somewhat doubtful. As noted above, very few of the metaphysical principles which philosophers have advanced or employed are completely divorced from empirical reality, and likewise, scientific theories are not entailed by empirical evidence. So either there are very few genuinely metaphysical statements in play, or else much of the theoretical side of science will be swept away along with the philosophical garbage. Even if there were some way to tread a verificationist line which puts, say, the principle of continuity on the metaphysical side and Einstein's principle of equivalence on the scientific side, the verificationist faces another problem: there can be no question but that all great scientists have appealed in one way or another to metaphysical principles, sometimes as central elements of their theories (e.g., Newton's postulation of absolute space; see NEWTON). The positivist adoration of science demands that such appeals be legitimated, hence the celebrated distinction between the "context of discovery" and the "context of justification." The former applies to the causal factors which lead someone to advance a scientific hypothesis (be they social, psychological, pathological, religious, or whatever), while the latter applies to the scientific verification of the hypothesis. The latter is the home of rationality, and within the former, irrational prejudices, hunches, or metaphysics can be safely contained.

The positivist theory of meaning has fallen (far) out of favor nowadays, but the distinction between the two contexts of scientific activity remains an important bulwark against metaphysics. Yet I believe that even a cursory examination of modern science will show that there is no easy way to separate metaphysical from "purely scientific" doctrine unless one is willing to swallow a radical (neo-empiricist) vision of science which itself embodies a metaphysical position. Furthermore, if we follow this route, the legitimate place of science within our culture is sadly diminished.

To begin, it is clear that metaphysical doctrines have a disconcerting habit of turning into scientific ones (and I suspect the reverse occurs as well). A well-known example concerns Newton's views of space. Newton believed in absolute space as a "container" for all material objects, against which *true motion* could be defined. He hypothesized that the Sun was at rest relative to absolute space, but admitted that, by the principle of inertial motion, there was no empirical test which would support his claim as against any of the infinite alternative velocities which one might assign to the

Sun (so in fact Newton's hypothesis met Ayer's extreme definition of a nonsensical metaphysical claim). Much later, Michelson and Morley attempted to measure the absolute velocity of the Earth (hence the Sun), but could find none of the expected effects. Their experiment was conceivable because, since the time of Newton, new theories had introduced novel empirical expectations, which Newton could not have anticipated. The results of Michelson and Morley were accommodated by Einstein in the special theory of relativity, which seems to refute the Newtonian metaphysical doctrine of absolute space (the status of space in the general theory of relativity is problematic, however (see EINSTEIN)). The application of non-Euclidean geometries in modern physics seems to refute the Kantian doctrine that certain geometrical propositions, such as that the shortest distance between two points is a straight line, are know a priori, and certain recent, if rather *outré*, physical theories posit that there are more than three spatial dimensions. The quantum theory is famous for having undercut the doctrine of causality that Kant believed to be an a priori truth, but it also appears to falsify Leibniz's principle of continuity.

An astonishing current example is the attack that quantum mechanics makes on a traditional view of substance and attribute (see QUANTUM MECHANICS). Normally, we think that the attributes a substance possesses are independent of our perception of them, and that it is the possession of certain attributes that *explains* why we have certain perceptions or obtain certain measurement results. In particular, if two substances, α and β, created together, are found always to have correlated properties, but are not in any kind of contact with each other, it would be natural to explain this in terms of their each having a fixed set of attributes which establish and maintain the correlation (i.e., the correlation is carried along with α and β via the possession of correlated attributes put into place at their creation). The theoretical work of John Bell (see Bell 1987 for this work and lots of quantum metaphysics) and a set of remarkable experiments have shown that this explanation is incorrect. The assumption that α and β possess definite, correlated attributes from their creation until their measurement is in conflict with quantum theory *and* observation (see d'Espagnet 1979).

One could certainly debate the truth or proper interpretation of all these claims, but it is clear that science has encroached on what was thought to be metaphysical territory. This is not strange if we take the view that the task of metaphysics is to produce a maximally comprehensive image of reality. Metaphysical principles have been put forth with one eye always on the empirical world which they are thought to underlie, and these principles have of course stemmed from human minds steeped in a certain view of the empirical world. Over the centuries, science has emerged out of "pure" speculation, and it seamlessly meshes with metaphysics in the task of presenting a coherent and comprehensive picture of the whole world. The development of science provides new data for the metaphysicians and new outlooks on the world which can lead to the revision of metaphysical principles. This is, roughly, the picture of Leibniz: a realist appreciation of science seen as the product of both metaphysical principle and empirical research.

I fear that the skeptical alternative ends up denying more than metaphysics. By relegating metaphysics to the "context of discovery," it prevents us from understanding science as helping to form a comprehensive view of the whole world. Of course, the skeptics have a point to make about how precarious our attempts to grapple with reality must be; but this point goes right through metaphysics into the heart of science. There are at least two ways that this skepticism plays itself out. One is instrumentalism (see REALISM AND INSTRUMENTALISM), which demands and permits of science no more than the attempt to produce empirical adequacy, because anything more ambitious, such as regarding a theory as an attempt to spell out the deep truth behind appearances, has to fall prey to the range of criticisms of metaphysics that we have glimpsed (such a view of science is presented in van Fraassen 1980). The second is to embrace what might be called the "metaphysics of surface." Here the tactic is to deflate scientific pretension by showing its content to be the contingent outcome of noncognitive forces. This is the programme of the sociology of science (variously represented, e.g., by Kuhn (1962), Bloor (1976), or Latour and Woolgar (1986)). Latour and Woolgar, for example, compare the development of science to the evolution of life, and state that: "if life itself arises from tinkering and chance, it is surely not necessary to imagine that we need more complex principles to account for science' (1986, p. 251). Unspoken is the further analogical inference that just as the pattern of life we find on earth today could have been entirely different, even given the same environmental constraints (see Gould 1989), so science could have been radically different, even given the same evidential constraints. In this case reality is not just a construction, but an accidental construction, whose driving forces are "surface phenomena" such as the ambition of scientists, the cultural *Zeitgeist*, capitalist industrialization, or some other "social force." As Latour and Woolgar see it, Plato's cave requires inversion: "reality . . . is the shadow of scientific practices" (1986, p. 186). Such a view sees the reality of social forces as the explanatory bottom (see SOCIAL FACTORS IN SCIENCE).

Fascinating as these sociological theories are, I doubt that the important issue here is the explanation of why a certain theory comes to be constructed at a certain time, but rather, what kind of world the theory is trying to tell us about. Figuring that out is difficult, bold, and speculative work always richly informed by substantial metaphysical as well as empirical labor. Give up that work, and science is mere technique.

References and further reading

Ayer, A. J. 1946: *Language, Truth, and Logic*, 2nd edn (London: V. Gollancz).

Bell, J. 1987: *Speakable and Unspeakable in Quantum Mechanics* (Cambridge: Cambridge University Press).

Bloor, D. 1976: *Knowledge and Social Imagery* (London: Routledge and Kegan Paul).

Brown, J. R. 1991: *The Laboratory of the Mind* (London: Routledge and Chapman Hall).

Burtt, E. A. 1964: *The Metaphysical Foundations of Modern Physical Science* (London: Routledge and Kegan Paul).

Descartes, R. 1644: *The Principles of Philosophy* (Amsterdam).

d'Espagnet, B. 1979: The quantum theory and reality. *Scientific American*, 241, 158–80.

Eddington, A. S. 1928: *The Nature of the Physical World* (Cambridge: Cambridge University Press).

Fine, A. 1986: *The Shaky Game: Einstein, Realism, and the Quantum Theory* (Chicago: University of Chicago Press).

Gould, S. J. 1989: *Wonderful Life: The Burgess Shale and Nature of History* (New York: W. W. Norton).

Hacking, I. 1983: *Representing and Intervening* (Cambridge: Cambridge University Press).

—— 1999: *The Social Construction of What?* (Cambridge, MA: Harvard University Press).

Hume, D. 1739: *A Treatise of Human Nature* (London).

—— 1748: *An Enquiry Concerning Human Understanding* (London).

Kant, I. 1781/7: *Critique of Pure Reason*.

Kuhn, T. S. 1962: *The Structure of Scientific Revolutions* (Chicago: University of Chicago Press).

Latour, B., and Woolgar, S. 1986: *Laboratory Life*, 2nd edn (Princeton: Princeton University Press).

Lindberg, D. C. 1992: *The Beginnings of Western Science* (Chicago: University of Chicago Press).

Redhead, M. 1995: *From Physics to Metaphysics (Tanner Lectures)* (Cambridge: Cambridge University Press).

Sellars, W. 1963: *Science, Perception and Reality* (London: Routledge and Kegan Paul).

van Fraassen, B. 1980: *The Scientific Image* (Oxford: Oxford University Press).

Weinberg, S. 1988: *The First Three Minutes: A Modern View of the Origin of the Universe* (New York: Basic Books).

43

Mill

GEOFFREY SCARRE

Son of the utilitarian James Mill and himself a major expositor of a utilitarian theory of ethics and politics, John Stuart Mill (1806–73) was also perhaps the greatest British empiricist philosopher of the nineteenth century. His massive work *A System of Logic* (1843 and several subsequent editions) was intended as a textbook of the doctrine which derives all knowledge from experience, including even our knowledge of the laws of mathematics and logic. Mill conceived his work as a sustained and careful polemic against the "German, or *a priori* view of human knowledge" (1981, p. 233), and defended a view of the scientific enterprise as a systematic inductive process of interrogating nature, unencumbered by "innate ideas," Kantian categories, or elaborate and speculative hypotheses.

Mill understood natural science to be concerned with the following tasks: (1) the explanation and classification of phenomena, distinguished by their observable properties; (2) the production of inductive generalizations descriptive of the causal principles of observable phenomena; (3) the arrangement of these causal principles into hierarchically structured systems of higher-level and lower-level laws; (4) the reduction of the more surprising or *recherché* features of nature to more familiar ones; (5) the attainment of theoretical closure in areas of research where careful application of inductive methods leaves nothing further to explain. Mill paid very little attention to the use of quantitative methods in science, and none at all to the construction of mathematical models in the development of theory. (Statistical reasoning receives a short chapter in the *Logic*; but statistical judgments are held to be "of little use . . . except as a stage on the road to something better" – namely, universal generalizations (1973, p. 592).)

The final book of *A System of Logic* surveys the "moral" or human sciences, though Mill's treatment of these lies beyond the scope of the present chapter. Mill sought to remove the study of man from "the uncertainties of vague and popular discussion" (1973, p. 833) and place it on a proper scientific footing. While social behavior could never be the subject of an exact science, Mill was hopeful that with the application of appropriate empirical methods interesting explanatory generalizations could be reached. Foundational to this program was "ethology," or the science of character, a relatively exact discipline which would establish "the kind of character produced in conformity to" the general laws of mind "by any set of circumstances, physical and moral" (1973, p. 869). Mill foresaw the social sciences developing out of ethology according to a methodological individualist principle that the workings of societies and their institutions are determined by the laws of individual human behavior, men in community having "no properties but those which are derived from, and may be resolved into, the laws of the nature of individual man" (1973, p. 879). Mill's interest in refining the methods of social-scientific inquiry was not, of course, wholly academic. As a utilitarian and radical reformer, he looked to the scientific study of how societies work to offer important clues as to how they could be made to work better, with enlargement of the prospects for human happiness.

Induction

Mill believed that induction was the only form of "real" inference, capable of leading to genuinely new knowledge. Since deductive processes enable us to do no more than "interpret" inductions, identifying the particular cases which fall under general propositions, it is induction alone "in which the investigation of nature essentially consists" (1973, p. 283). He did not clearly distinguish between induction as a mode of discovery and induction as a method of proof, proposing that "Induction may be defined, the operation of discovering and proving general propositions" (p. 284). Further, as the same evidence which entitles us to draw conclusions about classes of cases also enables us to draw a similar conclusion about a single unknown instance (which is frequently what we want to know about in everyday life), a common set of inductive rules serves us outside as well as inside science. On the Millian picture, science differs from everyday knowledge not in its methodology, but in its special subject matter – its directedness to the uncovering and proof of laws of nature.

Mill complained that the detailed study of inductive methods had been hitherto neglected; some of the "generalities of the subject" had been discussed, but previous analyses of the "inductive operations" had "not been specific enough to be made the foundation of practical rules, which might be for induction itself what the rules of the syllogism are for the interpretation of inductions" (1973, p. 283). Mill's interest was in identifying sound methods of inductive inquiry – a search which culminated in the famous canons of induction, which we shall examine shortly. Not all inductive extrapolations from examined instances are justified; and the difficult question is how we can establish which ones are legitimate, particularly in areas of research of which we have had little or no experience. The "fundamental axiom" of induction, Mill held, is the principle of the uniformity of nature, which depends on the law of causation that "there are such things in nature as parallel cases" and "that what happens once, will, under a sufficient degree of similarity of circumstances, happen again" (p. 306). That nature possesses a basic uniformity is attested by experience; it is a "universal fact," the law of causation standing "at the head of all observed uniformities, in point of universality, and therefore . . . in point of certainty" (p. 310). Yet not all apparently sound inductive reasonings, Mill realized, lead us to true conclusions. Not all swans are white, despite what Europeans once believed; nor can we safely infer that the succession of rain and fine weather will be the same every year, or that we will have the same dreams every night. Nature "is not only uniform, it is also infinitely various" (p. 311), and the task of a scientific theory of induction is to answer the question: "Why is a single instance, in some cases, sufficient for a complete induction, while in others, myriads of concurring instances, without a single exception known or presumed, go such a very little way towards establishing an universal proposition?" (p. 314).

The name "empirical laws" can be given, Mill wrote, to uniformities attested by observation or experiment, but which cannot be wholly relied on "in cases varying much from those which have been actually observed, for want of seeing any reason *why* such a law should exist" (1973, p. 516). "Ultimate laws" (such as Newton's laws of motion), by contrast, hold always and everywhere: thus we have "the warrant of a rigid induction for considering it probable, in a degree indistinguishable from certainty, that the known conditions for the sun's rising will exist to-morrow" (1973, p. 551). It is interesting that Mill never showed any awareness of the skeptical problem of induction raised by David Hume. Probably, like most philosophers before T. H. Green's revival of Hume studies in the 1870s, he failed to grasp the force of Hume's central skeptical contention, taking his problem to concern simply the *psychological explanation* of our belief in uniformity. Here was a question that Mill believed he could answer: the belief in uniformity is a higher-order induction from particular laws of causation, "not less certain, but on the contrary, more so, than any of those from which it was drawn" (1973, p. 570).

The eliminative methods

Mill firmly dismissed the idea that causal relations involve necessity and insisted that experience supports only a constant conjunction analysis of causation. The cause of a phenomenon is the sum total of contingent conditions "which being realized, the consequent invariably follows" (1973, p. 332). The main aim of science, in Mill's opinion, is to trace causal relationships, and a major role of inductive logic is to help it to do so. Despite his claim that effects normally depend not on a single factor but on a complex of factors acting together, Mill's famous methods of experimental inquiry are designed specifically to locate, by means of eliminative reasoning, a condition preceding or accompanying a phenomenon "with which it is really connected by an invariable law" (1973, p. 388). It has been fairly objected to this conception of science that the most interesting research is concerned much more with the discovery of novel entities and processes than with the identification of causes. But if causal explanation is not the whole of science, as Mill supposed, it is still a legitimate part of it; and the eliminative methods have also a useful role to play in everyday causal inquiry.

The most important of the experimental methods are those of Agreement and of Difference:

> *Method of Agreement*: If two or more instances of the phenomenon under investigation have only one circumstance in common, the circumstance in which alone all the instances agree, is the cause (or effect) of the given phenomenon. (1973, p. 390)

> *Method of Difference*: If an instance in which the phenomenon under investigation occurs, and an instance in which it does not occur, have every circumstance in common save one, that one occurring only in the former; the circumstance in which alone the two instances differ, is the effect, or the cause, or an indispensable part of the cause, of the phenomenon. (1973, p. 391)

The thought behind the Method of Agreement (MA) is that no feature *not* common to the instances in which the phenomenon occurs can be its cause, since the phenomenon is capable of occurring in its absence; so if there is a sole feature common to the different cases, this is the only remaining candidate to play the causal role. But this is problematic for two reasons: there is frequently great difficulty in obtaining different instances of a phenomenon coinciding in only one aspect, and there are often – as Mill himself reluctantly conceded – different causal routes to the same effect (as a man can be killed by shooting, stabbing, or poisoning). Strictly, MA establishes only that a condition not invariably present among the antecedents of a given phenomenon cannot be *necessary* for its occurrence. The Method of Difference (MD) corresponds to a familiar intuitive pattern of causal reasoning, but the difficulty of determining with certainty that all relevant differences between the instances in which a phenomenon occurs and those in which it does not have been taken into account leaves it unable to fulfill Mill's purpose for it of conclusively demonstrating nomological causal relationships. At most, MD can prove that a particular factor is not a *sufficient* condition of some phenomenon, where the factor occurs and the phenomenon does not. Nevertheless, as J. L. Mackie (1974) has pointed out, both MA and MD are suggestive and useful modes of causal investigation where we already have a good idea of the range of possible causes of the phenomenon at issue, though this implies that the methods will only be of much service in relatively well understood areas of inquiry, and will do little to advance more path-finding research.

What Mill calls the "Joint Method of Agreement and Difference" identifies as the cause of a phenomenon the only factor always present when the phenomenon occurs and always absent when it fails to occur. This is a particularly hard method to employ, involving the need to secure one pair of cases with a single similarity and another pair with a single difference; it will also locate a cause only in the uncommon cases where there is a *unique* cause to be found (where, that is, plurality of causes does not apply).

The remaining methods of inductive inquiry are those of Residues and of Concomitant Variation:

> *Method of Residues*: Subduct from any phenomenon such part as is known by previous inductions to be the effect of certain antecedents, and the residue of the phenomenon is the effect of the remaining antecedents. (1973, p. 398)

> *Method of Concomitant Variation*: Whatever phenomenon varies in any manner whenever another phenomenon varies in some particular manner, is either a cause or an effect of that phenomenon, or is connected with it through some fact of causation. (1973, p. 401)

Like the preceding methods, the Method of Residues (MR) can be useful in signaling causal possibilities, but as a mode of proof it fails because it falsely assumes that separate parts of a compound phenomenon always have separate causes, Finally, the Method of Concomitant Variation (MCV) properly, if vaguely, draws attention to the probability of some causal linkage between phenomena which vary in tandem.

Hypotheses

Mill's attitude to the use of hypotheses in science was ambivalent. On the one hand, he admitted the value of plausible conjectures suggestive of fruitful observations and experiments, and granted that "nearly everything which is now theory was once hypothesis" (1973, p. 496). On the other hand, his abhorrence of anything which smacked remotely of a priorism made him unwilling to countenance the admission to a process of reasoning of any proposition which could not be rigorously confirmed by observation or the inductive methods. In Mill's eyes, a hypothesis is always guilty until proved innocent – a vindication to be accomplished, where possible, by the Method of Difference. Mill conceded the legitimacy of two major species of hypothesis: those which posited a novel law of operation for a known cause and those which posited a novel cause operating according to a familiar law. Given the difficulty of calculating the effects of speculative causes (e.g., of Cartesian vortices or the luminiferous ether), Mill thought the former type of hypothesis more tolerable than the latter.

Mill's suspicious view of all but the least adventurous hypotheses contrasts strikingly with the enthusiasm for hypothetical method of the neo-Kantian William Whewell (see WHEWELL). In the latter's view, reality as we know it is in some part a construction of the human mind, and hypotheses are instruments for imposing form and order on the shapeless data of scientific inquiry. Mill strongly opposed Whewell's contention that no realm of facts exists independently of human mental activity, and maintained the reality of a wholly objective and epistemically accessible external world. Typifying their differences is their disagreement over the correct description of one of the milestones of Western science, Kepler's theory of the elliptical orbits of the planets. Whereas Mill asserted that Kepler, by a painstaking sequence of observations, had *discovered* the ellipticality of the orbits in the data, Whewell insisted that Kepler had *imposed* the idea of an ellipse on essentially formless data, in a brilliant example of creative hypothesizing. Irenically minded modern readers may perhaps consider that *both* Mill *and* Whewell had grasped something important about the scientific enterprise: Whewell the need for acts of constructive imagination in science, and Mill the requirement that scientists should represent the world as it really is.

Mill's philosophy of science can be criticized for its theoretical timidity and its limiting reduction of scientific methodology to a small number of rules for the determination of causes. Yet it is notable too for its single-minded devotion to a thoroughgoing empiricism, its subtle analysis of the notions of cause and of law, its attempt to probe the murky subject of the conditions of reliable inductive inferences, and its defense of the idea of science as a progressive program of ever more general and unified explanations of phenomena.

References and further reading

Works by Mill

1973: *A System of Logic: Ratiocinative and Inductive* (1st pub. 1843), ed. J. M. Robson, in *Collected Works*, vols 7 and 8 (Toronto: University of Toronto Press).

1981: *Autobiography*, ed. J. M. Robson and J. Stillinger, in *Collected Works*, vol. 1 (Toronto: University of Toronto Press).

Works by other authors

Buchdahl, G. 1971: Inductivist *versus* deductivist approaches in the philosophy of science as illustrated by some controversies between Whewell and Mill. Monist, 54, 343–67.

Mackie, J. L. 1974: Mill's methods of induction. In *The Cement of the Universe* (Oxford: Oxford University Press).

Ryan, A. 1974: *J. S. Mill* (London: Routledge and Kegan Paul).

Scarre, G. 1989: *Logic and Reality in the Philosophy of John Stuart Mill* (Dordrecht: Kluwer Academic Publishers).

Skorupski, J. 1989: *John Stuart Mill* (London and New York: Routledge).

Whewell, W. 1968: *William Whewell's Theory of Scientific Method*, ed. Robert E. Butts (Pittsburgh: University of Pittsburgh Press).

44

Models and Analogies

MARY HESSE

Models in classical physics

Questions about the structure and justification of theories, the interpretation of data, and the problem of realism have been in the forefront of debate in recent philosophy of science, and the topic of models and analogies is increasingly recognized as integral to this debate. Models of physical matter and motion – for example, models of atoms and planetary systems – were already familiar in Greek science, but serious analysis of "model" as a concept entered philosophy of science only in the nineteenth century. This was largely the result of proliferation in classical physics of theoretical entities such as "atom," "electro-magnetic wave," and "electron," for which there appeared to be no directly observable evidence (see THEORETICAL TERMS).

The senses of "model" discussed in classical physics were of two types, which may be distinguished as "material" and "formal" (Hesse 1966). A *material model* is, or describes, a physical entity – familiar examples are billiard balls, a fluid medium, a spring, or an attracting or repelling electric particle. A *formal model* is the expression of the form or structure of physical entities and processes, without any semantic content referring to specific objects or properties. For example, a "wave equation" in mathematical symbols may express the laws of a simple pendulum, of sound or light waves, of quantum wave functions, etc., while remaining neutral to any specific application. Another example is the formal structure of a computer program (the software), which may be realized in a number of different hardware setups, and has provided useful formal models of brain structure in Artificial Intelligence. Formal models are syntactic structures; material models are semantic, in that they introduce reference to real or imaginary entities.

"Analogy" will be taken here to refer to some relation of similarity and/or difference between a model and the world, or (less question begging) between a model and some theoretical description of the world, or between one model and another. Models are relata of analogy relations; that is, a model is an analogue. Analogy relations themselves may be formal or material: they may be merely analogies of structure, such as that between a light wave and a simple pendulum, or they may introduce material similarities, as when gas particles are held to be like billiard balls in all mechanical properties relevant to Newton's laws.

Analogy relations, like similarity, come in degrees and in different respects, and are therefore not generally transitive. This makes rigorous treatment difficult, but it is useful as a start to distinguish three types of material analogy relation: positive, negative, and neutral. A *positive (material) analogy* picks out those features of the analogues that are identical or strongly similar; a *negative analogy* picks out those known to be different or strongly dissimilar; a *neutral analogy* picks out those for which there is no evidence yet as to similarity or dissimilarity. For example, DNA models built of painted balls and metal struts are positively analogous to DNA molecules in spatial structure and connectedness, but negatively analogous in size, material, shape, and color of the constituents, etc. These models have a neutral analogy with molecules insofar as their further detailed properties are used to explore as yet unknown features of genetic material. The dividing line between these three sorts of analogy will of course shift as research goes forward – the better the model, the more of the neutral analogy will eventually be accepted as positive, whereas a poor model will become more and more negatively analogous.

So much for the somewhat rough definitions that sufficed for the description of models in classical physics and chemistry. Models served there to introduce unobservable entities and processes into physical theory by analogy with familiar observable entities and processes, thus providing pictures of the explanatory entities held to underlie phenomena. The problem of justifying these explanatory models led to a polarization of epistemological views. Realists held that successful models are positive analogues of the real world; positivists denied the reality of the theoretical entities referred to, and regarded models merely as working pictures to be dispensed with in accepted theories, having at best a formal analogy with the world (see REALISM AND INSTRUMENTALISM).

The philosophical debate about models was initiated by Norman Campbell (1920, ch. 6) in the course of a critique of the so-called hypothetico-deductive (HD) theory of theories (see THEORIES). According to this view, of which Maxwell's electromagnetic theory is a paradigm case, an explanatory theory in physics consists of a set of mathematical equations, some but not all the terms of which are interpreted by means of directly observable or measurable properties, such as shape, position, momentum, time interval, weight, texture, color, light intensity, temperature, etc. These interpretations were called "bridge principles" or (with Campbell) the "dictionary." A theory is confirmed if, with the bridge principles, laws and predictions can be deduced and shown to give a good fit with the experimental *explananda*; if the fit is poor, the theory is disconfirmed or refuted. In the positivist version of HD, models are used only as aids to discovery, and are not a logically essential part of the theory.

Campbell argued, on the contrary, that models, as interpretations of unobservable terms, are essential elements of theory, because a merely mathematical formalism gives no meaningful information other than that contained in the experimental laws and properties themselves. Taking the "billiard ball" model of gases as his main example, he showed how the experimental laws are "explained" (unified and made intelligible) by this model and, most importantly, how theoretical inference proceeds by modification and extension of the model to give new predictions. The logic of such inference is analogical argument from the properties of the model's familiar source (observable mechanical particles) to the *explanandum* (gases). For example, the original point model of particles that explained Boyle's and Charles's laws is extended to particles of finite size, thus predicting the corrections to Boyle's law which are necessary to obtain greater experimental range and accuracy for real gases. Thus models are shown to be essential to argument in physics, not merely dispensable heuristic devices.

Two kinds of issue arise from this analysis, one epistemological and the other ontological. Campbell has an implicit epistemological thesis in his argument for the essentiality of models: namely, that they *justify* reliance on predictions from models in virtue of the known positive analogy between model and *explanandum*. That there is such reliance was pointed out by Putnam (1963, p. 779) in a striking example from the construction of the first atomic bomb. Although laboratory-sized tests of the nuclear reactions involved had been performed successfully, large-scale tests had not, and their failure would have been catastrophic. Such tests would not have been carried out unless there had been *some* intuitive confidence that analogical extrapolations from evidence and theory justified expectation of success (Hesse 1974, ch. 9).

Underlying all such intuitions there is a metaphysics of the "analogy of nature," and this brings with it ontological questions about the status of models. If the kinetic model, for example, has no relation to real analogies in nature beyond those already observed, there is no basis for prediction to its analogical extrapolations. Does this imply, however, that there *are* molecules as described in the theory? Campbell's reply to this question was subtle. The model of molecules is not identical with the substructure of gases, only *materially analogous* to it. Models are entities which share the properties of mechanical particles insofar as these are required to explain already known phenomena (the positive analogy), and to predict phenomena yet to be examined (the neutral analogy). But analogies always have negative elements, and realistic identification of models with nature is therefore unjustified. Campbell, then, was anti-realist about theoretical entities and some of their first-order properties, but realist about their positive analogy relations.

The semantic conception of theories

Campbell's view anticipates more recent emphasis on the tentative and dynamical character of theory making, in contrast with the HD account, in which theories tend to be seen ahistorically, as static formal systems. Subsequently, however, there has been a greater concern with static ontology than with dynamic epistemology, and the analysis of models has become part of the general philosophical debate about realism. The syntactical HD account has been transformed into the so-called semantic conception of theories (SCT), in which emphasis shifts from formal theory structure to the set of semantic or metamathematical models for the theory (Suppe 1989, pp. 86ff). Each model of this set is an interpretation of the formal system that makes the axioms of the system true. The models may be real entities or, more often, imaginary idealizations of real entities, such as frictionless planes, point particles, or workshop mock-ups of the next mark of stretched limosine; or they may be mathematical entities such as geometrical spaces as models of some geometric axiom set. The semantic content of a theory is then said to be the whole class of its models – that is, all possible interpretations. If the theory is empirically acceptable, the real world will be (probably only approximately) among these models. This "family of models" is a highly abstract conception, carrying no information other than the structure of its parent formal system. Even if the models are conceived in some sense as real entities, the properties they have over and above their formal structure are irrelevant to the theory; as "models of

301

the theory," they are logically equivalent, and therefore do not compete with each other for "reality" or "truth."

The semantic conception makes a welcome move from talk about linguistic formulations to talk about things and processes, and thus comes nearer to talk of models as this actually occurs in science. But SCT adds little of philosophical interest to the topic of models itself, and nothing to the epistemological issues of the previous section (van Fraassen 1989, p. 216). Emphasis is still on the properties of a theory as frozen in a particular structural formulation. It is significant how many accounts of SCT refer to theories as expressed in "textbooks" (Cartwright 1983, p. 46; Giere 1988, p. 78). Like HD, SCT has nothing to say about theory change, or about general theory frameworks or "paradigms," because these are rarely formalizable in deductive axiom systems, and therefore do not define a set of semantic models (Suppe 1989, p. 269). The old problem of the "meaning of theoretical terms" gets pushed into the philosophy of language rather than philosophy of science. How models are thought up, and how their descriptive terminology is understood, becomes no different in this view from the introduction of any new terms into language, whether in new dialects, novels, science fiction, or literature in general (van Fraassen 1980, 221). But such distinctions between the philosophy of models and the philosophy of language are unjustified. Connections have already been found, for example, between the use of models as scientific metaphors and the linguistic analysis of metaphor in general (Black 1962, chs 3, 13; Hesse 1966, pp. 157ff) (see METAPHOR IN SCIENCE). Such comparisons have important implications for the philosophy of both science and language, and it makes no sense to exclude discussion of the development of scientific language from analysis of the structure of science.

An even greater weakness of the semantic conception lies in its tacit acceptance of the distinction made in classic HD between theoretical and observation terms. It is now generally accepted that this is a grave oversimplification (see OBSERVATION AND THEORY). As long ago as 1960, Suppes pointed out that the subject matter of science is not raw observation, but *models of data*. In the case of mathematical science, these come as sets of measurable quantities representing observable properties derived from idealizations of the real world, and not from raw experience. For example, theories of mechanics are related to experience by means of a set of variables interpreted as particles, time intervals, and space, mass, and force functions. These represent idealized mechanical entities and their measurable properties. Suppes himself did not go on to discuss *un*observable terms, but subsequently the much more general thesis of theory-ladenness of observation has blurred sharp distinctions between "observable" and "unobservable," and made his analysis relevant to theoretical models also. The question of what the particular sensory equipment of *Homo sapiens* can or cannot directly observe has lost most of its interest in relation to the nature and structure of theory. Scientific knowledge can now be conceived as a hierarchy of models, some of which are more particular and lie closer to the data, some of which are theoretical and more distantly related to the world.

What, though, *is* this theory–world relation? Answers to the question within SCT depend on how far it is construed as a realist or anti-realist theory of science. The generally received view is realist, at least in the sense that the real world is supposed to be (approximately) among the models of a good theory, and attempts have been made to specify criteria of "goodness" which will reduce the indefinitely large set of possible models to a manageable few. These criteria, by definition, have to be nonempirical, because it is assumed that the family of models which constitute a successful theory are all consistent with the data so far (or, rather, with models of these data). Different data define different theories. Nonempirical criteria that have been suggested include unification of phenomena, formal simplicity, and economy and non-*ad hoc*-ness of theory, but there has so far been little success in showing that these criteria are relevant to *truth*, or in showing that sequences of theories in a particular domain tend to converge upon a unique "best explanation" (see EXPLANATION, and INFERENCE TO THE BEST EXPLANATION).

Ronald Giere has suggested a more flexible realist version of SCT, which he calls "constructive realism." Here it is explicitly recognized that there is some looseness of fit between theoretical models, models of data, and the real world. Even in the HD conception, numerical approximation and statistical likelihood already disrupt the purely deductive character of theories. More generally, Giere identifies *similarity* as the primary relation between all types of models and the real world (Giere 1988, p. 81). This is logically an intransitive relation, and cannot yield "truth" or "correspondence." Giere declines to discuss it in logical terms at all, but regards the recognition of sufficient similarity in relevant respects as a wholly natural cognitive process, depending both on human biological capacities and on socially accepted conventions and paradigms (Giere 1988, pp. 94ff). In this type of realism there is no guarantee of convergence of finality in the process of theory making; it is an ontological analysis of what a theory *is*, not of how it is developed or justified.

Giere's constructive realism brings SCT closer to real science, and also to the type of anti-realism or "constructive empiricism" adopted by van Fraassen (1980). The difference between these two views seems to relate chiefly to the nature of the theory–observation distinction. Where Giere sees a seamless hierarchy of models of theory and data, van Fraassen makes a distinction (which cannot be more than pragmatic) between the empirical adequacy of a theory and the nonrealistic models whose relation to experience is mediated through the deductive apparatus of the theory and its bridging principles. Thus the relation of theory to world remains one of satisfaction of propositions, that is of "truth" or "correspondence," but at the empirical level only. Theory models are not held to carry truth-values in relation to the world in any interesting sense. Both Giere and van Fraassen, however, continue to neglect problems of theory change and model choice, preferring to refer these either to cognitive neurophysiology or to the general philosophy of perception and language. In other words, the ghosts of the formal, static, HD and SCT approaches still linger.

The analogical conception of theories

In order to address issues of meaning and justification, we need to abandon two dogmas still lurking in SCT. The first is the undue concentration on ontology and realism at the cost of banishing linguistic and epistemological questions from philosophy of science. The second is the emphasis on static, "textbook" formulations of theory, to the neglect of the ongoing process of theory making and the consequent problems of theory choice and theory change. Recent discussions have made a sharp break with both these dogmas, chiefly as a result of detailed studies of the historical and contemporary dynamics of theory and experiment.

The new approaches emphasize empirical study of science itself, rather than "logical reconstructions" of it. The theory–world relation is explicitly described in terms of physiological and cognitive science, instead of being regarded as a deep and intractable philosophical problem, and the attempts to find rigorous logical relations throughout scientific theory are replaced by various degrees of approximation, looseness of fit, similarities, and analogies. The new approaches have to face the objection that all this necessarily results in fuzzy thinking. They have to show that, although reality and science *are* irreducibly fuzzy, nevertheless philosophical talk about them can be conducted in rigorous, precise, and intelligible terms, but without falling into unrealistic and inapplicable logic. They have begun to do this by reintroducing similarity, analogy, and related concepts into serious philosophical discussion, thus releasing model talk from the metamathematical straitjacket in which SCT has encased it (e.g., Gooding 1990; Harré 1986, ch. 11).

These points emerge explicitly in the analysis put forward by Nancy Cartwright in what she calls the "simulacrum theory of explanation," described as follows: "To explain a phenomenon is to find a model that fits it into the basic framework of the theory and that allows us to derive *analogues* for the messy and complicated phenomenological laws which are true of it" (Cartwright 1983, p. 152, italics added). Here models cease to be abstract metamathematical entities, and are seen in a more historical light as what groups of scientists adopt as manageable paradigms. The indefinitely large "families of models" are thus reduced to very few working models with their empirical bridge principles. Cartwright argues that these have no claim to reality status or truth – they are fictions, used piecemeal, exploited, and superseded to suit convenience. So far, her conception is similar to that of van Fraassen, and, like him, she maintains a split between lower and higher levels of theorizing. But she differs from him in allowing that *un*observable causes and entities do exist, and that causal laws have truth-values, at least locally (Cartwright 1983, pp. 160f).

Cartwright's argument is bolstered by a wealth of detailed examples from physics. It remains unclear, however, just how the concept of "real cause" is distinguished from fictitious theory models. Cartwright does not seem to hold a strong *modal* concept of "natural kinds" or of laws (1983, p. 95); so it is not easy to see what the notion of "true causal relations" contributes that cannot equally be said (in local contexts) in terms of empirically adequate laws and models of the data (see NATURAL KINDS). It seems preferable to tell the same story all along the theory–observation spectrum. To talk in terms of propositions for a moment, models can then be regarded as satisfying theoretical propositions with truth-value throughout; but at higher levels of theory these are almost certainly *false*, whereas nearer the phenomenal level they are likely to be approximately and locally *true*, because they are subject to multiple sources of evidence and test. We then have a conception of theory as essentially an embodiment of analogies, both formal and material, which describe regularities among the data of a given domain (models of data and phenomenal laws), with analogies between these and models of data in other domains, and so on in a hierarchy of levels of a unifying theoretical system. The "meaning of theoretical terms" is given by analogies with familiar natural processes (e.g., mechanical systems), or by hypothetical models (e.g., Bohr's planetary atom). In either case, descriptive terms of the analogues are derived metaphorically from ordinary language.

A useful model, then, represents the real world, not by correspondence or isomorphism, but by analogy, and this may be strong or weak, depending on how much evidence there is from different analogous domains. The justification of predictions from models to new domains becomes a question of the strength of analogical argument within the whole theory–data network. That strong analogues justify prediction in turn depends on a metaphysical and inductive assumption of the analogy of nature; that is, past similarities, differences, and regularities are taken to indicate real and persisting structural regularities. This assumption is weaker than that of "natural kinds" related by universal and causally necessary laws, but in order to operate counterfactual predictions, it does of course have a modal component (see LAWS OF NATURE). This may be expressed as: "If a number of objects were found to be more similar than different in specific respects, they would justifiably be expected to be more similar than different in other respects." This is the basis of theorizing with idealized models when these are applied to the real world in a series of analogical steps. For example, analogy takes us from the initial state of a falling body in air to the concept of a sphere falling in a vacuum, with its lawlike initial and final states, and then by analogy to the (approximate) final state of the real body. Similar counterfactual arguments are required for exploration and application of all models which are hypothesized but not necessarily assumed to exist.

A good test of the analogical conception of theories (ACT) is provided by quantum physics. This has always been a difficult case for model theory, because it is generally accepted that no familiar mechanical (or any other) models are adequate interpretations of its formalism. The so-called Copenhagen Interpretation takes a robustly positivist view, according to which the essence of quantum theory is its mathematics, for which no consistent and comprehensive analogies with other physical processes can, or need, be found. Realists, on the other hand, continue to look for "hidden variable" models which will restore comprehensive dynamical reality to the theory, though so far without much success. Meanwhile "particle," "wave," and "field" language continues to be used, and physicists have learned to use these partial models piecemeal in appropriate experimental situations, without assuming anything other than analogical relations with reality.

In terms of ACT, none of this should be surprising. ACT argues only for the reality of certain formal and material analogies in nature. This does not imply any uniquely "true" models of reality, and the history of quantum theory shows that it need not imply that we can articulate any models at all that are adequate for a given theory and its data. It is ironic that SCT, with its abstract "family of models," was being developed at exactly the same time as it was found that in quantum theory there are insoluble problems in articulating even one comprehensive model for its mathematics. There are, however, piecemeal and mutually conflicting models at various levels of the theoretical hierarchy, and these can be seen to function like any other models in aiding intuition and manipulation, and permitting justified local extrapolations to novel data by analogical inference. Quantum theory therefore provides a strong argument for the adoption of ACT, rather than SCT.

305

A major problem for ACT remains. Structures of similarity and analogy within theories and between theory and world have been amply illustrated in the historical and philosophical literature, but they are still largely unanalyzed primitives within the new conception. Attempts by Carnap and others to formalize a logic of analogical argument have quickly got lost in a "combinatorial jungle" of relative similarities and differences (see EVIDENCE AND CONFIRMATION). Something other than standard logic is required to do justice to the new metaphysical position, but it must be something that has its own rigor, and preferably more general application than just to the philosophy of scientific theories.

Only one recent approach seems to offer hope along both these dimensions. This is the development in cognitive science of parallel distributive processing (PDP), which has been analyzed from the point of view of philosophy of science by Paul Churchland (1989) (see COGNITIVE APPROACHES TO SCIENCE). A PDP system is itself a model (though not yet a quite satisfactory one) of human and animal brains – indeed, of any system that learns economically from experience. To take Churchland's simplified example, suppose the problem is to discriminate between sonar reflections from mines and from rocks as received by ships at sea. The input terminals of the PDP system are presented with vector sets specifying discriminating features of mines and rocks. These pass in parallel to a level of "hidden units" along pathways which weight the input in variable ways. The system can be "taught," by a complex hidden network of feedbacks, to build up prototype profiles of "mine" and "rock" respectively, in such a way that differential responses are made and corrected at the output. Eventually these responses are triggered off appropriately without explicit correction from the teacher when new data are fed in. Even the learning phase does not necessarily require a human programmer, but can be conceived as the result of natural processes of feedback, such as conditioned response to danger, or Darwinian selection.

Tests of the system exhibit speedy and successful learning, but its philosophical interest lies rather in the learning principles presupposed. These show it to be an excellent model of models of scientific theorizing, as this is construed in ACT. The principal virtue of PDP is that it models the process of analogical classification much more faithfully than any previous models in logic or probability theory. It does this simply by building in the assumption (similar to that of Wittgenstein's family resemblances) that perception, discrimination, and successful extrapolation naturally take place by clustering objects and properties with sufficient similarities for our purposes, and distinguishing clusters from one another according to differences with respect to our purposes (Hesse 1988).

To summarize: models have been discussed in philosophy of science from two opposing points of view. The "standard" approach – for example, the semantic conception of theories, is formal and ahistorical, defining a model as one of the entities and processes that satisfy the formal axioms of a theory. The theory itself consists of its formal structure plus the family of all its models. Realist versions of SCT strive to define a "good" theory as one whose models can be taken approximately to represent the real world. Anti-realist versions regard the models as fictions having no direct relation to reality, but to be used purely heuristically for the discovery and explanation of phenomenal laws. Both realist and anti-realist versions of SCT tend to analyze theories as static "textbook" entities, and both tend to make a sharp distinction between a theory with its models and data derived from observation and experiment. SCT consequently neglects epistemological problems of theory development and theory choice.

The alternative approach has been called here the "analogical conception of theories." According to this view, theories are historically changing entities, and consist essentially of hypothetical models or analogues of reality, not primarily of formal systems. Theoretical models, models of data, and the real world are related in complex networks of analogy, which are continually being modified as new data are obtained and new models developed. Analogies with familiar entities and events introduce descriptive terms for theoretical concepts, by processes similar to the use of metaphor in language. Inferences within theories, and from theory to data and predictions, are analogical rather than propositional. Their justification must be sought in some metaphysical principle of the "analogy of nature," a principle that is weaker than the usual assumptions of "natural kinds" or "universal laws." It has been suggested that a suitable philosophical model for the difficult concept of "analogy" may be found in artificial learning systems such as parallel distributive processing.

References

Black, M. 1962: *Models and Metaphors* (Ithaca, NY: Cornell University Press).

Campbell, N. R. 1920: *Physics, the Elements* (Cambridge: Cambridge University Press; subsequently published as *Foundations of Science* (New York: Dover Publications, Inc., 1957).

Cartwright, N. 1983: *How the Laws of Physics Lie* (Oxford: Clarendon Press; New York: Oxford University Press).

Churchland, P. M. 1989: *A Neurocomputational Perspective, the Nature of Mind and the Structure of Science* (Cambridge, MA: MIT Press).

Giere, R. N. 1988: *Explaining Science, a Cognitive Approach* (Chicago and London: University of Chicago Press).

Gooding, D. 1990: *Experiment and the Making of Meaning* (Dordrecht: Kluwer Academic Publishers).

Harré, R. 1986: *Varieties of Realism* (Oxford: Blackwell).

Hesse, M. B. 1966: *Models and Analogies in Science* (Notre Dame, IN: University of Notre Dame Press).

—— 1974: *The Structure of Scientific Inference* (London: Macmillan).

—— 1988: Theories, family resemblances and analogy. In *Analogical Reasoning*, ed. D. H. Helman (Dordrecht: Kluwer Academic Publishers), 317–40.

Putnam, H. 1963: "Degree of confirmation" and inductive logic. In *The Philosophy of Rudolph Carnap*, ed. P. A. Schilpp (La Salle, IL: Open Court; London: Cambridge University Press), 761–83.

Suppe, F. 1989: *The Semantic Conception of Theories and Scientific Realism* (Urbana, IL, and Chicago: University of Illinois Press).

Suppes, P. 1960: A comparison of the meaning and uses of models in mathematics and the empirical sciences. *Synthese*, 12, 287–301.

van Fraassen, B. C. 1980: *The Scientific Image* (Oxford: Oxford University Press).

—— 1989: *Laws and Symmetry* (Oxford: Clarendon Press).

45

Naturalism

RONALD N. GIERE

Naturalism in the philosophy of science, and philosophy generally, is more an overall approach to the subject than a set of specific doctrines. In philosophy it may be characterized only by the most general ontological and epistemological principles, and then more by what it opposes than by what it proposes.

Ontologically, naturalism implies the rejection of supernaturalism. Traditionally this has meant primarily the rejection of any deity, such as the Judeo-Christian God, which stands outside nature as creator or actor. Positively, naturalists hold that reality, including human life and society, is exhausted by what exists in the causal order of nature. Some naturalists have embraced materialism, while others have struggled to avoid it.

Epistemologically, naturalism implies the rejection of all forms of a priori knowledge, including that of higher-level principles of epistemic validation. Positively, naturalists claim that all knowledge derives from human interactions with the natural world. This includes sense perception, but may also include both techniques and technologies of human origin, such as statistical hypothesis testing and microscopes.

Naturalists typically laud Aristotle, Hume, and Mill as supporting naturalism, while criticizing Plato, Leibniz, and Kant for their anti-naturalism (see HUME; MILL; and LEIBNIZ). Probably the single most important contributor to naturalism in the past century was Charles Darwin, who, while not a philosopher, was a naturalist both in the philosophical and the biological senses of the term. In *The Descent of Man* (1871), Darwin made clear the implications of natural selection for humans, including both their biology and psychology, thus undercutting forms of anti-naturalism which appealed not only to extra-natural vital forces in biology, but to human freedom, values, morality, etc. These supposed indicators of the extra-natural are all, for Darwin, merely products of natural selection (see DARWIN).

In the twentieth century, the last major, self-consciously naturalistic school of philosophy was American pragmatism, as exemplified particularly in the works of John Dewey. The pragmatists replaced traditional metaphysics and epistemology with the theories and methods of the sciences, and grounded their view of human life in Darwin's biology. Following the Second World War, pragmatism was eclipsed by logical positivism (see LOGICAL POSITIVISM) and more general analytic philosophy, both largely of European origin. Having naturalized Kant's fundamental categories of space, time, and causality in terms of the physics of Einstein (see EINSTEIN), particularly relativity and quantum theory, Reichenbach, Carnap, and other logical empiricists (see LOGICAL EMPIRICISM) saw the philosophy of science as consisting solely in the logical analysis of scientific concepts, theories, and methods, an a priori activity. Similarly, following Wittgenstein, analytic philosophers practiced "conceptual analysis," regarded as something clearly separate from natural science.

The recent rekindling of interest in a naturalistic approach to philosophy and the philosophy of science derives from several, mainly contemporary, sources. One was Quine's "Epistemology naturalized" (1969, repr. in Kornblith 1985), which argued from the failure of logical empiricist programs for reducing all knowledge to a set of observation statements to the replacement of epistemology by empirical psychology (see QUINE). More recent epistemological programs, such as Goldman's (1986), follow this lead (see Kitcher 1992). A cognitive approach to science also represents a development of Quine's program, with his own behaviorism replaced by contemporary cognitivism (see COGNITIVE APPROACHES TO SCIENCE).

Another inspiration for a naturalistic approach in the philosophy of science was Kuhn's *Structure of Scientific Revolutions* (1962) (scc KUHN). Although not explicitly naturalistic, Kuhn's account of how science develops in fact invokes only naturalistic factors grounded in either psychology (gestalt switches) or sociology (change in generations). Insofar as they attempt to provide explanations as well as descriptions of past scientific achievements, historians of science tend to be implicit naturalists.

Among recent naturalistic approaches to epistemology and philosophy of science, only *evolutionary* approaches appeal explicitly to earlier naturalistic traditions, in this case the Darwinian tradition. Donald Campbell's influential article, "Evolutionary epistemology" (1974), surveys both historical precedents and his own work of several decades. As Campbell makes clear, the application of evolutionary ideas to epistemology and the philosophy of science takes place at several different levels. The most basic level is the evolution of the biological mechanisms for human perception (eyes), cognition (brain), and motor activity (hands). Just as Descartes appealed to a beneficent God to vouchsafe the reliability of some judgments, so the evolutionary naturalist appeals to the fact of evolution to vouchsafe the general reliability of everyday judgments (see DESCARTES). Without a reasonable level of rough-and-ready reliability, the human species would not have evolved. Like Descartes's argument, this argument is circular if regarded as an attempt to establish an ultimate foundation for all knowledge. Not seeing any possible way of providing such foundations, naturalists are satisfied with an evolutionary explanation for the beginnings of knowledge. Others (Stroud 1981, repr. in Kornblith 1985) insist on an extra-naturalistic argument to show that traditional foundationist programs cannot be fulfilled.

At a higher level one finds sociobiological explanations for the evolution of inductive principles such as the consilience of inductions (Ruse 1986). More abstractly, David Hull (1988) has argued for a strict isomorphism between the evolution of animal populations, scientific communities, and scientific concepts. Many others (Toulmin 1972; Giere 1988) have been satisfied with a much looser application of selectionist models to the development of science.

Apart from the desire for an ultimate foundation for knowledge, a common objection to naturalistic approaches in the philosophy of science is that such approaches are limited to the mere description of what scientists do. The philosophy of science, it is claimed, is concerned to develop *normative* models of how science should be pursued – that is, model of scientific rationality. But there is a way in which naturalists can also be normative (Laudan 1987; Giere 1988). Naturalists are not limited merely to describing what scientists say and do. They can also develop *theoretical explanations* – for example, cognitive or evolutionary explanations – of how science works. Just as theoretical mechanics provides a basis for designing rockets, so a powerful theoretical account of how science works could provide a basis for sound advice on scientific practice and policies. So those (e.g., Putnam 1982) who use the desire for normative conclusions as an argument against naturalism beg the question by assuming a notion of rationality that goes beyond "conditional" or "instrumental" rationality.

References

Campbell, D. T. 1974: Evolutionary epistemology. In *The Philosophy of Karl Popper*, ed. P. A. Schilpp (La Salle, IL: Open Court), 413–63.

Giere, R. N. 1988: *Explaining Science: A Cognitive Approach* (Chicago: University of Chicago Press).

Goldman, A. I. 1986: *Epistemology and Cognition* (Cambridge, MA: Harvard University Press).

Hull, D. 1988: *Science as a Process: An Evolutionary Account of the Social and Conceptual Development of Science* (Chicago: University of Chicago Press).

Kitcher, P. 1992: The naturalists return. *Philosophical Review*, 101, 53–114.

Kornblith, H. (ed.) 1985: *Naturalizing Epistemology* (Cambridge, MA: MIT Press).

Kuhn, T. S. 1962: *The Structure of Scientific Revolutions* (Chicago: University of Chicago Press; 2nd edn 1970).

Laudan, L. 1987: Progress or rationality? The prospects for normative naturalism. *American Philosophical Quarterly*, 24, 19–31.

Putnam, H. 1982: Why reason can't be naturalized. *Synthese*, 52, 3–23.

Ruse, M. 1986: *Taking Darwin Seriously* (Dordrecht: Reidel).

Toulmin, S. 1972: *Human Knowledge* (Princeton: Princeton University Press).

46

Natural Kinds

JOHN DUPRÉ

A central aspect of science is the classification of natural phenomena. Not only is this to some extent an end in itself, an account of what kinds of things there are being an important part of the picture of the world that science aims to provide, but classification is also inextricably connected with the development of scientific theories. The change from phlogiston theory to atomic chemistry, for example, involved not just a different theory but an entirely new way of sorting the domain of chemistry into kinds. It is often supposed that a necessary condition for an adequate or correct scientific theory is that its generalizations be formulated in terms of *natural* kinds – those kinds, roughly speaking, that really exist in nature. Thus oxygen, but not dephlogisticated air, may be said to mark a natural kind. Natural kinds are sometimes conceived precisely as being those kinds to which true scientific laws apply. In addition, natural kinds have generally been thought of as defined by the common possession of an essence, a property both necessary and sufficient for an entity to be a member of the kind, and from which the further important properties of the kind flow.

I shall first give a brief account of the historical background to the issue, and then turn to a more detailed account of contemporary debates about natural kinds. (The historical section is indebted to an important paper by Ayers (1981); for a different view of Locke's contribution, see Mackie 1974.)

Historical background

Philosophical doctrines of natural kinds are generally traced historically to Aristotelian and Scholastic theories of substance. Substances were divided into individual substances, such as pigs or humans, and "homeomerous" substances, such as water or iron. Every individual pig or particular chunk of iron was held to share a common nature or essence. Although different individuals were distinguished one from another by the different matter from which they were constituted, matter by itself lacked any particular nature. Indeed, matter could not exist without instantiating some particular substantial form; and, conversely, a substance could not exist except as instantiated in some quantity of matter. The essence, which determines what kind of substance an individual is, can be specified in terms of a real definition, to be understood in terms of the five "predicables" genus, species, difference, properties, and accidents. To cite the most famous example, the *species* (or substance) man is a part of the *genus* animal, distinguished by the *difference*, rational. The essence of man is given by genus and difference, as rational animal. Certain *properties* – for example, language – were held to flow necessarily from this essence. Features not necessarily connected with the

311

essence, such as being tall, or bald, were *accidents*. Scientific investigation, finally, was taken to consist in discovering the properties of a substance inductively, and inferring the essence through philosophical reflection.

The most crucial sense in which this Aristotelian set of doctrines prefigures contemporary theories of natural kinds is that it presupposes that there are real boundaries – indeed, perfectly sharp boundaries – between naturally occurring kinds of things, and that the species distinguished by these boundaries are discoverable by some kind of investigation. Thus the Aristotelian account of scientific knowledge assumes the existence of natural kinds.

The Aristotelian view of science and its development by the Scholastics was the philosophical foil for the new philosophy of science developed in the seventeenth century. The idea of essences accessible to mere rational analysis, as famously parodied by Voltaire with the explanation of the soporific effects of opium by its *virtus dormitiva*, was seen as the antithesis of the empirical conception of science that was increasingly accepted at this time. The philosopher who attacked these Aristotelian ideas in the greatest detail and with the greatest effect was John Locke (see LOCKE).

A driving idea behind the philosophical developments of the seventeenth century was the commitment to the corpuscularian account of matter, according to which matter was ultimately composed of tiny elastic particles in motion through empty space. The first point to emphasize, in contrast to the formless matter of Aristotelian metaphysics, is that this is an *account* of matter at all. This lies at the heart of Locke's skepticism about natural kinds. Matter, having a form of its own, does not require a substantial form to be imposed on it. And indeed, the reductionist tendency of seventeenth-century thought suggested, rather, that the features of complex objects were to be understood as consequences of their underlying microstructural constitutions, the form of such objects, therefore, turning out to be a consequence of the properties of matter.

Locke's views on kinds begin with the distinction between real essence, which he defines as "the being of anything whereby it is what it is," and nominal essence, "the abstract idea which the general, or sortal . . . name stands for" (1975, p. 417) – that is, a set of properties associated with the term. Thus the nominal essence of gold might be malleable yellow metal. Although Aristotelians had recognized something like the latter as defining nonnatural kinds – for example, compound kinds, such as musician (a man can cease to be a musician without ceasing to exist) – and also as a possible, though less than ideal, way of referring to natural kinds, it is clear that the former notion, real essence, comes closer to the Aristotelian idea of a natural kind. Locke argued that all general terms referred to kinds defined only by nominal essences. For the members of a kind to share a real essence would be for them to have the same important microstructural properties. But lacking "microscopical eyes," we can never know this to be the case. Since this ignorance in no way prevents us from using and understanding general terms, it should be inferred that the extension of these terms is determined only by the nominal essence.

This claim reflects a deeper ontological view. Locke did not believe that nature was generally characterized by sharp boundaries. "The boundaries of the Species," he wrote, "whereby Men sort them, are made by Men" (1975, p. 462). Locke held that the boundaries between the kinds distinguished by our language were entirely permeable by nature, as, for example, by a creature he claimed to have observed "that was the

Issue of a Cat and Rat, and had the plain marks of both about it" (p. 451). It is easy to see how Locke's corpuscularianism makes this view reasonable. In principle, the corpuscles that compose matter might be capable of combining into innumerable and individually unique complex structures, or, as they were seen in the seventeenth century, machines. Perhaps the most crucial point that emerges from Locke's discussion is that the question of whether there are definite boundaries between the kinds of structures into which the basic constituents of matter can be arranged is a wholly contingent one.

Contemporary discussions

Recent philosophers have been much more optimistic than Locke about our ability to peer into the ultimate structure of things. Although we still lack microscopical eyes, scientific ingenuity has found methods of investigation that circumvent this disadvantage. Many philosophers, reflecting on the results of these investigations, have decided that Locke was too hasty in his rejection of natural boundaries, and have resurrected the Aristotelian notion of kinds determined by real essences. Chemistry and physics, they have noted, have found out a great deal about the inner structure of things, and many antecedently distinguished kinds have been found to share important internal properties. Thus they have argued that many natural kinds might, after all, be demarcated by real essences.

However, reconciling this aspect of Aristotelian essentialism with the seventeenth-century corpuscularian conception of matter, or its modern descendants, requires a very different epistemological perspective on essences. Essences are not to be discovered by philosophical reflection, but by scientific investigation. And essences will consist of fundamental microstructural properties. It will be recalled that for both Locke and Aristotle essences had to do both with the definition of general terms and with the natures of things, two functions that Locke saw as radically incompatible. The recent development that has rekindled interest in the topic of natural kinds can be seen in part as a move to reconcile these two functions of essences. The relevant ideas are due to Saul Kripke (1980) and Hilary Putnam (1975), and involve a radical suggestion about how the meanings of terms that refer to natural kinds should be understood.

Extending the idea promoted by Kripke for the meaning of proper names, Kripke and Putnam proposed that natural kind terms, rather than referring to the members of a kind by means of the familiar features by which kind terms are typically learned or explained – that is, by something similar to a Lockean nominal essence – should be seen as referring directly to the real members of the kind, the latter being demarcated by a real essence. Thus Putnam, in whose work this idea has been worked out in greatest detail, analyzed the meaning of a natural kind term into four components: a syntactic marker, a semantic marker, a stereotype, and an extension. To illustrate, the term "elephant" might have as syntactic marker "noun," as semantic marker "animal," as stereotype "large grey animal with a long nose and flapping ears," and as extension the kind determined by the real microstructural essence of elephants. The central point is that while the stereotype is crucial in facilitating the use of the term by people who may know nothing about the real essence of elephants, it does not fix the extension. This potentially ignorant talk is facilitated by what Putnam refers to as "the

313

division of linguistic labour" (1975, p. 227). When we really need to know whether something is a member of a particular kind, we appeal to the relevant experts. A compelling example of this phenomenon is that of gold. Few of us know how to tell whether some piece of metal is really gold or not. But the fact that there are experts who can, and who, we often suppose, have at some point validated the authentic status of the things we take to be gold, renders this incapacity generally harmless. And it is the criteria applied by experts, rather than the superficial stereotypic properties (soft, yellow, etc. metal) that we take as decisive as to whether something is really gold or not. Indeed, Putnam argues, by appeal to various hypothetical scenarios, that satisfaction of the stereotype is neither necessary nor sufficient for membership of the kind. Finally, Putnam and Kripke explain the introduction of such terms into language as indexical. A term is introduced in connection with an example, or a sample, of the extension, and is taken as applying to that and to anything else that shares the same essential nature.

Before evaluating this account in more detail, it is worth taking note of one reason why Putnam's analysis has been enormously attractive to many philosophers of science. Standard, broadly Fregean accounts of the meanings of terms in scientific theories played a central part in presenting the problem of incommensurability (see INCOMMENSURABILITY). Starting from the assumption that the meaning of a general term was to be understood in terms of beliefs that scientists had about their referents, it seemed that any significant change in scientific belief – on some accounts, any change in the relevant beliefs whatever – would change the meaning of the term. Thus what scientists said about, say, electrons at different times would turn out to be statements about different things. It would therefore be impossible to say that our knowledge of electrons had increased over time: we would have only a sequence of statements at different times, all with distinct subject matters. The idea of direct reference then seemed to solve this problem at a stroke. Whereas successive scientific theories may be seen as involving different stereotypes for the word "electron," the extension remains the same: whatever kind of things embody the real essence of the natural kind of electrons.

However, this example, it must now be noted, involves an important extension to the original theory of natural kinds provided by Putnam and Kripke. These philosophers presented their account as applying to natural kind terms in ordinary language, rather than to terms of theoretical science. Typical examples are "water," "gold," and "lemon." They claimed, on the basis of intuitions about hypothetical cases, that the intention to refer to a natural kind determined by a possibly unknown real essence is part of a correct account of the normal use of these terms in ordinary language. If this is right, then it is certainly reasonable to extend the account to the technical uses of theoretical terms in science. If the account cannot be sustained for the case of ordinary language kind terms, appeal to it as an account of scientific terms will be more problematic.

The most widely discussed argument for this view of natural kind terms involves Putnam's (1975) Twin Earth example. Putnam invites us to consider a place identical to Earth in all respects except one: the stuff that fills oceans, comes out of taps, and falls on occasion from the sky on Twin Earth is not in fact composed of the chemical H_2O, but of some quite different chemical which Putnam calls XYZ. Putnam then asks us to consider what would be said by scientists from Earth who discovered this anomalous fact. Surely, he claims, they would say that what looked like water (satisfied the stereotype) had proved after all not to be water but some different kind of liquid. This, and similar cases, lead him to argue that quite generally in the case of such natural kind terms it is the real nature or essence of the kind, as assessed by appropriate scientific methods, that determines the extension of the kind. There has been a great deal of discussion of this and related thought experiments (useful examples are Mellor 1977 and Zymach 1976). What most clearly emerges from this discussion is that intuitions differ markedly in these cases. The argument would seem a lot less plausible, for example, if we imagined that two or more quite different water-like substances had been discovered on Earth. Would we choose one of these to be the kind referred to by "water"? Or would we not rather have discovered that there was more than one kind of water? The latter interpretation seems possible even in Putnam's Twin Earth case. An argument based on such intuitions is on shaky ground if the intuitions are not widely shared.

In evaluating the Kripke–Putnam theory, the crucial point to note for present purposes is that it rests on a strong ontological presupposition: contrary to Locke, a large class of ordinary language kind terms must actually pick out (more or less) the requisite sort of natural kind. (Unless, at any rate, the theory of natural kind terms is based on massive metaphysical delusion.) And, in addition, we must have some sense, in advance of scientific illumination of the real essence of a kind, whether that kind is indeed a natural kind. Putnam claims explicitly, for example, that the stuff on Twin Earth would not have been water even if the explorers from Earth had arrived before any means had been discovered for distinguishing H_2O from XYZ. The intention to refer to the real nature preexists the characterization of that nature. I shall now suggest that a careful look at some of what we have discovered about the internal natures of things lends more support to Locke than it does to the neo-Aristotelianism of Putnam and Kripke.

Beginning with the second point from the last paragraph, that we must have some intuitive sense of which of our terms are intended to apply to natural kinds, it is certainly fair to say that there are many terms of ordinary language that refer to naturally occurring objects or stuffs that are relatively homogeneous and reliable in their properties. Water is usually wet, transparent, drinkable, etc.; and stuff with these properties is generally water. When you've seen one squirrel, you've seen them all. But though this observation suggests that we might have some use for the concept of a natural kind, it is some way from showing that our prescientific language contains terms for many kinds demarcated by real essences.

The main sources for examples of natural kinds in these discussions are biology and chemistry, and I shall now consider briefly the applicability of Putnam's theories to each of these domains. Beginning with biology, I suggest that the attempt to attribute real essences to familiar kinds is an unpromising one. (This argument is developed in more detail in Dupré 1993, chs 1–2.) For the purposes of scientific biology, the basic unit of classification is the species. Whether or not species form natural kinds with real essences, no biologist makes such a claim for higher-level groups (or "taxa") such as genera, families, etc. But familiar terms from ordinary language, when they can be correlated with any scientifically recognized taxon, refer most often to taxa at higher levels than the species. Many names of trees – oak, beech, elm, willow, pine – refer to genera. Among birds, kingbirds and cuckoos correspond to genera; ducks, wrens, and woodpeckers form families; and owls and pigeons make up whole orders. And so on. Moreover, many distinctions in ordinary language divide scientific taxa in ways which have no particular biological significance. Some examples are frogs and toads, rabbits and hares, and onions and garlic. The last case, being one in which the practical, though nonscientific, point of the distinction is obvious, points to the general problem. Ordinary language classifications are directed to the interests of ordinary life, whether these be the specialized purposes of the gastronome, the forester, or the furrier, or merely a general interest in the natural environment. These interests frequently, perhaps generally, do not coincide with the specialized interests of the scientific classifier. Indeed, there are cases where these barely overlap. One example is the term "cedar." Various species of tree, not even closely related, are referred to by this term. An obvious supposition is that the reason for this has to do with the particular uses of a kind of timber.

The difficulty so far is that if indeed our use of prima facie natural kind terms in ordinary language involves an intention to refer to a natural kind, members of which share a common essence, nature frequently frustrates this intention. There are, on the other hand, terms of natural language that do refer to a unique species, and if we count the esoteric vocabularies of bird-watchers, butterfly-collectors, and amateur botanists as parts of ordinary language, there are very many such terms. However, although this point has not always been appreciated by philosophers, contemporary scientific understanding of species lends little support to essentialism (see Hull 1965; Dupré 1993). Although the correct characterization of species remains highly controversial (see Ereshefsky 1992), one thing common to all post-Darwinian conceptions of species is recognition of the omnipresence of variation; and the idea of the members of species being demarcated by some necessary and sufficient internal property is quite antithetical to such conceptions. Indeed, a number of philosophers and biologists have argued that species are not kinds at all, but historically specific, though spatially discontinuous, individuals (Hull 1976). Ironically, perhaps, this would make the Kripke–Putnam approach much more plausible insofar as scientific species names could be seen as not merely analogous to, but literally, proper names. It would, on the other hand, lend no comfort to the account of *natural kind* terms in ordinary language.

Perhaps the heartland of contemporary essentialism is chemistry. Surely chemical formulae really do specify an essential property of being a particular chemical substance. Even if this is so, the Kripke–Putnam theory applies poorly to ordinary language terms for kinds of stuff. To begin with, there are examples with just the structure of the garlic and onions case. The same chemical, for example, is called topaz if it is yellow and sapphire if it is blue. Jade is an ordinary language term applied to two distinct chemicals, nephrite and jadeite. Moreover, pure chemicals – even nearly pure chemicals – are not often encountered in nature. Water is perhaps the most familiar case of one that is, though even here we may wonder whether seawater, polluted water, ditch water, etc. are not perfectly good kinds of water, even though far from pure. And finally, only the development of scientific knowledge has given us much insight into which of the many substances we encounter are in fact relatively pure chemical kinds. Surely, for example, there was no prescientific reason to suppose that water was a natural kind and air was not?

It appears, then, that if Putnam is right in supposing that we use terms for kinds of naturally occurring things and stuffs with the intention of referring directly to an Aristotelian natural kind, these intentions are consistently frustrated by nature. Given, in addition, the debatable status of the intuitions on which Putnam's account rests, it seems more plausible to return to a broadly Lockean account of the meaning of such terms as involving only some kind of nominal essence. Might we still think of scientific terms as designating natural kinds on this strong sense? In this case the argument would rest not on intuitions about pre-analytic linguistic intentions, but rather on a conception of the aims of science. Science has traditionally been conceived as concerned with the discovery of laws of nature, and natural kind terms, as noted above, are exactly suited to figure in such laws. (Quine (1969) provides an eloquent account of such a conception of the development of scientific classification.) The crucial point to note here is just that any conception of the aims of science should be subject to feedback from the results of scientific inquiry. Therefore, this is a plausible account of scientific terms just to the extent that we have reason to think that there exist such natural kinds for scientific terms to refer to.

With regard to this last issue, it now seems plausible that the elementary particles of physics, and the more or less precise structures of these defined by atomic physics and chemistry, realize something like an Aristotelian conception of natural kinds. We should, of course, take this judgment as only provisional: perhaps it is only the difficulty of observing these minute objects that prevents us from seeing that each electron is as different from the next as are two dogs or two oak trees. However that may be, it does also seem reasonable to impute to physical scientists the intention of referring to natural kinds, and thus to ground some argument of the sort sketched above against the incommensurability of consecutive uses of theoretical terms.

At levels of complexity greater than the chemical, however, it appears that Locke was broadly correct from the perspective of science as much as from that of everyday life. Although the basic constituents of matter may be quite limited in the kinds of structures they can form at the simplest levels, the greater the complexity of structure becomes, the less constraint there appears to be. For biological organisms there is no such canalization towards a definite range of possible forms. As mentioned above, there is great controversy about how best to define biological species. Although many biologists and philosophers remain convinced that we must decide on one unique criterion for distinguishing biological species, this is generally based on the idea that only thus can communication be maintained between different biologists, rather than on the claim that some particular system reflects the true order of nature. Defenders of natural kinds in biology will typically resort to a higher level of abstraction at which they suppose that real laws of nature might be formulated, suggesting such candidates as predator, population, or the species category (as opposed to any particular species). But whatever the merits of such proposals, these putative natural kinds are far removed from the set of individuals sharing a common internal essence that we have been considering.

I myself believe that the spectre of a Tower of Babel has been greatly exaggerated, and we have no reason to fear a multiplicity of classifications of the biological world. We have already seen that prescientific language has uses for classifications that cut across those requisite for scientific purposes. I see no reason why the same should not apply to different projects of inquiry within biology. The kinds distinguished by the genealogical investigations of evolutionists may well differ significantly from those suited to the investigations of contemporary interactions in ecology. And the criteria suitable for distinguishing kinds in slow-breeding, genetically largely isolated species such as mammals or birds may be quite inappropriate for bacteria or even flowering plants. If we take seriously the Lockean picture of potentially continuous variation, but note also that with respect to various properties of relevance to various concerns there is a good deal of clustering, we should acknowledge that both science and every-day life may require a variety of more or less cross-cutting classificatory systems. Since these classifications may represent perfectly real features of the objects classified, there is no reason to divorce such a pluralistic view from realism about the kinds distinguished. (This perspective on biology is defended in detail in Dupré 1993, under the rubric "promiscuous realism.") Such a position is even more attractive in the case of the human sciences when we reflect on the very different grounds on which people are sorted for the purposes of medicine, economics, psychology, anthropology, etc. I am sympathetic even to the rehabilitation of the term "natural kind" in the context of such a pluralistic view, though clearly in a sense stripped of its traditional essentialist connotations.

This position on natural kind has profound implications for various aspects of our understanding of science. If there are no Aristotelian natural kinds to be discovered in most parts of science, then we cannot expect to find laws of nature of the traditional universal form in these sciences. This lends support to the currently much discussed semantic view of theories, which considers theories as embodying sets of models, rather than universal laws (see THEORIES and MODELS AND ANALOGIES). It also presents major obstacles to traditional doctrines of reductionism and the unity of science, the latter generally having been articulated in terms of the former (see REDUCTIONISM and UNITY OF SCIENCE). Philosophers who have continued to insist that science requires universal laws applying to the members of natural kinds have been driven to defend eliminative or instrumentalist views of various domains. The most notable example of the former strategy is the eliminative view of the mental (see, e.g., Churchland 1986), which accounts for the lack of mental natural kinds by denying that there are any mental entities to form such kinds. Instrumentalist reactions are well illustrated by Rosenberg's (1994) account of biology (see REALISM AND INSTRUMENTALISM). At any rate, the presence or absence of natural kinds continues to be a focus of debate with respect to many areas of science, and is a question with major ramifications for how science should be understood.

References

Ayers, M. R. 1981: Locke versus Aristotle on natural kinds. *Journal of Philosophy*, 78, 247–72.
Churchland, P. S. 1986: *Neurophilosophy* (Cambridge, MA: Bradford Books/MIT Press).
Dupré, J. 1993: *The Disorder of Things: Metaphysical Foundations of the Disunity of Science* (Cambridge, MA: Harvard University Press).
Ereshefsky, M. (ed.) 1992: *The Units of Evolution: Essays on the Nature of Species* (Cambridge, MA: MIT Press).
Hull, D. L. 1965: The effect of essentialism on taxonomy: 2,000 years of stasis. *British Journal for the Philosophy of Science*, 15, 314–26; 16, 1–18.

——— 1976: Are species really individuals? *Systematic Zoology*, 25, 174–91.

Kripke, S. 1980: *Naming and Necessity* (Cambridge, MA: Harvard University Press).

Locke, J. 1975: *An Essay Concerning Human Understanding* (1689), ed. P. H. Nidditch (Oxford: Oxford University Press).

Mackie, J. L. 1974: Locke's anticipation of Kripke. *Analysis*, 34, 177–80.

Mellor, D. H. 1977: Natural kinds. *British Journal for the Philosophy of Science*, 28, 299–312.

Putnam, H. 1975: The meaning of "meaning." In *Mind, Language, and Reality. Philosophical Papers*, vol. 2 (Cambridge: Cambridge University Press), 215–71.

Quine, W. V. O. 1969: Natural kinds. In *Ontological Relativity and Other Essays* (New York: Columbia University Press), 114–38.

Rosenberg, A. 1994: *Instrumental Biology or the Disunity of Science* (Chicago: University of Chicago Press).

Zymach, E. 1976: Putnam's theory on the reference of substance terms. *Journal of Philosophy*, 73, 116–27.

47

Newton

RICHARD S. WESTFALL

Isaac Newton was born on 25 December 1642 in the hamlet of Colsterworth, Lincoln-shire, about six miles south of Grantham. The posthumous and only son of Isaac Newton, père, he found himself deposited with grandparents at the age of three when his mother married a second time; he remained with the grandparents for eight years until the death of his stepfather. After successfully resisting his mother's intention that he manage the considerable estate she had inherited from the two husbands, Newton graduated from the grammar school in Grantham and enrolled in Trinity College, Cambridge, in June 1661. Trinity was Newton's home for the following 35 years. He received his B.A. in 1665 and his M.A. three years later. Meanwhile he had been elected to a fellowship in the college, and in 1669, upon the resignation of Isaac Barrow, he was appointed Lucasian Professor of Mathematics. Newton did not strain himself with teaching. During his nearly 30 years as a fellow of Trinity he tutored only three students, all of them wealthy fellow commoners; he formed no perceptible bond with any of them. As Lucasian Professor, he lectured once a week during one term each year, at least early in his tenure. There is testimony that frequently no one attended the lectures, and even later, when he was famous and a connection with him was potentially valuable, only two men claimed to have heard them. He ceased to lecture altogether after 1687, and may have stopped before that.

However, Newton did not while away his time at Cambridge. On the contrary, he devoted himself to study with such intensity that the master of Trinity feared he might kill himself. Mathematics and physics, the studies we associate with Newton, were not the only ones in which he immersed himself. He also devoted extensive time to alchemy, which he pursued both in the study and in the laboratory, and to theology. In all these fields his creative intellectual work belonged to the years in Cambridge. In 1696, perhaps recognizing that his powers were waning, Newton abandoned the uni-versity for London, where he buried himself in administrative detail. He was appointed first warden, then master of the Mint, a position from which he derived sufficient income to die wealthy. He was elected president of the Royal Society in 1703. Newton died in London on 20 March 1727, the most celebrated intellectual in England, whose pall the Lord Chancellor, accompanied by two dukes and three earls, bore to a grave in Westminster Abbey.

As a scientist, Newton made a number of methodological pronouncements that remain of interest to philosophers of science. Against what he saw as the uncontrolled rationalism of Cartesian natural philosophy, he insisted on the primacy of empirical, experimental data in the establishment of scientific truth. Perhaps his most repeated words, "Hypotheses non fingo" (I do not feign hypotheses), come from a discussion of gravity in the General Scholium that he added to the second edition of the *Principia*. In this passage Newton asserted that natural phenomena establish the fact of universal gravitation. He refused, however, to compromise the solidity of that demonstration with speculations about the cause of gravity; *hypotheses non fingo*. In fact, Newton did entertain hypotheses about the cause of gravity and about much else, but in the work he published he isolated the speculations – for example, in the "Queries" that conclude the *Opticks* – from what he intended as demonstrations. His insistence on the distinction between matters he took to be mathematical demonstrations based on empirical evidence and speculations that went beyond such matters constituted the heart of his methodological stance.

Newton's primary significance for the philosophy of science, however, lies not in his pronouncements on methodology but in his scientific achievements. By 1661, the year when Newton arrived in Cambridge, the radical restructuring of natural philosophy that historians have labeled the "scientific revolution" was well under way. Central to it was heliocentric astronomy, which began with Copernicus in the sixteenth century – the realization that the Sun, rather than the Earth, stands at the center of our system and that the Earth is one planet among six (as far as astronomy then knew) circling the Sun. Early in the seventeenth century Johannes Kepler reformed heliocentric astronomy by replacing Copernicus's combinations of circles with elliptical orbits. Kepler's three laws of planetary motion not only summed up the observed phenomena of the heavens in generalizations of breathtaking simplicity; they also offered a new concept of natural law, a mathematic description of observed regularities, which became a central feature of modern science.

At much the same time Galileo Galilei was studying terrestrial motions (see GALILEO). He arrived at what was virtually the principle of inertia, that a body in uniform motion will persevere in that motion unless something external to the body acts to change its state. Galileo went on to define uniformly accelerated motion, which he identified with free fall, and to work out mathematically the kinematics of uniform and uniformly accelerated motion and their combination – a second exemplar of the new mathematical laws of nature.

The French philosopher René Descartes consciously set out to replace received Aristotelian philosophy with a philosophy constructed on a totally different foundation (see DESCARTES). Already in the work of Kepler, Galileo, and others, there were tendencies to treat questions in natural philosophy as problems in mechanics. Descartes generalized these tendencies. The physical world consists of nothing but inert matter in motion. Particles of matter, which are devoid of any source of spontaneous action, move where other particles that impinge upon them force them to go. The theory of vortices, which offered a physical basis for heliocentric astronomy, was only the best-known example of a mode of explanation that Descartes applied to all of nature. The qualities of Aristotelian philosophy are illusions. All that exists in the physical universe is matter defined by extension in three dimensions. Particles of matter that impinge on the nerve ends of sentient creatures provoke sensations that mankind has mistakenly projected onto reality. In fact, nothing exists in physical nature but particles of matter in motion. The world is a great machine. Together with other similar natural philosophies that appeared before the middle of the seventeenth century,

Descartes's philosophy came to be known as the "mechanical philosophy." By 1661 it had completely displaced Aristotelian philosophy among those active in the enterprise of science.

All of this Newton received, reworked, and fashioned into a consistent whole that became the enduring central structure of modern science. It was not what he encountered in the university's curriculum. There Aristotelian philosophy still reigned, and we know that Newton was duly initiated into it. But sometime during his undergraduate years, proceeding apparently on his own, he discovered the new science, and with it his own vocation.

The primary vehicle that promulgated Newton's vision of nature and science was *The Mathematical Principles of Natural Philosophy* (known as the *Principia* from the central word of its original Latin title), published in 1687. The *Principia* was, first of all, a work on the science of mechanics. Standing near the beginning of the book are the three laws of motion that continue to supply the foundation of physics. Together they define a mathematical science of dynamics, something students of mechanics before Newton had not successfully done. A number of the basic concepts of modern science first appeared here. The principle of inertia, the substance of the first law, was of course not Newton's, though it entered definitively into the structure of science with the *Principia*. He did develop the concept of force, rigorously defined in terms of the change in motion it causes, as well as a concept essential to the definition of force: that of mass, which the *Principia* first enunciated in satisfactory terms. The *Principia* also coined the term "centripetal force" and derived its quantitative measure, thus defining the mechanics of curvilinear motion, the central problem to which the book devoted itself. The ultimate genius of Newton's dynamics, however, lay in the fact that both Kepler's kinematics of celestial motion and Galileo's kinematics of terrestrial motion emerged as necessary consequences of it.

Newton also accepted the tradition of Descartes and the mechanical philosophers by treating the motions of celestial bodies as problems in mechanics. Like them, he considered that nature consists of material particles in motion. A large part of Newton's accomplishment in science, however, lay in the transformation he effected in the mechanical tradition. Half the goal of mechanical philosophers had been the elimination of insubstantial influences, such as sympathies and antipathies, from natural philosophy. The net result of Newton's application of his science of dynamics to the heavens was the predication of a force of universal gravitation whereby every body in the universe attracts every other body in precise inverse relation to the square of the distance between them. With forces that act at a distance went the elimination of matter; where Descartes's universe was a plenum, Newton's was a virtual void. To be sure, in the Queries that he added to the second English edition of the *Opticks* in 1716, Newton proposed a cosmic ether that could, among other things, cause the phenomenon of gravitational attraction. However, the ether of the Queries was composed of particles that repel each other at a distance, and even with that ether it is impossible to imagine the mechanism, to cite but one example, by which the Sun and the Moon, in proportion to their masses and inversely as the squares of the distances, "attract" the bulge of matter around the equator of the Earth, causing a conical motion of the axis. Newton's mechanical philosophy has been called dynamic, in contrast to the earlier kinematic one. Moreover, the dynamics in question was quantitative, and quantitative in a precise way.

In this respect it is significant that the first two words of Newton's title were "Mathematical Principles." Mathematics had been Newton's first great intellectual passion. About the same time, during his undergraduate years, when he discovered the new natural philosophy, he also discovered the new mathematics. For the next two years, from 1664 to 1666, mathematics dominated his life. During that time he absorbed the achievement of seventeenth-century mathematics before his time, what was referred to as "analysis," and went beyond early methods of finding tangents (or differentiation) and squaring curves (or integration) to understand the relation between the two procedures and to invent his method of fluxions, which we call the "calculus." A paper of October 1666 laid out the method in full generality.

Newton drew upon this expertise in composing the *Principia*. Nevertheless, the book looks strange to the modern reader; it appears to be a work of classical geometry rather than of calculus. In part this is an illusion. The concept of nascent and ultimate ratios that the book employs had no place in classical geometry. Rather, it belongs to the thought patterns of the calculus, which the *Principia* embraced, although it expressed them in a geometric rather than an algebraic idiom. What is essential is not the idiom but the rigor. Seizing on yet another of the crucial strands of the new science, which had characterized the work of such men as Kepler, Galileo, and Huygens, Newton extended it well beyond his predecessors, and by what he achieved with it insured that mathematical rigor would be one of the essential features of modern science.

Newton also did outstanding work in optics. He discovered the heterogeneity of white light, and as he investigated the rings that bear his name, he first demonstrated the periodicity of an optical phenomenon. Though quantitative, his *Opticks* was not mathematical in the manner of the *Principia*. Rather, it was experimental, and it furnished science with one of the most compelling examples of the power of experimental procedure to come out of the seventeenth century. In the 31 Queries that conclude the *Opticks*, Newton pointed toward many of the experimental investigations – for example, in electricity and chemistry – whereby the eighteenth century built upon the foundations set in place by the scientific revolution.

Without the work of earlier figures of the scientific revolution, on which he drew, Newton's own achievements would have been impossible. However, he raised almost everything he touched to a higher plane of generality and rigor, and in his major works, especially the *Principia*, he established the enduring framework of modern science.

References and further reading

Works by Newton

1934: *Mathematical Principles of Natural Philosophy*, trans. Andrew Motte, rev. Florian Cajori (Berkeley: University of California Press).

1952: *Opticks*, based on 4th edn (New York: Dover).

1959–77: *The Correspondence of Isaac Newton*, ed. H. W. Turnbull et al. 7 vols (Cambridge: Cambridge University Press).

1962: *Unpublished Scientific Papers of Isaac Newton*, ed. A. R. and M. B. Hall (Cambridge: Cambridge University Press).

1967–80: *The Mathematical Papers of Isaac Newton*, ed. D. T. Whiteside, 8 vols (Cambridge: Cambridge University Press).

1972: *Isaac Newton's Philosophiae naturalis principia mathematica*, 3rd edn with variant readings, ed. A. Koyré and I. B. Cohen, 2 vols (Cambridge: Cambridge University Press).

1984: *The Optical Papers of Isaac Newton*, ed. Alan E. Shapiro, 3 vols planned (Cambridge: Cambridge University Press).

Works by other authors

Burtt, E. A. 1952: *The Metaphysical Foundations of Modern Physical Science* (New York: Humanities Press).

Cohen, I. B. 1980: *The Newtonian Revolution* (Cambridge: Cambridge University Press).

Dobbs, B. J. T. 1992: *The Janus Faces of Genius* (New York: Cambridge University Press).

Herivel, J. 1965: *The Background to Newton's "Principia"* (Oxford: Oxford University Press).

Koyré, A. 1965: *Newtonian Studies* (Cambridge, MA: Harvard University Press).

McGuire, J. E., and Tamny, M. 1983: *Certain Philosophical Questions: Newton's Trinity Notebook* (Cambridge: Cambridge University Press).

Westfall, R. S. 1980: *Never at Rest: A Biography of Isaac Newton* (Cambridge: Cambridge University Press).

48

Observation and Theory

PETER ACHINSTEIN

During the first four decades of the nineteenth century a debate raged over the nature of light. Following proposals of Isaac Newton made early in the eighteenth century, many physicists accepted the theory that light is composed of tiny particles subject to mechanical forces (see NEWTON). At the beginning of the nineteenth century Thomas Young and Augustin Fresnel revived a competing theory originally suggested by Christiaan Huygens in the seventeenth century, according to which light consists not of particles, but of waves in a medium called the ether. In both cases theorists postulated entities – particles and waves in an ether – that could not be observed. Yet theorists on both sides gave arguments for the existence of these entities and the properties they ascribed to them. They did so on the basis of what they could observe. How was this possible?

A question of this sort, which leads to an examination of the relationship between observation and theory, has prompted philosophers of science to raise a series of more specific questions. What reasoning was in fact used to make inferences about light waves, which cannot be observed, from diffraction patterns that can be? Was such reasoning legitimate? How should claims about unobservables such as light waves be understood? Are they to be construed as postulating entities just as real as water waves only much smaller? Or should the wave theory be understood nonrealistically as an instrumental device for organizing and predicting observable optical phenomena such as the reflection, refraction, and diffraction of light? Such questions presuppose that there is a clear distinction between what can and cannot be observed. Is such a distinction clear? If so, how is it to be drawn?

These issues are among the central ones raised by philosophers of science about any theory that postulates unobservable entities.

How can theories about unobservables be inferred?

Two of the leading scientific methods provide an answer: the method of hypothesis – frequently called the hypothetico-deductive or H-D method – and inductivism. According to the former, the scientist begins with a theory, or hypothesis. This is a conjecture concerning what may be the case. It is produced in the scientist's mind by many factors, including observations, experiments, training, and even fortuitous events such as dreams. But the theory is not *inferred* by the scientist from any of these sources. The discovery of a new theory is a "happy guess" that may require the presence of certain causal conditions, but not a reasoning process subject to rules of inference. It is in the "context of discovery" (to use a term of Hans Reichenbach (1938)) in which theories are invented that the scientist will frequently introduce unobservables such as light waves.

Reasoning begins in the "context of justification" (again Reichenbach's term), in which the scientist attempts to defend or criticize his theory. This is accomplished by deriving conclusions deductively from the assumptions of the theory. Among these conclusions at least some will describe states of affairs capable of being established as true or false by observation. If these observational conclusions turn out to be true, the theory is shown to be empirically supported or probable. On a weaker version due to Karl Popper (1959), the theory is said to be "corroborated," meaning simply that it has been subjected to test and has not been falsified. Should any of the observational conclusions turn out to be false, the theory is refuted, and must be modified or replaced. So a hypothetico-deductivist can postulate any unobservable entities or events he or she wishes in the theory, so long as all the observational conclusions of the theory are true.

This simple account has advantages. In the context of discovery it allows the scientist to introduce unobservables unfettered by empirical constraints. Such constraints are introduced only in the context of justification: in order to conclude that the theory is true, probable, or supported, one needs to be able to deduce observational conclusions that are true. The account also has the advantage of reflecting in important ways how scientists frequently proceed: often they make a conjecture about what is unobservable, and test it by deriving consequences whose truth or falsity they can determine by observation and experiment.

The account also has its difficulties, however, as the enemies of the method of hypothesis, and even some of its friends, are happy to acknowledge. The most formidable is the "competing hypothesis objection," raised by many writers, most notably by Isaac Newton and John Stuart Mill, both avowed inductivists and opponents of the method of hypothesis (see MILL). Mill cites as an example the wave theory of light, which postulates an ether:

> The existence of the ether still rests on the possibility of deducing from its assumed laws a considerable number of actual phenomena. . . . Most thinkers of any degree of sobriety allow, that an hypothesis of this kind is not to be received as probably true because it accounts for all the known phenomena, since this is a condition sometimes fulfilled tolerably well by two conflicting hypotheses; while there are probably many others which are equally possible, but which, for want of anything analogous in our experience, our minds are unfitted to conceive. (Mill 1959, p. 328).

Indeed, if we consider various observed phenomena that were derived from the wave theory early in the nineteenth century, including the rectilinear propagation of light, reflection, refraction, and diffraction, Mill's objection is telling. These phenomena were also derived from the competing particle theory. So the wave theorist should not be able to claim truth, probability, or even very much support for his theory from the fact that it entails these observed phenomena.

This problem has led to more sophisticated versions of the H-D method. According to one, it is not sufficient that the theory entail simply known phenomena that have been observed. It must also entail new ones, particularly phenomena of a type of different from those that prompted the theory in the first place. William Whewell (who defended this more sophisticated version, which he called "induction," not the "method of hypothesis") speaks of this as the "consilience of inductions" (Whewell 1967; see also WHEWELL). He claimed that if a theory that entails known phenomena also satisfies consilience, then it is "certain." More weakly, a hypothetico-deductivist might say that under these conditions the theory is shown to be highly probable or strongly supported.

Mill, who engaged in an extensive debate with Whewell, remained unconvinced. Even if a theory T entails not only observed phenomena but also as yet unobserved ones of a different type, there may still be an incompatible theory T′ that entails the same phenomena. If so, Mill concluded, one cannot claim truth or high probability for the theory T. Indeed, there is a probability theorem, that Mill does not invoke, which seems to support his claim. Suppose that a theory T entails a set of observational conclusions O_1, O_2, ... If T has at least one incompatible competitor T′ that also entails these phenomena, and whose probability on information other than O_1, O_2, . . . is at least as great as that of T, then T's probability on O_1, O_2, . . . cannot rise above one-half, no matter how many observable consequences T entails (see Achinstein 1991, p. 126). Moreover, this is so even if T entails new observational predictions (interpreted here as observational predictions whose probability is less than 1). So even if your favorite theory entails all known phenomena that are supposed to be explained by that theory, and even if that theory makes new predictions that turn out to be true, this is not sufficient to establish its high probability. Moreover, there are some probabilistic interpretations of "consilience" under which the theorem remains applicable (see Achinstein 1991, essay 4). To avoid this consequence, a defender of the more sophisticated H-D position needs to provide some different, precisely defined concept of "consilience" which is such that if a theory T entails phenomena O_1, O_2, . . . , and satisfies "consilience," then its probability must rise above one-half as the number of entailed phenomena increases. This, however, has not been accomplished by hypothetico-deductivists.

Mill, by contrast to Whewell, proposes what he calls the "deductive method" for making inferences to scientific theories. It has three steps, the first of which is inductive and includes inferences to causes and laws governing them from the observation of their effects and similar causes in other instances. The second step Mill calls "ratiocination," which involves logical and mathematical computations to determine the consequences of the causal claims. The third step is verification: the "conclusions must be found, on careful comparison, to accord with the results of direct observation wherever it can be had" (1959, p. 303). What Whewell omits, says Mill, is the crucial inductive step at the beginning.

It is indeed possible to use Mill's "deductive method" to infer the existence of something unobservable from what is observable. Suppose we observe certain effects in a system 1, and we also observe that the same or very similar effects are produced in system 2 by a cause C; then we may infer that a cause of the same or a similar type is operating in system 1, even if that cause is not observable. Once such causal-inductive reasoning is employed, Mill's second step, ratiocination, can be used to generate other observational consequences that can be verified empirically.

327

The underlying idea can be given a simple probabilistic formulation. Suppose that on the basis of an observed phenomenon O_1 we inductively infer some causal hypothesis T with high probability. We may write this as $p(T/O_1) > k$, which means that the probability of T on O_1 is greater than some number k that represents a threshold for high probability (say $k = \frac{1}{2}$). Now suppose from T we deduce other observable phenomena $O_2, \ldots, O_n$. It is a theorem of probability that the probability of T on $O_1, \ldots, O_n$ must remain at least as high as it was on O_1 alone; that is, $p(T/O_1, \ldots, O_n)$ is at least as great as $p(T/O_1)$. So if $p(T/O_1) > k$, then $p(T/O_1, \ldots, O_n) > k$. Deriving additional observable phenomena from a theory sustains its high probability and may increase it. Under these circumstances there can be no incompatible theory T′ that entails the same observational conclusions whose probability is also high.

This type of reasoning, or some more complex variation of it, is frequently employed by scientists postulating unobservables. Nineteenth-century wave theorists, for example, typically begin with a range of observed optical phenomena, including rectilinear propagation, reflection, refraction, diffraction, the finite velocity of light, and many others. From a small subset S of these, particularly diffraction and the finite velocity of light, the wave theorist makes causal-inductive inferences to the high probability of the wave conception. These inferences involve two steps. First, from the observed fact that light travels from one point to another in a finite time, and that in nature one observes such motion occurring only by the transference of particles or by a wave movement, he inductively infers that it is probable that light is *either* a wave phenomenon *or* a stream of particles. Then he shows how in explaining observed diffraction patterns the particle theorist introduces forces that, on inductive grounds, are very unlikely to exist, since known forces in nature that have been observed are very dissimilar to the ones postulated. From this, together with the fact that if light were composed of particles, such forces would be very probable, it can be shown to follow that the particle theory is very improbable, given the subset S of observed optical phenomena and background information about known forces in nature. Since he has already argued that it is probable that light is composed either of waves or of particles, he concludes that the wave hypothesis is probable, given S. The steps so far involve inductive reasoning from similar observed effects to similar causes. Finally, he shows how the remaining known optical phenomena can be derived from the wave conception (without introducing any improbable hypotheses), thus at least sustaining the high probability of that theory on the basis of a large range of observations, not just S. It is in this manner that wave theorists argued from observed phenomena to unobservables. (For details see Achinstein 1991.)

Mill, indeed, was historically inaccurate in his criticism of the wave theory. Nineteenth-century wave theorists did not use the method of hypothesis to infer the existence of light waves and an ether. They used arguments of a sort that Mill himself could have accepted, since these arguments can be shown to be in accordance with his own "deductive method."

How should theories about unobservables be construed?

Suppose that Mill's "deductive method" or some other gives a basis for making inferences to theories about unobservables. This does not settle the issue of how such theories should be understood. Traditionally, two opposing views have been held, realism and anti-realism (or "instrumentalism"). These doctrines take many different forms. One prominent type of realism makes these two claims: (1) Entities exist that (for one reason or another – see next section) may be regarded as unobservable; they exist independently of us and of how we may describe them in our theories; they include light waves, electrons, quarks, genes, etc. (2) Scientific theories which purport to describe such entities are either true or false, and are so in virtue of their relationship to the entities themselves. Thus, the claim "Protons are positively charged" is true (we suppose) in virtue of the fact that the term "proton" denotes some real entity which exists independently of us and our theories, and which does have the property of being positively charged.

By contrast, one typical form of anti-realism is this: (A) There is no realm of independently existing unobservables. The only entities that have such existence are observable ones. (B) Our scientific theories which purport to describe unobservables should not be construed as making assertions that are either true or false. Instead, these theories should be treated as instruments for making inferences from some statements about observables to others. Statements about observables, by contrast, should be construed realistically. Another version of anti-realism accepts (A) but substitutes the following for (B): Theories about unobservables are true or false, but not because of their relationship to some independently existing realm of unobservables. "True" does not mean "corresponds with some independent reality," but something like "organizes our observations in the simplest, most coherent, manner." On both versions of anti-realism, there can be several conflicting theories about unobservables that are equally acceptable – that is, that organize our observations equally well – there being no realm of independently existing unobservables to establish or falsify any of these theories.

In recent years different formulations of realism and anti-realism (see REALISM AND INSTRUMENTALISM) have been proposed by van Fraassen (1980), who suggests that both viewpoints be construed as speaking not about what exists but about the aims of science. Realism should say that science in its theories aims to present a literally true story of what the world is like; and accepting such a theory involves believing it is true. By contrast, anti-realism, at least of the sort van Fraassen himself supports, proposes a different aim for science: to provide theories that are "empirically adequate." A theory satisfies this condition if, roughly, all observable events are derivable from it or at least compatible with it. Furthermore, accepting a theory does not involve believing that it is true, only that it is empirically adequate. This version of anti-realism, which van Fraassen calls "constructive empiricism," does not deny the existence of unobservables. It claims only that it is not the aim of science to produce true theories about them. A false theory can be acceptable so long as it "saves the phenomena."

Although the formulations of the method of hypothesis and inductivism given earlier might seem to presuppose realism, this is not necessarily the case. Both the H-D method and inductivism can, but need not, be construed realistically as allowing inferences to the truth (or the probability of the truth) of theories about an independently existing realm of unobservables. But an anti-realist can also use either method to make inferences about the usefulness of theories in organizing what is observable, or about the empirical adequacy of such theories. He can do so without a commitment to realism.

329

The realism/anti-realism debate has a long, honored history. Here it must suffice briefly to give just a few of the arguments used in favor of both positions without attempting to settle the issue. First, three arguments given by realists:

1. *An appeal to common sense.* Realism is an intuitive view that makes the best sense of what scientists in fact say. Scientists frequently speak of theories about atoms, electrons, photons, etc. as being true or false (e.g., J. J. Thomson's plum pudding theory of the atom is false; the contemporary quantum theory is true). And they speak of atoms, electrons, and photons as if they were independently existing constituents of the universe.

2. *The "miracle" argument, or "inference to the best explanation."* Suppose a theory T "saves the phenomena." The best explanation of this fact is realism: that is, the theory is true in the realist's sense: the entities it describes exist and have the properties the theory ascribes to them (see INFERENCE TO THE BEST EXPLANATION). If the theory is not true in this sense, it would be a miracle that it "saves the phenomena" (see Putnam 1975, p. 73).

3. *The "principle of common cause" argument.* Suppose there is a correlation between two observable facts or events A and B. Then either A causes B, or B causes A, or some third thing C causes both A and B. But sometimes there is a correlation between A and B where A does not cause B and B does not cause A and where no *observable* thing causes both A and B. For example, in geometrical optics there is a correlation of this sort between the fact that light can be reflected and the fact that it can be refracted; any ray of light with one property has the other. The principle of the common cause then allows us to postulate some unobservable C that causes both A and B – for example, light waves. This sanctions an inference to the real existence of such entities (see Salmon 1975).

Next, several arguments used by anti-realists:

1. *An appeal to empiricism.* Anti-realism is empirically more satisfying than realism. No mysterious, unobservable, unknowable world underlying what can be observed needs to be postulated. A world beyond the appearances is metaphysical and scientifically otiose. Or, if such a world exists, it is not one the scientist can know about (see Duhem 1982).

2. *An appeal to ontological simplicity.* Anti-realism (in many versions, but not van Fraassen's) is ontologically more satisfying than realism. For the realist, there is both a realm of unobservable entities and a realm of observable ones. According to many formulations of anti-realism, the world is much simpler, since it contains only observable entities. Talk about unobservables is legitimate, but only as a way of organizing our knowledge of observables.

3. *An appeal to scientific aims and practices.* Even if realism reflects how scientists speak on some occasions, anti-realism better represents their underlying aims and actual practice. What scientists really aim to do is not to produce true statements about an independently existing realm of unobservables, but to "save the phenomena." This is particularly evident from the fact that they frequently use theories and models that are incompatible. For example, some nineteenth-century physicists used

the wave theory to explain certain optical phenomena (e.g., interference) and the particle theory to explain others (e.g., dispersion). A realist, for whom truth is essential, must abhor such inconsistencies. An anti-realist, for whom "saving the phenomena" is all-important, can tolerate and even encourage this practice.

These are but a few of the arguments employed. Each side has sharp criticisms of, and responses to, its opponents' claims. Of the arguments not mentioned, one of the most powerful employed against the anti-realist is that he, but not the realist, must draw a certain type of questionable distinction between observables and unobservables. Let us turn finally to this issue.

What are "observables" and "unobservables" anyway?

In the twentieth century it was the logical positivists, particularly Rudolf Carnap and Carl G. Hempel, who first emphasized the importance of the distinction between observables and unobservables for understanding theories in science. (For a history of this topic in the twentieth century, see Kosso 1989.) The logical positivists, who defended a form of anti-realism, did not define the categories of "observables" and "unobservables," but they gave examples of what they meant. (Electrons are unobservable; tracks left by electrons in bubble chambers are observable.) They made various assumptions about this distinction: first, that although there are borderline cases that fit neither category, there is a clear distinction to be drawn, in the sense that there are obvious cases of unobservables (e.g., electrons) and obvious cases of observables (e.g., electron tracks); second, that what is observable does not depend upon, or vary with, what theory one holds about the world (the category of observables is "theory-neutral"); third, that the distinction does not depend on, or vary with, what questions scientists may be asking or different contrasts they may be making in disparate contexts of inquiry (it is "context-neutral"); fourth, that it would be drawn in the same way by different scientists and philosophers, including realists and anti-realists; fifth, that what is observable is describable in some unique or best way, using a special "observational vocabulary," by contrast with a "theoretical vocabulary" to be used for describing unobservables postulated by theories. All these assumptions but the last are also made by van Fraassen in his version of anti-realism. Are these assumptions reasonable?

Each of them has been challenged by some philosophers and defended by others. One of the best-known challenges is to the second assumption. The objection has been raised by Hanson (1958), Feyerabend (1962), and Kuhn (1962) that observation is "theory-laden": that what you see or observe depends crucially on what theories you hold. So, for example, when Aristotle and Copernicus both looked at the Sun, they literally saw different objects. The former saw a body in motion around the Earth; the latter a stationary body around which the Earth and other planets revolve. There is no theory-neutral category of observables.

Among the best replies to this objection is one due to Dretske (1969). A distinction needs to be drawn between "epistemic" and "non-epistemic" seeing or observing, between (roughly) seeing *that* X is P and seeing X. The epistemic sense involves beliefs or theories about what is seen or about the seeing of it; the non-epistemic sense does not.

331

On Dretske's view, in the non-epistemic sense, if person A sees X, then X looks some way to A, and this is such as to allow A to visually differentiate X from its immediate environment. So Aristotle and Copernicus could both see the Sun – the same Sun – if the Sun looked some way to each that would allow each to visually differentiate it from its immediate environment. Aristotle and Copernicus held quite different theories about the Sun, and the Sun could look different to each; yet we need not conclude that each saw something different. If the Sun was observable to one scientist, it was to the other as well, independently of the theories they held. (See Fodor 1984 for a more recent defense of the theory neutrality of observation.)

Another challenge questions both assumptions 1 and 3. One can observe X by observing its effects, even though in some sense X is "hidden from view." The forest ranger can observe a fire in the distance, even though all he can see is smoke; the artillery officer can observe enemy planes, even though all he can see are jet trails; the astronomer may be observing a distant star, though what he sees is its reflection in a telescopic mirror. In such cases, one observes X by attending to some Y associated with X in some way. In these cases, Y is produced by X, and attending to Y, given the present position of the observer, is the usual way – perhaps the only way – to attend to X from that position. (For more details, see Achinstein 1968, ch. 5.) These are standards for observation employed by scientists. Using them, many, if not most, of the items classified by positivists and others as unobservable are indeed observable. Electrons, to take the most often cited example, are observable in bubble chambers by visually attending to the tracks they leave. Physicists themselves speak in just such a manner. Moreover, as Shapere (1982) emphasizes, science itself is continually inventing new ways to observe, new types of receptors capable of receiving and transmitting information; the realm of what counts as observable in science is always expanding.

One response is to say that in the circumstances mentioned, electrons are not *directly* observable (Hempel 1965), or are not *themselves* observable (only their effects are) (see van Fraassen 1980, p. 17). The problem with such a response is that what counts as "directly" observing, or observing the thing "itself," depends on what contrast one seeks to draw, and this can vary contextually. If a physicist claims that electrons are not directly observable, he may mean simply that instruments such as bubble chambers and scintillation counters are necessary. On the other hand, physicists frequently say that electrons are directly observable in bubble chambers, but that neutrons are not. Here the contrast is not between observing with and without instruments, but between observing a charged particle such as an electron which leaves a track and observing a neutral particle such as a neutron which does not produce a track but must cause the ejection of charged particles – for example, alpha particles – that do leave tracks. Which contrast to draw depends on the interests and purposes of the scientists involved. Whether the entity is classified as "observable" (or "directly" observable, or "observable itself") shifts from one context to another. There is no unique, or preferred, distinction mandated either by what exists to be observed or by the use of the term "observable." These points directly challenge the first and third assumptions about the observable/unobservable distinction.

A related claim has been made by Churchland (1985) in questioning the first and fourth assumptions. There are, Churchland notes, many reasons why some entity or process may be unobserv*ed* by us, including that it is too far away in space or time, too small, too brief, too large, too protracted, too feeble, too isolated, etc. Now not all such unobserved entities and processes are deemed "unobservable" by proponents of the distinction. Yet is there any clear principle here for dividing the list into "items that are observable but have not been observed" and "items that are unobservable"? (See also Rynasiewicz 1984.) Is there any such principle that will bear the burden that anti-realists place on this distinction: namely, that while there is a realm of independently existing observables, no such realm of unobservables exists, or that the aim of science is to make true statements about observables, not about unobservables? Churchland finds no principle clear or powerful enough for the job. Distinctions can be made, but they are pragmatic and contextual.

Anti-realists need a fundamental distinction between observables and unobservables – one that is generated by some clear principle, that does not change from one context of inquiry to another, and that yields as "observable" and "unobservable" the sorts of entities that anti-realists cite. Realists are justified in claiming that such a distinction has not been drawn. Moreover, realists do not require a distinction between "observables" and "unobservables" *of the sort noted above*. They can recognize the contextual character of this distinction, that there are different respects in which, for example, electrons are or are not observable, some contexts calling for one, some for others. Realists, of course, are concerned with whether electrons exist, whether theories about them are true or false, and how one establishes that they are. But such questions can be answered without an observable/unobservable distinction of the sort that anti-realists require. In particular, the epistemic question discussed earlier – how to argue for a theory postulating entities such as light waves or electrons, whether inductively or hypothetico-deductively – does not require the anti-realist's categories.

References

Achinstein, P. 1968: *Concepts of Science* (Baltimore: Johns Hopkins University Press).

—— 1991: *Particles and Waves* (New York: Oxford University Press).

Churchland, P. 1985: The ontological status of observables: in praise of the superempirical virtues. In *Images of Science*, ed. P. Churchland and C. Hooker (Chicago: University of Chicago Press).

Dretske, F. 1969: *Seeing and Knowing* (London: Routledge and Kegan Paul).

Duhem, P. 1982: *The Aim and Structure of Physical Theory* (Princeton: Princeton University Press).

Feyerabend, P. 1962: Explanation, reduction, and empiricism. In *Minnesota Studies in the Philosophy of Science*, vol. 3, ed. H. Feigl and G. Maxwell (Minneapolis: University of Minnesota Press), 28–97.

Fodor, J. 1984: Observation reconsidered. *Philosophy of Science*, 51, 23–43.

Hanson, N. 1958: *Patterns of Discovery* (Cambridge: Cambridge University Press).

Hempel, C. 1965: The theoretician's dilemma. In *Aspects of Scientific Explanation* (New York: Free Press), 173–226.

Kosso, P. 1989: *Observability and Observation in Physical Science* (Dordrecht: Kluwer).

Kuhn, T. 1962: *The Structure of Scientific Revolutions* (Chicago: University of Chicago Press).

Mill, J. S. 1959: *A System of Logic* (London: Longmans).

Popper, K. 1959: *Logic of Scientific Discovery* (London: Hutchinson).

Putnam, H. 1975: *Mathematics, Matter, and Method* (Cambridge: Cambridge University Press).

Reichenbach, H. 1938: *Experience and Prediction* (Chicago: University of Chicago Press).

Rynasiewicz, R. 1984: Observability. In *PSA 1984*, vol. 1, 189–201.

Salmon, W. C. 1975: Theoretical explanation. In *Explanation*, ed. S. Korner (Oxford: Blackwell), 118–45.

Shapere, D. 1982: The concept of observation in science. *Philosophy of Science*, 49, 485–525.

van Fraassen, B. 1980: *The Scientific Image* (Oxford: Oxford University Press).

Whewell, W. 1967: *Philosophy of the Inductive Sciences* (New York: Johnson Reprint Corporation).

49

Peirce

CHERYL MISAK

Charles Sanders Peirce (1839–1914) is generally acknowledged to be America's greatest philosopher, although he was never able to secure a permanent academic position and died in poverty and obscurity (see Brent 1993). He founded pragmatism, the view that a philosophical theory must be connected to practice. The pragmatic account of truth, for which he is perhaps best known, thus has it that a true belief is one which the practice of inquiry, no matter how far it were to be pursued, would not improve. (See Misak 1991 for an elaboration and defense of this position.) He is also considered the founder of semiotics and was on the cutting edge of mathematical logic, probability theory, astronomy, and geodesy. (See Brent 1993 for an account of Peirce's achievements in science.)

Pragmatism is a going concern in philosophy of science today. It is often aligned with instrumentalism (see REALISM AND INSTRUMENTALISM), the view that scientific theories are not true or false, but are better or worse instruments for prediction and control. But you will not find in Peirce's work the typical instrumentalist distinction between theoretical statements about unobservable entities (mere instruments) and statements about observable entities (candidates for truth and falsity). For Peirce identifies truth itself with a kind of instrumentality. A true belief is the very best we could do by way of accounting for the experiences we have, predicting the future course of experience, etc.

A question about underdetermination arises immediately: what if two incompatible theories are equally good at accounting for and predicting experience? (see UNDERDETERMINATION OF THEORY BY DATA). Must Peirce say that they are both true? To understand Peirce's response here, we must focus first on the fact that his view of truth is set against those realist views which hold that the truth of a statement might well be unconnected with properties which make it worthy of belief. On this kind of view, a statement is true if and only if it "corresponds to" or "gets right" the world, and this relationship holds or fails to hold independently of human beings and what they find good in the way of belief. It is these realist views which insist that only empirical adequacy can count for or against the truth of a belief; hence the problem of underdetermination of theory by empirical data is especially pressing for them. (Peirce would argue that it is unclear how they are entitled to insist even upon empirical adequacy, for they are committed to the thought that, if a belief stands in a correspondence relationship with the world, then it is true, regardless of the evidence that might be for or against it.)

Peirce, on the other hand, explicitly links truth with what we find worthy to believe. Thus, factors such as simplicity, economy, etc. (see SIMPLICITY), if we find that they are useful in theory choice, can determine between theories underdetermined by the data; empirical adequacy is not the only thing relevant to the truth or to what would be the best theory to believe.

Indeed, Peirce was the first to put a name to, and discuss, a kind of inference that is important here, inference to the best explanation, or what Peirce called "abduction" (see INFERENCE TO THE BEST EXPLANATION). It is "the process of forming an explanatory hypothesis" (5.171), and its form is as follows (5.189):

> The surprising fact, C, is observed;
> But if A were true, C would be a matter of course.
> Hence, there is reason to suspect that A is true.

If we are faced with two hypotheses with the same empirical consequences, but one hypothesis has been actually inferred as the best explanation of what we have observed, then that hypothesis, Peirce holds, has something going for it over the other.

This thought has implications for the kind of paradoxes of confirmation pointed to by Nelson Goodman. (See CONFIRMATION, PARADOXES OF and Goodman 1955.) Suppose we define a predicate "grue": an object is grue if it is observed to be green and first examined before 1 January 1999 or if it is first examined after 1 January 1999 and is observed to be blue. The hypothesis "All emeralds are grue" has just as much inductive or empirical support as does the hypothesis "All emeralds are green," and Goodman asks why we should think that "green," not "grue," is the appropriate predicate to apply here. Since inductive inference rests solely on the enumeration of observed instances, we cannot, it seems, appeal to non-enumerative considerations to support the "green" hypothesis.

On Peirce's view of the scientific method, one arrives at hypotheses through abduction, derives predictions from them via deduction, and then tests the predictions via induction (see SCIENTIFIC METHODOLOGY). He would argue that the choice between the "green" and "grue" hypotheses is not a matter for induction, but rather, for abduction. We infer that all emeralds are green as the best explanation of the observed regularity that they have all thus far been green. If we are still interested in the matter, we will test this hypothesis, and the fact that it was the conclusion of an abductive inference will give it a weight that not just any inductively tested generalization would have.

But if pragmatic factors (see PRAGMATIC FACTORS IN THEORY ACCEPTANCE) such as inference to the best explanation, simplicity, and the like still fail to distinguish one empirically equivalent hypothesis or theory from another, then Peirce would advise us to hope or assume that inquiry will eventually resolve the matter. For he is a thoroughgoing fallibilist, arguing that, since we can never know that inquiry has been pushed as far as it could go on any matter, we can never know whether we have the truth in our hands. Similarly, we can never know whether two theories would be underdetermined, were we to have all the evidence and argument we could have.

The possibility remains, of course, that the theories in question would be underdetermined, no matter how far we pursued inquiry. In that case, Peirce would hold that either the theories are equivalent (the pragmatic maxim is, as William James once put it, that there is no difference which makes no difference) or that there is no truth of the matter at stake. Again, because Peirce's account of truth does not rest on the realist assumption that every statement either corresponds or fails to correspond to a bit of determinate reality, he has no qualms about admitting such a possibility. This, however, does not alter the fact that it is a regulative assumption of inquiry that, in any matter into which we inquire, there is a determinate answer to our question.

The pragmatic maxim also links Peirce to logical positivism (see LOGICAL POSITIVISM). For that maxim, like the verifiability principle, requires all hypotheses to have observable consequences. The logical positivists notoriously concluded that the only legitimate area of inquiry is physical science. Areas of inquiry such as psychology and ethics are viable only insofar as they restrict themselves to observable phenomena, such as the overt behaviour of their subjects. An exception is made for mathematics and logic, which are supposed to have a special kind of meaning, and thus be exempt from meeting the requirements of a criterion which they would, it seems, clearly fail. True mathematical and logical statements are supposed to be analytically true, or true come what may.

Talk of analytic statements has been out of fashion since Quine, and verificationism has been out of fashion since the logical positivists' attack on much of what is thought to be legitimate inquiry (see QUINE). Peirce, however, provides the materials with which to retain the thought that our hypotheses and theories must be connected to experience without the baggage of the analytic/synthetic distinction and without the consequence that we must abandon areas of inquiry which we have no intention of abandoning.

Peirce holds that mathematics is an area of inquiry which is open to experience and observation, for these notions are, for him, very broad. Experience is a shock or something that impinges upon one. It is not explicitly connected to the senses, for anything that we find undeniable, compelling, surprising, or brute is an experience. Thus Peirce holds that arguments and thought experiments can be compelling; experiences can be had in those contexts (see THOUGHT EXPERIMENTS). The mathematician or logician, for instance, carries out experiments upon diagrams of his own creation, imaginary or otherwise, and observes the results. He is sometimes surprised by what he sees; he feels the force of experience (see 3.363, 4.530).

This novel view of mathematics and logic (see Putnam 1992 for a discussion) as an area of diagrammatic thought leads naturally to an empiricism which requires all areas of inquiry, including mathematics and logic, to be connected to experience. We are presented with a picture of the unity of science, one which does not set up a dichotomy between different kinds of statements, and one which does not prejudice certain kinds of inquiry at the outset (see UNITY OF SCIENCE).

Peirce is also known for his work on induction (see INDUCTION AND THE UNIFORMITY OF NATURE) and the theory of probability (see PROBABILITY). Some see in his writing an anticipation of Reichenbach's probabilistic response to Hume's skepticism about induction (see Ayer 1968 and HUME). Peirce at times does put forward a view like Reichenbach's, where induction is self-correcting, in that probability measures attached to hypotheses mathematically converge upon a limit (see 2.758). But at other times he says, "it may be conceived . . . that induction lends a probability to its conclusion. Now that is not the way in which induction leads to the truth. It lends no definite probability to its conclusion. It is nonsense to talk of the probability of a law, as if we could pick universes out of a grab-bag and find in what proportion of them the law held good" (2.780). (Thus Peirce is also not best thought of as holding that science will get closer and closer to the truth. See VERISIMILITUDE.)

337

Peirce called the sort of inference which concludes that all *A*'s are *B*'s because there are no known instances to the contrary "crude induction." It assumes that future experience will not be "utterly at variance" with past experience (7.756). This is, Peirce says, the only kind of induction in which we are able to infer the truth of a universal generalization. Its flaw is that "it is liable at any moment to be utterly shattered by a single experience" (7.157).

The problem of induction, as Hume characterizes it, concerns crude induction; it is about the legitimacy of concluding that all *A*'s are *B*'s or that the next *A* will be a *B* from the fact that all observed *A*'s have been *B*'s. Peirce takes Hume's problem to be straightforwardly settled by fallibilism. We do, and should, believe that, say, the sun will rise tomorrow; yet it is by no means certain that it will. To show that induction is valid, we need not show that we can be certain about the correctness of the conclusion of a crude inductive inference. For the fallibilist holds that nothing is certain.

And against Mill, Peirce argues that induction does not need the "dubious support" of a premise stating the uniformity of nature (6.100; see MILL). What we have to show, rather, is that induction is a reliable method in inquiry. Thus, Hume's problem is bypassed, and not dealt with in Reichenbach's way.

Peirce holds that it is a mistake to think that all inductive reasoning is aimed at conclusions which are universal generalizations. The strongest sort of induction is "quantitative induction," and it deals with statistical ratios. For instance:

Case: These beans have been randomly taken from this bag.
Result: $^2/_3$ of these beans are white.
Rule: Therefore $^2/_3$ of the beans in the bag are white.

That is, one can argue that if, in a random sampling of some group of *S*'s, a certain proportion r/n has the character *P*, then the same proportion r/n of the *S*'s have *P*. From an observed relative frequency in a randomly drawn sample, one concludes a hypothesis about the relative frequency in the population as a whole.

Peirce is concerned with how inductive inference forms part of the scientific method, how inductive inferences can fulfill their role as the testing ground for hypotheses. Quantitative induction can be seen as a kind of experiment. We ask what the probability is that a member of the experimental class of *S*'s will have the character *P*. The experimenter then obtains a fair sample of *S*'s, and draws from it at random. The value of the proportion of *S*'s sampled that are *P* approximates the value of the probability in question. When we test, we infer that if a sample passes the test, the entire population would pass the test. Or we infer that if 10 percent of the sample has a certain feature, then 10 percent of the population has that feature.

It is in the context of the scientific method that some have seen Peirce's best insights regarding induction and probability. (For instance, Levi (1980) and Hacking (1980) have argued that Peirce anticipates the Neyman–Pearson confidence interval approach to testing statistical hypotheses.)

References and further reading

Works by Peirce

1931–58: *Collected Papers of Charles Sanders Peirce*, ed. C. Hartshorne and P. Weiss (vols 1–6) and A. Burks (vols 7 & 8) (Cambridge, MA: Belknap Press, Harvard University Press). [References in text refer, as is standard, first to volume number, then to paragraph number.]

1982: *Writings of Charles S. Peirce: A Chronological Edition*, ed. M. Fisch (Bloomington: Indiana University Press).

Works by other authors

Ayer, A. J. 1968: *The Origins of Pragmatism* (London: Macmillan).

Brent, J. 1993: *Charles Sanders Peirce: A Life* (Bloomington: Indiana University Press).

Goodman, N. 1955: *Fact, Fiction, and Forecast* (Cambridge, MA: Harvard University Press).

Hacking, I. 1980: The theory of probable inference: Neyman, Peirce and Braithwaite. In *Science, Belief and Behavior: Essays in Honour of R. B. Braithwaite*, ed. D. H. Mellor (Cambridge: Cambridge University Press), 141–60.

Levi, I. 1980: Induction as self-correcting according to Peirce. In *Science, Belief and Behavior: Essays in Honour of R. B. Braithwaite*, ed. D. H. Mellor (Cambridge: Cambridge University Press), 127–40.

Misak, C. J. 1991: *Truth and the End of Inquiry: A Peircean Account of Truth* (Oxford: Clarendon Press).

Putnam, H. 1992: Comments on the lectures. In *Reasoning and the Logic of Things: The Cambridge Conferences Lectures of 1898*, ed. Kenneth Laine Ketner (Cambridge, MA: Harvard University Press), 55–102.

50

Physicalism

WILLIAM SEAGER

The crudest formulation of physicalism is simply the claim that everything is physical, and perhaps that is all physicalism ought to imply. But in fact a large number of distinct versions of physicalism are currently in play, with very different commitments and implications. There is no agreement about the detailed formulation of the doctrine, even though a majority of philosophers would claim to be physicalists, and a vast majority of them *are* physicalists of one sort or another. There are several reasons for this lack of agreement: deep and imponderable questions about realism and the legitimacy of a narrowly defined notion of *reality*, questions about the scope of the term "physical," issues about explanation and reduction, and worries about the proper range of physicalism, particularly with respect to so-called abstract objects such as numbers, sets, or properties. And while the traditional challenge of psychological dualism poses a live threat only to a few iconoclasts, the project of specifying a physicalism which plausibly integrates mind into the physical world remains unfinished, despite the huge number and diversity of attempts.

It would be natural to begin the delineation of physicalism by contrasting it with the older doctrine of materialism, but this contrast is itself vague and murky. For a great many philosophers "physicalism" and "materialism" are interchangeable synonyms. But for some, "physicalism" carries an implication of reductionism, while materialism remains a "purely" ontological doctrine (one which perhaps better respects the ancient roots of the doctrine, which wholly predates science itself). Still others invoke a contrast between the traditional materialist view that all that exists is *matter* (and its modes) and the modern view in which the physical world contains far more than mere matter, and in which the concept of matter has been significantly transformed. This latter sense of the term also expresses the proper allegiance of the physicalist, which is unabashedly to the science of physics: physics will tell us what the fundamental components of the world are, from which all else is constructed.

Here is the rub. How are we to understand this relation of "construction" which holds between the "fundamental" features of the world discovered by physics and everything else? Ironically, it was the tremendous growth in scientific activity in the nineteenth and twentieth centuries that forced the issue. For the sciences grew not as offshoots of physics, but rather as more or less independent, autonomous disciplines with their own domains and methodologies (especially problematic were the most recent entrants to the fold: psychology and the social sciences). For those seeking to embrace a physicalist outlook – at first mostly the vigorous and philosophically evangelical *logical positivists* (see LOGICAL POSITIVISM) – the proliferation of scientific disciplines demanded a reductionist reply that combined, or perhaps confused, onto-logical and epistemological issues. The positivist slogan to be underwritten by reduction was "unity of science," in which scientific legitimacy was granted by tracing a dis-cipline's pedigree back to the patriarchal science of physics (see UNITY OF SCIENCE).

In a remarkable burst of philosophical energy, marshaling modern logical technique and results, as well as a good acquaintance with science itself (especially physics), the doctrine of reductionism was quickly worked out (see REDUCTIONISM). In essence, reduc-tion permitted the translation of any statement from a reduced theory into an exten-sionally equivalent statement of physics. Reduction being transitive, there was no need for all theories to reduce directly to physics. Reduction could proceed step by step through the scientific hierarchy. A possible example of such a reductive hierarchy might be sociology reducing to psychology, psychology to neuroscience, neuroscience to chem-istry, and chemistry to physics. The positivist theory of explanation also entailed that explanations would reduce along with their theories. Altogether, it was a pretty picture.

In part because of its virtues of logical precision and clarity, the faults of positivist reductive physicalism have become ever more apparent. Its demands on scientific theo-rizing are too stringent, its vision of explanation too restricted, its view of science overly logicist and formal, its treatment of the history of science naive, and its view of the culture of science hopelessly crude and rather blindly laudatory.

It is possible to take another tack, to view physicalism as a purely ontological doc-trine which asserts no more than that everything is at bottom physical. Such a view *leaves* questions of inter-theoretic relations *as* questions of inter-theoretic relations. Our theories represent a myriad of distinct, though methodologically related, approaches to a world that is through and through physical; but there is no need to demand, and little reason to expect, that this myriad will align itself along the narrow line envisioned by the positivists. Some theories may so align themselves, especially if their domains are significantly overlapping, and there may be some special cases where positivist reduction can almost be achieved, such as in thermodynamics (though the extent to which this case meets the model of reduction is moot). On the positive side, this mild physicalism replaces reduction with *supervenience*. Specifically, supervenient physicalism requires that everything supervene on the physical, by which is crudely meant: no difference, no alteration without a physical difference or alteration. For example, though there are no prospects for reducing economics to physics, it seems plausible to suppose that there could not be a change in the bank rate without there being some physical changes which "underlie" or "ground" or, better, *subvene* the economic change. This doctrine is *empirically* plausible – we have lots of evidence that all change is dependent upon the physical structure of things. By and large, we don't understand the details of this dependency; maybe, for various reasons, including mere complexity, we will never understand the details. This need not lead us to reject a proper physicalist view of the world. Such a physicalism also imposes some con-straints on scientific theorizing in that it forbids postulation of nonphysical agencies in sciences other than physics (and, in physics, the postulation of a nonphysical agency is impossible, though often enough new *physical* agencies are postulated which transcend the current catalog of the physical).

Of course, the extent to which the reductionist dream can be fulfilled and the limits of what could reasonably be labeled a "reduction" remain open questions. It is another virtue of supervenient physicalism that it applauds and encourages the attempts to explore inter-theory relationships; it frowns only on the absolutist dictate that all such relations must conform to a single model. Science, and the world as well, is too complicated to ground any confidence in reductionism. The primary virtue of supervenient physicalism is its acceptance of the complexity of the world and our theorizing about it, along with denial that this entails any ontological fracturing of our scientific image of the world.

Further reading

Beckermann, A., Flohr, H., and Kim, J. (eds) 1992: *Emergence or Reduction: Essays on the Prospects of Nonreductive Physicalism* (Berlin: de Gruyter).

Davidson, D. 1980: Mental events. In *Essays on Actions and Events* (Oxford: Oxford University Press). (The first presentation of a nonreductive, supervenient physicalism.)

Field, H. 1980: *Science without Numbers* (Princeton: Princeton University Press). (An attempt to present a purely physicalist account of the use of mathematics in science.)

Fodor, J. 1981: Special sciences. In *RePresentations* (Cambridge, MA: MIT Press). (Argues that anti-reductionism poses no threat to physicalism or to scientific practice.)

Hellman, G., and Thompson, F. 1977: Physicalism: ontology, determination and reduction. *Journal of Philosophy*, 72, 551–64. (A well worked-out version of supervenient physicalism.)

Kim, J. 1989: The myth of nonreductive materialism. *Proceedings of the American Philosophical Association*, 63, 31–47. (Disputes whether supervenience is compatible with nonreductionism.)

Nagel, E. 1961: *The Structure of Science* (New York: Harcourt, Brace and World). (Classic treatise on philosophy of science with a beautiful discussion of reductionism.)

Oppenheim, P., and Putnam, H. 1958: Unity of science as a working hypothesis. In *Minnesota Studies in the Philosophy of Science*, vol. 2, ed. H. Feigl, M. Scriven, and G. Maxwell (Minneapolis: University of Minnesota Press), 3–36. (Classic statement of reductionism.)

Post, J. 1987: *The Faces of Existence* (Ithaca, NY: Cornell University Press). (An extremely wide-ranging attempt to provide a nonreductive supervenient physicalism.)

Suppe, F. (ed.) 1977: *The Structure of Scientific Theories*, 2nd edn (Urbana, IL: University of Illinois Press). (Suppe's extended introduction is, among other things, an excellent survey of classical reductive physicalism and its problems.)

51

Popper

JOHN WATKINS

Karl Raimund Popper was born in Vienna on 18 July 1902. He enrolled at the University of Vienna in 1918 and at the Pedagogic Institute in Vienna in 1925. He was a secondary schoolteacher for several years from 1930 on. His *Logik der Forschung* was published in 1934. It was brought to Einstein's attention by Frieda Busch, wife of the founder of the Busch Quartet, at the suggestion of her son-in-law, the pianist Rudolf Serkin, who was a good friend of Popper's. Einstein told Popper that, purified of certain errors he had noted, the book "will be really splendid"; a considerable correspondence ensued between them. The book also brought Popper invitations to England, where he spent much of 1935–6, meeting Susan Stebbing, Ayer, Berlin, Russell, Ryle, Schrödinger, Woodger, and, at the London School of Economics, Hayek and Robbins. In 1937 he took up a lectureship in philosophy at Canterbury University College, in Christchurch, New Zealand. There he wrote "The poverty of historicism" (1944–5) and *The Open Society and its Enemies* (1945); he called the latter his "war work." Both these works contain interesting ideas about the special problems and methods of the social sciences. Later works, including an autobiography (1976), are listed at the end of the chapter.

He went to the London School of Economics as Reader in Logic and Scientific Method in 1946, and was given the title of Professor in 1949. In 1950 he gave the William James Lectures at Harvard. At Princeton he gave a talk, attended by Bohr and Einstein, on indeterminism; this was a summary of his 1950 paper "Indeterminism in quantum physics and in classical physics." Bohr went on talking so long that, in the end, only the three of them were left in the room. Popper had several meetings with Einstein, and may have shaken the latter's faith in determinism; see the editorial footnote on p. 2 of Popper (1982a).

Popper's 1934 book had tackled two main problems: that of demarcating science from nonscience (including pseudo-science and metaphysics), and the problem of induction. Against the then generally accepted view that the empirical sciences are distinguished by their use of an inductive method, Popper proposed a falsificationist criterion of demarcation: science advances unverifiable theories and tries to falsify them by deducing predictive consequences and by putting the more improbable of these to searching experimental tests. Surviving such severe testing provides no inductive support for the theory, which remains a conjecture, and may be overthrown subsequently. Popper's answer to Hume was that he was quite right about the invalidity of inductive inferences, but that this does not matter, because these play no role in science; the problem of induction drops out. (This is discussed under criticism 1 below.)

What are theories tested against? The opening paragraph of *Logik* says that they are tested against *experience*. But in chapter 5, on the empirical basis, this gets revised. It is there emphasized that a hypothesis can stand in logical relations only to other *statements*. Then, is a scientific hypothesis to be tested against protocol statements à la Carnap and Neurath, whereby an observer testifies to having had a certain perceptual experience, the possibility that it was hallucinatory not being ruled out? No; Popper advocated a strictly nonpsychologistic reading of the empirical basis of science. He required "basic" statements to report events that are "observable" only in that they involve relative positions and movements of macroscopic physical bodies in a certain space-time region, and which are relatively easy to test. Perceptual experience was denied an epistemological role (though allowed a causal one); basic statements are accepted as a result of a convention, or agreement, between scientific observers. Should such an agreement break down, the disputed basic statements would need to be tested against further statements that are still more "basic" and even easier to test. (See criticism 2 below).

If a theory yields a law statement of the form "All F's are G," henceforth $(x)\,(Fx \rightarrow Gx)$, while a basic statement says that the object a is F and not-G, henceforth $Fa \wedge \sim Ga$, then that basic statement is a potential falsifier, henceforth PF, of the theory. Popper took its class of PFs as (a measure of) a theory's empirical content, and he proposed two ways of comparing theories for empirical content. The first relies on the subclass relation: T_2 has more empirical content than T_1, henceforth $Ct(T_2) > (Ct(T_1)$, if the PFs of T_1 are a proper subclass of those of T_2. The other, which we may call the *dimension* criterion, says that $Ct(T_2) > Ct(T_1)$ if minimal PFs of T_1 are more composite than those of T_2. Unlike Wittgenstein, Popper did not postulate propositions that are atomic in any absolute sense. But he did suggest that statements may be regarded as atomic relative to various sorts of measurement: for instance, competing statements of equal precision about an object's length, or weight, or temperature. And then a non-atomic basic statement might be regarded as a conjunction of relatively atomic statements, the number of such conjuncts determining its degree of compositeness. Suppose that T_1 says $(x)\,((Fx \wedge Gx) \rightarrow Hx)$, while T_2 says $(x)\,(Fx \rightarrow Gx)$; then we can say that $Ct(T_2) > Ct(T_1)$ on the ground that T_1 is of higher dimension than T_2, a minimal PF of T_1 such as $Fa \wedge Ga \wedge \sim Ha$ being more composite than a minimal PF of T_2 such as $Fa \wedge \sim Ga$. (Later, Popper (1972) introduced the idea that $Ct(T_2) > Ct(T_1)$ if T_2 can answer with equal precision every question that T_1 can answer, but not vice versa.) (See criticism 3 below.)

Following Weyl (who, as Popper subsequently acknowledged, had been anticipated by Harold Jeffreys and Dorothy Wrinch), Popper adopted a "paucity of parameters" definition of simplicity: T_2 is *simpler* than a rival T_1 if T_2 has fewer adjustable parameters than T_1. This means that the simpler T_2 is of lower dimension, and hence more falsifiable, than T_1. Other things being equal, the simpler theory is to be preferred, not because it is more probable (being more falsifiable, it will be *less* probable, unless they both have zero probability anyway), but on the methodological ground that it is *easier to test* (see SIMPLICITY).

The longest chapter of *Logik* is on probability (see PROBABILITY). Popper emphasized that probabilistic hypotheses play a vitally important role in science and are strictly unfalsifiable; then do they not defeat his falsificationist methodology? He sought to bring them into the falsificationist fold in the following way. Let "P(A|B) = r" mean that the probability of getting outcome A from experiment B is r (where $0 \leq r \leq 1$), and let m and n be respectively the number of A's and of B's in a long run. If "P(A|B) = r" is true, then as n increases, the value of m/n converges on r. Let $1 - \varepsilon$ be the probability that m/n lies in the interval $r \pm \delta$. Fixing any two of the variables n, δ, ε determines the third. Popper's first step was to fix δ by confining it within the limits of experimental error. Now ε can be made arbitrarily close to 0 and virtually insensitive to increases, even large ones, in δ by choosing a large enough value N of n. Popper laid down as a methodological rule that a hypothesis of the form P(A|B) = r should be treated as falsified if, after N repetitions of the experiment, the value of m/n is found to lie unambiguously outside the interval $r \pm \delta$, and this effect is reproducible.

With his 1950 article, Popper launched a campaign, which he continued in his 1965 book *Of Clouds and Clocks* and *The Open Universe* (1982a), against determinism: he sought to undermine metaphysical determinism, according to which the future is as fixed and unalterable as the past, by refuting the epistemological doctrine of "scientific" determinism which says that all future events are in principle scientifically predictable. (See criticism 4 below.)

His 1956 article "Three views concerning human knowledge" took off from the claim that an instrumentalist view of science of the kind the Church put upon Galileo's world system had now become the prevalent view among physicists. In opposition to it, he upheld, not the essentialist view (to which Galileo himself inclined) which assumes that science can attain ultimate explanations, but a conjecturalist version of scientific realism (see REALISM AND INSTRUMENTALISM). According to this, science aims at an ever deeper understanding of a multileveled reality; there is no reason to suppose that science will ever get to the bottom of it.

In his 1957c article Popper shifted from a frequency to a propensity interpretation of probabilistic hypotheses. This makes the probability a characteristic, not of a sequence, but of the experimental setup's propensity to generate certain frequencies; and it makes it possible to speak of the probability of a possible outcome of a single experiment – say, throwing a die once and then destroying it – without invoking hypothetical sequences. In his 1983 book on quantum theory and in his 1990 book, he used the propensity interpretation in his attempt at a realistic and paradox-free understanding of quantum mechanics.

In his 1957 article "The aim of science," Popper set out sufficient conditions for a new theory to have greater depth than the theory or theories it supersedes, this relation being well exemplified by the superseding of Kepler's and Galileo's laws, call them K and G, by Newton's theory, N: as well as going beyond the conjunction of K and G because of its greater generality, N corrects them, providing numerical results that are usually close to, but different from, theirs. Popper insisted that since N is strictly inconsistent with K and G, it cannot be regarded as inductively inferred from them.

His *Logik* came out in English translation in 1959, with new appendices. In Appendix *ix he sharply distinguished his falsificationist idea of corroboration from Carnap's quasi-verificationist confirmation function. Confirmability varies *inversely* with falsifiable content, and falsifiable law statements in infinite domains always have zero confirmation. To get round this, Carnap had introduced (1950, pp. 571f) what he called the "qualified-instance confirmation" of a law. Popper pointed out in his 1963b paper "The demarcation between science and metaphysics" that an unambiguously refuted law statement may enjoy a high degree of this kind of "confirmation."

In a historical sequence of scientific theories, it typically happens that a superseding theory T_2 implies that its predecessor T_1, though in some ways a good approximation, was in fact false; and perhaps T_2 is destined to be similarly superseded. If tests have gone in its favor, we can say that T_2 is better corroborated than its predecessors; but can we say that it is better with respect to *truth*, and if so, just what would this mean? This is the problem of verisimilitude (see VERISIMILITUDE). In his book *Conjectures and Refutations* (1963a, ch. 10 and addenda) Popper sought a solution along the following lines. The objective content of a theory being the set of all its consequences, its truth content is the set of its true consequences, and its falsity content is the set (which may be empty) of whatever false consequences it has; this allows us to say that T_2 is more truthlike than T_1, or $Vs(T_2) > Vs(T_1)$, if the truth content of T_1 is contained in that of T_2, and the falsity content of T_2 is contained in that of T_1, at least one of these containments being strict (but see criticism 5 below).

Down the years Popper's objectivism became increasingly pronounced (1972, 1977): human minds are needed to create scientific theories and other intellectual structures, but, once created, the latter do not need "knowing subjects" to sustain them; they stand on their own as objective structures in a quasi-Platonic "World 3." (See criticism 6 below.)

Criticisms

1. *The problem of induction.* Wesley Salmon (1981) and many others – e.g., Lakatos (1974) and Worrall (1989) – have urged that if Popper-type corroboration appraisals had no inductive significance, then science could provide no guidance to aeronautical engineers and others who make practical use of well-tested scientific theory.

2. *Empirical content.* None of Popper's criteria for $Ct(T_2) > Ct(T_1)$ works for the condition he laid down in his 1957a paper for T_2 to be deeper than T_1. Here the empirical content of T_2 goes beyond and revises that of T_1; it is rather as if T_1 predicted that all mice have long tails, while T_2 predicts that all rats and all mice have long tails if male and short tails if female. The subclass relation does not hold; the dimension criterion is either inapplicable or tells against T_2; and T_1, but not T_2, can answer the question "Is the mouse in this mouse-trap short-tailed?" I tried to repair this situation in Watkins 1984, ch. 5.

3. *The empirical basis.* Popper said that in testing theories we stop at basic statements that are especially easy to test. Presumably we go on actually to test these statements? One would hardly accept "There is a hippopotamus in your garage" merely on the ground that it is easy to test, and without actually testing it. But how are we to test them if perceptual experience is allowed no epistemological role? Popper said that we test a basic statement by deducing further, and even more easily testable, basic statements from it (with the help of background theories). This means that the statements at which we eventually stop have not, after all, been tested; all we are getting is an ever-lengthening chain of derivations (see Watkins 1984, pp. 249f).

4. *Indeterminism.* John Earman (1986, pp. 8–10) has objected that to refute an epistemological doctrine about the knowability of future events does not disturb the metaphysical doctrine that the future is fixed.

5. *Verisimilitude.* Miller (1974) and Tichy (1974) independently showed that we cannot have $Vs(T_2) > Vs(T_1)$ as defined above when T_1 and T_2 are both false. Proof: if at least one of the two containments is strict, there will be (i) a false consequence of T_1, call it f_1, that is not a consequence of T_2, or (ii) a true consequence of T_2, call it t_2, that is not a consequence of T_1. Let f_2 be a false consequence of T_2. Then $f_2 \wedge t_2$ will be a *false* consequence of T_2 but not of T_1, and $f_1 \vee \sim f_2$ will be a *true* consequence of T_1 but not of T_2. Much effort has been invested in trying to surmount these difficulties and provide a viable account of truthlikeness; see, for example, Niiniluoto 1987 and Oddie 1986.

6. *World 3.* Popper's World 3 differs from a Platonic heaven in that its contents have not existed timelessly (though once in, nothing ever drops out). It is man-made; and though it transcends its makers because it includes all unintended and unnoticed by-products of their conscious thoughts, it is supposed to go on expanding as new objects are added to it. However, as L. Jonathan Cohen (1980) pointed out, we know that some of its contents – for instance, Frege's *Grundgesetze* – are inconsistent; and it would have been filled up and would have stopped expanding on the arrival of the first contradiction, since a contradiction entails every proposition (as Popper has often insisted).

Popper was knighted in 1965; he became a Fellow of the Royal Society in 1976, and a Companion of Honour in 1982. He died in 1994.

References and further reading

Works by Popper

1934: *Logik der Forschung* (Vienna: Springer); English translation in (1959).

1944–5: The poverty of historicism; revised version in (1957b).

1945: *The Open Society and its Enemies*, 2 vols (London: Routledge and Kegan Paul); 5th edn 1966.

1950: Indeterminism in quantum physics and in classical physics. *British Journal for the Philosophy of Science*, 1, 117–33, 173–95.

1956: Three views concerning human knowledge; repr. in 1963a, pp. 97–119.

1957a: The aim of science; repr. in 1972, pp. 191–205.

1957b: *The Poverty of Historicism* (London: Routledge and Kegan Paul).

1957c: The propensity interpretation of the calculus of probability, and the quantum theory. In *Observation and Interpretation in the Philosophy of Physics*, ed. S. Körner (London: Butterworth Scientific Publications), 65–70.

1959: *The Logic of Scientific Discovery* (London: Hutchinson).

1963a: *Conjectures and Refutations* (London: Routledge and Kegan Paul).

1963b: The demarcation between science and metaphysics; repr. in (1963a), pp. 253–92.

1965: *Of Clouds and Clocks*; repr. in 1972, pp. 206–55.

1972: *Objective Knowledge: An Evolutionary Approach* (Oxford: Clarendon Press).

1976: *Unended Quest: An Intellectual Autobiography* (Glasgow: Fontana).

1977: (with John Eccles) *The Self and its Brain* (Berlin: Springer International).

1982a: *The Open Universe: An Argument for Indeterminism*, ed. W. W. Bartley III (London: Hutchinson).

1982b: *Quantum Theory and the Schism in Physics*, ed. W. W. Bartley III (London: Hutchinson).

1982c: *Realism and the Aim of Science*, ed. W. W. Bartley III (London: Hutchinson).

1990: *A World of Propensities* (Bristol: Thoemmes).

1992: *In Search of a Better World* (London: Routledge and Kegan Paul).

1994a: *Knowledge and the Body–Mind Problem*, ed. M. A. Notturno (London: Routledge and Kegan Paul).

1994b: *The Myth of the Framework*, ed. M. A. Notturno (London: Routledge and Kegan Paul).

1996: *The Lesson of this Century* (London: Routledge and Kegan Paul).

1998: *The World of Parmenides*, ed. Arne F. Petersen and Jørgen Mejer (London: Routledge and Kegan Paul).

1999: *All Life is Problem Solving* (London: Routledge and Kegan Paul).

Works by other authors

Carnap, R. 1950: *Logical Foundations of Probability* (London: Routledge and Kegan Paul).

Cohen, L. J. 1980: Some comments on Third World epistemology. *British Journal for the Philosophy of Science*, 31, 175–80.

Earman, J. 1986: *A Primer on Determinism* (Dordrecht: Reidel).

Lakatos, I. 1974: Popper on demarcation and induction. In Schilpp (ed.) (1974), pp. 241–73.

Miller, D. 1974: Popper's qualitative theory of verisimilitude. *British Journal for the Philosophy of Science*, 25, 166–77.

Miller, D. (ed.) 1983: *A Pocket Popper* (London: Fontana Pocket Readers).

Niiniluoto, I. 1987: *Truthlikeness* (Dordrecht: Reidel).

Oddie, G. 1986: *Likeness to Truth* (Dordrecht: Reidel).

Salmon, W. C. 1981: Rational prediction. *British Journal for the Philosophy of Science*, 32, 115–25.

Schilpp, P. A. (ed.) 1974: *The Philosophy of Karl Popper*, The Library of Living Philosophers, 2 vols (La Salle, IL: Open Court).

Tichy, P. 1974: On Popper's definitions of verisimilitude. *British Journal for the Philosophy of Science*, 25, 155–60.

Watkins, J. 1984: *Science and Scepticism* (London: Hutchinson; Princeton: Princeton University Press).

—— 1997: Karl Raimund Popper, 1902–1994. *Proceedings of the British Academy*, 94, 645–84.

Worrall, J. 1989: Why both Popper and Watkins fail to solve the problem of induction. In *Freedom and Rationality*, ed. Fred D'Agostino and Ian Jarvie (Dordrecht: Kluwer), 257–96.

52

Pragmatic Factors in
Theory Acceptance

JOHN WORRALL

Theory acceptance

The state of science at any given time is characterized, in part at least, by the theories that are *accepted* at that time. Presently accepted theories include quantum theory, the general theory of relativity, and the modern synthesis of Darwin and Mendel, as well as lower-level (but still clearly theoretical) assertions such as that DNA has a double-helical structure, that the hydrogen atom contains a single electron, and so on. What precisely is involved in accepting a theory?

The commonsense answer might appear to be that given by the scientific realist: to accept a theory means, at root, to *believe it to be true* (or at any rate "approximately" or "essentially" true). Not surprisingly, the state of theoretical science at any time is in fact far too complex to be captured fully by any such simple notion.

For one thing, theories are often firmly accepted while being explicitly recognized to be idealizations (see IDEALIZATION). Newtonian particle mechanics was clearly, in some sense, firmly accepted in the eighteenth and nineteenth centuries; yet it was recognized that there might well be no such thing in nature, strictly speaking, as a Newtonian particle, and it was certainly recognized that none of the entities to which this theory was applied exactly fitted that description.

Again, theories may be accepted, not be regarded as idealizations, and yet be *known* not to be strictly true – for scientific, rather than abstruse philosophical, reasons. For example, quantum theory and relativity theory were (uncontroversially) listed as among those presently accepted in science. Yet it is known that the two theories cannot both be strictly true. Basically, quantum theory is not a covariant theory, yet relativity requires all theories to be covariant; while, conversely, the fields postulated by general relativity theory are not quantized, yet quantum theory says that fundamentally everything is. It is acknowledged that what is needed is a synthesis of the two theories, a synthesis which cannot of course (in view of their logical incompatibility) leave both theories, as presently understood, fully intact. (This synthesis is supposed to be supplied by quantum field theory, but it is not yet known how to articulate that theory fully.) None of this means, however, that the present quantum and relativistic theories are regarded as having an authentically conjectural character. Instead, the attitude seems to be that they are bound to survive in (slightly) modified form as limiting cases in the unifying theory of the future – this is why a synthesis is consciously sought.

In addition, there are theories that *are* regarded as actively conjectural while none-theless being accepted in some sense: it is implicitly allowed that *these* theories *might not* live on as approximations or limiting cases in further science, though they are certainly the best accounts we presently have of their related range of phenomena. This used to be (perhaps still is) the general view of the theory of quarks: few would put these on a par with electrons, say, but all regard them as more than simply inter-esting *possibilities*.

Finally, the phenomenon of *change* in accepted theory during the development of science must be taken into account. (See INCOMMENSURABILITY.)

For all these reasons, the story about what it seems reasonable to believe about the likely relationship between accepted theories and the world is more complicated than might initially be expected. Is it the *whole* story of theory acceptance in science? That is, can everything involved in theory acceptance be explained directly in terms of some belief connected with the truth or approximate truth of the theories concerned?

The (uncontroversial) answer is that it cannot. A scientist's acceptance of a theory usually involves her "working on it" – using it as the basis for developing theories in other areas, or aiming to develop it into a still better theory. Accepting a theory, in this pragmatic sense of using it as a basis for further theoretical work, is certainly a differ-ent enterprise from believing anything. Acceptance in *this* sense involves a commit-ment – and commitments are vindicated or not, rather than true or not. The acceptance of a theory as a basis for theoretical work need not, moreover, tie in at all straight-forwardly with belief in the theory. A scientist might perfectly well, for example, "work on" a theory, while believing it to be quite radically false, perhaps with the express aim of articulating it so sharply that its falsity become apparent (so that the need for an alternative is established). This seems to have been at least part of Newton's intent in working on Cartesian vortex theory (and on the wave theory of light) and of Maxwell's initial intent in working on the statistical-kinetic theory.

It is also uncontroversial that sometimes theories are "chosen" for explicitly prag-matic reasons not connected with heuristics. NASA scientists, for example, choose to build their rockets on the basis of calculations in Newtonian rather than relativistic mechanics. This is because of the relative ease of mathematical manipulation of the classical theory. Again, teachers choose to introduce students to optics by teaching the theories of geometrical optics – even though they know that there is no such thing in reality as a light ray in the sense of geometrical optics. This is because those geomet-rical theories are more easily assimilated than "truer" theories, while sharing a good degree of their empirical adequacy.

In such cases, however, it is easy to separate out uncontentiously the epistemic and pragmatic aspects of acceptance (which is not to say, of course, that they are not interestingly interrelated). Which theory is epistemically the best available (whatever this precisely may mean) is not affected by pragmatic decisions to use theories in certain ways. There is, obviously, no contradiction in using a theory (accepting it *for certain purposes*) while holding that there is a better, epistemically more acceptable, theory that either is immediately available or could readily be developed.

So, no one denies that there are senses in which a theory may be accepted which involve pragmatic issues; no one should deny that some theories are not even in-tended to be straightforward descriptions of reality; nor should anyone deny that even the most firmly entrenched theories in current science *might* be replaced as science progresses. Nonetheless, although the issue is thus more complicated than might at first be supposed, the question of which theory is currently best *in the epistemic sense* seems intuitively to amount to which theory is, in view of the evidence we have, most likely to link up with nature in the appropriate way. And that would seem to be an objective, semantic matter, not at all dependent on pragmatic factors. This has quite frequently been denied, however, by philosophers of a strongly empiricist bent, who see even the issue of which theory is the best epistemically as in part a pragmatic matter.

This empiricist thesis has recently been newly formulated and sharply argued by Bas van Fraassen (1980 and 1985), whose work has also been the target of some interesting critical responses (see, especially, Churchland and Hooker 1985). The heart of the issue between empiricists, like van Fraassen, and realists is the question of whether such factors as simplicity (including relative freedom from *ad hoc* assumptions), unity, coherence, and the like (sometimes classified together as the explanatory virtues) are evidential factors, factors which speak in favor of the likely truth (or approximate truth) of theories which have them, or whether they are, instead, pragmatic – that is, factors which simply reflect the sorts of features that *we* happen to like theories to have. One way in which to approach this issue is as follows.

The reasons for accepting a theory

What justifies the acceptance of a theory? Although particular versions of empiricism have met many criticisms, it still seems attractive to look for an answer in *some sort* of empiricist terms: in terms, that is, of support by the available evidence. How else could the objectivity of science be defended except by showing that its conclusions (and in particular its theoretical conclusions – those theories it presently accepts) are somehow legitimately based on agreed observational and experimental evidence? But, as is well known, theories in general pose a problem for empiricism.

Allowing the empiricist the assumption that there are observational statements whose truth-values can be intersubjectively agreed (see OBSERVATION AND THEORY), it is clear that most scientific claims are genuinely theoretical: neither themselves observational nor derivable deductively from observation statements (nor from inductive generalizations thereof). Accepting that there are phenomena that we have more or less direct access to, then, theories seem, at least when taken literally, to tell us about what's going on "underneath" the observable, directly accessible phenomena in order to produce those phenomena. The accounts given by such theories of this trans-empirical reality, simply because it is trans-empirical, can never be established by data, nor even by the "natural" inductive generalizations of our data. No amount of evidence about tracks in cloud chambers and the like can deductively establish that those tracks are produced by "trans-observational" electrons.

One response would, of course, be to invoke some strict empiricist account of meaning, insisting that talk of electrons and the like is in fact just shorthand for talk of tracks in cloud chambers and the like. This account, however, has few, if any, current defenders. But, if so, the empiricist must acknowledge that, if we take any presently accepted theory, then there must be alternative, different theories (indeed, indefinitely many of them) which treat the evidence equally well – *assuming that the only evidential criterion is the entailment of the correct observational results.*

Particular ruses for producing empirically equivalent alternatives to given theories are well known: the theory that "all observable appearances are *as if* electrons exist, but actually they don't" is bound to be co-empirically adequate with the theory that "electrons exist"; while the Creationist can always use the "Gosse dodge," and write the "fossils" and all the other apparent evidence for Darwinian evolution into his account of the creation.

But there is an easy *general* result as well: assuming that a theory is any deductively closed set of sentences, and assuming, with the empiricist, that the language in which these sentences are expressed has two sorts of predicates (observational and theoretical), and, finally, assuming that the entailment of the evidence is the *only* constraint on empirical adequacy, then there are *always* indefinitely many different theories which are equally empirically adequate as any given theory. Take a theory as the deductive closure of some set of sentences in a language in which the two sets of predicates are differentiated; consider the restriction of T to quantifier-free sentences expressed purely in the observational vocabulary; then *any* conservative extension of that restricted set of T's consequences back into the full vocabulary is a "theory" co-empirically adequate with – entailing the same singular observational statements as – T. Unless very special conditions apply (conditions which do *not* apply to any real scientific theory), then some of these empirically equivalent theories will formally contradict T. (A similarly straightforward demonstration works for the currently more fashionable account of theories as sets of models.)

How can an empiricist, who rejects the claim that two empirically equivalent theories are thereby fully equivalent, explain why the particular theory T that is, as a matter of fact, accepted in science is preferred to these other possible theories T′ with the same observational content? Obviously the answer must be "by bringing in further criteria beyond that of simply having the right observational consequences." Simplicity, coherence with other accepted theories, and unity are favorite contenders. There are notorious problems in formulating these criteria at all precisely; but suppose, for present purposes, that we have a strong enough intuitive grasp to operate usefully with them. What is the *status* of such further criteria?

The empiricist-instrumentalist position, newly adopted and sharply argued by van Fraassen, is that those further criteria are *pragmatic* – that is, involve essential reference to ourselves as "theory-users." We happen to prefer, for our own purposes, simple, coherent, unified theories – but this is only a reflection of our preferences: it would be a mistake to think of those features as supplying extra reasons to believe in the truth (or approximate truth) of the theory that has them. Van Fraassen's account differs from some standard instrumentalist-empiricist accounts in recognizing the extra content of a theory (beyond its directly observational content) as genuinely declarative, as consisting of true-or-false assertions about the hidden structure of the world. His account accepts that the extra content can *neither* be eliminated as a result of defining theoretical notions in observational terms, *nor* be properly regarded as only apparently declarative but in fact as simply a codification scheme. For van Fraassen, if a theory says that there are electrons, then the theory should be taken as meaning what it says – and this without any positivistic gerrymandering reinterpretation of the meaning that might make "There are electrons" mere shorthand for some complicated set of statements about tracks in cloud chambers or the like.

In the case of contradictory but empirically equivalent theories, such as the theory, T_1, that "there are electrons" and the theory, T_2, that "all the observable phenomena are as if there are electrons but there aren't," van Fraassen's account entails that each has a truth-value, at most one of which is "true." *But* his view is that science need not be concerned with which one (if either) *is* true; T_1 is reasonably accepted in preference to T_2, but this need not mean that it is rational to believe that it is more likely to be true (or otherwise appropriately connected with nature). So far as *belief* in the theory is concerned, the rational scientist believes nothing about T_1 that she does not equally believe about T_2. The only *belief* involved in the acceptance of a theory is belief in the theory's *empirical adequacy*. To accept the quantum theory, for example, entails believing that it "saves the phenomena" – *all* the (relevant) phenomena, but *only* the phenomena. Theories do "say more" than can be checked empirically even in principle; what more they say may indeed be true, but acceptance of the theory does not involve belief in the truth of the "more" that theories say.

Preferences between theories that are empirically equivalent are accounted for, because *acceptance involves more than belief*: as well as this epistemic dimension, acceptance also has a pragmatic dimension. Simplicity, (relative) freedom from *ad hoc* assumptions, "unity," and the like are genuine virtues that can supply good reason to accept one theory rather than another; but they are pragmatic virtues, reflecting the way we happen to like to do science, rather than anything about the world. Simplicity and unity are not virtues that increase likelihood of truth. Or, rather, there is no *reason* to think that they do so; the rationality of science and of scientific practices can be accounted for without supposing that it is necessary for a rational scientist to believe in the truth (or approximate truth) of accepted theories.

Van Fraassen's account conflicts with what many others see as very strong intuitions. This is best understood by approaching the problem from a somewhat different angle.

Factors in theory choice

Many critics have been scornful of the philosophical preoccupation with underdetermination (see UNDERDETERMINATION OF THEORY BY DATA), pointing out that a scientist seldom, if ever, in fact faces the problem of choosing between two different, but equally empirically adequate theories: it is usually difficult enough (often impossible) to find *one* theory to fit all the known data, let alone several, let alone indefinitely many. (Both the critics and the philosophers are right. It *is* easy to produce theories that are empirically equivalent to a given one. But it is also a fact that scientists cannot produce *satisfactory* alternatives at will, and would regard the alternatives that *can* be produced easily as philosophers' playthings. They would not even count these readily produced rivals as proper *theories*, let alone satisfactory ones. These considerations are not contradictory, however; instead, they just show how deeply ingrained in science are the further criteria – over and above simply entailing the data – for what counts as even a potentially acceptable theory. There is no easy proof that, given any accepted theory, we can produce another, inconsistent with the first, which not only entails the same data, *but also* satisfies equally well all the further criteria for what makes a good theory – for one thing, as already noted, there is no agreed articulation of what *exactly* those further criteria are!)

But, whatever may be the case with underdetermination, there is a (very closely related) problem which scientists certainly *do* face whenever two rival theories (or more encompassing theoretical frameworks) are vying for acceptance. This is the problem posed by the fact that one framework (usually the older, longer-established framework) can accommodate (that is, produce *post hoc* explanations of) *particular pieces* of evidence that seem intuitively to tell strongly in favor of the other (usually the new "revolutionary") framework. (Thomas Kuhn makes a great deal of this point in his celebrated book (1962).)

For example, the Newtonian particulate theory of light is often thought of as having been straightforwardly refuted by the outcomes of experiments – like Young's two-slit experiment – whose results were correctly predicted by the rival wave theory. Duhem's (1906) analysis of theories and theory testing already shows that this cannot logically have been the case. The bare theory that light consists of some sort of material particles has no empirical consequences in isolation from other assumptions; and it follows that there must always be assumptions that *could be* added to the bare corpuscular theory, such that the combined assumptions entail the correct result of any optical experiment. And indeed, a little historical research soon reveals eighteenth- and early nineteenth-century emissionists who suggested at least *outline* ways in which interference results could be accommodated within the corpuscular framework. Brewster, for example, suggested that interference might be a *physiological* phenomenon; while Biot and others worked on the idea that the so-called interference fringes are produced by the peculiarities of the "diffracting forces" that ordinary gross matter exerts on the light corpuscles.

Both suggestions ran into major conceptual problems. For example, the "diffracting force" suggestion would not even come close to working with any forces of kinds that were taken to operate in other cases. Often the failure was qualitative: given the properties of forces that were already known about, for example, it was expected that the diffracting force would depend in some way on the material properties of the diffracting object; but whatever the material of the double-slit screen in Young's experiment, and whatever its density, the outcome is the same. It could, of course, simply be assumed that the diffracting forces are of an entirely novel kind, and that their properties just had to be "read off" the phenomena – this is exactly the way that corpuscularists worked. But, given that this was simply a question of attempting to write the phenomena into a favored conceptual framework, and given that the writing-in produced complexities and incongruities *for which there was no independent evidence*, the majority view was that interference results strongly favor the wave theory, of which they are "natural" consequences. (For example, that the material making up the double slit and its density have no effect at all on the phenomenon is a straightforward consequence of the fact that, as the wave theory sees it, the *only* effect of the screen is to absorb those parts of the wave fronts that impinge on it.)

The natural methodological judgment (and the one that seems to have been made by the majority of competent scientists at the time) is that, even given that interference effects could be accommodated within the corpuscular theory, those effects nonetheless favor the wave account, and favor it *in the epistemic sense of showing that theory to be more likely to be true*. Of course, the account given by the wave theory of the interference phenomena is also, in certain senses, pragmatically simpler; but this seems generally to have been taken to be, not a virtue in itself, but a reflection of a deeper virtue connected with likely truth.

Consider a second, similar case: that of evolutionary theory and the fossil record. There are well-known disputes about which particular evolutionary account gets most support from fossils; but let us focus on the relative weight the fossil evidence carries for *some sort* of evolutionist account versus the special creationist account. Intuitively speaking, the fossil record surely does give strong support to evolutionary theory. Yet it is well known – indeed, entirely obvious – that the theory of special creation can accommodate fossils: a creationist just needs to claim that what the evolutionist thinks of as bones of animals belonging to extinct species, are, in fact, simply items that God chose to include in his catalog of the universe's contents at creation; what the evolutionist thinks of as imprints in the rocks of the skeletons of other such animals are simply pretty pictures that God chose to draw in the rocks in the process of creating them. It nonetheless surely still seems true intuitively that the fossil record continues to give us *better reason to believe* that species have evolved from earlier, now extinct ones, than that God created the universe much as it presently is in 4004 BC. An empiricist-instrumentalist approach seems committed to the view that, on the contrary, any preference that this evidence yields for the evolutionary account is a purely pragmatic matter.

Of course, intuitions, no matter how strong, cannot stand against strong counter-arguments. Van Fraassen and other strong empiricists have produced arguments that purport to show that these intuitions are indeed misguided.

Are the explanatory virtues pragmatic or epistemic? The chief arguments

Van Fraassen's central argument is from parsimony. We can, he claims (1985, p. 255) "have evidence for the truth of a theory only via evidential support for its empirical adequacy." Hence we can never have better evidence for the truth of a theory than we have for its empirical adequacy. (Certainly, the claim that a theory is empirically adequate, being logically strictly weaker than the claim that the theory is true, must always be at least as probable on any given evidence.) Hence parsimony dictates that we restrict rational *belief* to belief in empirical adequacy. The "additional belief [in truth] is supererogatory" (ibid.). It is not actually irrational to believe the theory to be true, but it is unnecessary for all scientific purposes, and therefore not rationally mandated. The belief involves unnecessary metaphysics – a vice from which van Fraassen seeks to "deliver" us.

There is another way in which he makes the same point. Any empirical test of the truth of a theory would be, at the same time, a test of its empirical adequacy. By professing the belief that a theory is true, the realist seems to "place [herself] in the position of being able to answer more questions, of having a richer, fuller picture of the world"; but "since the extra opinion is not additionally vulnerable, the [extra] risk is . . . illusory, and *therefore so is the wealth*" (emphasis original). The extra belief, the claim to extra knowledge, is "but empty strutting and posturing," for which we should feel only "disdain" (1985, p. 255).

There have been a variety of responses to this argument (some of them directed at earlier articulations of similar theses, so predating van Fraassen). There is, I think, a common core to all these responses. If we were to freeze theoretical science, and consider its state *now*, not worrying either about how it got there or about where it is going, then there is a clear sense in which the trans-empirical "extra" in theories, and any belief about how it attaches to reality, is indeed "supererogatory" – we could for *present practical* purposes do without it. But once we take a diachronic view of science, and in particular look at the ways in which presently accepted theories were developed and came to be accepted, then, so realist opponents of van Fraassen allege, we find a variety of ways in which taking a realist interpretation of accepted theories (that is, taking the explanatory virtues as signs of likely truth) has played an essential role. For one thing, a significant part of the *empirical power* of present science has been arrived at by taking seriously theories with those virtues, and showing that their theoretical claims lead to the prediction of *new types* of phenomena, which were subsequently observed. Richard Boyd (1984 and elsewhere) has argued in some detail that there are ways in which "background knowledge" is standardly used in science that *both* depend on taking a realist view of the accepted background theories *and* have, *on the whole*, led to a striking degree of further *empirical* success (that is, they have led to the development and acceptance of theories that have continued to be empirically successful).

This sort of response to van Fraassen's argument might be expressed as the view that, although it is true that the "theoretical extra" in theories undergoes no direct empirical test not undergone by the "purely observational part," this "theoretical extra" *has* been tested (in an admittedly more attenuated sense) by the development of science itself. By taking a realist interpretation of theories, so by taking the explanatory virtues as indications of likely truthlikeness, science has on the whole made progress (and progress made visible at the observational level) that there is reason to think would not have been made had scientists taken those virtues as merely pragmatic. Van Fraassen is surely correct that, by regarding theories that are simple, unified, and so on as, other things being equal, more likely to be true than theories that are not, science becomes committed to certain weak metaphysical assumptions. (After all, the world's blueprint *might* be describable in no universal terms, but be riddled with *ad hoc* exceptions. In resisting *ad hoc* theories, science is implicitly assuming that the blueprint does not have this character.) Although they don't themselves put it this way, Boyd and other realists seem in effect to be claiming – rather plausibly to my mind – that some (fairly minimal) metaphysics is essential to the progress of science, and can indeed be regarded as (in a loose sense) confirmed by the history of scientific progress.

Boyd (1984) and, for example, Leplin (1986) should be consulted for details of these heuristic arguments; and van Fraassen (1985) both for responses and for a range of further arguments for regarding the explanatory virtues as pragmatic. Skating over the many details and further considerations, the general situation seems to be in outline as follows. If van Fraassen's strict empiricist premises are correct, then simplicity, unity, and the rest must indeed be regarded as *pragmatic* virtues. The main counterargument is that not only have scientists as a matter of fact standardly regarded these virtues as epistemic, there are a variety of instrumentally successful practices in science that depend for their success on those virtues being regarded in that way. If so, then van Fraassen's version of empiricism would seem to stand refuted as too narrow to capture science adequately.

Would giving up van Fraassen's position and regarding the explanatory virtues as epistemic rather than pragmatic amount to the abandonment of empiricism in general? He seems to assume so (without any discernible argument). But suppose we accept Hempel's (1950) frequently adopted characterization: "The fundamental tenet of modern empiricism is the view that all non-analytic knowledge is based on experience" (p. 42). As the history of twentieth-century empiricism shows, there is a good deal of flexibility in the notion of being "based on" experience. It would not seem to be stretching that notion *too* far to maintain that theory acceptance in science *is* simply a matter of being appropriately "based on" the evidence – but that, in order to be properly based on evidence *e*, it is not enough for a theory to entail *e*; it must entail it "in the right way." (This position is frequently adopted implicitly, and is explicitly espoused in, for example, Glymour 1977.) So, for example, interference fringes are entailed in the right way by the wave theory of light (naturally, simply, in a way that coheres with the rest of what we accept), but not by the corpuscular theory, which must make special *ad hoc* assumptions in order to accommodate them. And that means that they support the wave theory, but not the corpuscular theory. If some metaphysics is implicit in this empiricist position, this may show only that the sensible empiricist should not aim to eliminate metaphysics, but rather to countenance only as much as is necessary to make sense of science.

References

Boyd, R. 1984: The current status of scientific realism. In *Scientific Realism*, ed. J. Leplin (Berkeley: University of California Press), 41–82.

Churchland, P. M., and Hooker, C. A. (eds) 1985: *Images of Science* (Chicago: University of Chicago Press).

Duhem, P. 1906: *La Théorie physique: son object, sa structure* (Paris: Marcel Rivière & Cie); trans. as *The Aim and Structure of Physical Theory* by P. P. Weiner (Princeton: Princeton University Press, 1954).

Glymour, C. 1977: The epistemology of geometry. *Nôus*, 11, 227–51.

Hempel, C. G. 1950: The empiricist criterion of meaning. *Revue internationale de philosophie*, 4, 41–63.

Kuhn, T. S. 1962: *The Structure of Scientific Revolutions* (Chicago: University of Chicago Press; 2nd edn 1970).

Leplin, J. 1986: Methodological realism and scientific rationality. *Philosophy of Science*, 53, 31–51.

van Fraassen, B. 1980: *The Scientific Image* (Oxford: Oxford University Press).

—— 1985: Empiricism in the philosophy of science. In Churchland and Hooker (eds), 245–308.

53

Probability

PHILIP PERCIVAL

Introduction

The mathematical study of probability originated in the seventeenth century, when mathematicians were invited to tackle problems arising in games of chance. In such games gamblers want to know which betting odds on unpredictable events are advantageous. This amounts to a concern with probability, because probability and fair betting odds appear linked by the principle that odds of m to n for a bet on a repeatable event E are fair if and only if the probability of E is $n/(m + n)$. For example, suppose E is the occurrence of double-six on a throw of two dice and that its probability is $1/36 = 1/(35 + 1)$. Because of the intuitive connection between the probability of an event and its long-run frequency in repeated trials, we would expect E to occur on average 1 time in $(35 + 1)$ throws. Since to set odds on E at 35 to 1 is to agree that for a stake B, the one who bets on E gains $35B/1$ if E occurs and loses B to his opponent if it does not, those must be the fair odds for a bet on E. At them, we can expect 1 win of 35B to be balanced in the long run by 35 losses of the stake B, so that neither party to the bet enjoys net gain.

If this connection explains a gambler's interest in probabilities, it is the difficulty of calculating them which explains why the aid of mathematicians was sought. Mathematical analysis of problems such as the one set Pascal by the Chevalier de Méré in 1654 – namely, how many times must two fair dice be thrown to have a better than even chance of throwing double-six? – has resulted in a core set of axioms from which a large number of theorems concerning probabilities have been derived. However, apparently slight variations in the formulation of these axioms reflect deep disagreements regarding the nature of probability. One set-theoretic formulation of them equates probability with a numerical function which assigns a real number to subsets of a set S containing all the possible events that might occur in some situation. In the previous example, this set contains the 36 possible results of tossing two dice: namely, $<1,1>, <1,2>, <2,1> \ldots <5,6>, <6,5>$, and $<6,6>$. Probabilities are assigned to subsets of this set, like $\{<1,6>, <6,1>, <2,5>, <5,2>, <3,4>, <4,3>\}$, because gamblers are concerned to bet not just on specific results such as $<4,4>$, but on less particular outcomes such as 'getting seven'. Where "$x \cup y$" signifies the union of the subsets x and y of S, and "$x \cap y$" signifies their intersection, in this formulation each probability function "p" is held to satisfy the axioms: (i) $p(S) = 1$, (ii) $0 \leq p(x) \leq 1$, (iii) $p(x \cup y) = p(x) + p(y)$ if x and y have no members in common, and (iv) $p(x|y) = p(x \cap y)/p(y)$ provided $p(y) \neq 0$. (See van Fraassen 1980, ch. 6, sec. 4.1) for refinements and complications.)

The point of the first three axioms is straightforward. What must occur and what can't occur are maximum and minimum probabilities which are marked arbitrarily as 1 and 0. Since *one* of the possible results of throwing two dice *must* ensue if the dice are thrown, the set S = {<1,1>, <1,2> . . . <6,5>, <6,6>} of these possibilities has the maximum probability 1, in accordance with the first axiom, while each of its subsets can have a probability no less than the minimum and no greater than the maximum, in accordance with the second. Moreover, the probability of getting either eleven *or* twelve must equal the probability of getting eleven plus the probability of getting twelve. That is, p({<6,5>, <5,6>, <6,6>}) = (p({<6,5>, <5,6>}) + p({<6,6>}) = 1/18 + 1/36 = 1/12 (assuming the dice are fair). As the third axiom indicates, one can only add probabilities in this way if the sets in question have no members in common. Clearly, the probability of getting less than eleven *or* an even number is not equal to the probability of getting less than eleven plus the probability of getting an even number: were one to add *these* probabilities, the probability of results like {4,2} which are both even and less than eleven would be counted twice, the resulting "probability" being greater than the maximum.

By contrast, the fourth axiom is more obscure. The probability "p(x|y)" it contains is read as "the probability of x *given* y." This so-called conditional probability has a familiar use in gambling situations. For example, one might want to know the probability of having thrown a total of six given that one of the two fair dice has landed even. The fourth axiom allows this probability to be determined as follows: p(the total is six|one of the dice has landed even = p(the total is six and one of the dice lands even)/p(one of the dice lands even) = p({<2,4>, <4,2>})/p({<even, even>, <odd, even>, <even, odd>}) = (2/36)/(3/4) = 2/27. The status of the conditional probability "p(|)" varies in different formulations of the probability axioms. In some it is "relegated" in being *defined* by the equation of the fourth axiom, while in others it is "promoted" via the supposition that probability is *essentially* "conditional" because all meaningful talk of the probability of something involves at least tacit reference to something else. (A formulation of the axioms corresponding to this conception occurs in the section "A methodological probability concept" below.)

Though surprisingly powerful, the axioms of the probability calculus permit the calculation of unknown probabilities only in terms of those that are known or hypothesized. For this reason they cannot be applied to a problem situation unless some relevant probabilities are given. But how can these initial probabilities be ascertained? A philosopher knows better than to tackle this fundamental epistemological problem without bearing in mind the question as to what probabilities *are*. The probability axioms themselves offer little guidance. One difficulty is that since we cannot speak of the probabilities of subsets of, for example, {the Moon, Tony Blair, the number of planets}, the set-theoretic formulation of the axioms prompts the question as to *which* sets probability functions are defined over. Reflection on this kind of issue has led some to favor an alternative, "propositional" formulation of the probability calculus, in which probabilities are conceived as functions to real numbers not from sets of "events" but from *propositions*. Then, the set-theoretical operations of union and intersection are replaced by the logical operations "∨" (disjunction) and "∧" (conjunction), so that the axioms are: (i*) p(q) = 1 if q is logically true, (ii*) 0 ≤ p(q) ≤ 1, (iii*) p(q ∨ r) = p(q) + p(r) if q and r are mutually inconsistent, and (iv*) p(q|r) = p(q ∧ r)/p(r) provided p(r) ≠ 0.

It might be objected that this reformulation is either uncalled for, because propositions are sets (e.g., of possible worlds), or trivial, because the sets appealed to in the previous formulation were at any rate *tantamount* to propositions. But one must be careful here. For the "possible worlds" conception of propositions is problematic, while the "events" previously grouped together into sets and assigned probabilities were such items as "getting two sixes on a throw of two particular dice." An item of this kind is not a proposition, and many theorists deny that there is any proposition for which it is doing duty. In particular, they deny that it is doing duty for such "single-case" propositions as "the next throw of these two dice will produce two sixes" or "were these dice to be tossed now the result would be two sixes."

The issue as to whether "single-case" propositions have probabilities is explained and discussed further in the penultimate section below. But once it is decided, the axioms still only specify which functions probabilities *might* be in a given situation, not which of them they *are*. Consider again the domain comprising all the subsets of the possible results of throwing two dice. There are an infinite number of numerical functions over this domain which satisfy the axioms. But unless the dice are loaded absurdly, functions like the one which assigns the value 1 to subsets including <6,6> and 0 to the rest, reflect nothing of significance. An account of what probabilities are must remedy this defect by clarifying a concept which, in a given situation, picks out from among all the functions which satisfy the probability axioms the function having the correct numerical values.

Many have viewed this as an invitation to investigate the meaning of "probability." However, in my view one should not require the results of a conceptual investigation into the nature of probability to coincide exactly with what agents mean by "the probability of such and such." The really important issue is what agents might *profitably* mean by it. Nor should one require such an investigation to unveil just one concept: the term "probability" might be ambiguous, or there might be various uses of it which might profitably be adopted.

In fact, various concepts of probability have been held crucial to the understanding, respectively, of scientific practice, scientific methods, and scientific results. Although monists like de Finetti (1937) and Salmon (1965) have tried to make do with just one of them, pluralists such as Carnap (1945), Lewis (1986, ch. 19) and Mellor (1971), have defended at least two. I will consider these concepts in turn.

A psychological probability concept

Philosophy of science ought to enhance our understanding of the behavior of scientists. Much of what a scientist gets up to is of no concern to the philosophy of science. But his choice of a research project, his dismissal of a theoretical model, his skepticism about the previous night's data, his confidence that he has hit upon the right solution, his advocacy of a theory, etc. are all psychological facts which the philosophy of science is obliged to make sense of.

The obvious way to approach such a task is to build a normative psychological model restricted to the practice of science: the best scientists embody this model, and the exemplary nature of their scientific practice is explained by their so doing. Many believe that the basic principles of such a model involve a probability function. This function is a so-called personal probability, a function specific to the agent which captures the degrees to which he believes the various propositions he can conceive. Employing the "propositional" formulation of the probability axioms given in the previous section, such theorists hold that for each agent A there is a probability function, p_A, such that, for each proposition q that he understands, $p_A(q)$ is the degree to which he believes q, the value of $p_A(q)$ being closer to 1 the more confident he is of q's truth, and closer to 0 the more confident he is of its falsity, with *certainty* of truth and falsehood at these two extremes. It follows that provided there are propositions two agents M and N don't agree over, their respective personal probability functions p_M and p_N will differ.

This doctrine of "personal" probabilities is known as "personalism." It involves various assumptions: that beliefs differ in strength, not just qualitatively but quantitatively, that the numerical functions specifying these quantities for rational agents satisfy the axioms of the probability calculus, and that the values of these functions are sufficiently well connected to the existing employment of the term "probability" to be deemed *probabilities*. The last of these assumptions seems relatively straightforward given the other two, since it only seems plausible to maintain that there is a probability function specifying the numerical degrees of an agent's partial beliefs if he tends to judge p more probable than q whenever he believes p more strongly than he believes q. But difficulties with the first two assumptions have led some to reject personalist models of scientific (and more generally rational) behavior (see Kyburg 1978).

While it is implicit in the first assumption that the strength of a belief can be measured numerically, the direct numerical measurement of a particular belief is a difficult task. To carry it out for an agent A and proposition q, a quantifiable property must be identified which varies as the strength of A's belief in q varies. Clearly, it does not suffice for this property to be merely an effect of the strength of A's belief; ideally, nothing but the latter must affect it. Now, one of the most striking effects of the strength of an agent's belief in q is his attitude to odds for a bet on q: Jones, who is almost certain that Nijinsky will win, might eagerly accept a bet on Nijinsky's winning at odds of 1 to 5, while Smith, who strongly believes that Nijnsky won't win, will refuse bets at odds of less than 50 to 1. This has prompted many personalists to try to measure the strengths of an agent's beliefs in terms of betting odds. Clearly, since such causes as risk aversion and disapproval of gambling can also affect an agent's inclination to accept a bet, one cannot just measure the strength of his belief in q by the lowest odds at which he will bet on it, and personalists have appealed to more complicated circumstances and attitudes in trying to screen such causes out. Thus the measure of an agent's strength of belief in q proposed by Mellor (1971) is the "betting quotient" $n/(m + n)$ corresponding to the betting odds m to n he prefers in circumstances in which he is forced to bet while knowing neither whether he is to bet on or against q , nor the stake; while the measure proposed by Howson and Urbach (1989, ch. 3) is the betting quotient corresponding to the odds for a bet on q which he judges fair. In my view, neither of these proposals succeeds in screening out all mental causes other than strength of belief in the relevant proposition (cf. Milne 1991). Nevertheless, they are not so unsatisfactory as to warrant dismissing altogether the supposition that beliefs come in measurable degrees, and the more that personalist psychological models are explored, the more likely it is that the measurement problem will be solved.

Suppose, then, that beliefs have measurable strengths, and call that function which captures an agent's degrees of belief his "credence" function. There are still those who deny that the degrees of belief of a rational agent are probabilities. In so doing, they resist two main lines of argument to the opposite conclusion.

The first kind are "Dutch Book" arguments, which exploit the following mathematical result. Let b be some numerical function over a set S of propositions q, r, . . . which is not a probability, say because we have $b(q \vee r) < [b(q) + b(r)]$ for some mutually inconsistent q and r (i.e., contradicting axiom (iii*) of the preceding section). As in the first paragraph of this chapter, let the value $n/(m + n)$ of this function for each proposition in S determine betting odds of m to n for a bet on that proposition. In that case, b determines odds on the propositions $(q \vee r)$, q, and r, such that if the stakes and directions of bets at these odds are suitably chosen, someone embracing the bets which result is *certain* to lose (see Howson and Urbach 1989, ch. 3, sec. d.2). Because a set of bets with this feature is known as a "Dutch-Book," this result is known as a "Dutch-Book Theorem." It is often claimed that this result shows that the credence functions of rational agents satisfy the axioms of the probability calculus. But the credibility of this claim depends on how degrees of belief are construed. For example, if Mellor's measurement technique were tenable, it would show that any agent having nonprobabilistic degrees of belief could himself recognize that were he forced to bet on certain propositions choosing only the odds, his choice would condemn him to certain loss against a sufficiently astute opponent. But so what? Prima facie, the realization that there could be circumstances in which the beliefs I now hold are bound to lead to disaster need not make it irrational of me to retain them in *current* circumstances. By contrast, the Dutch-Book Theorem seems to have more bite if degrees of belief are measurable by betting odds adjudged fair. For in that case, it shows that if an agent judges the fairness of bets in such a way that his credence function doesn't satisfy the axioms of the probability calculus, he is committed to judging fair a bet which one party *must* lose. (Cf. Howson 1997; but see too Hacking 1967; Maher 1993, sec. 4.6.)

The second kind of argument for the claim that rational credence functions satisfy the axioms of the probabilty calculus is provided by so-called representation theorems. These theorems are to the effect that provided an agent's preferences between various acts meet certain requirements, there is a probability function p and utility function u with respect to which those preferences maximize expected utility. As Maher (1993, p. 9) puts it, from this point of view "an attribution of probabilities and utilities is correct just in case it is part of an overall interpretation of the person's preferences that makes sufficiently good sense of them and better sense than any competing interpretation does." Representation theorems of this kind are surprisingly powerful, but they yield no route to personal probabilities unless the constraints on preferences to which they appeal can be vindicated as conditions of rationality. (See Fishburn 1981 for a survey of representation theorems; see Maher 1993, chs 2–3, for discussion of the claim that the constraints on preference in question are conditions of rationality.)

Even if personalist psychological models are legitimate, a methodological issue remains regarding what must be added to them to obtain an adequate normative model of scientific behavior. "Subjective Bayesians" insist that this can be settled without introducing any further probability concepts. Effectively, they believe that the normative role of probability in science is exhausted by a "principle of conditionalization" which specifies necessary and sufficient conditions for rational changes of personal probabilities in the light of new evidence. Consider an agent A whose beliefs are measured by the personal probability function p_A. Suppose A then learns that e. What values should his new personal probability p_A^* take if he is to respond rationally to what he has learnt? The principle of conditionalization gives a simple answer to this question: it says that his new degrees of belief rationally assimilate the evidence e if and only if, for each proposition q of which he can conceive, $p_A^*(q) = p_A(q \mid e)$. That is, his new degree of belief in a proposition q upon learning that e should equal his old degree of belief in q *given* e. (I have assumed that the evidence is certain. See Jeffrey 1983, ch. 11 for when it isn't.)

Among personalists, the status of the principle of conditionalization is disputed. Some deny that it is a *necessary* condition of rationality, and contest for example Dutch-Book arguments that have been advanced in favor of it (see Teller 1973; Horwich 1982; Christensen 1991). More frequently, however, it has been held to be too weak to provide a *sufficient* condition of rational belief. Surely, it is objected, there is more to rational belief than the injunction: make your degrees of belief conform to the probability calculus and then conditionalize as new evidence is acquired. If this were the only constraint on rational belief (and hence the beliefs of scientists), the permissibility of assigning different probabilities *prior* to evidence would result in all manner of wildly divergent belief systems counting as rational. There are personalists (like de Finetti) who try to resist this conditional by appealing to theorems to the effect that in the limiting case successive pieces of new evidence eventually "wash out" initial differences in prior probability (see Earman 1992, ch. 6, for discussion of such theorems). But these so-called convergence theorems place no constraints whatsoever on *our* partial beliefs, *now*, and many have held that an adequate account of rational belief must uphold normative constraints not merely on how probabilities of propositions should change in the light of evidence, but on what they should *be* before the change (see Jaynes 1973; Rosenkrantz 1977; Williamson 1998, sec. 1). Probability concepts which have been developed to this end can be called "methodological."

A methodological probability concept

Probability is explicitly central to scientific methodology insofar as it is crucial to the foundations of statistics. Most straightforwardly, statistical methods are employed as a shortcut to the determination of the frequency with which some attribute occurs in a population too large to examine in its entirety. For example, a social scientist wanting to know what proportion of the UK electorate intends to vote Labour won't have the resources to question each elector individually. Rather, he will ascertain the proportion intending to do so in some sample, and then use statistical methods to establish how reliable an indicator this is of the proportion intending to do so in the entire population.

More interestingly, however, statistical methods are also employed in the discovery of causal relationships. To this end, they are again applied to finite bodies of data, such as a specification of the incidence of lung cancer in two groups of men in their fifth decade. Clearly, the mere fact that the incidence of lung cancer among smokers is higher than it is among nonsmokers does not in itself establish any kind of causal relationship between smoking and lung cancer. Just as it is possible for 30 consecutive tosses of a fair coin to yield 30 heads, the idea that lung cancer is causally independent of smoking is consistent with substantially more smokers than nonsmokers getting lung cancer. For this reason, statistical methods are employed to determine the extent to which disparities in the distribution of an attribute like "having lung cancer" between two groups are at odds with distributions one might reasonably expect in the absence of a causal relationship.

However, *which* methods one should employ in evaluating the significance of such data is hotly debated. Suppose we want to test the hypotheses H1, that singing to tomato plants helps them grow, and H2, that they are helped to grow by a chemical XYZ which is known to enhance the growth of potatoes. To this end we compare the results of singing to tomato plants, on the one hand, and treating them with XYZ, on the other hand, against the growth of tomato plants in a control group. Suppose that after some period the average heights of the plants in the three groups are 4.5 inches, 4.5 inches, and 4.1 inches respectively. Provided the samples are sufficiently large and properly selected, it is generally agreed that this is very strong evidence for the hypothesis H2. However, disagreements will arise over whether it is similarly strong evidence for H1. "Classical" statistics must say that it is, since it holds that in such cases whether or not data is statistically significant depends simply on the size of the sample. By contrast, "Bayesian" statistics insists that the impact of such data on such hypotheses depends in addition on the *prior* probabilities of the hypotheses, as indicated by the theorem "$p(h|e) = [p(h) \times p(e|h)]/p(e)$." Since this theorem says that the "posterior" probability $p(h|e)$ of a hypothesis h in the light of evidence e is higher the higher its "prior" probability $p(h)$, the Bayesian statistician can discriminate between H1 and H2, assigning a higher posterior probability to H2 than to H1, on the grounds that H2 has a much higher prior probability than H1. (The theorem appealed to here is "Bayes's Theorem." There are different versions of it, depending in part on the way in which the probability calculus is formulated.)

There is something to be said for both classical and Bayesian statistics. Proponents of the former object that the prior probabilities to which the Bayesian appeals have no scientific standing, whereas the Bayesian accuses classical methods of ignoring information. One sympathizes with the Bayesian here: we have some prior evidence that chemical XYZ, but not song, is the sort of thing which could help tomato plants grow. But what does it mean to assign prior *probabilities* to such hypotheses as H1 and H2? Subjective Bayesians happily identify these probabilities with the personal probabilities of the previous section; but classical statisticians object that so doing introduces an unacceptable subjectivity into scientific methodology.

Because probability has also been thought to play an *implicit*, as well as an explicit, role in scientific methodology, this dispute resurfaces in many related areas. For example, it has been suggested that the relationship *any* evidence bears to *any* hypothesis is to be understood in probabilistic terms. This suggestion has led to a series of refinements of the idea of a "criterion of positive relevance": evidence e confirms hypothesis h just in case $p(h|e) > p(h)$ (i.e., just in case the probability of the hypothesis *given* the evidence is higher than the prior probability of the hypothesis) (see Salmon 1975). Again, however, the question arises as to which concept of probability this criterion should employ. While subjective Bayesians again invoke personal probabilities, their doing so once more has the disconcerting consequence that "e confirms h" appears a subjective affair: since the credences of Smith and Jones can differ, it might happen that $p_S(h|e) > p_S(h)$ whereas $p_J(h|e) < p_J(h)$, so that Smith asserts that e confirms h whereas Jones maintains that e *dis*confirms h, without either of them committing any kind of error. Clearly, if evidential support is subjective in this way, science itself is seriously threatened.

If the criterion of positive relevance is to capture an objective relation of evidential support which our esteem for science seems to presuppose, it must employ a concept of *objective* probability. Proponents of so-called logical probability like Keynes (1921) and Carnap (1945) view one such concept as a generalization of the concept of logical consequence. Roughly, they maintain that there is a probability function, p_L, which specifies, for each pair of propositions p, q, the degree to which p "probabilifies" q, the limiting cases being 1, when p entails q, and 0, when p entails ¬q. Clearly, in being essentially conditional, probability thus conceived is a numerical *relation* holding *between* propositions, and some reformulation of the probability axioms is again called for to reflect this fact. For example, the axioms might be reformulated as: (i') $p(q|r) = 1$ if r entails q, (ii') $0 \leq p(q|r) \leq 1$, (iii') $p((q \vee r)|s) = p(q|s) + p(r|s)$ if q and r are mutually inconsistent, and (iv') $p(r|(q \wedge s)) = p(r|q) \times p(r \wedge q)|s)$.

Logical probability readily yields a direct, objective account of theory evaluation. If the total evidence currently accepted is E, and T_1 and T_2 are competing theories, then T_1 is currently to be preferred to T_2 if and only if $p_L(T_1|E) > p_L(T_2|E)$, while a theory T is reasonably accepted into the body of scientific knowledge only if $p_L(T|E)$ is close to 1. These formulations presuppose that each evidence statement is known with certainty, and it is a little awkward to accommodate the more realistic assumption that evidence itself can be doubtful. But in any case, a concern remains as to whether there really are logical probabilities. Their status is clearly more tenuous than that of entailments, since widespread agreement regarding the entailment relations which hold between a large number of propositions contrasts markedly with the dearth of intuitions regarding the degree to which one proposition nontrivially probabilifies another (cf. Ramsey 1990).

Accordingly, the proponent of logical probabilities would do well to try to demonstrate how they can be *calculated*. Many attempts to do so have their origins in the "indifference principle" of the "classical" theory of probability, according to which mutually exclusive propositions have the same probability if there is no reason to prefer one to the other. The basic idea is to apply the indifference principle to hypotheses prior to *any* evidence to obtain an a priori measure m on hypotheses in terms of which logical probabilities can be defined by: $p_L(q|r) = m(q \wedge r)/m(r)$ (Carnap 1950 is the classical source; see too Weatherford (1982, ch. 3)). However, though the indifference principle seems plausible, it is deeply flawed. The main difficulty arises because hypotheses which are the exhaustive alternatives to a given hypothesis can be variously described. Consider the hypothesis that the next ball to be chosen from an urn will be red. One exhaustive division of the possibilities supposes that the ball is either red or not, while another has it that the ball is either red or blue or neither. If the indifference principle is applied to these alternatives, different probabilities of the ball being red result: in terms of the former, it is ½, whereas in terms of the latter it is ⅓. (See Kyburg 1970 for discussion of various paradoxes generated by the indifference principle.)

Reflections on the indifference principle, and on the riddles generated by Goodman's predicate "grue", suggest that logical probability cannot be characterized by the essentially syntactic methods which Carnap employed. Most theorists have given up on logical probability as a result (but Williamson (1998) defends logical probabilities while acknowledging that our knowledge of them remains obscure). One early response was an attempt to ground Bayesian methods in an alternative "frequency" concept of probability, the basic idea being that the prior probability of a hypothesis is the frequency with which hypotheses of this kind have been vindicated hitherto in the history of science. However, this proposal is somewhat fantastic, in part because of the extreme vagueness of the notion of a "kind" of hypothesis, and most advocates of a frequency concept of probability have held that it lacks useful application to theories, and that the attempts of, for example, Reichenbach (1949) and Salmon (1964, 1965) to accord the frequency concept of probability such methodological significance are misguided. Instead, advocates of frequency concepts of probability typically propose to employ their concept in the characterization of certain natural phenomena discovered by *applying* scientific methods. For this reason, the frequency concept of probability is best considered in the next section. Suffice it to say that nowadays most Bayesians are reconciled to the view that the prior probabilities to which they appeal are personal, although, typically, they assume that there must be further constraints on rational belief in addition to the principle of conditionalization, even if they cannot say what these constraints are (see Lewis 1986, pp. 87–9).

Physical probability concepts

An English speaker's use of the phrase "probability of E" is often sensitive to his knowledge of the relative frequency of E in some population. Indeed, the intuitive connection between the probability of an event such as obtaining two sixes on a throw of two dice and the long-run frequency of that event in a series of repeated trials was noted in the opening paragraph of this chapter. Furthermore, the relative frequencies of events or "attributes" in finite populations behave like probabilities. For example, if the attributes A and B are mutually exclusive, the frequency of the property (A or B) in a finite population P is equal to the sum of the frequencies of A in P and B in P. Frequentists exploit such facts to develop a "frequency" concept of probability. The distinctive feature of this concept is that the content of an assertion "the probability of E is r" is *defined* in terms of the frequency with which E occurs in some population (or "reference class") to which, implicitly or explicitly, the assertion alludes.

Thus defined, frequentist probabilities have in common with logical probabilities the fact that they are supposed to be objective. And they are also similarly relational, in a sense, in that one cannot speak of the frequentist probability of an event or attribute as such: one must speak of its probability in a given reference class or population. But they are usually supposed to be unlike logical probabilities in being a posteriori and contingent. Because of the precise manner in which they are relational, they are also undefined with respect to so-called single-case propositions like "Jones will die of lung cancer," since Jones belongs to many different classes, such as married men who have entered their fifth decade, people who have lived in Oxford, etc., and the frequency of death from lung cancer differs from one class to another. Accordingly, the frequency concept of probability calls for a refinement of the set-theoretic formulation of the first section above (see Salmon 1965, ch. 4).

Primarily, frequentists propose to apply their concept of probability to "statistical" phenomena. A phenomenon is a repeatable event, such as the boiling of ethanol at a certain temperature and pressure. "Statistical" phenomena are those in which, with certain qualifications, what is repeatable is the frequency (neither 1 nor 0) with which some result is obtained in an experiment in which a certain kind of trial on some "chance" setup is repeated a large number of times. For example, the setup might be a roulette wheel and ball, a trial on it spinning the wheel and dropping the ball into it without thought or effort, and the result "obtaining zero." The qualifications are these: whether or not the result is obtained in a particular repetition of the trial is not practically foreseeable, and its frequency in a run of trials need only be *weakly* repeatable in that its repetition is not invariable (even if the trials in each run are perfectly good) and is only approximate (though better approximated, on the whole, the larger the number of trials in a run). Thus defined, statistical phenomena include the frequency with which zero occurs upon spinning certain roulette wheels, and the frequency of decay during a given interval among the atoms of a given radioactive element.

Trying to characterize statistical phenomena by means of a frequency concept of probability is problematic once one attempts to give truth conditions for assertions concerning the probability of a result in some trial on a setup. Because the relative frequency displayed in some statistical phenomenon is only approximately repeatable, and even then not invariably so, we cannot say that this probability *is* the frequency with which the result consistently appears in experiments. But the fact that the experiments in which a statistical phenomenon is displayed are themselves repeatable and of unlimited size prevents us from saying that the probability of obtaining the result in question is its relative frequency in some larger population from which the trials in actual experiments are taken as samples. Moreover, the problem is exacerbated once we suppose with von Mises (1957) and Popper (1957) that the frequencies exhibited in statistical phenomena are the manifestation of some dispositional physical property of the experiment, setup, or objects experimented upon. For so doing prompts the thought that probabilities *are* dispositional properties with exist or not irrespective of whether the relevant trial is ever carried out.

When taken together with the fact that the relative frequency of a result appears to converge as the number of times the trial is repeated increases, such difficulties have led many frequentists (e.g., von Mises 1957; Howson and Urbach 1989) to identify the "true" probabilities associated with a statistical phenomenon with the *limiting* relative frequencies of the various possible results which *would* be obtained were the relevant trials repeated *indefinitely*. On this construal, the trials in actual experiments are thought of as samples taken, so to speak, from an infinite sequence of *possible* trials. The results of such trials constitute an infinite sequence, for example, of heads and tails

H, H, T, H, T, H, H, T, H, T, H, T, T, . . .

which generates in turn an infinite sequence of relative frequencies

1, 1, 2/3, 3/4, 3/5, 2/3, 5/7, 5/8, 2/3, 3/5, 7/11, 7/12, 7/13 . . .

with which, for example, heads occurs among these results. If this sequence tends to a limit in the mathematical sense, the limit is the "limiting relative frequency" with which heads occurs in the original sequence.

Whether or not there are facts of the form "Were the trial T conducted indefinitely often on the setup S, the limiting relative frequency with which the result A is obtained would be r" depends in part on how the subjunctive conditional is analyzed. On its possible worlds analysis in Lewis (1973), "Were this coin tossed indefinitely, heads would occur with a limiting relative frequency of $\frac{1}{2}$" is true if and only if, among all the possible worlds in which this coin is tossed indefinitely, the worlds that *are nearest* to the actual world are worlds in which heads is obtained with limiting relative frequency $\frac{1}{2}$. However, since we are considering all *possible* worlds in which the coin is tossed indefinitely often, for each real number r there will be a possible world in which r is the limiting relative frequency with which heads is obtained. So how is the question as to which possible worlds are nearest to the actual world to be decided? Were we dealing solely with setups on which numerous trials are in fact performed, as in the case of a coin which has in fact been tossed a large number of times, we might hold that the nearest worlds are those in which the limiting relative frequency of A is the relative frequency of A in the trials actually conducted. But this account is unavailable in the case of setups upon which no trials have been conducted.

In fact, it would seem that the only general account of the distance between the actual world and a possible world in this context is the intuitive one: if the possible results of a trial T on S are $<o_1 \ldots o_n>$, and the limiting relative frequencies with which those results are obtained in w are $<f_1 \ldots f_n>$, then w is nearer to the actual world the nearer these frequencies are to the *actual* probabilities $<p_1 \ldots p_n>$ of obtaining the results $<o_1 \ldots o_n>$ in each *particular* repetition of the trial T on S.

Because the probabilities just appealed to apply to particular repetitions of a trial, they are "single-case" and nonrelational, like the personal probabilities described in the second section above. They permit one to consider what the probability was of obtaining heads on *that* particular toss of some coin *per se*, as opposed to the probability "relative to the evidence," as in the case of logical probability, or the probability "in such and such reference class," as with the frequency concept. However, if a relation of distance between possible worlds is to ground *objective* facts about what the limiting relative frequencies of its possible results would be if a trial were conducted indefinitely, these single-case probabilities must contrast with personal probabilities in being objective.

The suggestion that such "subjunctive" limiting relative frequencies might be determined in this way is ironic, and of no use to frequentists, since they take themselves to have given an account of objective, a posteriori probability which avoids what they and others take to be the obscurity of the concept of an objective *single-case* probability defended in, for example, Giere (1973), Mellor (1969), and Lewis (1986, ch. 19). Often, the obscurity has seemed epistemological: if it really is a feature of the world that propositions like "This atom will decay in 10 days" have "chances," how could we know what that chance is? Indeed, how could we know even that such a proposition's "chance" of being true is a *probability*? Just as personalists need a justification for their claim that rational degrees to belief satisfy the probability axioms, so proponents of objective chance need a justification for their claim that chances satisfy those axioms as well. Indeed, they need more than that, since they have to combat arguments like those in Humphreys 1985 and Milne 1986 that chances *cannot* satisfy the probability axioms (cf. McCurdy 1996).

There have been other attempts to show that chances must be probabilities (e.g., Sapire 1991), but the most influential attempts to show as much hold that chances have to satisfy the axioms of the probability calculus because they run so closely parallel to partial beliefs as to share their formal properties: chances are probabilities because they are *credences objectified*. The classical development of this viewpoint is in Lewis 1986, ch. 19, where the claimed parallel between credence and chance is held to be exhibited by what Lewis calls the "Principal Principle": $c(A \mid p(A) = r) = r$; that is, the credence in A conditional upon the belief that the chance of A is r must itself be r.

The Principal Principle is a powerful one. Not only does it establish the formal properties of chance, but in linking hypotheses about the chance of A with credence in A, it also exhibits the way in which data comprising information about relative frequencies of events provides evidence for hypotheses about chances. However, how successful this treatment of the metaphysics and epistemology of chance is depends on the status of the Principal Principle, and some of Lewis's critics have argued that the principle is too shaky to play any such foundational role.

In any case, we can continue to ask *why* we should take the world to contain chances. If they play no *explanatory* role, aren't they best avoided? But what explanatory role could they play? There seem to be two suggestions in the literature.

First, Mellor (1969) appears to claim that (i) there are odds for a bet on a possible event such as the decay of a particular radioactive atom during 1999 which are *objectively* fair, and, hence, that a degree of belief corresponding to the thought that these odds are fair is objectively "appropriate," and (ii) that the best explanation of this fact is that there is an objective single-case probability of the proposition "This atom decays in 1995." But both steps in this argument are problematic. For the first seems to require the questionable assumption that the evidence available beforehand determines a precise rational degree of belief in the occurrence of the event, while the second appears entirely gratuitous once this supposition is made, since what then best explains the singling out of a rational degree of belief is not a single-case chance, but the supposition that evidence constrains rational degree of belief. Secondly, it is sometimes supposed that single-case chances are needed to explain statistical phenomena: that the best, and perhaps the only, explanation of the fact that the frequency of some result o_i in repetitions of a trial T on a setup S stabilizes around the value f_i as the number of repetitions increases is that for each repetition of T on S, there is an objective single-case probability $p_i = f_i$ of obtaining the result o_i. But is this supposition really the best available explanation? The postulated single-case probability has explanatory power because it has the consequence that it is highly probable that the frequency of o_i in large numbers or repetitions of T on S will indeed stabilize around $f_i = p_i$. However, we can deduce this without supposing that the probabilities in question are *objective*. Would supposing them to be otherwise diminish the explanatory power of the hypothesis? One ground for denying that it would is that one might try to explain, for example, the stabilizing frequency of heads upon tossing a coin by saying that each toss of the coin has a single-case probability $\frac{1}{2}$ of landing heads even if one thinks both that objective single-case probabilities cannot occur in a deterministic system and that tossing a coin is deterministic (cr. Giere 1973). But, on the other hand, there is something peculiar in the thought that something which is *not* objective can explain an objective occurrence like the stability of the frequencies exhibited by statistical phenomena.

In any case, even if nothing which is not objective can explain objective facts, it remains to be shown that objective single-case probabilities *can* explain statistical phenomena. Consider the nonprobabilistic case in which there is constant conjunction between two attributes. While mere certainty that *this* individual is B if it is A won't explain why every A is B, does the attempted objectification of this mental state via the thought that each A *has* to be B explain it? Isn't this like suggesting that what explains an event E is the fact that *necessarily*, E? One worry which consistently surfaces within the empiricist tradition regarding this proposal is the thought that it is an illusion to suppose that the words "necessarily, E" have clear and appropriate truth conditions in this context. Likewise, although many attempts have been made to clarify truth conditions for an objective single-case probability statement "This A has a probability p_i of being o_i," one cannot help feeling that proposals like those of Giere (1973) and Sapire (1992) (p_i measures the strength of A's disposition of "propensity" to cause B), or Mellor (1969, 1971) (p_i is a value some disposition or propensity displays when put to the test), are pseudo-scientific gesturings. The stabilizing relative frequencies of statistical phenomena do indeed cry out for explanation. But it is questionable whether explanations in terms of objective single-case probabilities meet the canons of scientific explanation.

Conclusion

Having shown some sympathy for the concept of personal probability, I examined concepts of objective probability and described problems arising for logical probability, for single-case probabilities, and for the attempt to give truth conditions for probability claims about statistical phenomena in terms of a frequency concept. Should the concept of objective probability be abandoned altogether as subjectivists in the tradition of de Finetti (1937) propose? To a working scientist utterly dependent on statistical methods for ascertaining unknown chances, this might seem like an absurd proposal. But, as with many such disputes in the history of philosophy, the subjectivist line here is not to *replace* an existing practice, but to retain it without embracing the objective features of the world it appears to postulate. Central to this end are various construals of unknown chances in personalist terms which fall short of objective probability (cf. Jeffrey 1983, ch. 12; Skyrms 1980), and representation theorems to the effect that under certain conditions an agent's partial beliefs will be structured as they would be *if* the agent believed in single-case objective probabilities (cf. de Finetti 1937). Many such strategies appear to avoid objective probabilities by giving statements about the truth conditions of chances that are context-dependent, with the result that while, for example, Jones and Smith can empirically investigate such questions as "What is the chance of getting heads with this coin?" in the ordinary way, each construes such questions relative to his own credences in such a way that the correct answer for Jones need not be the correct answer for Smith.

That is one line to take if one repudiates the concepts of objective probability with respect of which statements about the chance of radioactive decay and the like have context-independent truth conditions. But, rather than preserve truth conditions at the expense of context independence, an alternative strategy is to try to retain context independence at the expense of truth conditions. From this alternative point of view, traditional frequentists went wrong not in supposing that a frequency concept of probability can be used to describe statistical phenomena, but in endeavoring to give *truth* conditions for claims about chances in terms of relative frequencies. They should have pursued a "conventionalist" strategy whereby a frequency concept of probability is defined in terms of the conditions under which a judgement "The probabilities of obtaining results $<o_1 \ldots o_n>$ in a large number of repetitions of a trial T on a setup S are $<p_1 \ldots p_n>$ is *assertible*, or *scientifically acceptable*, or *empirically adequate* (Braithwaite 1953; Gillies 1973; and van Fraassen 1980).

References

Braithwaite, R. 1953: *Scientific Explanation* (Cambridge: Cambridge University Press).

Carnap, R. 1945: The two concepts of probability. *Philosophy and Phenomenological Research*, 5, 513–32; repr. in *Readings in Philosophical Analysis*, ed. H. Feigl and W. Sellars (New York: Apple-Century-Crofts), 438–55.

—— 1950: *Logical Foundations of Probability* (Chicago: University of Chicago Press).

Christensen, D. 1991: Clever bookies and coherent beliefs. *Philosophical Review*, 100, 229–47.

Earman, J. 1992: *Bayes or Bust?* (Cambridge, MA: MIT Press).

de Finetti, B. 1937: Foresight: its logical laws, its subjective sources. In *Studies in Subjective Probability*, ed. H. Kyburg and H. E. Smokler (New York: Wiley).

Fishburn, P. 1981: Subjective expected utility: a review of normative theories. *Theory and Decision*, 13, 129–99.

Giere, R. 1973: Objective single-case probabilities and the foundations of statistics. In *Logic, Methodology and Philosophy of Science*, vol. 4, ed. P. Suppes et al. (Amsterdam: North-Holland), 467–83.

Gillies, D. 1973: *An Objective Theory of Probability* (London: Methuen).

Hacking, I. 1967: Slightly more realistic personal probability. *Philosophy of Science*, 34, 311–25.

Horwich, P. 1982: *Probability and Evidence* (Cambridge: Cambridge University Press).

Howson, C. 1997: Logic and probability. *British Journal for the Philosophy of Science*, 48, 517–31.

Howson, C., and Urbach, P. 1989: *Scientific Reasoning: A Bayesian Approach* (La Salle, IL: Open Court).

Humphreys, P. 1985: Why propensities cannot be probabilities. *Philosophical Review*, 94, 557–70.

Jaynes, E. T. 1973: The well-posed problem. *Foundations of Physics*, 3, 477–93.

Jeffrey, R. C. 1983: *The Logic of Decision*, 2nd edn. (Chicago: University of Chicago Press).

Keynes, J. M. 1921: *A Treatise on Probability* (London: Macmillan).

Kyburg, H. E. 1970: *Probability and Inductive Logic* (London: Macmillan).

—— 1978: Subjective probability: criticisms, reflections, and problems. *Journal of Philosophical Logic*, 7, 157–80.

Lewis, D. 1973: *Counterfactuals* (Oxford: Blackwell).

—— 1986: *Philosophical Papers*, vol. 2 (Oxford: Oxford University Press).

Maher, P. 1993: *Betting on Theories* (Cambridge: Cambridge University Press).

McCurdy, C. 1996: Humphrey's Paradox and the interpretation of inverse conditional probabilities. *Synthese*, 108, 105–25.

Mellor, D. H. 1969: Chance. *Proceedings of the Aristotelian Society*, supp. vol. 63, 11–36.

—— 1971: *The Matter of Chance* (Cambridge: Cambridge University Press).

Milne, P. 1986: Can there be a realist single-case interpretation of probability? *Erkenntnis*, 25, 129–32.

—— 1991: Annabel and the bookmaker: an everyday tale of Bayesian folk. *Australasian Journal of Philosophy*, 69, 98–102.

Popper, K. 1957: The propensity interpretation of the calculus of probability and the quantum theory. In *Observation and Interpretation*, ed. S. Körner (London: Butterworth).

Ramsey, F. P. 1990: Truth and probability (1926). In *Philosophical Papers*, ed. D. H. Mellor (Cambridge: Cambridge University Press), 52–109.

Reichenbach, H. 1949: *The Theory of Probability* (Berkeley: University of California Press).

Rosenkrantz, R. 1977: *Inference, Method and Decision: Towards a Bayesian Philosophy of Science* (Dordrecht: Reidel).

Salmon, W. C. 1964: Bayes's theorem and the history of science. In *Minnesota Studies in the Philosophy of Science*, vol. 5, ed. R. H. Stuewer (Minneapolis: University of Minnesota Press), 68–86.

—— 1965: *The Foundations of Scientific Inference* (Pittsburgh: University of Pittsburgh Press).

—— 1975: Confirmation and relevance. In *Minnesota Studies in the Philosophy of Science*, vol. 6, ed. G. Maxwell and R. Anderson (Minneapolis: University of Minnesota Press), 5–36.

Sapire, D. 1991: General causation. *Synthese*, 86, 321–47.

—— 1992: General causal propensities, classical and quantum probabilities. *Philosophical Papers*, 21, 243–58.

Skyrms, B. 1980: *Causal Necessity* (New Haven: Yale University Press).

Teller, P. 1973: Conditionalisation and observation, *Synthese*, 26, 218–58.

van Fraassen, B. 1980: *The Scientific Image* (Oxford: Clarendon Press).

von Mises, R. 1957: *Probability, Statistics and Truth* (New York: Dover Publications).

Weatherford, R. 1982: *Philosophical Foundations of Probability Theory* (London: Routledge).

Williamson, T. 1998: Conditionalising on knowledge. *British Journal for the Philosophy of Science*, 49, 89–121.

54

Qualities, Primary and Secondary

G. A. J. ROGERS

Philosophers and natural scientists have often drawn a distinction between two kinds of properties that physical objects may have. It is particularly associated with atomistic accounts of matter, and is as old as the ancient Greek theories of Democritus and Epicurus. According to the atomists, matter consists of tiny particles – atoms – having no other properties than those such as shape, weight, solidity, and size. Other putative properties – for example, those of color, taste, and smell – were regarded as the names of experiences caused by the impact of particles on an observer, and had no independent existence as qualities in the object. Since the seventeenth century it has been normal to characterize the division between these two supposedly different kinds of properties as the primary–secondary quality distinction.

The rejection by Aristotle of atomistic theories of matter and the dominance achieved in medieval Europe by his alternative account in terms of substance and form were undermined in the early seventeenth century by the revival of ancient atomic theories and concurrent developments in mathematical physics. Thus Galileo, to take one important example, argued both that the "Book of Nature" was written in the language of mathematics, specifically geometry, and that while properties such as shape, number, and size were independent of any observer, colors, tastes, and smells exist only as experiences (see GALILEO).

The distinction between primary and secondary qualities may be reached by two rather different routes: either from the nature or essence of matter or from the nature and essence of experience, though in practice these have tended to be run together. The former considerations make the distinction seem like an a priori, or necessary, truth about the nature of matter, while the latter make it appear to be an empirical hypothesis.

Taking the first of these, it is possible to argue from the nature of matter that its essential properties are the primary ones. Thus, irrespective of the number of times we divide it, a piece of matter is bound to have such properties as size, shape, solidity, and location. But if it becomes too small, it becomes invisible and thus loses the property of color. Similar considerations lead us to see that other secondary qualities are not essential to its being a material object.

This argument is obviously vulnerable to the displacement of classical atomism – in Newton's words, "solid, massy, hard, impenetrable movable particles" – by the atomic theory of Rutherford and Bohr and, more generally, by any account which denies sense to the notion of the solidity of ultimate particles (see BOHR).

The epistemic argument for the primary–secondary quality distinction is not so easily overtaken by modern science, although it faces difficulties of its own. On the

373

epistemic view, we begin from the fact that we often discover that we make mistakes about, say, the colors of objects (e.g., when lighting conditions change), even though the object itself does not change any of its properties. Or, to take another example, the vibrating bell has the same inherent properties whether it is in a vacuum or in a medium such as air; but only in the latter case does it generate sound. The sound, then, is nothing other than the experience of the auditor caused by the bell's motion in the medium. Properties which change in this way cannot be properties inherent in the object, but must be a function of the interaction between the object, the mechanisms of perception, and the observer.

For Galileo, and for others, such as Descartes, who accepted the distinction in this form, primary qualities were in material objects objectively, and every material object must have the primary qualities (see DESCARTES). But the secondary qualities were denied any objective existence at all. They were, in that sense, an effect existing only in the mind of the observer, but caused by the physical impact of the particles of colorless, odorless matter on the sense organs.

The philosopher most famous for his espousal of the distinction in the seventeenth century was John Locke; but, in taking it over, he was only following in the wake of earlier natural philosophers who subscribed to a corpuscular theory (see LOCKE). Following Boyle, but perhaps here differing from Descartes and Galileo, Locke did not deny that there was a sense in which physical objects are colored, etc. But he expressed this by saying that primary qualities are actual qualities that an object must always have, whereas secondary qualities are nothing but *powers* in the object to cause in us the sensations which we call, say, seeing blue, hearing a bell, or feeling cold. Locke, in other words, gave a dispositional account of the secondary qualities. Furthermore, and for Locke very importantly, our ideas of the primary qualities resemble their causes; our idea of the roundness of an orange, for example, resembles the actual shape that exists in the orange itself. But our idea of its color, while it corresponds to the orange, does not *resemble* anything in it. For all there are "out there" are multitudes of tiny colorless, odorless, tasteless, soundless particles that interact with our perceptual systems.

Although the distinction of first appears plausible, it faces severe criticisms, not all of which can be dismissed easily. Its most famous critic is Berkeley, who argued that epistemically it was untenable; for our experience presents us only with ideas of the so-called secondary qualities: patches of color, specific sounds, and so on. We are never presented with the primary qualities as such. We have no right, therefore, to assume that they exist. Further, an idea can be like nothing but an idea. It is incoherent to suppose that our ideas of, say, shape, size, and solidity can resemble properties in an independent physical object (see BERKELEY).

Berkeley's rejection of the distinction was combined with an espousal of his brand of idealism: the claim that ultimately there are only minds and ideas. But others, aware that Berkeley had a substantial point in his claim that the supporters of the distinction can never infer to the existence of an independent physical world, argued instead that the crucial mistake lies in making the secondary qualities subjective. Instead, the direct realist holds that bodies have both sets of properties, and that in normal perception we directly perceive those properties as they are in the object.

The direct realist, however, still has to contend with the fact that there are other arguments for drawing a distinction. Thus the primary qualities may be apprehended by more than one sense (e.g., size by both sight and touch), whereas the secondary qualities are sense-specific. Moreover, and for seventeenth-century scientists this was very important, the primary qualities are amenable to direct measurement in ways in which the secondary qualities are not. They thus fitted well with developments in mathematical physics that attempted to explain phenomena, including the properties of bodies, in terms of underlying quantifiable structures.

Further reading

Alexander, P. 1985: *Ideas, Qualities and Corpuscles: Locke and Boyle on the External World* (Cambridge: Cambridge University Press).

Bennett, J. 1971: *Locke, Berkeley, Hume: Central Themes* (Oxford: Oxford University Press).

Berkeley, G. 1993: *Principles of Human Knowledge* (1710), in Berkeley, *Philosophical Works*, ed. M. R. Ayers (reissued London: J. M. Dent).

Locke, J. 1975: *An Essay Concerning Human Understanding* (1690), ed. P. H. Nidditch (Oxford: Oxford University Press).

Mackie, J. 1976: *Problems from Locke* (Oxford: Oxford University Press).

Van Melsen, A. G. 1960: *From Atomos to Atom: The History of the Concept Atom* (New York: Harper & Brothers; repr. of 1952 edn).

55

Quantum Mechanics

RICHARD HEALEY

The early twentieth century saw the development of two revolutionary physical theories: relativity (see SPACE, TIME AND RELATIVITY) and quantum mechanics. Relativity theory had an immediate impact on the rise of logical positivism, as philosophers like Carnap, Reichenbach, and Schlick struggled to come to terms with its content and implications (see LOGICAL POSITIVISM). By contrast, discussion of philosophical issues raised by quantum mechanics began among physicists who created the theory before being taken up by technically inclined philosophers of science.

But the philosophical issues raised by quantum mechanics are by no means esoteric. Does quantum mechanics imply the overthrow of causality, and if so, how (if at all) is science still possible? If the observer creates the result of his or her observation, can one consistently suppose that there is a single objective world accessible to our observations? Such questions strike at the heart not only of the philosophy of science, but also of metaphysics. Even after they have been more carefully formulated, philosophers of quantum mechanics continue to disagree on how best to answer them.

By the beginning of the twentieth century, it had already become apparent that serious difficulties arose when physicists attempted to apply the classical physics of Newton and Maxwell to atomic-sized objects (see NEWTON). But not until the 1920s did a theory emerge that seemed capable of replacing classical mechanics here and, at least in principle, in all other domains. This new quantum mechanics quickly proved its empirical success, both in accounting for general phenomena (such as the stability of atoms) and in predicting quantitative details (including the wavelengths and intensities of the light that atoms are observed to emit when excited). But the theory possessed certain features which set it apart from classical mechanics, and indeed all previous physical theories.

In classical mechanics, the state of a system at a particular time is completely specified by giving the precise position and momentum of each of its constituent particles. This state fixes the precise values of all other dynamical quantities (e.g., the system's kinetic energy and its angular momentum). The state typically changes as the system is acted on by various forces. The theory specifies how it changes by means of equations of motion. At least in the case of simple isolated systems, the solutions to these equations uniquely specify the state of a system at all later times, given both its initial state and the forces acting on it. In this sense, the theory is *deterministic*: the future behavior of the system is uniquely determined by its present state.

While any particular method of observing a system may disturb its state, there is no theoretical reason why such disturbances cannot be made arbitrarily small. An ideal observation of a system's state would then not only reveal the precise position and momentum of each of its constituent particles at a particular time, but also permit prediction of its exact future state (given the forces that will act on it, and setting aside computational difficulties).

Although it uses almost the same dynamical quantities, quantum mechanics does not describe a system to which it applies (such as an electron) by a state in which all these quantities have precise values. Instead, the state of an isolated system is represented by an abstract mathematical object – namely, a wave function or, more generally, a state vector. For as long as a system remains free of interaction with another system, and is not observed, its state vector evolves deterministically: the vector representing the system's state at later times is uniquely determined by its initial value. But this vector specifies only the probability that a measurement of any given dynamical quantity of the system would yield a particular result; and not all such probabilities can equal zero or one (see PROBABILITY). Moreover, no attempt to establish a system's initial state by measuring dynamical quantities can provide more information than can be represented by a state vector. It follows that no measurement, or even theoretical specification, of the present state of a system suffices within the theory to fix the value that would be revealed in a later measurement of an arbitrary dynamical quantity. In this deeper sense, the theory is *indeterministic*.

The famous two-slit experiment illustrates these features of quantum mechanics. If a suitable source of electrons is separated from a detection screen by a barrier in which two closely spaced horizontal slits have been cut, then impacts of individual electrons on different regions of the screen may be detected. Quantum mechanics is able to predict at most the probability that an electron will be observed in a given region of the screen. The resulting probability distribution is experimentally verified by noting the relative frequency of detection in different regions among a large collection of electrons. The resulting statistical pattern of hits is characteristic of phenomena involving interference between different parts of a wave, one part passing through the top slit and the other part through the bottom slit.

Now, according to quantum mechanics, the electrons have a wave function at all times between emission and detection. But the theory does not predict, and experiment does not allow, that any electron itself splits up, with part passing through each slit. The electron wave function specifies no path by which any particular electron travels from source to screen: it specifies only the probability that an electron will be detected in a given region of the screen. The theory neither predicts just where any electron will be detected on the screen, nor has anything to say about how individual electrons get through the slits.

After heated discussions at Bohr's institute in Copenhagen in the 1920s, there emerged a consensus among many physicists which became known as the "Copenhagen Interpretation" (see BOHR). A central tenet of this interpretation is that the quantum-mechanical description provided by the state vector is both predictively and descriptively *complete*. What this implies for the two-slit experiment is that it is both impossible in principle to predict just where on the screen an electron will be detected and also impossible to say anything true (or even meaningful) about an individual electron's path through the two slits in that experiment.

The Copenhagen Interpretation asserts more generally that the most complete description of a system at a given time typically permits only probabilistic predictions of its future behavior. It holds, moreover, that this description, though complete, must be more *indeterminate* than any classical description. That is, while a complete classical description assigns a single number to each dynamical quantity as its value, the quantum state vector at most assigns a single number to each of some severely restricted set of dynamical quantities, while any other quantity is assigned an extended range of numbers as its value in that state.

As an example, if a system's wave function makes it practically certain to be located within a tiny region of space, then the system's momentum must be very imprecise – that is, its value must be a very wide range of numbers. A quantitative measure of the reciprocal precision with which quantities such as position and momentum are simultaneously defined is provided by the *Heisenberg indeterminacy relations*. According to the Copenhagen Interpretation, rather than restricting our knowledge of an electron's precise simultaneous position and momentum, these relations specify how precise their simultaneous values can be.

The Copenhagen Interpretation of quantum mechanics implies that the world is indeterministic. Does this mean the overthrow of causality and the end of science? The continued flourishing of science since the quantum revolution merely dramatizes the falsity of certain claims about the presuppositions of science. Even if there can be no science unless natural events conform to laws, or at least manifest general, repeatable patterns, it does not follow that the laws must be deterministic rather than probabilistic, or that the patterns must be uniform rather than statistical. A probabilistic law can be established and then exploited in giving explanations and making predictions, irrespective of whether or not the phenomena it describes conform to some underlying deterministic laws. Probability theory began with applications to games of chance, and statistical mechanics was employing probabilistic laws to great effect well before quantum mechanics.

"Causality" is an ambiguous term. If causality holds just if there are repeatable phenomena that conform to general laws, then quantum mechanics does not overthrow causality, even if it does imply that the world is ultimately indeterministic. If it is simply equated to determinism, then causality fails trivially in such a world: but this failure has no untoward consequences for science. "Causality" may be treated as just a synonym of "causation," in which case a different issue arises. Can there be causation in an indeterministic world (see CAUSATION), and in particular in a world subject to quantum mechanics?

Intuitively, it seems clear that there can be. Surely opening the slits in the barrier in the two-slit experiment causes the detection of electrons at the screen. Bringing subcritical lumps of plutonium together to form a supercritical mass is clearly a cause of the subsequent nuclear explosion, even if the process of radioactive decay which occasioned it is itself an indeterministic quantum-mechanical process.

But quantum causation is not so easy to square with popular philosophical theories of causation. Effects of quantum causes often have neither necessary nor sufficient conditions of their occurrence. On the Copenhagen Interpretation, a quantum cause may be connected to its effect by no spatiotemporally continuous process. Some cases perplex causal intuitions as well as theories of causation (see Healey 1992). Philosophers who wish to understand causation have much to learn from quantum mechanics.

378

Some have objected to the Copenhagen Interpretation of quantum mechanics because of its rejection of determinism in physics. Einstein, for example, stated his belief that "God does not play dice" (see EINSTEIN). But Einstein's main objections to the Copenhagen Interpretation sprang from his conviction that it was incompatible with realism (see REALISM AND INSTRUMENTALISM). In its most general form, *realism* is the thesis that there is an objective, observer-independent reality which science attempts (with considerable success) to describe and understand. To see how the Copenhagen Interpretation seems to conflict with this thesis, consider once more the two-slit experiment.

If one performs an observation capable of telling through which slit each individual electron passes, then one will indeed observe each electron passing through one slit or the other. But performing this observation will alter the nature of the experiment itself, so that the pattern of detections on the screen will now look quite different. The characteristic interference pattern will no longer be observed: in its place will be a pattern which, ideally, corresponds to a simple sum of the patterns resulting from first closing one slit, then opening it, and closing the other.

Observation of the electrons passing through the slits therefore affects their subsequent behavior. The Copenhagen Interpretation further implies that it is only when this observation is made that each electron passes through one slit or the other! Observation reveals that an electron has a definite position at the barrier, even though the Copenhagen Interpretation maintains that it would not have had a definite position if it had not been observed. The observed phenomenon so depends on its being observed that its objective reality is threatened. Moreover, the quantum-mechanical probabilities explicitly concern results of just such observations. Quantum mechanics, on the Copenhagen Interpretation, appears, then, to be a theory not of an objective world, but merely of our observations. If there is an objective world somehow lying behind these observations, then quantum mechanics seems notably unsuccessful in describing and understanding it!

A proponent of instrumentalism could rest easy with this conclusion. For the instrumentalist, the task of a scientific theory is simply to order previous observations and predict new ones, and quantum mechanics succeeds admirably at this task. But if the Copenhagen Interpretation is correct, then the theory does so without even permitting a description of what lies behind these observations. Realists such as Einstein and Popper have therefore rejected the Copenhagen Interpretation, while attempting to accommodate the great success of quantum mechanics itself by offering an alternative interpretation of that theory (see EINSTEIN and POPPER).

According to the simplest realist alternative, a quantum system always has a precise value of every dynamical quantity like position and momentum. The state vector incompletely describes a large number of systems of the same kind that have been prepared in the same way, by specifying the fraction of these systems which may be expected to have any given value of any particular quantity. On this view, each electron follows its own definite path through one slit or the other in the two-slit experiment, and the wave function simply specifies the relative fractions that may be expected to take paths ending in each particular region of the screen.

Unfortunately, there are strong technical objections which appear to rule out this simple variety of realistic interpretation of quantum mechanics (see, e.g., Redhead

1987; Bub 1997). It proves to be inconsistent both with features of the mathematical representation of dynamical quantities in the theory and with the assumption that the state of a quantum system cannot be directly affected by anything which is done far away from where it is located. The second problem may be seen already in the two-slit experiment.

Suppose each electron follows a path through one slit or the other. An electron going through the top slit travels nowhere near the bottom slit, so its path would be expected to be the same whether or not the bottom slit is open. Similarly, electrons going through the bottom slit in the two-slit experiment should be unaffected by closing the top slit. It follows that the pattern on the screen in the two-slit experiment should be a simple sum of two patterns, the first generated when just the bottom slit is open, the second generated when just the top slit is open. But the actual pattern is quite different – it shows the periodic variations in intensity characteristic of an interference phenomenon. If each electron does follow a definite path through one slit or the other, then it seems that this path may be affected merely by opening or closing the slit through which it does not pass!

Einstein had two main arguments against the Copenhagen claim that the wave function provides the most complete description of a quantum system. He developed a version of the first argument in consultation with his colleagues Podolsky and Rosen, and their joint paper (1935, EPR hereafter) became a classic. In it they described a thought experiment in which quantum mechanics implies the possibility of establishing either the position or the momentum of one particle with arbitrary accuracy, solely by means of measurements performed on a second particle. They argued that neither measurement would affect the state of the first particle, given that the two particles are far apart and in no way physically connected. They concluded that the first particle *has* both a precise position and a precise momentum, despite the fact that no quantum-mechanical wave function describes these quantities as having such simultaneous precise values.

Einstein later clarified the assumption that a distant measurement could not affect the state of a particle. It may be decomposed into two assumptions, which I shall call "locality" and "separability." *Separability* requires that spatially separated systems have their own states, and that these wholly determine the state of the system they compose. *Locality* requires that if A and B are spatially separated, then the state of B cannot be immediately affected by anything that is done to A alone. The locality assumption may be justified in certain circumstances by appeal to the principle that causal influences cannot propagate faster than light. The separability assumption certainly holds for the states of classical systems; but, remarkably, it typically fails in quantum mechanics if one takes the state of a compound system to be wholly specified by its state vector.

Bohr (1935) rejected the conclusion of the EPR argument. His reply may be interpreted as maintaining the completeness of quantum mechanics by rejecting some of the argument's premises. While Bohr's justification for rejecting these assumptions was questionable, later work by Bell (1964) showed how they could be subjected to experimental test. In an experimental setup very similar to the one EPR had described, EPR's assumptions turn out to imply predictions which conflict with those of quantum mechanics itself! Subsequent verification of the quantum-mechanical predictions has provided strong evidence that either locality or separability is false (see, e.g., Redhead 1987).

Of course, it does not follow that the Copenhagen Interpretation is correct, and several rival interpretations have been proposed by physicists, philosophers, and mathematicians. An examination of Einstein's second main argument against the Copenhagen Interpretation may help explain why people have made such proposals.

If quantum mechanics is a universal theory, then it must apply not only to atoms and subatomic particles, but also to ordinary objects like beds, cats, and laboratory apparatus. Now while it may seem unobjectionable for an electron to have an indefinite position, it is surely ridiculous to suppose that my bed is nowhere in particular in my bedroom. The Copenhagen Interpretation seems committed to just such ridiculous suppositions.

Assume that a macroscopic object like a bed remains isolated, and represent its initial state by a wave function which is extremely small everywhere except in a certain bed-sized region where it is located. Because the bed is so heavy, it follows that this wave function will remain very small everywhere outside such a region for an enormously long time. It may seem that one can thus reconcile the determinate location of the bed with its quantum-mechanical description.

But it is possible to transfer the alleged indeterminateness of a microscopic object's state to that of a macroscopic object by means of an appropriate interaction between them. Indeed, this is exactly what happens when a macroscopic object is used to observe some property of a microscopic object.

In Schrödinger's (1935) famous example, a cat is used as an unconventional (and ethically questionable) apparatus to observe whether or not an atom of a radioactive substance has decayed. The cat is sealed in a box containing a sample of radioactive material. A Geiger counter is connected to a lethal device in such a way that if it detects a radioactive decay product, then the device is triggered and the cat dies. Otherwise, the cat lives. The size of the sample is chosen so that there is a 50 percent chance of detecting a decay within an hour. After one hour the box is opened.

The quantum-mechanical description couples the wave function describing the radioactive atoms to the wave function describing the cat. The fact that the atomic wave function is indeterminate between decay and no decay implies that the wave function describing the total coupled system after one hour is indeterminate between the cat's being alive and dead. If this wave function completely describes the state of the cat, it follows that the cat is then neither alive nor dead! This is hard to accept, since cats are never observed in such bizarre states. Indeed, when an observer opens the box, she will observe either a dead cat or a live cat. But if she finds a corpse, she is no mere innocent witness: rather, her curiosity has killed the cat!

Most proponents of the Copenhagen Interpretation would reject this conclusion. They would claim that an observation had already taken place as soon as the decay of a radioactive atom produced an irreversible change in a macroscopic object (such as the Geiger counter), thus removing any further indeterminateness and causing the death of the cat. But this response is satisfactory only if it can be backed up by a precise account of the circumstances in which an observation occurs, thereby leading to a determinate result.

The problem of explaining just why and when a measurement of a quantum-mechanical quantity yields some one determinate result has come to be known as the "measurement problem." The problem arises because if quantum mechanics is a universal theory, it must apply also to the physical interactions involved in performing quantum measurements. But if the Copenhagen Interpretation is correct, then a quantum-mechanical treatment of the measurement interaction is either excluded in principle or else leads to absurd or at least ambiguous results.

One radical response to the measurement problem is given by the *many worlds interpretation*, due originally to Everett (1957). It is to deny that a quantum measurement has a single result: rather, every possible result occurs in some actual world! This implies that every quantum measurement produces a splitting, or branching, of worlds. A measurement is just a physical interaction between the measured system and another quantum system (call this the "observer apparatus") which, in each world, correlates the result with the observer apparatus's record of it. One can show that the records built up by an observer apparatus in a world will display just that pattern which would have been expected if each measurement had actually had a single result.

While further technical discussion would be out of place here, two consequences of the many worlds interpretation deserve special mention. Since every possible result actually occurs in every quantum measurement, the evolution of the physical universe is *deterministic* on this interpretation: indeterminism is a kind of illusion resulting from the inevitably restricted perspective presented by the world of each observer apparatus. Second, while it may seem to each observer apparatus that locality is violated, this too turns out to be an illusion.

Unfortunately, the many worlds interpretation faces severe conceptual difficulties. It must distinguish between the physical universe and the "worlds" corresponding to each observer apparatus. But the status of these "worlds" is quite problematic. If they are objective, they can scarcely all coexist in the one space occupied by the physical universe. But if each world has its own space, then how are all these spaces related to one another?

Suppose instead that a "world" is just a mental representation of a sentient observer apparatus. Then the wave function of a typical system does not describe its physical state at all, but is rather a device for describing how a sentient observer's mental state represents that system, and predicting likely changes in that representation. A realist may well find this last view even less acceptable than the Copenhagen Interpretation!

After writing a classic textbook exposition of the Copenhagen Interpretation, Bohm (1952) rejected its claims of completeness, and proposed an influential alternative which sought to restore determinism. On this alternative, a particle always has a precise position. Changes in this position are produced by a physical force generated by a field described by the wave function of the entire system of which the particle is a component. Other dynamical quantities are of secondary significance: the particle's state does not specify all their values, and their measurement is analyzed into observations of some system's position.

Quantum mechanics is understood as offering probabilities for the results of these measurements. Since each result is actually determined, in part by the initial positions of measured system and apparatus (which are not described by the wave function), quantum indeterminism is just a consequence of the incompleteness of the quantum description.

Bohm's interpretation clearly involves interactions which violate locality. A measurement on one particle can instantaneously affect the behavior of a distant particle by altering the force acting on it. But even if the interpretation is correct, it turns out that this instantaneous action at a distance cannot be exploited to transmit signals instantaneously. Thus the nonlocality inherent in the interpretation remains hidden.

Yet other interpretations have subsequently been proposed by mathematicians (e.g., Kochen 1985) and philosophers (e.g., Healy 1989, van Fraassen 1991; Bub 1997). Such interpretations tend to be motivated more by the need to give a clear statement of the theory which resolves the measurement problem than by any desire to return to a classical or deterministic world view. Van Fraassen regards his interpretation as an explication of the Copenhagen view, and defends it from an anti-realist philosophical perspective. I defend my interpretation from a realist perspective. It is interesting that despite this philosophical difference, the two interpretations have a lot in common at the level of technical detail.

Quantum mechanics has stimulated philosophical reflection on several other topics which cannot be adequately discussed here. I shall briefly mention must two.

Putnam (1969) and other philosophers have been attracted to the view that the key to a realistic interpretation of quantum mechanics is to realize that it requires the rejection of classical logic in favor of a new "quantum" logic. The resulting quantum-logical interpretation of quantum mechanics has attracted few supporters. But as an instance of the general claim, that logic is itself a high-level empirical discipline whose laws are subject to revision in the light of experience, quantum logic has further stimulated philosophical debate on the grounds of logical truth.

The second topic is identity and individuation (see van Fraassen 1991, chs 11, 12; Redhead and Teller 1992). In classical mechanics, exchanging the roles of two particles of the same kind produces a numerically distinct, but qualitatively identical, state. This is reflected in the fact that the equilibrium probability of a qualitatively described state is proportional to the number of numerically distinct ways of realizing it. In quantum mechanics things are more interesting.

Physicists often call particles of the same kind "identical." Electrons are one kind of particle, while light consists of "particles" of another kind: namely, photons. These kinds come in two varieties: bosons and fermions. Electrons are fermions, photons are bosons.

The wave function of a system of "identical" bosons remains unchanged if the roles of any two bosons are exchanged. Although the wave function of a system of "identical" fermions changes sign under an analogous operation, this does not affect the probabilities it predicts. In both cases, exchanging two "identical" particles seems to produce a state that is not merely qualitatively, but also numerically, identical to the original state. And this is reflected in the equilibrium probabilities of states in quantum mechanics – what are called Bose or Fermi statistics respectively.

Here is a dilemma. If electrons, say, are genuine individuals, then surely exchanging a pair of such individuals must yield a numerically distinct state. However, both the quantum-mechanical description and the observed statistics appear to conflict with this conclusion. But if electrons are not individuals, then how is it possible to think of, or intelligibly refer to, them?

This brief review illustrates the following general conclusions. While quantum mechanics has proved to be of great philosophical interest, this is not because it has resolved any outstanding philosophical issues. No such resolution is to be expected in the absence of an agreed understanding of the theory itself. But there can be no doubt that philosophical reflection on quantum mechanics has reinvigorated and deepened debate on a host of issues central to metaphysics and the philosophy of science.

References and further reading

Bell, J. S. 1964: On the Einstein–Podolsky–Rosen paradox. *Physics*, 1, 195–200; repr. in Wheeler and Zurek 1983, pp. 403–8.

Bohm, D. 1952: A suggested interpretation of the quantum theory in terms of "hidden variables": parts I, II. *Physical Review*, 85, 166–93; repr. in Wheeler and Zurek 1983, pp. 369–96.

Bohr, N. 1935: Can quantum-mechanical description of physical reality be considered complete? *Physical Review*, 48, 696–702; repr. in Wheeler and Zurek 1983, pp. 145–51.

Bub, J. 1997: *Interpreting the Quantum World* (Cambridge: Cambridge University Press).

Einstein, A., Podolsky, B., and Rosen, N. 1935: Can quantum-mechanical description of physical reality be considered complete? *Physical Review*, 47, 777–80; repr. in Wheeler and Zurek 1983, pp. 138–41.

Everett, H., III 1957: "Relative state" formulation of quantum mechanics. *Reviews of Modern Physics*, 29, 454–62; repr. in Wheeler and Zurek 1983, pp. 315–23.

Healey, R. A. 1989: *The Philosophy of Quantum Mechanics: an Interactive Interpretation* (Cambridge: Cambridge University Press).

—— 1992: Chasing quantum causes: how wild is the goose? *Philosophical Topics*, 20, 181–204.

Kochen, S. 1985: A new interpretation of quantum mechanics. In *Symposium on the Foundations of Modern Physics*, ed. P. Lahti and P. Mittelstaedt (Singapore: World Scientific Publishing Co.), 151–70.

Popper, K. R. 1982: *Quantum Theory and the Schism in Physics* (London: Hutchinson).

Putnam, H. 1969: Is logic empirical? *Boston Studies in the Philosophy of Science*, 5, 216–41.

Redhead, M. L. G. 1987: *Incompleteness, Nonlocality and Realism: A Prolegomenon to the Philosophy of Quantum Mechanics*. (Oxford: Clarendon Press).

Redhead, M., and Teller, P. 1992: Particle labels and the theory of indistinguishable particles in quantum mechanics. *British Journal for the Philosophy of Science*, 43, 201–18.

Schrödinger, E. 1935: Die gegenwärtige Situation in der Quantenmechanik. *Naturwissenschaften*, 23, 807–12, 823–8; 844–9; trans. J. D. Trimmer as "The present situation in quantum mechanics: a translation of Schrödinger's 'cat paradox' paper," *Proceedings of the American Philosophical Society*, 124 (1980), 323–38; repr. in Wheeler and Zurek 1983, pp. 152–67.

van Fraassen, B. 1991: *Quantum Mechanics: An Empiricist View* (Oxford: Clarendon Press).

Wheeler, J. A., and Zurek, W. H. (eds) 1983: *Quantum Theory and Measurement* (Princeton: Princeton University Press).

56

Quine

LARS BERGSTRÖM

Willard Van Orman Quine was born on 25 June 1908 in Akron, Ohio. For many years he was a professor of philosophy at Harvard University and is now emeritus. To some extent his views are connected with the American pragmatist tradition, but a more important influence comes from the empiricist tradition and, in particular, from the logical positivism of the Vienna Circle (see LOGICAL POSITIVISM). Quine has always remained faithful to the spirit of empiricism, but he has also criticized and revised the empiricist doctrine in important ways. He has published 20 books and numerous articles, and he is perhaps the most influential analytical philosopher of the second half of the twentieth century.

Naturalism

A basic element in Quine's philosophy of science is his idea of a naturalized epistemology (see NATURALISM). Naturalism is the view that science is the only means we have of finding out the truth about the world. The aim of traditional epistemology was to defend science against skeptical doubts by showing how it can be derived from a secure foundation – for example, from clear and distinct ideas (rationalism) or from immediate sensory evidence (empiricism). This idea of a "first philosophy" – of a discipline which is different from and methodologically prior to science – is rejected by Quine. Traditional epistemology has been unsuccessful so far, and Quine believes that it cannot possibly succeed. Science cannot be justified by anything outside science. Instead, Quine claims that it is up to science itself to consider its own foundations and knowledge claims. There is no transcendental perspective from which we can question our own system of the world. Our system can only be questioned from within itself (1981, p. 72). Thus, epistemology is part of our total theory of nature.

The primary aim of a naturalized epistemology is to explain how our scientific theories are related, causally and inferentially, to sensory input. Quine is an *empiricist* in the sense that he holds that whatever evidence there is for science is sensory evidence (1969, p. 75). In accordance with naturalism, this doctrine is itself a scientific hypothesis, and as such it might conceivably turn out to be mistaken. So far, however, current science tells us that information about the external world can reach us through our sensory receptors, and that this is our only channel of information (1992, pp. 19–21).

Another consequence of Quine's naturalistic stance is *physicalism* (see PHYSICALISM). This means that every state or event in the world involves or is determined by some physical state or event (1981, p. 98). There can be no change without a change in

385

physical micro states. Furthermore, it seems that Quine is a *realist* about scientific theories (see REALISM AND INSTRUMENTALISM), and that his reason for this is again naturalism. There are indeed some passages in his writings which suggest that he has an instrumentalist attitude towards science, but as a naturalist he holds that questions of truth are to be settled within science, and this makes him, in his own judgment, a scientific realist (see, e.g., Barrett and Gibson 1990, p. 229).

Objectivity of science

According to empiricism, our evidence for scientific theories consists of observations. In order to study the relation between science and observation, Quine focuses on the corresponding linguistic formulations. Thus, he takes a *theory* to be a set or conjunction of sentences (1981, p. 24), and an *observation sentence* for a given speech community to be a sentence which is directly and firmly associated with sensory stimulations for every member of the community and on which all members give the same verdict when witnessing the same situation (1992, p. 3). Observation sentences are also the first sentences we master when we learn our first language as children. They can be learnt by ostension. Examples of observation sentences are "It's cold," "There is a dog," "This is a flower." Most observation sentences report physical things and events, but some – for example, "Tom perceives a dog" – are mentalistic (1992, p. 62).

Observation sentences are true on some occasions and false on others. Therefore, they cannot be implied by scientific theories, which are either true or false once and for all. However, two observation sentences can be combined into a general sentence of the form "Whenever this, that." An example would be "Whenever there is a raven, it is black," or simply "All ravens are black." Such sentences are called *observation categoricals*; they are true or false once and for all, and they can be implied by scientific theories. An observation categorical is synthetic for a given speaker if the stimulations associated with the antecedent are not completely included among the stimulations associated with the consequent. Synthetic observation categoricals can be tested in experiments. Two observation categoricals are synonymous for a speaker if their respective components are associated with the same stimulations. The empirical content of a theory for a given speaker consists of the set of synthetic observation categoricals implied by it. Two theories are empirically equivalent for a given community if they have the same empirical content for each member (1992, pp. 16–17). More generally, empirical equivalence obtains when "whatever observation would be counted for or against the one theory counts equally for or against the other" (1992, p. 96).

Observation sentences are very important in Quine's philosophy of science. They are causally connected to sensory stimulation, and they contain words which also occur in theoretical sentences. Thereby, they constitute the link between observation and theory. They are "the vehicle of evidence for objective science" (1993, p. 109).

In recent decades, several philosophers have criticized empiricism by questioning the objectivity of observation. It is often said that observation is theory-laden, that it involves interpretation, that different people observe differently, and that observation is therefore not sufficiently neutral and reliable to serve as the basis for scientific theories. Quine rejects this argument. In any case, he claims that its conclusion does not apply to observation sentences. He agrees that these are theory-laden taken analytically, word by word. But, as a response to stimulation, observation sentences should be seen holophrastically, as unanalyzed wholes, and as such they are not theory-laden and do not involve interpretation. Moreover, they are intersubjective by definition, and therefore they can indeed provide an objective basis for science. For the same reason, Quine disputes the popular idea that radically different theories in natural science may be incommensurable (see INCOMMENSURABILITY); he claims that observation sentences provide the shared reference points for comparing such theories (1993, p. 111).

Aims and methods of science

Some readers of Quine have supposed that he regards naturalized epistemology as a purely descriptive discipline, but this is not Quine's view. There is also normative epistemology, and Quine thinks of this as "the technology of anticipating sensory stimulation" (1992, p. 19).

Normative epistemology offers recommendations concerning the construction of hypotheses as well as good-making characteristics which should govern our preferences with respect to competing theories. Quine has listed six virtues which characterize a good hypothesis: conservatism, modesty, simplicity, generality, refutability, and precision (1978, pp. 66–79, 98). It may be asked whether he holds these traits to be desirable in themselves. It seems not. Rather, his claim appears to be that the six virtues distinguish hypotheses which "prove on the whole to be richest in their verifiable predictions" (1978, p. 135). This suggests that the ultimate aim of science is successful prediction, which in turn would provide a motivation for Quine's idea that normative epistemology is a "technology of anticipating sensory stimulation."

However, Quine also says that prediction is *not* the main purpose of science. The primary purposes of science are "technology and understanding" (1992, p. 20). But predictions are important, in that they are essential for testing. Quine's position can perhaps be stated as follows. Theories constitute good science to the extent that they imply true, synthetic observation categoricals, and good science is valuable to us because it enables us to understand the world and to construct efficient tools for practical purposes.

If understanding is an aim of science, scientific theories have to be taken realistically. If theories were mere instruments for prediction, they could not tell us much about the structure of the world. Moreover, if understanding is an aim, theories should be formulated in a language which is as clear and intelligible as possible. For Quine, this means that a scientific theory should be such that it can be formulated in a purely extensional language, which contains only individual variables and general terms combined by predication, quantification, and truth functions (1960, pp. 227–9).

However, Quine does not put any special restriction on what general terms should be tolerated. In particular, although he is a physicalist, he accepts the use of mentalistic idioms in science. He believes that mental events are physical (neural) events, but he does not believe that mentalistic predicates can be reduced to physicalistic ones, or that mentalistic predicates which cannot be so reduced should not be allowed in science (1992, pp. 71–3). In other words, theories in the social sciences and the humanities, which typically contain mentalistic predicates, can be scientifically respectable even though they cannot be reduced to natural science.

Underdetermination of science

A theory is supported by evidence only if it implies some observation categorical. However, following Pierre Duhem, Quine points out that observation categoricals can seldom if ever be deduced from a single scientific theory taken by itself; rather, the theory must be taken in conjunction with a whole lot of other hypotheses and background knowledge, which are usually not articulated in detail and may sometimes be quite difficult to specify. A theoretical sentence does not in general have any empirical content of its own. This doctrine is called *holism* (see HOLISM).

Holism is Quine's main reason for naturalism; if most theoretical sentences lack empirical content, a "first philosophy" along the lines of traditional empiricism cannot succeed. Holism also throws doubt on several other ideas which have been popular in the philosophy of science, such as the verifiability principle of logical positivism, Karl Popper's falsifiability criterion, and the distinction between analytic and synthetic sentences.

Moreover, holism suggests that the empirical basis for our science could in fact constitute the basis for many different theoretical superstructures. This is the idea of *underdetermination* (see UNDERDETERMINATION OF THEORY BY DATA). According to holism, no theoretical sentence is immune to revision. Therefore, we may replace any sentence in our total system of the world by its negation; this is possible as long as we make "drastic enough adjustments elsewhere in the system" (1953, p. 43). The result of such adjustments, if they are successful, would be another system of the world with the same empirical content as ours. The two systems would even be logically incompatible, since one contains the negation of a sentence in the other. This possibility is precisely what the underdetermination thesis tells us to take into account.

The underdetermination thesis does not say only that, at any given time, the available evidence is equally compatible with several rival theories. Rather, it says that drastically different scientific theories may be equally supported by all possible evidence. Therefore, one cannot hope to eliminate, even in the long run, all theories but one. In particular, the underdetermination thesis says that our global system of the world (at any given time) has some, probably unknown, empirically equivalent but irreducible rival, which is equally as good as our system.

Suppose, unrealistically, that we were to come across such a rival theory. Given that we regard our own theory as true, should we call the rival theory true, false, or meaningless? This question has worried Quine, and he has answered it in different ways in different writings. In *Pursuit of Truth* (1992) he suggested that it is "a question of words" (p. 101).

A more interesting thought, which may seem natural to many people, is that underdetermination leads to skepticism. There is no sign that Quine would accept this, but it is hard to see how it can be ruled out. Our system of the world may be supported by a lot of evidence, but if the underdetermination thesis is true, some other system of the world can account for all our evidence just as well, and this alternative system may even be incompatible with our own. If we believe this, we can hardly be justified in believing that our own system is true. Consequently, those of us who accept underdetermination are also committed to a kind of skepticism.

References and further reading

Works by Quine

1953: *From a Logical Point of View* (Cambridge, MA: Harvard University Press).

1960: *Word and Object* (Cambridge, MA: MIT Press).

1969: *Ontological Relativity and Other Essays* (New York: Columbia University Press).

1978 (with Ullian, J. S.): *The Web of Belief*, 2nd edn (New York: Random House; 1st pub. 1970).

1981: *Theories and Things* (Cambridge, MA: Belknap Press/Harvard University Press).

1992: *Pursuit of Truth*, rev. edn (Cambridge, MA: Harvard University Press; 1st pub. 1990).

1993: In praise of observation sentences. *Journal of Philosophy*, 90, 107–16.

1995: *From Stimulus to Science* (Cambridge, MA: Harvard University Press).

Works by other authors

Barrett, R. B., and Gibson, R. F. (eds) 1990: *Perspectives on Quine* (Cambridge, MA, and Oxford: Blackwell).

Gibson, R. F. 1988: *Enlightened Empiricism: An Examination of W. V. Quine's Theory of Knowledge* (Tampa, FL: University Press of Florida).

Gibson, R. F. (ed.) forthcoming: *The Cambridge Companion to Quine* (Cambridge: Cambridge University Press).

Hahn, L. E., and Schilpp, P. A. (ed.) 1986: *The Philosophy of W. V. Quine* (La Salle, IL: Open Court).

57

Ramsey Sentences

FREDERICK SUPPE

In what is known as the "*received view*" analysis, logical positivism construed scientific theories TC as being axiomatized in first-order predicate calculus using proper axioms T (the theoretical laws) and having distinct observational and theoretical vocabularies V_O and V_T which are related to each other via a *dictionary* of *correspondence rules C* (see THEORIES). Prior to 1936 the correspondence rules were required to be *equivalences* between V_T terms and simple or complex observational conditions expressible using just V_O terms that provided noncreative *explicit definitions* of the terms (see CRAIG'S THEOREM). The *empirical content* of the theory TC was identified with the set O of theorems which contained V_O but not V_T terms.

Frank Plumpton Ramsey explored some of the properties of such theories:

1. Can we say anything [empirically] in the [V_T] language of this theory that we could not say without it? . . .
2. Can we reproduce the structure of our theory by means of explicit definitions within the primary [V_O] system? (Ramsey 1960, pp. 219–20).

His answer to the first question is an early statement of the noncreativity of explicit definitions. The second explores the eliminability of V_T assertions, considering such explicit definition techniques as truth-functional expansions of monadic predicates. He concludes that while in principle "we can always reproduce the structure of our theory by means of explicit definitions" (p. 229), we need not, since "if we proceed by explicit definition we cannot add to our theory without changing the definitions" (p. 230).

Next he asks, "Taking it then that explicit definitions are not necessary, how are we to explain the functioning of our theory without them?" His answer is that theories express judgments, "the theory being simply a language in which they are clothed, and which we can use without working out the laws and consequences" (pp. 230–1). Immediately he concludes: "The best way to write our theory seems to be this" and introduces what has come to be known as the *Ramsey sentence* for a theory with laws or axioms $T(\phi_1, \ldots, \phi_n)$ and correspondence rules $C(\phi_1, \ldots, \phi_n)$ containing V_T terms $\phi_1, \ldots, \phi_n$:

$$(\exists \phi_1^*) \ldots (\exists \phi_n^*)[T^*(\phi_1^*, \ldots, \phi_n^*) \wedge C^*(\phi_1^*, \ldots, \phi_n^*)]$$

where the V_T predicate terms $\phi_1, \ldots, \phi_n$ in T and C have been replaced by the distinct predicate variables $\phi_1^*, \ldots, \phi_n^*$ thereby obtaining T^* and C^*. It can be shown that for any V_O sentence o,

$$[T(\phi_1, \ldots, \phi_n) \wedge C(\phi_1, \ldots, \phi_n)] \to o \text{ if and only if}$$
$$(\exists \phi_1^*) \ldots (\exists \phi_n^*)[T^*(\phi_1^*, \ldots, \phi_n^*) \wedge C^*(\phi_1^*, \ldots, \phi_n^*)] \to o.$$

Ramsey's concerns ceased to be important after 1936, when the positivists rejected the requirement that V_T terms be explicitly defined by weaker partial definition requirements (see CRAIG'S THEOREM and DEFINITIONS.) However, the Ramsey sentence construction can be applied to partial definition versions.

Instrumentalists maintained that theoretical laws and terms were simply computational devices that enabled predictions but did not purport to describe anything real (see REALISM AND INSTRUMENTALISM). Non-eliminable laws were legitimate in theories provided theoretical terms did not refer to anything nonobservational. Ramsey sentences were thought to provide an analysis of how this was possible. However, as Hempel (1958) noted, this avoids reference to nonobservational entities only by name, but still asserts the existence of entities of the kind postulated by T.

Long after many positivists had abandoned the analytic/synthetic distinction, Carnap continued to seek a satisfactory analysis of analyticity using *meaning postulates* to separate off the analytic from the empirical content of a theory. Using the Ramsey sentence, he proposed (1966) that the meaning postulate for theory TC be

$$(\exists \phi_1^*) \ldots (\exists \phi_n^*)[T^*(\phi_1^*, \ldots, \phi_n^*) \& C^*(\phi_1^*, \ldots, \phi_n^*)] \to [T(\phi_1, \ldots, \phi_n) \wedge C(\phi_1, \ldots, \phi_n)].$$

The proposal had little impact, since it presupposed the syntactical received view which was widely being abandoned in favor of more promising semantic analyses.

Sneed (1971) observed that applications of classical mechanics must measure values for force and mass, but that the very processes of measurement presuppose classical mechanics. The real problem of theoretical terms, he claimed, was showing how this practice avoided vicious circularity. Construing theories semantically (see AXIOMATIZATION), he proposed a controversial semantic solution that uses a Ramsey sentence applied to possible partial models of the theory T to characterize the empirical content of an application of T, thus making T and the measurement application distinct.

These proposals address versions of a more general problem:

Theories characterize phenomena in terms of descriptive variables $v_1, \ldots, v_n$ by laws characterizing how values of those variables change over time. Specification of laws often takes recourse to mathematical mechanisms m or entities e (e.g., phase spaces) that do not correspond to physical structures or entities. How is it possible to endow theoretical language containing v, m, and e with a full extensional semantic interpretation without requiring that terms m and e refer to physical existents?

Instrumentalism denies that it can be done. Ramsey sentences provide a syntactical approach liable to Hempel's objections but adaptable to simple semantic applications.

Van Fraassen's *theory of semi-interpreted languages* provides a general semantic solution. All the v, m, and e terms of T are interpreted as referring to points or regions of some logical space L, using standard extensional semantics. Individual physical entities or properties can be mapped from the real world to points or regions of logical space using distinct functions *loc*. All terms in T thus have full semantic properties, but only those terms which are images of *loc* functions refer to physical things, the references being mediated through *L*. Since one has free choice of *loc* functions, all ranges and intricacies of ontological commitment can be accommodated. By imposing transformation groups on the logical spaces or embedding the logical spaces into richer topological constructions, the analysis extends to both modal terms and empirical probabilities. Ontological commitments remain a function of choice of *loc* functions.

The philosophical importance of Ramsey sentences today is largely historical, having been eclipsed by van Fraassen's work.

References and further reading

Carnap, R. 1966: *Philosophical Foundations of Physics*, ed. M. Gardner (New York: Basic Books), chs 26–8.

Hempel, C. G. 1958: The theoretician's dilemma. In *Minnesota Studies in the Philosophy of Science*, vol. 2, ed. H. Feigl, M. Scriven, and G. Maxwell (Minneapolis: University of Minnesota Press), 37–98. Repr. in C. Hempel, *Aspects of Scientific Explanation and Other Essays in the Philosophy of Science* (New York: Free Press, 1965), 173–226.

Lewis, D. 1970: How to define theoretical terms. *Journal of Philosophy*, 67, 427–46. (Perhaps the most sophisticated version of the Ramsey approach to defining theoretical terms.)

Ramsey, F. P. 1960: Theories. In *The Foundations of Mathematics*, ed. R. B. Braithwaite (Paterson, NH: Littlefield, Adams & Co.), 212–36. (1st pub. London, 1931).

Sneed, J. D. 1971: *The Logical Structure of Mathematical Physics* (Dordrecht: D. Reidel Publishing Co.). See esp. chs 2–4.

Suppe, F. 1974: The search for philosophic understanding of scientific theories. In *The Structure of Scientific Theories*, ed. F. Suppe (Urbana, IL: University of Illinois Press), pp. 3–241 (2nd edn 1977). See esp. secs II–IV, V–C.

—— 1989: *The Semantic Conception of Theories and Scientific Realism* (Urbana, IL: University of Illinois Press, 1989). (Part II, esp. ch. 3, uses semi-interpreted languages to develop a semantic analysis of theories that does not presuppose an observational/theoretical distinction.)

van Fraassen, B. 1967: Meaning relations among predicates. *Noûs*, 1, 161–89. (Develops the basic theory of semi-interpreted languages.)

—— 1969: Meaning relations and modalities. *Noûs*, 3, 155–68. (Extends the theory of semi-interpreted languages to include the modalities.)

—— 1980: *The Scientific Image* (Oxford: Oxford University Press). (Largely an application of the theory of semi-interpreted languages (especially its treatment of modalities) which underlies his account of empirical adequacy (ch. 3), theories (ch. 4), and explanation (ch. 5). Unlike Suppe 1989, his applications invoke an observability/nonobservability distinction. Ch. 6 extends the theory to accommodate empirical probability statements.)

58

Realism and Instrumentalism

JARRETT LEPLIN

The main issue

The debate between realism and instrumentalism is at an impasse. That is the state of the art, and the competing positions and arguments are best understood by seeing how they have produced it. When scientists familiar with a common body of evidence, and with the resources of alternative theories for handling that evidence, nevertheless disagree as to which theory is best, something has gone wrong methodologically. Standards of evidential warrant, the criteria by which theories are to be judged, and not just the theories themselves, are in dispute. When philosophers disagree about theories of science, without disputing the evidence brought to bear for or against the contenders, the legitimacy of standards of philosophical argument is similarly unresolved. In the debate over realism the central bone of contention is abductive inference.

In abductive inference, the ability of a hypothesis to explain empirical facts counts in its assessment. The stronger its explanatory resources – the better the explanation it gives, the more diverse the phenomena it explains, the greater its role in larger explanatory systems – the more justified is the inference to it. A point may be reached at which explanatory success is so great, and so much greater than that which any rival hypothesis can claim, as to justify belief. On this basis, realists characteristically claim warrant for believing some hypotheses of theoretical science, hypotheses that, because they posit entities or processes inaccessible to experience, are not confirmable in other ways. What matters to realism as a philosophy of science is not so much which hypotheses are thus to be believed, or whether such belief must be hedged somehow to admit refinement or qualification in light of new information, but that the abductive mode of inference itself be legitimate. Realism minimally maintains – though not all self-proclaimed realists grant this – that our empirical knowledge *could* be such as to warrant belief in theory, on the basis of its explanatory success.

Anti-realism originated as the denial that explanation counts at all. Hypotheses whose credibility depended, because of inaccessibility of subject matter, on their explanatory power were considered unscientific. Thus the inductivist methodology of John Stuart Mill dismissed the wave theory of light in the nineteenth century, not because of any clash between the theory and experiment, but because the hypothesis of an all-pervasive ether medium, on which the wave theory depended, was of a kind that only explanatory achievements could credit (see MILL). No matter how many and important were the optical effects that the theory could explain, no matter how inadequate the rival corpuscular theory proved to these explanatory tasks, the wave theory simply was not to be an option. Science was limited to theories whose subject matter was experientially accessible.

By the end of the century it was clearly impossible to do science without theoretical hypotheses. The ether hypothesis in particular was indispensable in electromagnetic theory, which, with the equations of Maxwell, had reached a level of achievement comparable to the Newtonian synthesis in mechanics. So instrumentalism was born. Explanation would now count, but only pragmatically. If theoretical hypotheses were good and important science, they were nevertheless not to be believed. They would be evaluated as instruments, tools of research, not as claims about the world whose correctness could be at issue in experimental tests.

The real thesis of instrumentalism, arising in early twentieth-century positivism, is harsher: there is no autonomously theoretical language. Some theory can be reduced to observation by defining or translating theoretical terms into terms that describe observable conditions (see THEORETICAL TERMS). The remainder must be construed instrumentally, not as descriptive or referential at all. The function of this remainder is to predict and systematize observations, not to explain them. This move parallels the positivist construal of metaphysical, ethical, or religious sentences, which failed the verification criterion of meaningfulness, as having an emotive function, rather than as expressing propositions admitting of truth or falsity.

More recent versions of instrumentalism, despairing of translation or reduction schemes to rewrite theoretical sentences in observational terms, concede them propositional status, but deny them any possibility of warrant. This weakening is realist to the extent that it allows for truth and falsity in virtue of the way of the world, independent of language and observers. But that is mere metaphysical realism; it is non-epistemic and not at issue in contemporary philosophical debate about science. The instrumentalist theme of the irrelevance of the truth or falsity of a theoretical sentence to its scientific utility is sustained. Since belief in theoretical hypotheses is unwarrantable, truth, even if possessed, can be no part of what we value them for. This much instrumentalism is inevitable once the abductive mode of warrant is disallowed.

Arguments against abduction

There are four principal lines of attack. The first, best represented by W. V. Quine, is based on the multiplicity of explanatory options (see QUINE). Given any explanatory theory, another, incompatible theory can be constructed that equally accords with all relevant evidence. This is the Thesis of Empirical Equivalence (EE). From it is inferred the Underdetermination Thesis (UD), that no body of evidence supports any theory to the exclusion of all (actual or potential) rivals (see UNDERDETERMINATION OF THEORY BY DATA). If the evaluative situation is as these theses describe, then theory choice is at best pragmatic, and theoretical belief is unwarranted in principle.

Quine himself regards EE as a point of logic, grounded in the Löwenheim–Skolem theorem and the restriction of science to extensional language. But the applicability of the presuppositions of this framework to actual science is questionable. Also dubious on behalf of EE are logical algorithms that operate on an arbitrary theory to produce an observational equivalent by changes stipulated to leave the observational consequences of the given theory intact (see CRAIG'S THEOREM and RAMSEY SENTENCES). For the variations thus produced are not rival theories, and are parasitic on the original theory for the determination of their testable implications. More persuasive are examples from science in which equivalent alternatives to a given theory are generated by appeal to the theory's content. Some part of Newtonian theory is usually chosen, so that the mechanical equivalence of different states of absolute motion can be used as a kind of scientific algorithm for generating the alternatives.

Bas van Fraassen (1980) defends EE by such examples, and by a second argument that rejects abduction on the basis of a pragmatic analysis of explanation. A theory explanatory with respect to one purpose may not be so with respect to another purpose, and there is no independent perspective from which a theory may be pronounced explanatorily successful as such. In fact, theories as such are *not* explanatory; they may be used to provide explanations, but that requires a context to fix interests and purposes. Only these pragmatic considerations determine that for which an explanation is wanting and what will qualify as explanatory.

A symptom of this context relativity is that the asymmetry of the explanatory relation is not fundamental, not dictated by theory. A change of context can reverse the status of *explanans* and *explanandum*, and context is determined pragmatically. There are no ultimate facts as to what is explanatory. The result is that explanation has no epistemic import: a theory's explanatory achievements do not warrant belief, because it can claim those achievements only relative to choices made on pragmatic rather than evidential grounds. Whereas Quine makes preferences among equally explanatory theories pragmatic, van Fraassen makes explanatory status itself pragmatic.

Third is a historical argument that induces the likely falsehood of any theory, however well supported by evidence, from the failure of past theories that once were equally well supported. Larry Laudan (1981) argues that explanatory virtues cannot warrant belief because they have been possessed by theories that subsequent developments refuted. More generally, claims Laudan, any form of empirical success cited as justification for believing a theory can be found in equal or better measure to have been achieved by theories that are incompatible with current science. His conclusion is that the standards for accepting theories are incapable of sustaining realism. As we are unable to construct a priori different or higher standards than science at its best has come to judge theories by – and could not know such an a priori standard to be correct if we had one – we must instead acknowledge an inherent limitation on the potential scope of the knowledge that science can provide. Warranted belief is possible only at the observational level. Induction tells us what theories are empirically successful, and thereby what explanations are successful. But the success of an explanation cannot, for historical reasons, be taken as an indicator of its truth.

The fourth argument, due to Arthur Fine (1986a), discerns a vicious circularity in the attempt to defend realism abductively. Fine distinguishes a level of reasoning within science, whereby particular physical theories and hypotheses are proposed and evaluated in light of relevant experimental evidence, from a second level of philosophical reasoning about science, whereby general theories about the epistemic status of the conclusions which science reaches are proposed and evaluated. Realism and instrumentalism are philosophical theories at the second level. The reasoning used at each level may be evaluated, and as realism is a metascientific thesis, the evaluation of the reasoning used to defend or reject it is metaphilosophical.

Now abduction at the first level, used to ground the acceptance of particular physical theories, is suspect (see INFERENCE TO THE BEST EXPLANATION). That much, Fine thinks, is established by the historical argument just adumbrated. So we should not base belief in physical theories on their explanatory resources with respect to particular observations and experiments. But if abduction is suspect as a mode of inference at level 1, it is inappropriate to use it at level 2. For the whole point of reasoning at level 2 is to determine the epistemic status of the conclusions to which level 1 reasoning leads. To say that a mode of inference at level 1 is suspect is to say that the conclusions to which it leads have inadequate support, hence that their epistemic status is not such as to warrant belief. If, then, abduction were used at level 2, the conclusion that it there delivers, as to the status of level 1 conclusions, would be similarly suspect. Having faulted abduction at level 1, one is not in a position to invoke it at level 2.

Of course one may dispute the historical argument. But the burden of realism is to *defend* abduction at level 1, not simply to refute arguments against it. For as realism advocates a positive epistemic status for level 1 conclusions, it is bound to defend the reasoning used to reach those conclusions. Such a defense cannot *itself* be abductive, for then it would presuppose its own conclusion. And in general, any mode of scientific reasoning, which it is the burden of a realist view of the results of that reasoning to defend, is out of bounds for the defense of realism.

Yet, complains Fine, realist arguments typically do stress the explanatory benefits of realism in just this circular way. This seems particularly true of Hilary Putnam's influential defense of realism (1978) as the only philosophy that does not require scientific success to be miraculous. If theoretical science does not correctly identify or describe the mechanisms responsible for experience, then it is a mystery, thought Putnam, how it manages to predict experience so accurately. An instrumentalist view of theories cannot explain why they work so well. Only realism accounts for the predictive and technological achievements of science, *and it is to be accepted for that reason.* To do this, adds William Newton-Smith (1981), realism need claim, not that any fully correct or final account of theoretical mechanisms has been reached, but only that one is being approached. The convergence of the sequence of theories that scientific change produces on an unqualifiedly true theory explains the increasing successfulness of science at the level of observation. Richard Boyd (1984) argues similarly that the success of scientific methods is uniquely explained by a realist view of the theories that underwrite those methods.

Fine (1986b) disputes such explanatory advantages of realism as Putnam, Newton-Smith, and Boyd discern. But even granting them, Fine's original argument refuses them any role in the defense of realism. How much, how much better, and how well realism explains anything cannot bear on its credibility, on pain of circularity. Since philosophical theories of science judge the legitimacy of modes of inference used within science, Fine concludes that the evaluation of philosophical theories of science must be restricted to modes of inference not used within science. Indeed, adds Fine, the restriction must be to modes of inference stronger – less questionable – than those found within science.

Realist replies

The most widely influential criticism of abduction is the first, and it has been scrutinized by Larry Laudan and Jarrett Leplin (1991) and by Jarrett Leplin (1997). They dispute both EE and the inference from it to UD. The judgment that rival theories are identical in their observational commitments must be historically indexed to the auxiliary information available for drawing observational consequences from theories, and to the technology and supporting theory that determine the range of things accessible to observation. As a consequence, no final judgment of equivalence is pronounceable; it is always possible that new means of observation or revisions in auxiliary knowledge will afford an opportunity for observational discrimination of theories. The relations of theory and observation cannot be fixed on semantic or syntactic grounds, as Quine and van Fraassen suppose, but depend on the shifting epistemic fortunes of auxiliary assumptions and on technological advances effected by collateral theories. And if the *content* of a pair of theories determines for them the same class of observational consequences – by placing limits on observability, for example – a judgment of empirical equivalence then depends on the empirical evaluation of that common content, and is therefore defeasible.

For criticism of this response see André Kukla (1998). Naturally it does not preclude the possibility of permanently equivalent, rival theories. Its burden, to the contrary, is to make the existence of such cases an empirical question, devoid of the philosophical guarantee it would require to serve as a basis for rejecting realism. But even in such cases, should there be any, it cannot be assumed, as per UD, that the rival theories are identically confirmed or refuted by the same evidence. There are other ways for evidence to bear on theory than through logical relations of entailment. Observations can support a theory that does not entail them, for example by confirming a deeper, background theory or by analogy. Unentailed evidence can disconfirm a theory by supporting a rival. There is no reason to expect that theories entailing the same observations across all relevant historical changes, which is what empirical equivalence amounts to, will fare identically in the face of all relevant evidence, which is what underdetermination amounts to.

The Laudan–Leplin argument does not explicitly endorse realism, and, indeed, Laudan himself naturally denies that the possibility of uniquely warranted theory preferences supports realism. It is unclear, however, what intermediate epistemic stance, stronger than pragmatism but weaker than realism, the argument leaves open. For if it is not to be theoretical posits, as realism would have it, that the evidence uniquely warrants, it is unclear what it warrants instead.

To reject UD is to maintain the possibility – even in cases of empirical equivalence – of endorsing a theory over its rivals on evidentially probative grounds, not merely on the pragmatic or instrumental grounds that any pragmatist, Quine included, readily allows. What, however, is it for evidence to support a theory over its rivals, but for that evidence to establish that one set of theoretical posits is more likely or more worthy of credence than others? If this realist reading is to be avoided, then some other attribute of the supported theory, independent of its theoretical posits, will have to be identified as the object of support. This other attribute cannot be merely pragmatic; nor can it be any attribute that the empirical equivalence of theories would guarantee that they share equally. Either possibility would obviate the argument against UD.

If we canvass the theoretical desiderata favored by anti-realists, we quickly find that none meets these constraints. For example, predictive reliability is arguably an epistemic, rather than a merely pragmatic, trait, and it carries no realist commitments. But how can predictive reliability attach differentially to empirically equivalent theories? Predictive efficiency or economy – pragmatic virtues – can distinguish such theories, but not reliability. Another candidate, surely an epistemic trait, is well-testedness. The empirical equivalence of theories does not imply that they are equally well tested. But if what is tested for, and what passing tests well attests to, is not predictive reliability, then what can it be but the correctness of theoretical posits? It will not do to say that it is *the theory* that is being tested, and stop at that. Things are tested *for* something, some attribute or capacity; they are not just "tested." And one cannot very well test a theory for the attribute of being well tested.

The needed intermediate stance between its being the theory as a whole, theoretical posits and all, that the evidence uniquely or preferentially warrants, and a judgment of pragmatic superiority that is warranted, proves elusive. The conclusion, strongly suggested if yet tentative, is that the rejection of EE and UD is at once a defense of realism.

To the second criticism that explanation is pragmatic, it may be objected that a context-independent, asymmetric structure underlies the pragmatic variations that van Fraassen adduces. What makes him think that context is crucial to explanatory status is that events which it is natural to identify as causes are generally not sufficient conditions for their effects. Explanations leave many necessary factors unspecified, and what does seem salient varies with context. Science seeks to identify everything relevant, to describe a complete causal network leading to the event to be explained. But that is not explanatory. Explanation is achieved by singling out the causal factor that pragmatic considerations make salient.

Thus it is uncommon to cite atmospheric oxygen to explain the lighting of a match, although oxygen is necessary. One can imagine a context in which oxygen is salient, however, as when one lights a match to determine whether it is safe to remove one's oxygen mask. Since context determines not only what is explanatory but also what is to be explained, one can further imagine reversing this direction of explanation: suppose that the purpose in lighting the match is to reduce the oxygen supply.

John Worrall (1984) undercuts this context relativity. The full causal account that theory supplies, from which one selects to fit the occasion, shows the ignition of the match to be a deterministic process. Accordingly, it is not symmetric. Only by changing the event to be explained from the lighting of the match to the decision that the match be lit is the direction of explanation reversed. The theory is not about decision making. Oxygen explains ignition explains reduction in oxygen, not the reverse. Alternative causes and apparent reversals of explanatory direction are understandable without making the explanations that theories give pragmatic or contextual. It is not that the full scientific account fails to be explanatory; rather, ordinary explanations are incomplete.

The third argument against abduction restricts the cumulativity of scientific knowledge to the empirical level, denying that theories are reducible to, or recoverable from, their successors. If they were, there would be a respect in which they survive disconfirmation, and then evidence for current theory would be reason to judge current theory survivable – partially correct or "on the right track," at least. Instead, theories that once excelled as much as do current theories must be wholly rejected, so that such excellence is no basis for credence.

There is an obvious way in which rejected theories survive that is not merely pragmatic: their formal laws are often limiting cases of the laws of their successors. This is unavailing for two reasons. The actual, logical relationship is more complicated. New theories characteristically correct their predecessors, and do not merely extend them. Logically there is incompatibility, and so independence; the old is not logically retained in the new. More important, despite mathematical, limiting case relations, the theoretical interpretation of the formalism of successive theories is characteristically different. And that is what the realist supposes warranted by success.

The induction to the likely falsity of current theories from that of past theories is a powerful anti-realist argument. It can be neutralized somewhat by other inductions. Quantum mechanics has been tested repeatedly to incredible levels of accuracy and has never been found wanting. This is a rather embarrassing example for the realist, since quantum mechanics is widely held to defy realist interpretation, and its domain is restricted by scientists to the observable. More important, such a response feeds the inductivist strategy. Theories that have been tested repeatedly and not found wanting have, historically, eventually been found wanting. We have been in the position of grasping final theory before and been wrong. (It seems to be an end-of-century sort of thing.) It is unsatisfactory to argue that the situation is fundamentally different today, that science really began in the 1930s, and that history does not matter. Maybe it will begin again (see QUANTUM MECHANICS).

More promising is to let the historical argument restrict the kinds of success that warrant a realist interpretation of theory and to acknowledge the achievement of such success only where some substantial theoretical commitment has survived theory change. Not all explanatory achievements support abductive inference; not all scientific success demands a realist explanation. It is sufficient that some does, for realism's burden is no sweeping endorsement of science, but only a defense of the possibility of warranted theoretical belief. That is all it takes to refute prevailing anti-realist epistemologies.

One form of scientific success that seems to require realism for an explanation is *novel predictive success*. This has not been evident, because philosophical accounts of novelty focus on what the theorist knows in advance or on what empirical results he uses in constructing his theory. But a result not used might exemplify an empirical law that was used and was antecedently affirmed on the strength of other exemplifications. In this situation, there is a kind of circularity in treating the result as support for a realist interpretation of the theory, even though it was not used and may, for good measure, have been unknown. In general, results that qualify as novel on such grounds can be interchangeable, in reconstructions of the reasoning that produces theories, with results that do not so qualify. And then the fact that the resultant theory successfully predicts results that are novel can be explained by the involvement of results of the same kind in the foundation of the theory. The theory need not be true to generate them.

399

A conception of novelty requiring a more robust independence from the theory's provenance will identify a form of success that false theories could not be expected to achieve, and that historically successful but completely superseded theories have not achieved. Let a result be novel for a theory only if no generalization of it is essential to the theory's provenance, where the standard for being essential is that omission vitiates the rationale that provenance supplies. An analysis of novelty along these lines promises the best answer to the historical challenge to realism. For a detailed development of this approach, see Leplin (1997).

The dilemma which Fine constructs for realism, although subtle and persuasive, comes to less than the historical objection. The argument is inspired by Hilbert's program of finite constructivism in metamathematics. That program rested on the admonition that it would be question begging to employ in metatheory a means of proof used in the theory whose credentials are at issue. Fine acknowledges but dismisses the obvious rejoinder that Hilbert's program was proved to be too stringent. It is not true of standardly accepted metamathematical reasoning that proof procedures are more narrowly restricted and more rigorous than those of the theories under investigation. It should be all the more apparent, in connection with the application that Fine makes to science, that the constructivist ideal is unrealizable. For it will be impossible to identify *any* rationally cogent form of argument without instances in scientific reasoning. Nor does it help to disqualify only "suspect" forms of argument. Every kind of argument found in science is sometimes involved in the drawing of false conclusions, so is "suspect" to some extent. If forms of reasoning used in science are banished from reasoning about science, then science *cannot be* reasoned about. Perhaps Fine approves of this result; it sounds much like his own position, the "natural ontological attitude." But an objection to realism that reduces to an objection to the very idea of philosophizing about science does not engage the issue.

Even if the circularity of which realism stands accused is vicious, it does not follow that realism is defeated. It follows not that an abductive defense of realism is unavailing, but only that the legitimacy of such a defense may not be *presupposed*. The legitimacy of abduction is to some extent underwritten by comparison with other standards of ampliative justification. Evidently Fine has no objection to ordinary, enumerative induction, since he endorses the historical argument. But induction encounters well-known paradoxes and inconsistencies that it takes abduction to resolve. The mere fact that certain properties have accompanied one another in experience is not, by itself, reason to project further concomitance, and ordinary reasoning does not so treat it, absent some explanation of why there should be such a regularity. Ordinarily, one is loath to infer, simply by enumerative induction, a hypothesis that proposes no explanation of the facts from which one infers. Ordinary justificatory reasoning ineliminably incorporates both inductive and abductive modes of ampliative inference, as well it must. And appeal to ordinary reasoning is only fair. Where else can one find neutral ground to acquit a demand for justification of one's justificatory standards?

It is also possible to respond to anti-abductive arguments by defending realism without appeal to its explanatory virtues. Two such attempts deserve notice. They depend on restricting realism to entities, eschewing it for theories. Nancy Cartwright (1983) and Ian Hacking (1983) develop a form of realism that involves no endorsement of theory, and appeals, therefore, to no explanation of its success. It is theoretical entities that they wish to be realist about, not the theories that posit such entities. Thus their realism claims not truth, however qualified, but existence: electrons, for example, exist, even if no theory of them is true.

The arguments they offer are different but complementary. Cartwright appeals to the causal agency of electrons in producing empirical regularities; Hacking, to their technological use in investigating yet conjectural aspects of nature. Cartwright thinks that citing electrons as causes of the phenomena we codify in empirical laws is different from using hypotheses about electrons to explain such laws. Hacking thinks that we can believe that electrons exist without believing any theoretical statement about them.

These contentions are surely problematic. It is not clear that inference to causes can succeed if abduction fails. And Hacking's realism seems to imply a causal theory of reference, which he is happy to deploy as motivation, even if he stops short of endorsement. But Jarrett Leplin (1979, 1988) has shown that whatever the benefits of the causal theory for understanding reference to observable entities, it is not feasible to extend it to entities initially posited by theory.

References

Boyd, R. 1984: The current status of scientific realism. In *Scientific Realism*, ed. J. Leplin (Berkeley and Los Angeles: University of California Press), 41–83.

Cartwright, N. 1983: *How the Laws of Physics Lie* (Oxford: Clarendon Press).

Fine, A. 1986a: *The Shaky Game: Einstein, Realism, and the Quantum Theory* (Chicago: University of Chicago Press).

—— 1986b: Unnatural attitudes: realist and instrumentalist attachments to science. *Mind*, 95, 149–79.

Hacking, I. 1983: *Representing and Intervening* (Cambridge: Cambridge University Press).

Kukla, A. 1998: *Studies in Scientific Realism* (New York and Oxford: Oxford University Press).

Laudan, L. 1981: A confutation of convergent realism. *Philosophy of Science*, 48, 19–50.

Laudan, L., and Leplin, J. 1991: Empirical equivalence and underdetermination. *Journal of Philosophy*, 88, 449–73.

Leplin, J. 1979: Reference and scientific realism. *Studies in History and Philosophy of Science*, 10, 265–84.

—— 1988: Is essentialism unscientific?. *Philosophy of Science*, 55, 493–510.

—— 1997: *A Novel Defense of Scientific Realism* (New York and Oxford: Oxford University Press).

Newton-Smith, W. 1981: *The Rationality of Science* (Boston: Routledge and Kegan Paul).

Putnam, H. 1978: *Meaning and the Moral Sciences* (Boston: Routledge and Kegan Paul).

van Fraassen, B. 1980: *The Scientific Image* (Oxford: Clarendon Press).

Worrall, J. 1984: An unreal image. *British Journal for the Philosophy of Science*, 35, 65–80.

59

Reductionism

JOHN DUPRÉ

The term "reductionism" is used broadly for any claim that some range of phenomena can be fully assimilated to some other, apparently distinct range of phenomena. The logical positivist thesis that scientific truth could be fully analyzed into reports of immediate experience was a reductionistic thesis of great significance in the history of the philosophy of science (see LOGICAL POSITIVISM). In recent philosophy of science, "reductionism" is generally used more specifically to refer to the thesis that all scientific truth should ultimately be explicable, in principle at least, by appeal to fundamental laws governing the behavior of microphysical particles.

A classic exposition of this kind of reductionism is that of Oppenheim and Putnam (1958). Oppenheim and Putnam propose a hierarchical classification of objects, the objects at each level being composed entirely of entities from the next lower level. They suggest the following levels: elementary particles, atoms, molecules, living cells, multicellular organisms, and social groups. The investigation of each level is the task of a particular domain of science, which aims to discern the laws governing the behavior of objects at that level. Reduction consists in deriving the laws at each higher (reduced) level from the laws governing the objects at the next lower (reducing) level. Such reduction will also require so-called bridge principles identifying the objects at the reduced level with particular structures of objects at the reducing level. Since such deductive derivation would be transitive, the end point of this program will reveal the whole of science to have been derived from nothing but the laws of the lowest level, or the physics of elementary particles, and the bridge principles. (Another classic source on reductionism is Nagel 1961; for critical discussion see Dupré 1993, ch. 4.)

Two major motivations can be discerned for such a reductionist conception of science. The first of these is the contention that the history of science in fact exemplifies instances of reduction as among its major achievements. Second, a commitment to reductionism may be based on a priori philosophical argument: given the belief common to most scientists and philosophers that everything there is, is composed entirely of elementary physical particles, and the idea that the behavior of elementary particles is fully described by laws at the most basic physical level, it seems that the behavior of structures of physical particles (objects higher in the Oppenheim–Putnam hierarchy) must ultimately be fully determined by the behavior of their constituent physical parts.

The first ground for reductionism has become more problematic in the light of a good deal of recent work in the history and philosophy of science. Even paradigm cases of reduction, such as that of thermodynamics to statistical mechanics, have come to be seen as much more complex and debatable than was once thought. A great deal of work on the relation of Mendelian genetics, the study of the transmission of traits from organisms to their descendants, to molecular genetics, has led a majority of philosophers of biology to conclude that this promising case for reduction cannot be subsumed under anything like the Oppenheim–Putnam model (Hull 1974; Kitcher 1984; but see Waters 1990).

One response to this has been to move to a very different conception of reduction. A number of philosophers have suggested that higher-level theories cannot be reduced to lower levels because, strictly speaking, the former are false. Thus they propose that rather than higher-level theories being deduced from lower-level theories, we should anticipate the ultimate rejection of the former in favor of expanded versions of the latter. A notorious example of this strategy is the claim that our commonsense understanding of the mind, in terms of sensations, beliefs, feelings, etc., as well as the psychological theories that have attempted to study and refine these commonsense concepts, should ultimately be rendered obsolete by some future science of neurophysiology (Churchland 1986). Such eliminative reductionism, and the derivational reductionism described above, can be seen as forming a spectrum of possible cases. In between these extremes we can imagine that the theories ultimately to be derived from the reducing science will, in the process, be more or less substantially altered by developments in the reducing science.

A difficulty with the eliminative conception is that it is much more difficult to provide plausible examples of eliminative reduction from the history of science. Scientific theories have often been replaced, but by rival theories at the same structural level. The putative redundancy of traditional mental concepts has yet to be demonstrated. Eliminativists tend, therefore, to base their claims on philosophical arguments of the kind sketched above, and on claims about the limited prospects of the science to be eliminated. Other philosophers, persuaded that the practical failure of reductionism reveals deep difficulties with the program, have retreated to an insistence on the possibility of reduction in principle, but have denied its actual feasibility in the practice of science. One common such position is referred to as "supervenience" (though the relation of this position to reductionism remains controversial). A domain of phenomena is said to supervene on another domain – in this case higher-level phenomena are said to supervene on lower-level phenomena – when no difference at the supervening level is possible without some difference at the lower level supervened on (but not vice versa) (Kim 1978). Thus the supervenient phenomena are fully determined by the state of the phenomena on which they supervene. States of the mental are often said to supervene on states of the brain, or biological phenomena on underlying chemical processes.

More radical rejections of the philosophical argument for reductionism are also possible. In particular, the assumption that the behavior of microphysical particles is completely described by microphysical laws could be questioned. Evidence for microphysical laws is derived from extremely unusual and specialized experimental setups, and the extension of these to particles embedded in complex structures requires a debatable inductive leap. In the absence of this assumption it is possible to see science as consisting of equally autonomous, though incomplete, generalizations at many levels. I defend such a position in Dupré 1993, pt. 2. (See also Cartwright 1983.)

References

Cartwright, N. 1983: *How the Laws of Physics Lie* (Oxford: Oxford University Press).

Churchland, P. S. 1986: *Neurophilosophy* (Cambridge, MA: Bradford Books/MIT Press).

Dupré, J. 1993: *The Disorder of Things: Metaphysical Foundations of the Disunity of Science* (Cambridge, MA: Harvard University Press).

Hull, D. L. 1974: *The Philosophy of Biological Science* (Englewood Cliffs, NJ: Prentice-Hall).

Kim, J. 1978: Supervenience and nomological incommensurables, *American Philosophical Quarterly*, 15, 149–56.

Kitcher, P. 1984: 1953 and all that: a tale of two sciences. *Philosophical Review*, 93, 335–76.

Nagel, E. 1961: *The Structure of Science: Problems in the Logic of Scientific Explanation* (New York: Harcourt, Brace, and World).

Oppenheim, P., and Putnam, H. 1958: The unity of science as a working hypothesis. In *Minnesota Studies in the Philosophy of Science*, vol. 2, ed. H. Feigl, M. Scriven, and G. Maxwell (Minneapolis: University of Minnesota Press), 3–36.

Waters, K. 1990: Why the anti-reductionist consensus won't survive: the case of classical Mendelian genetics. In *PSA 1990*, vol. 1, ed. A. Fine, M. Forbes, and L. Wessels (East Lansing, MI: Philosophy of Science Association, 1990), 125–39.

60

Relativism

JAMES W. MCALLISTER

Relativism about a property P is the thesis that any statement of the form "Entity E has P" is ill formed, while statements of the form "E has P relative to S" are well formed, and true for appropriate E and S. Relativism about P therefore entails the claim that P is a relation rather than a one-place predicate. In the principal forms of relativism, the variable S ranges over cultures, world views, conceptual schemes, practices, disciplines, paradigms, styles, standpoints, or goals.

Innumerably many forms of relativism are entirely unobjectionable. An obvious example is relativism about the property "utility," since a tool or instrument is useful only relative to a particular goal. Even relativism about the merits of conceptual systems and taxonomic schemes is reasonably uncontentious, since practically everyone finds different concepts and taxonomies preferable in different circumstances. Among more controversial forms of relativism, the following are those of greatest interest in philosophy of science.

Relativism about truth, or epistemological relativism, is the thesis that propositions have truth-value not absolutely, but only from a particular standpoint, and that in general a proposition has different truth-values from different standpoints. Unrestricted relativism about truth – that is, a relativism that counts its own statement among the propositions to which it applies – is self-subverting. This flaw can be evaded by, for instance, locating the statement of relativism on a level different from that of the propositions to which it applies, but such a maneuver requires a piecemeal theory of truth. Introducing a predicate "subjective truth," for which relativism holds, fails to establish relativism about truth: subjective truth is a property different from truth, as we may establish by reasoning that a proposition's having subjective truth from a particular standpoint is compatible with its being also absolutely true or false. Those who espouse relativism about truth would in many cases better achieve their aims by endorsing the anti-realist's denial of truth-value to the propositions of a specified set, and applying to these propositions a distinct and truly standpoint-relative predicate, such as "acceptability" (Meiland and Krausz 1982; Siegel 1987; see also REALISM AND INSTRUMENTALISM).

Relativism about rationality is the thesis that none of the many possible canons of reasoning has a privileged relationship with reality, and thus none has special efficacy in establishing truths or intervening in the world. For instance, Paul K. Feyerabend sees many different patterns of reasoning followed opportunistically by scientists, and none to which special status may be attributed. Similarly, Thomas S. Kuhn suggests that what communities take to be rational reasoning is prescribed by their current paradigm; he is sometimes understood as advancing also a form of relativism about reality, according to which the users of different patterns of reasoning inhabit different worlds. A conclusion drawn by many relativists about rationality is that mutually incompatible criteria for theory choice have equivalent status, and therefore that science cannot be said to achieve any absolute progress. Anthropologists and historians often recommend versions of relativism about rationality as aiding the reconstruction and understanding of alien modes of thought (Hollis and Lukes 1982; Laudan 1990; see also FEYERABEND and KUHN).

One of the most interesting forms of relativism formulated in recent years in science studies is relativism about the evidential weight of empirical findings. It was assumed by logical positivism that empirical data would have equal evidential weight in all contexts. More sensitive historiography has established that evidential weight is in fact attributed to sets of empirical findings within particular domains. The procedure of vesting evidential weight in a set of findings is frequently arduous and contested, and a finding that is accredited in one context may fail to carry evidential weight in another (Shapin and Schaffer 1985, pp. 3–21; McAllister 1996; see also LOGICAL POSITIVISM and EXPERIMENT).

A fruitful language in which to express relativist views is offered by the notion of style. The foundationalist tradition inaugurated by René Descartes holds that philosophy and the sciences possess method, a unique procedure permitting the demonstration of truths. An alternative view suggests that, rather than a unique truth-yielding method, there exist alternative styles of inquiry among which a choice is open. There is abundant evidence that scientists' practice is governed to some extent by style: even reasoning in axiomatic mode is the outcome of a choice among styles. Since it may occur that an intervention that is well formed, well founded, persuasive, or otherwise successful within one style is not so in another, this view of scientific practice readily accommodates forms of relativism (Hacking 1992).

Many of the most lively recent discussions of relativism have occurred in social studies of science. The approach known as "sociology of scientific knowledge," formulated by H. M. Collins and others, holds that the work of scientists is most fruitfully studied from an attitude that combines realism about models of social phenomena with relativism about the merits of the scientists' own knowledge claims. On this approach, a scientific community's settling on particular matters of fact is to be explained by references to the constitution of society, which are regarded as relatively unproblematic. An alternative approach advocated by Bruno Latour and others, actor-network theory, denies that social structure predates the outcomes of scientific controversies, and therefore that it can explain them. It portrays scientific practice as the construction of networks in which the resources of human and nonhuman actors are harnessed to establish knowledge claims. Parts of the network still under construction are best understood from a relativist attitude, while parts that are settled lend themselves to analysis in realist terms (Pickering 1992, pp. 301–89; see also SOCIAL FACTORS IN SCIENCE).

Relativism is one of the central issues in the science wars, an acrimonious dispute involving natural scientists, sociologists of science, and others, that arose in the mid-1990s. Some natural scientists have claimed that sociologists of science espouse relativism about the status of scientific findings in an attempt to discredit science, and have pointed out that much work in sociology of science shows ignorance of natural science. Sociologists have responded by reasserting the legitimacy of critical analysis of the concepts of truth and scientific rationality (Jardine and Frasca-Spada 1997; Brown 2000).

References

Brown, J. R. 2000: *Who Should Rule? A Guide to the Epistemology and Politics of the Science Wars* (Berlin: Springer).

Hacking, I. 1992: "Style" for historians and philosophers. *Studies in History and Philosophy of Science,* 23, 1–20.

Hollis, M., and Lukes, S. (eds) 1982: *Rationality and Relativism* (Oxford: Blackwell).

Jardine, N., and Frasca-Spada, M. 1997: Splendours and miseries of the Science Wars. *Studies in History and Philosophy of Science,* 28, 219–35.

Laudan, L. 1990: *Science and Relativism: Some Key Controversies in the Philosophy of Science* (Chicago: University of Chicago Press).

McAllister, J. W. 1996: The evidential significance of thought experiment in science. *Studies in History and Philosophy of Science,* 27, 233–50.

Meiland, J. W., and Krausz, M. (eds) 1982: *Relativism: Cognitive and Moral* (Notre Dame, IN: University of Notre Dame Press).

Pickering, A. (ed.) 1992: *Science as Practice and Culture* (Chicago: University of Chicago Press).

Shapin, S., and Schaffer, S. 1985: *Leviathan and the Air-Pump: Hobbes, Boyle, and the Experimental Life* (Princeton: Princeton University Press).

Siegel, H. 1987: *Relativism Refuted: A Critique of Contemporary Epistemological Relativism* (Dordrecht: Reidel).

61

Russell

PAUL J. HAGER

The eminent British philosopher Bertrand Arthur William Russell (born 18 May 1872, died 2 February 1970) studied philosophy and mathematics at Trinity College, Cambridge, and subsequently held posts at Cambridge and various other major universities, interspersed with periods devoted to political, educational, and literary pursuits. He was author of numerous influential books and papers on philosophy, politics, and education. Few philosophers of science have had as strong a scientific background as Russell. His mathematical training at Cambridge was almost entirely in applied mathematics which was largely physics.

A main thrust of Russell's work was the development of a method of philosophizing, which he consistently applied throughout his career, and which led him to a distinctive philosophy of science. However, because the method has not been well understood, his philosophy of science has not attracted wide notice. Accordingly, I will here first outline major characteristics of Russell's method of philosophical analysis and, second, show how this method underpins a distinctive philosophy of science.

Russell's method of philosophical analysis

Russell developed a method of philosophical analysis, the beginnings of which are clear in the work of his idealist phase. This method was central to his revolt against idealism and was employed throughout his subsequent career. Its main distinctive feature is that it has two parts. First, it proceeds backwards from a given body of knowledge (the "results") to its premises, and second, it proceeds forwards from the premises to a reconstruction of the original body of knowledge. Russell often referred to the first stage of philosophical analysis simply as "analysis," in contrast to the second stage, which he called "synthesis." While the first stage was seen as being the most philosophical, both were nonetheless essential to philosophical analysis. Russell consistently adhered to this two-directional view of analysis throughout his career (Hager 1994).

His initial major applications of this method of philosophical analysis were to mathematics in *Principles of Mathematics* and *Principia Mathematica*. However, at that time, he held that this mathematical work was in principle no different from work in the foundations of any science. In all cases, philosophical analysis was a nonempirical intellectual discovery of propositions and concepts from which could be fashioned premises for the basic data with which the analysis had begun. The links that Russell saw between philosophical analysis and the sciences become clear from considering some important characteristics that he ascribes to his method of analysis:

408

(i) *Analysis is unlikely to be final.* This applies in several ways. Not only is analysis never final in the sense that new premises may be discovered in relation to which existing premises are results, but also there is the ever present possibility of alternative sets of premises for the same results. In the former case, further stages of analysis in no way invalidate earlier ones. As Russell repeatedly emphasizes, no error will flow from taking complex objects to be simple at one level of analysis, as long as it is not assumed that such objects are incapable of further analysis. In the latter case, to ask what are the minimum premises for a given set of results "is a technical question and it has no unique answer" (Russell 1975, p. 162). Hence, one important task for philosophy is to devise alternative sets of premises.

(ii) *Analysis enlarges the domains of particular subjects.* The current science on which analysis is practiced changes as the subject itself evolves. Formerly tentative premises for a science later become a part of that science. As the frontier is extended, territory that once belonged to philosophy becomes exact enough for incorporation into science. Thus "every advance in knowledge robs philosophy of some problems which formerly it had" (Russell 1986, p. 243). In terms of Russellian analysis, yesterday's premises become tomorrow's results from which a new generation of philosophers will start the backwards journey of analysis. Thus the philosophy/science distinction "is one, not in the subject matter, but in the state of mind of the investigator." (Russell 1970a, p. 1). It remains for philosophy to move to the new frontier. Hence Russell's description of philosophy as occupying the "No Man's Land" between "theology and science" (Russell 1971, p. 13) and the maxim that "science is what you more or less know and philosophy is what you do not know" (Russell 1986, p. 243).

(iii) *Analysis leads to premises that are decreasingly self-evident.* Russell made this point emphatically:

> When pure mathematics is organized as a deductive system . . . it becomes obvious that, if we are to believe in the truth of pure mathematics, it cannot be solely because we believe in the truth of the set of premises. Some of the premises are much less obvious than some of their consequences, and are believed chiefly because of their consequences. This will be found to be always the case when a science is arranged as a deductive system. It is not the logically simplest propositions of the system that are the most obvious, or that provide the chief part of our reasons for believing in the system. With the empirical sciences this is evident. Electro-dynamics, for example, can be concentrated into Maxwell's equations, but these equations are believed because of the observed truth of certain of their logical consequences. Exactly the same thing happens in the pure realm of logic; the logically first principles of logic, – at least some of them – are to be believed, not on their own account, but on account of their consequences. (Russell 1988, pp. 163–4)

Likewise, "[i]n mathematics, the greatest degree of self-evidence is usually not to be found quite at the beginning, but at some later point; hence the early deductions, until they reach this point, give reasons rather for believing the premises because true consequences follow from them, than for believing the consequences because they follow from the premisses" (Russell 1925–7, vol. 1, p. v).

The decreasing self-evidence of the premises can have ontological implications. According to Russell, the current premises provide our best guide to the nature of the most fundamental entities – hence, for example, his replacement of commonsense physical objects by sense-data and events. The decreasing self-evidence of the premises was also the basis of Russell's vintage statement that "the point of philosophy is to start with something so simple as not to seem worth stating, and to end up with something so paradoxical that no one will believe it" (Russell 1986, p. 172). This decreasing self-evidence of the premises, coupled with the earlier claim that there may be alternative premises from which the same given set of results is deducible, is the basis of Russell's characteristic open-mindedness about the finality or otherwise of his philosophical views at any given stage.

Russell's philosophy of science

Given that a philosophy of science should centrally concern itself with such matters as the rational basis of science, the distinctive nature of its theories and modes of explanation, and the nature of the reality it describes, Russellian philosophical analysis, as outlined above, clearly implies a philosophy of science. Since, according to Russell, the method of philosophy is like that of science, the three features of Russellian philosophical analysis that have been outlined so far together incorporate the main elements of Russell's philosophy of science.

First, because *analysis is unlikely to be final*, science is at any stage deducible from premises, which may be themselves deducible from still further premises, and so on. Moving backwards in this chain, some premises are eventually reached that currently belong to philosophy rather than to science. These philosophical premises are likely to become themselves the "results" for which a search for still further suitable premises will be initiated. Thus, however advanced a particular science might be, it will still be based on premises which are not themselves a part of that science.

Likewise, however well founded might be the premises for a particular science, they are never immune from drastic revision, since the issue of "the minimum hypotheses" from which that science is deducible "is a technical question and it has no unique answer" (Russell 1975, p. 162). Thus, there is always the possibility that, as in the case of Einsteinian relativity displacing Newtonian mechanics, an entirely new structure of premises will come to ground that science.

Second, because *analysis enlarges the domains of particular subjects*, there is a gradual shift in the border between philosophy and science. Science expands to incorporate what was once philosophy; however, at each new border there is further work for philosophy. It must be stressed, nevertheless, that Russell does not view philosophy as the main source of knowledge advance. Rather, this comes from within the special sciences. One role of philosophy is that of "suggesting hypotheses as to the universe which science is not yet in a position to confirm or confute. But these should always be presented *as* hypotheses" (Russell 1988, p. 176). However, this is not philosophy's major role:

> Philosophy is more concerned than any special science with relations of different sciences and possible conflicts between them; in particular, it cannot acquiesce in a conflict between

physics and psychology, or bctwccn psychology and logic. . . . The most important part . . . [of the business of philosophy] . . . consists in criticising and clarifying notions which are apt to be regarded as fundamental and accepted uncritically. As instances I might mention: mind, matter, consciousness, knowledge, experience, causality, will, time. (Russell 1988, pp. 176–7)

On this view, rather than aiming to provide overarching theories of the universe, as attempted by most of his great predecessors, Russell limits philosophy to attacking in a piecemeal way problems posed by the sciences. Philosophy's task is "piecemeal" because, according to Russell, it does not enjoy an independent standpoint from which it can judge science as a whole:

> The philosophic scrutiny . . . though sceptical in regard to every detail, is not sceptical as regards the whole. That is to say, its criticism of details will only be based upon their relation to other details, not upon some external criterion which can be applied to all the details equally. (Russell 1969, p. 74)

By making all sound philosophy parasitic on science, Russell is in grave danger of excluding ethical, political, and social questions from the philosophical agenda. Although many would see this as a too narrowly rigid conception of philosophy, Russell accepts this consequence:

> There remains, however, a vast field, traditionally included in philosophy, where scientific methods are inadequate. This field includes ultimate questions of value; science alone, for example, cannot prove that it is bad to enjoy the infliction of cruelty. Whatever can be known, can be known by means of science; but things which are legitimately matters of feelings lie outside its province . . . philosophers who make logical analysis the main business of philosophy . . . confess frankly that the human intellect is unable to find conclusive answers to many questions of profound importance to mankind, but they refuse to believe that there is some "higher" way of knowing, by which we can discover truths hidden from science and the intellect. (Russell 1971, pp. 788–9)

Thus, for Russell, not only is philosophy parasitic on science, but, in an important sense, all philosophy is philosophy of science.

Third, because *analysis leads to premises that are decreasingly self-evident*, Russell is committed to a nonfoundationalist philosophy of science as that term is usually understood. Rejecting all aspirations to reach certain premises, Russell insists that though the "demand for certainty is . . . natural . . . , [it] is nevertheless an intellectual vice. . . . What philosophy should dissipate is *certainty*, whether of knowledge or of ignorance . . . all our knowledge is, in a greater or less degree, uncertain and vague" (Russell 1970b, pp. 32–3). Because, firstly, the premises become decreasingly self-evident as knowledge advances, and, secondly, alternative sets of premises are always a possibility, Russell holds that both science and philosophy offer "successive approximations to the truth," rather than certainty (Russell 1971, p. 789). (We are inevitably reminded here of Popper's swamp analogy. (see POPPER))

Space limitations have permitted a consideration of only the main aspects of Russell's philosophy of science. Further details would emerge from a study of works such as *The*

411

Analysis of Matter. Likewise, an examination of Russell's inductive skepticism would show it to have anticipated Goodman's "new riddle of induction." However, the key point that philosophy of science was the focus of Russell's philosophizing should by now be evident.

References and further reading

Works by Russell

1925–7: *Principia Mathematica*, with A. N. Whitehead, 3 vols (Cambridge: Cambridge University Press; 1st pub. 1910–13).

1948: *Human Knowledge: Its Scope and Limits* (London: Allen & Unwin).

1969: *Our Knowledge of the External World as a Field for Scientific Method in Philosophy*, rev. edn (London: Allen & Unwin; 1st pub. 1914).

1970a: *An Introduction to Mathematical Philosophy* (London: Allen & Unwin; 1st pub. 1919).

1970b: Philosophy for laymen (1950). In *Unpopular Essays* (London: Allen & Unwin), 27–37.

1971: *History of Western Philosophy*, 2nd edn (London: Allen & Unwin; 1st pub. 1946).

1975: *My Philosophical Development*, Unwin Books edn (London: Allen & Unwin; 1st pub. 1959).

1986: The philosophy of logical atomism (1918). In *The Philosophy of Logical Atomism and Other Essays: 1914–19*, *The Collected Papers of Bertrand Russell*, vol. 8, ed. J. Slater (London: Allen & Unwin), 157–244.

1988: Logical atomism (1924). In *Essays on Language, Mind and Matter: 1916–26*, *The Collected Papers of Bertrand Russell*, vol. 9, ed. J. Slater (London: Unwin Hyman), 162–79.

Works by other authors

Hager, P. J. 1994: *Continuity and Change in the Development of Russell's Philosophy* (Dordrecht: Kluwer).

62

Scientific Change

DUDLEY SHAPERE

General description of the topic

Broadly, the problem of scientific change is to give an account of how scientific theories, propositions, concepts, and/or activities alter over history. Must such changes be accepted as brute products of guesses, blind conjectures, and genius? Or are there rules according to which at least some new ideas are introduced and ultimately accepted or rejected? Would such rules be codifiable into a coherent system, a *theory* of "the scientific method"? Are they more like rules of thumb, subject to exceptions whose character may not be specifiable, not necessarily leading to desired results? Do these supposed rules themselves change over time? If so, do they change in the light of the same factors as more substantive scientific beliefs, or independently of such factors? Does science "progress"? And if so, is its goal the attainment of truth, or a simple or coherent account (true or not) of experience, or something else?

Controversy exists about what a theory of scientific change should be a theory of the change *of*. Philosophers long assumed that the fundamental objects of study are the acceptance or rejection of individual beliefs or propositions, change of concepts, propositions, and theories being derivative from that. More recently, some have maintained that the fundamental units of change are theories or larger coherent bodies of scientific belief, or concepts or problems. Again, the kinds of *causal factors* which an adequate theory of scientific change should consider are far from evident. Among the various factors said to be relevant are observational data; the accepted background of theory; higher-level methodological constraints; psychological, sociological, religious, metaphysical, or aesthetic factors influencing decisions made by scientists about what to accept and what to do.

These issues affect the very delineation of the field of philosophy of science: in what ways, if any, does it, in its search for a theory of scientific change, differ from and rely on other areas, particularly the history and sociology of science? One traditional view was that those others are not relevant at all, at least in any fundamental way. Even if they are, exactly how do they relate to the interests peculiar to the philosophy of science? In defining their subject, many philosophers have distinguished matters *internal* to scientific development – ones relevant to the discovery and/or justification of scientific claims – from ones *external* thereto – psychological, sociological, religious, metaphysical, etc., not directly (scientifically) relevant but frequently having a causal influence. A *line of demarcation* is thus drawn between science and non-science, and simultaneously between philosophy of science, concerned with the internal factors

413

which function as *reasons* (or count as *reasoning*), and other disciplines, to which the external, nonrational factors are relegated.

This array of issues is closely related to that of whether a proper theory of scientific change is normative or descriptive. Is philosophy of science confined to description of what scientists do and how science proceeds? Insofar as it is descriptive, to what extent must scientific cases be described with complete accuracy? Can the history of internal factors be a "rational reconstruction," a retelling that partially distorts what actually happened in order to bring out the essential reasoning involved?

Or should a theory of scientific change be *normative*, prescribing how science *ought* to proceed? Should it counsel scientists about how to improve their procedures? Or would it be presumptuous of philosophers to advise them about how to do what they are far better prepared to do? Most advocates of normative philosophy of science agree that their theories are accountable somehow to the actual conduct of science. Perhaps philosophy should clarify what is done in the *best* science; but can what qualifies as "best science" be specified without bias? Feyerabend objects to taking certain developments as paradigmatic of good science. With others, he accepts the "pessimistic induction," according to which, since all past theories have proved incorrect, present ones can be expected to do so also: what we consider good science, even the methodological rules we rely on, may be rejected in the future.

Historical background

The seventeenth-century founders of modern science viewed their work as discontinuous with previous thought in employing a new *method* of gaining knowledge. For Descartes the basic laws, if not the specific facts, of nature are deducible by reason (see DESCARTES). Ultimately Cartesian deductivism was almost universally rejected, though Laudan (1981) claims that its influence persisted in the turn to the hypothetico-deductive method.

By contrast, early inductivists held that (1) science begins with data collection; (2) rules of inference are applied to the data to obtain a theoretical conclusion (or, at least, to eliminate alternatives); and (3) that conclusion is established with high confidence or even proved conclusively by the rules. Rules of inductive reasoning were proposed by Francis Bacon and by Newton in the second edition of the *Principia* ("Rules of Reasoning in Philosophy") (see NEWTON). Such procedures were allegedly applied in Newton's *Opticks* and in many eighteenth-century experimental studies of heat, light, electricity, and chemistry.

According to Laudan (1981), two gradual realizations led to rejection of this conception of scientific method: first, that inferences from facts to generalizations are not establishable with certainty (hence scientists were more willing to consider hypotheses with little prior empirical grounding); second, that explanatory concepts often go beyond sense experience (and that such trans-empirical concepts as "atom" and "field" can be introduced in the formation of such hypotheses). Thus, after the middle of the eighteenth century, the inductive conception began to be replaced by the *method of hypothesis*, or the *hypothetico-deductive method*. On this view, the order of events in science is seen as, first, introduction of a hypothesis, and second, testing of observational predictions of that hypothesis against observational and experimental results.

More may be involved than Laudan suggests. Newton's third rule allowed generalizations from sense experience employing the same concepts as the individual cases upon which the generalizations are based; the concept of "atom," at least, was conceived as a straightforward generalization from the universal, invariant properties of the observed whole (solidity, extension, mass, etc.) to the properties of the smallest parts. Even for concepts harder to define empirically ("field," perhaps), history might have taken a course different from the introduction of the method of hypothesis: namely, denying (3) above (the certainty or near-certainty of scientific conclusions) and replacing it with a stage of inference, beyond generalization, by which trans-empirical concepts could be introduced. Perhaps a rule for making analogies – for fields, the analogy of patterns formed by iron filings near a magnet – might have been tried. Hesse, Schaffner, and others have highlighted the role of analogies in the introduction of new ideas in science. Others argue that analogy alone cannot account for the introduction of ideas as radically nonempirical as those fundamental to contemporary quantum theories and cosmology.

In any event, with growing realization that scientific change was pervasive, and particularly that radically new concepts foreign to sense experience could be introduced, interest arose in the sources of new ideas. Among the pioneers developing such theories of scientific method in the nineteenth century were Herschel and Whewell (see WHEWELL).

Twentieth-century relativity and quantum mechanics alerted scientists even more to the potential depths of departures from common sense and earlier scientific ideas (see QUANTUM MECHANICS). Bohr especially emphasized the pragmatic character of scientific innovation: scientists should be willing to try whatever might help solve a problem, no matter how "crazy" the suggestion might seem. Yet, at the same time, many philosophers (Reichenbach was an exception) saw those developments in the light of their even greater awe of mathematical logic. Their attention was called away from scientific change and toward analyses of atemporal "formal" characteristics of science; the dynamical character of science, emphasized by physicists, was lost in a quest for unchanging characteristics definitory of science and its major components (theory, confirmation, lawlikeness, etc.). So, while Laudan is correct in claiming that some investigations of scientific change are initiated by new developments in science, this was not the case in the first half of the twentieth century, at least among logical empiricists. "The logic of science" being seen as the study of the "form" rather than the "content" of thought, the "meanings" of fundamental "metascientific" concepts and methodological strictures were divorced from the specific content of science. The hypothetico-deductive conception of method, endorsed by logical empiricists, was likewise construed in these terms: "discovery," the introduction of new ideas, was grist for historians, psychologists, or sociologists, whereas the "justification" of scientific ideas was the application of logic and thus the proper object of philosophy of science (see LOGICAL EMPIRICISM). Logical empiricists like Carnap sought an "inductive logic," no longer conceived as beginning with collected data and inducing theoretical conclusions, but rather as assessing the relations between an already proposed hypothesis and its relevant evidence. Despite his opposition to logical empiricism, Popper shared this general "logical" approach. For him, there is no Carnapian inductive logic; the logic of justification lies in falsification, the negation of deductive conclusions from scientific "conjectures."

Logical empiricism faced increasing difficulties in the 1950s, just when new interpretations of major historical episodes were proliferating among historians of science. The view that there are no sources of scientific ideas worthy of philosophical investigation – that there is nothing that can be called reasoning in "the context of discovery" – became increasingly suspect. Consequently, philosophical interest in the rational sources of discovery arose, initiated in the writings of Hanson, Toulmin, Feyerabend, and Kuhn, and followed by a broadening interest in scientific change in such writers as Laudan, McMullin, Shapere, and subsequently an increasing host of others.

Discovery versus justification

Much discussion of scientific change since Hanson centers on the distinction between contexts of discovery and justification (see DISCOVERY). The distinction is usually ascribed to Reichenbach (1938), and, as generally interpreted, reflects the attitude of the logical empiricist movement and of Popper, that discovery is of no philosophical concern, being a matter of purely psychological or sociological factors affecting the discoverer (see POPPER). Some doubt that what either Reichenbach or Popper had in mind corresponds to the interpretations of the distinction and its implications among more recent philosophers of science.

Hanson (1958), following Peirce, proposed that there is a logic of discovery, an *abductive logic*, as opposed to both deduction and induction which are used in justification, in the sense of a formal rule by application of which hypotheses are generated. (This view of abduction as a "logic" does not depart in principle from the logical empiricist program.) By "abduction" Hanson had in mind the acceptance of a hypothesis on the basis of the fact that a certain phenomenon could be understood if that hypothesis were true. Presupposing prior formulation of the hypothesis, this is not a "logic of *discovery*" after all.

The promise of a "logic" of discovery, in the sense of a set of algorithmic, content-neutral rules of reasoning distinct from justification, remains unfulfilled. Upholding the distinction between discovery and justification, but claiming nonetheless that discovery is philosophically relevant, many recent writers propose that discovery is a matter of a "methodology," "rationale," or "heuristic" rather than a "logic." That is, rather than there being formal rules for mechanically generating discoveries, there is only a loose body of strategies or rules of thumb – still formulable independently of the content of scientific belief – which one has some reason to hope will lead to the discovery of a hypothesis.

Details of the discovery/justification distinction and its adequacy are disputed in three major respects.

(a) *Whether we must distinguish between several phases of scientific change*: a "preparation" or "incubation" stage before discovery, a "discovery" or "hypothesis-generating" stage, a "pursuit" stage, a "testing" or "assessment" stage, and a "clarification" stage in which the hypothesis is clarified and developed. Laudan (1977) distinguishes the discovery (generation) stage, which he views as a matter of psychology, etc., from the pursuit stage, in which reasons exist for pursuing the hypothesis further.

(b) *Whether any rationale (much less any "logic") is involved in some or all of the stages*, particularly in the early ones, or whether considerations involved in, say, the discovery phase, carry any epistemic weight at all.

(c) *Whether, if a rationale is involved, it is different in fundamental ways from that involved in the testing phase*, or whether the rationale involved in, say, the pursuit phase consists in the same sorts of considerations later used as part of a fuller justification. Laudan considers that the kinds of reasons involved in pursuit are those also employed in the final assessment stage, the only difference being that when they are given prior to final testing, they are matters of "preliminary evaluation." Others argue that there are rules, or at least heuristics, of (say) discovery or pursuit which differ from those used in justification.

Paradigms and cores of scientific thought

In the enthusiasm over the problem of scientific change in the 1960s and 1970s, the most influential theories were based on holistic viewpoints within which scientific "traditions" or "communities" allegedly worked. Kuhn (1962) suggested that the defining characteristic of a scientific tradition is its "commitment" to a shared "paradigm" (see KUHN). A paradigm is "the source of the methods, problem-field, and standards of solution accepted by any mature scientific community at any given time" (1962, p. 102). *Normal science*, the working-out of the paradigm, gives way to *scientific revolution* when "anomalies" in it precipitate a crisis leading to adoption of a new paradigm. Besides many studies contending that Kuhn's model fails for some particular historical case, three major criticisms of Kuhn's view are as follows. First, ambiguities exist in his notion of a paradigm (Masterman 1970; Shapere 1964). Thus a paradigm includes a cluster of components, including "conceptual, theoretical, instrumental, and methodological" (Kuhn 1962, p. 42) commitments; it involves more than is capturable in a single theory, or even in words. Second, how can a paradigm fail, since it determines what count as facts, problems, and anomalies? Third, since what counts as a "reason" is paradigm-dependent, there remain no trans-paradigmatic reasons for accepting a new paradigm upon the failure of an older one.

Such radical relativism is exacerbated by the "incommensurability" thesis shared by Kuhn (1962) and Feyerabend (1975) (see INCOMMENSURABILITY). Even the meanings of terms are paradigm-dependent, so that a paradigm tradition is "not only incompatible but often actually incommensurable with that which has gone before" (Kuhn 1962, p. 102). Different paradigms cannot even be compared, for both standards of comparison and meanings are paradigm-dependent.

Besides confronting such objections, philosophers of science gradually assimilated important lessons from Kuhn's work. The influence of presuppositions on the conduct of science became a major theme. Lakatos (1970) proposed that the fundamental units for the analysis of scientific change are "research programmes," a succession of theories having in common a *hard core* of suppositions (see LAKATOS). The "methodological decision of its protagonists" (p. 133) ordains that this core be irrefutable. What are to be rejected in the light of empirical evidence are the auxiliary hypotheses forming the *protective belt*. A *positive heuristic*, a set of partially articulated suggestions or

417

hints, provides the methodology for changing and developing the refutable protective belt. For a research program to be "progressive," each step in it must be a *consistently progressive theoretical problem shift* ("consistently content-increasing"), and, "at least every now and then the increase in content should be seen to be retrospectively cor-roborated: the programme as a whole should also display an *intermittently progressive empirical shift*" (p. 134). Otherwise it is degenerative and open to rejection.

Lakatos's view exposes certain Popperian themes which had been implicit in Kuhn's view: only falsification of a core/paradigm is possible; while a research program/paradigm can be corroborated, this means only that it has not failed its tests up to now, there being no such thing as evidence increasing degree of confirmation. Many objections to Kuhn's views arise with Lakatos's also. The content of the hard core is as difficult to specify as that for paradigms. Do scientists simply decree that the core be irrefutable, or do they have good reasons to maintain it at least until objections arise? To what degree is there commitment to the irrefutability of the hard core? Are not a multitude of competing interpretations of its propositions constantly examined, and alternatives considered? Is there *no* rationale by which new research programs are introduced?

Feyerabend (1970) objects that Lakatos's directives for judging degeneracy are vacu-ous (see FEYERABEND). Such judgments are possible only with hindsight, foreseeing none of the future revisions that might eventuate in triumph of the rejected method. Hence no theory or methodology should ever be ruled out. Feyerabend's own (1975) view is that scientists *ought* not to adhere to any single theory or method: in science "anything goes." Adherence is dogmatism; to avoid dogmatism, science ought to be "irrational." Furthermore, adherence to a single methodology, theory, paradigm, or core blinds scientists to facts which an alternative method might expose. Like Kuhn, Feyerabend has modified some of his earlier views significantly, but he still maintains that there are no rules common to all scientific viewpoints, and no valid distinction between scientific and nonscientific viewpoints. All points view (even witchcraft) and all methodologies (even astrology) must be encouraged.

Scientific progress

The predominant view has been that scientific progress consists in advancement toward some goal; but much disagreement exists as to what that goal is. Major candidates for the (primary) goal of science include truth, simplicity, coherence, and explanatory power – none of the latter three necessarily entailing truth. An air of arbitrariness frequently pervades such contentions, it often being claimed that, what-ever they are alleged to be, the goals are simply those we set up at the outset – as "founding intentions" (Gutting 1973, p. 226) definitory of the scientific enterprise.

Views of the relation between progress and advance toward *truth* typify controver-sies regarding scientific progress. Popper maintained that science progresses by in-creasing approximation to truth, or "verisimilitude," this being a function of the relative truth and falsity contents of the theories being compared. Severe technical flaws under-mine his conception of verisimilitude (see VERISIMILITUDE), and in any case we have no way of counting the number of true and false statements in a theory. Those who retain the general notion are hard pressed to analyze truth approximation (Newton-Smith

1981, ch. 8). Some writers speak of "convergence toward truth," though Laudan argues that the notion that science converges toward truth is contradicted by the history of science.

The issue of progress is complicated by Kuhn's contention that scientific change is not cumulative, in either the empirical sense (of a linear accumulation of facts) or the theoretical (that every later theory contains earlier ones as approximations). Kuhn describes revolutionary scientific developments as frequently regressive, answering fewer problems than their predecessors, and holds that we may have to abandon the notion that paradigm changes bring scientists closer to the truth. While there is "a sort of progress," its ultimate criterion is the decision of the scientific group.

Like Toulmin and Kuhn earlier, Laudan claims that science (and being rational) is essentially a problem-solving activity. Progress is measured not by approximation to truth (which is improbable and uncertifiable in any case), but by problem-solving capability. Objections arise concerning individuation of problems, their countability. Also, Laudan's classification of problems as empirical or conceptual remains at a quite general level, conceptual problems arising from almost any source at all. Some problems are more significant for science than others in the light of what science has (putatively) learned. Yet, if significant scientific problems do arise in this way, can we, with Laudan, dispense with the notion that truth, or at least acceptability, is more fundamental than rationality defined in terms of problem-solving capacity? Laudan's view also fails to deal adequately with criteria of progress toward solving a problem, or with ways in which problems change over the history of science. Nickles, also espousing a problem-solving approach, writes of the role of "constraints" in determining the formulation and significance of problems. But do such constraints arise solely from substantive accepted scientific conclusions, or are some "metascientific"? Should attention be focused, not on problems, but on the substantive scientific conclusions, and perhaps science-independent factors, which give rise not only to problems, but also to other concepts and activities including observation, evidence, possible and acceptable answers to the problems?

As opposed to those who interpret scientific progress as advancement *toward* some goal, some argue that progress consists moving *away from* some kind of status, for instance in correcting old errors rather than arriving at new truths. The view that the primary characteristic of science is its self-correctiveness (Peirce, Reichenbach, Salmon) is one example (see PEIRCE).

Related issues

Questions about scientific change embody wider issues in the interpretation of the knowledge-seeking enterprise. The entire body of problems regarding beliefs and their acceptability lies in the background of this topic, including the nature of knowledge and explanation, questions about verification and confirmation, refutation, evidence and whether it underdetermines theory, whether there are crucial experiments, and issues about meaning and reference.

Other topics besides those described here are widely discussed. Darden, Nersessian, and Thagard employ tools of cognitive and computer science in considering questions about the introduction of new ideas in science. Salmon applies Bayes's theorem in analyzing scientific discovery. Much is being done regarding innovation in particular fields outside the physical sciences, too long the focus of attention. Many, like Hull and Kitcher, while defending rationality in science, attempt to include social influences as integral parts of that rationality.

Concluding remarks

Many problems concerning scientific change have been clarified, and many new answers suggested. Nevertheless, concepts central to it (like "paradigm," "core," "problem," "constraint," "verisimilitude") still remain formulated in highly general, even programmatic, ways. Many devastating criticisms of the doctrines based on them have not been answered satisfactorily.

Problems centrally important for the analysis of scientific change have been neglected. There are, for instance, lingering echoes of logical empiricism in claims that the methods and goals of science are unchanging, and thus are independent of scientific change itself, or that if they do change, they do so for reasons independent of those involved in substantive scientific change itself. By their very nature, such approaches fail to address the changes that actually occur in science. For example, even supposing that science ultimately seeks the general and unalterable goal of "truth" or "verisimilitude," that injunction itself gives no guidance as to what scientists should seek or how they should go about seeking it. More specific scientific goals do provide guidance, and, as the transition from mechanistic to gauge-theoretic goals illustrates, those goals are often altered in light of discoveries about what is achievable, or about what kinds of theories are promising. A theory of scientific change should account for these kinds of goal changes, and for how, once accepted, they alter the rest of the patterns of scientific reasoning and change, including ways in which more general goals and methods may be reconceived.

To declare scientific changes to be consequences of "observation" or "experimental evidence" is again to overstress the superficially unchanging aspect of science. We must ask how what count as observations, experiments, and evidence themselves alter in the light of newly accepted scientific beliefs (Shapere 1982). (This also applies to concepts, methods, explanations, problems, etc.) On the other hand, it is now clear that scientific change cannot be understood in terms of dogmatically embraced holistic cores; the factors guiding scientific change are by no means the monolithic structures which they have been portrayed as being. Some writers prefer to speak of "background knowledge" (or "information") as shaping scientific change, the suggestion being that there are a variety of ways in which a variety of prior ideas influence scientific research in a variety of circumstances. But it is essential that any such complexity of influences be fully detailed, not left, as by Popper, with cursory treatment of a few functions selected to bolster a prior theory (in this case, falsificationism). Similarly, focus on "constraints" can mislead, suggesting too negative a concept to do justice to the positive roles of the information utilized. Insofar as constraints are scientific and not trans-scientific, they are usually *functions*, not *types*, of scientific propositions.

Traditionally, philosophy has concerned itself with relations between propositions which are specifically relevant to one another in form or content. So viewed, a philosophical explanation of scientific change should appeal to factors which are clearly more scientifically relevant in their content to the specific directions of new scientific research and conclusions than are social factors whose overt relevance lies elsewhere. However, in recent years many writers, especially in the "strong programme" in the sociology of science, have maintained that all purportedly "rational" practices must be assimilated to social influences.

Such claims are excessive. Despite allegations that even what is counted as evidence is a matter of mere negotiated agreement, many consider that the last word has not been said on the idea that there is in some deeply important sense a "given" in experience in terms of which we can, at least partially, judge theories. Again, studies continue to document the role of reasonably accepted prior beliefs ("background information") which can help guide those and other judgments. Even if we can no longer naively affirm the sufficiency of "internal" givens and background scientific information to account for what science should and can be, and certainly not for what it often is in human practice, neither should we take the criticisms of it for granted, accepting that scientific change is explainable only by appeal to external factors.

Equally, we cannot accept too readily the assumption (another logical empiricist legacy) that our task is to explain science and its evolution by appeal to metascientific rules or goals, or metaphysical principles, arrived at in the light of purely philosophical analysis, and altered (if at all) by factors independent of substantive science. For such trans-scientific analyses, even while claiming to explain "what science is," do so in terms "external" to the processes by which science actually changes.

Externalist claims are premature: not enough is yet understood about the roles of indisputably scientific considerations in shaping scientific change, including changes of methods and goals. Even if we ultimately cannot accept the traditional "internalist" approach to philosophy of science, as philosophers concerned with the form and content of reasoning we must determine accurately how far it can be carried. For that task, historical and contemporary case studies are necessary but insufficient: too often the positive implications of such studies are left unclear, and their too hasty assumption is often that whatever lessons are generated therefrom apply equally to later science. Larger lessons need to be extracted from concrete studies. Further, such lessons must, where possible, be given a systematic account, integrating the revealed patterns of scientific reasoning and the ways they are altered into a coherent interpretation of the knowledge-seeking enterprise – a theory of scientific change. Whether such efforts are successful or not, it will only be through attempting to give such a coherent account in scientific terms, or through understanding our failure to do so, that it will be possible to assess precisely the extent to which trans-scientific factors (metascientific, social, or otherwise) must be included in accounts of scientific change.

References and further reading

Feyerabend, P. K. 1970: Consolations for the specialist. In *Criticism and the Growth of Knowledge*, ed. I. Lakatos and A. Musgrave (Cambridge: Cambridge University Press), 197–203.

—— 1975: *Against Method* (London: Verso).

Gutting, G. 1973: Conceptual structures and scientific change. *Studies in History and Philosophy of Science*, 4/3, 209–30.

Gutting, G. (ed.) 1980: *Paradigms and Revolutions: Applications and Appraisals of Thomas Kuhn's Philosophy of Science* (Notre Dame, IN: Notre Dame University Press).

Hanson, N. R. 1958: *Patterns of Discovery* (Cambridge: Cambridge University Press).

Kuhn, T. S. 1962: *The Structure of Scientific Revolutions* (Chicago: University of Chicago Press); 2nd edn, with new "Postscript," 1970.

Lakatos, I. 1970: Falsification and the methodology of scientific research programmes. In *Criticism and the Growth of Knowledge*, ed. I. Lakatos and A. Musgrave (Cambridge: Cambridge University Press), 91–195.

Laudan, L. 1977: *Progress and Its Problems: Towards a Theory of Scientific Growth* (Berkeley: University of California Press).

—— 1981: *Science and Hypothesis* (Dordrecht: Reidel).

Masterman, M. 1970: The nature of a paradigm. In *Criticism and the Growth of Knowledge*, ed. I. Lakatos and A. Musgrave (Cambridge: Cambridge University Press), 59–89.

McMullin, E. 1979: Discussion review: Laudan's "Progress and Its Problems," *Philosophy of Science*, 46(4), 623–44.

Newton-Smith, W. H. 1981: *The Rationality of Science* (Boston: Routledge and Kegan Paul).

Nickles, T. (ed.) 1980: *Scientific Discovery, Logic, and Rationality* (Dordrecht: Reidel).

Reichenbach, H. 1938: *Experience and Prediction* (Chicago: University of Chicago Press).

Shapere, D. 1964: The structure of scientific revolutions. *Philosophical Review*, 73, 383–94. Repr. in Gutting 1980, 27–38.

—— 1982: The concept of observation in science and philosophy. *Philosophy of Science*, 49(4), 485–525.

63

Scientific Methodology

GARY GUTTING

Generically, "scientific methodology" denotes whatever generalized and systematically formulable procedures may be behind the successful pursuit of science. Since the ancient Greeks, people reflecting on science have been strongly attracted to the idea that there is a single comprehensive method employed in any genuinely scientific work. We will begin with this idealizing assumption, although we will later encounter ways in which it might be doubted.

Questions about the nature of scientific methodology have arisen in three contexts. Most frequently, philosophers have raised them and have tried to provide answers on the basis of general metaphysical and epistemological theories. This sort of reflection has been the historically dominant approach from Plato through the logical empiricists (see LOGICAL EMPIRICISM). But the issue of scientific methodology has also sometimes been a serious part of debates among scientists themselves. This has been particularly so at times of major theoretical upheaval. Since the middle of the nineteenth century, a third approach to scientific methodology has become very prominent. This is the effort to construct an account of scientific methodology from a study of the history of scientific practice (see HISTORY, ROLE IN THE PHILOSOPHY OF SCIENCE).

We should not expect to find sharp separations of these three contexts. Philosophical theories of scientific methodology are always guided by some paradigm extracted from successful practice; and methodological discussions among scientists, even when they arise directly from scientific problems, are never free from (often rather naive) philosophical presuppositions. And no historian approaches the scientific past without some substantial orientation from available philosophical and scientific accounts of scientific methodology.

This threefold division is, nonetheless, a useful tool for organizing the history of reflection on scientific methodology. For the most part, ancient and medieval discussions fall into the first category of treatments based on general philosophical principles; such treatments are also prominent in early modern philosophy (Bacon, Descartes, Locke, Hume, Kant) and in the work of twentieth-century logical empiricists. Methodological reflection has been a major part of science itself during periods of revolution such as the seventeenth-century revolt against Aristotelian science and the twentieth-century relativistic and quantum revolutions. Methodological questions have also been particularly prominent among those attempting to establish scientific approaches to the study of psychology and society. The historical approach to scientific methodology emerged in the nineteenth century with Whewell and Herschel (see WHEWELL) and became prominent once again in the wake of the decline of logical empiricism and of

Thomas Kuhn's seminal work (see KUHN). The following chronological survey will cover major examples of each approach to scientific methodology.

Aristotle

We inevitably begin with Aristotle, whose work was seminal in both science and the philosophy of science. His scientific ideas and practice, as well as his philosophical account of scientific methodology, dominated Western and Islamic science until the seventeenth century. In fact, although it is customary to accept the self-description of the sixteenth- and seventeenth-century founders of modern science as anti-Aristotelians, their methodology remained Aristotelian in some fundamental ways – as does all contemporary scientific methodology.

The Aristotelian core of all subsequent thought about scientific methodology is the dual insistence that science is *experiential* in origin and *necessary* in content. For Aristotle, the ultimate root of science is not divine authority, human imagination, or speculative reasoning, but our sense experience of material objects. However, his conception of the nature of this experience and its role in science was very different from that of modern methodologists. For Aristotle, our sensory encounters with individual material things provide a basis for intellectual judgments (*epagogai*) about their essences (the ontological structures that make them members of a natural kind). So, for example, observations of heavy objects (those predominantly composed of earth) show that they have an innate tendency to fall toward the ground. *Epagōgē* takes us beyond the mere contingent empirical generalization that such objects behave this way to the intellectual judgment that it is the essence of heavy objects to seek the center of the Earth.

In this way, Aristotle sees sense experience as leading to necessary truths (truths valid in virtue of the essential natures of things) about the natural world. Such truths obtained through *epagōgē* constitute the axiomatic basis of scientific knowledge. (Recall that Aristotle had before him the developing model of axiomatic geometry.) Given a body of such necessary truths, the goal of Aristotelian science is to explain natural phenomena by logically deriving statements expressing them from the intuited first truths. This logical derivation is carried out by the methods of syllogistic deductive inference, as first explicated in Aristotle's own writings on logic. The logical necessity of such deductions transmits the natural necessity of the premises to the conclusions, thereby assuring that the entire body of scientific truths, fundamental and derived, is an expression not of contingent juxtapositions but of essential truth.

Aristotle was never very clear as to the nature of the process whereby sense experiences of material objects lead to the crucial *epagōgē* yielding necessary scientific first principles. His own scientific practice – for example, the analysis of motion in his *Physics*, the description and classification of organisms in his biological writings – is hard to reconcile with the philosophical analyses of the *Posterior Analytics*. Even early Hellenistic commentators on his work, such as Alexander of Aphrodisias, puzzled over these questions. Following the rediscovery of Aristotle's works in Western Europe during the twelfth century, similar questions perplexed medieval commentators. As a result, there was a continuous line of philosophical reflection on scientific methodology throughout the later Middle Ages. By the sixteenth century, this increasingly critical and revisionary reflection converged with similar lines of development in scientific theorizing to constitute the methodological dimension of what is called the "scientific revolution."

The scientific revolution

The two greatest philosophers of the scientific revolution, Francis Bacon and René Descartes, were utterly explicit and emphatic in their rejection of Aristotelian methodology. The new science, they agreed, had no place for essences and substantial forms as the objects of knowledge, or for the mysterious *epagōgē* as the source of knowledge. There was, however, much less agreement in their positive accounts of scientific methodology. Descartes broke with Aristotle at two crucial points. First, he replaced the intellectual intuition of essences with the clear and distinct perception of ideas. For him, science was grounded not in the mind's dubious vision of external essences but in its indubitable awareness of its own contents. Second, he rejected Aristotle's syllogistic logic in favor of a new style of mathematical reasoning (see DESCARTES).

Bacon, by contrast, proposed a method that was much more empirical, but much less mathematical than Descartes's. He offered an elaborate account of how hypotheses should be constructed from, and tested in the light of, experiential data, emphasizing both the guiding role of facts and the active power of intellectual understanding. His metaphor of the scientist as neither an ant (merely gathering data) nor a spider (spinning theories from purely internal resources), but rather a bee (building hypotheses out of the facts) deftly catches the spirit of the new science. On the other hand, his failure to appreciate the scientific power of mathematical analysis and reasoning made his account inadequate to the celestial and terrestrial mechanics that, after Galileo and Newton, became the paradigmatic achievements of the new science (see GALILEO and NEWTON).

But seventeenth-century discussions of methodology were not limited to formal treatises, such as those of Descartes and Bacon. Treatments of methodology were also integral parts of first-order scientific work. This was because the new science – as developed, for example, by Galileo, Newton, and Boyle – required not just a new description of the behavior of natural objects, but a fundamental reconception of how they should be studied. At the heart of this reconception was a new understanding of both scientific experience and the conclusions based on it. Scientific experience was now regarded as primarily a matter of carefully controlled laboratory experiments, and conclusions from it as probable hypotheses, not necessary truths.

The reconception developed slowly and with difficulty. Galileo made extensive use of a new instrument of observation, the telescope, to overthrow the Aristotelian cosmology. He also made some (there is still controversy as to how much) use of experiments in the development of his law of falling bodies. But he remained close enough to the Aristotelian ideal of scientific truths as necessary and totally certain to refrain from frankly presenting his results as hypotheses made probable, but not proved, by observation. Similarly, Newton sharply distinguished the results (which he regarded as definite) of his combined deductive and inductive method from the various explanatory hypotheses, tentative in principle, that he put forward in, for example, the "Queries" appended to his *Opticks*. The former method consisted of a precise mathematical description of a given set of phenomena (which description Newton called a "deduction from the phenomena"), followed by a straight inductive generalization from these cases to a law governing all similar phenomena. Further scientific conclusions could be drawn by mathematical reasoning from the laws so derived. As the "Queries" indicate, Newton was willing to propose hypotheses (about, for example, unobservable corpuscles or hidden forces) that were not directly derived from phenomena but were postulated to explain them. But he did not regard these products of what we now call hypothetico-deductive (or retroductive) method as having the full scientific status of, for example, his laws of motion and of gravitation.

One major seventeenth-century scientist who did wholeheartedly embrace the hypothetico-deductive method was Robert Boyle. He regarded scientific hypotheses (for him, conjectures about the invisible corpuscular makeup of physical bodies) as conjectures more or less probable, depending on the degree of support given them by observation. He further required that scientific observation not take the form (as it often had with Bacon and his followers) of a casual gathering of random data; it was, rather, to be a matter of the disciplined execution of controlled experiments specifically designed to test a given hypothesis. Boyle even provides a detailed list of criteria (such as novelty, testability, and empirical superiority to alternatives) for evaluating the worthiness of a hypothesis. With Boyle we for the first time encounter something recognizable as a formulation of the experimental practice of modern science.

The thinkers we have been discussing promulgated and promoted methods for modern science *avant le lettre*. Their efforts were centrally important in the formation and eventual triumph of the new, non-Aristotelian sciences. After the triumph of Newtonian science, philosophical reflection on methodology (from Locke on) has been in the very different position of starting from the unquestionable success of a scientific paradigm. The question is no longer how to build an engine of scientific progress, but how to understand and justify the one we have.

Immanuel Kant's project for a "critique of pure reason" was the culmination of more than a century of modern philosophy's epistemological and metaphysical probing of the new science. Kant provided a philosophical explication and grounding for Newtonian science by presenting its fundamental concepts and principles (space, time, substance, causality, etc.) as necessary forms and categories of human experience and understanding. Instead of Aristotle's dubious effort to extract necessary scientific truths from particular sense experiences, Kant offered an account of scientifically necessary truths as essential preconditions for our experience of the natural world.

Logical empiricism

The dominant philosophy of science of the twentieth century, logical empiricism, was strongly rooted in the relativistic and quantum revolutions in physics. The logical empiricists saw these revolutions as undermining philosophical efforts (from Aristotle to Kant) to ground the necessity of science in essential truths about the natural world. Conceptions such as Aristotle's essences and Kant's categories had been put forward as defining a priori boundaries that had to be respected by all scientific conceptions. As the logical empiricists saw it, relativity and quantum theory required revisions in even these fundamental concepts, thereby showing that there were no a priori limits on the content of scientific accounts of the world (in Kant's terminology, no synthetic a priori truths). According to their neo-Humean view, the necessity of science resided only in the mathematical formalism used to formulate its laws and in the deductive logical relations between axioms and theorems.

This conception of science left no room for the traditional philosophy of nature (originated by Plato and Aristotle, but still vital in Descartes, Leibniz, Kant, and Hegel), which grounded and limited scientific theorizing through substantive philosophical conclusions about the essential nature of the physical world. Philosophers were restricted to employing the techniques of formal logic to analyze the content of already established scientific theories by, for example, giving them a logically rigorous axiomatic formulation. But such analyses were of limited significance because of their derivative nature and were extremely difficult to carry out because of their logical complexity. Although there was (and continues to be) a great deal of valuable work done in this area – particularly on the foundations of relativity and of quantum theory – the primary focus of logical empiricist philosophy of science was on scientific methodology itself.

Given the recent upheavals in physics, the fundamental methodological question which the logical empiricists faced was how to make sense of science as a permanent body of knowledge. Hadn't the work of Einstein, Bohr, and others (see EINSTEIN and BOHR) simply overturned Newtonian science (just as Newton had overturned Aristotelian science) and replaced it with a new fundamental view of the physical world, which, it was reasonable to expect, would eventually be itself replaced? If so, how could science pretend to be a cumulative, progressive epistemic enterprise? The logical empiricists' response to this challenge was based on a sharp distinction between the *observational* facts of science and their *theoretical* interpretations. The former (e.g., the existence of a planet, the observed correlation between pressure, volume, and temperature in a gas) were unchanging truths. The latter (e.g., Newton's postulation of a gravitational force to explain a planet's motion, the kinetic theory that regarded the gas as composed of unobservable molecules) were variable constructions that could be discarded in the light of new observational data.

Given this distinction, a primary goal of logical empiricist methodology was to show how all the essential assertions of science could be formulated in an entirely observational (nontheoretical) language. Rudolf Carnap, in particular, introduced the logical technique of reduction sentences, which was intended to extract the observational core of any scientific statement. Carnap and others also developed an elaborate system of inductive logic to show how the most basic observation statements (e.g., expressions of immediate sense-data) supported other such statements (e.g., empirical generalizations). These two enterprises were the heart of the logical empiricist account of the experiential dimension of science.

The necessity of science was treated in logical empiricist accounts of the logic of explanation. Here the most important work was done by Carl Hempel, who elaborated and defended the covering law model of explanation. According to this model, statements expressing natural phenomena are explained by being logically derived (deductively or probabilistically) from laws of nature. The empirical claims of science are necessary in the relative sense that they can be derived via formal logical rules from higher-order empirical premises (i.e., laws of nature). The question of the necessity of

427

the laws of nature themselves remained a disputed one among logical empiricists, but the standard view was that the necessity of any scientific statement was merely a matter of its derivability within a logical system.

Although logical empiricist philosophy of science still sets much of the agenda for contemporary discussions, its basic viewpoint has been generally rejected. The sharp distinction between observational and theoretical terms proved untenable, and Carnapian reduction projects failed. Moreover, neither formal systems of inductive logic nor the covering law model of explanation proved adequate to the realities of scientific practice. Even more important, critics, beginning with Thomas Kuhn, have rejected the logical empiricist effort to develop a methodology of science primarily from a priori philosophical principles (with contemporary mathematical physics as a privileged example) and have emphasized the need to understand science through the history and sociology of its practice (see KUHN and FEYERABEND).

The idea of constructing an account of scientific methodology on the basis of reflections on the history of science originated in the nineteenth century. It arose from two fundamental beliefs: that physics (and other related disciplines such as astronomy and chemistry) had, since Newton, arrived at what Kant called "the sure path of a science"; and that the nature of a phenomenon could (as Hegel and Darwin had, in different contexts, urged) be best understood in terms of its historical development. The two most important early proponents of this approach were the Englishmen William Whewell and John Herschel. Later in the century, Ernst Mach in Germany and Pierre Duhem in France likewise deployed historical erudition as a basis for important accounts of scientific methodology (see WHEWELL and MACH). The methodological views of Duhem and, especially, of Mach had an important influence on logical empiricist accounts of science; but the logical empiricists detached these views from their origin in detailed historical investigation. For Carnap and his followers, the new analytic tools of symbolic logic were much more important for understanding science than was its history, in which they had little interest beyond the recent revolutions in physics. The achievements of the pioneers of the historical approach to scientific methodology were once again appreciated only with the decline of logical empiricism in the 1960s.

Kuhn

The challenge to logical empiricism on historical grounds originated in Thomas Kuhn's immensely influential book *The Structure of Scientific Revolutions*. Kuhn's project was much more tentative and much less revolutionary than his proponents and opponents have usually suggested. His leading idea was that looking at science through the lens provided by the historiography of science (particularly since the work of Koyré) would offer a perspective superior to that of formal logical analysis. The main upshot of his discussion was that the rationality of science was not reducible to any set of explicit methodological rules, but ultimately resided in the informed judgment of the scientific community.

Kuhn developed this view through his three key concepts of *paradigm, incommensurability,* and *revolution*. A paradigm was a signal scientific achievement (e.g., what Newton accomplished in his *Principia*) that came to serve as a model for an entire domain of scientific activity. Initiation into a given scientific discipline, Kuhn maintained, is not a matter of knowing that certain theoretical and methodological principles are true but of knowing how to think and act in terms of the techniques, values, and world view implicit in the paradigm. Incommensurability referred to the fact that there is no methodological algorithm for resolving disagreements between proponents of competing paradigms (see INCOMMENSURABILITY). Kuhn fully accepted the relevance to scientific disputes of standard criteria (values) such as empirical adequacy, explanatory power, and simplicity; and he had no inclination to introduce passions, prejudices, unconscious drives, or other "irrational" factors as determinants of scientific conclusions. But he did insist that methodological rules expressing the standard criteria could not by themselves resolve disputes between proponents of rival paradigms. The resolution had to come from a consensus in judgment within the scientific community. A scientific revolution occurs when the community consensus rejects a reigning paradigm in favor of a rival. Kuhn never said that revolutions were irrational, but he maintained that their rationality was far removed from that of the proof of a mathematical theorem. A change in paradigm was, rather, he said, in a comparison bound to mislead most philosophers of science, similar to a religious conversion.

Kuhn's work (along with that of other important critics of logical empiricism such as Popper, Quine, Hanson, and Feyerabend) led to a variety of new approaches to scientific methodology (see POPPER; QUINE; FEYERABEND). Philosophers of science became preoccupied with analyzing the nature of scientific change and drawing consequences from their analyses about the rationality of science. This led to a number of alternative models of scientific change, most prominently from Lakatos, Hesse, Laudan, and Shapere, all based on their authors' knowledge of the history of science and all designed to ward off the purported irrationalism of Kuhn's own account (see LAKATOS). Here a vigorous minority voice was that of Paul Feyerabend, who argued with wit, erudition, and persistence for the "anarchist" view that there are no inviolable principles of scientific method and that, for any proposed methodological rule, there are cases where scientists have been successful precisely because they ignored or even contradicted it. These discussions of scientific change and rationality flourished throughout the 1970s but after that dwindled away with few substantive results. Some of their themes were fruitfully revived in the 1980s, however, by philosophers of language (e.g., Hilary Putnam), who applied to them the resources of new accounts of meaning and reference.

Another sequel, particularly to Kuhn's work, was a seemingly endless series of efforts to apply his account to the social sciences. This was odd, since, as Kuhn explicitly pointed out, he had introduced his notion of a paradigm as precisely that which distinguished the natural sciences, where there was consensus and some form of progress, from the social sciences, where there was not. Nonetheless, the literature is crammed with attempts to identify the "paradigms" employed in various social-scientific inquiries. The standard mistake of such discussions is to confuse a general framework of inquiry, which has attracted a significant body of followers, with a paradigm in Kuhn's sense of an achievement that has been endorsed by all the competent investigators in a given domain.

Perhaps the most important development following from Kuhn's work was the emergence of a rich and challenging group of studies in the sociology of science, both past and present. In one way, this development was quite alien to Kuhn, who, for all

429

the questions he raised about scientific rationality, remained firmly "internalist," avoiding explanations of scientific thought in terms of external factors such as economic forces or political ideology. On the other hand, many plausibly read Kuhn's subordination of methodological rules to communal judgment as an invitation to scrutinize the sociological sources of this judgment. Here, the work of various European schools (e.g., in Edinburgh, Bath, and Paris) has been particularly important. The Edinburgh school, for example, became famous for its "strong program" in the sociology of science, which challenged the standard assumption that only scientific errors admitted of explanation by social, economic, or political causes. Instead, the program proposed a symmetry thesis asserting the relevance of sociological explanation to every sort of scientific endeavor. Philosophers of science have often been uneasy with what they see as the philosophical naiveté and confusion (often tied to an all-too-facile relativism) of sociological approaches. But it must be admitted that, at a minimum, these approaches have given us an invaluable awareness of the highly complex social contexts in which even the most abstruse scientific enterprises take place. What has not yet been achieved, by either philosophers or sociologists of science, is an account of scientific thinking that does justice to both its undeniably strong social origin and its ability to achieve a viewpoint (in some sense objectively true) that does not merely reflect that origin. (See also SOCIAL FACTORS IN SCIENCE.)

The future of methodology

When we compare current discussions of scientific methodology with the long history of such discussions, perhaps the most striking feature of the current scene is its distance from the actual practice of science. Through the nineteenth century, most of the major methodologists were either themselves major scientific figures (e.g., Aristotle, Galileo, Descartes, Boyle, Newton, Herschel, Mach) or else philosophers (such as Bacon, Locke, and Kant) with close ties to contemporary working scientists, who took their methodological views very seriously. Particularly since the days of the logical empiricists, reflection on methodology has itself become a highly technical subdiscipline requiring so much specialization of its own that practicing scientists are no longer able to make significant contributions to it. Correspondingly, the intense demands of scientific specialization in the twentieth century made it impossible for scientists to look much beyond their disciplinary boundaries. As a result, the study of scientific method became autonomous in a way that it had never been before.

Contemporary philosophers of science have tried to overcome this distance by increasingly requiring of themselves and their students a detailed knowledge of particular scientific disciplines and/or of the history (and, more recently, the sociology) of science. This demand has greatly enriched philosophy of science, but there is no denying the fundamental difficulty: practicing scientists no longer make important contributions to discussions of scientific methodology; nor do they have any particular interest as scientists in philosophers' treatments of the subject. This fact raises serious questions about the value of an essentially autonomous study of scientific methodology.

Some philosophers seem inclined to admit that there is in fact little or no point to purely philosophical reflections, no matter how well informed historically or sociologically, about the proper way for scientists to proceed. This is, for example, a plausible way of reading Feyerbend's epistemological anarchism, which denies the normative status of any methodological principles. It is also a natural conclusion from Rorty's rejection of any privileged epistemological position for philosophy over other disciplines. (A similar conclusion might also, more arguably, find support in Arthur Fine's "natural ontological attitude," which can be read as counseling philosophers to leave science to scientists.) Certainly, many of the most talented young philosophers of science (particularly philosophers of physics and philosophers of biology) seem uneasy with methodological issues and much more comfortable with technical clarifications of fundamental scientific concepts.

There is no doubt that philosophical accounts of scientific methodology aimed at telling scientists how to proceed with their work are today otiose. Such accounts made sense in the seventeenth century, for example, when there were serious rivals to modern science's methodology and when central aspects of it (e.g., the use of hypotheses, the propriety of postulating unobservable entities) were still in dispute. But over the last 300 years, science has arrived at an effective understanding of how, in general, to proceed in investigating the natural world. This understanding is embodied in well-established theoretical and experimental procedures, and does not need endorsement or promulgation by philosophers. On the level of practice, scientists know, as well as any of us can, how they should proceed and need no advice from philosophers or other outsiders. But it is important to realize that (as Polanyi particularly has emphasized) scientists' practical knowledge of their methodology is implicit, not explicit, more a knowing *how* than a knowing *that*. Philosophical accounts of methodology have at least the advantage of explicitness, something that may be of value even to working scientists but is certainly important for nonscientists who need to understand and appreciate what scientists are doing.

Philosophical explicitness is particularly important for making informed judgments about the extra-scientific significance of scientific (and allegedly scientific) achievements. Evolutionary biology gives us authoritative information about the processes whereby humans evolved over millions of years from lower life forms. But what does this tell us about the essential nature of human beings? Does it follow that they are just material mechanisms, that all the higher functions of consciousness can be reduced to the motions of molecules? Contemporary cosmology provides fascinating and increasingly well-supported theories about the origin of the universe in the big bang. What is their significance for religious accounts of divine creation? Do they show that the universe originated simply by chance? Or do they require (or make highly probable) a creator God? Psychologists of various schools claim to have made important discoveries about the cognitive and emotional development of children. But are their claims scientific in the same sense as those of physics and chemistry? Should we rely on them in raising our children in something like the way that engineers rely on Newton's laws in designing a space shuttle? Such questions are obviously not purely scientific ones. But answering them requires a very clear reflective understanding of the methods whereby scientific results are obtained. It is here that philosophical reflection on scientific methodology still plays an essential role in the human understanding of science.

References and further reading

Asquith, P., and Kyburg, Henry Jr. (eds) 1979: *Current Research in Philosophy of Science* (East Lansing, MI: Philosophy of Science Association).

Blake, R. M., Ducasse, C. J., and Madden, E. H. 1960: *Theories of Scientific Method: The Renaissance through the Nineteenth Century* (Seattle: University of Washington Press).

Butts, R., and Davis, John (eds) 1970: *The Methodological Heritage of Newton* (Toronto: University of Toronto Press).

Carnap, R. 1966: *Philosophical Foundations of Physics: An Introduction to the Philosophy of Science* (New York: Basic Books).

Cohen, I. B. 1985: *Revolutions in Science* (Cambridge, MA: Harvard University Press).

Feyerabend, P. 1975: *Against Method: Outline of an Anarchistic Theory of Knowledge* (London: NLB).

Friedman, M. 1992: *Kant and the Exact Sciences* (Cambridge, MA: Harvard University Press).

Giere, R., and Westfall, R. S. (eds) 1973: *Foundations of Scientific Method: The Nineteenth Century* (Bloomington, IN: Indiana University Press).

Hempel, C. G. 1966: *The Philosophy of Natural Science* (Englewood Cliffs, NJ: Prentice-Hall).

Jardine, N. 1984: *The Birth of History and Philosophy of Science* (Cambridge: Cambridge University Press).

Kuhn, T. 1970: *The Structure of Scientific Revolutions*, 2nd edn (Chicago: University of Chicago Press).

Lakatos, I., and Musgrave, A. (eds) 1970: *Criticism and the Growth of Knowledge* (Cambridge: Cambridge University Press).

Laudan, L. 1968: Theories of scientific method from Plato to Mach: a bibliographical review. *History of Science*, 7, 1–63.

McKirahan, Richard D. 1992: *Principles and Proofs: Aristotle's Theory of Demonstrative Science* (Princeton: Princeton University Press).

McMullin, E. 1990: Philosophy of science, 1600–1900. In *Companion to the History of Modern Science*, ed. R. C. Olby et al. (London: Routledge), 816–37.

64

Simplicity

ELLIOTT SOBER

Scientists often appeal to a criterion of simplicity as a consideration that helps them decide which hypotheses are most plausible. Some such principle seems to be essential; the data, all by themselves, apparently cannot single out as best one hypothesis among the set of competitors.

A standard setting in which considerations of simplicity are brought to bear is the *curve-fitting problem*, depicted in figure 64.1. Suppose a scientist wishes to discover what general relationship obtains between two quantitative characteristics – for example, the temperature of the gas in a closed chamber and the pressure that the gas exerts on the sides of the chamber. The scientist might gather data on this question by observing a system that displays some value of the independent variable, x, and seeing what value of the dependent variable, y, the system exhibits. By making several observations of this sort, the scientist accumulates a set of <x, y> values, each of which may be represented as a point in the coordinate system depicted in the figure.

What role might these data points play in the task of evaluating different competing hypotheses, each of which corresponds to some curve drawn in the x–y plane? If one

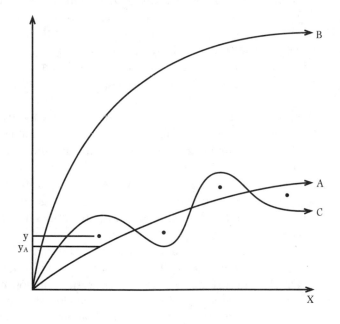

Figure 64.1

could assume that the observations were entirely free from error, one could conclude that any curve that fails to pass exactly through the data points must be false. However, observations are never entirely free from error. This means that the true curve may fail to pass exactly through the data points. How, then, are the data to be used?

The standard scientific procedure for measuring what the data say about the plausibility of hypotheses is by way of the concept of *goodness of fit*. Consider curve A. For each data point, one computes the distance from the observed y value to the y value that curve A predicts. For the leftmost data point, this is the quantity $(y - y_A)$. One then squares this quantity and does the same thing for each data point. The *sum of squares* (SOS) measures how far the curve is from the data. If SOS measures how well the data support the curve in question, we may conclude that the data in the figure support curve A better than they support curve B.

The rationale for SOS as a measure of evidential support is not far to seek. Curves that are close to the data are more likely, in the technical sense of likelihood defined by R. A. Fisher (1925). Curve A is more likely than curve B (given standard assumptions about the probability of errors) because Pr(data | curve A) > Pr(data | curve B). Of course, it is important not to confuse the likelihood of a hypothesis with its probability. The conclusion is not being drawn that Pr(curve A | data) > Pr(curve B | data).

Although likelihood as measured by SOS helps explain why the data seem to favor curve A over curve B, it says nothing about how we should compare curve A and curve C. They are about the same with respect to their SOS values. It is at this point that simplicity is said to enter the fray. Curve A is smooth, and curve C is bumpy. Curve A seems to be simpler. If we allow simplicity to guide our judgments about which curve is more plausible, we can conclude that curve A is superior to curve C, given the data at hand.

If curve fitting decomposes into the use of the SOS measure and the use of a simplicity criterion, three philosophical problems must be addressed. The first concerns how the simplicity of a curve should be measured. Vague reference to smoothness and bumpiness is obviously not enough. The second question is how the use of simplicity is to be justified. SOS reduces to likelihood, but what rationale can be given for the use of a simplicity criterion? The third question concerns how SOS and simplicity are to be traded off against each other. How much of the one quantity should be sacrificed for how much of a gain in the other?

This last question is quite fundamental. The goodness of fit of a hypothesis can usually be improved by making it more complicated. In fact, with n data points, perfect goodness of fit (an SOS of zero) can always be secured by a polynomial of degree n (i.e., an equation of the form $y = a_1 + a_2x + a_3x^2 + \ldots + a_nx^{n-1}$). If there are just two data points, a straight line will fit them perfectly; if there are three, a parabola can be found with an SOS of zero, and so on. The practice of science is to view goodness of fit as one consideration relevant to evaluating hypotheses, but not the only one. The problem is to understand that practice.

Even though curve fitting is a pervasive task in scientific inference, simplicity is an issue in other inferential contexts as well. The idea of minimizing the complexity of curves can be thought of as an instance of Occam's razor, the principle of parsimony, which concerns hypotheses of every sort. This venerable principle says that hypotheses that postulate fewer entities, causes, or processes are to be preferred over hypotheses that postulate more. Occam's razor expresses the idea that a theory that provides a unified, common treatment of different phenomena is preferable to a theory that treats each phenomenon as a separate and independent problem. It is to be hoped that an adequate account of the role of simplicity would also provide an understanding of the value that science seems to place on unified, general, and parsimonious theories. These considerations are sometimes dismissed as merely "pragmatic" in character, the idea being that they make a theory more or less useful, but have nothing to do with whether a theory ought to be regarded as true. However, time and again, scientists seem to use simplicity and kindred considerations as a guide to formulating their views concerning the way the world is.

As important as the problem of simplicity is to the task of understanding scientific inference, it has additional philosophical ramifications as well. Empiricists have long held that science is, and ought to be, an enterprise that is driven by observational data; to the degree that simplicity is an extra-empirical consideration that is integral to scientific deliberation, to that degree does empiricism fail to be adequate to the subject it seeks to characterize. In similar fashion, debates over the rationality of science continually return to the role of allegedly "aesthetic" considerations in scientific decision making. If simplicity is merely aesthetic, but intrinsic to the scientific outlook for all that, how can the so-called scientific method be viewed as the paradigm of rationality, which all other types of human thought should emulate?

In some ways, attempts to justify the use of simplicity as a criterion in inference have recapitulated patterns of argumentation that have been deployed in answer to David Hume's problem of induction. Just as Peter Strawson (1952) argued that "induction is rational" is an analytic truth, others have been inclined to suggest that "it is rational to use a simplicity criterion in scientific inference" is analytic as well. Just as Max Black (1954) tried to defend an inductive justification of induction, others have thought that simplicity considerations deserve to be taken seriously because simplicity has been a reliable guide to true hypotheses in the past. And just as John Stuart Mill (1843) said that induction relies on the assumption that nature is uniform (see MILL), so the idea has been considered that Occam's razor is justified by the fact that nature itself is simple.

These suggestions inherit defects in the approaches to the problem of induction that they imitate. Even if it is analytic that "using a simplicity criterion is part of the scientific method," it cannot be analytic that reliance on simplicity will be a reliable guide to what is true. The Strawsonian strategy severs the connection between rationality and the tendency of a method to lead to the truth. Likewise, the suggestion that simplicity is justified by its track record in the past raises the question of how the method's past success allows us to infer success in the future. It would appear that this extrapolation is preferable to other predictions in part because it is simpler. If so, it is hard to see how a simplicity criterion can be justified inductively without a vitiating circularity. And finally, the suggestion that nature itself is simple faces much the same problem; it is difficult to defend this proposition without begging the question. But perhaps more importantly, the use of simplicity considerations in scientific inference does not seem to depend on this dubious postulate; for no matter how complicated the hypotheses are that the data force us to consider, we still seem to give weight to simplicity considerations when we compare these hypotheses with each other. Even if the evidence

435

amply demonstrates that nature is quite complex, simplicity continues to serve as a guide to inference.

Attempts to characterize and justify the use of simplicity considerations in hypothesis evaluation usually derive from more general conceptions of how scientific inference works. For example, consider how a Bayesian might go about trying to understand the role of a simplicity criterion. Bayes's theorem shows how the probability of a hypothesis, after an observation is made, is related to its probability prior to the observation: $Pr(H|O) = Pr(O|H)Pr(H)|Pr(O)$. If two hypotheses, H_1 and H_2, are to be evaluated in the light of the observations O, then the theorem entails that $Pr(H_1|O) > Pr(H_2|O)$ if and only if $Pr(O|H_1)Pr(H_1) > Pr(O|H_2)Pr(H_2)$. In other words, if H_1 has the higher posterior probability, then it must have the higher likelihood or the higher prior probability (or both).

Bayes's theorem is a mathematical fact; as such, it is not controversial philosophically. However, Bayesians interpret this mathematical fact as having epistemic significance; they propose that the mathematical concept of probability be used to explicate the epistemological concept of plausibility. It is this application of the mathematics that is a matter of debate.

Let us apply the Bayesian framework to the curve-fitting problem represented in figure 64.1. How could curve A have a higher posterior probability than curve C, relative to the data depicted? There are just two possibilities to explore, and one of them will not help. We have already noted that the two curves have the same likelihoods. This means that a Bayesian must argue that curve A has the higher prior probability.

This is the approach that Harold Jeffreys (1957) adopts. He proposes that simpler laws have higher prior probabilities. Jeffreys develops this idea within the assumption that hypotheses in physics must be differential equations of finite order and degree. This guarantees that there is a denumerable infinity of hypotheses to which prior probabilities must be assigned.

Jeffreys suggests that we measure the complexity of a law by summing the number of adjustable parameters it contains with the absolute values of its integers (powers and derivatives). For example, $y^3 = ax^{-2}$ has a complexity value of 6, while $y = 2 + 3x$ has a complexity of 2.

Jeffreys's proposal has been criticized on a variety of fronts (Hesse 1967). The proposal appears to have some counterintuitive consequences; it is difficult or impossible to apply the proposal to simple transcendental functions such as $y = \sin x$; and it is arguable that the set of hypotheses that science contemplates goes beyond the set of differential equations and, moreover, may not be denumerable.

More fundamental problems become visible when we consider the Bayesian philosophy on which it relies. If prior probabilities merely report an agent's subjective degrees of belief, then there is no reason why we should accept Jeffreys's proposal rather than inventing one of our own. If these prior probabilities are to have probative force, they must reflect objective features of reality that rational agents can be expected to recognize. It is entirely mysterious how a prior probability ordering of this type can be defended.

In addition, there is a difficulty in Jeffreys's proposal that goes beyond this standard general complaint about Bayesianism. To see it clearly, one must keep clearly in mind the distinction between an *adjustable* and an *adjusted* parameter. The equation

$y = a_1 + a_2x$ has two adjustable parameters. It defines an infinite set of curves – the family LIN of straight lines. In contrast, the equation $y = 1.2 + 3.7x$ contains no adjustable parameters. The values of a_1 and a_2 have been adjusted; the resulting equation picks out a unique straight line that is a member of LIN.

Jeffreys's proposal covers both equations with adjustable parameters and equations in which all parameters have been adjusted. His counting procedure has the curious consequence that LIN is only two units greater in complexity than $y = 1.2 + 3.7x$. However, LIN is in fact an infinite disjunction; if we somehow wrote it out as a disjunction, it would be infinitely complex, according to Jeffreys's proposal. The problem, so far, is that it is difficult to obtain a language-invariant measure of a hypothesis's complexity.

But there is a further difficulty. Let us consider LIN and a second equation in which the parameters are adjustable: $y = a_1 + a_2x + a_3x^2$. This is PAR, the family of parabolic curves. Jeffreys says quite reasonably that PAR is more complex than LIN. His proposal is that PAR should therefore be assigned a lower prior probability than LIN. But this is impossible, since LIN *entails* PAR. Any specific curve that is a member of LIN is also a member of PAR, but not conversely. It is a theorem of the probability calculus that when one hypothesis entails another, the first cannot have the higher probability, regardless of what the evidence is on which one conditionalizes. The general point is that with nested families of curves, it is impossible for simpler families to have higher probabilities. Paradoxically, probability favors complexity in this instance.

The Bayesian may consider several responses. One would be to abandon the goal of analyzing the complexity of families of curves and to focus exclusively on curves with no adjustable parameters. Setting aside the question of whether this or other responses can be made to work, let us consider another, quite different framework for thinking about the role of simplicity in hypothesis evaluation.

Karl Popper (1959) turns the Bayesian approach on its head (see POPPER). For him, the goal of science is to find highly *im*probable hypotheses. Hypotheses are to be valued for their falsifiability – for their saying more, rather than less. Popper conjoins this requirement with the idea that we should reject hypotheses that have been falsified. So the two-part epistemology he proposes is that we should prefer hypotheses that are unfalsified though highly falsifiable. Within the context of the curve-fitting problem, this means that if neither LIN nor PAR is falsified by the observed data points, then we should prefer the former over the latter. LIN is more falsifiable, says Popper, because at least three data points are needed to falsify it; for PAR to be falsified, however, at least four observations are required.

Whereas Bayesianism encounters problems when equations with adjustable parameters are compared, Popper's approach has difficulty with equations in which all parameters are adjusted. If we consider, not LIN and PAR, but some specific straight line and some specific parabola, then each can be falsified by a single observation. All specific curves are equally falsifiable.

A further limitation of the Popperian approach should be noted. In point of fact, a purely deductivist methodology cannot explain why scientists do not reject specific curves that fail to pass exactly through the data points. The three curves in figure 64.1 are all "falsified" by the data. However, as Popper sometimes intimates, a more adequate assessment of the fit of hypotheses to data is afforded by the concept of

437

likelihood; rather than considering the yes/no question of whether a hypothesis has been falsified, we need to consider the quantitative issue of the degree to which a hypothesis says that the observations were to be expected. It is arbitrary to impose on that continuum a cutoff point that separates the stipulated categories of "falsified" and "unfalsified."

Two further questions are worth posing about the Popperian approach. It has always been unclear why scientists should have more confidence (stronger belief) in more falsifiable hypotheses than in less falsifiable ones. For all the difficulties that arise in connection with Bayesianism, the Popperian still needs to explain why we should be more confident in hypotheses that are less probable. Popper's account of simplicity inherits this difficulty; even if simplicity could be equated with falsifiability, the question remains as to why simplicity should be regarded as a guide to truth.

A second problem that Popper's approach faces is a problem that pertains also to Jeffreys's. One must show how simplicity is related to other considerations in hypothesis evaluation. How much fit of hypotheses to data should be sacrificed for how much of a gain in simplicity? The tasks of defining simplicity and justifying its epistemic relevance do not exhaust what there is to this philosophical problem.

These critical remarks about Bayesianism and the Popperian approach set the stage for the third theory to be discussed here. The statistician H. Akaike (1973) and his school have developed a set of theorems concerning how the predictive accuracy of a family of curves may be estimated. These theorems show how simplicity, as measured by the number of adjustable parameters in an equation, and goodness of fit are relevant to estimating predictive accuracy.

Working scientists are often aware of the risk involved in adopting a curve that *over-fits* the data. As noted earlier, given any body of data, it is possible to make the SOS value as small as one likes by making one's hypothesis sufficiently complex. The problem of over-fitting arises when the resulting hypothesis is asked to predict *new* data. Although a complex hypothesis fits the old data quite well, it usually does a poor job of predicting new data. Scientists find that hypotheses that are quite close to the data are often quite distant from the truth. Simpler hypotheses often do worse at fitting the data at hand, but do a better job at predicting new data.

This common experience suggests how the predictive accuracy of a family of curves should be defined. Consider LIN. Suppose we confront a given data set, D_1, and find the member of LIN that best fits this data set. This straight line in LIN is the likeliest member of that family; call it $L(\text{LIN})$. Suppose we then obtain a new set of data, D_2, and measure how well $L(\text{LIN})$ fits D_2. Now imagine repeating this process over and over, first finding the best-fitting member of LIN with respect to one data set and then using that best member to predict a new data set. The average performance of LIN defines its predictive accuracy.

It should be evident that the predictive accuracy of a family of curves depends on which specific curve happens to be true. If the truth is a member of LIN, then LIN will do quite well in the prediction problem just described. But if the truth is some highly nonlinear curve, then LIN will perform poorly.

The predictive accuracy of a family might also be termed its closeness to the truth. It isn't that LIN will fit the data perfectly when the true curve is a member of LIN; errors in observation may produce non-colinear data points, and the $L(\text{LIN})$ curve constructed

438

to fit one data set can easily fail to perfectly fit a new one. However, over the long run, LIN will do better than PAR if the true curve is in fact a member of LIN.

It is an interesting property of the concept of predictive accuracy that LIN can be closer to the truth (more predictively accurate) than PAR even though LIN entails PAR. If the truth happens to be a straight line, L(PAR) will fit a single data set at least as well as L(LIN) will; but, on average, L(LIN) will predict new data with greater accuracy than L(PAR) will.

Predictive accuracy is obviously a desirable feature of hypotheses; however, for all that, predictive accuracy seems to be epistemologically inaccessible. It seems that we can't tell, from the data at hand, how predictively accurate a family of curves will be. To be sure, it is easy to tell how close a family is *to the data*; what seems inaccessible is how close the family is *to the truth*.

Akaike's remarkable theorem shows that predictive accuracy is, in fact, epistemologically accessible. The predictive accuracy of a family can be estimated from the data at hand. The theorem may be stated as follows:

Estimate[distance from the truth of family F] = SOS[$L(F)$] + k + constant.

Here, $L(F)$ is the likeliest (best-fitting) member of the family F relative to the data at hand, k is the number of adjustable parameters in F.

The first term on the right-hand side, SOS[$L(F)$], is the quantity that has traditionally been taken to exhaust the testimony of the evidence. For any family with at least one free parameter, the second term will be positive, as long as observation is at least somewhat prone to error. This means that the SOS of $L(F)$ *under*estimates how far family F is from the truth; in other words, $L(F)$ is always guilty of some degree of overfitting. The third term disappears when hypotheses are compared with each other, and so may be ignored.

The second term in Akaike's theorem gives simplicity its due; the complexity of a family is measured by how many adjustable parameters it contains. It is important to see that the number of adjustable parameters is not a syntactic feature of an equation. Although the equations "y = ax + bx" and "y = ax + bz" may each seem to contain two free parameters (a and b), this is not the case. The former equation can be *reparameterized*; let $a' = a + b$, in which case the first equation can be restated as "y = a'x. For this reason, the first equation in fact contains one free parameter, while the second contains two.

Let us apply Akaike's theorem to LIN and PAR. Suppose the data at hand fall fairly evenly around a straight line. In this case, the best-fitting straight line will be very close to the best-fitting parabola. So L(LIN) and L(PAR) will have almost the same SOS values. In this circumstance, Akaike's theorem says that the family with the smaller number of adjustable parameters is the one we should estimate to be closer to the truth. A simpler family is preferable if it fits the data about as well as a more complex family. Akaike's theorem describes how much improvement in goodness of fit a more complicated family must provide for it to make sense to prefer the complex family.

Another feature of Akaike's theorem is that the relative weight given to simplicity declines as the number of data points increases. Suppose there is a slight parabolic bend in the data, reflected in the fact that the SOS value of L(PAR) is slightly lower than the SOS value of L(LIN). It is a property of the SOS measure that the SOS of a

439

hypothesis almost certainly increases as the number of data points increases. With a large number of data points, the estimate of a family's distance from the truth will be determined largely by goodness of fit and only slightly by simplicity. But with a smaller amount of data, simplicity plays a more determining role. This is as it should be: as the number of data points increases, their parabolic bend should be taken more seriously.

Akaike's theorem has a number of interesting applications, only one of which can be mentioned here. Perhaps the most famous example of the role of simplicity considerations in hypothesis evaluation is the dispute between Copernican and Ptolemaic astronomy. In Ptolemy's geocentric model, the relative motion of the Earth and the Sun is replicated within the model for each planet; the result is a system containing a very large number of adjustable parameters. Copernicus decomposed the apparent motion of the planets into their individual motions around the Sun together with a *common* Sun–Earth component, thereby drastically reducing the number of adjustable parameters. In *De Revolutionibus*, Copernicus argued that the weakness of Ptolemaic astronomy traces back to its failure to impose any principled constraints on the separate planetary models. Thomas Kuhn (1957, p. 181) and others have claimed that the greater unification and "harmony" of the Copernican system is a merely aesthetic feature. The Akaike framework offers a quite different diagnosis; since the Copernican system fit the data then available about as well as the Ptolemaic system did, Akaike's theorem entails that the former had an *astronomically* superior degree of estimated predictive accuracy.

Akaike's theorem is a theorem, so it is essential to ask what the assumptions are from which it derives. Akaike assumes that the true curve, whatever it is, remains the same for both the old and new data sets considered in the definition of predictive accuracy. He also assumes that the likelihood function is "asymptotically normal." And finally, he assumes that the sample size is large, in the sense that enough data are available that the value of each parameter can be estimated.

Akaike's result is that the formula described above provides an *unbiased* estimate of a family's distance from the truth. This means that the estimator's average behavior centers on the correct value for the distance; it leaves open that individual estimates may stray quite considerably from this mean value. Akaike's procedure is not the only one that is unbiased, so it is a matter of continuing statistical investigation how one best estimates a family's distance from the truth. In addition, there are other desirable statistical properties of an estimator besides unbiasedness, so there is a genuine question as to how various optimality properties ought to be traded off against each other. However, the fact that important details remain unsettled should not obscure the fact that Akaike's approach has made significant headway in the task of explaining the role of simplicity in hypothesis evaluation.

References and further reading

Akaike, H. 1973: Information theory and an extension of the maximum likelihood principle. In *Second International Symposium on Information Theory*, ed. B. Petrov and F. Csaki (Budapest: Akademiai Kiado), 267–81.

Black, M. 1954: Inductive support of inductive rules. In *Problems of Analysis* (Ithaca, NY: Cornell University Press), 191–208.

Burnham, K., and Anderson, D. 1998: *Model Selection and Inference: A Practical Information-theoretic Approach* (New York: Springer Verlag). (A useful exposition of the theory and practice of Akaike's approach to model selection.)

Fisher, R. 1925: *Statistical Methods for Research Workers* (Edinburgh: Oliver and Boyd).

Forster, M., and Sober, E. 1994: How to tell when simpler, more unified, or less *ad hoc* theories will provide more accurate predictions. *British Journal for the Philosophy of Science*, 45, 1–36. (An exposition and application of Akaike's ideas, as well as criticisms of some other proposals.)

Glymour, C. 1980: *Theory and Evidence* [Princeton: Princeton University Press). (A discussion of curve fitting within the context of Glymour's bootstrapping account of confirmation.)

Hesse, M. 1967: Simplicity. In *The Encyclopedia of Philosophy*, ed. P. Edwards (New York: Macmillan), vol. 7, pp. 445–8.

Jeffreys, H. 1957: *Scientific Inference*, 2nd edn (Cambridge: Cambridge University Press).

Kuhn, T. 1957: *The Copernican Revolution* (Cambridge, MA: Princeton University Press).

Mill, J. S. 1843: *A System of Logic: Ratiocinative and Inductive* (London: Longmans Green and Co.).

Popper, K. 1959: *The Logic of Scientific Discovery* (London: Hutchinson).

Rosenkrantz, R. 1977: *Inference, Method, and Decision* (Dordrecht: D. Reidel). (A Bayesian proposal in which simplicity is connected with the idea of average likelihood.)

Sober, E. 1988: *Reconstructing the Past: Parsimony, Evolution and Inference* (Cambridge, MA: MIT Press). (An exploration of why common cause explanations are superior to separate cause explanations within the context of the evolutionary problem of phylogenetic inference.)

—— 1990: Let's razor Ockham's razor. In *Explanation and Its Limits*, ed. D. Knowles (Cambridge: Cambridge University Press), 73–94. (Argues that invocation of the principle of parsimony is a surrogate for an assumed background theory, which, once stated explicitly, renders appeal to parsimony superfluous.)

Strawson, P. 1952: *Introduction to Logical Theory* (London: Methuen).

65

Social Factors in Science

JAMES ROBERT BROWN

Although there has long been an interest in how social factors play a role in science, recent years have seen a remarkable growth of attention to the issue. There are quite different ways in which social influences might function, some of which are more controversial than others.

1. *Setting the goals of research.* This is often done with an eye to technology, and the major directions are usually set by the various sponsoring governments. That a large number of scientists work in solid state physics and in cancer research, while relatively few work in astrophysics, simply reflects social policy. And sometimes direct public pressure can have an influence, as is evident in the case of AIDS research.

2. *Setting ethical standards for research procedures.* Some pretty grizzly things have been done in the pursuit of knowledge. In response, social pressure has been brought to bear on the methods of research. The use of animals, for example, is strictly regulated in most countries. The use of human subjects is similarly constrained in diverse ways. Informed consent, for instance, must typically be sought, even though telling an experimental subject what an experiment is about may diminish the value of the experiment. Certain sorts of research are prohibited altogether.

These two types of social factor in science concern aims and methods, respectively. Even though there is massive social involvement with science in both of these senses, still, there need be no social factor in the *content* of the resulting theories. The military had an interest in knowing how cannon balls move, so it funded research in this field. Still, it was Galileo's objective discovery that they move in parabolic paths (see GALILEO). Though Nazis committed a horrendous crime by immersing people in ice water to see how long they would survive, still it is an objective fact that they died within such and such a time. Social factors in the above senses are of the utmost philosophical concern, but they are mainly the concern of political and moral philosophy. It is a different sense from these, one focused on epistemology, that has been the subject of much (often heated) debate among philosophers of science.

3. *Social factors influencing the content of belief.* Perhaps our scientific beliefs are the result of various interests and biases we possess that stem from our social situations. Instead of "evidence" and "good "reasons" leading us to our theoretical choices, it may be class or professional interest, gender bias, or other such "noncognitive" factors.

Most philosophers of science and most traditional historians of ideas have tried to account for science in rational terms. That is, they try to show that scientific theories have been accepted or rejected because of the evidence available at the time. Of course, there are many cases in which scientists did not do the rational thing, but such episodes are interesting anomalies; the bulk of science has been a rational process. The Popperians, the Bayesians, and others may differ sharply on what scientific rationality is, but they are firmly agreed on this one point: that scientific belief is largely determined by the available evidence.

This rational picture is upheld by many, but it has been largely rejected over the past three decades by a variety of commentators who stress the dominating presence of social factors in the formation of scientific belief. Even if we allow – as we surely must – that social factors abound in science, it is far from clear how they manage to get in and what their effect is. I will briefly describe some of the variations and trends in current thinking about this issue.

A large number of case studies with social factors at their very heart were produced starting in the early 1970s. One of the most famous of these is Paul Forman's (1971) study of the rise of quantum mechanics in Germany between the wars. Following the First World War and Germany's collapse, the public was quite disillusioned with science and technology. The German public adopted a mystical, anti-rational attitude. The sense of crisis was epitomized by Spengler's *Decline of the West*, in which he attacked causality and mechanism and linked them to death and the declining fortunes of Germany. Salvation, according to Spengler, could come only by turning away from this outlook. Deprived of the prestige they had enjoyed before and during the war, German scientists were impelled to change their ideology so as to improve their public image. They resolved, says Forman, to rid themselves of the "albatross of causality." Acausal quantum mechanics was the result. Thus, far from being driven to the new physics by what it normally thought of as the evidence, they adopted the new physics, according to Forman, because of a desire to regain their lost stature.

Forman's account is typical of an "interest" explanation of belief. The scientists' interest (in regaining lost prestige) caused their belief. It is a much cited, paradigm example of a way of understanding science. For sociologists of knowledge such as David Bloor, interests and similar causes are the only way to explain scientific beliefs; "evidence" as normally conceived is just a mythical entity. Bloor's extremely influential views (Bloor 1991) are known as the "strong programme." It is to be contrasted with what Bloor calls "the weak programme" in the sociology of science, which attempts to explain rational and irrational episodes in quite different ways. Sociologists (such as Robert Merton) and traditional intellectual historians might explain the acceptance of Newton's universal gravitation as due to the evidence that was then available; but they would explain adherence to Lysenko's biological theories as due to the peculiar social situation in the USSR at the time. The strong programme disavows this asymmetry, and calls for the same type of explanation of belief (citing social causes) in every case.

Bloor claims to be doing "the science of science"; we should explain science the way science itself explains things. It is a species of naturalism. To this end, he lists four tenets: (1) *Causality*: Any belief must be explained in terms of what caused that belief. (2) *Impartiality*: Every belief must be explained, regardless of its truth or falsehood,

443

rationality or irrationality. (3) *Symmetry*: The style of explanation must be the same for true and false, and for rational and irrational beliefs. (4) *Reflexivity*: This theory applies to itself.

Each of these claims has considerable plausibility. In fact, traditionalists are probably happy to embrace all but (3). If reasons are thought of as causes (and this is orthodoxy in most philosophical circles) then to cite reasons for, say, Newton's belief in universal gravitation is to cite a cause for his belief. And to explain by citing either reasons (for rational beliefs) or social factors (for nonrational beliefs) is to explain. So all beliefs are being explained, and they are being explained causally. As for reflexivity, who can deny that it is rational to be rational?

The real heart of the strong programme (or of any serious sociology of knowledge) is the symmetry principle. Bloor is persuasive when he insists that biologists explain both healthy bodies and sick ones; engineers are interested in why some bridges stand and others fall down. Similarly, he claims, sociologists should use their tools on all scientific beliefs, not just the "false" or "irrational" ones. However, the initial plausibility of the symmetry principle may crumble upon closer inspection. Just what does "same style of explanation" mean? It cannot mean exactly the same explanation, since the presence of germs explains A's sickness, but not B's good health. But once the principle is appropriately loosened, it is very hard not to allow social factors in some explanations and reasons in others.

All along, Bloor and others have insisted that they are not denying "reason" and "evidence"; it's just that these things must be understood in a social context. However, what this usually means is that they allow the rhetorical force of talk about reason and evidence; but they certainly do not give them causal powers, in the sense that traditional intellectual historians do.

Another favorite argument for getting social factors into the very content of science is based on the problem of underdetermination (see UNDERDETERMINATION OF THEORY BY DATA). There are many versions of the problem, but a simple one will here suffice. Consider a theory T that does justice to the available evidence, which we'll take to be a body of observation sentences, $\{O_i\}$. And we will take "doing justice" to mean explaining the data by entailing it; that is, $T \Rightarrow \{O_i\}$. The problem of underdetermination arises from the fact that there will always be another theory (indeed, infinitely many others) that will do equal justice to the very same data; that is, there is always another theory, T′, such that $T' \Rightarrow \{O_i\}$. So how do we choose? The available evidence does not help us decide; it is not sufficient to determine a unique choice.

Let $\{O_i\}$ be our body of data, the evidence, and let T, T′, T″, . . . all equally account for it. We shall further suppose that T is believed, and we want to know why. At this point the sociologist of knowledge gleefully jumps in and posits an interest, I, which is the cause of T being adopted; that is, T serves interest I, and that is why T was accepted. The interest explains the belief.

Thus, as well as accounting for $\{O_i\}$, a theory must do justice to the interest I. However, underdetermination works here, too. There are indefinitely many theories, T, T*, T**, . . . which will do equal justice to $\{O_i, I\}$. So why was T selected? We don't yet have an explanation.

Now the sociologist of knowledge can take either of two routes. The first is to propose a second interest, I_2, to explain why T was chosen over its rivals; but this leads to an obvious regress. Undoubtedly there are only a finite number of interests a person could possibly have; so eventually the explanatory power of interests will stop, leaving it unexplained why T was chosen over indefinitely many alternatives which could do equal justice to $\{O_i, I, I_2, I_3, \ldots, I_n\}$.

The second route is to deny that there are really indefinitely many alternatives. The sense in which indefinitely many alternative theories *exist* is a logician's Platonic sense, not a practical or realistic sense. In reality, there are usually only a very small handful of rival theories from which to choose, and from these one will best serve the scientist's interests.

The second route would indeed be sufficient for an explanation. But if there is only a very small number of theories to choose from, the evidence $\{O_i\}$ will almost always suffice to decide among them. In other words, there is no practical problem of under-determination, so there is no need to appeal to something social, an interest, to explain the adoption of the theory T. Underdetermination, consequently, cannot be used to justify a sociological approach to science.

Laboratory studies have stimulated a large and interesting literature. A classic in this genre is Bruno Latour and Steve Woolgar's *Laboratory Life* (1979). This work is in the form of field notes by an "anthropologist in the lab," and it has an amazing story to tell about the creation/discovery of TRF(H) (Thyrotropin Releasing Factor, or Hormone). The accepted view (whether created or discovered) is that this is a very rare substance produced by the hypothalamus, which plays a major role in the endocrine system. TRF(H) triggers the release of the hormone thyrotropin by the pituitary gland; this hormone in turn governs the thyroid gland, which controls growth, maturation, and metabolism.

The work on TRF(H) was done by Andrew Schally and Roger Guillemin, independently; they shared the Nobel Prize in 1977 as co-discoverers. The amount of physical labor involved in isolating TRF(H) is mind boggling. Guillemin, for example, had 500 tons of pig's brains shipped to his lab in Texas; Schally worked with a comparable amount of sheep's brains. Yet, the quantity of TRF(H) extracted in each case was tiny.

The lack of any significant amount of the hormone leads to an identification problem. As the existence of the stuff is somewhat precarious, any test for its presence is highly problematic. The philosophical claims about facts by Latour and Woolgar largely turn on this point. Consider gold, for a moment. We have lots of this stuff; it's observable, easily recognized by ordinary people, and paradigm samples abound. To protect ourselves from "fool's gold" and from outright fraud, tests (assays) have been developed. How do we know that a particular assay is a good test? Simple. We use standard samples of gold and non-gold. An assay is a good one insofar as it can distinguish between them.

But such a procedure is not possible in the TRF(H) case. We simply do not have recognizable samples that we can use to "test the test." Different bioassays for TRF(H) were developed, but without a standard sample of TRF(H) there is no independent check on the bioassay; that is, there is no way to be sure that the bioassay is really "true to the facts." The relevant fact is this: there is a substance in the hypothalamus that releases the hormone thyrotropin from the pituitary, and its chemical structure is pyroGlu-His-Pro-NH$_2$. The *existence of the fact* rests on *acceptance of some particular bioassay*; they stand or fall together. Since there is no *direct* test of the bioassay, say

445

Latour and Woolgar, it must have been adopted as a result of *social negotiation*. Schematically:

> TRF(H) exists if and only if bioassay B is accepted.
> B is accepted as a result of social negotiation.
> ∴ TRF(H) is not discovered; it is a social construction.

The argument is interesting, and the story behind it sufficiently engrossing, that the conclusion is plausible. But after a bit of reflection we can see that neither premise is acceptable. For example, the first means that there was no gold until there was an assay for it. Perhaps this is an uncharitable reading of the authors' words. The exact claim is: "Without a bioassay a substance could not be said to exist." This may just mean that without a test we have no grounds for *asserting* the existence of the stuff. But this leaves it entirely open whether the stuff in question (gold, TRF(H)) actually exists. Consequently, facts need not be social constructions after all, contrary to Latour and Woolgar.

What about the second premise of the argument? The picture painted by Latour and Woolgar is reminiscent of Quine's "web of belief," a conception popular with social constructivists. Propositions are connected to one another in a network. Sometimes the connections are strong, at other times weak; but in any case the network is huge. Latour and Woolgar's picture is again initially plausible; much of our belief about any substance is intimately connected to whatever bioassays we have adopted. However, Latour and Woolgar's network consists of only the two propositions: "This is TRF(H)" and "This is the bioassay that works." They are linked in such a way that they stand or fall together. Such simplemindedness is easily countered. In the bioassay which has been adopted, rats are used instead of mice, because mice are believed to have more sensitive thyroids; males are used instead of females, because the female reproductive cycle might interfere; 80-day-old rats are used, since that is the age when the thyrotropin content of the pituitary is greatest; and so on. Of course, these are fallible considerations, and social factors may have played a role in their adoption (e.g., why does the female reproductive cycle "interfere"?). But the crucial thing is that they are *independent* reasons for thinking that the particular bioassay adopted is the right one for detecting TRF(H). So, *contra* Latour and Woolgar, we do not have a little circle consisting of only two propositions which will stand or fall together; we have a very much larger network, and the bioassay is supported by numerous far-reaching strands. It may be a social construction that rats have more sensitive thyroids than mice, but it was not constructed by the TRF(H) scientists. For them, it functions as a kind of external constraint. The claim that the bioassay is accepted through social negotiation will now be much harder to sustain.

The highly interesting and influential works of Harry Collins (1985), Andrew Pickering (1984), Steven Shapin and Simon Schaffer (1985), and others also focus on the laboratory, and each stresses social factors as well. For example, in their study of the rise of experimentation with Boyle and its relation to the politics of Hobbes's *Leviathan*, Shapin and Schaffer remark, "Solutions to the problem of knowledge are solutions to the problem of social order." The sociology of scientific knowledge has been a very fast-moving field recently. What I have described so far might be considered *passé* by many adherents to the general outlook. In the recent past much attention has focused on Latour's work (1987, 1988). Among other things, Latour makes the interesting claim that we cannot automatically suppose that the social should explain the scientific; science, he says, makes society just as society makes science. Debates between Latour and others have arisen on several key issues. An excellent survey of different points of view can be found in Pickering (1992).

Sociologists of knowledge have not been the only ones concerned with social factors in science. There is also a very significant literature by feminists (see FEMINIST ACCOUNTS OF SCIENCE). All branches of the sciences have been discussed, but, as might be expected, those parts which directly touch on humans have received more attention than the physical sciences. Biology, for example, has been especially important, since this is a main source of deterministic arguments that "explain" women's role in society. That is, they explain why it is natural for women to play a subordinate role, why men are "smarter" and more aggressive, and so on (see Hubbard 1990; Harding 1986; Keller 1985). The spectrum of feminist views is very wide, some of it quite dismissive of science. However, the most interesting and important are reform-minded.

Typical of the reformist approach is Helen Longino's *Science as Social Knowledge* (1989), which has the aim of reconciling "the objectivity of science with its social and cultural construction." The search for human origins – both anatomical and social – has enormous social ramifications. It informs our picture of ourselves, and so plays a role in the determination of social policy and civil life. It's a major case study for Longino.

One prominent hypothesis is the "man-the-hunter" view. The development of tools, on this account, is a direct result of hunting by males. When tools are used for killing animals and for threatening or even killing other humans, the canine tooth (which had played a major role in aggressive behavior) loses its importance, and so there will be evolutionary pressure favoring more effective molar functioning, for example. Thus human morphology is linked to male behavior. Male aggression, in the form of hunting behavior, is linked to intelligence, in the form of tool making. Notice that on this account women play no role in evolution. We are what we are today because of our male ancestors' activities.

But this is not the only account of our origin. A view of more recent vintage is the "woman-the-gatherer" hypothesis. This account sees the development of tool use to be a function of female behavior. As humans moved from the plentiful forests to the less abundant grasslands, the need for gathering food over a wide territory increased. Moreover, women are always under greater stress than men, since they need to feed both themselves and their young. Thus, there was greater selective pressure on females to be inventive. So innovations with tools were due mainly to females. Why, on this account, should males lose their large canine teeth? The answer is sexual selection. Females preferred males who were more sociable, less prone to bare their fangs and to other displays of aggression.

So, on the "woman-the-gatherer" account of our origins, our anatomical and social evolution is based on women's activities. On this account, we are what we are today largely because of the endeavor of our female ancestors.

The kinds of evidential consideration thought relevant in deciding this issue include fossils, object identified as tools, the behavior of contemporary primates, and the activities of contemporary gatherer-hunter peoples. Obviously, each of these is somewhat problematic. Fossils, for example, are few and far between, and are little more than fragments; tools such as sticks will not last the way stone tools do, so we may have a very misleading sample of primitive artifacts. Moreover, it is often debatable whether any alleged tool was really used for hunting an animal rather than preparing some vegetation for consumption. Finally, an inference from the behavior of contemporary primates and gatherer-hunter humans to the nature of our ancestors who lived from two to twelve million years ago is a leap that only a Kierkegaard would relish.

None of these considerations should be dismissed out of hand; each provides evidence of some sort, but it is abysmally weak. The moral of this example is that it displays how values can affect choices. If one is already inclined to think of males as the inventors of tools, then some chipped stone will be interpreted as, say, a tool for hunting. This will then become powerful evidence in the man-the-hunter account of our origin. On the other hand, if one is a feminist, then one might be inclined to see some alleged tool as an implement for the preparation of food. On this interpretation the tool becomes strong evidence for the woman-the-gatherer account of our evolution.

There is a traditional view of science which allows that social factors may play a role in the generation of theories. But it holds that such "corruptions" are filtered out in the process of testing. This is the well-known distinction between "the logic of discovery" and "the logic of justification." This view is now largely untenable. There is simply too much contrary evidence produced by sociologists, historians, and feminist critics of science to allow us to think that science works this way. Science seems to be shot through with social factors. There is even an argument (Okruhlik 1994) that this is inevitable. It is based on two assumptions: first, social factors play a role in the generation of theories; second, rational theory choice is comparative. This second point means that theories are chosen on the basis of how they fare compared to one another with respect to nature, not by being directly held up to nature.

On this view, there need not be any social factor involved in choosing the best theory from among the available rivals. However, the pool of rivals is itself shot through with social factors. If all anthropologists were males, then perhaps origin stories would tend to be variations on man-the-hunter. Objective evaluation would (on the comparative account) result in selecting the best of these. The gender bias that is present could not be filtered out.

The only way to improve this situation appears to be to insure that the most diverse set of rival theories possible is made available. And the only way to insure this is to have the most diverse set of theorizers possible. This will lessen the chances of one social factor being systematically present in all the rival theories.

This last point raises an interesting question about social policy, which I'll list as a fourth sense of social involvement in science.

4. *Social intervention to improve science.* There is a significant difference between sociologists of science like Bloor, Collins, and Latour, on the one hand, and feminist critics, on the other. The former think of themselves as reporting on a kind of natural phenomenon. The latter do that, too, but they are also interested in making science better. They see women and various racial groups, for example, as all too often victims of science, and – quite rightly – they want to put an end to it. They see women as contributing a different perspective, sometimes to our great advantage. Evelyn Fox

Keller in her biography of Barbara McClintock (1983) describes a classically trained scientist who tends to work in a way quite different from her male colleagues. Keller suggests that McClintock's "feeling for the organism" was typically female, in that it was much more "holistic." Because of this she was able to imagine processes of inheritance in the corn she was studying that her more reductionistically minded male colleagues could not so readily conceive (see FEMINIST ACCOUNTS OF SCIENCE).

If the analysis above (by Longino and by Okruhlik) is on the right track, then the problem stems from not having enough of the right sort of theories to choose from. Expanding the pool of competing theories seems to be the solution. This, however, requires making the pool of theorizers less homogeneous. Scientific institutions could achieve such diversity in their ranks only by hiring with an eye to diversity as a goal.

No one would deny the right of funding agencies to say how their money is to be spent (Those who pay the piper call the tune). Similarly, no one would deny the right of outsiders to insist that laboratory animals not be abused. This kind of social intervention in science does not affect the *content* of the theories we accept, and that perhaps, is why we do not instinctively resist it. On the other hand, affirmative action proposals are often viewed as "undermining excellence"; yet exactly the opposite may be the case. Greater diversity within the ranks of science – even if forced on it by the larger society – may actually better promote its primary aim: to achieve the best science possible.

What role, if any, social factors play in science has been hotly debated in recent years. The level of vitriol is greater than in other academic debates, perhaps because a great deal hangs on it politically. One of the most interesting developments stems from the "Sokal hoax." Physicist Alan Sokal concocted an article in the worst postmodern gibberish, which was then unwittingly published in a major journal of cultural studies (1996a). He then revealed the hoax (1996b) to applause and derision. But Sokal's aim in the Science Wars, as they are often called, was not to defend science from postmodern critics and other so-called social constructivists, but to defend the political Left, which he feared was succumbing to a form of self-defeating anti-intellectualism. The hoax has sparked a great deal of comment (especially on the Internet), some of it useful, some not. Naturally, there are fears that his prank will only further polarize the warring camps. On the other hand, it has forced many to look seriously at the work of rivals, so that critiques are less frequently based on caricatures. More importantly, perhaps, the Sokal affair has prompted people to think more deeply about the role of science in politics and the attitude to science that might best serve various social goals.

References and further reading

Bloor, D. 1991: *Knowledge and Social Imagery*, 2nd edn (Chicago: University of Chicago Press).

Brown, J. R. 1989: *The Rational and the Social* (London and New York: Routledge).

—— 2000: *Who Should Rule? A Guide to the Epistemology and Politics of the Science Wars* (New York and Berlin: Springer-Verlag).

Brown, J. R. (ed.) 1984: *Scientific Rationality: The Sociological Turn* (Dordrecht: Reidel).

Collins, H., 1985: *Changing Order: Replication and Induction in Scientific Practice* (Beverly Hills, CA: Sage).

Collins, H., and Pinch, T. 1993: *The Golem: What Everyone Should Know about Science* (Cambridge: Cambridge University Press).

Forman, P. 1971: Weimar culture, causality and quantum theory. *Historical Studies in the Physical Sciences*, 3, 1–115.

Fuller, S. 1988: *Social Epistemology* (Bloomington, IN: Indiana University Press).

Hacking, I. 1999: *The Social Construction of What?* (Cambridge, MA: Harvard University Press).

Harding, S. 1986: *The Science Question in Feminism* (Ithaca, NY: Cornell University Press).

Hubbard, R. 1990: *The Politics of Women's Biology* (New Brunswick, NJ: Rutgers University Press).

Keller, E. F. 1983: *A Feeling for the Organism: The Life and Work of Barbara McClintock* (New York: Freeman).

—— 1985: *Reflections on Gender and Science* (New Haven: Yale University Press).

Latour, B. 1987: *Science in Action: How to Follow Scientists and Engineers Through Society* (Cambridge, MA: Harvard University Press).

—— 1988: *The Pasteurization of France*, trans. A. Sheridan and J. Law from *Les Microbes: Guerre et Paix suivi de Irréductions* (Paris, 1984) (Cambridge, MA: Harvard University Press).

Latour, B., and S. Woolgar 1979: *Laboratory Life: The Social Construction of Scientific Facts* (Beverly Hills and London: Sage).

Longino, H. 1989: *Science as Social Knowledge: Values and Objectivity in Scientific Inquiry* (Princeton: Princeton University Press).

Okruhlik, K. 1994: Gender and the biological sciences. *Canadian Journal of Philosophy*, suppl. vol. 20, 21–42.

Pickering, A. 1984: *Constructing Quarks* (Chicago: University of Chicago Press).

—— 1995: *The Mangle of Practice* (Chicago: University of Chicago Press).

Pickering, A. (ed.) 1992: *Science as Practice and Culture* (Chicago: University of Chicago Press).

Ruse, M. 1999: *Mystery of Mysteries: Is Science a Social Construction?* (Cambridge, MA: Harvard University Press).

Schiebinger, L. 1999: *Has Feminism Changed Science?* (Cambridge, MA: Harvard University Press).

Shapin, S., and Schaffer, S. 1985: *Leviathan and the Air Pump* (Princeton: Princeton University Press).

Sokal, A. 1996a: Transgressing the boundaries: toward a transformative hermeneutics of quantum gravity. *Social Text*, nos 46–7, 217–52.

—— 1996b: A physicist experiments with cultural studies. *Lingua Franca*, May–June, 62–4.

66

Social Science, Philosophy of

ALEX ROSENBERG

Do the social sciences employ the same methods as the natural sciences? If not, can they do so? And should they do so, given their aims? These central questions of the philosophy of social science presuppose an accurate identification of the methods of natural science. For much of the twentieth century this presupposition was supplied by the logical positivist philosophy of physical science. The adoption of methods from natural science by many social scientists raised another central question: why had these methods so apparently successful in natural science been apparently far less successful when self-consciously adapted to the research agendas of the several social sciences? Alternative answers to this last question reflect competing philosophies of social science. On one view, the social sciences have not progressed because social scientists have not yet applied the methods of natural science well enough. Another answer has it that the positivists got the methods of natural science wrong, and that social scientists aping wrong methods have produced sterility in their disciplines. Still another response to this question argues that the social sciences are using the right methods and are succeeding, but that the difficulties they face are so daunting that rapid progress is not to be expected. Finally, a fourth answer which has attracted many philosophers of social science has it that social science has made much progress, but that its progress must be measured by different standards from those applied in natural science. This view reflects the belief that social science ought not to employ the methods of natural science, and that many of its problems stem precisely from the employment of an inappropriate method, reflecting profound philosophical confusion. These differing philosophies of social science must be assessed along the dimensions outlined below.

Causation, laws, and intentionality

The thesis that the social sciences should and can employ the methods of natural science is traditionally labeled "naturalism." The issue on which naturalism and its contrary – anti-naturalism – have longest contended is that of the nature of human action and its explanation. All parties agree that social sciences share a presumption, along with history and common sense, that the subject of social science is action, its consequences, and its aggregation into social institutions and processes; moreover, there is agreement that action is explained by the joint operation of desires and beliefs. This is a theory which, on the one hand, is so obvious that in history and biography it goes unmentioned, and, on the other hand, has been transformed into the economist's formalization as rational choice theory. If explanation of human action in common sense and all the social sciences appeals to this principle, formalized or not, then the vindication of naturalism or its refutation turns on whether some version of the theory of rational choice is a *causal* law or not. Similarly, the adequacy of anti-naturalism hinges on whether such explanations' force lies in some noncausal power to illuminate their *explanans* (see NATURALISM).

Why? Because scientific explanation is held crucially to require derivation of the *explananda* from causal laws and initial or boundary conditions (Hempel 1965). Though this analysis of explanation in science has been significantly qualified and circumscribed over the years, the notion that explanation involves systematizing laws and/or causality remains a fixed point in the philosophy of natural science (Kitcher and Salmon 1989). Thus, something like the following statement will have to be a causal law, or an empirical generalization capable of being improved in the direction of a law:

L   (x) (if x desires d, and x believes that all things considered doing action a is
      the most efficient means of attaining desire d, then x does a)

The debate on whether a principle like [L] embodies a law, or an approximation to one, has been a lively one in the philosophy of psychology.

For [L] to be a law, beliefs, desires, and actions must behave in the way that causes and effects behave: they must be logically independent of one another: it must be possible to establish that each of a particular package of belief, desire, and consequent action obtain without having to establish that the other two obtain. However, the existence of a logical connection between descriptions of these states is something that has long been recognized in the philosophy of psychology. Beliefs and desires can be unmetaphorically described as "aboutness," or content or intentionality. Beliefs are about actual or (in the case of false beliefs) non-actual states of the world, as are desires; actions are distinct from mere bodily motion only because they reflect sets of desires and beliefs. Thus, action too is imbued with intentionality.

Intentionality is an obstacle to a naturalistic treatment of the psychological, for two reasons. First, beliefs and desires are identified and distinguished from one another by their intentional content; but the only way to establish the content of a person's beliefs from observation of the person's actions is to have *prior* knowledge of the person's desires and all other beliefs, and vice versa. Moreover, in order to infer desire and belief from action, one must be able to distinguish the action governed by the combination of the desire and belief from mere behavior, and this cannot be done unless we can establish that the body's motion constitutes action – that is, that it is movement caused by belief and desire. We are thus caught in an intentional circle. There is no way independent of observing an action to establish that its causes obtain, and vice versa. But causes and effects must be logically distinguishable from each other; more than that, they must be methodologically distinguishable. Otherwise there will be no way to *test* the claim that some combination of desire and belief causes some particular action, and consequently no way to systematically apply our theory of rational choice in the prediction of actions with any hope of improvement on common sense.

Of course, if we could establish when a certain combination of desires and beliefs obtain independent of the actions they cause – by, for example, reading mental contents off the neurological states of the brain – we could in principle identify causes and effects independently. Alas, philosophers and psychologists from Descartes to Skinner (1953) have rejected this course as unavailing, either in principle or in practice. Despite the scientific conviction that the mind is the brain, no one has yet overcome Descartes's arguments in *The Meditations* against this claim, or shown to general satisfaction how intentional states could be physical states (see DESCARTES). And even were this obstacle overcome, as Skinner (1953) long argued, neuroscience can provide no real-time help in establishing the boundary conditions to which intentional generalizations like those of rational choice are applied for explanation and prediction.

Regularities versus rules

In the latter half of the twentieth century the philosophy of social science was dominated by a dispute about whether the intentional explanations of common sense, history, cultural anthropology, a good deal of political science and sociology, or their economic and psychological formalizations are, or could be, causal laws. Naturalists proclaimed that they were, but with the coming of Wittgenstein's influence in the philosophy of mind, this view came increasingly under attack (Wittgenstein 1953). The gist of this attack on naturalism was that since intentional states and actions are logically connected with one another, explanations that appealed to them could not be causal. Instead, the explanatory force of such explanations had to have different nonnaturalistic foundations. It was held to be a conceptual confusion to treat belief and desire as causes of action and to approach the generalization which links them as an empirical generalization that might be improved and refined so as to become a law. Instead, many anti-naturalists held that beliefs and desires are logically linked to actions as their *reasons*, and that the linkage is established by *rules*. On their view, much of the sterility and vacuity of social science is attributable to the confusion of rules with regularities. Rules are learned by asking those who already understand them the right questions, not by making experimental observations of behavior. When we attempt to apply empirical methods to what is in essence a conceptual inquiry, the result is bound to be unsatisfactory (Winch 1958).

Opponents of naturalism before and since Wittgenstein have been animated by the notion that the aims of social science are not causal explanation and improving prediction, but uncovering rules that make social life intelligible to its participants. For these purposes there is no alternative to "folk psychology," the account of action and its sources implicit in our everyday beliefs and ubiquitous in all cultures. And folk psychology is what enables us to "interpret" the behavior of others, to show it rational or reasonable by our own lights. If we fail so to understand the actions of others, then by and large the fault is not in our "theory" but in our application of it. We have misdiagnosed the beliefs and the desires of those we seek to understand. On this view, the goal of social inquiry is to be understood on the model of the cultural anthropologist "going native" in order to learn the meaning of day-to-day life for the participant of the culture under study. These meanings are embodied in rules. Indeed, principles of rationality like [L] are but rules which may characterize only Western culture. Note that rules may be broken, and that when they are broken, there are further rules which dictate the sanctions to be imposed, and so forth. The social scientist's objective is to uncover these rules which render what happens in a society intelligible, though they never make social life predictable beyond the limits set by commonsense folk psychology.

That social science is a search for intelligibility will explain why its theories ought not to be construed causally and why it neither embodies nor needs explanatory and predictive laws. Thus, the failure of empirically inspired social science to uncover generalizations about human action that seem reasonable candidates for laws or even their rough and improvable precursors is explained not by the complexity of human behavior and its intentional causes, but by the fact that a search for such laws misconstrues the aim of social science and the role which concepts like rationality sometimes play in attaining this aim. This aim is an interpretation of actions and events which assigns them meaning – sometimes the participant's meaning, sometimes a "deeper" meaning, but always one which presupposes a significant overlap in most of the broader beliefs and desires of the participants and observers as well. Anti-naturalists point out that the sort of interpretation sought is inevitably underdetermined by the evidence, is defeasible, and is a sort of construction subject to negotiation among interested parties. For this reason, the sort of explanation social science is suited to provide is no basis for predicting human action.

Between naturalism and anti-naturalism there is third alternative, which combines the most controversial parts of each of these doctrines: eliminativism adopts the interpretationalist conclusion that under their intentional characterizations beliefs and desires cannot be linked to behavior via laws. It also adopts the naturalistic claim that our goal in social science is causal explanation and improving precision and scope of prediction. Accordingly, the eliminativist argues that we should surrender any hope of explaining behavior as action caused by intentionally characterized psychological attitudes. Instead, in behavioral science we should adopt a neuroscientific or some other sort of non-intentional perspective on individual behavior, and in social science we should seek aggregate generalizations about large-scale processes while remaining agnostic on their psychological foundations (Churchland 1981; Rosenberg 1988).

Eliminativism has found few defenders in either social science or its philosophy. Most naturalists among social scientists and philosophers participating in this debate have challenged a presumption shared by both eliminativists and interpretationalists: the claim that, on its naturalistic interpretation, rational choice theory cannot reasonably be construed as embodying empirical generalizations with any prospect of being improved into causal laws (Fodor 1991). Instead, naturalists point with pride to relatively controversial and certainly limited successes, in economics, political science, and sociology. Naturalists also reject the philosophical theses from which the interpretationalists and the eliminativists draw their pessimistic conclusions about the prospect for intentional laws. Some naturalists dispute the requirement that the descriptions of causes be logically distinct from those of their effects (Davidson 1980). Others argue that, despite ineliminable *ceteris paribus* clauses, the status of the rationality principle as a general law is unimpeachable, or at least no more problematical than *ceteris paribus* laws elsewhere – in biology, for example (Fodor). These philosophers of social science continue to seek the obstacles to predictive power in the social sciences in the complexity and interdependence of human behavior, and in the openness of social processes to interference by exogenous forces. Others seek an explanation of the limitations on predictive strength of social science in the reflexivity of social phenomena.

Ontology in social science and biology

Over the last quarter of the twentieth century the philosophy of social science has made common cause with the philosophy of biology. The affinities are easy to explain. Theories in both biology and the social sciences appear to quantify over entities at a variety of levels of organization, and to appeal to the purposes, goal, ends, or functions that systems serve in order to explain their presence and operation.

Biology employs generalizations that obtain at the level of the molecule, the cell, the tissue, the individual organ, the kin group, the interbreeding population, the species, etc. So too, some theories in social science are committed to the existence of entities beyond the individual: groups and institutions whose existence and behavior cannot be exhaustively analyzed into or explained by appeal to the existence and behavior of the individual agents who participate in them. Such hypostasis has been based on the explanatory role and apparently autonomous coercive powers which such wholes seem to have.

To those who are skeptical about the notion that there are, in Durkheim's terms, social facts above and beyond individual ones, the slogan that the whole is greater than the sum of its parts is simply mystery-mongering provocation. Economicsts have been foremost among social-scientific opponents of holism. Their own theory rigorously rejects any dispensation from the obligation to explain all economic phenomena by derivation from the rational choices of individual agents. Empiricist philosophers of social science have argued that since social wholes, facts, institutions, etc. cannot be directly observed, and since, ontologically, their existence depends wholly on the existence of the agents who compose them, such wholes must be reducible by definition or by explanation to the behavior of individuals. Allowing such wholes autonomous existence is not only metaphysically bootless, but epistemically gratuitous (Elster 1990).

Given a willingness throughout science – natural and social – to postulate a variety of entities not amenable to observation, much of the force of the purely epistemological argument against holism must fade. Thus contemporary debates about methodological individualism versus holism turn on the explanatory *indispensability* of the appeal to trans-individual entities within social science and biology. The fact that evolutionary biology seems to countenance *holistic* theories, ones which postulate the existence and causal role of family groups, interbreeding populations, species and ecosystems independent of the individual organisms that compose them, provides grist for the holist mill. If such postulation is legitimate in biology, it should be equally legitimate in social science.

But this argument seems doubtful to individualists, both in biology and in the social sciences. For, despite appearances, evolutionary explanation in biology is implicitly or explicitly individualist. Therefore, it must be so in social science as well, holds the individualist (see HOLISM).

Teleology and function

Whether the argument from biology tells in favor of or against the methodological holist in social science hinges on the correct understanding of teleological explanation and functional analysis in biology. And this is still another reason why philosophers of social science closely follow or participate in debates among philosophers of biology.

Durkheim was among the earliest to explicitly attribute functions to social institutions above and beyond ones that individuals ordain or recognize. On this functionalist view, social wholes exist *in order* to fulfill certain functions. The fact that they fulfill such functions can actually explain their existence. For example, the presence of the kidney is explained by citing its function, the removal of soluble wastes. By contrast, the presence of the pineal gland in humans is problematical, just because it has no identifiable function. Similarly, certain social structures exist in order to fulfill some function vital to the society's survival or well-being.

Empiricist philosophers have been dubious about the cognitive value of functional and other forms of purposive explanation, just because they seem to reverse the order of causation (Hempel 1965). To explain the presence of the kidneys by appeal to their ability to remove waste is to cite an effect to explain its cause. Moreover, there are many things with functions that do not fulfill their purposes. For instance, a chameleon's skin color functions to camouflage the chameleon, even when the camouflage doesn't work, and the chameleon is eaten by a bird. Can we explain the presence of a particular strategy for evading predatory by citing a function it fails to perform? Add to this the difficulty of identifying functions, both in the biological case and the social one: what exactly is the function of the peacock's plumage, or the function of puberty rites in a hunter-gatherer society?

Some philosophers of biology have sought to legitimize, eliminate, and minimize the metaphysical excesses of functional explanation by appeal to the explanatory role of the theory of natural selection. The theory of natural selection is sometimes invoked to legitimate functional explanation and sometimes to show that teleology is a mere appearance – an overlay we impose on purely causal processes. Either way the analysis is roughly the same: the appearance or the reality of adaptation is the result of a long, slow process of random variation in hereditary traits being subjected to the culling of nature in the struggle for survival. In a relatively constant environment over a long period of successive generations, variants that are fortuitously advantageous will be selected and will increase their proportion in the whole of a population until the variants become ubiquitous. Thus, to say animals have kidneys in order to remove soluble wastes is explanatory, because the kidney exists in animals now as the result of a long process of selection for soluble waste-removers over the evolutionary past (Wright 1976).

Once teleology is legitimated in biology, it is mere caviling to withhold it from the social sciences. Functions can be accorded to properties of social wholes on the basis of their selection in the struggle among groups for survival. We thus apparently underwrite holism and functionalism by the same token.

The trouble for the holists is that appeals to evolutionary biology seem to underwrite methodological individualism. Most evolutionary biologists agree that selection does not operate at the level of the group, that the mechanism of selection and hereditary transmission operates exclusively at the level of the individual, and that group properties must be fully explained by appeal to the properties and interactions of individuals. Functional explanation in biology turns out to be resolutely individualist (see TELEOLOGICAL EXPLANATION).

This result is disturbing for holists. Their best argument for the existence of social wholes and their properties is the explanatory role of social roles and their properties. But these properties – say preferential marriage rules – are usually accorded a functional role – for example, that of enhancing social integration – and their existence is explained by citing these functions. If functions have explanatory force only to the extent that their existence can in turn be given a thoroughly individualist foundation, then the holist argument for such wholes and their functional properties from irreducible explanatory indispensability collapses.

Reflexive knowledge in social science

The subjects of social science are themselves epistemic agents, who can be influenced by their own beliefs about the generalizations, the input, and the output of social theory. Thus, the publication of an economic prediction can result in its being falsified by the action of agents who act on the prediction; broadcasting polling results can make predictions based on them self-fulfilling.

Theories and predictions whose dissemination can effect their confirmation are known as "reflexive" ones. Some philosophers and social scientists have held that theories in social science must be different *in kind* from those of natural science, because the former are potentially reflexive, whereas the latter are not. One extreme version of this view denies to social theory the potential for predictive improvement beyond certain narrow limits, and so constitutes a serious obstacle to a naturalistic social science that proceeds by uncovering a succession of improvable causal generalizations (McCloskey 1986). Among advocates of Critical Theory the doctrine is held not so much to limit knowledge, but to burden it with a special responsibility (Habermas 1971).

Critical Theorists adopt an empiricist account of natural science, but they insist that the reflexive character and normative aims of social science make for important differences in its methods and epistemology: social science is a search for emancipative intelligibility, not laws, or even rules. Its methods are social criticism of ideologies masquerading as fixed truths, mere interpretations which are to be unmasked as social constructs, not the inevitable result of natural forces. Critical Theorists hold that since theory *can* influence social behavior, the social scientist has the responsibility of framing and disseminating theory in a way that will emancipate people by showing them the real meaning of social institutions and freeing people from false and enslaving beliefs about themselves and their communities (Habermas 1971).

Reflexivity raises important methodological questions: Can we minimize the obstacles to empirical testing that self-fulfilling prophecies and suicidal predictions raise? If not, how do we assess the cognitive status of explanations embodying reflexive theories? Even more radically anti-naturalistic doctrines start from similar premises but deny the empirical character even of natural science. These doctrines infer from the non-empirical character of social science to the ideological, negotiated noncumulative character of natural science. Heavily influenced by a radical interpretation of Kuhn's (1962) theory of scientific change, post-modernist thought not only denies that the social sciences have a methodology or an epistemology, it denies that natural science constitutes objective information about the way the world is or even the way observable phenomena are (McCloskey 1986).

457

From interpretation to historicism

The natural sciences have traditionally been viewed as *a*historical in a number of different senses. Natural laws are typically time-symmetrical: given laws and initial conditions for a closed deterministic system, we can retrodict past events as well as future ones; there is no causal action at a temporal distance: a past event can only influence a future one transitively through its intermediate effects on each intervening event between the cause and its ultimate effect. If there are causal laws, their writ runs across all places and all times: a regularity confirmed in one spatiotemporal region and disconfirmed in another is not a law in either region; it is at best a local empirical regularity to be explained by some exceptionless universal law.

Historicism in social science is an anti-naturalistic doctrine which involves the denial that one or another of the three theses discussed above characterizes the social sciences. More specifically Marxian and Freudian historicists sometimes argue that social processes reflect the operation of asymmetrical principles which mandate a fixed order of events: thus capitalism could not have come into existence without the prior appearance of feudalism; adult neurosis could not have come into existence without the prior frustration of infantile sexuality. Historicism sometimes also embodies the thesis that each historical epoch operates in accordance with its own distinct explanatory laws, and that sometimes the discovery of such laws can usher in a new era with new laws – whence the connection between historicism and reflexivity.

Historicism raises metaphysical questions of a fundamental sort. What is it about human action, social institutions, and large-scale historical events which make them so different from causal processes governed by symmetrical causal laws? These questions either go unanswered in contemporary debates or lead us back to problems on the intersection of the philosophy of social science and the philosophy of psychology: problems about the nature of intentionality and the mind. For these are the sources, if any, of historically conditioned action.

Dangerous knowledge, ideology, and value freedom

The social sciences have special relevance to normative questions about individual and social policy. Well-confirmed theories of human behavior provide us with the means to ameliorate or to worsen human life. This fact raises moral questions about how this knowledge should be employed. In addition, in choosing which research questions to examine and which hypotheses to test, the social scientist makes value judgments about the interest, importance, and significance of alternatives. However, these questions are in principle no different from those raised by advances in theoretical physics. Different questions are raised by the fact that well-confirmed social theories would enable us to control and manipulate individual and aggregate human behavior. A further distinctive problem is raised by the need to employ human subjects in order to test some theories of human behavior. Whether it is permissible to treat humans in the way we treat laboratory animals, or even animals in field studies, is an issue that social scientists face along with medical researchers.

Additionally, there is the problem of potentially dangerous knowledge. Some inquiries are best left unmade, or so it is alleged. Studies of the correlation of criminality and chromosomal abnormality, the heritability of intelligence, or the dynamics of jury deliberation, will be condemned on the ground that disseminating the findings may be harmful whether they are scientifically significant or not. Over the last half-century, this claim has been made with special force in regard to studies of the statistical heritability of IQ among various groups. Some philosophers have attacked such studies on cognitive grounds, arguing that IQ is not a measure of intelligence, and that heritability shows little about genetic inheritance. But they have also held that the examination of such questions should be foresworn because merely pursuing them is socially inflammatory, and that the results, even of well-conceived and executed studies, are likely to be misused. Whether this prospect raises special questions beyond those faced in research on toxic chemical substances or harmful disease vectors which might be inadvertently released is one that philosophers of social science need to debate.

But beyond the normative issues to which social science is relevant, there is the further debate among philosophers and social scientists about whether claims in the social sciences are themselves explicitly, implicitly, or inevitably evaluative, normative, prescriptive, or otherwise "value-laden" (Taylor 1985). Arguments for this claim often turn on the allegedly evaluative aspect of certain descriptive and explanatory concepts indispensable to varying social theories. For example, the term "rational" has a positive connotation, as do expressions like "functional," or "adaptational." No matter what meaning is stipulated for these terms, they may nevertheless convey attitudes or obstruct certain questions without the social scientist recognizing their normative role. More fundamentally, the very treatment by social science of institutions and norms as subjects of objective scientific exploration may unwittingly suggest the naturalness, inevitability, and immutability of social arrangements which are in fact artificial, constructed, and subject to negotiation. As noted above, it is the responsibility of social scientists, some will hold, to unmask the character of some oppressive or exploitative institutions. This is a responsibility made more difficult to discharge by the failure to recognize the implicit normative dimension of social theories and methods.

On a traditional view of the normative dimension of social science, there is a distinction between facts and values, and by scrupulousness about this distinction the social scientist *qua scientist* can and should avoid making normative claims implicitly or explicitly. The normative force of this claim rests on the view that scientists should be as objective as possible. For the value of scientific knowledge and its further accumulation is jeopardized by the appearance of partiality. Radically opposed to this view is the claim that objectivity is impossible in any science, social or natural; that "knowledge" is a coercive term that obscures the partiality, the negotiated social constructions of physical science; and that in any case, as Kuhn is taken to have shown, there is no real accumulation or progress in the history of science. Ironically, on this radical view, the problem of normative versus positive theory is not a distinctive one for philosophy of social science at all. Either it is a problem for the philosophy of all the sciences or for none of them (see VALUES IN SCIENCE).

References and further reading

Churchland, P. 1981: Eliminative materialism and the propositional attitudes. *Journal of Philosophy*, 78, 67–90.

Davidson, D. 1980: *Essays on Actions and Events* (Oxford: Oxford University Press).

Durkheim, E. 1965: *The Rules of the Sociological Method* (New York: Free Press).

Elster, J. 1990: *The Cement of Society* (Cambridge: Cambridge University Press).

Fodor, J. 1991: *A Theory of Content* (Cambridge, MA: MIT Press).

Habermas, J. 1971: *Knowledge and Human Interests*, trans. T. McCarthy (Boston: Beacon Press).

Hempel, C. G. 1965: *Aspects of Scientific Explanation, and Other Essays* (New York: Free Press).

Kitcher, P., and Salmon, W. 1989: *Scientific Explanation: Minnesota Studies in the Philosophy of Science* (Minneapolis, University of Minnesota Press).

Kuhn, T. 1962: *The Structure of Scientific Revolutions* (Chicago: University of Chicago Press).

McCloskey, D., 1986: *The Rhetoric of Economics* (Madison, University of Wisconsin Press).

Rosenberg, A. 1988: *Philosophy of Social Science* (Oxford: Clarendon Press and Boulder, CO: Westview).

Skinner, B. F. 1953: *Science and Human Behavior* (New York: Macmillan).

Taylor, C. 1985: *Collected Papers* (Cambridge: Cambridge University Press).

Winch, P. 1958: *The Idea of a Social Science* (London: Routledge and Kegan Paul).

Wittgenstein, L. 1953: *Philosophical Investigations* (Oxford: Blackwell).

Wright, L. 1976: *Teleological Explanation* (Berkeley: University of California Press).

67

Space, Time, and Relativity

LAWRENCE SKLAR

Given the central role played by space and time both in our ordinary experience and in our attempts to understand the world by means of scientific theory, it is no surprise that attempts to understand space and time form a central locus of the interaction of philosophy and the physical sciences.

In the long history of the discussions of the central topics in the philosophy of space and time, there has been a recurrent dynamic. Central issues in the philosophical problematic in this area are isolated. A number of alternative solutions to the core questions are outlined by the philosophers. At the same time, however, new results are obtained by those developing the mathematical and physical theories of space and time. Reflection on these results forces upon the philosophers a radical rethinking of the very framework in which the philosophical exploration takes place. This dynamic can be seen in three cases: the impact of Greek geometry and physics on Greek philosophy of space and time; the impact of Newtonian physics on the philosophy of space and time in the period of the great scientific revolution of the seventeenth century; and the impact of the development of the theories of special and general relativity on contemporary philosophy of space and time.

Greek beginnings

Anticipations of what later became the doctrine of substantivalism with regard to space can be found in Plato's characterization of the "receptacle" in his *Timaeus* and in the atomist's notion of atoms as motion in the void. In Aristotle's notion of "place," and even more trenchantly in his characterization of time as "the measure of motion," one can see anticipations of later relationism. Furthermore, Aristotle's exploration of motion and its causes, particularly his concept of "natural" as opposed to "forced" motions, was the beginning of the concepts that lie behind the Newtonian argument for absolute space.

Most important of all, however, was the Greek discovery of deductive, axiomatic geometry. It was this that gave rise to the epistemic account of knowledge as founded on the deduction of conclusions from self-evident first principles, which became the centerpiece of the rationalist theory of knowledge. This was the approach to knowledge that from then on took our knowledge of the structure of space, formalized in geometry, as the paradigm of a priori knowledge of the world.

461

The scientific revolution

The great metaphysical debate over the nature of space and time has its roots in the scientific revolution of the sixteenth and seventeenth centuries. An early contribution to the debate was Descartes's identification of matter with extension, and his concomitant theory of all of space as filled by a plenum of matter (see DESCARTES).

Far more profound was Leibniz's characterization of a full-blooded theory of relationism with regard to space and time (see LEIBNIZ). Space was taken to be a set of relations among material objects. The latter, the deeper monadological view to the side, were the substantival entities. No room was provided for space itself as a substance over and above the material substance of the world. All motion was then merely relative motion of one material thing in the reference frame fixed by another. The Leibnizian theory was one of great subtlety. In particular, the need for a modalized relationism to allow for "empty space" was clearly recognized. An unoccupied spatial location was taken to be a spatial relation that could be realized but that was not realized in actuality. Leibniz also offered trenchant arguments against substantivalism. All of these rested upon some variant of the claim that a substantival picture of space allows for the theoretical toleration of alternative world models that are identical as far as any observable consequence is concerned.

Contending with Leibnizian relationism was the "substantivalism" of Newton and his disciple S. Clarke (see NEWTON). Actually Newton was cautious about thinking of space as a "substance." Sometimes he suggested that it be thought of, rather, as a property – in particular as a property of the Deity. But what was essential to his doctrine was his denial that a relationist theory, with its idea of motion as the relative change of position of one material object with respect to another, can do justice to the facts about motion made evident by empirical science and by the theory that does justice to those facts.

The Newtonian account of motion, like Aristotle's, has a concept of natural or unforced motion. This is motion with uniform speed in a constant direction, so-called inertial motion. There is, then, in this theory an absolute notion of constant velocity motion. Such constant velocity motions cannot be characterized as merely relative to some material reference frame, for an inertially moving object will have many motions relative to other material objects, some of which will be non-inertial. Space itself, according to Newton, must exist as an entity over and above the material objects of the world, in order to provide the standard of rest relative to which uniform motion is genuine inertial motion.

Such absolute uniform motions can be empirically discriminated from absolutely accelerated motions by the absence of inertial forces felt when the test object is moving genuinely inertially. Furthermore, the application of force to an object is correlated with the object's change of absolute motion. Only uniform motions relative to space itself are natural motions requiring no forces as explanatory. Newton also clearly saw that the notion of absolute constant speed requires a notion of absolute time, for, relative to an arbitrary cyclic process as defining the time scale, any motion can be made uniform or not, as we choose. But genuine uniform motions are of constant speed in the absolute time scale fixed by "time itself." Periodic processes can be at best good indicators or measurers of this flow of absolute time.

462

Newton's refutation of relationism by means of the argument from absolute acceleration is one of the most distinctive examples of the way in which the results of empirical experiment and of the theoretical efforts to explain these results impinge upon what had been initially a purely philosophical debate of metaphysics. Although philosophical objections to Leibnizian relationism exist – for example, in the claim that one must posit a substantival space to make sense of Leibniz's modalities of possible position – it is the scientific objection to relationism that causes the greatest problems for that philosophical doctrine.

Between Newton and Einstein

A number of scientists and philosophers continued to defend the relationist account of space in the face of Newton's arguments for substantivalism. Among them were Leibniz, Huygens, and Berkeley (see BERKELEY). The empirical distinction between absolute uniform motion and motion that is absolutely accelerated continued, however, to frustrate their efforts.

In the nineteenth century, Mach made the audacious proposal that absolute acceleration might be viewed as acceleration relative not to a substantival space, but to the material reference frame of what he called the "fixed stars" – that is, relative to a reference frame fixed by what might now be called the "average smeared-out mass of the universe" (see MACH). As far as observational data went, he argued, the fixed stars could be taken to be the frame relative to which uniform motion was absolutely uniform. Mach's suggestion continues to play an important role in debates up to the present day.

The nature of geometry as an apparently a priori science also continued to receive attention. Geometry served as the paradigm of knowledge for rationalist philosophers, especially for Descartes and Spinoza. Kant's attempt to account for the ability of geometry to be both a priori and "synthetic" – that is, to have full-blooded factual content and go beyond the analytic truths of logic extended by definition – was especially important. His explanation of the a priori nature of geometry by its "transcendentally psychological" nature – that is, as descriptive of a portion of mind's organizing structure imposed on the world of experience – served as his paradigm for legitimate a priori knowledge in general.

The theories of relativity

A peculiarity of Newton's theory, of which Newton was well aware, was that whereas acceleration with respect to space itself had empirical consequences, uniform velocity with respect to space itself had none. The theory of light, particularly in J. C. Maxwell's theory of electromagnetic waves, suggested, however, that there was only one reference frame in which the velocity of light would be the same in all directions, and that this might be taken to be the frame at rest in "space itself." Experiments designed to find this frame seem to show, however, that light velocity is isotropic and has its standard value in all frames that are in uniform motion in the Newtonian sense. All these experiments, however, measured only the average velocity of the light relative to the reference frame over a round-trip path.

463

It was Einstein's insight to take the apparent equivalence of all inertial frames with respect to the velocity of light to be a genuine equivalence (see EINSTEIN). His deepest insight was to see that this required that we relativize the notion of the simultaneity of events spatially separated from one another to the state of motion of an inertial reference frame. For any relationist, the distance between nonsimultaneous events is frame-relative. Einstein proposed the symmetrical claim that for noncoincident events simultaneity is relative as well. This theory of Einstein's later became known as the Special Theory of Relativity.

Einstein's proposal accounts for the empirical undetectability of the absolute rest frame by optical experiments, because in his account the velocity of light is isotropic and has its standard value in all inertial frames. The theory had immediate kinematic consequences, among them the fact that spatial separation (lengths) and time intervals are frame-of-motion-relative. A new dynamics was needed if dynamics was to be, as it was for Newton, equivalent in all inertial frames.

Einstein's novel understanding of space and time was given an elegant framework by H. Minkowski in the form of Minkowski space-time. The primitive elements of the theory were pointlike locations in both space and time of unextended happenings. These were called the "event locations" or the "events" of a four-dimensional manifold. There is a frame-invariant separation of event from event called the "interval." But the spatial separation between two noncoincident events, as well as their temporal separation, are well defined only relative to a chosen inertial reference frame. In a sense, then, space and time are integrated into a single absolute structure. Space and time by themselves have only a derivative and relativized existence.

Whereas the geometry of this space-time bore some analogies to a Euclidean geometry of a four-dimensional space, the transition from space and time by themselves to integrated space-time required a subtle rethinking of the very subject matter of geometry. "Straight lines" are the straightest curves of this "flat" space-time, but they include "null straight lines," interpreted as the events in the life history of a light ray in a vacuum, and "timelike straight lines," interpreted as the collection of events in the life history of a free, inertial, material particle, as well as purely spatial straight lines.

Einstein's second great contribution to the revolution in scientific thinking about space and time arose from the problem of fitting the theory of gravity into the new relativistic framework. The result of his thinking was the theory known as the general theory of relativity.

The heuristic basis for the theory rested upon an empirical fact known to Galileo (see GALILEO) and Newton, but whose importance was made clear only by Einstein. Gravity, unlike other forces such as the electromagnetic force, acts on all objects independently of their material constitution or of their size. The path through space-time followed by an object under the influence of gravity is determined only by its initial position and velocity. Reflection upon the fact that in a curved space the path of minimal curvature from a point, the so-called geodesic, is uniquely determined by the point and by a direction from it, suggested to Einstein that the path of an object acted upon by gravity can be thought of as a geodesic followed by that path in a curved space-time. The addition of gravity to the space-time of special relativity can then be thought of as changing the "flat" space-time of Minkowski into a new, "curved" space-time.

The kind of curvature implied by the theory is that explored by B. Riemann in his theory of intrinsically curved spaces of arbitrary dimension. No assumption is made that the curved space exists in some higher-dimensional flat embedding space. Curvature is a feature of the space that shows up observationally to those in the space. One of its manifestations is the fact that the least-curved curves in the space are no longer straight lines, just as the paths of light rays or particles acted upon only by gravity are not straight lines. Curvature also shows up in metric effects, just as the shortest distances between points on the Earth's surface cannot be reconciled with putting those points on a flat surface. Einstein (and others) offered other heuristic arguments to suggest that gravity might indeed have an effect on relativistic interval separations as determined by measurements using measuring tapes (for spatial separations) and clocks (to determine time intervals).

The actual theory offered by Einstein posits that space-time has a structure analogous to that of an intrinsically curved four-dimensional Riemannian space. Once again one must be cautious in using the analogy, since it is a curved version of Minkowski's space-time that is posited, not a curved spatial manifold. The core of the theory is a law associating the curvature of space-time with the distribution within that space-time of mass-energy. This takes the place of the older Newtonian law that had a gravitational field of force generated by the presence of matter. The equation expressing this law is a local partial differential equation. To solve it, one must usually impose "boundary conditions" on the space-time, just as one normally does to solve any local differential equation. These play an important role in the discussions about possible relationistic interpretations of the theory.

Space-times and pre-relativistic physics

Once the concept of space-time had been developed to serve as a framework for relativistic theories, it became evident that space-time notions could play an important role in understanding pre-relativistic theories as well.

We noted above that one difficulty with the Newtonian theory was that in positing space itself as the reference frame relative to which accelerations were absolute accelerations, it also posited the existence of the absolute velocity of an object relative to space itself. But such absolute velocities had no observational consequences in Newton's own theory. By using space-time notions, one can construct a concept of space-time, called Galilean space-time or neo-Newtonian space-time, that is, in some ways, more appropriate to the Newtonian theory than Newton's own "space itself" and "absolute time." In this new space-time absolute time intervals are retained, so the theory is nonrelativistic. But while absolute acceleration is definable in the space-time (and empirically observable), the empirically unobservable absolute velocity simply does not exist as a structure in it.

The Newtonian theory of gravity also posited structures in the world with no observable consequences. Consider, for example, a universe in which all the matter is embedded in a gravitational force field of constant strength and direction. All this matter would be undergoing uniform acceleration with respect to space itself. But this particular absolute acceleration would have no detectable consequences. This is again due to the fact that the accelerating influence of the gravitational force is universal and independent of the constitution or size of the object on which the force acts. Awareness of this fact was implicit in Newton and was emphasized by Maxwell.

Using the new space-time notions, a "curved space-time" theory of Newtonian gravitation can be constructed. In this space-time time is absolute, as in Newton, Furthermore, space remains flat Euclidean space. This is unlike the general theory of relativity, where the space-time curvature can induce spatial curvature as well. But the space-time curvature of this "curved neo-Newtonian space-time" shows up in the fact that particles under the influence of gravity do not follow straight line paths. Their paths become, as in general relativity, the curved timelike geodesics of the space-time. In this curved space-time account of Newtonian gravity, as in the general theory of relativity, the indistinguishable alternative worlds of theories that take gravity as a force superimposed on a flat space-time collapse to a single world model.

Epistemological issues

The strongest impetus to rethink epistemological issues in the theory of space and time came from the introduction of curvature and of non-Euclidean geometries in the general theory of relativity. The claim that a unique geometry could be known to hold true of the world a priori seemed unviable, at least in its naive form, in a situation where our best available physical theory allowed for a wide diversity of possible geometries for the world and in which the geometry of space-time was one more dynamical element joining the other "variable" features of the world. Of course, skepticism toward an a priori account of geometry could already have been induced by the change from space and time to space-time in the special theory, even though the space of that world remained Euclidean.

The natural response to these changes in physics was to suggest that geometry was, like all other physical theories, believable only on the basis of some kind of generalizing inference from the lawlike regularities among the observed observational data – that is, to become an empiricist with regard to geometry.

But a defense of a kind of a priori account had already been suggested by Poincaré, even before the invention of the relativistic theories. He suggested that the limitation of observational data to the domain of what was both material and local (i.e., concerned with relations among material things at a point), along with the need to invoke laws connecting the behaviour of material objects with the geometry of space (or space-time) in order to derive any observable predictions from a geometric postulate, made the matter of choosing a geometry for the world a matter of convention or decision on the part of the scientific community. If any geometric posit could be made compatible with any set of observational data, Euclidean geometry could remain a priori in the sense that we could, conventionally, decide to hold to it as the geometry of the world in the face of any data that apparently refuted it.

The central epistemological issue in the philosophy of space and time remains that of theoretical underdetermination, stemming from the Poincaré argument. In the case of the special theory of relativity the question is the rational basis for choosing Einstein's theory over, for example, one of the "aether reference frame plus modification of rods and clocks when they are in motion with respect to the aether" theories that it displaced. Among the claims alleged to be true merely by convention (see CONVENTION,

ROLE OF) in the theory are those asserting the simultaneity of distant events, those asserting the isotropy of the velocity of light in inertial frames of reference, and those asserting the "flatness" of the chosen space-time. Crucial to the discussion are the fact that Einstein's arguments themselves presuppose a strictly delimited local observation basis for the theories and that in fixing upon the special theory of relativity, one must make posits about the space-time structure that outrun the facts given strictly by observation. In the case of the general theory of relativity, the issue becomes one of justifying the choice of general relativity over, for example, a flat space-time theory that treats gravity, as it was treated by Newton, as a "field of force" over and above the space-time structure.

In both the cases of special and general relativity, important structural features pick out the standard Einstein theories as superior to their alternatives. In particular, the standard relativistic models eliminate some of the problems of observationally equivalent but distinguishable worlds countenanced by the alternative theories. But the epistemologist must still be concerned with the question as to why these features constitute grounds for accepting the theories as the "true" alternatives.

Other deep epistemological issues remain, having to do with the relationship between the structures of space and time posited in our theories of relativity and the spatiotemporal structures we use to characterize our "direct perceptual experience." These issues continue in the contemporary scientific context the old philosophical debates on the relationship between the realm of the directly perceived and the realm of posited physical nature.

Metaphysical issues

A first reaction on the part of some philosophers was to take it that the special theory of relativity provided a replacement for the Newtonian theory of absolute space that would be compatible with a relationist account of the nature of space and time. This was soon seen to be false. The absolute distinction between uniformly moving frames and frames not in uniform motion, invoked by Newton in his crucial argument against relationism, remains in the special theory of relativity. In fact, it becomes an even deeper distinction than it was in the Newtonian account, since the absolutely uniformly moving frames, the inertial frames, now become not only the frames of natural (unforced) motion, but also the only frames in which the velocity of light is isotropic.

At least part of the motivation behind Einstein's development of the general theory of relativity was the hope that in this new theory all reference frames, uniformly moving or accelerated, would be "equivalent" to one another physically. It was also his hope that the theory would conform to the Machian idea of absolute acceleration as merely acceleration relative to the smoothed-out matter of the universe.

Further exploration of the theory, however, showed that it had many features uncongenial to Machianism. Some of these are connected with the necessity of imposing boundary conditions for the equation connecting the matter distribution to the space-time structure. General relativity certainly allows as solutions model universes of a non-Machian sort – for example, those which are aptly described as having the smoothed-out matter of the universe itself in "absolute rotation." There are strong arguments to suggest that general relativity, like Newton's theory and like special relativity, requires the positing of a structure of "space-time itself" and of motion relative to that structure, in order to account for the needed distinctions of kinds of motion in dynamics. Whereas in Newtonian theory it was "space itself" that provided the absolute reference frame for motion, in the special theory of relativity it was the set of inertial reference frames. In general relativity it is the structure of the null and timelike geodesics that perform this task. The compatibility of general relativity with Machian ideas is, however, a subtle matter and one still open to debate.

Other aspects of the world described by the general theory of relativity argue for a substantivalist reading of the theory as well. Space-time has become a dynamic element of the world, one that might be thought of as "causally interacting" with the ordinary matter of the world. In some sense one can even attribute energy (and hence mass) to the space-time (although this is a subtle matter in the theory), making the very distinction between "matter" and "space-time itself" much more dubious than such a distinction would have been in the early days of the debate between substantivalists and relationists.

On the other hand, a naive reading of general relativity as a substantivalist theory has its problems as well. One problem was noted by Einstein himself in the early days of the theory. If a region of space-time is devoid of nongravitational mass-energy, alternative solutions to the equation of the theory connecting mass-energy with the space-time structure will agree in all regions outside the matterless "hole," but will offer distinct space-time structures within it. This suggests a local version of the old Leibniz arguments against substantivalism. The argument now takes the form of a claim that a substantival reading of the theory forces it into a strong version of indeterminism, since the space-time structure outside the hold fails to fix the structure of space-time in the hole. Einstein's own response to this problem has a very relationistic cast, taking the "real facts" of the world to be intersections of paths of particles and light rays with one another and not the structure of "space-time itself." Needless to say, there are substantivalist attempts to deal with the "hole" argument as well, which try to reconcile a substantival reading of the theory with determinism.

Space-time in science and in experience

Although most of the debate between substantivalists and relationists has hinged on the ability or inability of the relationist to find in his account the structures needed to ground the kinds of "absolute" motions needed for our dynamical theories, it may be that the debate ultimately has deeper roots and must be decided on deeper philosophical grounds.

There are arguments on the part of the relationist to the effect that any substantivalist theory, even one with a distinction between absolute acceleration and mere relative acceleration, can be given a relationistic formulation. These relationisitic reformulations of the standard theories lack the standard theories' ability to explain why non-inertial motion has the features that it does. But the relationist counters by arguing that the explanations forthcoming from the substantivalist account are too "thin" to have genuine explanatory value anyway.

Relationist theories are founded, as are conventionalist theses in the epistemology of space-time, on the desire to restrict ontology to that which is present in experience, this taken to be coincidences of material events at a point. Such relationist-conventionalist accounts suffer, however, from a strong pressure to slide into full-fledged phenomenalism.

As science progresses, our posited physical space-times become more and more remote from the space-time we think of as characterizing immediate experience. This will become even more true as we move from the classical space-times of the relativity theories into fully quantized physical accounts of space-time. There is strong pressure from this growing divergence of the space-time of physics from the space and time of our "immediate experience' to completely dissociate the two and, perhaps, to stop thinking of the space-time of physics as being anything like our ordinary notions of space and time. Whether such a radical dissociation of posited nature from phenomenological experience can be sustained, however, without giving up our grasp entirely on what it is to think of a physical theory "realistically" is an open question.

Further reading

Alexander, H. (ed.) 1956: *The Leibniz–Clarke Correspondence* (Manchester: Manchester University Press).

Earman, J. 1989: *World Enough and Space-Time* (Cambridge, MA: MIT Press).

Friedman, M. 1983: *Foundations of Space-Time Theories* (Princeton: Princeton University Press).

Grünbaum, A. 1973: *Philosophical Problems of Space and Time* (Dordrecht: Reidel).

Nehrlich, G. 1976: *The Shape of Space* (Cambridge: Cambridge University Press).

Newton-Smith, W. H. 1980: *The Structure of Time* (London: Routledge and Kegan Paul).

Sklar, L. 1985a: *Philosophy and Spacetime Physics* (Berkeley: University of California Press).

—— 1985b: *Space, Time, and Spacetime* (Berkeley: University of California Press).

Toretti, R. 1978: *Philosophy of Geometry from Riemann to Poincaré* (Dordrecht: Reidel).

—— 1983: *Relativity and Geometry* (Oxford: Pergamon).

469

68

Statistical Explanation

CHRISTOPHER READ HITCHCOCK AND
WESLEY C. SALMON

Generally speaking, scientific explanation has been a topic of lively discussion in twentieth-century philosophy of science; philosophers of science have endeavored to characterize rigorously a number of different types of explanation to be found in the various fields of scientific research. Given the indispensability of statistical concepts and techniques in virtually every branch of modern science, it is natural to ask whether some scientific explanations are essentially statistical or probabilistic in character. The answer would seem to be yes. For example, we can explain why atoms of carbon 14 have a $\frac{1}{4}$ probability of surviving for 11,460 years because the half-life of that species is 5,730 years. As we shall see, explanations of this type are not especially problematic. As another example, we might explain why a particular weed withered by citing the fact that it received a dose of a herbicide, even though we know that the herbicide is not invariably effective. This means that the withering is related probabilistically to the herbicide treatment but is not necessitated by it. Explanations of this kind, by contrast, lead to severe difficulties.

The standard view

Carl G. Hempel's (1965) is the *locus classicus* for discussions of scientific explanation in the latter part of the twentieth century. This work contains a profound investigation of the existence and nature of statistical explanation (also known as probabilistic explanation). In this essay Hempel distinguished two types of statistical explanation, represented by the deductive-statistical (D-S) and the inductive-statistical (I-S) models respectively. The case of the decay of C^{14} atoms illustrates the D-S model; the case of the withered weed exemplifies the I-S model.

According to Hempel, all legitimate scientific explanations are arguments, either deductive or inductive. The conclusion (the *explanandum*) states the fact to be explained; the premises (collectively, the *explanans*) furnish the explanatory facts. Furthermore, in every case, at least one of the premises in the *explanans* – essential to the validity of the argument – must state a law of nature (see LAWS OF NATURE). This requirement characterizes the covering law conception, which has been the focus of enormous controversy in discussions of scientific explanation.

In his *nonstatistical* deductive-nomological model, Hempel distinguished between explanations of particular facts and explanations of general regularities (see EXPLANATION). Newton, for example, could explain the falling of a particular apple in his

470

garden on the basis of the law of universal gravitation. Using the same general law, he could explain the regular behavior of the tides. The same distinction is required for Hempel's theory of statistical explanation. In the example of the withered weed, we explained the occurrence of withering in one particular instance. In the case of C^{14} decay, we explained a statistical generalization about all C^{14} atoms by deducing it from another statistical law: namely, that any C^{14} atom has a probability of $\frac{1}{2}$ of decaying within a period of 5,730 years (and it has that probability regardless of its age).

There seems, however, to be little need to distinguish deductive-statistical explanations from deductive explanations of universal laws on the basis of universal laws. Although, as Hempel realized, there are problems with D-N explanations of universal generalizations (see Salmon 1990, pp. 9–10), there are no other special difficulties for deductive explanation of statistical generalizations on the basis of statistical laws. Since it is widely agreed that, certain philosophical difficulties notwithstanding, there are genuine scientific explanations of general phenomena by deduction from more fundamental general laws, there is little opposition to the idea that there are genuine deductive explanations of statistical laws on the basis of more basic statistical laws. The special problems of statistical explanation arise in the context of statistical explanations of particular facts. We shall focus mainly on this type of explanation here.

If, along with Hempel and many other philosophers, one regards explanations as arguments, then it is natural to think that inductive-statistical explanations of particular facts are arguments that render the *explanandum* (the fact to be explained) highly probable given the *explanans* (which contains at least one statistical law and some statements of initial conditions). Indeed, an explanation of a particular fact, whether it is D-N or I-S, is an argument to the effect that the fact to be explained was to be expected on the basis of the explanatory facts. This suggests a strong analogy between these two types of explanation, but there is a fundamental point of difference. While D-N explanations of particular facts or general laws and D-S explanations of statistical laws all qualify as deductive arguments, I-S explanations are inductive arguments. The fundamental schema can be represented as follows:

$$(\text{I-S}) \quad \begin{array}{l} Pr(G \,|\, F) = r \\ Fi \\ \overline{\overline{}} \; [r] \\ Gi \end{array}$$

The double line separating the premises from the conclusion in this schema signifies an inductive argument; the bracketed r is the degree of inductive support given to the conclusion by the premises. The first premise is taken to be a statistical law (Hempel, 1965, p. 390).

As Hempel realized, certain basic disanalogies between deductive and inductive arguments raise crucial difficulties for statistical explanation. Deductive logic is *monotonic*; that is, given a valid deductive argument, you may add whatever premises you wish (without taking away any of the original premises), and the argument remains valid. For example, given that

> All men are mortal.
> Socrates is a man.
> ∴ Socrates is mortal.

471

is valid, we can obviously add the premise "Xantippe (wife of Socrates) is a woman," without destroying its validity.

Inductive logic clearly lacks this characteristic of monotonicity. From the fact that Yamamoto is a Japanese man, we may conclude with high probability that he weighs less than 300 pounds; but if we add that he is a sumo wrestler, it becomes highly *im*probable that he weighs less than 300 pounds. By adding a premise to a strong probabilistic argument which concludes that Yamamoto weighs less than 300 pounds, we completely undermine it, with the result that we have a strong probabilistic argument for the opposite conclusion. This is a phenomenon that cannot happen with deductive arguments. If two valid deductive arguments have conclusions that contradict one another, the premises of each of these must contradict the premises of the other. In the inductive case, the premises of the two arguments are completely consistent with one another. This leads to what Hempel called *the ambiguity of inductive-statistical explanation.*

To cope with the problem of ambiguity, Hempel devised a *requirement of maximal specificity*, which demanded the inclusion of all relevant evidence (while excluding information – such as the *explanandum* itself – that would render the explanation circular) (Hempel 1965, p. 400). This requirement involved reference to a given knowledge situation in which an explanation is sought, and therefore led to the *essential epistemic relativization of inductive-statistical explanation*. (See Coffa 1974 for a penetrating discussion of problems arising from epistemic relativization.)

One misgiving often felt regarding I-S explanations is that they are not *bona fide* explanations in their own right but, rather, incomplete D-N explanations. This is not to deny that incomplete explanations are valuable when we cannot construct complete ones. Consider the explanation, given above, of the withering of the weed. Even though we claimed only high probability that a plant treated with the herbicide would wither and die, we may well believe that a careful study would reveal a specifiable subclass of cases in which the plant, so treated, would certainly succumb. This specification would include such factors as the type of plant; its age, size, and health; the precise dose of herbicide; and so on. Our sketchy "explanation" simply failed to take account of the details. Whether our present knowledge is complete enough to provide a deductive explanation is beside the point; it seems in principle possible to acquire such knowledge, and therefore our example was incomplete. At least, it is tempting to think so.

The claim that every I-S explanation can be regarded as an incomplete D-N explanation is tantamount to determinism, the doctrine that events are completely determined by antecedent conditions. Such conditions, if known, could be used to construct D-N explanations. It would, however, be a mistake to assume that the universe is completely deterministic. Modern physics, especially quantum mechanics, strongly suggests just the opposite: namely, that there are events that occur by chance – their outcomes are not completely determined by previous conditions. We referred above to the spontaneous radioactive decay of C^{14} atoms; such events have definite probabilities of occurrence, but there are no certainties. Given two such atoms, one of which decays during a specific period of 5,370 years while the other does not, there are no antecedent physical conditions which determine that the one will decay while the other will not. Each one had a 50–50 chance of decaying during that time; one did, the other did not, and that is the whole story. (It is, of course, possible that both or neither would decay.) Our theory of statistical explanation should at least leave open the possibility that the world is actually indeterministic. In that case, there might be statistical explanations that are complete – not merely explanations that, on account of our ignorance, fail to achieve full D-N status. If, for example, we have a fairly large number of C^{14} atoms at one particular time, we can explain why there are only about half that many in that sample 5,730 years later, because there is a very high probability that this result will occur; but it is only a probability, not a certainty.

Alternative views

Hempel's claim that statistical explanations of particular facts are arguments whose conclusions are highly probable on the basis of the premises has been seriously criticized on two scores. The first criticism focuses on the requirement of high probability, arguing that it is neither sufficient nor necessary for a satisfactory statistical explanation. Consider an example. A patient with a troubling psychological problem undergoes psychotherapy. During treatment, or shortly thereafter, the symptom vanishes. Suppose that there is quite a high probability that this particular type of symptom will disappear when treated by the type of therapy this patient has undergone. One can construct an inductive argument in which the patient's improvement follows with high probability from the premises stating that this particular patient has the type of symptom in question, and that a particular type of therapy was used. Does the psychotherapy explain the patient's improvement? Not necessarily. To answer that question, it is essential to consider the spontaneous remission rate – that is, the probability that the symptom would go away without any sort of treatment whatever. Spontaneous remission rates can be quite high. If the probability of a "cure" with a given kind of treatment is high, but no higher than the spontaneous remission rate, then the psychotherapy does not explain the patient's improvement. This shows that the high probability requirement is not *sufficient* for a satisfactory statistical explanation.

Another example shows that a high probability is not *necessary* for explanatory success. Consider a patient who has an illness that is almost certainly fatal unless a particular operation is performed. The operation is, however, extremely risky; suppose, for example, that it succeeds in only 25 percent of cases. This particular patient undergoes the operation, survives, and returns to good health. We would unhesitatingly say that the operation explains the cure.

What is crucial in both examples is the question of whether there is a *difference* in the probability of recovery given the treatment and the probability of recovery without it. The second example shows what we mean by statistical or probabilistic relevance: an event E is *relevant* if it makes a difference to the probability of an outcome O. In the first example relevance is lacking, and for that reason it does not constitute a legitimate explanation. The second example exhibits an appropriate sort of relevance, and for that reason it qualifies as a genuine explanation. Thus, *statistical or probabilistic relevance* – rather than high probability – is the key explanatory relation. Notice that in the second example we cannot construct any inductive argument in which the recovery is highly probable relative to some set of premises; nevertheless, the explanation is legitimate.

473

The second major criticism of the I-S model, offered by Richard Jeffrey (1969), is that the probability value attached to a given event is no gauge of our degree of understanding of it. Suppose we have some stochastic process that yields one outcome with high probability and another with low probability. A biased coin with probability 0.9 of heads will sometimes yield tails when tossed. Even though tails occurs relatively infrequently, Jeffrey argued, we understand that outcome just as well as we understand the outcome heads. Each outcome is the result of a process that results in heads most of the time and in tails less often; there is no ground for the supposition that we understand why heads comes up better than we understand the outcome tails.

It was Jeffrey who first expressed the preceding criticism and, in connection with it, drew the conclusion that statistical explanations *need not be* arguments; it was Salmon who first argued that statistical relevance, rather than high probability, is the key explanatory relation, and drew the stronger conclusion that statistical explanations *are not* arguments. A new model of scientific explanation – the statistical-relevance (S-R) model – was introduced and was contrasted with Hempel's I-S model (Salmon et al. 1971, p. 11). According to this model, an explanation is an *assembly of facts statistically relevant* to the *explanandum*, along with the associated probability values, regardless of whether the degree of probability of the *explanandum* under the given conditions is high, middling, or low.

The S-R model led to a result that many philosophers found shocking – namely, that the very same type of circumstance C could sometimes explain one outcome and on other occasions explain just the opposite. Recall the heavily biased coin and assume that the process is genuinely indeterministic; in that case the circumstances of the flip sometimes explain heads and sometimes explain tails. In response to this result, many philosophers have argued that *there are no statistical explanations of particular events*; the best we can do is to explain why a given type of event has a particular probability of occurring. An explanation of this sort would be deductive, and it would explain a probability value; for example, from the laws of quantum mechanics we could presumably calculate the probability that an atom of C^{14} will decay during a specified period of time. It should be carefully noted that such an explanation does not explain the decay of any C^{14} atom, for the atom has a certain probability of disintegrating whether it decays or remains intact.

Peter Railton (1978) attempted to escape the defects of Hempel's I-S model by offering a deductive-nomological account of probabilistic explanation of particular facts (D-N-P model). "What I take to be the two most bothersome features of I-S arguments as models for statistical explanation – the requirement of high probability and the explicit relativization to our present epistemic situation . . . – derive from the inductive character of such inferences, not from the nature of statistical explanation itself," he writes (pp. 211–12). He escapes the problems associated with inductive inferences by offering an explanatory *account* that is *not* an argument. The example he chooses is the quantum-mechanical explanation of spontaneous alpha decay of a particular nucleus u of U^{238} during a specified time span (say, an hour), an extraordinarily improbable event given that the half-life of U^{238} is 4,500 million years. The first stage of a D-N-P explanation is a deduction (1) of a probabilistic law (2) from our fundamental theory; we then add the premise (3) that u is an atom of U^{238}. We then *deduce* (4) the probability of decay of u during the hour just passed; for this argument to be valid, the probability in (2) must be constructed as a *single case* probability (see PROBABILITY). If that were the whole story, we would have explained, not the decay of *u*, but the probability of its decay. Railton's account contains one more item, a parenthetic addendum (5) stating that *u* did decay. Here is a simplified version of the schema:

(D-N-P) (1) Derivation of (2) from theoretical principles
$\qquad$ (2) $(x)\ (Gx \supset Pr(Fx) = r)$
$\qquad$ (3) Gu

$\qquad$ (4) $Pr(Fu) = r$
$\qquad$ (5) u is, in fact, F

The result is that a complete D-N-P explanation is not an argument, though it contains deductive arguments as essential components. It escapes Hempel's problem of ambiguity of I-S explanation, because this model contains no inductive arguments – only deductive. In this way it escapes the problem of epistemic relativization, because, like Hempel's D-N model, it requires that the statements that constitute the explanation be true (not just believed to be true).

Railton was not the first philosopher to employ single case probabilities in statistical explanations. James H. Fetzer (1974) first introduced a single case propensity interpretation into the discussion and offered an account of statistical explanation in which this interpretation played a crucial role. According to this approach, each trial on a chance setup has a particular disposition to yield a given outcome; for example, each flip of the above-mentioned biased coin has a disposition of 0.9 to yield heads (whether or not heads comes up on that trial). Interpreting universal laws as representations of universal dispositions, he draws a strong parallel between D-N and I-S explanations of particular occurrences. Formulating a revised version of Hempel's requirement of maximal specificity, he applies it to D-N as well as I-S explanations. Like Hempel (and in contrast to Jeffrey, Railton, and Salmon) he regards statistical explanations as inductive arguments, but his revised form of maximal specificity takes care of the problems of statistical relevance that gave rise to the S-R model. He also relinquishes the high probability requirement.

An interesting feature of Fetzer's account is that he regards his probabilistic dispositions as causal. Other authors (e.g., Salmon 1984, 1998) have claimed that an appeal to causality is an important component of many scientific explanations, although Hempel was not among them. Salmon explicitly recognized the basic role of causal relations in statistical as well as other kinds of explanations (Salmon et al. 1971). Although he did not try to offer an explication of causality in strictly statistical terms, he thought it would be possible to do so; but he no longer retains that hope. Nevertheless, if causality is to play a basic role in statistical explanation, we need an account of causality that is compatible with indeterminism.

Probabilistic causality

Traditional accounts of causation have taken causes to be necessary and/or sufficient conditions for their effects (see CAUSATION). (Hume proposed the most famous example (see HUME).) However, the same pressures that have led philosophers to countenance statistical explanations have led others to formulate probabilistic theories of causation. The central idea behind such theories is that causes *raise the probabilities of their effects*. Thus, smoking is a cause of lung cancer, not because all or only smokers develop lung cancer, but because smokers are *more likely* to suffer from lung cancer than are non-smokers; that is, smoking is *positively relevant* to lung cancer.

The determinist, who argued that I-S explanations are nothing more than incomplete D-N explanations, is likely to repeat her argument here. To characterize an individual as a smoker or a nonsmoker is to provide only a partial description, whereas it is a rich constellation of factors that determines whether an individual will suffer from lung cancer. At this point in time, however, belief in the existence of constellations of factors that are causally sufficient for lung cancer (or its absence) can be nothing more than an article of faith. In particular, the sort of statistical evidence that has led researchers to conclude that smoking causes lung cancer seems to have little bearing on the existence of such constellations. By contrast, this statistical evidence provides direct confirmation of the claim that smoking *raises the probability* of lung cancer.

There is, however, some justification for the determinist's concern with complexes of factors. In the nineteenth century, it was believed that malaria was caused by the "bad air" produced in swamps, and malaria is indeed more prevalent in regions where the air is bad, so bad air raises the probability of malaria. But we now know that the disease is caused by mosquitoes bearing the *Plasmodium* virus, and not by bad air at all. In the presence (or absence) of virus-bearing mosquitoes, breathing bad air does *not* increase the probability of malaria; bad air is *screened off* from malaria by the presence of the mosquitoes. By contrast, the presence of mosquitoes is *not* screened off from malaria by bad air. Both Reichenbach (1956) and Suppes (1970) explicitly required that causes not be screened off from their effects by other factors. A recent account of probabilistic causality (Eells 1991) modifies this requirement by evaluating probabilistic relevance relative to causally homogeneous background contexts. C is the cause of E when the conditional probability $Pr(E \mid C.B)$ is greater than the conditional probability $Pr(E \mid \sim C.B)$ for a range of homogeneous causal backgrounds B. C prevents E, or is a negative cause of E, if this inequality is reversed. (See PROBABILITY for the definition of conditional probability.)

We will mention briefly two issues that are currently being debated by researchers in this area. First, how are the homogeneous causal backgrounds to be characterized? In particular, can they be specified without reference to some sort of causal relation? If not, then the probabilistic theory cannot provide a reductive *analysis* of causation, but can only impose constraints upon the antecedently given causal relation. Second, must causes raise the probabilities of their effects in *all* causally homogeneous backgrounds? Eells (1991) argues that an affirmative answer is necessary if the theory is to enable us to make precise causal claims. Unfortunately, an affirmative answer would also seem to render many of our ordinary causal judgments false. For example, most people would continue to accept the claim that high cholesterol causes heart disease, even though there is evidence for a rare gene which protects its bearers from the dangers of high cholesterol.

Since probabilistic theories of causation characterize causation in terms of probabilistic relevance relations, they would seem to have a clear affinity with statistical theories of explanation. Surprisingly, the connection has not been well explored. An important exception to this generalization is Humphreys's (1989) theory of "aleatory explanation." (Fetzer (1974) is another exception, although his account of probabilistic causation is quite different from that sketched above.) An aleatory explanation provides a partial list of probabilistic causes (both positive and negative) of the phenomenon to be explained. An interesting feature of these explanations is that, unlike Railton's D-N-P explanations and Salmon's S-R explanations, they do not cite the probability of the *explanandum*. Humphreys argues that to require that an explanation cite the probability of the *explanandum* is to require that it include *all* factors that are statistically relevant to the *explanandum*, for the omission of any relevant factor will result in an *explanans* that confers the wrong probability upon the *explanandum* (barring fortuitous cancelation). This completeness requirement seems too strict; Humphreys's account allows for statistical explanations that are partial, but nonetheless correct.

In Humphreys's account, the theory of probabilistic causation is primary, and statistical explanation is then characterized in terms of causation. Another approach, advocated by Hitchcock (1993), is to reverse the order of precedence. Explanation would then be characterized in terms of statistical-relevance relations, much as in Salmon et al. (1971), and causal terminology would emerge as a useful tool for presenting encapsulated information about the statistical-relevance relations.

Common cause explanation

One type of causal explanation that appears frequently both in science and in everyday life is the *common cause explanation*. This pattern is used, not to explain particular events, but to explain correlations between phenomena. Such correlations are often explained by citing a common cause. Suppose there is a geyser that spouts rather irregularly, and that another, not too distant from the first, also spouts irregularly. However, for the most part, they spout more or less simultaneously. We do not consider the correlation a chance coincidence; we attribute it to a subterranean connection – a common aquifer that feeds both.

Within probability theory it is possible to give a precise characterization to the intuitive notion that two phenomena A and B are correlated; that is, they are correlated if $Pr(A.B) > Pr(A) \, Pr(B)$. If $Pr(A.B) = Pr(A) \, Pr(B)$, then A and B are said to be independent. If A and B are correlated, then A is more likely to occur when B is present, and vice versa. According to Reichenbach (1956, §19), common cause explanations fit the following schema:

$$(CC) \quad 1. \quad Pr(A \mid C) > Pr(A \mid {\sim}C)$$
$$2. \quad Pr(B \mid C) > Pr(B \mid {\sim}C)$$
$$3. \quad Pr(A{\wedge}B \mid C) = Pr(A \mid C) \, Pr(B \mid C)$$
$$4. \quad Pr(A{\wedge}B \mid {\sim}C) = Pr(A \mid {\sim}C) \, Pr(B \mid {\sim}C)$$

Together, these conditions entail that $Pr(A.B) > Pr(A) \, Pr(B)$ (on the assumption that none of the probabilities is zero or one). The first two conditions state, in effect, that C is a probabilistic cause of both A and B – hence it is a common cause. But C is more; by (3) and (4) it renders A and B probabilistically independent of one another; once it has been determined that C either has or has not occurred, there is no residual correlation between A and B. C explains the correlation between A and B in the following sense: sometimes C occurs, in which case both A and B are relatively likely; sometimes C does not occur, in which case they are relatively unlikely. A and B are correlated, because they both have a greater tendency to occur at those times when C is present. Note that a common cause explanation of the correlation between A and B does not merely cite a probabilistic cause of the conjunction $A.B$. C may be a probabilistic cause of the conjunction $A.B$ (according to the theory sketched above) without satisfying *any* of conditions (1)–(4). Thus common cause explanations form an independent class of statistical explanations.

The schema for common cause explanations was elevated by Reichenbach to the status of a principle, aptly named the *Common Cause Principle*. This principle states that when there is a probabilistic correlation between two phenomena (and neither is the cause of the other), then the two phenomena have a common cause satisfying conditions (1)–(4). As plausible as this principle may seem in the realm of macroscopic objects, it is belied by certain phenomena in quantum mechanics (see QUANTUM MECHANICS). These distant correlation phenomena are similar in spirit to a thought experiment described by Einstein, Podolsky, and Rosen (1935) (see EINSTEIN). Pairs of microscopic particles are prepared in the so-called *singlet state*. The particles are then fired at a pair of detectors labeled "left" and "right." Each detector measures the spin (along some particular axis) of its respective particle; the measurement will yield one of two possible results – "plus" or "minus." For each detector, the two results "plus" and "minus" occur with equal frequency; yet, whenever the left detector registers "plus," the right detector registers "minus," and vice versa. There is, therefore, a correlation between the results of the measurements on the two particles. The detectors can be positioned sufficiently far apart to preclude any causal connection between the measurement outcomes. According to the Common Cause Principle, then, we should look for some common cause, perhaps in the preparation of the particles. It would seem plausible to suppose, for example, that the particle pairs are prepared in two different states: "left-plus/right-minus" and "left-minus/right-plus." Theorems proved by John Bell and others, however, have shown that no such common cause explanation is possible.

Indeed, distant correlations and other phenomena within the realm of the quantum raise important questions about the nature and even the possibility of statistical explanation. No genuinely causal explanations of these phenomena seem possible, and for many, this is tantamount to saying that there is no explanation at all. Some (e.g., Fine 1989) have argued that quantum correlations simply do not demand explanation. Others (e.g., Hughes 1989) have argued that their explanation is to be found in the nonclassical structure of the probability spaces that represent such phenomena. If this is right, then we have a statistical explanation that is radically different from any of those characterized above (see QUANTUM MACHANICS).

The success of quantum mechanics in the twentieth century provides strong evidence that the world we inhabit is not a deterministic one, and that if our world is to admit of scientific understanding, we must countenance the possibility of explanations that are irreducibly statistical. It is ironic, then, that quantum mechanics poses the greatest challenge currently facing the theory of statistical explanation.

References

Coffa, J. A. 1974: Hempel's ambiguity. *Synthese*, 28, 145–64.

Eells, E. 1991: *Probabilistic Causality* (Cambridge: Cambridge University Press).

Einstein, A., Podolsky, B., and Rosen, N. 1935: Can quantum-mechanical description of physical reality be considered complete? *Physical Review*, 47, 777–80.

Fetzer, J. H. 1974: A single case propensity theory of explanation. *Synthese*, 28, 171–98.

Fine, A. 1989: Do correlations need to be explained? In *Philosophical Consequences of Quantum Theory: Reflections on Bell's Theorem*, ed. J. T. Cushing and E. McMullin (Notre Dame, IN: University of Notre Dame Press), 175–94.

Hempel, C. G. 1965: Aspects of scientific explanation. In *Aspects of Scientific Explanation and Other Essays in the Philosophy of Science* (New York: Free Press), 331–496.

Hitchcock, C. R. 1993: A generalized probabilistic theory of causal relevance. *Synthese*, 97, 335–64.

Hughes, R. I. G. 1989: Bell's theorem, ideology, and structural explanation. In *Philosophical Consequences of Quantum Theory: Reflections on Bell's Theorem*, ed. J. T. Cushing and E. McMullin (Notre Dame, IN: University of Notre Dame Press), 195–207.

Humphreys, P. 1989: *The Chances of Explanation: Causal Explanations in the Social, Medical, and Physical Sciences* (Princeton: Princeton University Press).

Jeffrey, R. C. 1969: Statistical explanation vs. statistical inference. In *Essays in Honor of Carl G. Hempel*, ed. Nicholas Rescher (Dordrecht: D. Reidel), 104–13; repr. in Salmon et al. 1971, 19–28.

Railton, P. 1978: A deductive-nomological model of probabilistic explanation. *Philosophy of Science*, 45, 206–26.

Reichenbach, H. 1956: *The Direction of Time* (Berkeley and Los Angeles: University of California Press).

Salmon, W. C. 1984: *Scientific Explanation and the Causal Structure of the World* (Princeton: Princeton University Press).

—— 1990: *Four Decades of Scientific Explanation* (Minneapolis: University of Minnesota Press). (A comprehensive survey of the period 1948–87.)

—— 1998: *Causality and Explanation* (New York and Oxford: Oxford University Press).

Salmon, W. C., et al. 1971: *Statistical Explanation and Statistical Relevance* (Pittsburgh: University of Pittsburgh Press).

Suppes, P. 1970: *A Probabilistic Theory of Causality* (Amsterdam: North-Holland).

69

Supervenience and Determination

WILLIAM SEAGER

In the mid-part of the twentieth century, the union of youthful science and the ancient philosophical dream of metaphysical completion begot a visionary doctrine known as the unity of science (see UNITY OF SCIENCE). This view of the relationship among scientific theories maintained that any theory aspiring to be truly "scientific" must fit into a hierarchy in which every theory was *reducible* to the theory immediately below it, save for the foundational theory of physics. Reduction would be accomplished by establishing relations of coextension between the predicates of the reduced theory, T_1, and the reducing theory, T_2, sufficient to allow the mathematical deduction of the laws of T_1 expressed entirely in terms drawn from T_2 (see REDUCTIONISM). The classic example is the reduction of phenomenological to statistical thermodynamics. The example suffers from the apparently serious defects of not meeting the conditions laid down by the unity of science and of being pretty much unique in achieving the degree of approximation to these conditions which it enjoys. It has become clear in the later stages of the century that despite the rich and complex interrelationships that prevail among scientific theories, there is little or no prospect of even roughly fulfilling the dream of the grand unification of all theories into a complete hierarchy of reduction. (One philosopher of science, Ian Hacking, bluntly assesses unity of science to be just another of philosophy's "idle pipedreams.")

Still, the metaphysical dream of completion remains a strong motivator. Perhaps there is an alternative to the reductionism of unity of science which can yet satisfy philosophical desire? The notion of *supervenience* (equivalently, *determination*) has recently emerged in attempts to give an affirmative answer to this question.

Strange to say, the original home of the idea is in the realms of *value*: aesthetics and ethics. The idea was first broached by G. E. Moore in his famous *Principia Ethica* (though not under the title of "supervenience") and was formally introduced by R. M. Hare in *The Language of Morals* (1952). An example used by Hare makes the basic idea clear. Suppose that we possess two oil paintings that are physically qualitatively indistinguishable. Could we nonetheless say that they are aesthetically distinguishable? If not, we say that aesthetic qualities *supervene* upon physical qualities. To speak more abstractly, one domain, α, supervenes on a second domain, β iff any change in α requires a change in β (in our example, if aesthetic quality supervenes on physical state, then some physical change must be made to ground any change in aesthetic quality).

The intricacy of the idea of supervenience has been vigorously explored in recent work, and certain varieties of supervenience deserve note. A primary distinction marks off *global* from *local* supervenience. The former is the weaker notion, maintaining that α supervenes globally upon β iff no two possible worlds that are indistinguishable with respect to β properties are distinguishable with respect to α properties. The latter notion applies supervenience to individuals: α supervenes locally upon β iff any two β-indistinguishable *objects* must be α-indistinguishable. The difference is illuminated by the property of being an uncle, which supervenes upon the physical, but not locally. I can become an uncle without physical change, yet, of course, the *world* must change physically. By contrast, the property of being alive supervenes locally, for I cannot go from life to death without suffering an internal physical change. Within the category of local supervenience, one can distinguish *weak local* from *strong local* supervenience. Weak supervenience requires only that the supervening domain posses some subvening base in each possible world, whereas strong supervenience requires the very same subvening base across worlds (as example of weak supervenience is that of linguistic truth (at a time) upon sentencehood). Varieties of supervenience are also distinguished by their "modal force" – that is, by whether they hold by logical, metaphysical, or mere physical necessity.

The differences between supervenience and reduction are suggestive. The relation of supervenience is not, in the first instance, a relation between *theories*, but rather between ontological domains. Thus one can maintain, for example, that the chemical supervenes upon the physical without espousing any doctrine of reduction concerning the corresponding theories. On the other hand, where we actually possess a reduction of one theory to another, supervenience of the associated reduced domain upon the reducing one will automatically follow. Supervenience is neutral about reductionism. The grounds for maintaining a claim of supervenience are generally quite distinct from those supporting a claim of reduction. The latter involve a "meta-enterprise" primarily concerned with issues of the syntactical structure of the relevant theories. The former essentially involve empirical data about the workings of an ontological domain. For example, the grounds for maintaining the supervenience of the chemical upon the physical stem from our understanding of how chemical processes actually proceed and how chemical kinds are assembled from basic *physical* parts.

The advantages of replacing a reductionist understanding of the scientific enterprise with one based upon supervenience are many. The supervenience-based approach espouses a judicious naturalism, which allows that the physical is ontologically basic, but does not implausibly demand that this primacy be directly reflected in our theories. This sophistication permits other virtues to emerge: maintenance of a supervenience thesis between two domains can coexist with the recognition of mystery and ignorance. We can have grounds for believing in supervenience even when we cannot see any secure links between the theories involved (e.g., consciousness appears to supervene on the physical, but we have no clue as to how this works). Different theories may embody quite radically distinct systems of description or classification (an instance might be the essential appeal to *rationality* in psychology). While this would preclude reduction, it does not prevent the associated domains from entering the supervenience relation. The idea of supervenience makes us recognize that the old model of reductionism missed the point by putting forward a syntactical doctrine (about theory structure), when what was required was a semantical doctrine (about the nature of the objects to which theories are applied). Finally, if we employ the notion of supervenience, we can happily accept the full autonomy of separate scientific disciplines without fear of a corresponding ontological fracture.

Supervenience has its vices as well as its virtues, and perhaps the primary one is the encouragement of indolence. The doctrine of reductionism at least enjoined constant effort to bring theories into proper interrelation. This in turn brought an increase in explanatory power as the deeper theories became applicable to new, higher-level domains. And the virtue of recognizing the autonomy of disciplines can easily slip into vice if the autonomy ends up leaving a discipline too distant from the scientific mainstream (so, a rationality-based psychology may lack connection with physiology, despite an accepted supervenience – surely a result that would leave psychology perilously isolated and could threaten its status as a *science*).

Nonetheless, the deployment of supervenience has been, and promises to remain, a liberating movement, bringing us closer to the empirical phenomena which are science's true concern and forming in us a deeper appreciation of the complexity of the natural world and the multitudes of legitimate scientific approaches to it.

References and further reading

Beckermann, A., Flohr, H., and Kim, J. (eds) 1992: *Emergence or Reduction: Essays on the Prospects of Nonreductive Physicalism* (Berlin: de Gruyter).

Blackburn, S. 1985: Supervenience revisited. In *Exercises in Analysis: Essays by Students of Casimir Lewy*, ed. Ian Hacking (Cambridge: Cambridge University Press), 47–67.

Davidson, D. 1970: Mental events. In *Experience and Theory*, ed. L. Foster and J. Swanson (Amherst, MA: University of Massachusetts Press); repr. in Davidson, *Essays on Actions and Events* (Oxford: Oxford University Press, 1980).

Hare, R. M. 1952: *The Language of Morals* (Oxford: Clarendon Press).

Haugeland, J. 1982: Weak supervenience. *American Philosophical Quarterly*, 19, 93–103.

Hellman, G., and Thompson, F. 1977: Physicalism: ontology, determination and reduction. *Journal of Philosophy*, 72, 551–64.

Horgan, T. 1982: Supervenience and micro-physics. *Pacific Philosophical Quarterly*, 63, 27–43.

Kim, J. 1978: Supervenience and nomological incommensurables. *American Philosophical Quarterly*, 15, 149–56.

—— 1984: Concepts of supervenience. *Philosophy and Phenomenological Research*, 45, 153–76.

—— 1989: The myth of nonreductive materialism. *Proceedings of the American Philosophical Association*, 63, 31–47.

—— 1993: *Supervenience and Mind* (Cambridge: Cambridge University Press).

Moore, G. E. 1903: *Principia Ethica* (Cambridge: Cambridge University Press).

Seager, W. 1992: *Metaphysics of Consciousness* (London: Routledge), esp. chs 1 and 4.

70

Technology, Philosophy of

MARY TILES

Philosophy of technology is a relatively new philosophical subdiscipline, and some would argue that it does not even now have that status. Reasons for philosophy's tendency to ignore technology will be considered below; but first it may be instructive to see why it has been difficult to stake out a territory for "philosophy of technology."

Analytic philosophy has a strong tradition in philosophy of science, whose concern has been mostly with the nature of scientific knowledge, the kind of rational justification that can be afforded for scientific claims and for making choices between competing theories, and with the mechanisms of scientific change (or progress). Since, in the public mind, science and technology have become inextricably linked, it was not unnatural that philosophers of science should think of extending their domain of inquiry to include technology. (See, e.g., Laudan 1984.) There are nonphilosophical reasons why this kind of research should be given support in the form of research funding. Many governments and large multinational companies have an interest in determining what are the methods by which technological advances are made. Science policy units have been established with the aim of trying to determine what scientific and technological research should be funded in order to keep up with, and hopefully steal a march on, industrial and military competitors. New, "advanced" technology is seen as the key to industrial competitiveness and military supremacy, and hence to national prosperity and security. At least some philosophers of science have suggested that there are methods by which scientific progress can be secured. Would a philosophical study of technology yield a theoretical understanding of the processes of technological change, one which could be used to provide a rational basis for policy decisions aimed at securing technological dominance?

At first sight it might seem that sophisticated technology is the product of applying scientific insights in practical, problem-solving contexts. In this case the answer is simple: technological advance depends on, and will flow from, scientific research. Indeed, generous government and company funding for fundamental scientific research has in the past been based on this premise. But, as budgets get tight and as fundamental scientific research and development get more expensive, questions are asked about the effectiveness of such a strategy. Historical studies of past technological developments suggest that there is no consistent pattern (for discussion see, e.g., Volti 1992, pp. 56–7; Gibbons 1984a; Webster 1991; Mowery and Rosenberg 1989). Although some technologies have arisen out of fundamental scientific work, many did not, and even in those cases where the path from pure theory to practical outcome was most direct, as in the case of the development of the atomic bomb in the Manhattan

Project, much more than pure science was involved (Gowing and Arnold 1979). In other cases, such as the steam engine, widespread adoption of a new technology preceded and seems to have spurred scientific research in the direction of developing the theory of thermodynamics (Kuhn 1959; Musson and Robinson 1969, ch. 12).

In short, technical inventions, like the first occurrences of scientific ideas, may be the product of all sorts of circumstances. It was for this reason that philosophers of science did not focus on how scientists come by their ideas, but on the justifications that can be offered for accepting or rejecting those ideas once they arise. What, in the case of technology, would be the counterparts of acceptance, rejection, and justification? Possible candidates would be adoption (widespread use), non-adoption, and the justifications offered for making technological choices – the rational choice being that technology which best performs the desired task or best solves the practical problem (this being the analogue of truth, or best explanation). A significant disanalogy is that whereas theories are presumed to be true or false, technologies can often fall between being widely adopted and totally ignored.

This disanalogy is indicative of further differences between the two contexts. Analytic philosophers of science have assumed that scientific theories are assessed in terms of either their truth or their predictive success as an indicator of truth, and that there are unambiguous empirical standards which form the basis of such assessments (see THEORIES). But it is quite clear that technology choices are never made by determining which device performs a task best in a narrowly technical sense of "best." The "best" chosen is relative to all sorts of constraints other than those of merely performing a certain technical task as well as possible. This is another way of saying that the problems which technology is introduced to solve are never purely technical problems. Technologies, to be widely adopted, have to be mass-manufactured, marketed, and purchased. Adoption thus depends in the end on manufacturing and on actual or projected purchasing decisions. It is well known that purchasing decisions, even at their most rational, are the result of trade-offs between many factors: cost, appearance, size, operating costs, availability of parts, after sales service, material inputs required and their availability, etc. Frequently, as advertisers know, and help to insure, purchasing decisions are less than rational in this narrow utility-maximizing sense.

This means that technological change, in the sense of change in widely adopted technology, whether military or civilian, cannot be understood by looking simply at technical problem solving and its methods, those practical-cognitive contexts of engineering research and development which are most closely analogous to theoretical science. Instead, it requires understanding the complex economic, political, sociocultural, and infrastructural contexts of technological decision making (see, e.g., Mowery and Rosenberg 1989, ch. 11). For this reason, science and technology studies have tended to become science, technology, and society interdisciplinary programs, requiring input from historians, economists, sociologists, engineers, and even occasionally philosophers. From this point of view it would seem more pertinent to ask what philosophy has to contribute to STS programs, rather than to continue to try to carve out an autonomous domain called "philosophy of technology." Latour (1987) suggests that philosophy has nothing to contribute at present. He proposes a methodology, based on methods drawn from sociology, for investigating techno-science in its social contexts, presenting powerful and challenging analyses.

On the other hand, it could be argued that there is already a "philosophy of technology" (a theoretical view and set of attitudes towards technology) implicit in the long-standing philosophical tradition of ignoring technology. This omission is not a mere oversight, but a consequence of not regarding technology as a topic relevant to the concerns of a philosopher (in this respect, it is much like gender, and the human body). When introducing a discussion of artificial intelligence, Steve Woolgar made the following, perceptive remark: "Discussions about technology – its capacity, what it can and cannot do, what it should and should not do – are the reverse side of the coin to debates on the capacity, ability, and moral entitlements of humans" (Woolgar 1987, p. 312). If this is correct, there are views of technology implicit in debates about human beings and in the philosophical practice of separating these from discussions of technology. To the extent that philosophical debates about what it is to be human are being thrown open, technology should be playing a role in the discourse at the heart of philosophy, not in a peripheral subdiscipline. Technology may even be contributing to the opening of these debates, as its role in the problems posed in medical and legal ethics becomes more prominent. If there is a view of technology implicit in a long tradition of philosophical practice, it will be very difficult for that tradition to explicitly incorporate critical philosophical debate concerning technology. This would explain why philosophy of technology has not achieved anything like the status of philosophy of science, even though technology figures more prominently, and often more problematically, in the lives of most people than pure research science. It also indicates why the definition of "technology" is itself highly contested and is indeed one of the fundamental debates for the philosophy of technology. If this is the reverse side of debates about human nature, any definition of technology carries with it implications for conceptions of human nature.

Mesthene (1969), for example, says that "Whether modern technology and its effects constitute a subject deserving of special attention is largely a matter of how technology is defined" (p. 73) and Ihde (1993, p. 47) reminds us that definitions are not neutral. Some definitions that have been offered are: (1) "the organisation of knowledge for practical purposes" (Mesthene 1969, p. 74); (2) "systems humans create to carry out tasks they otherwise could not accomplish" (Kline and Kash 1992); (3) "a system based on the application of knowledge, manifested in physical objects and organisational forms, for the attainment of specific goals" (Volti 1992, p. 6); (4) "systems of rationalised control over large groups of men, events and machines by small groups of technically skilled men operating through an organised hierarchy" (McDermott 1969, p. 95); (5) "forms of life in which humans and inanimate objects are linked in various kinds of relationships" (Winner 1991, p. 61); (6) "the totality of methods rationally arrived at and having absolute efficiency (for a given stage of development) in every field of human activity" (Ellul 1964, p. xxv); (7) "a social construct and a social practice" (Stamp 1989, p. 1).

To introduce some order into this prolific field of definitions, it may be useful to adopt and adapt a categorization proposed by Feenburg (1991). He suggests that most established theories of technology belong to one of two major types; they are

485

either *instrumental theories* or *substantive theories*. He himself advocates a third style of approach which he calls a *critical theory* of technology.

Instrumental theories are based on the commonsense idea that technologies consist of tools, designed by potential users and available to be used by them and others to suit their purposes. In other words, technologies are artifacts which do not themselves embody any values, but are subservient to the values established in other spheres (politics, culture, religion, economics). Instrumental theories thus tend to see technology either as consisting merely of artifacts and devices (tools and machines), or more broadly as applied science, and they define it accordingly. Thus definitions (1) and (2) above incorporate this kind of view.

Substantive theories, on the other hand, attribute an autonomous cultural force to technologies, a force which overrides all traditional or competing values. They claim that what the very employment of technology does to humanity and nature is more significant than its ostensible goals. As Marcuse (1964, p. 131) argues: "The liberating force of technology – the instrumentalisation of things – turns into a fetter of liberation; the instrumentalisation of man." Ellul's definition (6) was designed to capture this vision of technology.

Critical theories don't see technologies as instruments or as autonomous technical systems, but as nonneutral social constructs, because social values enter into the design, and not merely the use, of technical systems (see definitions (4) and (7) above). Technological development is thus viewed as an ambivalent political process. "Technology is not a destiny but a scene of struggle. It is a social battlefield . . . on which civilizational alternatives are debated and decided" (Feenberg 1991, p. 14).

One advocate of an instrumental view is Mesthene. As he sees it, the primary effect of new technology is to create new possibilities. New value issues arise because there are more possibilities to choose from. There are choices to be made which did not need to be made before (such as whether to abort a genetically defective fetus). However, technology itself, as creator of possibilities, in value-neutral. This value neutrality is just a special case of the general neutrality of instrumental means, which, being only contingently related to the ends they serve, are also independent of the values implicit in the adoption of such ends. This is the reverse side of the view of human beings as having the autonomy to determine ends to be achieved in the light of values they adopt and as rational beings capable not only of making rational choices about means for achieving their adopted ends but also of devising new means for achieving them. The contingency of the means–end relation mirrors the autonomy of reason and desire vis-à-vis purely material determinations; that is, it is a reflection of the conception of man as endowed with reason and the capacity (freedom) to use this to direct action.

Already in Plato and Aristotle we find clearly expressed the idea that it is reason which marks humans off from animals. For humans to fully realize their human potential is thus for them to rise above their merely animal nature by cultivating and employing their rational capacities, allowing these, rather than animal instincts and appetites, to direct their actions. This means that only those who have the time (the leisure) to cultivate their rational faculties, freed from concern with the material necessities of biological life, can live truly fulfilled human lives. Practical work, whether of the craftsman or the agricultural laborer, is thus devalued; it is something that humans need to be freed from if they are to become fulfilled beings. Moreover, it is activity whose function and direction should ideally be the product of knowledge, of understanding the nature of the goals to be achieved as well as the means to achieve them. The bridle maker (weapons manufacturer) needs to receive design specifications from those who will use it (armed forces personnel), and these in turn are directed by generals (military strategists), who determine the role of cavalry (bombs and artillery). Generals in turn look to statesmen for their military objectives. In a well-ordered state the means–end hierarchy maps onto a social-political hierarchy, an authority structure, in which direction flows from those who are qualified, in virtue of their theoretical and practical wisdom, to deliberate about ends and the best means of achieving them, to the craftsmen and laborers who must bring them about (Plato, *Republic* 601c; Aristotle, *Nicomachean Ethics*, bk 1, 1094a10–15).

For the Greeks, the freedom of some to lead fulfilled human lives was contingent upon the labor of others (slaves and females) providing for the material necessities of life (production and reproduction). Marx and Engles dreamed of the possibility of overcoming the need for this division of labor through the development of industrial technology. Technology, provided it belonged to and was managed by the whole community for the communal good, was envisioned as replacing slaves, freeing people from the necessity of labor, and so making available to all the possibility of a fulfilled human life (Marx and Engels 1970, p. 61). (This is a dream already outlined, in somewhat different technological terms, by Francis Bacon in *New Atlantis*.) In other words, both the program of modern science and Marxist revolutionary politics are founded on an instrumental view of technology and on a vision of science as that which delivers the rational tools for controlling nature and freeing humans from enslavement to it. It is an instrumental vision founded on a separation of the distinctively human from the natural, and hence on the conception that humans can realize their full potential only when freed from the practical demands of the work and labor necessary to insure their biological well-being. The fact that this view of technology has transcended the political divide of the cold war years has lent credibility to the view that technology is value-neutral – it seems to be neutral between the very different value frameworks of democratic individualism with free-market capitalism and totalitarianism with state capitalism.

On this view, the way to advance technology is to advance science. Technology is applied science: that is, the application, via rational problem-solving techniques, of rationally acquired understanding to material situations to achieve freely chosen ends. The assumed independence of both science and material problem situations from social determination gives a double-edged objectivity to technology. It is a product of rational, acquired, universal knowledge of the laws of nature, laws which hold no matter where one is in the universe. This knowledge is applied to material situations and could be similarly applied, with similar results, to such situations wherever they arise. Success or failure is evident; either the goal is or is not achieved. Technological progress consists in making more possibilities available (being able to do more things), so that more desired ends can be achieved and can be achieved more efficiently.

It is this scheme which is implicit in the philosophical tradition, and which also underlies the decision-making practices of many contemporary institutions. Stamp (1989) illustrates this in the case of development agencies. A vision of development is founded on the belief that lack of development is a result merely of lack of financial resources to acquire available technology. This is to assume that a machine or process which works in one place will work when transferred to another. Development aid then takes the form of financing technology transfer. Stamp also illustrates the fallacies of this approach. Most poignantly, these are demonstrated in the failures of development policies and the consequent human suffering and social disruption. It is precisely with regard to the problems of technology transfer that the limitations of viewing technology in purely instrumental terms have become most evident. Technologies, by their very specification, are introduced not into purely material contexts, but into social contexts. They are to be used by human beings, to perform tasks previously done in other ways by other means, possibly by other people, or to do wholly new things. Their introduction is bound to have social effects.

Most philosophical writing on technology has for this reason tended to be critical of instrumental theories – hence the perceived need to come up with a better definition of technology, one in which its social embedding is acknowledged. Substantive theories basically concur with the equation of the technical with the rational and instrumental, but they take into account the fact that tools, or instruments, if they are to achieve any ends, have to be used. An instrument rationally designed for a specific purpose must be envisioned as being used in specific ways. Instrument design and use thus carry implications for the working practices and social organization of those who will use them (as well as for those who may be displaced by their use). The instrumental approach to technology has tended to seek to secure intended uses for technology devices basically by treating those who must use them as part of the instrument, part of the natural material which must be dominated and controlled if the desired end is to be realized. In this it is aided by approaches to the human sciences which have modeled themselves on the natural sciences and which, in their quest for laws linking behavior to initial conditions, have treated human beings as objects of study just like any other natural objects. Such laws can be used to predict and manipulate behavior by modifying the conditions in which people are placed.

Substantive theories thus talk about technological systems and technical practices (techniques), rather than about devices. They see these systems as embodying values beyond those which are evident in the selection of the ends intended to be achieved by technological means. The instrumental criterion of "efficiency" masks the presence of these values. If efficiency is a measure of the ratio of costs to benefits, how costs and benefits are counted becomes crucial – costs to whom, benefits to whom, and of what type? The purely instrumental approach, because founded in a bifurcated vision of the world into natural and individual human, tends to overlook the social costs of implementing a technology by not according any reality to the social or to the socially constructed. Substantive theories count social structures as part of reality and thus see technology as acting not only to control nature for the benefit of individual human beings but as acting on the reality which is at once material and social – the environment in which people live and work. Technological systems then become part of this reality to the extent that they constitute this environment by creating and sustaining it.

So substantive theories concur in the identification of technology as a product of the exercise of instrumental rationality devoted to securing material well-being. Their rejection of the instrumentalists' positive endorsement of technology as the vehicle for human progress is founded on an explicit rejection of the conception of human fulfillment as consisting in either the exercise of reason or material satisfaction. Instead, they tend to emphasize other routes to human fulfillment, whether through religion, artistic creativity, or the development of interpersonal relationships. These are the values that they see being overridden by the implementation of technical systems. Because destruction of ways of life needed to sustain and realize these values is not counted as a cost when evaluating technological efficiency, the technology itself cannot be regarded as value-neutral. Its introduction elevates one set of values at the expense of others, not merely at a level of ideological preference but at the real level of making choice of alternative values unavailable. In this sense the values are destroyed, and technology, far from creating human possibilities, destroys them. Technological systems turn human beings into mere natural objects by leaving them no alternative to be anything else. Some proponents of substantive theory, such as Ellul and Heidegger, see wholesale rejection of technology and a return to the primitive, as the only route to preserving the possibilities for distinctively human ways of leading a fulfilling life. Others, such as Winner (1986) and Borgmann (1984) have investigated the possibility of putting limits on technology. Borgmann calls for a "two-sector" economy in which an expanding craft sector takes up the slack in employment from an increasingly automated economic core.

Critical theories question the identification of technology with instrumental rationality. This rejection is founded on a conception of human beings which owes much to Marx's materialism and to existentialist rejections of the idea that humans have a fixed essence. Marx's rejection of idealist philosophy and its inherently dualist conception of man as having a (higher) spiritual, intellectual, or mental aspect, which is problematically and contingently associated with a (lower) physical, material aspect is one route which has paved the way for opening up debates about technology. There is, however, a deep tension in Marx's own work between the instrumentalist and the dualist values inherent in the formulation of his political ideals and his dialectical materialism. The former predicates human freedom on mastery over nature, with freedom from labor achieved via an industrial technology of mass production. The latter sees humans as biological beings who must fulfill their biological needs in order to survive, but who distinguished themselves from other animals once they began to produce their own means of subsistence. Human intellectual development is a product of social organization, the relations formed in order to produce means of subsistence. These relations are in turn conditioned by material circumstances and by the technologies and techniques employed (modes of production). Technology, then, has a crucial role to play in forming a society and its ideology at any given period of history. "Technology reveals the active relation of man to nature, the direct process of the production of his life, and thereby it also lays bare the process of the production of the social relations of his life, and of the mental conceptions which flow from those relations" (Marx 1867, p. 406). Historically determinist forms of Marxism would make this into another variety of substantive theory, but one where there is not even the possibility of rejecting or limiting technology. Changes in technology and society occur in a historically determined sequence independent of the actions of individuals. A more dialectical Marxism, on the other hand, sees each generation as formed by the economic and social structures into which it was born, but as exploiting and modifying what has been handed down to it as it adapts to changing circumstances.

489

Emphasis on the contextual, the particular, and the situational is not, however, confined to Marxism. Existentialists, by focusing on lived experience have also approached the "phenomenon of technology," examining the way it is experienced and the many ways in which it frames human lives. By not presuming a fixed human nature, existentialism explicitly refuses to endorse any universal scheme of values. This makes room for a reevaluation of work which questions the assumption that it is a necessary evil. Arendt (1958), for example, explores the distinction between work and labor, to reveal inherently different schemes of human values associated with them. A similar distinction is picked up by Feenburg (1991), who argues that our present and inherited traditions of work practice make available an alternative conception of a role for technology in the enhancement of work and of providing fulfillment in work, rather than merely as a means of eliminating work. This was also a vision propounded by Schumacher, who looked to Buddhism for a conception of human fulfillment which did not dismiss technological and economic development, but which nevertheless provided a basis for a critique of their currently dominant forms, a critique which is urgent for developing countries. He argued that in Buddhism the function of work is at least threefold: to give man a chance to utilize and develop his faculties, to enable him to overcome his ego-centeredness by joining with others in a communal task, and to bring forth goods and services needed for a becoming existence (Schumacher 1973, p. 45). Ihde (1990) also looks to other cultures for concrete grounding of critiques of Western technological development.

It is from the basis of its traditional concern with conceptions of human nature and human fulfillment that philosophy has a role to play in studies of, and debates about, technology; but it can participate fully only once it has opened up those debates by becoming explicitly aware of the conceptions inherent in much of its traditional practice. This opening has already begun at the urgings of feminists and post-modernists. Their critiques make a space for the insertion of technology into philosophical discourse. It is one which needs to be more extensively explored if philosophy is to contribute to ongoing public debates such as those over the environment, health care, unemployment, development and cultural diversity, or education.

References and further reading

Arendt, H. 1958: *The Human Condition* (Chicago: University of Chicago Press).

Bijker, W. E., Hughes, Thomas P., and Pinch, T. (eds) 1987: *The Social Construction of Technological Systems* (Cambridge MA: MIT Press).

Borgmann, A. 1984: *Technology and the Character of Contemporary Life* (Chicago: University of Chicago Press).

Ellul, J. 1964: *The Technological Society*, trans. John Wilkinson (New York: Knopf).

Feenberg, A. 1991: *Critical Theory of Technology* (Oxford and New York: Oxford University Press).

Gibbons, M. 1984: Is science industrially relevant? The interaction between science and technology. In Gibbons and Gummett 1984, 96–116.

Gibbons, M., and Gummett, P. (eds) 1984: *Science, Technology and Society* (Manchester: Manchester University Press).

Gowing, M., and Arnold, L. 1979: *The Atomic Bomb* (London: Butterworth).

Ihde, D. 1990: *Technology and the Life World: From Garden to Earth* (Bloomington, IN: Indiana University Press).

—— 1993: *Philosophy of Technology: An Introduction* (New York: Paragon House).

Kline, S. J., and Kash, D. E. 1992: Technology policy: what should it do? *Technology and Society Magazine*, 11(2); repr. in Teich 1993.

Kuhn, T. S. 1959: Energy conservation and an example of simultaneous discovery. In *Critical Problems in the History of Science*, ed. Marshal Clagett (Madison: University of Wisconsin Press), 321–56; repr. in Kuhn 1977, 66–104.

—— 1977: *The Essential Tension* (Chicago: University of Chicago Press).

Latour, B. 1987: *Science in Action* (Cambridge, MA: Harvard University Press).

Laudan, R. 1984: *The Nature of Technological Knowledge: Are Models of Scientific Change Relevant?* (Dordrecht and Boston: Reidel).

Marcuse, H. 1964: *One-Dimensional Man* (London: Routledge and Kegan Paul); repr. London: Sphere Books, 1968.

Marx, K. 1867: *Capital*, vol. 1 (New York: Vintage Books, 1977).

Marx, K., and Engels, F. 1970: *The German Ideology*, ed. C. J. Arthur (London: Lawrence and Wishart).

McDermott, J. 1969: Technology: the opiate of the intellectuals. *New York Review of Books*, July; repr. in Shrader-Frechette and Westra 1997, 87–104.

Mesthene, E. G. 1969: The role of technology in society. *Technology and Culture*, 10/4, 489–536; repr. in Shrader-Frechette and Westra 1997, 71–84.

Mowery, D. C., and Rosenberg, N. 1989: *Technology and the Pursuit of Economic Growth* (Cambridge: Cambridge University Press).

Musson, A. E., and Robinson, E. 1969: *Science and Technology in the Industrial Revolution* (London: Curtis Brown Ltd. and New York: Gordon and Breach).

Shrader-Frechette, K., and Westra, L. 1997: *Technology and Values* (Lanham, MD, and Oxford: Rowman & Littlefield).

Schumacher, E. F. 1973: *Small is Beautiful: A Study of Economics as if People Mattered* (London: Blond and Briggs); repr. New York: Sphere Books, 1974.

Stamp, P. 1989: *Technology, Gender and Power in Africa* (Ottawa: International Development Research Centre).

Teich, A. H. (ed.) 1993: *Technology and the Future*, 6th edn (New York: St Martins Press).

Volti, R. 1992: *Society and Technological Change* (New York: St Martin's Press).

Webster, A. 1991: *Science, Technology and Society* (New Brunswick, NJ: Rutgers University Press and London: Macmillan Education Ltd.).

Winner, L. 1986: *The Whale and the Reactor* (Chicago: University of Chicago Press).

—— 1991: Artifact/ideas and political culture. *Whole Earth Review*, 73, 18–24; repr. in Shrader-Frechette and Westra 1997, 55–68.

Woolgar, S. 1987: Reconstructing man and machine. In Bijker et al. 1989, 311–28.

71

Teleological Explanation

ANDREW WOODFIELD

Human curiosity leads people to ask what things are *for*. Teleological explanations answer "What for?" questions by appealing to forward-looking reasons.

Children learn that certain things – a snapped twig, a pattern of pebbles washed up by the tide – are *not* thought to be explainable in this way. But there is a vast range of phenomena which adults do try to explain teleologically. Children accept these explanations and learn rules for constructing them. Folk acceptability is no guarantee of scientific acceptability, however. Many thinkers have held that the study of purposes is no business of science.

Let us distinguish between supernatural purposes and natural purposes. Even if mainstream science eschews the former, it smiles on the latter. There are four main domains where natural teleology is found:

1 goal-directed behavior;
2 artifacts endowed with functions by organisms that design, manufacture, and use them to serve their goals;
3 features and parts of living things which have natural functions for their possessors;
4 social phenomena which have functions within social organizations.

Goal explanations are widely employed in psychology, ethology, and Artificial Intelligence; natural function explanations figure in biology; social function explanations occur in anthropology, sociology, and sociobiology. Contemporary philosophy has investigated the meanings, the logical forms, and the truth conditions of such explanations, their evidential bases, and their predictive utility. Although they may disagree over details, most philosophers of science believe that natural teleological explanations are empirically respectable (Nagel 1961; Taylor 1964; Woodfield 1976; Wright 1976).

Yet there are some whose idealized conception of good science challenges the legitimacy of goal talk, and others who believe that natural function attributions are not wholly objective. Their worries will be briefly sketched.

Goals

An organism has a goal G if and only if it is motivated toward G. Explanations of goal-directed action depend upon a theory which imputes intentional states to the agent. Such states are individuated by their representational contents. In recent philosophy of psychology, eliminativists have argued that such states probably do not correspond to any physiological or physical state types. They predict that as science advances,

492

content-individuated states will cease to play any serious explanatory role. Future scientists will not hypothesize that an organism intends G, because they will have concluded that intentions are not real states. If eliminativists are right, the term "goal" belongs to a primitive folk theory which cannot be grafted onto science. That this concept still features in cognitive psychology shows that what currently passes for science is not finished science.

Admittedly, few philosophers adopt such a radical view. But the arguments in favour of eliminating mentalistic idioms from science threaten teleological explanations of types (1) and (2).

Functions

Typically, a functional explanation in biology says that an organ x is present in an animal because x has function F. What does this mean?

Some philosophers maintain that an activity of an organ counts as a function only if the ancestors of the organ's owner were naturally selected partly because they had similar organs that performed the same activity. Thus the historical-causal property, *having conferred a selective advantage*, is not just evidence that F is a function; it is constitutive of F's being a function.

If this reductive analysis is right, a functional explanation turns out to be a sketchy causal explanation of the origin of x. It makes the explanation scientifically respectable. The "because" indicates a weak relation of partial causal contribution.

However, this construal is not satisfying intuitively. To say that x is present because it has a function is normally taken to mean, roughly, that x is present because it is supposed to do something useful. This looks like the right sort of answer to a "What for?" question. Unfortunately, this normal interpretation immediately makes the explanation scientifically problematic, because the claim that x is supposed to do something useful appears to be normative and non-objective.

One possible ground for such a claim is that a designer meant x to do F. If the designer is held to be a supernatural being, the claim is not testable. If the designer is held to be Nature, the claim involves a metaphorical personification. Dennett (1987) argued that discerning natural functions always involves tacitly conceiving Nature as a designer.

By contrast, Millikan (1984) argued that evolution establishes norms within nature; hence biological function statements can be both normative and objective. This move was seized upon by philosophers of mind, who saw in it a way to "naturalize intentionality" – that is, establish that the representational properties of minds are explicable by natural science. Just as the content of a representation may be true or false, depending on how the world is, so the function of a device may be performed or not performed, depending on conditions outside the device. The parallelism inspired a hope of explicating the semantic properties of natural representations in terms of specialized natural functions (Millikan 1984; Dretske 1988; Papineau 1987; McGinn 1989; but see Woodfield 1990 for reservations). If this strategy works, it undercuts the eliminativist challenge to intentionality and saves the notion of goal teleology. But natural functions – the ultimate foundation – have to be sufficiently objective to withstand the skeptics.

In the nineteenth and twentieth centuries, philosophical attitudes to teleological explanation had a great impact upon the theories and methods espoused by working scientists. This interaction can be expected to continue into the twenty-first century on a higher plane of sophistication.

References

Dennett, D. C. 1987: Evolution, error and intentionality. In *The Intentional Stance* (Cambridge, MA: MIT Press), 287–321.

Dretske, F. 1988: *Explaining Behavior* (Cambridge, MA: MIT Press).

McGinn, C. 1989: *Mental Content* (Oxford: Blackwell).

Millikan, R. 1984: *Language, Thought, and Other Biological Categories* (Cambridge, MA: MIT Press).

Nagel, E. 1961: *The Structure of Science* (London: Routledge and Kegan Paul).

Papineau, D. 1987: *Reality and Representation* (Oxford: Blackwell).

Taylor, C. 1964: *The Explanation of Behaviour* (London: Routledge and Kegan Paul).

Woodfield, A. 1976: *Teleology* (Cambridge: Cambridge University Press).

—— 1990: The emergence of natural representations. *Philosophical Topics*, 18, 187–213.

Wright, L. 1976: *Teleological Explanations* (Berkeley: University of California Press).

72

Theoretical Terms:
Meaning and Reference

PHILIP PERCIVAL

Introduction

It is one thing for a scientist to speak a language in which he can conduct and communicate his investigations, another for him to possess a reflective understanding enabling him to explain the nature and workings of that language. Many who have sought such an understanding have held that the concepts of "meaning," "reference," and "theoretical term" play a crucial role in developing it. But others – instrumentalist skeptics about reference, Quinean skeptics about meaning, and skeptics about the theory/observation distinction – have denied this.

Reference: semantic instrumentalism versus semantic realism

"Reference" has been variously construed. The three most important ways of understanding it hold that to state the reference of an expression is to state either (i) the contribution the expression makes to the truth-values of the sentences in which it occurs (the expression's *semantic role*), or (ii) the entity to which the expression bears that one–*one* relation – *designating* – which holds between, for example, the particular utterance of "that" and the object demonstrated when an agent utters the words "That is my pen" with an accompanying demonstrative gesture (the expression's *designation*), or (iii) the entities to which the expression bears the one–*many* relation – *denotation* which holds between, for example, the word "goose" and each goose. Some hold that general (theoretical) terms like "electron" have references in each of these senses. But within the "instrumentalist" tradition, others insist that such terms cannot have references in any of them.

Whereas the influential new brand of instrumentalism in van Fraassen 1980 exploits a distinction between theoretical and observable *entities*, instrumentalism traditionally employs some such epistemological criterion as "not applicable on the basis of observation alone" to distinguish "theoretical" from "observational" *terms* (see below, section on the theory/observation distinction). Two kinds of instrumentalism then emerge, depending on the significance which a term's satisfaction of this criterion is claimed to have. For the *epistemological* instrumentalist, it is merely epistemological: one cannot know whether or not the world is as contingent claims involving theoretical terms suggest. But for the *semantic* instrumentalist it is more radical, in that it restricts the kind of linguistic function that the term can have. In the broadest sense, this is a matter of rendering discourse comprising (contingent) sentences in which theoretical terms occur *nonfactual* (just as, for example, emotivists hold ethical discourse to be nonfactual). In a narrower sense, semantic instrumentalism maintains that theoretical terms cannot have either denotations (so that theoretical *objects* cannot exist), designations (so that theoretical *properties* or *natural kinds* or *universals* cannot exist), or semantic roles (so that (contingent) theoretical *truths* and *falsehoods* cannot exist). (See Horwich 1993 and Field 1994 for the view that supposing S is truth-apt (a bearer or truth and falsity) falls short of supposing that it is factual.) Semantic instrumentalism can be traced back at least as far as Berkeley (see Newton-Smith 1985), the basic motivation being a brute empiricism about conceivability (to the effect that an agent can stand in a cognitive relation to an entity only if that entity is observable), a very strict notion of observability, and convictions that a language cannot outstrip thought and that a sentence can only be factual/truth-apt if its subject terms and predicates relate to actual entities.

Semantic instrumentalism opposes a doctrine I call "semantic realism." Semantic realism about a language for science holds that the parts of the language that the instrumentalist deems theoretical function in the same way as the parts he deems observational. In particular, it holds that theoretical claims *are* either true or false – thus that terms of a language for science have semantic roles – and that general terms like "electron" and "has mass r kg" have "extensions," in that they *aspire to have* denotations. (A term aspires to denote when its linguistic function is such that it *does* denote if the world is a certain way; the extension of such a term is the set of entities it denotes.) But as I shall employ the phrase, "semantic realism" is neutral as to whether general theoretical terms designate "intensional" entities like natural kinds or properties, etc. (Semantic realists who deny that general terms like "electron" and "has mass r kg" designate such entities in addition to denoting electrons and bodies having mass r kg might still hold that these terms designate something else – namely, their extensions.)

Semantic realism holds that, for example, "There are electrons, and they have rest mass 9.11×10^{-31} kg" represents the world as being a certain way, that the terms "electrons" and "have rest mass 9.11×10^{-31} kg" aspire to denote, and that this sentence is true or false depending on how the world is. Yet two related considerations have been taken to suggest that there is no conceptual room for semantic instrumentalism to deny this. The first is an equivalence claim about truth: someone upholding, in the sense of assertively uttering, *any* sentence "S," is *obliged* to uphold the sentence " 'S' is true." The second is the claim that the English quantifier "there are" carries ontological commitment. Once one upholds the theoretical sentences that science advances, as the semantic instrumentalist does, the first of these claims commits one to thinking those sentences true, while the second of them commits one to thinking that, for example, electrons *really* exist, and that "electron" denotes them. However, few now hold that an English sentence of the form "There are F's" admits of only one construal, and that on it this sentence claims that F's *really* exist. And while the equivalence claim about truth is currently received more sympathetically (Horwich 1990), I think there is more to be said about truth than this claim allows. The main route to the equivalence claim about truth is a "deflationism" whereby the *only* concept of truth is a *thin* one given by the disquotational schema " 'S' is true iff S" (for all sentences S meeting minimal conditions of susceptibility to assertion). But the semantic instrumentalist can retort that whether or not there is *a* thin concept of truth – $true_m$ – about which this much is correct, there is nevertheless a "thick" concept – perhaps involving the notion of correspondence with the facts – for which the disquotational schema fails when S involves theoretical terms. Similarly, to preserve an appealing connection between truth and reference, namely " 'Fb' is true iff 'F' denotes b," the instrumentalist will maintain that whether or not there is a "thin" concept of denotation – $denotes_m$ – which supports the analogous disquotational schema " 'F' denotes b iff Fb," there is a "thick" concept for which this schema fails when "F" is a theoretical term. (Cf. Jackson et al. 1994, which argues, further, that even if there is only one concept of truth and deflationism is right about it, it still doesn't follow that all sentences meeting minimal conditions of susceptibility to assertion are truth-apt.)

Whereas semantic realism employs concepts of truth and reference to explain the linguistic function of terms the instrumentalist deems "theoretical," semantic instrumentalism narrowly construed rejects this explanation on the grounds that it employs (thick) concepts of truth and reference which are inappropriate to the theoretical realm. In so doing, it must offer an alternative explanation. Perhaps one alternative is readily available to it on the cheap if the thin concepts "$truth_m$" and "$reference_m$" are substituted for the thick concepts which occur in the explanation it rejects. But in its earlier manifestations semantic instrumentalism involved the more radical suggestion that *no* concepts of truth or reference are appropriate in the theoretical realm. To see how the linguistic function of theoretical terms might be explained within the constraints which this suggestion imposes, we must clarify, in a manner partly neutral between realism and instrumentalism, the notion of "definition."

Definitions: explicit, operational, and implicit

A term *t* is "explicitly" definable iff it functions merely as an alternative (typically a shorthand) for some other expression *e*. If *t* is explicitly definable, substituting *e* for *t* (or vice versa) in any sentence leaves the semantic status of the sentence unaffected (with the exception of quotational contexts like " 'electron' has three vowels," a complication I shall ignore), so that the rule for *t*'s use, in effect, is this: employ *e* in accordance with the rules for *e*, and then substitute *t* for *e*. Accordingly, explicitly defining a term defers problems attending the term's function to the expression by which it is defined, and were all the terms of a language for science explicitly definable, a straightforward account of that language would ensue: it would function as a shorthand for something else. However, since it cannot be the case that all the terms of a language are explicitly definable in that language – that would involve circularity – the most one could hope for from explicit definition would be for all *theoretical* terms to be explicitly definable. Yet, even if the distinction between theoretical and observational terms can be sustained, at first glance the prospects of finding explicit definitions for all theoretical terms appear poor. Some theoretical terms – particularly those involved in functional identities – do seem to invite somewhat trivial explicit definitions. For example, "linear momentum" looks like shorthand for "product of mass and velocity." But others are not explicitly definable in so brisk a manner. Though "has mass r kg" is a phrase not of everyday (observational) language, but of physics, it cannot be treated as "momentum" was treated – that is, as expressing a quantity mathematically related to more fundamental magnitudes. There aren't any.

P. W. Bridgman once argued that terms like "has mass r kg" should be definitionally associated with operations – paradigmatically, measurement procedures – by which one can determine whether or not the terms apply in a particular case. But although this doctrine has an obvious appeal, in that there is *some* connection between what it means to say of, for example, a brick that it has mass 10 kg and getting the reading "10" upon implementing certain measurement procedures, Bridgman never succeeded in refining his concept of "operational" definition into a precise theory (see Feigl 1945). It won't do for the "operationalist" to say that "x has mass r kg" can be explicitly defined by some such expression as "x is subjected to measurement procedure M using device D → D indicates r." For if an entity b – my watch, say – is not subjected to M, the sentence "b is subjected to M using D → D indicates 150" will be true in virtue of the falsity of the antecedent of its truth-functional conditional "→." So the operationalist must try another tack. There are two alternatives (see Hempel 1965, ch. 5).

The first alternative strengthens the definition just rejected, either by prefacing it with an intensional operator "It is a law of nature that," or else by replacing "→" with a subjunctive conditional. But this strategy suffers one main defect and one striking limitation. It is defective because these operators are obscure enough to suggest that they cannot serve to clarify linguistic function. And it is limited because it has no obvious extension to terms like "x is an electron." In particular, procedures for detecting the presence of electrons cannot give rise to the kind of explicit definition under consideration: they are not operations by which to determine whether the expression "x is an electron" applies to some independently identified item.

By contrast, the second alternative promises a unified account of the function of all manner of terms while avoiding the obscurities of intensional operators. Its trick is to relax the requirement of explicit definition in favor of definition which is merely "implicit." Unlike explicit definition, the "implicit" definition of a term t need not provide some other expression e which is everywhere substitutable for t. Instead, it purports to confer (or capture) the linguistic function of t by imposing certain constraints on sentences in which t occurs. Classically, this is a matter of *stipulating* that certain sentences containing t are what I will call "unassailable." Recourse to implicit definition in this sense has two advantages. First, it permits the operationalist's account of a term like "x has mass r kg" to be made more precise while remaining within the confines of extensional language. (For example, this term might be taken to be implicitly defined by what Carnap (1953, sec. 4) calls a "bilateral reduction sentence" of the form "x is subjected to M using D → [x has mass r kg ↔ D indicates r]".) Second, there is no reason to confine stipulations of unassailability to sentences having the logical forms so far considered, and, as just noted, every reason not to in the case of terms like "electron." Indeed, such stipulations need not be confined to sentences in which just one term to be defined occurs: stipulating the unassailability of sentences $S_1 \ldots S_n$ each of which contains at least two terms $t_1 \ldots t_m$ might be taken jointly to define these terms *en masse*. It is upon implicit definition in this wider sense that the account of theoretical terms that has come to be known as the "received" or "standard" or "orthodox" view of theories is built. I shall call it the "classical" view of scientific theories.

The classical view of theories

The classical view of theories holds that the linguistic function of theoretical terms is entirely dependent upon stipulative connections which these terms bear to observational terms. There are different versions of this view, but each of them contains four core claims about a transparently formulated scientific theory: (i) the theory comprises a deductively closed set of sentences (one specified by means of an axiomatic system comprising axioms and derivation rules); (ii) the theory is formulated in a language the terms of which can be classified into "observational," "theoretical," and "logico-mathematical"; (iii) the theoretical terms are implicitly defined by stipulations of the unassailability of certain sentences that are not logical truths; and (iv) these stipulations must concern some sentences containing both theoretical and observational terms, but may also concern some others devoid of observational terms. Call any sentence containing both observational and theoretical terms which is not a logical truth a "bridge principle," and call bridge principles that are stipulated to be unassailable "coordinating definitions." On the classical view it is coordinating definitions that are crucial. Only via them can the independent meaningfulness of observation terms seep up to theoretical terms. Stipulations concerning sentences devoid of observation terms can serve only to *refine* linguistic functions already conferred (see Carnap 1939, secs 23–5; Feigl 1970).

Different versions of the classical view of theories arise because the neutral term "unassailable" has realist and instrumentalist readings. For the semantic realist, to stipulate that a sentence is "unassailable" is to stipulate that the sentence is (thickly) true. But semantic instrumentalism thinks otherwise. For the modest version of this doctrine, it is to stipulate that the sentence is (thinly) true$_m$; for the radical version, it is *merely* to license the use of the sentence when deriving consequences (especially observational consequences) from sentences which involve theoretical terms.

Semantic realist and modest semantic instrumentalist construals of early formulations of the classical view encounter certain objections (see section below on the Carnap–Lewis refinement), and even Carnap's 1936 doctrine of "partial" definition can be viewed as contending that *some* concession must be made to the radical semantic instrumentalist. As previously noted, Carnap here argues that the linguistic function of a term like "water soluble" is given by stipulative bilateral reduction sentences. But while he supposes that such stipulations ensure that some entities fall within the extension of this term (namely, those which have dissolved in water), and that some entities fall outside it (namely, those which have been placed in water but which have not dissolved), he contends that these stipulations leave indeterminate whether other entities – the ones never placed in water – fall within the extension of the term. For Carnap, greater determinacy in the reference of a term implicitly defined by bilateral reduction sentences can be achieved by adding further stipulations. (When a general term t is "completely" defined, for all entities e it is determinate whether e falls within the extension of t.)

Critique of semantic instrumentalism

As construed by the semantic instrumentalist, the classical view of theories holds that a language for science merely embodies rules connecting its theoretical sentences with observation sentences: neither the purely theoretical sentences nor the bridge principles themselves effect representations of the world inexpressible by means of observation sentences. The *point* of theoretical language thus conceived is *pragmatic*: theoretical language allows one to systematize in a convenient and fruitful way the *observational* sentences one believes (thickly) true.

It is sometimes claimed that this is an essentially holistic account of the meanings of the expressions of a theoretical language which engenders a "meaning variance" thesis to the effect that the meanings of theoretical terms change whenever there is a change in theory. But this is a mistake. The observational consequences of a given theoretical sentence depend on the further theoretical sentences and bridge principles with which that sentence is allied in a theory. So, if the point of a theoretical sentence is its contribution to observational content, the point of *asserting* such a sentence must derive from the contribution to observational content that the sentence makes within the *speaker's* theory. It follows that the ultimate point of such an assertion cannot be grasped unless one knows which theoretical sentences and bridge principles the speaker accepts, so that any change in the speaker's theory will affect the point of his asserting some theoretical sentences. However, this is only to say that the *meaning* of some theoretical sentence (and hence of some theoretical term) will be altered upon such a change if this meaning is identified with the contribution the sentence makes to the observational content of a theory in which it happens to be embedded – and, hence, with the point of asserting it. But although this identification appealed to some instrumentalistically minded proponents of the classical view of theories (under the influence of logical positivism), it should not have done. It is as misguided as a proposal to identify the meaning of the sentence "That man is drinking martini" with the content "Jones is drinking martini" expressed when this sentence is uttered while pointing to Jones. As construed by the semantic instrumentalist, the classical view should hold that context dependence of content is more rife than is generally supposed, and that in the case of sentences containing theoretical terms one content-determining feature of context amounts to the nonstipulative nonobservational sentences embraced by the speaker. So a speaker produces changes in the *meaning* of his terms only by adding or excising a given sentence from the theory *if that sentence is stipulative*.

While semantic instrumentalist versions of the classical view can avoid the meaning-variance thesis in this way via a holistic view of content, some will think the latter just as unpalatable. But in any case the semantic instrumentalist's contention that theories are of simply *pragmatic* utility remains untenable. Suppose the classical view is right about the logical connection between theories and observation sentences, so that theories do indeed facilitate the axiomatization of observational sentences. Even so, the semantic instrumentalist is wrong to maintain that theories are thereby dispensable in principle. On the contrary, theories *realistically construed*, are *essential* to the scientific enterprise. It is only by aspiring to talk, *literally*, about theoretical entities, that theories can aspire to the explanations for which science aims. Admittedly, like their epistemological cousins, semantic instrumentalists have played down the explanatory pretensions of science (see van Fraassen 1980). But their doing so has been in the main *ad hoc*: they downplayed them only because their account of theoretical language seemed not to allow them to do otherwise. And this was to embrace the wrong alternative. They should have kept faith with the explanatory pretensions of science and rejected their own account of theoretical language. Even if theories would acquire *some* explanatory potential on the mere assumption that they systematize observation sentences, their explanatory power is more substantial and stems from a different source. A theory which aspires to talk, *literally*, of molecules, offers an explanation, for example, of Brownian motion, not just because it systematizes certain observational claims – if that is what it does – but because it (aspires to) *denote* (thickly) entities which *cause* this phenomenon (see Putnam 1975a, chs 13–14).

This, then, is the main consideration in support of semantic realism. Unless the theoretical terms of languages for science *aspire* to refer (thickly), so as to *have* extensions and ensure the (thick) truth or falsity of theoretical claims, science cannot perform one of its central tasks: it cannot explain natural phenomena. Of course, in the light of this consideration, not only is it incumbent upon philosophers of language to try to frame semantic realist accounts of languages for science: they must also expose semantic instrumentalist arguments to the effect that such accounts are *impossible*. Suffice it to say here that while the arguments in Dummett 1978, ch. 21, underlie the most sophisticated arguments around for semantic instrumentalism (although Dummett himself seems wary of putting them to exactly *this* purpose), McGinn (1980), Craig (1982), and McDowell (1981) make compelling objections.

Incommensurability

That even terms like "electron" (and "phlogiston"!) aspire to refer is obscured by Kuhn (1962), and Feyerabend (1975, ch. 17; 1981, chs 1–6). Invoking the holistic thesis that a term has no meaning independently of the theory it helps to express, rejecting the theory/observation distinction, and not always giving the distinction between meaning and reference its due, these authors are led to a (semantic) incommensurability thesis: since the meaning of *all* terms depends on theory, (at least!) theories which are genuinely *revolutionary* do not engage the theories they supersede – there can be no (contingent) sentence of either the truth of which requires the falsity of any (contingent) sentence of the other (see INCOMMENSURABILITY).

A striking objection to this thesis in Davidson 1984, ch. 13, appeals to a principle of charity when interpreting other speakers: by one's own lights, the beliefs one assigns to other speakers must mostly come out true, and, *a fortiori*, commensurable with one's own. But even when, under pressure from other principles operating in the theory of interpretation (see Lewis 1983, ch. 8), one feels uncharitable, the thesis of semantic incommensurability has extremely implausible presuppositions. To formulate competing claims about, say, quasars, two theorists need not agree on the definition or meaning of "quasar"; truth depends on *reference*, not *meaning* (Sheffler 1967, ch. 3). Nor, even, need they give the term "quasar" the same extension. If T_1 includes "All quasars$_1$ are F's" while T_2 asserts that all quasars$_2$ are non-F's, they disagree provided *some* quasars$_1$ are quasars$_2$ (see Martin 1971). And it is hard to see how T_2 and T_1 can even aspire to compete – be theories of *the same phenomena* – if T_2 is couched in a language such that *none* of the references of its terms overlap *any* of the references of the terms of the language of T_1. Yet, even in this extreme case T_2 and T_1 might turn out to be commensurable: even if *no* predicate of T_1 denotes *anything* denoted by any predicate of T_2, they might still disagree. Arguably, there is nothing in virtue of which the Newtonian term "mass 1.2 kg" can be said to denote entities with relativistic mass 1.2 kg (in the standard frame of reference) rather than entities with proper mass 1.2 kg, in that both alternatives result in central Newtonian claims about mass coming out false. (So the denotation of "mass" is not invariant across Newtonian and relativistic mechanics.) Yet, since other Newtonian claims come out true on both alternatives, instead of just saying, simply, that this term lacks denotation, we do better to hold that it "partially" denotes entities of both kinds. A theorem "Some Q's and some S's are both R and V" of T_1 might then contradict a theorem "All G's are H" of T_2 even if G and H do not denote. For if G partially denotes Q's and partially denotes S's, while H partially denotes ~R's and partially denotes ~V's, the two theorems contradict one another on each resolution of this indeterminacy (see Field 1973; Devitt 1979).

The supposition that the references of *all* terms are determined by the theories in which they are embedded does not warrant the claim that the references of common terms cannot remain invariant when a theory T_1 is replaced by a genuinely revolutionary theory T_2. Nor does the semantic incommensurability of T_1 and T_2 follow from the supposition that these references change when T_1 is replaced by T_2. However, to be persuaded that the semantic incommensurability thesis is misguided is not to understand why it is wrong. To understand this, we need to know *how* the terms of different theories concerning the same phenomenon can have references ensuring commensurability: we need to solve *the problem of the denotations of (theoretical) terms*. Solving this problem involves more than answering questions like "Which entities do 'electron' and 'hydrogen peroxide' denote?" For such questions can be answered all too readily: respectively, these terms denote electrons and bits of hydrogen peroxide. Solving it demands *informative* (nontrivial) answers to such questions.

The Carnap–Lewis refinement of the classical view

As construed by the semantic realist, the classical view of theories offers a nontrivial solution to the problem of denotation: theoretical terms purport to denote entities that stand in certain relations to each other and to the entities that observation terms denote. But this answer must be refined so as to take account of difficulties which the section on the classical view forewarned. As stated so far, the classical view says nothing which (i) prevents the set of stipulations governing a term t from having observational consequences (yet how can a set of sentences having observational consequences be *stipulated* to be true?), or (ii) prevents theoretical terms from denoting only *mathematical* entities, or (iii) guarantees *the existence* of a semantic role such that were t to have that role, the sentences containing t which have been stipulated to be true would indeed be true, or (iv) ensures that any such role is *unique* (see Winnie 1967; Horwich 1997) (see THEORIES).

These difficulties are tackled in Carnap 1974, part V, and, in effect, in Lewis 1983, ch. 6. Carnap proposes to solve the first difficulty as follows. Suppose that T is a theory which is to implicitly define theoretical terms $t_1 \ldots t_n$, and which is axiomatized by sentences $S_1 \ldots S_m$. Conjoin these sentences into a single sentence $F(t_1 \ldots t_n)$, the "theoretical postulate" of T. Typically, the postulate $F(t_1 \ldots t_n)$ will have (contingent) observational consequences; yet the classical view of theories insists that it also harbours stipulations which implicitly define the terms $t_1 \ldots t_n$. To dispel the tension, Carnap suggests that these features are respectively possessed by two formulae, one synthetic, the other stipulative, that can be factored out from $F(t_1 \ldots t_n)$ because jointly equivalent to it. The (same) observational consequences are borne by T's "Ramsey sentence," $\exists x_1 \ldots \exists x_n(Fx_1 \ldots x_n)$, while the stipulations govern T's "Carnap sentence," $\exists x_1 \ldots \exists x_n F(x_1 \ldots x_n) \rightarrow F(t_1 \ldots t_n)$ (which has no observational consequences at all).

Though concerned with the first difficulty, Carnap's proposal seems also to solve the third: for the Carnap sentence $\exists x_1 \ldots \exists x_n F(x_1 \ldots x_n) \rightarrow F(t_1 \ldots t_n)$ *shows* the semantic roles to be conferred on the terms $t_1 \ldots t_n$ – they are to purport to designate entities belonging to a sequence denoted by the theory's "realization formula," $F(x_1 \ldots x_n)$. Equally, however, Carnap's proposal highlights the fourth difficulty: while the Carnap sentence says, in effect, that if the realization formula $F(x_1 \ldots x_n)$ denotes *any* sequences of entities $<e_1 \ldots e_n>$, the terms $t_1 \ldots t_n$ respectively designate the entities of *one* of those sequences, the Carnap sentence doesn't say *which* sequence is to be selected if the realization formula denotes *more* than one sequence. However, as Lewis observes, this problem is solved if the Carnap sentence is sharpened to the "Lewis sentence," $\exists !x_1 \ldots \exists !x_n(Fx_1 \ldots x_n) \leftrightarrow F(t_1 \ldots t_n)$, which expresses a *uniqueness* requirement to the effect that the terms $t_1 \ldots t_n$ respectively designate the entities of the sequence $<e_1 \ldots e_n>$ if *and only if* $<e_1 \ldots e_n>$ is the *only* sequence denoted by $F(x_1 \ldots x_n)$.

In two respects, Lewis's refinements leave the classical view unduly severe. First, while the theoretical terms of a defining theoretical postulate $F(t_1 \ldots t_n)$ are said to lack designations in the case in which more than one sequence of entities is denoted by the realization formula $F(x_1 \ldots x_n)$, these terms might be held to designate the entities of that sequence which does best with respect to certain additional criteria (O'Leary-Hawthorne 1994), or they might be held to *partially* designate the members of these sequences if none does (Lewis 1994). Second, theoretical terms are said to lack designations unless there is a sequence of entities which is *exactly* as the relevant realization formula characterizes a sequence of entities as being. But it is absurd, for example, to suppose that "electron" lacks denotation if current theory has the mass of an electron wrong at the second decimal place. Hence – as in effect Lewis (1972) concedes – the classical view *must* be amended to allow the possibility that theoretical terms designate even if their defining theory isn't *exactly* right. (Perhaps this can be done simply by prefixing some such expression as "mostly" or "approximately" to the theory's realization formula.)

Another problem concerns the inflated ontology – reflected in the care with which I spoke of the *designation* of theoretical terms – which the Carnap–Lewis refinement brings to the classical view. As Horwich (1997) complains, Carnap follows Ramsey in replacing theoretical terms by bound variables, which, being *second-order*, range over some such "intensional" entities as properties or natural kinds. For example, consider a theoretical sentence "For all x, if x is an electron, then x has negative charge." The second-order Ramsey sentence corresponding to this is $\exists \phi \exists \phi(x)(\phi x \rightarrow \phi x)$, where the second-order variables range over properties two of which are respectively designated by the predicates "x is an electron" and "x has negative charge." Admittedly, Lewis observes that if predicates like "x is an electron" are reparsed as "x *has* electron-hood," then theoretical properties can be thought of as being designated by such *singular terms* as "electron-hood," etc., and the trick can be pulled using only first-order quantification:

replacing these terms by *first-order* variables then yields a first-order formulation of the Ramsey sentence, $\exists y \exists z(x)(x\ has\ y \rightarrow x\ has\ z)$, and hence a first-order formulation of the Lewis sentence. But whether first- or second-order quantifiers are employed, the fact is that the Carnap–Lewis refinement of the classical view embraces an ontology of intensional entities to which the classical view as originally conceived was not committed. (Having complained of this ontology, Horwich (1997) advocates a novel conception of implicit definition in an attempt to avoid it. On the question as to whether the Carnap–Lewis refinement of the classical view can be reconstructed without it, see Hintikka 1998.) (see RAMSEY SENTENCES.)

The theory/observation distinction

But what of the second difficulty raised in the previous section? The realization formula $F(x_1 \ldots x_n)$ denotes sequences of *mathematical* entities unless a requirement such as the following is imposed: roughly, there must be at least one observational term which shares some part of its extension with the extension of some theoretical term. But is this requirement legitimate? Isn't it of *the essence* of the distinction between theoretical and observational terms that nothing falls within the extension of both?

In asking these questions, we can no longer ignore a criticism that has been repeatedly voiced against the classical view of theories – namely, that its distinction between theoretical and observational terms is untenable. But in fact most of the objections to this distinction have been overplayed. What is at issue is whether or not there is an epistemologically *significant* distinction among the nonlogical terms of languages for science, and all that such a distinction requires is the existence of an epistemological problem, not the intractability of an *insoluble* problem. As van Fraassen (1980, ch. 2) points out, arguments to the effect that theory/observation distinctions are vague are inconclusive. The fact that a distinction is vague does not preclude a matter of substance from turning on it; indeed, *contra* Bird (1998, ch. 4), the fact that a distinction is vague does not prevent a *semantic* matter from turning on it. Furthermore, features which create epistemological problems are easily identified. Here are three (see van Fraassen 1980, ch. 2; Ulises Moulines 1985; Fodor 1984): (i) not referring to any observable entity, (ii) not (always?) being applicable unless the truth of some theory in which the term occurs is presupposed, (iii) not occurring in any atomic sentence which an agent's senses can represent to the agent as being true. Each of these criteria for being a theoretical term is compatible with the requirement that there are observational terms the extensions of which overlap the extension of some theoretical term. (In particular, it should not be supposed that no observational *term* refers to any *unobservable* entity. Perhaps an observational term of which this is true is "is a proper part of.")

Still, even if some such criterion can be sustained, the classical view might have misconceived the relationship between the two kinds of term. Once theoretical terms are held to (purport to) refer (thickly), the possibility arises of their *having* those references *independently* of any stipulative, conventional connection with observational terms. This possibility, which is anathema to the semantic instrumentalist tradition, need not involve the quite general Quinean skepticism about meanings, stipulations, and analyticity considered below (see QUINE). Even if there are *some* conventionally stipulated sentences in (natural) languages for science, the coordinating definitions invoked by the classical view may not be among them (see Putnam 1975b, chs 2 and 13; Hempel 1970, 1973).

More recently, the rejection of the classical view's presumption that theory and observation are linked by *linguistic* convention has helped to foster a "semantic" or "structuralist" account of theories, according to which the primary function of theoretical language is the characterization of a set of abstract models, the minimal intention being that some one among these models should "fit" the phenomena falling within the scope of the theory (see van Fraassen 1980, ch. 3, 1987; Bauer 1990; da Costa and French 1990). The crucial idea here is that the relationship between the models of the theory and the relevant phenomena is mediated not by linguistic rules or conventions but by *pragmatic* or *theoretical* considerations. For example, consider how Newton's theory of gravity fits the solar system. The connection which the theoretical terms of Newton's theory bear to observable features of the solar system does not appear to be stipulative. Rather, it is pragmatic in two respects. First, masses, distances, and velocities are measured *as best we can* for the purposes at hand. Second, predictions are obtained from the theory by ignoring features of the solar system which are too subtle to be accommodated, such as the gravitational pull of the planets on each other, the size of the bodies involved, etc. The theory of masses, forces, accelerations, etc. is mathematically precise: we know exactly how an abstract Newtonian gravitational system behaves. But the way in which the theory "fits" an actual system is messy. Contrary to the classical view of theories, it is not a matter of (prior) linguistic stipulations; it is a matter of brute science (see Giere 1988, ch. 3).

Although the semantic view of theories has *something* to say about the function of theories, it does not answer the questions the classical view sought to answer: it does not constitute a comprehensive account of the linguistic function of theoretical terms. And it may well be that the classical view has much to teach us even if its distinction between observational and theoretical terms cannot be sustained, or its presumption that terms of the former kind are invariably connected with terms of the latter kind by linguistic stipulation is misguided. In particular, a "neo-classical" view can reconceptualize the distinction between theoretical and observational terms in a non-epistemological way. Neither the appeal which the classical view makes to conventional stipulations nor its idea that linguistic function is transmitted from some terms to others presupposes any *epistemological* distinction between the terms on which linguistic function is conferred, and the terms which confer it. So a neo-classical view can simply *equate* the classical view's theoretical/observational distinction with the defined/defining distinction. This seems to be the import of Lewis's (1983, ch. 6) suggestion that "observational" term should be read as *old* term and "theoretical" term as *new* term. To be sure, on this reading, terms like "electron," "quark," etc. are classified as old terms, so that the neo-classical view would then have nothing to say about the function of such terms. But Lewis thinks that his account of how new terms can be defined by theories which introduce them could be extended to *old* terms if those terms were defined by theories which attended them when they were first introduced. On the other hand, as Lewis recognizes, this proposal encounters a dilemma. Our theories change over time: the current theory of the electron is not Rutherford's. Thus, after the initial introduction of theoretical terms $t_1 \ldots t_n$ at time T_0 via a postulate

505

$F_0(t_1 \ldots t_n)$, each subsequent time T_i is associated with a theory $F_i(t_1 \ldots t_n)$, which might well be different from, and even contradict, earlier theories in the series. Clearly, saying that for each time T_j the references of the terms $t_1 \ldots t_n$ at T_j are to be determined by the realization formula $F_j(x_1 \ldots x_n)$ engenders the meaning-variance thesis: every change of theory brings a change in the meanings of the terms by which it is expressed (though not necessarily a change in their reference). As Lewis observes, that seems wrong. So it seems we should say that at T_j the references of the terms $t_1 \ldots t_n$ are determined with respect to some *earlier* theory $F_i(t_1 \ldots t_n)$ and somehow subsequently inherited. Saying this involves certain difficulties. But I will only be able to consider them in the section after next. For saying it also involves a historical chain theory of reference a version of which is one of the main rivals to the classical and neo-classical views (see OBSERVATION AND THEORY).

Historical chain theory

A "historical chain theory" of the reference of a token n of a term "N" holds that n has the reference that it does in virtue of the fact that (i) n stands in a historical chain comprising tokens of "N" linked by a certain relation (e.g., a certain *causal* relation); (ii) the chain begins with uses of "N" whereby reference is *fixed*; and (iii) tokens of "N" later in the chain inherit their reference from earlier tokens of "N." So historical chain theory has two components: a theory of reference *fixing* and a theory of reference *transmission*. The latter amounts to a "division of linguistic labor." In one respect, this division is undeniable, since a term as used by a speaker on a particular occasion can acquire its meaning and reference from the meaning and reference that other contemporaneous speakers assign to it. (For example, I can assert that my watch contains molybdenum even though all I know about the use of the term "molybdenum" is that it refers to a metal.) But whereas this *synchronic* division of labor merely *allows* a (strictly) incompetent speaker to exploit the linguistic resources of his *contemporaries*, the *diachronic* division of labor claimed by historical chain theory *obliges* even competent speakers to defer to the linguistic resources of their *predecessors*. Specifying exactly what such deference involves – that is, identifying the relation which links tokens in the chain – is a subtle business, not least because reference can fail to be inherited when speakers intend that it should be (see Evans 1973; Hacking 1983, ch. 6). Nevertheless, it is the historical chain theory's account of reference *fixing* that has occasioned most debate.

Many accounts of reference fixing have been canvassed. *Stipulative* theories hold that the original reference of a term "N" is fixed (at least tacitly!) by a stipulation of the form " 'N' refers to an entity e iff e is D," where "D" picks out the entity or entities to which the token of "N" employed in the stipulation thereby refers. (If "D" picks out an entity or entities at least in part by picking out some entity *nondescriptively* – as when it achieves *demonstrative* reference to certain entities via expressions like "this cat," "that," "those rocks," etc. – then the stipulation is *indexical*. Otherwise the stipulation is *purely descriptive*.) But *nonstipulative* theories hold that no stipulations are involved even implicitly: rather, the reference of a term is fixed in virtue of certain relations (typically causal) the referent(s) bear(s) to original uses of the term.

Historical chain theory came to prominence when Kripke (1980, ch. 2) recommended it for ordinary proper names like "Aristotle," and Putnam (1975b, chs 8, 11, 12) joined him in extending it to (so-called!) natural kind terms like "tiger" and "gold" (Kripke 1980, ch. 3). But their accounts of how the reference of a natural kind term of this ilk is fixed is unclear: while they denied that it was fixed by a purely descriptive stipulation, their account is somewhat ambiguous between an indexical stipulative theory and a nonstipulative theory. The indexical stipulative theory that is at least hinted at is along the following lines. An original user wanted to introduce the term "tiger" as a general term denoting entities of the same kind as the apparently striped cat-like animals which confronted him. So he did one of two things. Either he fixed the extension of "tiger" directly by stipulating that it is to comprise entities of the same natural kind as *these demonstrated entities*, or else he identified the natural kind these entities belonged to and then stipulated that "tiger" designates *that*, thereby fixing the extension of this term indirectly as the entities belonging to the term's designation. (In holding that terms like "tiger'" and "gold" are "rigid" designators, Kripke is committed to the second alternative and its implicit ontology of natural kinds. The first alternative avoids this ontology.)

On this account of reference fixing, "tiger" has *no* reference unless the demonstrated entities *did* exemplify a natural kind – which they wouldn't have if, say, half of them had been biological entities and the rest had been look-alike robots from Mars. But that seems wrong: like jade, tigers might not form a natural kind. Similarly, the account's appeal to the *concept* of a natural kind is problematic. At best, the "same natural kind" relation involves different considerations in different sciences: in physics and chemistry it is a matter of the same inner constitution, whereas in biology it is a matter of common ancestry and capacity for interbreeding (see Dupré 1981). In any case, there is no *one* kind of thing that a particular exemplifies – a tiger is a cat, a mammal, an animal, etc. – while entities which exemplify kinds like gold, ruby, water, etc. are typically impure (cf. Miller 1992; Brown 1998). These are difficulties enough, but even greater difficulties arise when Kripke and Putnam try to extend this kind of theory to theoretical terms like "electron," "quark," and "electricity."

To begin with, the reference of such terms cannot be fixed in a *strictly* analogous manner. Whereas "tiger" and "gold" denote entities at least some of which are perceptible, the entities denoted by "election" and "quark" are imperceptible. (see van Fraassen 1980, ch. 2); I will mark this contrast by speaking of "O-terms" and "non-O terms" respectively. Still, although no entities denoted by non-O terms can be demonstratively identified, entities to which they bear some relation might be. And since the relation between imperceptible and perceptible entities of most obvious interest to science is the causal relation between *events*, the most straightforward extension of the indexical theory of reference fixing to non-O terms is this: the extension of a term like "electron" was fixed by means of an indexical stipulation involving an impure description of the form "the entities which are of the same kind as the entities causing events of the same kind as *this* (observable event)." (Reference is made here to events of *the same kind* as the event(s) demonstrated to accommodate the point that science is primarily concerned with *regularities*, not *token* events.)

There are two reasons why this account cannot work for all theoretical terms, reasons which also militate against purely descriptive analogues of it which employ pure descriptions of the form "the entities which cause events with such and such observable properties." First, such accounts make reference to imperceptible entities too easy and, correlatively, *failure* of reference too hard. If the reference of a non-O term could be fixed in this manner, even Thales could have referred to electricity simply as a result of observing that pieces of straw are picked up by rubbed amber, while "electric effluvia" and "phlogiston" – concepts most certainly introduced in an attempt to give causal explanations of observable phenomena – would not have the null extensions which historians of science have often taken them to have (see Enç 1976; Nola 1980; Kroon 1985). Second, such a theory is a nonstarter in the case of those non-O terms which, like "black hole" and "positron," were originally introduced not in an attempt to explain previously known phenomena, but in the context of deducing the theoretical possibility of a novel kind of entity. (To mark this contrast, I will speak of "E-terms" and "non-E terms.") It may be years after a non-E term is introduced before any causal explanatory role for the entities it purports to denote is so much as conjectured.

Accordingly, accounts of reference fixing are needed which make reference by non-O E terms harder and reference by non-O non-E terms possible. The first of these requirements could be met by supposing that the extension of a non-O term is fixed by stipulating not just *what* phenomena its denotations are supposed to cause, but *how* its denotations bring those phenomena about (see Nola 1980; Kroon 1985), while both requirements could be met if we suppose that the extension of a non-O term is fixed by stipulating the nonrelational features of a new *kind* of entity which is *postulated* (not stipulated) to have certain causal powers (see Enç 1976). But such alternatives are far removed from the radical departures from earlier thinking that Kripke and Putnam seemed to offer us. In effect, each of them is a variant of the reorientated classical or neo-classical view of theories, whereby those views are seen not as theories of reference, but as theories of reference *fixing*.

Quine's challenge

The section on the theory/observation distinction ended by alluding to difficulties which arise if one tries to escape the first horn of Lewis's dilemma by pursuing this reorientation. Here they are. If the classical/neo-classical view escapes the meaning-variance thesis by supposing that at each time T_k the references of the terms $t_1 \ldots t_n$ is determined by a stipulation $\exists!x_1 \ldots \exists!x_n F_i(x_1 \ldots x_n) \leftrightarrow F_i(t_1 \ldots t_n)$ corresponding to some *earlier* theory $F_i(t_1 \ldots t_n)$, it encounters a difficulty which seems to undermine *any* attempt to implicate theoretical terms in historical chain theory. Scientists are not historians of language. Hence, they would be ignorant of any stipulations by which the references of theoretical terms were originally fixed. But in that case they would be entirely in the dark about whether their own speculations were even consistent with those stipulations. Since that methodology would be irresponsible – absurd even – historical chain theory cannot be generally correct for theoretical terms. (This point does not undermine historical chain theory for other terms: in special cases scientists may *want* to talk about entities standing at the end of a historical chain.)

Lewis's dilemma and the dispute between, for example, Enç and Nola (previous section) over how *much* theory should be employed in reference fixing both involve issues brought into sharp relief by Quine's philosophy of language. Quine's fundamental challenge, issued initially to Carnap, but equally pertinent to subsequent formulations of the classical and neo-classical view, is this: any attempt to implicitly define theoretical terms via a theory they help formulate must decide *which* theory involves stipulations (Lewis's dilemma) and *which* parts of that theory are stipulative (Enç versus Nola). But whereas these decisions presuppose distinctions between sentences that are analytic or stipulative and those which are synthetic or nonstipulative, Quine argues that no (noncircular) definitions can be given of these notions (1953, ch. 2), and that no behavioral criterion can be given for when a sentence as used by a speaker or community is analytic or stipulative (1970). (Correlatively, none can be given for determining whether a change in linguistic practice – say, coming to accept that radiation can seep from black holes, or coming to equate the momentum of a photon with $h\lambda$ – involves change in a term's meaning or, merely, a change in belief.) Quine concludes that attempts to illuminate the linguistic function of theoretical terms in a classical or neo-classical manner are utterly misguided. Linguistic competence consists in assimilating appropriate linguistic *dispositions*, not in grasping *meanings* reflected in a set of rules or definitions.

Quine's skepticism about the analytic/synthetic distinction has sometimes been frostily received (Grice and Strawson 1956), while others (Lewis 1969) have tried to reconstruct the distinction in elaborate terms (see Boghossian 1997). But in any case, by his own lights, Quine's critique may have less significance for the classical or neo-classical views than is often supposed. Papineau (1996) observes that reference might be fixed (determinately) in accordance with these views even if it is indeterminate which bits of theory are stipulative. And these views might survive the demise of the analytic/synthetic distinction by exploiting Quine's own distinction between "analysis" and "explication," so as to insist that they aim not at *capturing* prior distinctions embedded in languages of science, but at *making* distinctions in a language which can serve as a rational *alternative* to the obscure "languages" that scientists have hitherto employed. From such a perspective, the project is no longer the one with which this essay started. The aim is not a reflective understanding of natural languages of science, but their rational reconstruction. Still, even if this latter project could be successfully pursued in the classical/neo-classical manner, one would want to know what it is about natural languages of science that *permits* them to be thus reconstructed. For example, one would want to know what is it about "electron" which permits its replacement by a term "electron*" which *does* denote electrons via some stipulation. Since the natural answer (at least in part) is that "electron" denotes electrons, the problem we have been addressing would remain. All that would have been achieved is the negative result that *that* problem cannot be answered by identifying a reference-determining rule or stipulation which governs linguistic behavior. And this result might ensue by a different route even if the stipulative/nonstipulative distinction is retained in the face of Quine's skepticism. Arguably, the moral of the repeated failure of attempts to define theoretical terms is not Quine's skeptical claim to the effect that there are *no* stipulations/rules which competent speakers implicitly grasp, but the despairing claim that those stipulations/rules take such nonilluminating forms as " 'electron' denotes electrons" (see QUINE).

The theory of interpretation

Suppose that the project of solving the problem of denotation within an explicit theory of linguistic understanding is misguided, either because it is not the case that competent speakers follow linguistic rules, or because the only rules which competent speakers follow are trivial rules to the effect, for example, that "electron" denotes electrons. In that case we might hope to illuminate denotation via the theory of *interpretation*, the theory of what it is about an L-speaker in virtue of which he speaks L. (For the contrast between these approaches see Lewis 1983, ch. 11, and Peacocke 1976.) To this end, we might seek informative necessary and sufficient conditions for the claim that in the language spoken by population P, for example, a term "N" denotes electrons. However, as yet the theory of interpretation has done little to illuminate the problem of denotation. Stemming from Quine (1960) and Davidson (1984), the dominant approach to it has been holistic in two respects. First, it is held that there is no way of telling whether, for example, a single term "N" has a given extension without determining the linguistic function of *all* the terms in the language to which it belongs. Second, it is held that the principles constraining correct interpretation operate on *sentences*, not on individual terms, the conclusion drawn from this, for example, by Davidson being that the conditions restricting the reference of a term are much too weak to determine reference uniquely. The upshot is a thesis of the "inscrutability of reference": *different* accounts of the references of the terms employed by a speaker can be equally correct. This thesis is related to Field's conception of "partial" denotation discussed earlier, and could be viewed as an extension of it. It is also related to Quine's doctrine of the "relativity of ontology" – and is best viewed as a less paradoxical articulation of Quine's point (see Quine 1969, ch. 2; Davidson 1984, ch. 16).

Both respects in which the dominant tradition in the theory of interpretation is holistic have found their critics. Dummett (1993, ch. 1) criticizes holism in the first respect, on the grounds that it makes piecemeal language acquisition impossible; while Field (1972) hopes for a theory of interpretation which eschews holism in the second respect. But in view of the degree of holism and vagueness in both the theory of interpretation and the best available account framed within the theory of linguistic understanding – some version of the classical/neo-classical view – it is not entirely unfair to say that this century's explosion of interest in the philosophy of language has yielded little more than the negative result that attempts to pass beyond banalities like " 'electron' denotes electrons" have failed. (The extent of the mess to which the problem of denotation has given rise is evident in Kitcher's (1993, chs. 3–5) attempt to reconcile competing intuitions in a notion of "reference potential" which allows different tokens of one and the same theoretical term *as used by a single scientist* to have different references.) (See HOLISM.)

The reference of theoretical terms: semantic role and designation

If electrons exist, "electron" denotes them. Is the semantic role of "electron" determined by what this term denotes? A language is called "extensional" only if the semantic role of its general terms is determined by their denotations in the following sense: terms which denote the same objects can be substituted *salva veritate* in all the sentences of the language. But there are two reasons why a language for science appears not to be extensional in this sense. First, expressions like "it is a law of nature that . . . ," and ". . . caused ——," and "the explanation of . . . is that ——" appear to be non-extensional contexts the use of which is central to the practice of *all* science. Second, expressions like "—— believes that . . ." and "—— intended to . . ." appear to be non-extensional contexts the use of which is unavoidable in the pursuit of social sciences.

If it is admitted that a language for science includes non-extensional contexts, it cannot be the case, generally, that the denotations of its theoretical terms determine their semantic roles. What, then, might the semantic role of a general term otherwise be? The answer to this question that is given by an intensional entity theory holds that general terms *designate* "intensional" entities (like "universals," "properties," "natural kinds," etc.) having the following features: (i) the extension of a general term is determined by the intensional entity it designates and that entity's relations to other entities, and (ii) co-designating general terms are inter-substitutable *salva veritate* in all the sentences of the language in which those terms occur, So, for example, the extension of a term "x is an electron" might comprise all those entities which *possess* or *instantiate* the property which "x is an electron" designates. Clearly, the first feature of intensional entity theory is relevant to the problems that were addressed in the main body of this essay. Equally clearly, however, it does not solve them: it just sets them one stage further back: if "x is an electron" designates a property such that an entity is an electron if and only if it has that property, the questions arise as to *which* property "electron" designates, and in virtue of *what* it designates that property. Nor does intensional entity theory promise an explanation of how it is possible to use a general term. It might be suggested that one is able to apply a general term like "crustacean" to new cases because something remains constant as one applies it to one crustacean after another – for example, the property of being a crustacean. However, why should it be easier to recognize that *this* intensional entity is quantitatively or qualitatively *identical* to *that* one than it is to recognize, for example, that this non-intensional entity is qualitatively *similar* to that one?

Still, one might suppose that the ability of intensional entity theory to explain how co-denoting general terms can have different semantic roles is justification enough. Yet there are alternative responses to the apparent existence of non-extensional contexts in languages for science. One of them, pursued by Davidson (1984, ch. 7), is syntactic revisionism: what seems to be a non-extensional sentence-forming operation isn't a sentence-forming operation at all. Failing this, a second alternative is to deny that science has any use for the non-extensional contexts, and while eliminativists like Churchland (1987) argue for the inappropriateness of the folk-psychological language of belief and desire, etc. to a genuinely scientific account of human behavior, van Fraassen (1980), for example, argues that the fact that science *affords* causal explanations of phenomena isn't to say that it aims to provide them.

The meanings of theoretical terms

We have reason to think that "electron" denotes electrons. Maybe, in addition, it designates the property of being an electron, or the natural kind or universal *electron*.

511

Arguably, we need look no further than these two features for the semantic roles of theoretical terms which occur in the natural sciences. Suppose this is so. Would this be to say that we need look no further for the *meanings* of such terms?

Suppose we specify the extensions of all the terms of some theoretical language so as to capture their semantic roles. Will the result serve as a theory of meaning for the language? Not necessarily. Intuitively, different languages can include terms which in fact have the same extensions. (It might be a rule of L that "electron" denotes electrons, and a rule of L* that "electron" denotes electrons if $2 + 2 = 4$, and nothing otherwise. The extension of "electron" in L is identical to its extension in L*. Yet, surely, L and L* are different languages in virtue of the different ways in this common extension is specified.) Can we get round this difficulty if we suppose that theoretical terms designate as well as denote? Not if two natural kind terms can designate one and the same natural kind in different ways – and a case can be made for supposing that, for example, "tiger" and "*Felis Tigris*" do so (see Wiggins 1993).

This shows, I think, that even if a theory of reference is taken as the core of a theory of meaning, there is more to meaning than reference. Whether reference is denotation, designation, or semantic role, there can be different *ways* of stipulating one and the same reference, and the identity of a language is sensitive to these ways of so doing. This is a Fregean viewpoint on meaning, since Frege's *Sinn* (mostly translated "sense") is a way of having a reference. On the other hand, the thought that a theory of meaning is a theory of reference which *shows* the senses (and hence the meanings) of the expressions of the language has been thought to be fraught with difficulties. In particular, Dummett complains that such a theory offers no insight into what it is to have mastered a language thus construed. But perhaps there is none to be had.

References

Bauer, M. 1990: On the aim of scientific theories in relating to the world: a defence of the semantic account. *Dialogue*, 29, 323–33.

Bird, A. 1998: *Philosophy of Science* (London: UCL Press).

Boghossian, P. 1997: Analyticity. In *A Companion to the Philosophy of Language*, ed. B. Hale and C. Wright (Oxford: Blackwell), 331–68.

Brown, J. 1998: Natural kind terms and recognitional capacities. *Mind*, 107, 275–303.

Carnap, R. 1939: *Foundations of Logic and Mathematics* (Chicago: University of Chicago Press), repr. in *Readings in the Philosophy of Science*, ed. B. Brody (Englewood Cliffs, NJ: Prentice-Hall).

—— 1953: Testability and meaning (1936). In *Readings in the Philosophy of Science*, ed. H. Feigl and M. Brodbeck (New York: Apple-Century-Crofts), 47–92.

—— 1974: *An Introduction to the Philosophy of Science* (New York: Basic Books).

Churchland, P. 1987: Eliminative materialism and the propositional attitudes. *Journal of Philosophy*, 78, 67–89.

Craig, E. 1982: Meaning, use and privacy. *Mind*, 91, 541–64.

da Costa, N. C. A., and French, S. 1990: The model-theoretic approach to the philosophy of science. *Philosophy of Science*, 57, 248–65.

Davidson, D. 1984: *Inquiries into Truth and Interpretation* (Oxford: Oxford University Press).

Devitt, M. 1979: Against incommensurability. *Australasian Journal of Philosophy*, 57, 29–50.

Dummett, M. 1978: *Truth and other Enigmas* (London: Duckworth).

—— 1993: *The Seas of Language* (Oxford: Clarendon Press).

Dupré, J. 1981: Natural kinds and biological taxa. *Philosophical Review*, 90, 66–90.

Enç, B. 1976: Reference of theoretical terms. *Nous*, 10, 261–82.

Evans, G. 1973: The causal theory of names. *Proceedings of the Aristotelian Society*, supp. vol. 47, 187–208.

Feigl, H. 1945: Operationism and scientific method. *Psychological Review*, 52, 250–9.

—— 1970: The "orthodox" view of theories: remarks in defense as well as critique. In *Minnesota Studies in the Philosophy of Science*, vol. 4, ed. M. Radner and S. Winokur (Minneapolis: University of Minnesota Press), 3–15.

Feyerabend, P. 1975: *Against Method* (London: Verso).

—— 1981: *Realism, Rationalism and Scientific Method* (Cambridge: Cambridge University Press).

Field, H. 1972: Tarski's theory of truth. *Journal of Philosophy*, 13, 347–75.

—— 1973: Theory change and the indeterminacy of reference. *Journal of Philosophy*, 70, 462–81.

—— 1994: Disquotational truth and factually defective discourse. *Philosophical Review*, 103, 405–52.

Fodor, J. 1984: Observation reconsidered. *Philosophy of Science*, 51, 23–43.

Giere, R. 1988: *Explaining Science* (Chicago: University of Chicago Press).

Grice, H. P., and Strawson, P. F. 1956: In defence of a dogma. *Philosophical Review*, 65, 141–58.

Hacking, I. 1983: *Representing and Intervening* (Cambridge: Cambridge University Press).

Hempel, C. 1965: *Aspects of Scientific Explanation* (New York: Free Press).

—— 1970: On the "standard conception" of scientific theories. In *Minnesota Studies in the Philosophy of Science*, vol. 4, ed. M. Radner and S. Winokur (Minneapolis: University of Minnesota Press), 142–63.

—— 1973: The meaning of theoretical terms: a critique of the standard empiricist construal. In *Logic, Methodology and Philosophy of Science*, vol. 4, ed. P. Suppes et al. (Amsterdam: North-Holland).

Hintikka, J. 1998: Ramsey sentences and the meaning of quantifiers. *Philosophy of Science*, 65, 289–305.

Horwich, P. 1990: *Truth* (Oxford: Blackwell).

—— 1993: The essence of expressivism. *Analysis*, 54, 19–20.

—— 1997: Implicit definition, analytic truth, and a priori knowledge. *Nous*, 31, 423–40.

Jackson, F., Oppy, G., and Smith, M. 1994: Minimalism and truth aptness. *Mind*, 103, 287–302.

Kitcher, P. 1993: *The Advancement of Science* (Oxford: Oxford University Press).

Kripke, S. 1980: *Naming and Necessity* (Oxford: Blackwell).

Kroon, R. 1985: Theoretical terms and the causal view of reference. *Australasian Journal of Philosophy*, 63, 143–66.

Kuhn, T. S. 1962: *The Structure of Scientific Revolutions* (Chicago: University of Chicago Press).

Lewis, D. 1969: *Convention* (Oxford: Blackwell).

—— 1972: Psychophysical and theoretical identifications. *Australasian Journal of Philosophy*, 50, 249–58.

—— 1983: *Philosophical Papers*, vol. 1 (Oxford: Oxford University Press).

—— 1994: Reduction of mind. In *Companion to the Philosophy of Mind*, ed. S. Guttenplan (Oxford: Blackwell).

Martin, M. 1971: Referential variance and scientific objectivity. *British Journal for the Philosophy of Science*, 22, 17–26.

McDowell, J. 1981: Anti-realism and the epistemology of understanding. In *Meaning and Understanding*, ed. H. Parret and J. Bouveresse (Berlin and New York: de Gruyter), 225–48.

McGinn, C. 1980: Truth and use. In *Reference, Truth, and Reality*, ed. M. Platts (London: Routledge and Kegan Paul), 2–40.

Miller, R. 1992: A purely causal solution to one of the qua problems. *Australasian Journal of Philosophy*, 70, 425–34.

Newton-Smith, W. H. 1985: Berkeley's philosophy of science. In *Essays on Berkeley*, ed. J. Foster and H. Robinson (Oxford: Clarendon Press), 149–62.

Nola, R. 1980: Fixing the reference of theoretical terms. *Philosophy of Science*, 47, 505–31.

O'Leary-Hawthorne, J. 1994: A corrective to the Ramsey–Lewis account of theoretical terms. *Analysis*, 54, 105–10.

Papineau, D. 1996: Theory-dependent terms. *Philosophy of Science*, 63, 1–20.

Peacocke, C. 1976: Truth definitions and actual languages. In *Truth and Meaning*, ed. G. Evans and J. McDowell (Oxford: Oxford University Press), 162–88.

Putnam, H. 1975a: *Mathematics, Matter, and Method* (Cambridge: Cambridge University Press).

—— 1975b: *Mind, Language, and Reality* (Cambridge: Cambridge University Press).

Quine, W. V. O. 1953: *From a Logical Point of View* (New York: Harper and Row).

—— 1960: *Word and Object* (Cambridge, MA: MIT Press).

—— 1969: *Ontological Relativity and Other Essays* (New York: Columbia University Press).

—— 1970: On the reasons for indeterminacy of translation. *Journal of Philosophy*, 67, 178–83.

Scheffler, I. 1967: *Science and Subjectivity* (New York: Bobbs-Merrill).

van Fraassen, B. 1980: *The Scientific Image* (Oxford: Clarendon Press).

—— 1987: The semantic approach to scientific theories. In *The Process of Science*, ed. N. Nersessian (Dordrecht: Martinus Nijhof), 105–24.

Ulises Moulines, C. 1985: Theoretical terms and bridge principles: a critique of Hempel's (self-)criticisms. *Erkenntnis*, 22, 97–117.

Wiggins, D. 1993: Putnam's doctrine of natural kind words and Frege's doctrines of sense, reference, and extension: can they cohere? In *Meaning and Reference*, ed. A. Moore (Oxford: Oxford University Press), 192–207.

Winnie, J. 1967: The implicit definition of theoretical terms. *British Journal for the Philosophy of Science*, 18, 223–9.

73

Theories

RONALD N. GIERE

Some decades ago, Fred Suppe (1974, p. 3) remarked that "it is only a slight exaggeration to claim that a philosophy of science is little more than an analysis of theories and their roles in the scientific enterprise." The truth of this remark is attested by the fact that so many topics in contemporary philosophy of science continue to be framed in terms of theories. The issue of realism and instrumentalism, for example, is typically understood as the question of whether various terms in statements making up scientific theories refer to real objects or merely serve the role of facilitating inferences among claims about observations (see REALISM AND INSTRUMENTALISM). Again, reduction has often been seen as depending on whether statements in one theory can be logically deduced from those in another theory (see REDUCTIONISM). Similarly, scientific change has been understood as the replacement of one theory by another (see SCIENTIFIC CHANGE). Finally, relativism is typically portrayed as the view that the choice of one theory over another has no "objective" or "rational" basis, but depends merely on the interests of those with the power to enforce their decision (see RELATIVISM). In fact, the framing of these and many other issues not only centers on relationships involving theories, it often presumes a particular account of the general nature of theories.

So ingrained has been the "theory centrism" of contemporary philosophy of science that it often seems difficult even to imagine that things could be otherwise. Yet it need not be so and, indeed, has not always been so. The writings of both Mill and Whewell, for example, exhibit no preoccupation with the nature of scientific theories (see MILL and WHEWELL). Of course, these and other earlier philosophers of science talked about particular scientific theories, but their analyses of philosophical issues were not framed in terms of any particular account of the nature of scientific theories. They were much more concerned with the status of laws of nature (see LAWS OF NATURE). So the idea that an analysis of scientific theories lies at the core of a philosophy of science does not arise directly from a philosophical concern with science itself, but must be part of a particular interpretation of the nature of science. One must inquire how this interpretation arose.

The development of the classical view of scientific theories

Here the term "classical view" of scientific theories will be used in place of the expression "received view" popularized in the 1970s by philosophers such as Suppe. By whatever name, this view had its origins in Europe, particularly the German-speaking regions, in the early decades of the twentieth century. Its original proponents were self-proclaimed "scientific philosophers," such as Rudolf Carnap and Hans Reichenbach, who founded what later became logical empiricism (see LOGICAL EMPIRICISM). Their original idea was to use Einstein's theory of relativity and quantum mechanics as the basis for an analysis of the nature of space, time, and causality (see EINSTEIN; QUANTUM MECHANICS; SPACE, TIME AND RELATIVITY; and CAUSATION). These basic concepts, they argued, were to be understood *scientifically*, through an analysis of the appropriate scientific theories, and not through any process of extra-scientific reasoning, as advocated by the neo-Kantian philosophers who then dominated German philosophy.

If understanding basic ontological categories such as space and causality is the province of natural science, what is left for philosophy? Is there a role for a philosophy of science that is distinct from the sciences? Yes. Philosophy of science, they said, becomes the logical analysis of scientific concepts and theories. As Carnap (1937, p. xiii) put it: "Philosophy is to be replaced by the logic of science – that is to say, by the logical analysis of the concepts and sentences of the sciences, for the logic of science is nothing other than the logical syntax of the language of science." This is not an empirical, but a logical (analytic) task. Here talk of "logical analysis" has a very specific meaning. The background was provided by Hilbert's formalization of geometry, Peano's axiomatization of arithmetic, and the attempted reduction of mathematics to logic by Russell and Whitehead (see RUSSELL). It thus came to be assumed that, for purposes of philosophical analysis, any scientific theory could ideally be reconstructed as an axiomatic system formulated within the framework of Russell's logic. Further analysis of a particular theory could then proceed as the logical investigation of its ideal logical reconstruction. Claims about theories in general were couched as claims about such logical systems.

In both Hilbert's geometry and Russell's logic an attempt was made to distinguish between logical and nonlogical terms. Thus the symbol "&" might be used to indicate the logical relationship of conjunction between two statements, while "P" is supposed to stand for a nonlogical predicate. As in the case of geometry, the idea was that underlying any scientific theory is a purely formal logical structure captured in a set of axioms formulated in the appropriate formal language. A theory of geometry, for example, might include an axiom stating that for any two distinct Ps (points), p and q, there exists a unique L (line) such that $O(p, l)$ and $O(q, l)$, where O is a two-place relationship between Ps and Ls (p lies on l). Such axioms, taken all together, were said to provide an *implicit definition* of the meaning of the nonlogical predicates. Whatever Ps and Ls might be, they must satisfy the formal relationships given by the axioms.

The logical empiricists were not primarily logicians; they were empiricists first. From an empiricist point of view, it is not enough that the nonlogical terms of a theory be implicitly defined; they also require an empirical interpretation. This was provided by "correspondence rules" which explicitly linked some of the nonlogical terms of a theory with terms whose meaning was presumed to be given directly through "experience" or "observation." The simplest sort of correspondence rule would be one that takes the application of an observationally meaningful term, such as "dissolves," as being both necessary and sufficient for the applicability of a theoretical term, such as "soluble" (see THEORETICAL TERMS). Such a correspondence rule would provide a *complete* empirical interpretation of the theoretical term.

A definitive formulation of the classical view was finally provided by Carnap (1956), who divided the nonlogical vocabulary of theories into theoretical and observational components. The observational terms were presumed to be given a complete empirical interpretation, which left the theoretical terms with only an *indirect* empirical interpretation provided by their implicit definition within an axiom system in which some of the terms possessed a complete empirical interpretation.

Among the issues generated by Carnap's formulation was the viability of "the theory–observation distinction." Of course, one could always arbitrarily designate some subset of nonlogical terms as belonging to the observational vocabulary, but that would compromise the relevance of the philosophical analysis for any understanding of the original scientific theory. But what could be the philosophical basis for drawing the distinction? Take the predicate "spherical," for example. Anyone can observe that a billiard ball is spherical. But what about the moon, on the one hand, or an invisible speck of sand, on the other? Is the application of the term "spherical" to these objects "observational"?

Another problem was more formal. Craig's theorem seemed to show that a theory reconstructed in the recommended fashion could be re-axiomatized in such a way as to dispense with all theoretical terms, while retaining all logical consequences involving only observational terms (see CRAIG'S THEOREM). Thus, as far as the "empirical" content of a theory is concerned, it seems that we can do without the theoretical terms. Carnap's version of the classical view seemed to imply a form of instrumentalism, a problem which Hempel christened "the theoretician's dilemma."

Even the above brief summary conveys a strong sense that the development of the classical view of theories was driven more by concerns with logical structure and empiricist conceptions of meaningful language than by the examination of any genuine scientific theories. These aspects of logical empiricism provided a basis for general criticisms by philosophers of science who rejected any approach based on the logical reconstruction of theories. But the foundations of the classical view were also criticized by many who objected not to formalism as such, but to what they regarded as the wrong sort of formalism.

Formal alternatives to the classical view

In the late 1940s, the Dutch philosopher and logician Evert Beth published an alternative formalism for the philosophical analysis of scientific theories. He drew inspiration from the work of Alfred Tarski (and also Carnap) on formal semantics, but also from von Neumann's work on the foundations of quantum mechanics. (See Suppe 1989, p. 6, for more details and references.) Twenty years later, Beth's approach was developed in North America by Bas van Fraassen (1970), a Dutch emigrant who left Holland around the time Beth's works were first published. Here we follow van Fraassen (1980, 1989).

To elaborate the difference between the "syntactic" approach of the classical view and the "semantic" approach of Beth and van Fraassen, consider the following simple geometrical theory (van Fraassen 1989, pp. 218–20), presented first in the form of three axioms.

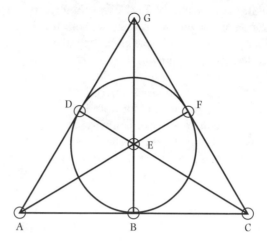

Figure 73.1

A1: For any two lines, at most one point lies on both.
A2: For any two points, exactly one line lies on both.
A3: On every line there are at least two points.

Note first that these axioms are stated in more or less everyday language. On the classical view one would have first to reconstruct these axioms in some appropriate formal language, thus introducing quantifiers and other logical symbols. And one would have to attach appropriate correspondence rules. Contrary to common connotations of the word "semantic," the semantic approach downplays concerns with language as such. Any language will do, so long as it is clear enough to make reliable discriminations between the objects which satisfy the axioms and those which do not. The concern is not so much with what can be deduced from these axioms, valid deduction being a matter of syntax alone. Rather, the focus is on "satisfaction," what satisfies the axioms – a semantic notion. These objects are, in the technical, logical sense of the term, *models* of the axioms. So, on the semantic approach, the focus shifts from the axioms, as linguistic entities, to the models, which are nonlinguistic entities.

Among the objects satisfying the three axioms above are a single line with only two points lying on it. A more interesting model is the set of seven lines (including the circle) and seven points pictured in figure 73.1. Simple inspection reveals that this figure satisfies each of the three axioms. Here the expression "this figure" may be taken to refer to the actual paper and ink material on the page, provided one adopts an appropriate *interpretation* for the terms "point," "line," and "lies on." Another more obviously physical model could be constructed by driving seven nails into a board following the configuration in figure 73.1. The nails could then be connected by wires looped around each nail respectively. One could also consider a completely abstract interpretation in which a point is understood as something that has location only and no size whatsoever, while a line is an ideal object that has length but no breadth.

It is not enough to be in possession of a general interpretation for the terms used to characterize the models, one must also be able to *identify* particular instances – for example, a particular nail in a particular board. In real science much effort and sophisticated equipment may be required to make the required identifications – for example, of a star as a white dwarf or of a formation in the ocean floor as a transform fault. On a semantic approach, these complex processes of interpretation and identification, while essential to being able to use a theory, have no place within the theory itself. This is in sharp contrast to the classical view, which has the very awkward consequence that various innovations in instrumentation automatically require changes in our philosophical analysis of the theory itself. The semantic approach better captures the scientists' own understanding of the difference between theory and instrumentation.

On the classical view the question "What is a scientific theory?" receives a straightforward answer. A theory is (i) a set of uninterpreted axioms in a specified formal language plus (ii) a set of correspondence rules that provide a partial empirical interpretation in terms of observable entities and processes. A theory is thus true if and only if the interpreted axioms are all true. To obtain a similarly straightforward answer within a semantic approach requires looking at the axioms, in whatever language, a little differently. Return to the axioms for seven-point geometry displayed above. Rather than regarding them as free-standing statements, consider them to be part of a *theoretical definition*, a definition of seven-point geometry. The definition could be formulated as follows: Any set of points and lines constitutes a seven-point geometry if and only if A1, A2, and A3. Since a definition makes no claims about anything and is not even a candidate for truth or falsity, one can hardly identify a theory with a definition. But claims to the effect that various things satisfy the definition may be true or false of the world. Call these claims *theoretical hypotheses*. So we may say that, on the semantic approach, a theory consists of (i) a theoretical definition plus (ii) a number of theoretical hypotheses. The theory may be said to be true just in case all its associated theoretical hypotheses are true. (See Giere 1988, ch. 3.)

Adopting a semantic approach to theories still leaves wide latitude in the choice of specific techniques for formulating particular scientific theories. Following Beth, van Fraassen adopts a *state space* representation which closely mirrors techniques developed in theoretical physics during the nineteenth century – techniques which were carried over into the development of quantum and relativistic mechanics. The technique can be illustrated most simply for classical mechanics.

Consider a simple harmonic oscillator, which consists of a mass constrained to move in one dimension subject to a linear restoring force – a weight bouncing gently while hanging from a spring provides a rough example of such a system. Let "x" represent the single spatial dimension, "t" the time, "p" the momentum, "k" the strength of the restoring force, and "m" the mass. Then a linear harmonic oscillator may be *defined* as a system which satisfies the following differential equations of motion:

$$dx/dt = DH/Dp, \quad dp/dt = -DH/Dx, \text{ where } H = (k/2)x^2 + (1/2m)p^2.$$

The Hamiltonian, H, represents the sum of the kinetic and potential energy of the system. The state of the system at any instant of time is a point in a two-dimensional position–momentum space. The history of any such system in this state space is given by an ellipse, as illustrated in figure 73.2. In time the system repeatedly traces out the ellipse in state space. Its motion in real one-dimensional space is the projection of the ellipse onto the x axis. (For more details see Giere 1988, ch. 3, or any physics text

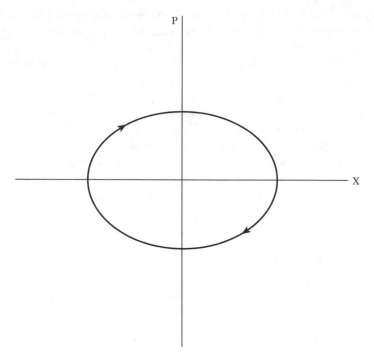

Figure 73.2

covering classical mechanics.) It remains to be discussed how well any real-world system, such as a bouncing spring, satisfies this definition.

Other advocates of a semantic approach differ from the Beth–van Fraassen point of view in the type of formalism they would employ in reconstructing actual scientific theories. One influential approach derives from the work of Patrick Suppes during the 1950s and 1960s (some of which is reprinted in Suppes 1969). Suppes was inspired by the logicians J. C. C. McKinsey and Alfred Tarski. In its original form, Suppes's view was that theoretical definitions should be formulated in the language of set theory. To characterize a theory, on this view, is to define a set-theoretical predicate – that is, a predicate formulated using the formalism of set theory. Suppes's approach, as developed by his student Joseph Sneed (1971), has been adopted widely in Europe, and particularly in Germany, by the late Wolfgang Stegmüller (1976) and his students. Frederick Suppe has for several decades developed a related approach (elaborated in Suppe 1989) that shares features of both the state space and the set-theoretical approaches.

Realism, causal necessity, and laws of nature

There are few if any issues in the philosophy of science that can be resolved simply by changing one's understanding of the nature of scientific theories. Nevertheless, changing one's view of scientific theories restructures many issues. Philosophically important differences between the classical view and the semantic view of theories may thus be better appreciated by examining how some standard issues in the philosophy of science are structured by a semantic approach.

Advocates of a semantic approach span the full range of alternatives on the issue of scientific realism. Sneed advocates a form of instrumentalism; van Fraassen is an anti-realist empiricist, but not an instrumentalist; while Giere and Suppe are (qualified) realists. The difference between van Fraassen's empiricism and the realism of Giere and Suppe can be illustrated by reconsidering van Fraassen's example of seven-point geometry. This geometry is obviously not Euclidian, because its space is not continuous – it has only a finite number of points. But it may be regarded as being *embedded* in a Euclidian space. The seven-point structure is *isomorphic* with a substructure of a Euclidian structure and thus may be regarded as a *sub-model* of a more inclusive Euclidian model. These notions are clearest when both the seven-point space and the enclosing Euclidian space are purely abstract mathematical spaces. But the notions of embedding and sub-models are still fairly clear if we move to physical models such as lines on a page or nails with attached wires pounded into a flat board.

Van Fraassen extends these ideas to physical theories exemplified by our theory of simple harmonic oscillators. He would distinguish between the state variables, position and momentum, and the theoretical variables, kinetic and potential energy. The behavior of the state variables constitutes a sub-model of the full model which includes values for the total energy of the system. Identifying the state variables as *observable phenomena* and the energy as a *theoretical quantity*, van Fraassen would allow that hypotheses about both may be true or false. But, he argues, only claims about the embedded, observable sub-models may legitimately be believed. Science, he argues, aims only for *empirical adequacy* (truth regarding the observable phenomena), not full correspondence between theoretical definitions and empirical models.

Van Fraassen's arguments for these views are acknowledged to be independent of his commitment to a semantic view of theories. And so they are. But the semantic framework clarifies the issues. It makes clear, for example, that van Fraassen requires an identification of the distinction between state variables and theoretical variables with a traditional epistemological distinction between what is observable and what is not. There seems little basis for such an identification. Neither the position nor the momentum of a rapidly oscillating, but otherwise clearly observable object, such as a billiard ball, is observable to the unaided human observer. On the other hand, physicists clearly wish to regard the position of atomic particles, such as Rutherford's α-rays, as observable.

One advantage of a state space representation is that physical modalities (e.g., natural necessity) are so easily represented. For example, the trajectory of the state of a harmonic oscillator in position–momentum space represents all the *physically possible* states of that system. Moreover, the theoretical definition permits one to calculate what the physically possible states *would be* if the parameters of the system were different in ways in which they in fact never will be. Empiricists, such as van Fraassen, insist that only the actual sequence of states is physically real. The attribution of necessity or counterfactual truth to states in any real system is a mistake. These aspects of the state space are merely artifacts of the form of representation. Realists, such as Giere or Suppe, insist that the state space represents the underlying causal structure of such systems, so that counterfactual claims about what the state might be under some unrealized condition are true if they correctly reflect that causal structure. Either way, the issue is clarified by being framed in terms of state space representations.

In the framework of the classical view of theories, questions about causal necessity are typically framed as being about the status of *scientific laws of nature* (see LAWS OF NATURE). Laws play a major role in that account, because they constitute the nonlogical axioms that give theories most of their empirical content or instrumental value. Within this framework, the minimal empiricist view is that laws are true statements of universal scope. Realists typically argue that the laws also express some sort of necessity. Philosophers inclined toward a semantic account of theories (including Cartwright (1983)) have recently begun to insist that there simply are no laws, not even just true universal generalizations, that can play the role which the classical view of scientific theories requires.

Take the equations in the definition of a simple harmonic oscillator given above to constitute a set of putative universal laws intended to describe the actual behavior of a designated class of real objects. The problem is that either the intended class must be empty, or the supposed laws false, and thus no laws at all. There are in the real world no systems answering to these laws. The reason is that the simple harmonic oscillator, as described by these laws, is a perpetual motion machine. Its total energy is a constant, which means that there could be no friction of any kind anywhere in the system. Such systems do not exist in the real world. Nor can this difficulty be eliminated by complicating the supposed laws in any straightforward way. Frictional forces are too complex to be captured by any even moderately complex formulae. Although such problems are obvious even to beginning students of classical physics, they tended to be overlooked by a tradition that focused attention on simple qualitative predicates like "red" or "soluble."

A generalized model-based picture of scientific theories

Most of those who have developed "semantic" alternatives to the classical "syntactic" approach to the nature of scientific theories were inspired by the goal of reconstructing scientific theories – a goal shared by advocates of the classical view. Many philosophers of science now question whether there is any point in creating *philosophical* reconstructions of scientific theories. Rather, insofar as the philosophy of science focuses on theories at all, it is the scientific versions, in their own terms, that should be of primary concern. But many now argue that the major concern should be directed toward the whole practice of science, in which theories are but a part. In these latter pursuits what is needed is not a technical framework for reconstructing scientific theories, but merely a general *interpretive* framework for talking about theories and their various roles in the practice of science. This becomes especially important when considering sciences such as biology, in which mathematical models play less of a role than in physics. (See Beatty 1981; Lloyd 1988; and Thompson 1989 for applications of a model-based approach to biology.)

Here there are strong reasons for adopting a generalized model-based understanding of scientific theories which makes no commitments to any particular formalism

– for example, state spaces or set-theoretical predicates. In fact, one can even drop the distinction between "syntactic" and "semantic" as a leftover from an old debate. The important distinction is between an account of theories that takes *models* as fundamental versus one that takes *statements*, particularly laws, as fundamental. A major argument for a model-based approach is that just given. There seem in fact to be few, if any, universal statements that might even plausibly be true, let alone known to be true, and thus available to play the role which laws have been thought to play in the classical account of theories. Rather, what have often been taken to be universal generalizations should be interpreted as parts of definitions. Here it may be helpful to introduce explicitly the notion of an idealized, *theoretical model*, an abstract entity which answers precisely to the corresponding theoretical definition. Theoretical models thus provide, though only by fiat, something of which theoretical definitions may be true. This makes it possible to interpret much of scientists' theoretical discourse as being about theoretical models rather than directly about the world. What have traditionally been interpreted as laws of nature thus turn out to be merely statements describing the behavior of theoretical models.

If one adopts such a generalized model-based understanding of scientific theories, one must characterize the relationship between theoretical models and real systems. Van Fraassen (1980) suggests that it should be one of *isomorphism*. But the same considerations that count against there being true laws in the classical sense also count against there being anything in the real world strictly isomorphic to any theoretical model, or even isomorphic to an "empirical" sub-model. What is needed is a weaker notion of similarity, for which it must be specified both in which respects the theoretical model and the real system are similar, and to what degree. These specifications, however, like the interpretation of terms used in characterizing the model and the identification of relevant aspects of real systems, are not part of the model itself. They are part of a complex practice in which models are constructed and tested against the world in an attempt to determine how well they "fit."

Divorced from its formal background, a model-based understanding of theories is easily incorporated into a general framework of naturalism in the philosophy of science (see NATURALISM). It is particularly well-suited to a cognitive approach to science (see COGNITIVE APPROACHES TO SCIENCE). Many forms of representation now discussed in the cognitive sciences utilize models of some sort. Many of these are presumed to be embodied as mental states in the brains of real people, unlike the abstract theoretical models discussed above. But it is very plausible to suppose that physicists, for example, possess mental models, or at least partial mental models, for harmonic oscillators and other staples of both classical and contemporary physics. It is the possession of such mental models that makes it possible for them to recognize a new situation as one for which a particular sort of theoretical model is appropriate.

Finally, a generalized model-based understanding of scientific theories makes contact with the Kuhnian notion of theoretical science as grounded in exemplary problem solutions rather than abstract laws or theories (see KUHN). Exemplary problem solutions may be seen as based on exemplary models. One could even follow Kuhn into the social dimensions of science. Contemporary theories are notoriously difficult to isolate. Scientists who ostensibly share fundamental principles nevertheless disagree on how they are to be deployed in constructing models of particular phenomena (see Hull

1988, ch. 6, on evolutionary theory). The family of models one would identify with a high-level theory, such as evolutionary theory or quantum mechanics, turns out not to be anywhere localized, but distributed among many scientists operating in diverse specialities. The idea that theories are well-defined entities seems to have been an artifact of the classical view of theories that led philosophers to identify a theory such as Newtonian mechanics with a definite set of propositions – for example, Newton's three laws plus the law of universal gravitation. A model-based understanding of scientific theories provides resources for appreciating the illusiveness of theories in the practice of science.

References

Beatty, J. 1981: What's wrong with the received view of evolutionary theory? In *PSA 1980*, vol. 2, ed. P. D. Asquith and R. N. Giere (East Lansing, MI: Philosophy of Science Association), 397–426.

Carnap, R. 1937: *The Logical Syntax of Language* (London: Routledge and Kegan Paul).

—— 1956: The methodological character of theoretical concepts. In *Minnesota Studies in the Philosophy of Science*, vol. 1: *The Foundations of Science and the Concepts of Psychology and Psychoanalysis*, ed. H. Feigl and M. Scriven (Minneapolis: University of Minnesota Press), 38–76.

Cartwright, N. 1983: *How the Laws of Physics Lie* (Oxford: Clarendon Press).

Giere, R. N. 1988: *Explaining Science: A Cognitive Approach* (Chicago: University of Chicago Press).

Hull, D. 1988: *Science as a Process: An Evolutionary Account of the Social and Conceptual Development of Science* (Chicago: University of Chicago Press).

Lloyd, E. 1988: *The Structure and Confirmation of Evolutionary Theory* (New York: Greenwood Press).

Sneed, J. D. 1971: *The Logical Structure of Mathematical Physics* (Dordrecht: Reidel).

Stegmüller, W. 1976: *The Structure and Dynamics of Theories* (Berlin: Springer).

Suppe, F. 1989: *The Semantic Conception of Theories and Scientific Realism* (Urbana, IL: University of Illinois Press).

Suppe, F. (ed.) 1974: *The Structure of Scientific Theories* (Urbana, IL: University of Illinois Press; 2nd edn, 1977).

Suppes, P. 1969: *Studies in the Methodology and Foundations of Science: Selected Papers from 1951 to 1969* (Dordrecht: Reidel).

Thompson, P. 1989: *The Structure of Biological Theories* (Albany, NY: SUNY Press).

van Fraassen, B. 1970: On the extension of Beth's semantics of physical theories. *Philosophy of Science*, 37, 325–39.

—— 1980: *The Scientific Image* (Oxford: Oxford University Press).

—— 1989: *Laws and Symmetry* (Oxford: Oxford University Press).

74

Theory Identity

FREDERICK SUPPE

In 1925 the *old quantum mechanics* of Planck, Einstein, and Bohr was replaced by the *new (matrix) quantum mechanics* of Born, Heisenberg, Jordan, and Dirac. In 1926 Schrödinger developed *wave mechanics*, which proved to be equivalent to matrix mechanics in the sense that they led to the same energy levels. Dirac and Jordan joined the two theories into one *transformation quantum theory*. In 1932 von Neumann presented his Hilbert space formulation of quantum mechanics and proved a *representation theorem* showing that sequences in transformation theory were isomorphic to sequences in the Hilbert space formulation (see QUANTUM MECHANICS). Three different notions of theory identity are involved here: theory individuation, theoretical equivalence, and empirical equivalence.

Individuation of theories

What determines whether theories T_1 and T_2 are instances of the same theory or distinct theories? By construing scientific theories as partially interpreted syntactical axiom systems TC, positivism made specifics of the axiomatization individuating features of the theory. Thus different choices of axioms T or alterations in the correspondence rules – say, to accommodate a new measurement procedure – resulted in a new scientific theory. Positivists also held that axioms and correspondence rules implicitly defined the meanings of the theory's descriptive terms τ. Thus significant alterations in the axiomatization would result not only in a new theory $T'C'$ but one with changed meanings τ'. Kuhn and Feyerabend maintained that the resulting changes could make TC and $T'C'$ noncomparable, or *incommensurable*. Attempts to explore individuation issues for theories via meaning change or incommensurability proved unsuccessful and have been largely abandoned (see INCOMMENSURABILITY).

Individuation of theories in actual scientific practice is at odds with the positivistic analysis. For example, difference equation, differential equation, and Hamiltonian versions of classical mechanics (CM) are all formulations of one theory, though they differ in how fully they characterize CM. It follows that syntactical specifics of theory formulations cannot be individuating features, which is to say that scientific theories are not *linguistic entities*. Rather, theories must be some sort of *extra-linguisitic structure* which can be referred to via alternative and even inequivalent formulations (as with CM). Also, the various experimental designs, etc., incorporated into positivistic correspondence rules cannot be individuating features of theories. For improved instrumentation or experimental technique does not automatically produce a new theory. Accommodating these individuation features was a main motivation for *the semantic conception of theories* where theories are state spaces or other extra-linguistic structures standing in mapping relations to phenomena (see THEORIES).

Scientific theories undergo development, are refined, and change. Both syntactic and semantic analyses of theories concentrate on theories at mature stages of development, and it is an open question whether either approach adequately individuates theories undergoing active development.

Theoretical equivalence

Under what circumstances are two theories equivalent? On syntactical approaches, axiomatizations T_1 and T_2 having a common definitional extension would be sufficient. Robinson's theorem (see CRAIG'S THEOREM) says that T_1 and T_2 must have a model in common to be compatible. They will be equivalent if they have precisely the same (or equivalent) sets of models. On the semantic conception the theories will be two distinct sets of structures (models) M_1 and M_2. The theories will be equivalent just in case we can prove a *representation theorem* showing that M_1 and M_2 are *isomorphic* (structurally equivalent). In this way von Neumann showed that transformation quantum theory and the Hilbert space formulation were equivalent.

Empirical equivalence

Many philosophers contend that only part of the structure or content of theories is descriptive of empirical reality. Under what circumstances are two theories identical or equivalent in empirical content? Positivists viewed theories as having a separable observational or empirical component, O, which could be described in a theory-neutral observation language. Let O_1 and O_2 be the observational content of two theories. The two theories are empirically equivalent just in case O_1 and O_2 meet appropriate requirements for theoretical equivalence. The notion of a theory-independent observation language was challenged by the view that observation and empirical facts were *theory-dependent*. Thus, syntactically equivalent O_1 and O_2 might be not be empirically equivalent. (See OBSERVATION AND THEORY and INCOMMENSURABILITY.)

In van Fraassen's version of the semantic conception, a theory formulation T is given a semantic interpretation in terms of a logical space into which lots of models can be mapped. (This presupposes his theory of semi-interpreted languages, for which see RAMSEY SENTENCES.) A theory is *empirically adequate* if the actual world A is among those models. Two theories are empirically equivalent if for each model M of T_1 there is a model M' of T_2 such that all empirical substructures of M are isomorphic to empirical substructures of M', and vice versa with T_1 and T_2 exchanged. In this sense wave and matrix mechanics are equivalent. Van Fraassen assumes an observability/nonobservability distinction in presenting his empirical adequacy account, but this notion and his formal account can be divorced from such distinctions, the generalized empirical adequacy notion being applicable whenever not all of the structure in a theory (e.g., the dimensionality of the state space) corresponds to reality.

For all the above accounts the underlying idea is that two theories are empirically equivalent if those sub-portions of the theories making empirically ascertainable claims are consistent (in the sense of Robinson's theorem) and assert the same facts. Suppe's quasi-realistic version of the semantic conception (which employs no observability/ nonobservability distinction) maintains that theories with variables v purport to describe only how the world A would be if phenomena were isolated from all influences other than v. Thus the theory structure M stands in a *counterfactual mapping relation* to the actual world A. Typically, neither A nor its v portion will be among the sub-models of empirically true theory M, so true theories will not be empirically adequate. Further, two closely related theories M_1 and M_2 with variables v_1 and v_2 could both be counterfactually true of the actual world without their formulations having extensional models in common (violating Robinson's theorem). Thus issues of empirical equivalence largely become preempted by questions of empirical truth for the theories.

Further reading

Ellis, B. 1965: The origin and nature of Newton's laws of motion. In *Beyond the Edge of Certainty*, ed. R. Colodny (Englewood Cliffs, NJ: Prentice-Hall), 29–68.

Glymour, C. 1973: Theoretical realism and theoretical equivalence. In *Boston Studies in the Philosophy of Science*, Vol. 8, ed. R. Buck and R. Cohen (Dordrecht: D. Reidel), 275–88.

Schrödinger, E. 1926: "Quantisierung als Eigenwertproblem," *Annalen der Physik*, ser. 4: 79, pp. 361–76, 489–527; 80, pp. 437–90; 81, pp. 109–39. Translated as "Quantitization as an eigenvalue problem" by J. F. Shearer and W. W. Deans in E. Schrödinger, *Collected Papers on Wave Mechanics* (London: Glasgow, 1928).

Stegmüller, W. 1976: *The Structure and Dynamics of Theories*, trans. W. Wohlhueter (New York: Springer-Verlag); 1st pub. 1973.

Suppe, F. 1977: *The Structure of Scientific Theories*, 2nd edn (Urbana, IL: University of Illinois Press), esp. 16–86, 125–208, 624–46.

—— 1989: *The Semantic Conception of Theories and Scientific Realism* (Urbana, IL: University of Illinois Press), ch. 1; pt II; chs 10, 14.

van Fraassen, B. 1980: *The Scientific Image* (Oxford: Oxford University Press), esp. ch. 3.

von Neumann, J. 1932: *Mathematische Gründlagen der Quantenmechanik* (Berlin), trans. R. T. Beyer as *Mathematical Foundations of Quantum Mechanics* (Princeton: Princeton University Press, 1955).

75

Thought Experiments

JAMES ROBERT BROWN

We need only list a few of the well-known thought experiments to be reminded of their enormous influence and importance in the sciences: Newton's bucket, Maxwell's demon, Einstein's elevator, Heisenberg's gamma-ray microscope, Schrödinger's cat. The seventeenth century saw some of its most brilliant practitioners in Galileo, Descartes, Newton, and Leibniz. And in our own time, the creation of quantum mechanics and relativity are almost unthinkable without the crucial role played by thought experiments.

Galileo and Einstein were, arguably, the most impressive thought experimenters, but they were by no means the first. Thought experiments existed throughout the Middle Ages and can be found in antiquity too. One of the most beautiful early examples (from Lucretius, *The Nature of Things*) attempts to show that space is infinite: If there is a boundary to the universe, we can toss a spear at it. If the spear flies through, it isn't a boundary after all; if the spear bounces back, then there must be something beyond the supposed edge of space – a cosmic wall which is itself in space – that stopped the spear. Either way, there is no edge of the universe; space is infinite.

This example nicely illustrates many of the common features of thought experiments: We visualize some situation; we carry out an operation; we see what happens. It also illustrates their fallibility: in this case we've learned how to conceptualize space so that it is both finite and unbounded.

Often a real experiment that is the analogue of a thought experiment is impossible for physical, technological, or just plain practical reasons; but this need not be a defining condition of thought experiments. The main point is that we seem able to get a grip on nature just by thinking, and therein lies the great interest for philosophy. How is it possible to learn (apparently) new things about nature *without* new empirical data?

Ernst Mach (who seems to have coined the expression *Gedankenexperiment*) developed an interesting empiricist view in his classic, *The Science of Mechanics* (1960). We possess, he says, a great store of "instinctive knowledge" picked up from experience. This need not be articulated at all, but comes to the fore when we consider certain situations. One of his favorite examples is due to Simon Stevin. When a chain is draped over a double frictionless plane, as in figure 75.1a, how will it move? Add some links, as in figure 75.1b. Now it is obvious. The initial setup must have been in static equilibrium. Otherwise, we would have a perpetual motion machine; and, according to our experience-based "instinctive knowledge," says Mach, this is impossible (see MACH).

Thomas Kuhn's "A function for thought experiments" (1964) employs many of the concepts (but not the terminology) of his well-known *Structure of Scientific Revolutions*. On his view a well-conceived thought experiment can bring on a crisis or at least

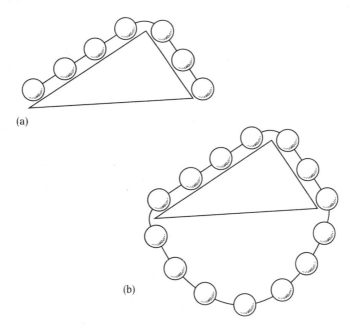

(a)

(b)

Figure 75.1

create an anomaly in the reigning theory and so contribute to paradigm change. So thought experiments can teach us something new about the world, even though we have no new data, by helping us to reconceptualize the world in a better way.

Recent years have seen a sudden growth of interest in thought experiments. The views of Brown (1991) and Norton (1991) represent the extremes of Platonic rationalism and classic empiricism, respectively. Norton claims that any thought experiment is really a (possibly disguised) argument; it starts with premises grounded in experience and follows deductive or inductive rules of inference in arriving at its conclusion. The picturesque features of any thought experiment which give it an experimental flavor may be psychologically helpful, but are strictly redundant. Thus, says Norton, we never go beyond the empirical premises in a way to which any empiricist would object. (For criticisms see Brown 1991, 1993, or Gendler 1998, and for a defense see Norton 1996.)

By contrast, Brown holds that in a few special cases we do go well beyond the old data to acquire a priori knowledge of nature. Galileo showed that all bodies fall at the same speed with a brilliant thought experiment that started by destroying the then reigning Aristotelian account. The latter held that heavy bodies fall faster than light ones ($H > L$). But consider a heavy cannon ball attached to a light musket ball ($H + L$); it must fall faster than the cannon ball alone (fig. 75.2). Yet the compound object must also fall more slowly, since the light part will act as a drag on the heavy part. Now we have a contradiction ($H + L > H$ and $H > H + L$). That's the end of Aristotle's theory; but there is a bonus, since the right account is now obvious: they all fall at the same speed ($H = L = H + L$).

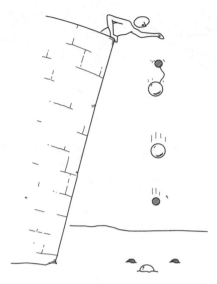

Figure 75.2

This is said to be a priori (though still fallible) knowledge of nature, since no new data are involved, nor is the conclusion derived from old data, nor is it some sort of logical truth. This account of thought experiments is further developed by linking the a priori epistemology to a recent account of laws of nature which holds that laws are relations between objectively existing abstract entities. It is thus a rather Platonistic view, not unlike Platonistic accounts of mathematics such as that urged by Gödel. (For details see Brown 1991.)

The two views just sketched might occupy the opposite ends of a spectrum of positions on thought experiments. Some of the promising new alternative views include those of Sorensen (somewhat in the spirit of Mach), who holds that thought experiments are a "limiting case" of ordinary experiments; they can achieve their aim, he says, without being executed. (Sorensen's book (1992) is also valuable for its extensive discussion of thought experiments in philosophy of mind, ethics, and other areas of philosophy, as well as the sciences.) Other promising views include those of Gooding (who stresses the similar procedural nature of thought experiments and real experiments), Miščević, and Nersessian (each of whom tie thought experiments to "mental models"), several of the accounts in Horowitz and Massey 1991, and recent works by Humphreys (1994) and Gendler (1998).

References and further reading

Brown, J. R. 1991: *Laboratory of the Mind: Thought Experiments in the Natural Sciences* (London: Routledge).

——— 1993: Why empiricism won't work. In *PSA 1992*, vol. 2, ed. D. Hull, M. Forbes, and K. Okruhlik (East Lansing, MI: Philosophy of Science Association).

Gendler, T. 1998: Galileo and the indispensability of scientific thought experiments. *British Journal for the Philosophy of Science*, 49 (Sept.), 397–424.

Gooding, D. 1993: What is *experimental* about thought experiments? In *PSA 1992*, vol. 2, ed. D. Hull, M. Forbes, and K. Okruhlik (East Lansing, MI: Philosophy of Science Association), 280–90.

Hacking, I. 1993: Do thought experiments have a life of their own? In *PSA 1992*, vol. 2, ed. D. Hull, M. Forbes, and K. Okruhlik (East Lansing, MI: Philosophy of Science Association), 302–10.

Horowitz, T., and Massey, G. (eds) 1991: *Thought Experiments in Science and Philosophy* (Savage, MD: Rowman and Littlefield).

Humphreys, P. 1994: Seven theses on thought experiments. In *Philosophical Problems of the Internal and External Worlds*, ed. J. Earman et al. (Pittsburgh: University of Pittsburgh Press), 205–27.

Kuhn, T. 1964: A function for thought experiments. Repr. in Kuhn, *The Essential Tension* (Chicago: University of Chicago Press, 1977), 240–65.

—— 1970: *The Structure of Scientific Revolutions*, 2nd edn (Chicago: University of Chicago Press).

Mach, E. 1960: *The Science of Mechanics*, trans. J. McCormack, 6th edn (LaSalle, IL: Open Court).

—— 1976: On thought experiments. In *Knowledge and Error* (Dordrecht: Reidel), 79–91.

Miščević, N. 1992: Mental models and thought experiments. *International Studies in the Philosophy of Science*, 6(3), 215–26.

Nersessian, N. 1993: In the theoretician's laboratory: thought experimenting as mental modeling. In *PSA 1992*, vol. 2, ed. D. Hull, M. Forbes, and K. Okruhlik (East Lansing, MI: Philosophy of Science Association), 291–301.

Norton, J. 1991: Thought experiments in Einstein's work. In Horowitz and Massey 1991, 129–48.

—— 1996: Are thought experiments just what you thought? *Canadian Journal of Philosophy*, 26, 333–66.

Sorensen, R. 1992: *Thought Experiments* (Oxford: Oxford University Press).

76

Underdetermination of Theory by Data

W. H. NEWTON-SMITH

It is a familiar fact in the practice of science that the available observational evidence may not decide between rival hypotheses or theories. For instance, at the time of Copernicus it was widely held that his theory and the Ptolemaic theory did not differ in their predictions in regard to the available astronomical data. This situation can be illustrated by an analogy. Imagine a finite number of dots on a page of paper representing the available evidence. It will always be possible to draw more than one curve connecting the points. How do we decide which curve or theory to adopt? A standard response is to look to some area where the theories make different predictions. We then seek new evidence, perhaps by conducting an experiment, to determine which theory's prediction is correct and opt tentatively for that one. If it is not feasible to gather this evidence, we can be agnostic for the moment on the question of which it is best to adopt. Or, more likely, we will look to factors other than fit with the data to assist us in resolving the matter (see PRAGMATIC FACTORS IN THEORY ACCEPTANCE). We will call this phenomena *weak underdetermination of theories*, or WUT.

Some philosophers of science have entertained a much more exciting thesis of underdetermination. Perhaps there could be rival theories that no data could decide between; perhaps all theories are underdetermined by all actual and possible observational evidence. According to this thesis of the *strong underdetermination of theories*, or SUT, any scientific theory has an incompatible rival theory to which it is empirically equivalent. That is, for any theory T_1, there is another theory T_2 with which it is inconsistent but which makes exactly the same observational predictions. In this case even an observationally omniscient God who knew the observational states of the entire universe past, present, and future, would not be able to decide on that basis alone between T_1 and T_2. This thesis is both outlandish and controversial. Some have denied that it is even intelligible. Some who think it intelligible hold it to be obviously true. Others argue that there is not a shred of evidence in its favor. The attention it has received is not surprising in light of its consequences for the realist perspective on science.

A popular version of realism (see REALISM AND INSTRUMENTALISM) holds that the diligent application of the scientific method provides us with theories which give ever more approximately true accounts of the world, in particular of the underlying theoretical entities and structures which explain our observations. On this realist picture, the history of a mature science like physics is a sequence of theories $T_1, T_2, \ldots, T_n$ which give ever better accounts of how the world is. However, given SUT, there is some other sequence of possible theories $T'_1, T'_2, \ldots, T'_n$, where each T_i and T'_i are incompatible but empirically equivalent. That being so, we can entertain the fantasy of a community of physicists on Alpha Centauri whose history was represented by the latter sequence. These physicists would have achieved precisely the same predictive and manipulative power over the world as we have. If this is possible, we have no grounds for thinking that it is our sequence of theories which has been getting nearer the truth. Perhaps it is our scientific colleagues on Alpha Centauri who have the truer picture of the underlying theoretical structures in the world. SUT, *if* true, represents a decisive blow to realism. On the other hand, the instrumentalist, for whom theories are mere tools for predicting and manipulating the observable world, is not threatened by SUT. Indeed, instrumentalists positively welcome SUT as a primary weapon in their struggle with the realists.

Not surprisingly, realists have been quick to argue against SUT. Some hold that the thesis is not even intelligible, on the grounds that it presupposes an untenable dichotomy between the observable and the theoretical. SUT just is the thesis that two theories could give the same results at the observable level while making incompatible claims about the world at the theoretical level. So if there is no difference in kind between the observable and nonobservable or theoretical, the thesis cannot even be given content. For arguments that this cannot be done see OBSERVATION AND THEORY.

Even setting aside this difficulty, further problems for SUT emerge if we consider how we would respond to an apparent case of underdetermination. Icabod and Isabel, let us imagine, have theories which seem to be empirically equivalent. The theories have been used by each of them to produce the same impressive predictions and technological spin-offs. In Isabel's theory there are important items, electrons and positrons. According to her, electrons are negatively charged, and positrons are positively charged. This looks incompatible with Icabod's theory, according to which electrons are positively charged and positrons negatively charged. This is not of interest. The words "electron" and "positron" have simply been swapped, giving the superficial impression of an incompatibility, whereas in fact all we have is an equivocation: a simple case of notional variance. Critics of SUT (Dummett 1973, p. 617n) have maintained that any apparent case of underdetermination can be treated similarly as a case of equivocation. Those who take SUT seriously need to provide a systematic way of meeting the challenge that in any case of empirical equivalence we really have two notional variants of the same theory, not two incompatible theories. (For further discussion see THEORY IDENTITY.)

The challenge of equivocation would arise if we had two empirically equivalent theories. But there would be difficulties in showing that two actual scientific theories really were empirically equivalent. To derive observations from a theory such as Newtonian mechanics, we need to make a host of auxiliary assumptions. In this case we need to assume among many other things that a certain device provides a good clock. It T_1 and T_2 seem to be empirically equivalent using a common set of background assumptions B, we have no guarantee that they will be empirically equivalent if, with scientific progress, we change to a different set of background assumptions B'. It may be that T_1 taken with B' leads to different observational predictions than T_2 taken with B' (see Laudan and Leplin 1991). To show convincingly that T_1 and T_2 are genuinely empirically equivalent, we would need some algorithm that showed that for any set of background assumptions B, T_1 and T_2 would lead to precisely the same observational predictions. But if we have such an algorithm, the possibility that T_1 and T_2 are mere notational variants of one another is even more worrying.

It has also been argued that strong underdetermination is only a pseudo-problem generated by taking too narrow an empiricist view of evidence. Fit with the observational data is not the only sort of evidence to be taken into account in choosing between scientific theories. For instance, we have a preference for simpler theories, all things being equal. Suppose that simplicity is not merely a pragmatic factor in theory choice (see PRAGMATIC FACTORS IN THEORY ACCEPTANCE) but is genuinely evidential. That is, suppose that T_1's being simpler than T_2 provides fallible evidence that T_1 is more likely to be true or more likely to be more approximately true than T_2. In this case, even if T_1 and T_2 were empirically equivalent and genuinely incompatible, we would still have rational grounds for preferring T_1 to T_2 as a better representation of how the world is if T_1 is the simpler theory. The appeal to simplicity is problematic (see SIMPLICITY). For it may be that simplicity only renders theories more likeable, not more likely to be true. There are a host of other allegedly evidential factors that need to be considered in this context. SUT will be worrying only if it can be shown that no factors other than fit with the data are evidentially relevant to theory choice. For a discussion of the difficulties involved in seeking to justify appeal to other factors see CONVENTION, ROLE OF.

In order to sidestep this debate, some of those who take SUT seriously have considered the stronger thesis: For any theory T_1, there is an incompatible rival theory T_2 to which it is not only empirically equivalent but also evidentially equivalent. That is, both theories fare equally well on all epistemically viable principles covering theory choice. But if the thesis is strengthened in this way, the charge of equivocation becomes still more worrying.

Are there reasons for thinking that SUT is true? Critics frequently remark that there are no cases in the actual history of the physical sciences. Proponents of SUT retort that this is not surprising. For it is difficult enough to find even one decent theory for a particular subject matter, and having found one theory, there are no incentives to develop an empirically equivalent rival theory. Still, if underdetermination is a serious matter, the absence of any convincing examples does give pause for thought. Critics of SUT think that its proponents have been misled by the analogy (see above) of the points on a page through which more than one line can be drawn. They counter that if we consider all actual and possible data, this would be like having a point on the page for each point along the x-axis, in which case there is only one curve connecting the points: namely, the curve constituted by the set of points.

Significant figures in the history of the philosophy of science have held SUT or something similar, including Duhem and van Fraassen. Quine (see QUINE) at one stage wrote that he expected general agreement that all theories were underdetermined. However, he abandoned this view on consideration of the problem of equivocation, remarking that insofar as the thesis was intelligible, there were no reasons for thinking it was true (Quine 1975). Duhem's reflections on the alleged holistic (see HOLISM) character of science remain one of the primary sources of inspiration for those who take SUT seriously. As noted above, the experimental testing of a scientific theory takes place against a background of a host of auxiliary assumptions. Should the experiment or observation not bear out the prediction derived from the theory together with the background, there is no algorithm determining what should be blamed. One might hold that the theory is at fault and seek to revise it. Or one might stick by the original theory and consequently reject or modify one or other of the auxiliary assumptions. If there is indeed always this freedom to maneuver in the face of recalcitrant results, two scientists could maintain different theories in the face of any observations by suitable revisions of their auxiliaries. But as Quine (1975) noted, this holistic assumption that theories face the tribunal of observations as a whole, and that any aspect of a theory can be maintained by making suitable adjustments elsewhere, is question begging in the context of discussions of underdetermination. For unless we already assume underdetermination, there is no reason to think that our two scientists can pull off this trick. What we find in practice is that sooner or later one avenue becomes blocked. Those proceeding down it eventually are simply unable to find reasonable revisions to make in the face of the data. Or, if they can, the combination of their theory and auxiliaries becomes transparently *ad hoc*. Only on the assumption of SUT can we be assured that our rival scientists will arrive at equally reasonable theories following different revisionary routes.

It has been argued on formal grounds that, given a theory, one can always construct an empirically equivalent rival (see PRAGMATIC FACTORS IN THEORY ACCEPTANCE for details). However, critics regard the rival constructs as artifices and not real theories and argue that the originating theory is always to be preferred (see REALISM AND INSTRUMENTALISM).

The only reasonable conclusion we can draw is that SUT is *at the very best* a highly speculative, unsubstantiated conjecture. Even if the thesis can be expressed intelligibly in an interesting form, there are no good reasons for thinking that it is true. Even if it is granted that there is the abstract possibility that SUT holds, the realist may well argue that it remains rational to believe that our theories do approximate the truth. We will only need to rethink this matter when we actually meet our hypothetical scientific colleagues on Alpha Centauri. The idea that there might be totally different stories to tell about the nonobservational world, stories that not even our observationally omniscient God can adjudicate between, is sufficiently exciting that we can anticipate that SUT will continue to be a subject of speculative discussion in the philosophy of science.

Even if there are no convincing arguments for SUT, there have been intriguing conjectures that some aspects of space and time (or more accurately space-time) are subject to strong underdetermination. For instance, science standardly represents space and time as being not merely dense but also continuous. That is, the points along an interval of time or space are mapped onto the real numbers, not the rational numbers. However, as all measurements are limited in accuracy to a finite number of decimal places, we use only rational numbers in recording the results of observation. This gives rise to the conjecture that we have a choice between regarding space and time as continuous or as merely dense. In the former case we would regard our rationally valued measurements of, say, position as approximating real values. In offering this as a sort of underdetermination, we are interpreting the notion of the observable as the directly measurable. The conjecture is then that different hypotheses about space and time (mere density versus continuity) are compatible with all actual and possible measurements. While it is no doubt simpler to represent space and time as continuous,

535

rather than merely dense, it might be that this is merely a matter of convenience, and that no measurement data can decide the matter (Newton-Smith 1978).

It may be that this story cannot be elaborated in a convincing manner. For there may be indirect arguments favoring the assumption of continuity. Even if the story can be made convincing, it will not support a general conjecture of strong underdetermination. Space and time are special; consequently, it would be rash to generalize from underdetermination of aspects of their structure to underdetermination in theories about things other than space and time. (See CONVENTION, ROLE OF, for further possible examples.)

If some aspects of space-time are underdetermined, we face the interesting issue as to how to respond. In this case one of our theories will entail some hypothesis h, and the other theory will entail not-h. One possible response is a form of limited skepticism (see QUINE). Either h is true or h is not true in virtue of how the world is. Due to underdetermination, we will never know which. Not even our observationally omniscient God will be able to form a rational belief about the matter. This is at odds with the general orientation of the empiricist, who takes it as objectionably metaphysical to posit aspects of the world totally beyond our powers of investigation. Consequently, some maintain that we should not think that there is a matter of fact at stake with regard to h. They hold instead that whether to accept h or not is a matter of adopting a convention, rather than a matter of conjecturing as to the facts.

While most discussion of underdetermination has focused pro and con on SUT, there has been some interest in WUT. Some sociologists have argued that there are rival theories for any given finite body of data; that scientists do, nonetheless, make choices; and that those choices can be explained only by reference to psychological and/or sociological factors. For exposition and criticism of this move see SOCIAL FACTORS IN SCIENCE.

References and further reading

Bergstrom, L. 1984: Underdetermination and realism. *Erkenntinis*, 21, 349–65.

Boyd, R. 1973: Realism, underdetermination and a causal theory of evidence. *Noûs*, 7, 1–12.

Dummett, M. A. E. 1973: *Frege: The Philosophy of Language* (London: Duckworth).

Earman, J. 1993: Underdetermination, realism and reason. *Midwest Studies in Philosophy*, 18, 19–38.

Glymour, C. 1971: Theoretical realism and theoretical equivalence. *Boston Studies in the Philosophy of Science*, 8, 275–88.

Laudan, L. 1990: Demystifying underdetermination. *Minnesota Studies in the Philosophy of Science*, 14, 267–97.

Laudan, L., and Leplin, J. 1991: Empirical equivalence and underdetermination. *Journal of Philosophy*, 88, 449–72.

Newton-Smith, W. H. 1978: The underdetermination of theory by data. *Aristotelian Society*, suppl. vol. 52, 71–91.

Quine, W. V. O. 1975: On empirically equivalent theories of the world. *Erkenntnis*, 9, 313–28.

Sklar, L. 1982: Saving the noumena. *Philosophical Topics*, 13, 49–72.

van Fraassen, B. 1980: *The Scientific Image* (Oxford: Oxford University Press).

77

Unification of Theories

JAMES W. MCALLISTER

Unification of theories is achieved when several theories $T_1, T_2, \ldots, T_n$ previously regarded as distinct are subsumed into a theory of broader scope T^*. Classic examples are the unification of theories of electricity, magnetism, and light into Maxwell's theory of electrodynamics, and the unification of evolutionary and genetic theory in the modern synthesis (Mayr and Provine 1980).

In some instances of unification, T^* logically entails $T_1, T_2, \ldots, T_n$ under particular assumptions. This is the sense in which the equation of state for ideal gases, $pV = nRT$, is a unification of Boyle's law, $pV =$ constant for constant temperature, and Charles's law, $V/T =$ constant for constant pressure. Frequently, however, the logical relations between theories involved in unification are less straightforward. In some cases, the claims of T^* strictly contradict the claims of $T_1, T_2, \ldots, T_n$. For instance, Newton's inverse-square law of gravitation is inconsistent with Kepler's laws of planetary motion and Galileo's law of free fall, which it is often said to have unified. Calling such an achievement "unification" may be justified by saying that T^* accounts on its own for the domains of phenomena that had previously been treated by $T_1, T_2, \ldots, T_n$. In other cases described as unifications, T^* uses fundamental concepts different from those of $T_1, T_2, \ldots, T_n$, so the logical relations among them are unclear. For instance, the wave and corpuscular theories of light are said to have been unified in quantum theory, but the concept of the quantum particle is alien to classical theories. Some authors view such cases not as a unification of the original $T_1, T_2, \ldots, T_n$, but as their abandonment and replacement by a wholly new theory T^* that is incommensurable with them (see INCOMMENSURABILITY).

Standard techniques for the unification of theories involve isomorphism and reduction. The realization that particular theories attribute isomorphic structures to a number of different physical systems may point the way to a unified theory that attributes the same structure to all such systems. For example, all instances of wave propagation are described by the wave equation, $\partial^2 y/\partial x^2 = (\partial^2 y/\partial t^2)/v^2$, where the displacement y is given different physical interpretations in different instances (see MODELS AND ANALOGIES). The reduction of some theories to a lower-level theory, perhaps through uncovering the microstructure of phenomena, may enable the former to be unified into the latter. For instance, Newtonian mechanics represents a unification of many classical physical theories, extending from statistical thermodynamics to celestial mechanics, which portray physical phenomena as systems of classical particles in motion (see REDUCTIONISM).

Alternative forms of theory unification may be achieved on alternative principles. A good example is provided by the Newtonian and Leibnizian programs for theory unification. The Newtonian program involves analyzing all physical phenomena as the effects of forces between particles. Each force is described by a causal law, modeled on the law of gravitation. The repeated application of these laws is expected to solve all physical problems, unifying celestial mechanics with terrestrial dynamics and the sciences of solids and of fluids. By contrast, the Leibnizian program proposes to unify physical science on the basis of abstract and fundamental principles governing all phenomena, such as principles of continuity, conservation, and relativity. In the Newtonian program, unification derives from the fact that causal laws of the same form apply to every event in the universe; in the Leibnizian program, it derives from the fact that a few universal principles apply to the universe as a whole. The Newtonian approach was dominant in the eighteenth and nineteenth centuries, but more recent strategies to unify physical science have hinged on formulating universal conservation and symmetry principles reminiscent of the Leibnizian program (McAllister 1996, pp. 109–11).

There are several accounts of why theory unification is a desirable aim. Many hinge on simplicity considerations; the following is a brief selection. A theory of greater generality is more informative than a set of restricted theories, since we need to gather less information about a state of affairs in order to apply the theory to it. Theories of broader scope are preferable to theories of narrower scope in virtue of being more vulnerable to refutation. Bayesian principles suggest that simpler theories yielding the same predictions as more complex ones derive stronger support from common favorable evidence: on this view, a single general theory may be better confirmed than several theories of narrower scope that are equally consistent with the available data (see SIMPLICITY; EVIDENCE AND CONFIRMATION).

Theory unification has provided the basis for influential accounts of explanation. According to many authors, explanation is largely a matter of unifying seemingly independent instances under a generalization. As the explanation of individual physical occurrences is achieved by bringing them within the scope of a scientific theory, so the explanation of individual theories is achieved by deriving them from a theory of wider domain (see EXPLANATION). On this view, T_1, T_2, . . . , T_n are explained by being unified into T^* (Friedman 1974; Kitcher 1989; for a contrary view, see Morrison 2000).

The question of what theory unification reveals about the world arises in the debate between scientific realism and instrumentalism. According to scientific realists, the unification of theories reveals common causes or mechanisms underlying apparently unconnected phenomena. The comparative ease with which scientists have been able to unify theories of different domains would be fortuitous on an instrumentalist interpretation, realists maintain, but can be explained if there exists a substrate underlying all phenomena composed of real observable and unobservable entities (Salmon 1984, pp. 213–27; Forster 1986). Instrumentalists provide a methodological account of theory unification which rejects these ontological claims (see REALISM AND INSTRUMENTALISM).

Theory unification is at least as much a matter of ethos as of achievement. The commitment to unification is especially strong among particle physicists and cosmologists, who aspire to the construction of a grand unified theory and a theory of everything (Maudlin 1996). Other authors emphasize and celebrate the methodological and ontological diversity of scientific disciplines and theories, and cast doubt on the viability of reductionism (Galison and Stump 1996; see also UNITY OF SCIENCE). Recent work on emergence, complexity, and self-organization in physics, chemistry, and biology may open new perspectives on the subject, and identify forms of theory unification that do not depend on isomorphism or reduction (Cat 1998).

References

Cat, J. 1998: The physicists' debates on unification in physics at the end of the twentieth century. *Historical Studies in the Physical and Biological Sciences*, 28, 253–99.

Forster, M. R. 1986: Unification and scientific realism revisited. In *PSA 1986: Proceedings of the 1986 Biennial Meeting of the Philosophy of Science Association*, ed. A. I. Fine and P. K. Machamer (East Lansing, MI: Philosophy of Science Association), 1, 394–405.

Friedman, M. 1974: Explanation and scientific understanding. *Journal of Philosophy*, 71, 5–19.

Galison, P., and Stump, D. J. (eds) 1996: *The Disunity of Science: Boundaries, Contexts, and Power* (Stanford, CA: Stanford University Press).

Kitcher, P. 1989: Explanatory unification and the causal structure of the world. In *Scientific Explanation*, ed. P. Kitcher and W. C. Salmon (Minneapolis: University of Minnesota Press), 410–505.

Maudlin, T. 1996: On the unification of physics. *Journal of Philosophy*, 93, 129–44.

Mayr, E., and Provine, W. B. (eds) 1980: *The Evolutionary Synthesis: Perspectives on the Unification of Biology* (Cambridge, MA: Harvard University Press).

McAllister, J. W. 1996: *Beauty and Revolution in Science* (Ithaca, NY: Cornell University Press).

Morrison, M. 2000: *Unifying Scientific Theories: Physical Concepts and Mathematical Structure* (Cambridge: Cambridge University Press).

Salmon, W. C. 1984: *Scientific Explanation and the Causal Structure of the World* (Princeton: Princeton University Press).

78

The Unity of Science

C. A. HOOKER

The problem of unity

We live together in one natural, if complex, world, and our scientific knowledge of it ought to be correspondingly unified. But currently the sciences collectively form a very complex structure, partly interrelated, partly unrelated, and partly incompatible. How is this condition explained, and what may we expect of unity in science?

In times past it was common to assume that there must be a unique universal scientific method generating a unique, universal scientific knowledge. Both methodological and descriptive unity were tacitly assumed.

But how can the social sciences, with their normative and cultural content, be unified with descriptive, a-cultural natural science? Even cognitive science has been claimed a priori independent of natural science in both method and content, because it is confined to logical symbol manipulation, while the latter deals in material causes. And within natural science, biological science, with its focus on complex adaptive systems and irreversible, dissipative processes, has been asserted to be independent of, or even to contradict, physics. Even within physics, there are the deep differences between field and particle method and theory (see below). How are these differences to be understood/resolved?

The present condition of science might have one or more of three distinct explanations. It might reflect (i) the incompleteness and partial error of contemporary science, (ii) a real disunity to our world, (iii) the nature of any description, constructible by finite creatures, of a unified but complex world. What are the natures of these explanations? To what extent do they or must they, apply?

There is further issue, the role of unity in good scientific method. Newton made a certain kind of unity central to his method (see below; see also NEWTON). Is unity in science, then, an epistemic virtue? If so, how does it relate to the other epistemic virtues: empirical adequacy, explanatory power, simplicity, and so on? And how does its virtuousness translate into prescription for scientific method? More particularly, how dependent should that prescription be on the actual nature of the world and of ourselves as knowers?

There is no hope of addressing these complex questions adequately in the brief space available here. Rather, the following discussion aims only to open up the issues and provide some salient points of reference from which to proceed.

The unity of science in the historical philosophical tradition

While scientists have often demanded unity, and it is instructive to consider their disciplines in detail (confined here to Newton, see below), philosophers have as often sought to provide a general, a priori guarantee of unity in a universal, unified knowledge.

Among those mathematical sciences which had not only utility but some true apprehension of being, Plato distinguished a superior science, dialectic, which takes the basic principles of others not as given but as hypothetical, ascending from them to a single universal categorical principle, the Form of the Good, from which all sciences and all true knowledge logically follow. For Aristotle, on the contrary, though there is a science of all being, metaphysics, this provides only the generic principles of logic and those general constraints on being that render logic unambiguously applicable (e.g., that no two things can be in the same place at the same time). While each particular science is an instance of these principles, it is a unique instance, distinctively different from the other sciences, and, unlike in metaphysics, the fundamental principles of these sciences are not demonstrable (McRae 1973).

For Kant, reason demanded the methodological search for unity, the grounds offered ranging from the pragmatic convenience of simplicity to the necessity of a unified reality. According to the latter reading, it is a methodological requirement of reason that science is unified, because its basic principles are presupposed by the empirical application of the Kantian categories, thereby bestowing both necessity and unity on them, since only then can a coherent world of objects be expressed. According to the former reading, empirical generalizations achieve the status of laws only when they become organizable into a unified science. The two approaches are complementary; either way, to build a science is to build a unified system (Kitcher 1994).

Whewell followed in essentially this tradition, his colligation of facts and then consilience of inductions leading to greater unification through elimination of parameters and/or laws and of distinct ontologies. For example, Newton shows that diverse heavenly orbits are colligated by the gravitation law which also subsumes hitherto quite distinct terrestrial dynamical phenomena; the inference to the universal law of gravitation then represents a strong consilient induction, revealing a universal constant. Inspired by Newton, as Kant was, he held that science should aim at deep unification marked by the revealing of fundamental constants characterizing the forms of universal laws (Butts 1994; see also WHEWELL).

For logical positivists/empiricists, responsible for the *Encyclopedia of Unified Science* project, the unity of science was to be given a quasi-Kantian guarantee through choice of philosophical framework (Neurath et al. 1971). As a condition of intelligibility, all of science was to be expressed in a single formal language (commonly, the predicate calculus) within which science was to be deductively systematized as axiomatized theories. Each theory was to be inductively derived from a foundation of pure, incorrigible observation statements that were held to be mutually logically compatible (Humean logical compatibility of facts), first physics, then chemistry, biology, then the social sciences. Conversely, each theory logically reduced to that preceding it, thus guaranteeing a formal unity to science. (Note: This is the reduction of objects, involving both substantival and property reduction (Causey 1977); the reduction of complex systems also involves functional reduction (Hooker 1981).) Thus here, and in the subsequent analytic tradition, reduction was central to the conception of the unity of science (see LOGICAL POSITIVISM; REDUCTIONISM; UNIFICATION OF THEORIES).

More recently there has been a revival of the Kantian–Whewellian systemization tradition in which empiricism's logical induction is replaced by inference to a unified explanatory system (see INFERENCE TO THE BEST EXPLANATION). Empiricism's model of explanation as simple deduction from a law is replaced by derivation from a unified system (Butts 1994; Kitcher 1989). Reduction may, but typically does not, play a significant role.

There is also a historical reading of dialectic that leads to quite different conceptions of unity. Hegelian idealism and subsequently historical materialism (Marx, Engels, Lenin) assert simultaneously the fundamentalness of unity and of property opposites (e.g., sameness, difference), the latter providing the intrinsic dynamics to historical development. This conception can be traced back to an ancient religious and meta-physical tradition asserting the unity of opposites as the fundamental nature of the world and the grasp of that unity as the fundamental measure of ideally completed knowledge. Science has a unified method and metaphysics in historical dialectic, but its specific contents concern the complex imperfect unities of opposites in finite "organic" wholes (see HOLISM).

The logical empiricist, Kant–Whewell, and historical dialectic traditions present a spectrum ranging respectively from a purely formal conception of the unity of science to a substantive conception corresponding to dynamical unity in nature. Thus the historical discussion of unity has raised the same issues as the opening systematic discussion: truth and the unity of being, and their relation to the unity of content and method required for our knowledge of being. So let us return to the systematic discussion.

Truth and unity

The truth is one, but error many. Parmenidean argument aside, this is the oneness of uniqueness, not necessarily of unity. Truth does not suffice for unity unless the world is unified, and knowably so, but error can suffice for disunity even in a unified world, or (with less likelihood) for unified but erroneous beliefs. Also, incompleteness can characterize a disunity of missing connections; like the elephant described by the seven blind men, any complex reality can be given very different partial descriptions – hence the character of explanation (i) for disunity in science.

Ontological unity

Turning to explanation (ii), what we ought to expect of scientific unity depends on, but is not determined by, the ontological unity of this world, since it is the world that is the object of scientific description.

Parmenides argued that all being must be one because being could not coherently share a boundary with nonbeing. This is a unity compatible with property diversity and even with atomism if space-time is included in being. Parmenides did not accept the latter; for him change, and so time, was an illusion (no "degrees of being"), and unoccupied space an impossible vacuum of being itself. Was this a more unified position? Whatever one makes of these arguments, they show how unity is complex. The world's unity, such as it is, has four different aspects: its spatiotemporal unity, substantival unity, the unified structure of its properties, and its dynamical unity.

Spatio-temporal unity   Though not philosophically uncontroversial, the world is evidently a single spatial piece; even if it is as structurally complex as quantum "foam," it displays kinematical connectivity. Similar considerations apply to time, *a fortiori*, even if time should prove to be two-dimensional. However, the passage from Newtonian to relativistic mechanics represents an increase in unity, since space-time, though structurally more complex, now becomes a single piece, expressed formally in increased symmetries; the relativistic symmetry groups have the classical groups as degenerate special cases (though only exactly at the nonrelativistic limit).

Substantival unity   There are at least two distinct substance ontologies in science. A field fills all space and time (perhaps zero field locations aside for non-Parmenidean fields). There is only variegation. Motion is successive reproduction of a spatial variegation pattern, and change generally is a more complex version of the same. Atoms (corpuscles, particles) are separate, local objects with their identities guaranteed by the nonintersection of their space-time trajectories, hence independently of their properties. Where atomic properties are temporally stable, all change is reduced to spatial rearrangement. These are two very different kinds of unity. Quantum theories, especially quantum field theory, show stronger symmetries than nonquantal theories and, like relativistic vis-à-vis classical space-time, may on that account be said to be more unified. It is an open question whether, and if so how, quantum theory forms a third kind of ontological unity, especially (as has been claimed) one that transcends the field/particle distinction (see Hooker 1994a).

Property unity   Both field and atomic ontologies may show diversity of properties. There may be many fields simultaneously present, independent or interacting. The former case presents a fundamental disunity of properties. The latter case, which is equivalent to one field with many linked field properties, offers the unity of interdependence, with every interaction requiring a force law. Similarly, the atoms of an atomic world may show different properties, both across atoms and through time for any one atom. The division of atoms into natural kinds, each characterized by a distinctive, stable property cluster, is only one alternative, if the commonly assumed one. Atoms may be independent (mutually or not), a property disunity, or they may interact, again offering the unity of interdependence under force laws.

In both cases a decrease in the number of basic properties must be compensated for by increased demand on reduction, in order to explain (away) macroscopic property diversity. Spatial rearrangement of fewer natural kinds of atoms places severe constraints on reduction, but has proved successful, for example in structural chemistry, given only three natural kinds (protons, neutrons, and electrons). There is, however, no comparable program to that of the geometrization of all field properties begun by Einstein, who unified the mass and energy properties and identified both with a geometrical property of space-time curvature, and subsequently extended this to the unification of all known basic properties as aspects of a single eleven-dimensional geometrical

543

structure. Here the unity of properties and that of space-time profoundly relate in a way that cannot happen with an atomic ontology. In the quantum version there is a further, dynamical aspect to this geometric unity: all force laws reduce to a single law under suitable conditions (that of the big bang), only differentiating into those we commonly recognize (strong and weak nuclear, electromagnetic, gravitational) as space-time unfolds, and temperatures and densities decrease.

The Australo-British naturalists, (chronologically) Alexander, Anderson, and Armstrong, developed philosophies which conjoin these aspects of unity in instructively different ways (see NATURALISM). The naturalism of all three derives from their insistence that there is no being outside space-time, the final reality and unifying framework. The diversity comes in the treatment of properties. Alexander's world was populated by organized patterns of basic physical properties, and these in turn could support genuine emergence, the appearance of new properties of complex wholes. Consciousness, mind, and even natural deity are part of the rich diversity thrown up as complexity increases (Alexander 1920). Anderson took the idea of complexes still further, insisting that everything (more precisely, every "proposition") was a complex of activities in a spatiotemporal region. But he rejected emergence (and most reduction also), while supporting an a priori argument for atheism (Anderson 1962). Armstrong holds that properties themselves are real and complex, interrelating to form complexes which are instantiated by material things. At the same time he accepts a strong materialist program of both property and substance reduction to those of physics (Armstrong 1978). All three philosophies share the same basic naturalist spatio-temporal unity, yet within that they show quite different conceptions of the property unity of a complex world.

Dynamical unity If the world is as quantum field theory says it is, then there is a strong dynamical unity underlying the macroscopic diversity of dynamical laws. There is an equally strong, but evidently different, sense of dynamical unity underlying macroscopic diversity for the dialecticians (and different again as between their Platonic, Hegelian, and Marxist versions). Lying perhaps between them (whether the gap is bridged will be left unresolved) is the contemporary conception of complex adaptive, self-organizing, self-reproducing systems founded in nonlinear, far-from-equilibrium irreversible dissipative systems, a construct currently revolutionizing the sciences (Hooker 1994b). Here a rich diversity of properties and processes results in a specific dynamical unity strong enough to underpin internal reorganization for self-repair and increased adaptation (learning). The property diversity is essential to the dynamical unity.

Individuals generically are systems whose dynamics show some suitable set of invariances sufficient to identify and characterize them uniquely. Rigid object dynamics leaves mass and geometry invariant, and the nonintersecting joint space-time trajectories of the constituents provides a sufficient basis for individuating and counting these objects. Structure supporting relevant dynamical invariances is necessary for individuality, but not sufficient; though structured, waves on a pond are not individuals, because their capacity for superposition destroys too many invariances; neither are photons. While living individuals show similar unique space-time trajectories, their much more complex dynamics does not exhibit all the invariances of a rigid object, instead leaving a greater range of functions invariant – for example, digestion and immune response – together with whatever structure that may require (e.g., approximate organ location and constitution). In these learning, adaptive systems, operational completeness or universality, autonomy, and objectivity under epistemic norms, all co-develop as the mark of a flourishing regulatory system. Nothing less complex suffices to express their distinctive dynamical unity. This discussion hardly exhausts the subtlety of the notion of individuality, but it does establish the variety of unities that are involved.

Descriptive unity

Science aims to represent accurately actual ontological unity/diversity. The wholeness of the spatiotemporal framework and the existence of physics (i.e., of laws invariant across all the states of matter) do represent ontological unities which must be reflected in some unification of content. However, there is no simple relation between ontological and descriptive unity/diversity. A variety of approaches to representing unity are available (cf. the formal-substantive spectrum vis-à-vis the range of naturalisms above). Anything complex will support many different partial descriptions, and, conversely, different kinds of things may all obey the laws of a unified theory (e.g., quantum field theory of fundamental particles) or collectively be ascribed dynamical unity (e.g., as self-organizing systems).

It is reasonable to eliminate gratuitous duplication from description – that is, to apply some principle of simplicity (see SIMPLICITY). However, this is not necessarily the same as demanding that its content satisfy some further methodological requirement for formal unification. Elucidating explanation (iii), there is again no reason to limit the account to simple logical systemization; the unity of science might instead be complex, reflecting our multiple epistemic access to a complex reality.

Biology provides a useful analogy. The many diverse species in an ecology nonetheless each map, genetically and cognitively, interrelatable aspects of a single environment and share exploitation of the properties of gravity, light, etc. Though the somatic expression is somewhat idiosyncratic to each species, and the representation incomplete, together they form an interrelatable unity, a multidimensional functional representation of their collective world. Similarly, there are many scientific disciplines, each with its distinctive domains, theories, and methods specialized to the conditions under which it accesses our world. Each discipline may exhibit growing internal metaphysical and nomological unities. On occasion, disciplines, or components thereof, may also formally unite under logical reduction (cf. Hooker 1981 vis-à-vis Fodor 1974). But a more substantive unity may also be manifested: though content may be somewhat idiosyncratic to each discipline, and the representation incomplete, together the disciplinary contents form an interrelatable unity, a multidimensional functional representation of their collective world. Correlatively, a key strength of scientific activity lies, not in formal monolithicity, but in its forming a complex unity of diverse, interacting processes of experimentation, theorizing, instrumentation, and the like (Galison and Stump 1996; Hooker 1994b).

While this complex unity may be all that finite cognizers in a complex world can achieve, the accurate representation of a single world is still a central aim. Throughout the history of physics, significant advances are marked by the introduction of new

545

representation (state) spaces in which different descriptions (reference frames) are embedded as some interrelatable perspectives among many – thus Newtonian to relativistic space-time and to quantum Hilbert spaces – and real quantities are those invariant across perspectives (Hooker 1994a). Analogously, young children learn to embed two-dimensional visual perspectives in a three-dimensional space in which object constancy is achieved and their own bodies are but some among many. In both cases the process creates constant methodological pressure for greater formal unity within complex unity.

Unity and scientific method

The role of unity in the intimate relation between metaphysics and method in the investigation of nature is well illustrated by the prelude to Newtonian science. In the millennial Greco-Christian religion preceding Kepler, nature was conceived as essentially a unified mystical order, because suffused with the divine reason and intelligence. The pattern of nature was not obvious, but a hidden ordered unity which revealed itself to diligent search as a luminous necessity. In his *Mysterium Cosmographicum* Kepler tried to construct a model of planetary motion based on the five Pythagorean regular or perfect solids. These were to be inscribed within the Aristotelian perfect spherical planetary orbits in order, and so determine them. Even the fact that space is a three-dimensional unity was a reflection of the one triune God. And when the observational facts proved too awkward for this scheme, Kepler tried instead, in his *Harmonice Mundi*, to build his unified model on the harmonies of the Pythagorean musical scale.

Subsequently Kepler trod a difficult and reluctant path to the extraction of his famous three empirical laws of planetary motion; laws that made the Newtonian revolution possible, but had none of the elegantly simple symmetries that mathematical mysticism required. Thus we find in Kepler both the medieval methods and theories of metaphysically unified religio-mathematical mysticism and those of modern empirical observation and model fitting, a Janus-faced, transitional figure in the passage to modern science.

To appreciate both the historical tradition discussed earlier and the role of unity in modern scientific method, consider Newton's methodology, focusing just on Newton's derivation of the law of universal gravitation in *Principia Mathematica*, book III. The essential steps are these: (1) The experimental work of Kepler and Galileo is appealed to, so as to establish certain phenomena, principally Kepler's laws of celestial planetary motion and Galileo's terrestrial law of free fall. (2) Newton's basic laws of motion are applied to the idealized system of an object small in size and mass moving with respect to a much larger mass under the action of a force whose features are purely geometrically determined. The assumed linear vector nature of the force allows construction of the centre of mass frame, which separates out relative from common motions; it is an inertial frame (one for which Newton's first law of motion holds), and the construction can be extended to encompass all solar system objects.

(3) A sensitive equivalence is obtained between Kepler's laws and the geometrical properties of the force: namely, that it be directed always along the line of centres between the masses, and that it vary inversely as the square of the distance between

them. (4) Various instances of this force law are obtained for various bodies in the heavens – for example, the individual planets and the moons of Jupiter. From these one can obtain several interconnected mass ratios – in particular, several mass estimates for the Sun, which can be shown to mutually cohere. (5) The value of this force for the Moon is shown to be identical to the force required by Galileo's law of free fall at the Earth's surface. (6) Appeal is made again to the laws of motion (especially the third law) to argue that all satellites and falling bodies are equally themselves sources of gravitational force. (7) The force is then generalized to a universal gravitation and is shown to explain various other phenomena – for example, Galileo's law for pendulum motion, the tides, etc. (8) The corrections to the step 2 model for interplanetary interaction are shown suitably small, thus leaving the original conclusions drawn from Kepler's laws intact while providing explanations for the deviations (Friedman 1992; Harper 1993; see also NEWTON).

Newton's construction represents a great methodological, as well as theoretical, achievement. Many other methodological components besides unity deserve study in their own right. The sense of unification here is that of a deep systemization. Given the laws of motion, the geometrical form of the gravitational force and all its significant parameters needed for a complete dynamical description – that is, the components G, n of the geometrical form of gravity Gm_1m_2/r^n – are uniquely determined from phenomena and, after the law of universal gravitation has been derived, it plus the laws of motion determine the space and time frames and a set of self-consistent attributions of mass. For example, the coherent mass attributions ground the construction of the locally inertial centre of mass frame, and Newton's first law then enables us to consider time as a magnitude: equal times are those during which a freely moving body traverses equal distances. The space and time frames in turn ground use of the laws of motion, completing the constructive circle. This construction has a profound unity to it, expressed by the multiple interdependence of its components, the convergence of its approximations, and the coherence of its multiply determined quantities. The methodological demand for this kind of constructive unity was effectively expressed in Newton's Rule IV: (loosely) don't introduce a rival theory unless it provides an equal or superior unified construction – in particular, unless it is able to measure its parameters in terms of empirical phenomena at least as thoroughly and as cross-situationally invariantly (Rule III) as does the current theory. This gives unity a central place in scientific method.

Kant and Whewell seized on this feature as a key reason for believing that the Newtonian account had a privileged intelligibility and necessity. Significantly, the requirement to explain deviations from Kepler's laws through gravitational perturbations has its limits, especially in the cases of the Moon and Mercury; these need explanations, the former through the complexities of n-body dynamics (which may even show chaos) and the latter through relativistic theory. Today we no longer accept the truth, let alone the necessity, of Newton's theory. Nonetheless, it remains a standard of intelligibility. It is in this role that it functioned, not just for Kant, but also for Reichenbach, and later for Einstein and even Bohr: their sense of crisis with regard to modern physics and their efforts to reconstruct it are best seen as stemming from their acceptance of an essentially Newtonian ideal of intelligibility as complete, unified construction and their recognition of the falsification of this ideal by quantum theory (Hooker

1994a; see BOHR and EINSTEIN). Nonetheless, quantum theory represents a highly unified, because symmetry-preserving, dynamics, reveals universal constants, and satisfies the requirement of coherent and invariant parameter determination.

Newtonian method provides a central, simple example of the claim that increased unification brings increased explanatory power (see EXPLANATION). The law of universal gravitation unifies dynamics by reducing to one the number of force laws required to describe celestial and most terrestrial motions. The result is increased explanatory power for Newton's theory because of the increased scope and robustness of its laws, since the data pool which now supports them is the largest and most widely accessible, and it brings its support to bear on a single force law with only two adjustable, multiply determined parameters (the masses). Call this kind of unification (simpler than full constructive unification) "coherent unification." As noted earlier, much has been made of these ideas in recent philosophy of method, representing something of a resurgence of the Kant–Whewell tradition.

Coherent unification is not, however, the only weighty scientific consideration; scientists regularly trade off considerable increases in the diversity and complexity of laws and/or parameters for deeper insight into ontology and causal structure, as virtually every shift from macroscopic to microscopic theory attests – thus molecular biochemistry in place of macroscopic medical symptomology. But coherent unification and ontological depth are not unconnected: improvements along both dimensions go via ontological identifications which achieve unifications. Thus identifications within a given ontology allow coherent unification, and identifications across ontologies allow a new, more systematic underlying ontology to be introduced (e.g., the reductive identifications involved in the medical symptoms–molecular cause case). Coherent unification can be achieved without any increase in ontological depth. On the other hand, achieving ontological depth typically provides the basis for coherent unification across domains by reduction. In addition, the underlying ontology is related to the laws of other domains; for example, microbiological theory of disease relates medicine to various biological domains, such as molecular genetics. Thus unification stands at the heart of explanatory power and method.

References and further reading

To satisfy length constraints, references have sometimes been selected as much for their capacity to refer readers to yet further literature as for their own content, and sometimes as but one among several to illustrate a position. It has also not been possible to include specific references to the Thomist-Christian and Hegelian-Marxist traditions.

Alexander, S. 1920: *Space, Time and Deity*, 2 vols (London: Macmillan).

Anderson, J. 1962: *Studies in Empirical Philosophy* (Sydney: Angus and Robertson).

Armstrong, D. M. 1978: *Universals and Scientific Realism*, 2 vols (Cambridge: Cambridge University Press).

Butts, R. E. 1994: Induction as unification: Kant, Whewell and recent development. In *Kant and Contemporary Epistemology*, ed. P. Parrini (Boston: Kluwer), 273–89.

Causey, R. 1977: *The Unity of Science* (Dordrecht: Reidel).

Fodor, J. A. 1974: Special sciences (or: The disunity of science as a working hypothesis). *Synthese*, 28, 97–115.

Friedman, M. 1992: *Kant and the Exact Sciences* (Cambridge, MA: Harvard University Press).

Galison, P., and Stump, D. J. (eds) 1996: *The Disunity of Science* (Stanford, CA: Stanford University Press).

Harper, W. 1993: Reasoning from phenomena: Newton's argument for universal gravitation and the practice of science. In *Action and Reaction: Proceedings of a Symposium to Commemorate the Tercentenary of Newton's Principia*, ed. P. Theerman and A. F. Seeff (London: Associated University Presses).

Hooker, C. A. 1981: Towards a general theory of reduction. *Dialogue*, 20, part I, Historical framework, pp. 38–59; part II, Identity and reduction, pp. 201–36; part III, Cross-categorial reduction, pp. 496–529.

—— 1994a: Bohr and the crisis of empirical intelligibility. In *Niels Bohr and Contemporary Philosophy*, ed. J. Faye and H. Folse (Boston: Kluwer), 155–99.

—— 1994b: *Reason, Regulation and Realism* (Albany, NY: SUNY Press).

Kitcher, P. 1989: Unification and the causal structure of the world. In *Scientific Explanation*, ed. P. Kitcher and W. C. Salmon (Minneapolis: University of Minnesota Press), 410–505.

—— 1994: The unity of science and the unity of nature. In *Kant and Contemporary Epistemology*, ed. P. Parrini (Boston: Kluwer), 253–72.

McRae, R. 1948: The unity of the sciences: Bacon, Descartes and Leibnitz. *Journal of the History of Ideas*, 18, 27–48.

—— 1973: Unity of science from Plato to Kant. In *Dictionary of the History of Ideas*, ed. P. Edwards (New York: Charles Scribner), 431–7.

Neurath, O., Carnap, R., and Morris, C. (eds) 1971: *Encyclopedia of Unified Science* (Chicago: University of Chicago Press).

79

Values in Science

ERNAN MCMULLIN

A century ago, nearly all of those who wrote about the nature of science would have been in agreement that science ought to be "value-free." This had been a particular emphasis on the part of the first positivists, as it would later be on the part of their twentieth-century successors. Science, so it was said, deals with *facts*, and facts and values are irreducibly distinct. Facts are objective; they are what we seek in our knowledge of the world. Values are subjective; they bear the mark of human interest; they are the radically individual products of feeling and desire. Fact and value cannot, therefore, mix. Value cannot be inferred from fact; fact ought not be influenced by value. There were philosophers, notably some in the Kantian tradition, who viewed the relation of the human individual to the universalist aspirations of science rather differently. But the legacy of three centuries of largely empiricist reflection on the "new" sciences ushered in by Galileo and his successors was as unqualified in its distrust of value as in its extolling of the virtues of fact (see GALILEO).

A century later, the maxim that scientific knowledge is "value-laden" seems almost as entrenched as its opposite was earlier. The supposed wall between fact and value has been breached, and philosophers of science seem quite at home with the thought that science and values may be closely intertwined after all. What has happened to bring about such an apparently radical change? What are its implications for the objectivity of science, the prized characteristic that, from Plato's time onwards, has been assumed to set off *real* knowledge (*epistēmē*) from mere opinion (*doxa*)? To answer these questions adequately, one would first have to know something of the reasons behind the decline of logical positivism, as well as of the diversity of the philosophies of science that have succeeded it. For this, the reader can be referred to other essays in this collection. The shift in regard to the role of values in science is, as we shall see, as good a barometer as any of the changes that have occurred in the philosophy of science over recent decades.

The notion of value

Something should be said first about the amorphous notion of value itself. "Value" derives originally from *valoir*, to be of worth. To "value" something is to ascribe worth to it, to have it serve as a goal of effort, to regard it as desirable, to have a positive attitude towards it. And, correlatively, "value" is the characteristic that leads something to be so regarded. At this point there is a division. Something may be a value because of its relation to a valuer or a community of valuers. The acquisition of property is a value for some and not for others; the value of a particular property will depend on the estimate on the part of a particular community of its desirability relative to other possible goods. On the other hand, a characteristic may count as a value in an entity of a particular kind because it is objectively desirable for an entity of that kind. Here the emphasis is not on the relation to a valuer but on the part played by the characteristic in the proper functioning of its possessor. Thus, keenness of hearing would be a value for many types of animal. A particular range of temperature and humidity would be a value for an art museum. Sharpness of image would be a value in a TV set. Proper functioning in cases such as these might refer to the intrinsic needs of an organism or artifact or institution because of its constitution or the environment in which it is set; it might also involve the utility of some further agent.

Since value in many of these senses can be at least roughly quantified, a further derivative sense of the term refers simply to a quantity, apart from any reference to desirability, goals, proper functioning, or the like. Thus, we speak of the value of a variable in mathematics or the measured value of a particular physical property. This neutral sense of the term is commonplace in scientific usage; mathematicians and experimental scientists in their everyday practice use the term "value" incessantly, simply because they are dealing with quantities, and need generic terms to describe them. Though one might conceivably want to inquire into this focus on quantification on the part of scientists in the Galilean tradition, this sense of "value" will be disregarded in our further discussion.

Scanning across the other usages of the term, one can see why it is so readily assumed that values invariably involve some degree of subjectivity. First of all, something may be of value precisely *because* it is such for a particular subject or set of subjects. The nineteenth-century founders of value theory took "value" in this sense. Values then corresponded to such features of human experience as attraction, emotion, and feeling, and their reality lay primarily in the feelings of the subject, not in characteristics of the object. It was this sense of the term that the logical positivists usually had in mind when banning values from science. A second, connected sense would once again link value to a subject when it is (like the sharpness of a TV image) a necessary condition for the proper functioning of something whose operation is of interest to a subject.

There is another more general reason why values carry the connotation of subjectivity: their *estimation* often depends on the judgment of the individual, on possibly idiosyncratic likes and dislikes, rather than on intersubjective norms. It is, if you like, the value of the value that is subjective in such cases; it is the value judgment that is suspect, rather than the value itself. Once again, it is not hard to recognize here the source of positivist unease: subjective estimation would presumably undermine the sort of universality to which science has traditionally laid claim. Value judgment cannot substitute for the impartial working of logical rule. On the face of it, this seems like a good argument. I shall return to it later.

Science and values

Science and values can interact in a variety of ways. One of these is our primary concern: the manner in which values influence the actual making of science. But there are others, and first among them is the growing impact of science on human life and on the world in which that life is lived. When books are written or courses are offered under titles like "Science and Human Values," it is this that is usually intended. Nearly four centuries ago, Francis Bacon hailed the transformative potential of natural science; a deeper understanding of the capacities of matter would, he was sure, allow a broken world to be mended. Though he still retained the older metaphor of science as a kind of light, he was more interested in the metaphor of power. And for him, the test of the quality of light was the measure of power it afforded.

The link between light and power proved a good deal more elusive than he anticipated. It was a full two centuries before an understanding of what he had called the "latent processes" of matter began to show fruit in the practical order. In the meantime, technological innovation had accelerated; the steam engine became the symbol of a new kind of large-scale industry that would rapidly transform almost every aspect of daily life in Western society. As the nineteenth century progressed, the basic sciences gradually came to inform technological development more and more. Chemistry showed the way to new dyes; physics helped to harness the powers of electromagnetism; geology aided in the search for deposits of coal and oil. But the real explosion of science-driven technological change is recent; most would associate it with the Second World War, with its enormous expenditures on such developments as radar and atomic power.

When science is utilized for technological ends, values may be implicated in a variety of ways. The application of scientific knowledge may confer a straightforward benefit, like the eradication of an infectious disease in a population. More often, there are possible losses as well as uncertainties about the gains. A course of chemotherapy may extend the life of a particular cancer sufferer but may also reduce its quality. Building a nuclear plant of a particular design may lessen the cost of electricity but make the threat of nuclear terrorism more real. Estimation of the values involved in issues like these can be very difficult; different people may judge very differently. Furthermore, the relative likelihood of the different outcomes is also a matter of estimation – in this case, of probability.

What has come to be called "decision theory" brings together the two sets of variables involved in decisions of this kind, gains and losses on the one hand and probabilities on the other, to construct a lattice that in principle should enable a rational decision to be made. In games of chance, where both sets of variables can easily be quantified, this sort of approach works well. But where the proposal is to apply a scientific hypothesis to transform a situation that involves human welfare, it works only under special circumstances. Often, such decisions are almost intractably complex, both because the probability estimates are themselves disputed and because the goods involved are very differently estimated by different parties. Do we ban chlorofluorocarbons? Do we divert resources into fusion energy research? Decisions such as these involve large-scale issues of public policy, but in our technology-driven society analogues of them (Ought I to be screened regularly for breast cancer? Ought I to drop eggs from my diet?) face the individual at every turn.

The application of scientific theory to matters of human utility can involve values in a second, and often even less tractable, way when the values concerned are ethical ones. The question then is: Is it moral? rather than: Is it rational? Decision theory no longer helps, unless perhaps one is a utilitarian in ethics. The explosive growth in the field of medical ethics over the past few decades is an indication of where the most difficult issues seem to lie. The difficulty here is not for the most part with the science, but in deciding what is the moral course to follow, given the technological possibilities. Whether it be the prolongation of life of the terminally ill, or genetic manipulation to bring about the desired kind of offspring, it is at least in part the novelty of the questions that makes ethical response so hesitant or so divided. Ethical norms developed in human societies that changed only very slowly, and where genuinely new sorts of ethical problems rarely arose. The medical technologies that are pouring out of research laboratories today in such profusion give moralists little respite for reflection, little chance to discover the long-run moral consequences of a particular response to a new medical possibility.

In an article that gave rise to a continuing controversy, Richard Rudner (1953) argued, on the basis of examples such as these, that "value judgments are essentially involved in the procedures of science." Critics pointed out that the "procedures" he described were not those of science proper, but those involved in the *application* of a scientific insight or a technological advance to matters of human utility. That this latter would necessarily involve value judgment has never been disputed. But it would not establish the role of values in the day-to-day work of the scientist. Rudner's claim for the centrality of value judgments *in* science was vindicated by later philosophers of science, as we shall see, but on very different grounds.

Values of science

What are the goals of science? What values are the sciences supposed to secure? If one goes back to the origins of the ambition to discover an *epistēmē* of the natural world, the answer is relatively straightforward. For Aristotle, the goal of natural science was an understanding of the world around us in terms of its causes. From another perspective, its goal was truth, understood as a conformity between mind and world. The value of science was thus basically contemplative. Noting the distinction between the nobler, but less accessible, celestial domain of star and planet, and the perishable, but much more accessible, domain of plant and animal, Aristotle remarks: "The scanty conceptions to which we can attain of celestial things give us, from their excellence, more pleasure than all our knowledge of the world in which we live." On the other hand, our natural affinity to the animal world leads us, rather, to investigate the wonders of the living world around us: "For if some [animals] have no graces to charm the sense, yet nature which fashioned them, gives amazing pleasure in their study to all who can trace links of causation and are inclined to philosophy" (*On the Parts of Animals*, I, 5; revised Oxford translation).

A quite different goal was suggested at the outset of the seventeenth century, as we have already noted, by Francis Bacon and René Descartes (see DESCARTES). A genuine knowledge of the natural world ought to permit its powers to be utilized for human benefit. Though science still seeks to understand the causes of things, it is not merely to give pleasure or to satisfy curiosity but to transform, to put nature to work. And as that century wore on, natural philosophers were brought to admit that the discovery of underlying causes was a far more arduous and tentative affair than earlier accounts of science had allowed for. Hypothesis may be the best we can do, and hypotheses are to be tested by their consequences. Thus, the goal of accurate prediction came to the fore, the tracing of likely consequences, not only as a means of testing the validity of hypothesis but as a goal in itself.

In our own time, as scientists have reached further and further into latent process, the structures they have created have seemed more and more remote from the intuitions that guide us through our middle-sized world. Whether quantum theory affords a genuine understanding of basic physical process or merely a convenient and powerful means of prediction is endlessly debated. What drives (and funds) scientific effort in some domains is clearly not the values associated with understanding nearly so much as the practical values of technological utility. Yet in many other domains, like cosmology or paleontology, the lure of technological profit is absent, and one can only suppose that the more ancient goal still rules. People spend their lives tracing the life and death of distant stars or ancient species just because they, and those who fund their researches, derive satisfaction from the knowledge itself.

The ethos of science

So much for the values that scientific advance might in general be expected to achieve. What of the norms that guide the daily work of the scientist? Sociologists have labored to disentangle some general norms from the complex of motivations that historians discover within even the simplest of human actions. Robert Merton (1957) characterized what he called the "ethos" of science by a set of values that he proposed as distinctive of scientific activity: *universalism* (employing impersonal criteria in order to arrive at knowledge that holds universally); *communalism* (working as a community in research and sharing in discoveries made); *disinterestedness* (placing the search for truth ahead of personal advancement); *organized skepticism* (willingness to challenge every step of an argument); and *purity* (insisting on the autonomy of science in the face of political, economic, and other external demands). Merton knew very well that these values are by no means always characteristic of the way in which scientists conduct their affairs, but he clearly regarded them as norms which are on the whole observed. Later sociologists of science challenged him on this, suggesting that he was imposing his own ideals of how scientific activity should be guided, rather than reporting the empirical results of sociological inquiry. Some questioned, in effect, whether some of his proposed norms should even be counted as ideals.

Kuhn (1977), for example, emphasized the role played by what he called "dogma" in science, maintaining that scientists working within a paradigm ordinarily do not, and should not, question the authority of the paradigm (see KUHN). Lakatos and Feyerabend note that refusal on the part of its defenders to concede defeat when a theory seems to all intents and purposes refuted may serve the long-range benefits of science (see LAKATOS and FEYERABEND). Hull (1988) argues that disinterestedness is about as unlikely a description of the motivation of the average scientist as one could imagine; nevertheless, the single-minded and aggressive pursuit of recognition and credit on the part of individuals works well for science in the long run. Sociologists of scientific knowledge of the Edinburgh school urge the primacy of interests, personal, political, and economic, in scientific decision making, and hence the inadequacy of the traditional historiography of science because of its focus on "internal" considerations, like evidence and proof. These writers would question the entire idea of an "ethos" of science; the particularities of social context are so various that no general account of "the" norms of science can be given.

Fifty years ago, admirers of the ways of science were wont to say that the ethos of scientists could serve as a model for ethical behavior generally. Nowadays that encomium would be greeted with a degree of skepticism. Yet scientists do give one another credit; they do, for the most part, share their results; they very rarely cheat in reporting observations. There *are* norms; scientists are well aware of them and are careful to abide by them. Hull's somewhat cynical construal of this state of affairs is that they abide by the institutional goals of science because, by and large, these coincide with their own goals: "Scientists adhere to the norms of science because it is their own self-interest to do so" (1988, p. 394).

Values and rules

One of the most important shifts in recent philosophy of science concerns the role played by values in shaping the epistemic character of science. Earlier philosophies of science stressed the importance of rule following as the guarantee of *epistemic* propriety. Aristotle and Descartes saw deductive logic as the means of transferring truth, as it were, from premises to conclusion. Since the logical rules themselves could be seen to hold universally, the only problem was to insure the truth of the premises that served as starting points. Defenders of induction likewise relied on rules ("Mill's methods"), though they were rather more guarded about the truth of the resulting inductive generalizations from particulars to universal "laws." The advantage of formal rules is that they can be applied in a more or less automatic way, and there is an effective procedure to tell whether or not they have been applied properly.

Doubts about this attractively simple picture of what it is that makes science go back a long way. Already in the seventeenth century there was a growing realization that the tracing of an effect to one of the (usually many) possible causes could not be a simple matter of rule. It led Kepler, Boyle, Huygens, Locke, and others to shift attention to the evaluation of hypotheses. What sorts of criteria are appropriate? And what kind of assurance do they give? Kepler already urged that it could not be a matter of simply "saving the phenomena," of hitting on a hypothesis from which the desired consequences would follow. It had to *explain* them, to give a properly causal account. This would set further constraints on the hypothesis, enough ultimately (Kepler hoped) to allow its truth to be established. Evaluating an explanatory hypothesis was not at all like applying a rule; several different criteria might be involved (Boyle listed ten), each of which might, separately, be satisfied to a greater or lesser degree. Among the criteria usually mentioned were consistency with accepted physical theory and success in predicting novel results.

Newton's reluctance to allow hypothesis as part of science proper (see NEWTON) was a step backwards, looked on from the perspective of a later day. The difficulty he encountered in construing gravity in explanatory terms led him to bracket the methodological issues raised by causal explanation; he could then represent the derivation of the fundamentals of his mechanics as a "deduction" from the phenomena, subsequently made general by "induction." Such a science would consist of "laws," empirically determined regularities; it would advance by making these laws a more and more exact rendering of the empirical data. It was an attractively simple picture, pleasing to logicist and empiricist alike, and it commanded respect over the next two centuries among theorists of science as diverse as Reid, Mill, and Mach (see MILL and MACH). But, as structural explanations became the norm in fields as diverse as chemistry and geology, Whewell and, later, Peirce underlined the fundamental logical difference between an induction arriving at an empirical generalization like Boyle's law and the construction and evaluation of an explanatory hypothesis like the kinetic theory of gases (see WHEWELL and PEIRCE). Peirce gave the latter its own designation ("retroduction" or "abduction") in order to distinguish it from the simpler inference form for which he retained the traditional term, "induction." The criteria governing retroduction are obviously much more complex than those applicable to induction. The assessment of the relative merits of rival explanatory theories is a far more complicated affair than the evaluation, most often in statistical terms, of an empirical generalization drawn from a limited set of data.

A realization of the importance of this distinction was surprisingly slow in coming. The logical positivists, initially at least, focused, as one would expect, on induction; they distrusted the unobservable entities on which retroduction frequently depends. Carnap's inductive logic relied on a single criterion in the assessment of hypotheses, the extent to which a hypothesis is substantiated by the observable consequences deductively derivable from it. These consequences are reported in protocol sentences that are treated as foundational, as unproblematic. Much effort was expended in developing this idealized logical schema.

But from the beginning, the distance between idealization and the actual practice of science was evident. Popper noted that an element of decision is required in determining what constitutes a "good" observation (see POPPER). Carnap conceded that "external" factors are involved in the decision to adopt a particular mathematical or linguistic formalism: would the same not be true of the decision to favor one physical theory over another? Influenced by the earlier conventionalist tradition of writers like Poincaré, Popper described the decision as a matter of "convention," an unfortunate choice of term because of its overtone of arbitrariness. Carnap stressed the pragmatic character of the criteria governing the choice. For both of them, the point to be conceded was that individual decision was involved; instead of the impersonal application of rule, the scientist had to rely on value judgment, on a personal estimate of the relevant factors. How was this to be reconciled with the prized objectivity and universality of science?

Kuhn (1977) carried the analysis a stage further. He noted that, typically, there are many different considerations involved in theory choice, and that these operate as values to be maximized, not as rules to be satisfied. But this does not necessarily undermine the objectivity of the choices made: though pragmatic values like self-interest may intrude, the standard values recognized as appropriate to theory choice (accuracy, consistency, scope, simplicity, fertility) are such as to promote the objectivity of the choice made; indeed, they serve to define what is meant by "objectivity" in this context. In keeping with his earlier work, however, Kuhn argued that the transfer of allegiance from theory to theory is often better described as "conversion" rather than "choice." He also denied that the values sought in a "good" theory testify to that theory's likely truth or to the reality of the theoretical entities it postulates. In the eyes of his critics, this was to attenuate the objectivity of theory to an unacceptable degree (McMullin 1993).

Discussion of the additional values (or "virtues" as they have been called) guiding theory choice continues. Different lists have been proposed: unifying power, fertility, and explanatory success have come in for special emphasis. But the larger issue is their status: are they epistemic? That is, do they bear on the truth of the theory under evaluation? Or are they merely pragmatic, a matter of practical utility? Are they independent criteria, or are they satisfied simply by the persistent application of the primary inductive criterion: "saving" the phenomena? The logical positivists focused on the criterion of simplicity. It was important for them not to allow epistemic status to criteria other than the standard inductive one, and simplicity was easy to dismiss as merely pragmatic in character. Van Fraassen (1980) has argued that the "super-empirical" virtues mentioned in connection with theory choice – in practice, all virtues other than empirical adequacy and logical consistency – ought to be regarded as pragmatic.

The issue itself has become entwined with the issue of scientific realism. Critics of realism, like van Fraassen, are loath to allow epistemic status to such values as explanatory power or fertility, if these are taken to apply over and above the ground-level value of empirical adequacy. Many defenders of scientific realism rely on the fact that successful scientific theories *do* exhibit explanatory virtues of various sorts; they argue that this warrants giving such theories a realist interpretation under certain circumstances. Anti-realists retort that these virtues reduce, in evidential terms, to empirical adequacy; the validation of a novel prediction does no more for a theory (they urge) than if this datum were part of the data on which the theory was originally based. (Mill and Whewell already debated this issue, and with the same larger questions in mind.) Realists, in turn, respond that if a theory does no more than save the phenomena it was originally designed to save, this goes but a little way to warrant the existence of the entities it postulates. But when it does *more* than that, when it shows the sort of virtues one would expect it to show if, in fact, the kinds of underlying structures it postulates do exist, a realist construal of these structures may become the only reasonable one. The debate goes on.

Epistemic versus non-epistemic values

It would by now be fairly generally admitted that scientists rely on value judgment at the most critical moments in their researches. There would be less agreement, however, regarding the values that guide these judgments. Most would say that these are predominantly *epistemic* – that is, that reliance on them tends to improve the chances that the judgments based on them are (at least approximately) true. That other sorts of values may also play a role has never been denied, but it has been assumed that this role can be progressively limited by continued reliance on epistemic considerations. Francis Bacon described the "Idols" that could divert scientific inquiry from its proper path; they derived, he argued, primarily from the limitations imposed on individual scientists by their milieu. If they are recognized, they can, he was sure, be overcome. Nineteenth-century writers often used the term "ideology" as a sort of code word to underline the threat to the objectivity of science posed by the intrusion of values alien to science proper. But this notion of "science proper" has come under increasing attack in recent decades.

Sociologists of science of various stripes argue that social, political, economic, and personal values permeate scientific work at all levels. The history of past controversies in science, as well as the application to the current practice of science of the techniques of the social scientist, testify to this, they claim. Nor is it enough to distinguish between "internal" and "external" historiography of science and banish all mention of the sociopolitical to the latter. The distinction itself has been challenged on several grounds. Theory choice in science is, by general admission, underdetermined in conventional epistemic terms; closure can be arrived at, therefore, only by turning to other values. These latter are thus ineliminable from the actual practice of science. Putting this in a different way, such central features of science as theory choice and the decision as to when experimental results can be reported inevitably reflect interests of all sorts, personal and political (Barnes 1977). Furthermore, even the supposed epistemic values themselves *are* epistemic only because they are accepted as such by the scientific community. They are as socially constituted as any other. There is thus neither internal nor external; even the distinction between epistemic and non-epistemic becomes suspect. Science is a fundamentally social construction, with all of the contextual contingencies that this conveys. So the arguments go. (See, e.g., Pickering 1992.)

Neo-Marxist philosophers of science, seconded by some recent feminist theorists, focus specifically on the proper role of values in science. Their target is the "value-free" science preached by the positivists. Modern science, they claim, is dependent on values, including sociopolitical values, to a far greater extent than is generally admitted. It aims at power, at control, at domination, frequently with disastrous effects for society. The reductionism which underlies theory construction is a further evidence of this goal. The aim of these thinkers is not to work for the elimination of sociopolitical values, as Bacon recommended. Such a goal is, in their view, unattainable; and, more to the point, it is undesirable. What is needed, rather, is to substitute the *right* values for the ones now prevailing: "In order to practice science as a feminist, as a radical, or as a Marxist, one must deliberately adopt a framework expressive of that commitment . . . [The advice these give is:] if you share our political beliefs, here is a way to do science that expresses these beliefs" (Longino 1990, p. 197).

The obvious objection from the Baconian side is: Why should one think that reliance in theory choice in physics or chemistry, say, on a particular emancipatory ideal is likely to lead to truth? Isn't there a danger of anthropomorphism here, the fallacy to which Bacon referred? "The human understanding is no dry light, but receives an infusion from the will and affections; whence proceed sciences which may be called 'sciences as, one would'" (1960, aph. 49). The distinction between the epistemic and the non-epistemic separates those values that experience has shown *do* guide choice well (lead to theories that in the long run are more empirically adequate or postulate entities that are better verified directly) from those that do not. This leaves room for an intermediate possibility: values (and background assumptions) whose epistemic credentials, in the context of a particular domain of theory choice, have not yet been either established or discounted.

Feminist philosophers are divided as to whether their response should be to claim that the emancipatory values which they defend are likely to lead to better science (in the sense of leading science closer to theory truth), or simply to attack the notion of "closer to truth." Is a biological theory that conforms to feminist ideals more likely to be true than a rival theory that does not? Or ought one simply to say that science has more than one goal, and that each has to be weighted against the others? Philosophers of science in this tradition characteristically reject scientific realism, both because it seems to them to be associated with reductionism and because the sort of objectivity it claims for successful theory appears inhospitable to the infusion of political values into the realism of theory choice, in the natural sciences, at least.

Longino sees contemporary philosophy of science as falling under one of three heads: positivism (too narrow), holism of a Kuhnian variety (too relativistic), and realism (undermined, she claims, by the criticisms of Laudan, Fine, and van Fraassen). Rejecting all three, she attempts to show that an empiricism sensitive to feminist values can still retain objectivity, if objectivity be defined in communitarian terms. Critics find this notion of community too broad and argue that the epistemic character of certain values, as well as the non-epistemic character of others, is not just a matter of consensus, but is rooted in an intricate story of how scientific inquiry took shape over many millennia. Longino realizes that an appeal to community will not in any event serve her purpose unless that community embodies a different set of values from those that characterize Western society today. Radical science requires first, therefore, a radical political transformation of society.

Arguments of this sort have been going on in the more restricted area of social theory for a long time. Their foundations were already laid in the writings of Hegel, Comte, and Marx. They are to be found, in particular, in the debates surrounding the Frankfurt school from the 1930s onwards. On the one side were philosophers like Husserl and Schutz, who, although they repudiated positivism, shared with it the ideal of a neutral self-corrective form of research which, to begin with, brackets social and political values, though ultimately it can help to validate them. Against this, Horkheimer argued that although this idea served the natural sciences well in the past, this was because these sciences are concerned with the manipulation of nature and for this require only technologically applicable hypotheses. A properly *critical* theory, on the other hand, is guided by a practical interest in improving human existence, allowing mankind for the first time to determine its own way of life. Critical Theorists are concerned primarily with *social* realities, and do not disguise the values that guide their research (Bernstein 1978).

Habermas pressed this Critical Theory of society further. He claimed that social scientists have confused the practical with the technical, effectively reducing their science to matters of technical control. Knowledge is constituted by interests (values) of various sorts. The three principal ones, technical (associated with the "empirical-analytic" sciences), historical-hermeneutic, and emancipatory, have a quasi-transcendental status; each is grounded in one dimension of social existence: work, interaction, and power. Each is legitimate; what is not legitimate is that the technical interest should predominate and lay claim to being the only source of scientific knowledge. The emancipatory interest requires free and open communication and thus a critique of conditions that militate against this. Communication of this self-critical sort is basic in the long run, even to the empirical-analytic sciences; the emancipatory interest cannot therefore be excluded even in these latter. Habermas's analysis is an extremely dense one and has prompted a range of responses. At the heart of the debate is the proper role of values in the constitution of knowledge, and how these values themselves are to be validated in a noncircular way.

These two linked questions have quite evidently prompted some of the liveliest discussion in recent philosophy of science. They bear directly on the issue that has always intrigued philosophers about the activities loosely grouped under the label "science": what is it about these activities, collectively, that warrants the confidence they inspire in the knowledge they produce?

References and further reading

Bacon, F. 1960: *The New Organon* (1620), ed. F. Anderson (Indianapolis: Bobbs-Merrill).

Barnes, B. 1977: *Interests and the Growth of Knowledge* (London: Routledge).

Bernstein, R. 1978: *The Restructuring of Social and Political Theory* (Philadelphia: University of Pennsylvania Press).

Habermas, J. 1971: *Erkenntnis und Interesse* (Frankfurt: Suhrkamp, 1968), trans. J. Shapiro as *Knowledge and Human Interests* (Boston: Beacon Press).

Hempel, C. G. 1983: Valuation and objectivity in science. In *Physics, Philosophy, and Psychoanalysis*, ed. R. S. Cohen and L. Laudan (Dordrecht: Reidel), 73–100.

Hull, D. L. 1988: *Science as a Process* (Chicago: University of Chicago Press).

Kuhn, T. S. 1977: *The Essential Tension* (Chicago: University of Chicago Press).

Laudan, L. 1984: *Science and Values* (Berkeley: University of California Press).

Lowrance, W. W. 1984: *Modern Science and Human Values* (New York: Oxford University Press).

Longino, H. 1990: *Science as Social Knowledge: Values and Objectivity in Scientific Inquiry* (Princeton: Princeton University Press).

McMullin, E. 1983: Values in science. In *PSA 1982*, ed. P. Asquith and T. Nickles (East Lansing, MI: Philosophy of Science Association), 3–28.

—— 1993: Rationality and paradigm change in science. In *World Changes: Thomas Kuhn and the Nature of Science*, ed. P. Horwich (Cambridge, MA: MIT Press), 55–78.

Merton, R. K. 1957: *Social Theory and Social Structure* (New York: Free Press).

Pickering, A. (ed.) 1992: *Science as Practice and Culture* (Chicago: University of Chicago Press).

Proctor, R. 1991: *Value-free Science? Purity and Power in Modern Knowledge* (Cambridge, MA: Harvard University Press).

Rudner, R. 1953: The scientist qua scientist makes value judgments. *Philosophy of Science*, 20, 1–6.

van Fraassen, B. C. 1980: *The Scientific Image* (Oxford: Oxford University Press).

Verisimilitude

CHRIS BRINK

At the 1960 International Congress of Logic, Methodology, and Philosophy of Science, and again in his books *Conjectures and Refutations* (1963) and *Objective Knowledge* (1972), Karl Popper proposed a formal definition of what it means for one scientific theory to be "closer to the truth" than another (see POPPER). Such a concept was a necessary ingredient in Popper's philosophy of science, in which all our scientific theories are not only false, but *bound* to be false. We can never, according to Popper, arrive at *the truth*: a complete and adequate description of reality. (That such a reality exists, outside us, is the basic tenet of scientific realism.) Nonetheless, Popper holds, scientists do make progress – namely, when they replace one false theory by another which, though still false, is closer to the truth, or, as we shall say, has greater verisimilitude.

Popper's definition came out in terms of the notion of logical consequence: theory 2 has greater verisimilitude than theory 1 if theory 2 has more true consequences and fewer false ones than theory 1. (Here we use "more" and "fewer" in the sense of set-theoretic inclusion, and for convenience tacitly include also the possibility "or equal.") However, David Miller (1974) and Pavel Tichý (1974) simultaneously, though independently, showed that Popper's definition is untenable: no false theory could, on Popper's definition, have strictly greater verisimilitude than any other false theory. Thus we are left with the *problem of verisimilitude*: to specify rigorously what it means for one theory to be closer to the truth than another.

In its initial stage the debate was conducted mostly in the pages of the *British Journal for the Philosophy of Science* and culminated in the publication of three books: by Oddie (1986), Niiniluoto (1987), and Kuipers (1987), all reviewed by Brink (1989). The main thrust was to view the question of "What is closer-to-the-truth?" as a quest for an acceptable notion of "distance from the truth," and hence to investigate various ways of defining a metric over theories. Consider the following toy example. Reality is described by whether or not it is hot, whether or not it is rainy, and whether or not it is windy. The truth is that it is hot and rainy and windy; the toy scientist can have any one of sixteen (propositional) theories about reality, and the verisimilitude theorist seeks a definition that will, in this case as in others, order the possible theories with respect to closeness to the truth. The proposal to view verisimilitude as a notion of distance from the truth would then further involve insisting that the order sought must be a *linear* ordering.

Also very prominent in the initial stage of the debate was an objection of David Miller, who argued that any verisimilar ordering of theories will be inextricably linked to the language in which those theories are expressed. Thus, in the toy example, the ordering of theories is linked up with the choice of basic predicates expressing the atomic facts about the weather. For suppose we introduce two other predicates, "Minnesotan" and "Arizonan," stipulating that we call the weather "Minnesotan" if it is either hot and wet or cold and dry and "Arizonan" if it is either hot and windy or cold and still. It is then a small exercise in propositional logic to show that the sixteen theories expressed in terms of the basic predicates hot, rainy, and windy translate one–one into the sixteen theories expressed in terms of the basic predicates hot, Minnesotan, and Arizonan. But – and this is Miller's point – such a translation seems to disturb any reasonable verisimilar ordering, which contradicts the apparently innocent assumption that verisimilitude should be invariant under linguistic translation.

Current interest in verisimilitude covers a wide range – a trend already evident in the "parade of approaches" of Kuipers (1987). Thus, for example, Brink and Heidema (1987) propose a notion of verisimilitude based on power orderings; Van Benthem (1987) explores the analogy between the study of verisimilitude in the philosophy of science and that of conditionals in philosophical logic; Schurz and Weingartner (1987) propose to rescue the original Popperian idea by filtering out irrelevant logical consequences, and Orlowska (1990) proposes a notion of verisimilitude based on concept analysis. Interesting generalizations can be obtained by relativizing the notion of a verisimilar ordering. For example, we may wish to relativize the truth to some arbitrary world, or to order theories with respect to some given theory. Another perspective is pointed out by Brink, Vermeulen, and Pretorius (1992): it has been argued that the theory of domains and power domains in denotational semantics of programming languages may naturally be considered in terms of growth of information, and the so-called Egli–Milner ordering used to obtain power domains is exactly the power ordering proposed for verisimilitude by Brink and Heidema (1987). This also brings in a topological approach to the notion of verisimilitude. Verisimilitude has also been linked with topics related to artificial intelligence, such as belief revision and theory change – see, for example, Ryan and Schobbens (1995) and Britz and Brink (1995). Recent surveys of verisimilitude theory can be found in Niiniluoto (1998) and Zwart (1998).

References

Brink, C. 1989: Verisimilitude: views and reviews. *History and Philosophy of Logic*, 10, 181–201.

Brink, C., and Heidema, J. 1987: A verisimilar ordering of theories phrased in a propositional language. *British Journal for the Philosophy of Science*, 38, 533–49.

Brink, C., Vermeulen, J. C. C., and Pretorius, J. P. G. 1992: Verisimilitude via Vietoris. *Journal of Logic and Computation*, 2, 709–18.

Britz, K., and Brink, C. 1995: Computing verisimilitude. *Notre Dame Journal of Formal Logic*, 36, 30–43.

Kuipers, T. A. F. (ed.) 1987: *What is Closer-to-the-truth?* (Amsterdam: Rodopi).

Miller, D. 1974: Popper's qualitative theory of verisimilitude. *British Journal for the Philosophy of Science*, 25, 166–77.

Niiniluoto, I. 1987: *Truthlikeness* (Dordrecht: D. Reidel).

—— 1998: Verisimilitude: the third period. *British Journal for the Philosophy of Science*, 49, 1–29.

Oddie, G. 1986: *Likeness to Truth* (Dordrecht: D. Reidel).

Orlowska, E. 1990: Verisimilitude based on concept analysis. *Studia Logica*, 49, 307–19.

Popper, K. R. 1963: *Conjectures and Refutations: The Growth of Scientific Knowledge* (London: Routledge and Kegan Paul).

—— 1972: *Objective Knowledge*, 1st edn (Oxford: Oxford University Press; 2nd edn, 1979).

Ryan, M. D., and Schobbens, P. 1995: Belief revision and verisimilitude. *Notre Dame Journal of Formal Logic*, 36, 15–29.

Schurz, G., and Weingartner, P. 1987: Verisimilitude defined by relevant consequence-elements. In Kuipers 1987, pp. 47–77.

Tichý, P. 1974: On Popper's definition of verisimilitude. *British Journal for the Philosophy of Science*, 25, 155–60.

Van Benthem, J. 1987: Verisimilitude and conditionals. In Kuipers 1987, pp. 103–28.

Zwart, S. 1998: Approach to the truth: verisimilitude and truthlikeness (Ph.D. thesis, Rijksaniversiteit Groningen, The Netherlands).

81

Whewell

JOHN WETTERSTEN

William Whewell was born in 1794. He was the son of a carpenter. In spite of a rather sickly childhood, he was intellectually precocious. At Cambridge his talent was quickly recognized, and the expectations for him were high. He fully overcame the sickliness of his youth to become imposing and robust as a man, adventurous and rambunctious in his intellectual life. He helped to introduce the newer French mathematical techniques as a substitute for the outdated Newtonian ones taught at Cambridge and advocated mathematics as a foundation for good thinking in his research on pedagogy. For a while he was a mineralogist, a subject he studied under Mohs in Germany. He studied the tides and sought to determine the mean density of the Earth by comparing the motions of pendulums on the surface and in the interior of the Earth, in mines. He became a specialist on the architecture of cathedrals, which he inspected while traveling in Europe. He wrote poetry and studied languages. But all these activities brought him no great accomplishment or success. It was only when he turned his hand to the history and philosophy of science that he became a figure of major importance. His place in the history of the philosophy of science has by no means been secure, but his views have always reemerged, even though at various times they have been thoroughly rejected as not even serious.

He was reputed to be rough in argument, but de Morgan came to his defense, explaining that, if he was impatient with obtuseness, he was always graceful in defeat. He was elected Master of Trinity when he was 41 and remained in this position until his death in 1866. Whewell's ambitious intellectual life was matched by his deep religious belief. He was a preacher, who, it is said, had a poor delivery. He sought to integrate his religious devotion and his intellectual curiosity, arguing that the results of science can be seen to support only the view that the world was designed by God. The Earth is unique, and life will not be found elsewhere.

He married in 1841 and suffered intensely when his wife died after a long illness in 1855. He married again in 1858. His second wife died in 1865. He had no children but was consoled by his nieces. By the end of his life his philosophy had been rejected. He was not popular at Cambridge, where the views of Mill dominated, and where he was deemed oppressive and old hat. When the students cheered instead of booing, as they had been accustomed to do, after his return to Cambridge following a period of mourning and depression brought on by the death of his first wife, he wept.

In his path-breaking *History of the Inductive Sciences* and *The Philosophy of the Inductive Sciences, Founded upon their History*, Whewell developed a philosophy of science which dispensed with inductive methods of discovery as well as inductive proof. On the basis of new physiological discoveries such as those of rapid eye movement by Charles Bell, Whewell argued that perception is active, that we do not have given, observed, surefire facts on the basis of which we construct our theories. Before we can construct theories, we need, he said, to impose ideas on the world to organize and even to define the facts. We start with vague ideas which need to be empirically tested to make them clear. We make conjectures, and some of these conjectures become so refined that we know they are true. The scientifically educated know they are true on the basis of intuition. Whewell calls such ideas "fundamental." They describe the world as it is; Whewell was a realist. The discovery of each fundamental idea leaves open problems of refining and extending the fundamental idea and opens up new areas of research. We can, for example, unify two fundamental ideas – say, Kepler's description of the motions of the planets and Galileo's theory of falling bodies – with a new theory, such as Newton's mechanics (see GALILEO and NEWTON). Whewell calls this a "consilience of inductions." The theory of the consilience of inductions explains the continued progress of science. Consiliences of inductions provide for an ever deepening of our knowledge and the discovery of ever more general principles. Science is incomplete and will remain so. It will, we hope, progress, but its limits are not known to us.

All knowledge, Whewell said, possesses two aspects, which are ideas and senses, thoughts and things, theory and fact, necessary and experiential truth. The unity of these two aspects of knowledge Whewell called the "fundamental antithesis of philosophy." If we had no sensations, we would have no knowledge of the world. If we had no ideas, we would have no knowledge. In the process of obtaining knowledge, we find a movement, however, from ideas to facts, or, we can also say, to the idealization of facts – that is, to their clear and distinct formulation. As ideas become clearer and clearer, they then become facts: Kepler's theory of the motion of the planets becomes a fact, which in turn needs to be explained. But the idealization is never complete. If it were to be complete, we would no longer have knowledge. We can prove that the fundamental antithesis is present in all knowledge, but we cannot explain it within the bounds of reason. If we seek to do that, Whewell said, we will fall into the trap which led the Germans to create useless, speculative systems. The existence of this limit to knowledge is unsatisfying, but, for believers at any rate, theology can provide a more satisfying picture.

Whewell's philosophy of science is not designed to explain why science works in the way it does. Rather, it is designed to explain how science in fact works, and to do so better than any competitor. Whewell devised a new method of proof for his theory of science. He claimed that a theory of science which could explain how scientific knowledge was in fact obtained, and could do so clearly and distinctly, when no other theory could, was in fact true. His own theory, he claimed, could do that. It explained science in the same way that scientific theories explained some particular field of inquiry and was proved in the same way.

This theory posed the most serious challenge to the standard inductivist views of the nineteenth century. John Herschel, John Stuart Mill, David Brewster, Henry Mansel, and de Morgan raised objections to it (see MILL). These objections came from different perspectives but led to a unified judgment which was summarized in the *Dictionary of National Biography*: Whewell's theory was eclectic and confused. He was hardly forgiven for dispensing with inductive methods of discovery, even though no less a thinker than de Morgan praised him for it. He was damned for dispensing with inductive proof. It was his old friend from Cambridge schooldays and the then most influential physicist of the day – John Herschel – who supported the inductivist view of John Stuart Mill. The debate between Whewell and Mill, in which Mill presented his criticism in *A System of Logic*, especially in the third edition of 1851, was long and hard. Whewell could win whenever it came to describing science, which he, in contrast to Mill, knew quite intimately. But he failed to gain any support for his philosophy.

The theory of the existence of fundamental ideas which are not derived from facts but are known by intuition was rejected as excessive. It was seen as a path which must lead to the excesses of the German philosophers. Even though de Morgan had kind words to say about Whewell's theory of the growth of necessary truth, this view was hardly taken seriously. Whewell's theory of scientific method was, perhaps, a very good theory of scientific discovery and a very good psychology of scientific research. But no one accepted Whewell's claim that intuition was the basis of scientific proof.

Antagonism to Whewell's theory did not arise merely because it was too adventurous. It arose just as much because his intent was seen as too conservative. The idea that there were two sources of knowledge fitted very well with Whewell's theological ideas. But he did not use his theology to support his theory of science. Rather, he based his theory of science on the facts of science and found support for his theology in his theory of science. But this did not fit with the new attempts to find a moral and social theory which was based on the facts alone, such as that of John Stuart Mill.

The sharp, antagonistic, politically successful reaction to Whewell has often led to the impression that he is a forgotten figure, a man who impressed his colleagues for a while but whose lasting influence is minimal. But just when it appears that he has finally been consigned to the archives as a man of interest to only a few specialists in nineteenth-century thought, he emerges once again with new strength. This is due not only to the depth of Whewell's critique of the more established views, but also to the fact that he developed a powerful alternative, a theory which showed how at least some of these problems might be overcome. When the old problems reappear, he proves once again to be the thinker who thought seriously about how they could be solved, and his insight is sought.

The hidden reaction to Whewell has been quite different from the public one. For the problems he pointed to were not openly acknowledged as such even when they came to dominate the agendas in psychology, methodology, and epistemology. In the newly self-conscious scientific psychology of the nineteenth century, more strenuous efforts were made to carry out the reduction of higher thought processes to sensations, which Whewell claimed could not be done. In methodology, attempts were made to reconcile old-fashioned inductivist views with Whewell's view of scientific practice, as Mill tried so strenuously to do. Whewellian views appeared on the scene with those of Claude Bernard in France. The success of Bernard's view led to new research on Whewell and the puzzling relationship between them. The views of Duhem and others show his imprint. In Germany, when the views of the associationist psychologists could no longer be upheld, Külpe sought a new, dualistic scientific psychology and a new methodology to complement it, one which presumed that there are two sources of knowledge – that is, ideas and sensations. Külpe knew Whewell's work through his

History, which had been translated into German. This led in turn to the work of Karl Popper, who had to make vast improvements on Külpe's work in order to render it a plausible twentieth-century philosophy of science (see POPPER). Popper did not know of Whewell when he first developed his philosophy of science. But Whewell's influence had worked its way through the work of the Würzburg school to Popper.

Since the 1950s the literature on Whewell has been dominated by attempts to carry out three tasks. The first and foremost among these is the rehashing of old complaints against Whewell so as to stop the reemergence of his or any view like it. The second task has been to incorporate elements of Whewell's view into modern inductivist theories. His theory of the concilience of inductions and his theory of the discovery of a fact, for example, have been, subjects of inquiry. The third task, undertaken more recently, is to explain his social and political background, whereby one concern seems to be to find new justifications for his rejection in the nineteenth century.

Whewell's philosophy cannot be defended as a live option. But it is difficult to understand the contemporary debate without seeing how powerfully his views have been working behind the scene. Even today his views of the development of science and of the relation between the history and the philosophy of science remain modern and hardly absorbed. A new, deeper appreciation of this revolutionary thinker and his central role in the development of the philosophy of science is gradually being developed. For an understanding of how the fundamental problems of the philosophy of science over the last 150 years have arisen, an appreciation of Whewell and the reaction to him is indispensable.

References and further reading

Works by Whewell

1830: Review of John Herschel's "Preliminary Discourse on the Study of Natural Philosophy," *Quarterly Review*, 90, 374–407.

1858: *Novum Organon Renovatun* (London: John W. Parker and Son).

1967a: *History of the Inductive Sciences*, 3rd edn (London: Frank Cass & Co.), repr. of 1857 edn.

1967b: *The Philosophy of the Inductive Sciences, Founded upon their History*, 2nd edn (London: Frank Cass & Co.), repr. of 1847 edn; 1st pub. 1840.

1971: *On the Philosophy of Discovery* (New York: Burt Franklin), repr. of 1866 edn.

Index

a priori knowledge 227, 286, 373, 395, 426, 461, 463, 466, 530
abduction 184, 336; arguments against 394–6, 397; defense of 396
Abraham, R. H. 259
absolutism 104
abstraction 255
Achinstein, P. 130, 327, 332
aesthetics 435, 480
Akaike, H. 438–40
Akaike theorem 438–40
Alexander, H. G. 6
Alexander, S. 544
Algorithor 4–5
ambiguity 472, 533
analogical conception of theories (ACT) 303–7
analogy 301, 415; neutral 300; positive 299–300; see also models and analogies
analysis 408–10
analytic philosophy 483, 484
Anderson, J. 544
antecedent conditions 472
anthropology 136
anti-metaphysics 252–5
anti-naturalism 453–4
anti-realism see realism/anti-realism debate
Arbib, M. 279
Arendt, H. 490
argument 117, 337, 446
argument from economy see economy principle
Aristotelianism 149–51
Aristotle 117, 175, 205, 214, 215, 285, 332, 373, 424–5, 486–7, 529, 541, 553
Armstrong, D. M. 544
Arnold, L. 484
Aronson, J. L. 221, 281
artificial intelligence (AI) 45–6, 86, 87–90, 116, 562
artificial life 22–4
astronomy see cosmology
atomism 373
atoms 415, 543
Augustine, St 5
Austin, J. L. 244
auxiliary hypothesis 155–6
axiomatization 500; defined 9; in science 10–11; semantic/syntactic approaches to 9–10
axioms 9–10, 204, 427, 516
Ayer, A. J. 235, 244, 247–8, 249, 289, 290, 311, 337

background 356, 420, 421
Bacon, F. 3, 6, 118, 217, 414, 425, 552, 553, 557
Barnes, B. 558
Barrett, P. H. 72
Barrett, R. B. 386
Barrow, I. 12–13
Barwise, J. 46

Bauer, M. 505
Bayes, T. 111
Bayesian confirmation theory 111–12, 198–9, 238, 362, 364–5, 366, 419–20, 436–7
Beatty, J. 522
behavior 23–4
behavioral science 38
beliefs 118, 162–3, 314, 353, 355, 362, 452–3; nonadditive function 116; social factors 442–8
Bell, J. 37, 478
Bentley, J. L. 51
Bergson, H. 103
Berkeley, G. 12–15, 213, 373
Berlin, I. 244, 247, 248
Berlin School 244
Bernard, C. 566
Bernoulli, J. 110
Bernstein, R. 559
Beth, E. 517
Beth's theorem 66, 76
Bhaskar, R. 218, 221
biology 16, 72, 73, 315–16, 317–18, 522, 545; artificial life 22–4; contemporary issues 17–20; explanation/prediction 18–19; fitness/selection 19–20, 71, 72; gender bias 135, 447; laws 17–18; new directions 20–4; ontology in 455; organization 21–2; received view 17; reductionism 19; semantic conception 20–1
Bird, A. 504
Black, M. 277, 278, 279, 280, 302, 435
Bloor, D. 443–4
Boden, M. A. 44
Bohr, N. 26–9, 106, 246, 377, 380
Boids 23–4
bootstrapping 115, 238
Borgmann, A. 489
Born, M. 105
Boyd, R. 277, 279, 356, 396
Boyle, R. 230, 555
Braithwaite, R. 371
Brannigan, A. 91
Brent, J. 335
bridge principle 300, 499
Bridgman, P. W. 77, 249–50
Brink, C. 562
Broad, C. D. 182
Brown, J. R. 287, 507, 529, 530
Bub, J. 380, 383
Butts, R. E. 150, 541, 542

C-R model see causal-relevance (C-R) model
calculus 224–5, 323
Callebaut, W. 73
Campbell, D. T. 38, 73, 92, 94, 309
Campbell, N. 259, 300
Campbell, R. 39

Carnap, R. 10, 66, 76, 111, 118, 182, 203, 233, 234–5, 237, 238, 240, 243, 244–5, 248–9, 250, 306, 331, 360, 415, 427, 428, 498, 499, 502–4, 516, 517, 556
Carnot, S. 169
Cartwright, N. 4, 130, 140, 170, 218, 302, 304, 400–1, 403
Cat, J. 539
categories 426, 541
causal analysis 272–5
causal explanation 295, 511
causal necessity 34–5, 521, 572
causal relations 304, 476
causal theory of reference 178–9
causal-inductive reasoning 327–8
causal-relevance (C-R) model 129–30, 131
causation 18–19, 31, 129, 166–7, 168, 214, 217, 241, 253, 378, 401, 443, 451–2, 456; discovery 36; dispensability argument 37–8; generalization/singular claims distinction 32; Hume's account 32–3; manipulability/activity approach 38; mental 39; preliminaries 31–2; probabilistic accounts 35–6, 475–7; scientific considerations 37–9; skepticism about 36–7; standard problems 33; sufficiency accounts 33–4
cause–effect relationship 34
Causey, R. 541
change 22, 350; see also scientific change
chaos 21–2
chemistry 316–17
Cherniak, C. 50
Christensen, D. 363
Church, A. 248
Church–Turing thesis 47–9
Churchland, P. M. 42, 179, 332, 351, 454, 511
Churchland, P. S. 318, 403
circularity 395–6, 400
Clarke, S. 225–6, 462
classical mechanics (CM) 376, 525
classical view 499, 521; Carnap–Lewis refinement 502–4; development of 515–17; formal alternatives to 517–20
classification 231, 317–18
clock paradox 103–4
cognitive approach 41–3, 119, 179, 306, 309, 419
cognitive skills 195
Cohen, L. J. 347
Collins, H. M. 122, 125, 406, 446
common cause 330, 477–8
common sense 330, 403
complementarity 27
complexity 23–4, 545
computing 44, 419; definition of a computer 46–8; extravagant claims 45–6; findings 48–51; and other sciences 51
Comte, A. 246
concept 5; and Laws of Nature 215–16
conceptual analysis 309
conditionalization 362–3
confirmation 108–9, 163, 169, 240, 336, 345; entrenched projectability 109, 182; nonprobabilistic theories 113–15; paradoxes of 53–5; probabilistic theories 109–13
connectionism 89
consilience 327
constructive realism 303
context 144–5, 177, 395, 398
context of discovery/context of justification 87, 89–90, 289, 290, 326, 416–17, 448
contingent statements 18
continuity principle 255
control 22, 23–4

conventionalism 271–2; alleged underdetermination of theory 56–63; conventionality of geometry 63; denying the problem 58; geometry and general theory 57–8; holism about meaning 60–2; meaning of theoretical terms 60; Poincaré on geometry 56–7; Quine's approach 62–3; realist responses 58–60
convergence 170
Cook, T. 38
Copernicanism 149, 150
Copernicus, N. 321, 332, 440, 532
corpuscularianism 312, 313, 354, 394
correspondence rules 17, 26, 66
cosmology 149–52, 200, 297, 321, 322, 427, 440, 547
counterfactuals 18, 35, 215, 221, 274, 521–2
covering law model 127
Craig, E. 501
Craig, W. 65
Craig's theorem 65–6, 76, 268, 517
Crick, F. 16, 72
critical theories 486, 488–9
Critical Theory 457, 559
cultural relativism 147
Currie, G. P. 209, 211
curve-fitting problem 433–6; Akaike theorem 438–40; Bayes theorem 436–7; Popperian approach 437–8
Cziko, G. 92

D-N model see deductive-nomological (D-N) model
da Costa, N. C. A. 505
Darwin, C. 16, 68–74, 92, 184, 308
Davidson, D. 204, 454, 501, 510, 511
Dawkins, R. 72–3
de Finetti, B. 111, 360, 363, 370
decision theory 552–3
deduction 83, 125, 238, 327, 351, 352, 427, 555
deductive logic 471–2
deductive-nomological (D-N) model 127–9, 186, 470–3, 474–5, 476; counterexamples 129
definitions 76–8, 215; admissable forms 77; Aristotelian 77; of concepts 77–8; creative 76–7; disjunctive polytypic 77; explicit 76, 77; full 76; nominal 77; operational 77; partial 76; real 77
demarcation problem 155–61
democratic relativism 147
Dennett, D. C. 74, 493
denotation 496, 502, 510–11, 512
Descartes, R. 79–83, 229–30, 286, 309, 321, 322, 462, 553; Discourse and Essays 79–81; Meditations 81–2
descriptive facts 253
descriptive unity 545–6
designation 510 11
d'Espagnet, B. 290
determinism 105, 112, 261–2, 315, 378, 382, 476, 478; quantum 28–9
Devitt, M. 502
diffracting force 354
Dirac, P. 26
discovery 42, 85–95, 415; and AI 86, 87–90; and coupling 91–2; criticisms of 88–95; historical background 86–7; logic of 87, 91–5, 416; neglect of 120; questions concerning 85; relevance to epistemology 91; revival of interest in 87–8; see also context of discovery/context of justification
dispositions: affordances 98; bare/grounded 97–8; conditional properties 97; doctrine of qualities 99; dynamicist metaphysics of physical science 99–100; forms/varieties 97–9; historical development of concepts 99–100; liabilities/power 98–9; and scientific realism 100
Doell, R. 136

domain 266–8
Donovan, A. 88
double-slit experiment 354, 377, 378, 379–80
Dretske, F. 219, 331, 332, 493
Duhem, P. 3, 61, 120, 249, 388, 428, 534, 566
Dummett, M. A. E. 1–2, 163, 501, 510, 533
Dupré, J. 315, 318, 402, 507
Dutch-Book theorem 111, 362, 363
dynamical unity 544–5
dynamicism 99–100

Earman, J. S. 105, 250, 346, 363
economy principle 91–2, 254, 335
Eddington, A. S. 60, 283
Edinburgh School 430, 554
Eells, E. 35, 476
efficiency 488
Egli–Milner ordering 562
Einstein, A. 28–9, 102–6, 239, 255, 289, 380, 464–5
Eldredge, N. 71, 72
elements 252–3
eliminativism 295–6, 403, 454
Ellul, J. 485, 486
Elster, J. 455
Empirical Equivalence (EE) 394–5, 397–8
empiricism 330, 346, 357, 385, 455, 456; contextual
    140–1; feminist 138; Lockean 229–32; logical
    see logical empiricism; and theory acceptance 351–3
Enç, B. 508
Engels, F. 487
epistemic/non-epistemic difference 331–2
epistemology 308, 309, 385; feminist 138
equivalence 53–4, 61–2, 289, 336, 394–5, 397–8,
    467, 496, 526; empirical 526–7
equivocation see ambiguity
Ereshefsky, M. 316
error 273–4
essence 215, 230–1, 311–12, 313, 426
essentialism 311–12, 313
ethics 442, 553
Evans, G. 506
Everett, H. III 382
evidence 232, 357, 443, 444; and confirmation
    108–16
evolution 16, 309, 318, 355, 456; fitness/selection
    19–20; gender bias 135–6
existence 401
existentialism 490
experience 165–6, 168, 248, 344, 357, 373, 424, 425,
    516; and Laws of Nature 216–18
experiment 117; criticism of 122; as handmaiden of
    theory 117–19; implications of 124–6; and
    knowledge 122–3; neglect of 120; observation/
    instrumental practice 121; regress 121–2; and
    testing of theories 120–1
explanation 3, 5, 34, 127–32, 238–9, 241, 279, 341,
    369, 394, 538, 542–3, 548; biological 18–19;
    causal-relevance model 129–30; deductive-
    nomological model 127–9; as pragmatic 395,
    397–8; pragmatic aspect 131–2; types of 130–1;
    see also Inference to the Best Explanation
explanation, statistical see statistical explanation
explanation, teleological see teleological explanation
explanatory virtues 351; pragmatic or epistemic 355–7
explications 77–8
externalism 421

Fairbanks, 125
falsification 72, 118, 124, 145–6, 156, 199–200, 208,
    209, 343, 345, 415, 437–8
Fausto-Sterling, A. 135
Faye, J. 28

Feenburg, A. 485, 490
Feigenbaum, E. 89
Feigl, H. 234
feminist epistemology see epistemology, feminist
feminist science 134–5; biological evolution 135–6;
    contextual values 140, 141; empiricism 138; gender
    ideology 136–7; holism 140–1; integrated theory
    140; objectivity 139, 141; positivism 140–1;
    postmodernism 138–9; primatology 136; research
    139; scientific revolution 136–8; standpoint
    epistemology 138, 139
feminist theory 558–9; social factors 447–8
Fetzer, J. H. 46, 475, 477
Feyerabend, P. K. 2, 88, 143–7, 150, 154, 203, 331,
    405, 418, 429, 431, 501; notion of incommensurability
    176–8
Field, H. 260, 262, 496, 502, 510
Fine, A. 106, 395–6, 400, 478
Fisher, R. A. 113–14, 362, 434
Fodor, J. A. 163, 454, 504, 545
folk psychology 453, 511
Folse, H. 28
formalism 10
Forman, P. 443
forms 14
Forster, M. R. 538
Frankfurt School 559
Franklin, A. 120, 123
Frasca-Spada, M. 407
Freedman, D. 38
Frege, G. 247, 512
French, S. 505
Friedman, M. 104, 130, 260, 538, 547
function 455–7, 493–4
fundamental ideas 565–6

Galileo Galilei 18, 117, 146, 149–52, 197, 200, 205,
    259, 321, 373, 374, 425, 529, 546, 547
Galison, P. 123, 125, 126, 200, 545
gender ideology 134–42
Gendler, T. 529, 530
generalizations 196–7
generalized model-based understanding of theories
    522–4
genetics 16, 19–20, 71–2, 300, 403; Mendelian 20–1
geometry 57–8, 463, 466, 516, 517–20, 521, 547;
    see also conventionalism
Gibbons, M. 483
Gibson, J. J. 98
Gibson, R. F. 386
Giere, R. N. 41, 42, 119, 179, 303, 309, 310, 368,
    370, 505, 519
Gillies, D. 371
Glymour, C. 38, 44, 93, 115–16, 130, 357
goals 492–3
God 12, 13, 15, 79, 82, 225–6, 286–7
Gödel, K. 243
Goldman, A. I. 6, 309
Goldstein, I. 88
Goodfield, J. 119, 121
Gooding, D. 42, 119, 120, 123
Goodman, N. 54, 108, 182, 197, 215, 336
goodness of fit 434
Gould, S. J. 71, 72, 291
Gowing, M. 484
grammatical rules 215–16
gravity 464, 465–6, 470–1, 505, 538, 547, 555
Grice, H. P. 509
grue, problem of 54, 108–9, 112–13, 114, 182, 336, 366
Grünbaum, A. 63, 150
Guth, A. 105
Gutting, G. 418

H-D method *see* hypothetico-deductive method
Habermas, J. 457, 559–60
Hacking, I. 119, 120, 124, 140, 250, 338, 400–1, 406, 480, 506
Hager, P. J. 408
Hahn, H. 243
Haldane, J. B. S. 16
Hamilton, W. 73
Hampshire, S. 244
Hanfling, O. 243, 249, 250
Hanson, N. R. 88, 203, 245, 331, 416
Haraway, D. 136
Harding, S. 138, 139–40, 447
Hardy–Weinberg law 18
Hare, R. M. 480
Harper, W. 547
Harré, H. R. 123, 218
Hartley, D. 232
Hawking, S. 105
Healey, R. A. 106, 378, 383
Hegel, G. W. F. 154
Heidema, J. 562
Heil, J. 39
Heisenberg, W. 26, 27, 378
Hempel, C. G. 18, 51, 66, 90, 109, 127, 128, 131, 203, 234, 237–9, 241, 244, 249, 279, 331, 332, 357, 427, 452, 456, 470, 471, 473, 474–5, 498, 505
Heraclitus 284
Herodotus 284
Hesse, M. 103, 278, 279, 280, 299, 301, 306, 436
heuristics 89, 92, 93, 208–10, 209, 356, 417–18
Hintikka, J. 504
historical chain theory of reference 506–8
history 254, 458
history, role in philosophy of science 154–61; demarcation problem 155–61; interest in 154–5
Hitchcock, C. R. 477
holism 60–2, 140–1, 162–3, 456–7, 500
Hollis, M. 406
Holton, G. 106
Hooker, C. A. 351, 542, 543, 544, 545
Horkheimer, M. 559
Horwich, P. 363, 496, 502, 503, 504
Howson, C. 110, 112, 115, 362, 367
Hubbard, R. 135, 447
Hubble, E. P. 105
Hughes, R. I. G. 37, 478
Hull, D. L. 17, 309, 316, 403, 554, 555
Hume, D. 32–3, 36, 130, 165–8, 181, 190, 196, 197, 214, 216–18, 246, 287–8, 337, 435
Humphreys, P. 38, 369, 477
Husserl, E. 170
hypotheses 108–9, 113, 117, 253, 296–7, 336, 387, 394–5, 414, 437–8, 440, 519, 553–4; acceptance of 194, 196–8; attitude toward 201; gender bias 135; as true 198–9; as worthy of consideration 201
hypothetico-deductive (H-D) method 87, 118, 151–2, 186, 240, 300, 302, 303, 325–8, 329, 414, 415, 426

idealism 542
idealization 169–70, 349
ideas 165–6, 229–30, 565; simple/complex 230
identification fallacy 46
identity 383; gendered 137, 138; *see also* theory identity
ideology 458–9, 557
Ignatieff, M. 244
Ihde, D. 485, 490
impartiality 443–4
impetus 176
implicit term 498

incommensurability 77, 145, 146, 147, 172, 314, 417, 429, 501–2; Feyerabend's notion 176–7; Kuhn/Feyerabend comparison 177–8; Kuhn's notion 172–6; responses to 178–9
indeterminism 34–5, 346, 377, 378, 381, 382; quantum 28–9
indifference principle 365–6
individuation 525–6
induction 108, 118, 166, 167, 181–2, 185, 190–1, 217, 238, 294–5, 327, 328, 329, 337–8, 343–4, 351, 399, 414, 435, 555, 564–7; criticism of 346
inductive confirmation 195–9
inductive logic 472
inductive reason 116
inductive-statistical (I-S) explanation 470–3; criticisms of 473–5
inference 325–8, 336, 338, 435–6, 542
Inference to the Best Explanation 184–92, 303; abduction model 184–5; articulating 186–7; contrastive 188–9; descriptive problem 185–6; guiding challenge 189–90; identification 187–8; and inductive justification 190–1; justificatory applications 191–2; likeliest/loveliest model 187, 188, 190; and miracle argument 191; recession hypothesis 185; as true 191–2; vertical 186
instrumentalism 3–4, 14, 28, 41, 318, 329, 335, 352, 379, 394, 515; and technology 486, 487; *see also* realism/instrumentalism debate
intensional entity theory 511
intentionality 452–3
interaction 278–80
interpretation 458, 510
interventionism 140
intuition 301, 315, 317, 355, 477
Irvine, A. 261

Jackson, F. 498
Jardine, N. 407
Jaynes, E. T. 112, 363
Jeffrey, R. C. 370, 473
Jeffreys, H. 182, 436–7
Jevons, W. S. 87
Johnson-Laird, P. N. 42
judgment 42, 194; described 194–5; and falsification 199–200; and inductive confirmation 195–9; and observation 200; summary of 201–2
justification 87, 185, 190–2, 240, 400, 415; *see also* context of discovery/context of justification

Kant, I. 288, 290, 308, 426, 541
Kantorovich, A. 92
Kash, D. E. 485
Katz, M. 169
Kauffman, S. 22
Keller, E. F. 137, 140, 447, 448–9
Kelly, K. 94
Kepler, J. 149, 297, 321, 546, 547, 555, 565
Kepler's laws 128, 130
Kim, J. 403
Kitcher, P. 73, 130, 262–3, 309, 452, 510, 538, 541, 542
Kleene, S. C. 48, 49
Kline, S. J. 485
knowledge 160, 229–32, 458–9; and experiment 122–3
knowledge elicitation problem 89
Kochen, S. 383
Kornblith, H. 309
Kosso, P. 331
Koyré, A. 203
Koza, J. 92
Krantz, D. H. 259, 266

Kripke, S.   313, 507
Kroon, R.   508
Kuhn, T. S.   4, 5, 6, 91, 117, 118, 120, 122, 144–5,
    154–5, 157–8, 159, 198, 203–5, 207, 210, 240,
    280, 291, 309, 354, 405, 417, 419, 426–8, 440,
    457, 484, 501, 523, 528, 554, 556; notion of
    incommensurability   172–8
Kuipers, T. A. F.   561
Kukla, A.   397
Külpe, 566–7
Kyburg, H. E.   361

laboratory studies   445–7
Lakatos, I.   41, 156–7, 158, 159, 203, 207–11, 346,
    417–18
Lanczos, C.   102
Langley, P.   42, 90, 93
language   173–5, 204, 205, 497, 509; semi-interpreted
    391
Latour, B.   121, 122, 124, 125, 291, 445–6, 447, 484
Laudan, L.   87, 310, 395, 397, 406, 414–15, 419
Laudan, R.   483
laws   5, 452, 555; biological   17–18; explanatory
    128–9
Laws of Nature   230, 317, 427–8, 515, 522, 523;
    described   213–15; as descriptions of natural
    tendencies   218–21; as expressing relations among
    concepts   215–16; Hume's account   216–18;
    philosophical problems   214; as summaries of
    experience   216–18; summary of   221–2
Laymon, R.   169
Leibniz, G. W.   6, 224–8, 286–7, 462, 463, 538;
    contribution to science   224–6
Leplin, J.   356, 397, 400, 401
Lepore, E.   163
Levi, I.   338
Lewis, D.   35, 274, 366, 368, 501, 502–4, 505, 510
likelihood   438
linguistic analysis   239–40
Lloyd, B. B.   279
Lloyd, E.   522
Locke, J.   13, 99, 197, 229–32, 317, 373
logic   154, 248, 337, 383, 516
logic, deductive see deductive logic
logic of discovery see discovery
logic, inductive see inductive logic
logical empiricism   146, 163, 233, 308–9, 415–16,
    426–8, 516, 541, 542; critique of   143; current
    trends   237–9; historical background   233–7;
    opposition to   239–41
logical positivism   4, 41, 66, 87, 118, 203–4, 209, 231,
    233–4, 239, 308, 337, 340, 500, 541, 550, 556;
    origins   245–7; verification, meaning, truth   247–50;
    Vienna Circle   243–5
Longino, H.   136, 140–1, 447, 558, 559
Löwenhein-Skolem theorem   394
Lukes, S.   406

McAllister, J. W.   406, 538
McCloskey, D.   457
McCurdy, C.   369
McDowell, J.   501
McGinn, C.   493, 501
Mach, E. W. J. W.   104–5, 214, 218, 221, 245–6,
    252–5, 428, 528
Mackie, J. L.   35, 296, 311
McKinsey, J. C. C.   520
McMullin, E.   151, 557
McRae, R.   541
Madden, E. H.   218
magic   15
Maher, P.   362

Malament, D.   261
Malthus, T. R.   68–9
man-the-hunter/woman-the-gatherer   447–8
Marcuse, H.   486
Martin, M.   501
Marx, K.   487, 489
Marxism   489–90
material hindrance   151
materialism   12
mathematics   14–15, 207–8, 231, 257–63, 323, 337,
    408; application of   262–3; artifacts   259–60;
    measurement theory   257–9; need for   260–2;
    structuralist view   263
matrix mechanics   27
matter   312, 313, 373, 462, 468
Maudlin, T.   538
maximal specificity   472
Maxwell, G.   66
Maxwell, J. C.   463
Mayr, E.   537
meaning   163, 177, 231, 247–50, 304, 313–14, 351,
    500, 511–12
measurement   265–6, 382–3; mathematical theory
    266–70; philosophical theories   270–5
mechanical philosophy   322
mechanization   137
Mele, A.   39
Mellor, D. H.   315, 361, 368, 370
Mendel, G. J.   18, 20–1, 71–2, 91
Merchant, C.   136–7
Merton, R.   554
Mesthene, E. G.   485, 486
metaphor   277; comparison view   277–8; interactive
    view   278–80; scientific/literary contrast   280–1;
    type hierarchies   281
metaphysics   233, 283–4, 346, 357, 467–8; history of
    attitudes   284–8; in modern science   288–91
Method of Agreement (MA)   295–6
Method of Difference (MD)   296
methodological appraisal   209
methodology   146, 320–1, 363–6; see also scientific
    methodology
methodology of scientific research programs (MSRP)
    207–11, 417–18
Mill, J. S.   93, 145, 262, 293–7, 326, 338, 394, 435,
    566
Miller, D.   347, 561
Miller, R.   507
Millikan, R.   74, 493
Milne, P.   369
mind   229–30, 403
Minkowski, H.   102, 464
miracle argument   191–2, 330
Misak, C.   335
Miščević, N.   530
modalities   125
model-based understanding of theories see generalized
    model-based understanding of theories
models and analogies   42; analogical conception of
    theories   303–7; in classical physics   299–301;
    semantic conception of theories   301–3
Modern Synthetic Theory of Evolution (MST)   16
molecular genetics (MG)   19
Moor, J. H.   45
Moore, G. E.   480
More of the Same principle   185, 190
Morrison, M.   538
motion   14, 130, 146, 462–3, 543, 565
Mowery, D. C.   483
multiplicity of causes   34
Mundy, B.   259
Murdoch, D.   28, 29

Musgrave, A. 208
Musson, A. E. 484
MYCIN theory of confidence 116
Myhill, J. 49

Nagel, E. 4, 245, 259, 402, 492
names 313, 507
narrative 159, 160
natural kinds 304, 305, 311; Aristotelian 311–12;
   contemporary discussions 313–18; historical
   background 311–13; Kripke–Putnam theory
   313–17; Lockean 231–2, 312–13
natural selection 69–74, 184
naturalism 41, 308–10, 385–6, 453–4; epistemological
   308; evolutionary approaches 309; interest in 309;
   objections to 309–10; ontological 308; and
   pragmatism 308–9
nature, laws of see Laws of Nature
necessity see causation
neo-classical view 505
Neo-Darwinism 72
Nersessian, N. J. 42, 530
Neurath, O. 243, 248, 541
neutral monism 253
Newell, A. 44, 45, 88, 89
Newton, I. 6, 14, 104, 118, 130, 156, 197, 225, 232,
   259, 289–90, 320–3, 414, 415, 425–6, 462–3,
   470–1, 540, 546, 547–8, 555
Newton-Smith, W. H. 6, 103, 146, 396, 418, 496, 536
Neyman, J. 114, 115
Nickles, T. 87, 210
Nicod's condition 53–4
Niiniluoto, I. 347, 561, 562
Nola, R. 508
nonlogical terms 516
nonprobabilistic theories of confirmation: Fisher–Popper
   theory 113–14; Neyman–Pearson theory 115
Norton, J. 529
novel predictive success 399–400

O-terms 507–8
objectivity 163, 386–7, 556; gendered 139
observation 117–19, 143–4, 200, 245, 254, 325,
   344, 351, 386, 397, 420, 427, 496, 500, 515, 516,
   533–4; defined 331–3; and instrumental practice
   121; unobservables as constructed 328–31;
   unobservables as inferred 325–8; see also theory/
   observation distinction
observation sentences 386–7
Occam's razor 355, 435
Oddie, G. 347, 561
Okruhlik, K. 448
ontological unity 542–3
ontology 31–2, 455
operationalism 249–50, 270–1, 498
Oppenheim, P. 18, 239, 402
optics 12–13, 323, 394
order 268–70
ordinary language 239–40, 248, 305, 314, 315–16
organization, biological 21–2
Orlowska, E. 562
overdetermination 34

Papert, S. 88
Papineau, D. 493, 509
paradigm 172–8, 417–18, 428–9
paradox 53–4, 103, 108–9, 112–13, 114, 336
parallel distributive processing (PDP) 306
Parmenides 542
parsimony see Occam's razor
Pauli, W. 26
Peacocke, C. 510

Pearce Williams, L. 211
Pearl, J. 38
Pearson, E. S. 115
Peirce, C. S. 184, 335–8, 556
perceptions 12, 15, 80
perfect knowledge 82
personalism 361, 362–3, 364, 368
phenomenalism 235–6, 248
philosophical psychology 39
philosophy 247; definitions of 2; impact of Darwinism
   on 73–4; Russellian analysis of 408–10; tasks of 5
philosophy of science 516; aims, methods, tools, products
   5–7; Cartesian 79–83; disappointments/expectations
   of 6–7; Leibniz contribution to 226–8; naturalistic
   approach 308–10; role of history in 154–61;
   Russellian 410–12
physicalism 340–2, 385–6
Pickering, A. 119, 123, 125, 406, 446, 447, 558
Pinxten, R. 73
Planck, M. 91
Plato 285, 461, 486–7, 541
pluralism 144–5
Podolsky, B. 28, 106
Poincaré, H. 56–7, 253, 466
Polanyi, M. 203, 207, 431
Pólya, G. 208
Popper, K. R. 72, 87, 92, 113–14, 120, 143, 154, 155,
   168, 182, 203, 207, 208, 209, 217, 237, 244–5, 326,
   343–7, 367, 388, 415, 416, 418, 437–8, 556, 561,
   567
population genetics (PG) 19, 72
positive relevance, criterion of 364–5
positivism 140, 144, 246–7, 289; and axiomatization
   9–10
postmodernism, feminist 138–9
power 98–9, 219–21
pragmatism 308–9, 335–8, 395, 397–8, 415, 500,
   534, 556, 557; in theory acceptance 349–57
pre-relativistic theories 465–6
prediction 327; biological 18–19
Pretorius, J. P. G. 562
primatology 136
Principal Principle 369
principles 253
probability 108–9, 237, 238, 337, 338, 344–5,
   358–60, 377, 378, 379, 472–3, 474, 477; based on
   "logical" metric 110; bootstrap theory 115, 238;
   calculus 110; classical 110, 365–6; conditional
   110; Dempster–Shafer theory 115; fiducial 114;
   frequency concept 366–8, 371; grue, ravens, etc.
   112–13; logical 365; methodological concept
   363–6; objective 112, 365, 371; physical concepts
   366–70; positive relevance 364–5; posterior 111;
   prior 111, 364; psychological concept 360–3;
   single-case 368, 369–70; subjective 111–12;
   see also causation
problem solving 88–91, 415, 419, 510
production rules 42
progress 418–19
projection 237; entrenched 109, 182
proper names see names
properties 97, 219, 230, 301, 312, 361, 373, 543–4
property unity 543–4
propositions 118, 119, 163, 231, 344, 360, 361
Provine, W. B. 537
pseudo-problems 253
psychoanalysis 137
psychology 232, 309, 360–3, 431
Ptolemy 440, 532
Putnam, H. 17, 119, 163, 178, 191, 204, 261, 301,
   310, 313–15, 330, 337, 383, 396, 402, 501, 505
Pylyshyn, Z. W. 45

qualities, primary/secondary 13, 99, 230, 373–5
quantum mechanics 26–9, 105–6, 262, 290, 305, 349, 376–84, 399, 415, 472, 478, 516, 519, 525, 543, 544, 547–8; and causation 37–8, 378; Copenhagen Interpretation 106, 305, 377–83; EPR argument 28–9, 106, 380; indeterminate 377, 378, 381, 382; locality/separability 380–1, 383; locality 383; measurement problem 382–3; realist 379–80, 383; two-slit experiment 377, 378, 379–80; wave function 377, 380, 383
Quine, W. V. O. 61, 62–3, 120, 162, 204, 244, 245, 248, 250, 261, 263, 309, 317, 337, 385–8, 394–5, 504, 508–9, 510, 534, 535

Railton, P. 474, 475
Ramsey, F. P. 111, 365, 390
Ramsey sentences 62, 268, 390–2, 503
ratiocination 327
rationalism 41, 204, 289, 363, 385, 420, 454, 541
raven paradox 53–4, 112–13, 114, 185, 196
Ray, C. 105
realism 3–4, 28, 105–6, 132, 143–4, 237, 272–5, 300, 303, 306, 345, 351, 386, 515, 521–2, 532–3; as circular 395–6, 400; and conventionalism 58–60, 61–2; defense of 396, 400–1; see also scientific realism
realism/anti-realism debate 329–31, 333, 394
realism/instrumentalism debate: arguments against abduction 394–6; main issue 393–4; realist reply 397–401; see also semantic instrumentalism/semantic realism debate
recalcitrance 125–6
Redhead, M. L. G. 379, 381, 383
reductionism 19, 237, 250, 312, 318, 341–2, 402–3, 480, 481, 541–2
reference 495–7, 510–11
reflexivity 444, 457
Reichenbach, H. 60, 87, 93, 105, 233, 235–7, 238, 239, 240, 244, 326, 366, 416, 476, 477, 478, 547
relationism 225–6, 462–3, 468–9
relativism 145, 147, 405–7, 417, 515
relativity theory 6, 177, 239, 349, 376, 415, 463–5, 516, 519; general 104–5, 466–8; Newtonian 463, 467; special 102–4, 466–7
relevance 131–2
representation 42
representation of the domain 266–8
representation of order/scale 268–70
representation theorems 269, 362
research programs see methodology of scientific research programs (MSRP)
Resnik, M. 263
revolution 429
Reynolds, C. 23
Richards, R. J. 73
Robinson, E. 484
Robinson's theorem 66
Rorty, R. 2, 431
Rosche, E. 279
Rosen, N. 28, 106
Rosenberg, A. 17, 318, 454
Rosenberg, N. 483
Rosenkrantz, R. 113, 363
Ruben, D.-H. 130, 132
Rudner, R. 553
rules 89; and values 555–7; vs. regularities 453–4
Ruse, M. 19, 73, 309
Russell, B. A. W. 181, 234, 259, 408–12, 516; method of philosophical analysis 408–10; philosophy of science 410–12
Rutherford, E. 26

Ryan, M.D. 562
Rynasiewicz, R. 333

Salmon, W. C. 35, 130, 132, 240, 241, 330, 346, 360, 364, 366, 419, 452, 474, 475, 477, 538
Sapire, D. 370
Sarton, G. 203
scale 268–70
Schaffer, S. 95, 117, 121, 406, 446
Scheffler, I. 501
Scheines, R. 38
Schiebinger, L. 137
Schilpp, P. 106
Schlick, M. 60, 243, 247, 248, 249
Schobbeus, P. 562
Schrödinger, E. 27, 381
Schrödinger's cat 381
Schumacher, E. F. 490
Schurtz, G. 562
Schuster, J. 94
science: aims/methods/practices 3–4, 5, 317, 330, 345, 387; and balancing errors 15; as biological endeavor 254; cognitive approaches 41–3; as costly 7–8; definitions of 2–7; detractors of 1–2; ethos of 554–5; feminist accounts 134–42; goals 44; impact of 1–2; instrumentalist 14; narrow image 120; products of 5–6; questions concerning 7; as rational/objective 204; role of history in philosophy of 154–61; role of judgment in 194–202; role of mathematics in 257–63; role of metaphysics in 283–91; social character of 6; texts 120–1; tools 5; undetermination of 388; values of 553–4
scientific change: discovery vs. justification 416–17; general description 413–14; historical background 414–16; paradigms/cores of scientific thought 417–18; related issues 419–20; scientific progress 418–19; summary 420–1
scientific methodology 4, 94–5, 435, 546–8; Aristotelian 424–5; defined 423–4; future of 430–1; Kuhnian 428–30; logical empiricism 426–8; scientific revolution 425–6
scientific realism 100; see also realism
scientific revolution 321, 323, 425–6, 462–3
Scriven, M. 128–9
Seager, W. 227
Searle, J. R. 45
self-definition, gendered 137
self-organization 22, 23
semantic (in)stability 144
semantic instrumentalism 495–7; critique 499–501
semantic realism 495–7
semantic role 510–11
semantical conception of theories (SCT) 301–3, 306–7, 526
semantics 27–8, 318, 391, 521; approach to axioms 9, 10; biological 20–1; see also syntactic/semantic difference
Semmelweiss, I. 188–9
set-theoretical approach 520
Shaffer, S. 115
Shapere, D. 332, 420
Shapin, S. 117, 121, 406, 446
Shapiro, S. 261, 263
Shaw, C. D. 259
Shea, W. 152
Shrager, J. 42, 93
significance tests 114
Simon, H. 44, 88, 89
simplicity 112, 182, 330, 335, 344, 353, 356, 433–40, 534, 545; Akaikerian 438–40; Bayesian 436–7; and conventionalism 59–60; curve-fitting problem 433–5; Popperian 437–8

simulacrum theory of explanation   304
skepticism   312, 388, 410, 504, 509, 554
skills   194–5
Skinner, B. F.   453
Sklar, L.   33, 106
Skyrms, B.   370
Smart, J. C. C.   17–18
Sneed, J.   520
Sober, E.   17, 20, 73
social factors in science   41, 442–9; content of belief
   442–8; ethical standards   442; goals of research
   442; intervention   448–9
social influences   420, 421
social sciences   38, 232, 293–4, 429, 451; causation,
   laws, intentionality   451–3; dangerous knowledge,
   ideology, value freedom   458–9; interpretation to
   historicism   458; ontology in   455; reflexive
   knowledge in   457; regularities vs. rules   453–4;
   teleology and function   455–7
sociology   6, 291, 536, 554, 559
sociology of science   429–30
sociology of scientific knowledge   406
Sokal, A.   449
Sorensen, R.   530
Sowden, L.   39
space   104–5
space-time   31, 60, 102–3, 217, 220, 225, 461,
   535–6, 543–4, 546; epistemological issues   466–7;
   metaphysical issues   467–8; Newton to Einstein   463;
   pre-relativist physics   465–6; relativity   463–5; in
   science/experience   468–9; scientific revolution
   462–3
spatio-temporal unity   543
Spirtes, P.   38
Stamp, P.   485, 487
standpoint theory   138, 139, 140
state space approach   520, 521–2, 546
statements   248, 344; singular/universal   218–19
statistical explanation   127, 470; alternative views
   473–5; common cause   477–8; probabilistic causality
   475–7; standard view   470–3
statistical phenomena   367
statistical-relevance (S-R) model   474
Stegmüller, W.   520
Steiner, M.   763
Stevens, S. S.   265
Stevin, S.   528
stipulations   499, 500, 506–7, 508
straight-rule of inductive inference   93
Strawson, P. F.   181, 435, 509
"strong programme"   421, 443
strong underdetermination of theories (SUT)   532–6
structuralism   263
Stump, D. J.   545
substances   311
substantivism   461, 468; and technology   486, 487–8,
   489
sufficiency see causation
sum of squares (SOS)   434
supervenience   403, 480–2; global/local   481
Suppe, F.   10, 17, 66, 77, 301, 515, 517, 526–7
Suppes, P.   9, 302, 476, 520
supposition   80–1
SUT see strong underdetermination of theories
Swiss cheese theory   177
Swoyer, C.   259
symmetry   444
syntactic/semantic difference   517–20, 522–3
syntax   9–10

Tarski, A.   517, 520
Taylor, C.   492

technology, philosophy of   483–90; adoption/change
   484, 489–90; categorization   485–6; critical   486,
   488–9; instrumental   486, 487; omission of   485;
   substantive   486, 487–8, 489; as value neutral
   486–7
teleological explanation   455–7, 492; functions
   493–4; goals   492–3
Teller, P.   363, 383
testability, principle of   145
Thagard, P.   42, 89, 179, 419
theorems   9, 10, 269, 427
theoretical terms, meaning/reference   60, 144; Carnap–
   Lewis refinement of classical view   502–4; classical
   view   499; critique of semantic instrumentalism
   499–501; definitions   497–8; historical chain theory
   506–8; incommensurability   501–2; interpretation
   510; meanings   511–12; Quine's challenge   508–9;
   semantic instrumentalism vs. semantic realism
   495–7; semantic role/designation   510–11; theory/
   observation distinction   504–6
theories   6, 117–19, 253, 254, 386, 453, 484, 515;
   axiomatic-deductive system   17; classical view   499;
   construed   328–31; development of classical view
   515–17; formal alternatives to classical view   517–20;
   generalized model-based   522–4; historical argument
   395, 398–9; inferred   325–8; interpretation of   26,
   27–9; nonlinguistic   179; realism, causal necessity,
   laws of nature   520–2; structure of scientific   20–1;
   survival/success of   399; testing by experiment
   120–1
theory acceptance   349–51; factors in choice   353–5;
   pragmatic factors in   349–57; reasons for   351–3
theory choice   556–7, 558
theory identity   66, 525; empirical equivalence   526–7;
   equivalence   526; individuation   525–6; see also
   identity
theory/observation distinction   504–6, 517
theory–world relation   302–3, 304
thick/thin concept   497, 500, 501
Thomson, Sir J. J.   26
thought experiments   117, 337, 528–30
Tichý, P.   347, 561
Tiles, J. E.   117
Tiles, M.   261
time   33; see also space-time
Toulmin, S.   203
translation   173–5
TRF(H), creation/discovery   445–6
truth   247–50, 335–7, 353, 355, 394, 418–19, 420,
   496–7, 542
truth conditions   35
truth-function   498
truth-value   169–70, 351, 405
Turing complexity thesis   48–51
Turing imitation test   45
Turing machine   467
Twin Earth   314–15
twin paradox   103
type hierarchies   281

Ulises Moulines, C.   504
underdetermination   119, 121–2, 124, 335, 353–4,
   394, 397–8, 444–5
underdetermination of theory by data   388, 532–6
unification of theories   537–9; alternative forms   538;
   as desirable   538; and explanation   538; standard
   techniques   537
uniformity of nature   181–2
uniqueness theorems   269
unity   353, 356
unity of science   250, 480; historical philosophical tradition
   541–6; problem   540; and scientific method   546–8

universals   218–19, 307, 318, 381
unobservables *see* observation
Urbach, P. M.   110, 112, 115, 362, 367

value freedom   459
value neutrality   486–7
values: epistemic vs. non-epistemic   557–60; notion
  of   550–1; and rules   555–7; and science   551–2
values of science   550, 553–4
Van Benthem, J.   562
van Fraassen, B. C.   5, 131–2, 245, 291, 302, 303, 329,
  332, 351, 352, 355–7, 358, 371, 383, 391, 395, 495,
  500, 504, 505, 507, 511, 517, 521, 523, 526, 557
verification   29, 236–7, 247–50, 327, 337, 345
verisimilitude   170, 347, 418–19, 420, 561–2
Vermeulen, J. C. C.   562
Vienna Circle   234, 243–5, 246, 247, 248
Volti, R.   483, 485
von Mises, R.   367
von Neumann, J.   10

Waismann, F.   243, 247
Wartofsky, M.   266
Waters, K.   403
Watkins, J.   346
Watson, J.   16, 72
wave theory   27, 28, 305, 328, 354–5, 357, 377, 380,
  383, 394–5
"weak programme"   443

weak underdetermination of theories (WUT)   532, 536
Webster, A.   483
Weingartner, P.   562
Wheeler, J. A.   104
Whewell, W.   297, 327, 547, 564–7
Wiggins, D.   512
Wigner, E.   763
Williams, G. C.   72–3
Williams, M. B.   10
Williamson, T.   363, 366
Wilson, E. O.   73
Winch, P.   453
Winner, L.   485, 489
Winnie, J.   502
Wittgenstein, L.   77, 143, 204, 247, 453
Woodfield, A.   492, 493
Woolgar, S.   121, 122, 124, 125, 291, 445–6, 485
"World 3"   347
world   382
Worrall, J.   209, 211, 346, 398
Wright, L.   456, 492
WUT *see* weak underdetermination of theories (WUT)
  532, 536

Yeo, R.   94

Zahar, E.   209
Zwart, S.   562
Zymach, E.   315